ENCYCLOPEDIA OF PHYSICS

EDITED BY

S. FLÜGGE

VOLUME LII

ASTROPHYSICS III: THE SOLAR SYSTEM

WITH 225 FIGURES

SPRINGER-VERLAG
BERLIN · GÖTTINGEN · HEIDELBERG
1959

HANDBUCH DER PHYSIK

HERAUSGEGEBEN VON

S. FLÜGGE

BAND LII

ASTROPHYSIK III: DAS SONNENSYSTEM

MIT 225 FIGUREN

SPRINGER-VERLAG
BERLIN · GÖTTINGEN · HEIDELBERG
1959

ISBN-13: 978-3-642-45931-3 e-ISBN-13: 978-3-642-45929-0
DOI: 10.1007/978-3-642-45929-0

Softcover reprint of the hardcover 1st edition 1959

Inhaltsverzeichnis.

The Photosphere of the Sun.

By

LEO GOLDBERG and A. KEITH PIERCE[1].

With 27 Figures.

I. The continuous spectrum.

1. Introduction. An empirical model of the photosphere rests on observations of: (1) the solar constant—which fixes the temperature scale, (2) limb darkening—from which the variation of temperature with depth throughout the atmosphere is determined, and (3) the energy distribution—which, together with limb darkening, reveals the opacity at each depth and for each wavelength.

All the basic observational data for a satisfactory interpretation of the physics of the solar atmosphere were available shortly after the turn of the century. By 1905 ABBOT had clarified the causes of the disagreement among different observers in the determination of the value of the solar constant—values which varied all the way from 1.75 to 4.0 calories cm^{-2} min^{-1}. He showed that the diversity of solar constant values was not due to the use of inadequate instruments but was caused by a lack of an international scale of pyrheliometry and most particularly by failure of the extrapolation by empirical formula to zero air mass. ABBOT'S early value of 2.1 cal cm^{-2} min^{-1} is within a few percent of today's accepted figure.

The first monochromatic limb darkening measures were obtained by VOGEL (1873) and by VERY (1902). As so often happens in astronomy, the construction of new and much more sensitive detectors permitted ABBOT, FOWLE and ALDRICH (1900 to 1922) to extend and improve greatly upon VOGEL and VERY'S results. The Smithsonian observers, in an attempt to detect a variation of limb darkening with time, obtained thousands of center-to-limb traces at 22 wavelengths, distributed between $\lambda\lambda$ 3230—20970 Å. SCHUSTER in 1903 and SCHWARZSCHILD in 1906 were the first to provide an explanation of the limb darkening observations, based on the concept of a solar atmosphere absorbing and reemitting radiation. Parenthetically, the existence of this atmosphere was known at a very early date from its Fraunhofer spectrum (KIRCHHOFF and BUNSEN, 1859 to 1862), but it was not considered to affect the continuum.

An accurate determination of the energy distribution in the Sun's spectrum resulted from the work started by WILSING (1905) at the Astrophysical Observatory in Potsdam, and by LANGLEY (1883) at the Allegheny Observatory, and at the Smithsonian Institution, Washington. WILSING compared the ratio of the Sun's intensity to that of a black body in the interval $\lambda\lambda$ 4500—21000 Å. In the visible portion of the Sun's spectrum he found that the energy distribution corresponded to a color temperature of 6300°. The Smithsonian work followed as a byproduct of the solar constant determinations. To carry out LANGLEY'S program ABBOT, ALDRICH and FOWLE (1908) obtained bolograms ($\lambda\lambda$ 3400—

[1] Chap. I: The continuous spectrum was written by PIERCE; Chap. II: The Fraunhofer spectrum by GOLDBERG.

24000 Å) at different air masses, and from these computed the extraterrestrial energy curve of the Sun.

It is the purpose of this chapter to outline in a rather brief manner the present status of the observational work on photospheric radiation and to discuss a few of the observational models of the solar atmosphere. For the reader interested in the theoretical models, Refs. [1] to [8] and the paper by D. BARBIER in Vol. L of this Encyclopedia may be consulted.

a) Solar constant[1].

2. LANGLEY's method; Smithsonian results. Langley considered the observation of the amount of heat the Sun sends to the Earth as one of the most important and difficult problems in astronomical physics, and also the fundamental problem of meteorology. The measurement of the solar constant requires special instruments: one that measures the total radiation in absolute units and one that measures the atmospheric losses[2]. In 1838 POUILLET devised a radiation calorimeter which he called a pyrheliometer. In its modern version by ABBOT, it consists of a blackened silver disk 28 mm in diameter and 7 mm thick, drilled to receive a sensitive thermometer. The temperature rise during a 2-minute interval is taken as a measure of the incident energy. While the silver disk pyrheliometer is one of the most accurate and reliable of available instruments, it is not in itself an absolute instrument. Each instrument, however, is calibrated and is supplied with a constant that converts the observations to the absolute Smithsonian scale of 1913. The latter was established by use of the standard water-flow pyrheliometer of ABBOT and ALDRICH.

RADAU and LANGLEY showed that BOUGUER's law of absorption applied to observations of high and low Sun, and that linear extrapolation to zero air mass was possible if limited only to a single wavelength. Thus it became evident that the solar constant could not be determined from pyrheliometer observations alone; separate measurements at each wavelength are required in order to evaluate the atmospheric losses. The equation,

$$\ln I_\lambda = k_\lambda H_\lambda \sec Z + \ln I_{\lambda,0}, \tag{2.1}$$

relates the incident intensity $I_{\lambda,0}$ to the observed intensity I_λ. The optical path length is the equivalent height of the atmosphere H, times the secant of the Sun's zenith distance—corrected for refraction and curvature of the atmosphere at low solar altitudes ($<10°$). The coefficient of $\sec z$ is variable with time, but at excellent observing locations the use of Eq. (2.1) for the reduction of a day's observations can give $I_{\lambda,0}$ with an error of less than $\frac{1}{4}$ of 1%.

LANGLEY employed the spectrobolometer to observe I_λ. Since the spectrobolometer was not sufficiently stable, and could not be calibrated with sufficient accuracy, LANGLEY normalized each bologram through concurrent pyrheliometer observations. Thus, the spectrobolometer has served only to give the relative energy at each wavelength, to perform the extrapolation to zero air mass, and to define the areas of the water vapor, ozone, and CO_2 absorptions. The Smithsonian bolograms extend from 0.346 to 2.4 μ while the pyrheliometer observations cover the region receivable through the atmosphere: 0.295 to about 25 μ. Hence, in

[1] Treated also by F. MÖLLER, Vol. XLVIII, p. 182 of this Encyclopedia.

[2] E.g. the detailed accounts and references of B. STRÖMGREN: Handbuch der Experimentalphysik, Vol. 26. Leipzig: Akademische Verlagsgesellschaft 1937. — C.W. ALLEN: The Sun, ed. KUIPER. Chicago: Univ. of Chicago Press 1953. — A discussion by R.R. MCMATH and O.C. MOHLER of solar instruments and accessories is to be found in Vol. LIV of this Encyclopedia.

order to equate the integrated spectrobolograms obtained at each solar altitude to the pyrheliometer values, small corrections are added to the integrated curves to account for the unobserved ultraviolet and infrared regions. These corrections depend upon the solar altitude and the water vapor content of the atmosphere at the time of observation. The band absorption and the effects of dust and Rayleigh extinction are removed by the extrapolation of the normalized bolograms to zero air mass. To the integrated zero air mass bolograms the Smithsonian observers add theoretical estimates of the energy radiated by the Sun to the violet of 0.346 μ and the red of 2.4 μ. These estimates amount to 3.1 and 2.0% of the value of the total solar constant for the ultraviolet and infrared respectively.

The Smithsonian observers[1] conclude from their extensive observations and intercomparisons between instruments that the most probable absolute value of the mean solar constant is:

$$1.940 \text{ cal cm}^{-2} \text{ min}^{-1}.$$

3. Solar constant based on new solar spectral irradiance data. A critique of the Smithsonian reductions has been given by JOHNSON[2]. From the ground, radiation from the Sun at wavelengths less than 0.295 μ cannot be observed, and we are unable to predict the energy in this region of the spectrum from astrophysical theory. Only direct measures from high altitude rockets can supply the necessary data. By use of V-2 rockets, N.L. WILSON *et al.*[3] obtained calibrated ultraviolet spectra which, on a relative scale, were considered accurate to $\pm 5\%$. A relative spectral irradiance curve above the atmosphere and for the region 0.22 to 0.60 μ was obtained by fitting the rocket data to the spectral irradiance curve at 0.318 μ of DUNKELMAN and SCOLNIK[4]. Finally, MOON's[5] data were used to extend the curve to 2.4 μ. The resulting relative energy distribution, integrated over the restricted interval 0.346 to 2.4 μ, was then put on an absolute scale by using the Smithsonian results. JOHNSON's values for the solar energy distribution are given in Table 4. The first column gives the wavelength in microns; the second lists H_μ, the mean zero air mass spectral irradiance in watts cm^{-2} micron^{-1}; P is the percentage of the solar constant short of wavelength μ.

In the ultraviolet, the U.S. Naval Research Laboratory curve led to a 1.24% larger zero air mass correction than that used by the Smithsonian Institution. In the infrared, JOHNSON concluded that the Smithsonian observers failed to recognize the great absorbing power of small amounts of water, which would produce a sharp upturn in the extrapolation to zero air mass. Further he also assumed that the Sun radiates as a gray body in the region beyond 2.4 μ, with temperature 6000° K, emissivity 0.990. In all, a value 0.038 cal cm^{-2} min^{-1} greater than the Smithsonian Institution zero air mass correction was derived.

JOHNSON's final value is:

$$(2.00 \pm 0.04) \text{ cal cm}^{-2} \text{ min}^{-1}.$$

4. Rocket and balloon results. Observations from high altitudes offer the greatest possibility of determining the precise value of the solar constant. A

[1] L.B. ALDRICH and W.H. HOOVER: Ann. Astrophys. Obs. Smithsonian Inst. **7**, 176 (1954).

[2] F.S. JOHNSON: J. Meteorology **11**, 431 (1954). — Rapport de la Commission pour l'étude des Relations entre les Phénomènes Solaires et Terrestres. Paris: J. & R. SENNAC 1957.

[3] N.L. WILSON, R. TOUSEY, J.D. PURCELL, F.S. JOHNSON and C.E. MOORE: Astrophys. Journ. **119**, 590 (1954).

[4] L. DUNKELMAN and R. SCOLNIK: J. Opt. Soc. Amer. **42**, 876 (1952).

[5] P. MOON: J. Franklin Inst. **230**, 583 (1940).

determination of the atmospheric extinction is not required, and the infrared and ultraviolet corrections are zero, if the observations are made above about 50 km. Experiments made by A.L. QUIRK, based on telemeter information from semi-conductor bolometers recessed in the nose of an Aerobee rocket flown May 25, 1950 at an altitude of 70 km, gave maximum readings between 1.99 and 2.02 cal cm^{-2} min^{-1}.

Because of the failure of rockets in later flights, attempts were made to determine the solar constant with weather bureau global pyrheliometers suspended by a long cable below plastic balloons 90 feet in diameter at 90000 feet. Three flights in 1954 gave the values[1] 2.02, 1.995 and 2.01 cal cm^{-2} min^{-1} based on the Smithsonian scale of pyrheliometry of 1913. When reduced by 2.4% to bring the values to the true scale, the mean balloon value is 1.96 ± 0.03 (estimated accuracy) cal cm^{-2} min^{-1} or 1% greater than the most probable value 1.940 cal cm^{-2} min^{-1} adopted by the Smithsonian observers. It is clear that further work, both at the ground and at high altitude is desired.

5. Variations in the solar constant. A correlation between sunspot numbers and the solar constant has been suggested by ALDRICH and HOOVER[2]. From their tabulation the solar constant increased by 0.6% at sunspot maximum (1948) over the value at sunspot minimum (1944).

If the solar constant should vary, either because of changing temperature or because of sunspots and their associated plage areas, one would expect the range to be greater in the ultraviolet than at longer wavelengths. Of the several attempts to measure the ultraviolet variability we mention PETTIT's[3] observations extending from June 1924 to October 1931. PETTIT compared the zero air-mass ratio of intensities, $I(0.32\,\mu)/I(0.50\,\mu)$, observed through thin silver and gold films. This ratio varied from 0.95 to 1.57 during the period and, except for one year, correlated well with the sunspot curve.

The possibility of detecting long range variations in the solar constant through differential measures of the brightness of planets or their satellites versus stars has long been known to astronomers. In 1927 modern photoelectric techniques were used by STEBBINS and JACOBSEN[4] in their observations of Jupiter's satellites and the planet Uranus. From observations made on 47 nights, they conclude that there was no change in the solar constant large enough to be detected: about ± 0.33%. A more recent attempt by HARDIE and GICLAS[5] showed that, in the interval 1949 to 1954, any variation in the solar constant, as reflected by the measures of Uranus and Neptune, was probably smaller than 0.4%. The extension of this work by MITCHELL, who compared the observations obtained in the interval October 1956 to June 1957, with those obtained in October 1954 to June 1955, and in October 1955 to June 1956, indicated a variation in the energy of the Sun, in blue light, of about 0.4% per year; but this was only a little greater than the probable error of the measurements[6]. One is impressed by the very small variability in the Sun's energy output. Probably in the monochromatic wavelengths of Hα or Lα the variation is large, but the fundamental work of ABBOT certainly shows that the variation in the solar constant is small—so small that it approaches the limit of detection.

---

[1] A.L. QUIRK: Final Report-contract AF19 (122)—249 of Geophys. Res. Directorate Univ. of Rhode Island, Dept. of Physics 1955.

[2] L.B. ALDRICH and W.H. HOOVER: Ann. Astrophys. Obs. Smithsonian Inst. **7** (1954).

[3] E. PETTIT: Astrophys. Journ. **75**, 185 (1932).

[4] J. STEBBINS and T.S. JACOBSEN: Lick Obs. Bull. **8**, 180 (1928).

[5] R.H. HARDIE and H.L. GICLAS: Astrophys. Journ. **122**, 460 (1955).

[6] E.C. SLIPHER: Astronom. J. **62**, 296 (1957).

b) Observed limb darkening.

6. The integral relating the source function and the observed intensity. As MILNE[1] emphasized, the darkening of the Sun's disk toward the limb "is a consequence simply and solely of the temperature gradient in the outer layers". The intensity at the limb is less than the intensity measured at the center of the solar disk because of the inwardly increasing temperature and because radiation from a semi-transparent atmosphere originates, in the mean, at the same optical depth along the line of sight. Consider radiation leaving the surface of an atmosphere consisting of plane parallel layers. The observed brightness $I_\lambda(0,\mu)$ in the direction arc cos μ, in units of the brightness observed in the normal direction, is given by a summation of the source functions for the elements at depths τ_i, diminished by extinction of overlying layers, namely:

$$\frac{I_\lambda(0,\mu)}{I_\lambda(0,1)} \equiv I_\lambda^*(0,\mu) = \int_0^\infty B_\lambda^*(\tau_\lambda)\, e^{-\frac{\tau_\lambda}{\mu}} \frac{d\tau_\lambda}{\mu}. \tag{6.1}$$

Here we have expressed the source function

$$B_\lambda^*(\tau_\lambda) = B_\lambda(\tau_\lambda)/I_\lambda(0,1)$$

in units relative to the surface brightness in the normal direction. The use of the optical depth τ_λ in place of geometrical depth x conceals a large amount of information. But if the absorption coefficient $k_\lambda(x)$ be known then the geometrical depth x can be found from

$$\tau_\lambda = \int_x^0 k_\lambda(x)\, dx. \tag{6.2}$$

The many solutions of these equations, for the source function representing the solar atmosphere, are discussed later.

7. Observations outside of eclipse. The earlist extensive work on limb darkening was that of ABBOT, FOWLE, and ALDRICH[2]. Their Washington observations (1906 to 1908) were obtained with a horizontal telescope of 20 inches aperture and 140 feet focal length fed by light from a 17-inch aperture coelostat, and a 12-inch aperture fixed mirror. In their work at Mount Wilson, a tower served as mounting for coelostat and telescope. The image was formed by a 12-inch diameter, 75-foot focal length mirror used off-axis. In 1913, 577 drift curves were obtained at 7 different wavelengths. These observations were extended through 1920. By stopping the coelostat drive, the Sun's image was made to drift along a diameter and over the spectrobolometer slit which had a length of 2.1 cm and widths from 0.2 to 2.0 mm. The drift technique has great advantages: it eliminates errors due to unequal transmission in different parts of the optical apparatus; also, observations are made at equal solar altitudes, an important gain if they happen to be made in the early morning when the extinction is large, but when the seeing is best. There is, however, one disadvantage: since the time of traversal is 2.3 min, the response of the recorder must be quick, in order to minimize distortion due to its time constant.

Both the time constant of the receiving system and the finite size of the scanning slit distort the observations of the true center-to-limb variation. RAUDENBUSCH[3] attempted to minimize these effects by recording with a wire electrometer

[1] E.A. MILNE: Handbuch der Astrophysik, Vol. 3. Berlin: Springer 1930.

[2] C.G. ABBOT, F.E. FOWLE and L.B. ALDRICH: Ann. Astrophys. Obs. Smithsonian Inst. **2** (1912); **3** (1913); **4** (1914).

[3] H. RAUDENBUSCH: Astronom. Nachr. **266**, 301 (1938).

with a very short time constant, and by using a slit width $^1/_{900}$ of the Sun's diameter. His horizontal telescope gave a solar image 18 cm in diameter. Measurements were obtained at $\lambda\lambda$ 0.426, 0.506, 0.596, and 0.670 μ with a small prism apparatus and Cs and K photocells. A very marked asymmetry in brightness was found between east and west limbs.

Canavaggia and Chalonge[1] employed photography and a high dispersion quartz spectrograph to obtain values of the limb darkening in the ultraviolet ($\lambda\lambda$ 3149—3692 Å). They were able to select 25 regions of the continuum relatively free of absorption lines. The exposures were obtained with the spectrograph placed at the focus of an off-axis Cassegrain telescope which gave a solar image 50 mm in diameter.

An attempt to observe the brightness distribution at the extreme limb was made by Minnaert *et al.*[2] using the 40-inch refractor of the Yerkes Observatory. Photographs were taken only at moments of good seeing. By means of filters two effective wavelengths were isolated—λ 5593 and λ 6245 Å. The results, covering the interval 0.995 to 0.90 R were fitted at 0.90 R to the mean intensity of other observers. This important work should be repeated and extended to more wavelengths.

In the far infrared limb darkening observations were secured in atmospheric windows at 3.5, 8.3, and 10.2 μ by Pierce, McMath, Goldberg and Mohler[3]. A 24-inch Cassegrain reflector served to form a 5.4-inch image of the Sun on the slit (7 mm by 0.3 mm) of a Perkin-Elmer spectrometer. The radiation was chopped before reaching the thermo-couple. The resulting AC signal, amplified and rectified by a breaker-type amplifier, was recorded on a Brown balancing potentiometer.

Peyturaux[4] obtained in 1952 to 1953 an extensive and homogeneous set of limb-darkening measures at 25 wavelengths extending from λ 3190 to λ 23120 Å. His observations in the ultraviolet are of particular value because they cover the region of the Balmer discontinuity. A small heliostat directed light to a concave mirror. The image of the Sun, diameter 30 mm, was scanned in 4 minutes with a monochromator slit of aperture 1.5×0.03 mm. A photo-multiplier and a lead sulfide cell were used as detectors, together with a galvanometer for photographic recording and a Wilson-type amplifier.

The Snow telescope at Mount Wilson was used by Pierce[5] in 1952 to obtain drift curves at 43 wavelengths in the region λ 3811 to λ 24388 Å. A linear amplifier was used in conjunction with a 1P21 photomultiplier and a Wilson-type amplifier with a PbS cell. Both systems were carefully calibrated.

Fig. 1 shows the limb darkening as a function of wavelength for fixed points on the solar disk, namely $\cos\vartheta = 0.9$ to 0.2, where ϑ is the angle between the normal and the line of sight to the observer. There is excellent agreement between Peyturaux and Pierce in the interval λ 4000 to λ 12000 Å, but at longer wavelengths Peyturaux's results are systematically below Pierce's by about 1% for the limb points.

8. Corrections to observed limb-darkening curves. The true limb darkening differs from the observed limb darkening because of: (1) seeing, (2) scattered

[1] R. Canavaggia and D. Chalonge: Ann. d'Astrophys. **9**, 143 (1946).

[2] M. Minnaert, E. v. d. Hoven van Genderen and J. v. Diggelen: Bull. Astronom. Netherl. **11**, 55 (1949).

[3] A. K. Pierce, R. R. McMath, L. Goldberg and O. C. Mohler: Astrophys. Journ. **112**, 289 (1950).

[4] R. Peyturaux: Ann. d'Astrophys. **18**, 34 (1955).

[5] A. K. Pierce: Astrophys. Journ. **120**, 221 (1954). Only the observations at 26 wavelengths to the red of λ 7793 have been published.

light, (3) imperfections of mirror surfaces, (4) diffraction, (5) optical errors, and (6) the use of a finite scanning aperture. One observes limb darkening with a complex two-dimensional apparatus function. The atmospheric part consists

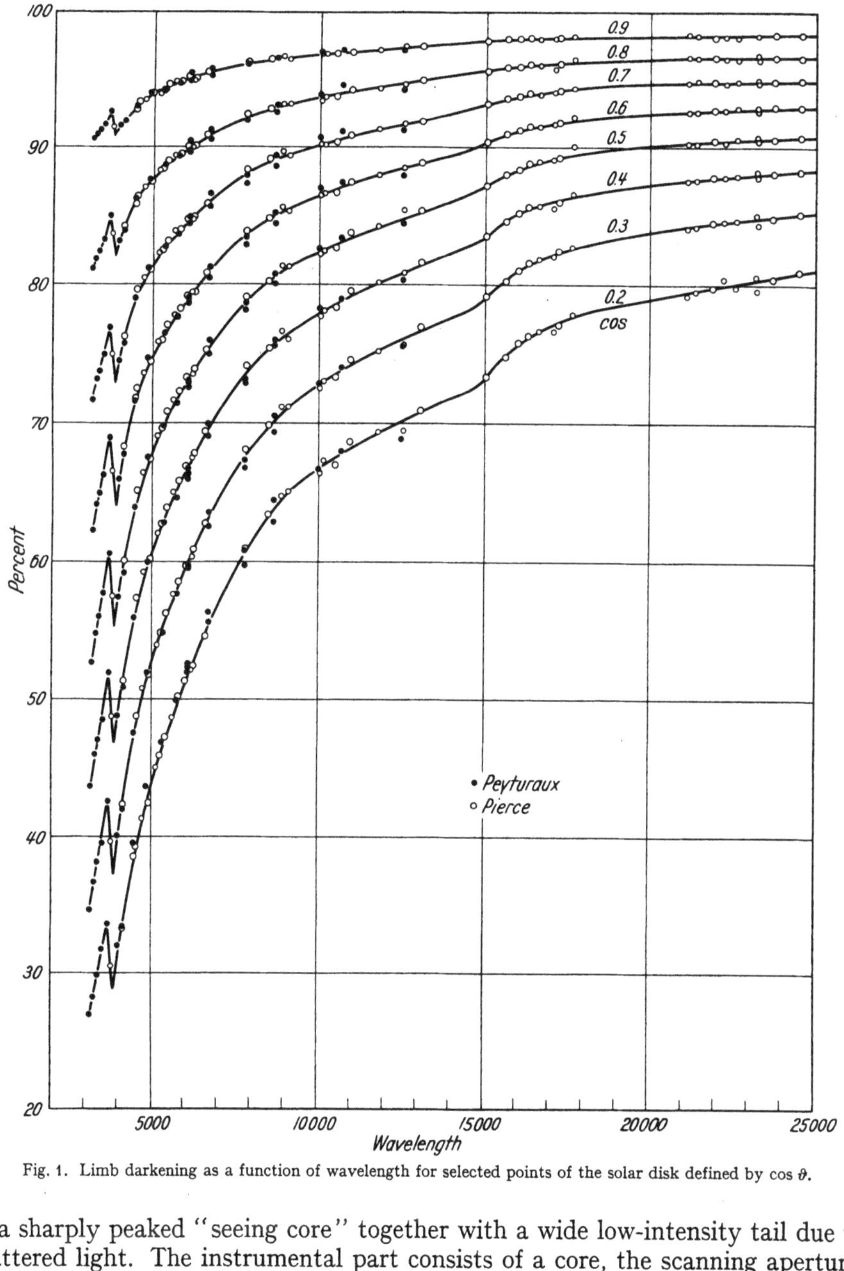

Fig. 1. Limb darkening as a function of wavelength for selected points of the solar disk defined by cos ϑ.

of a sharply peaked "seeing core" together with a wide low-intensity tail due to scattered light. The instrumental part consists of a core, the scanning aperture, and broad wings resulting from diffraction and instrumental scattered light. Let the intensity at different points of the Sun's surface be $I(x, y)$. Then the observed intensity is

$$i(x_0, y_0) = \frac{1}{N} \iint I(x, y)\, \Phi(x - x_0, y - y_0)\, dx\, dy, \tag{8.1}$$

where N is a normalization factor and Φ the diffusion function. We suppose that Φ can be written as a sum or product of two or more independent distributions. It is generally assumed that the central seeing image can be represented by the two-dimensional surface of revolution:

$$\Phi_1 = \frac{a}{\pi}\, e^{-a(x^2+y^2)} = \frac{a}{\pi}\, e^{-a\varrho^2}, \tag{8.2}$$

while in the scattered light region, $30' < \varrho < 5°$,

$$\Phi_2 = s\,\varrho^{-n} \tag{8.3}$$

where, according to VAN DE HULST[1], the value of n varies between 1.2 and 1.8. The proper formula for the intermediate region $1' < \varrho < 30'$ is not known. Probably expression (8.3) can be used with n equal to about 2.0.

α) Correction for seeing. The most satisfactory treatment of the effect of seeing on limb darkening is that of MINNAERT, VAN DER HOVEN VAN GENDEREN, and VAN DIGGELEN[2]. They assume that the intensity distribution can be represented by

$$\frac{I_\lambda(0,\vartheta)}{I_\lambda(0,0)} = \frac{1+\beta_0\cos\vartheta}{1+\beta_0}. \tag{8.4}$$

By substituting $\cos\vartheta = \sqrt{1-r^2}$, and by transferring to the limb with the expression $x = 1 - r$, an approximation yields

$$I_\lambda(0) = A_\lambda + B_\lambda\sqrt{x}. \tag{8.5}$$

Following WANDERS, they assume that seeing blurs a point according to a Gaussian distribution:

$$j = \frac{a}{\pi}\, e^{-a(\xi^2+\eta^2)}. \tag{8.6}$$

Then, if the limb be considered a straight edge, one has for the intensity distribution, modified by the seeing,

$$i(x) = \frac{a}{\pi}\int\limits_0^\infty (A + B\sqrt{\xi})\, d\xi \int\limits_{-\infty}^{+\infty} e^{-a[(x-\xi)^2+\eta^2]}\, d\eta, \tag{8.7}$$

which may be integrated over η to give

$$\left.\begin{aligned} i(x) &= \sqrt{\frac{a}{\pi}}\int\limits_0^\infty (A + B\sqrt{\xi})\, e^{-a(x-\xi)^2}\, d\xi \\ &= \frac{A}{\sqrt{\pi}}\int\limits_{-x\sqrt{a}}^\infty e^{-z^2}\, dz + \frac{B}{a^{\frac{1}{4}}\sqrt{\pi}}\int\limits_0^\infty \sqrt{w}\, e^{-(x\sqrt{a}-w)^2}\, dw. \end{aligned}\right\} \tag{8.8}$$

Values of the second integral are given in Table 1. A typical plot of this equation, which illustrates the effect of blurring of the limb by seeing, is shown in Fig. 2. It is found in practice that this system of curves fits the observed limb profile very well and thus yields a reliable value of the diffusion constant a.

[1] H. C. VAN DE HULST: Atmospheres of Earth and Planets, ed. KUIPER, 2nd ed. Chicago, Ill.: Chicago University Press 1952.

[2] M. MINNAERT, E. VAN DER HOVEN VAN GENDEREN and J. VAN DIGGELEN: Bull. Astronom. Netherl. **11**, 55 (1949).

Table 1. (The distance from the limb is $x \cdot R$, where R is the radius.)

$x\sqrt{a}$	$\int_0^\infty \sqrt{w}\, e^{-(x\sqrt{a}-w)^2} dw$	$x\sqrt{a}$	$\int_0^\infty \sqrt{w}\, e^{-(x\sqrt{a}-w)^2} dw$
−1.5	0.014	+0.5	1.120
−1.25	0.033	+0.75	1.388
−1.0	0.071	+1.0	1.644
−0.75	0.140	+1.25	1.879
−0.5	0.249	+1.5	2.093
−0.25	0.407	+1.75	2.285
0.0	0.612	+2.0	2.464
+0.25	0.856		

The difference between the observed limb darkening and the adopted approximation [Eq. (8.8)] is the error curve c:

$$j_{\text{obs}} = i(x) + c(x). \tag{8.9}$$

When a is known, the methods of BURGER and VAN CITTERT[1] or BRACEWELL[2] may be used to correct $c(x)$ for diffusion. The true limb darkening, J, is then given by

$$J = I + C \equiv A_\lambda + B_\lambda \sqrt{x} + C(x), \tag{8.10}$$

where C is the corrected error curve.

An accurate estimate of the diffusion constant can be obtained from observations of the limb of the Moon at the time of a partial eclipse. If a is known, the integral equation which relates the true intensity distribution $I(x)$ with the observed intensity is

$$i(x) = \int_{-\infty}^{+\infty} D(y)\, I(x-y)\, dy, \tag{8.11}$$

where D is the diffusion or weighting function. Eq. (8.11) can be solved by Fourier transforms[3], by BRACEWELL'S[4] graphical method, or by any of the numerous methods for solving an integral equation of the first kind[5].

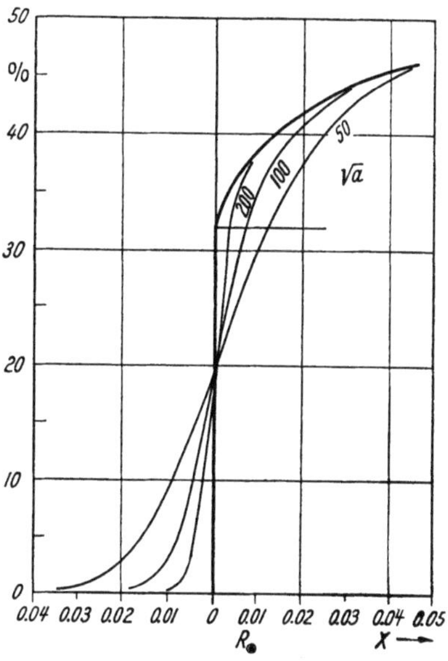

Fig. 2. The effect of seeing on the limb profile, λ 5000 Å.

β) Correction for scattered light. ABBOT carefully checked for the effects of stray light and sky radiation in his limb-darkening measures and concluded that they were of no measurable importance. He states that at a point $1.015R$ (14″ from the limb) the intensity is certainly less than 1% of the intensity at the center of the disk. RAUDENBUSCH included the short-range scattering with the seeing factor by increasing the dispersion of the latter by $\sqrt{2}$. PEYTURAUX made no mention of scattered light in his observations of limb darkening, possibly

[1] H.C. BURGER and P.H. VAN CITTERT: Z. Physik **79**, 722 (1932). — Z. Physik **81**, 428 (1933).

[2] R.N. BRACEWELL: J. Opt. Soc. Amer. **45**, 873 (1955).

[3] P.B. FELLGETT and F.B. SCHMEIDLER: Monthly Notices Roy. Astronom. Soc. London **112**, 445 (1952).

[4] R.N. BRACEWELL: loc. cit.

[5] H.C. VAN DE HULST: Bull. Astronom. Netherl. **10**, 75 (1946).

because a Wilson-type amplifier is non-linear for very small signals and tends to conceal the effects of light scattered beyond the limb. PIERCE *et al.* ignored the scattered light in their Lake Angelus observations. At shorter wavelengths the scattering becomes of greater importance, particularly when the altitude of the Sun is low, and in later work by PIERCE, at Mt. Wilson, a careful determination was made of the scattered light.

The magnitude of the scattered light, from instrument and from sky, has been determined by a number of observers. For the sky alone VAN DE HULST[1] finds that the true mean intensity of the aureole 4′ from the Sun's limb (at 1.25 R) is $\frac{1}{20}$ of one percent. But observations obtained with various solar instruments, in which no particular pains have been taken to reduce the scattered light within the instrument, give much larger results. DE JAGER[2], PETTIT and NICHOLSON[3], KINMAN[4], MICHARD[5], and PIERCE[6] are in fair agreement. For normal days, the observed scattered light is about 2% at 1.010 R (10″ from limb) and about 1% at 1.062 R (60″ from limb). MICHARD and PIERCE find, however, that there is a strong dependence on wavelength.

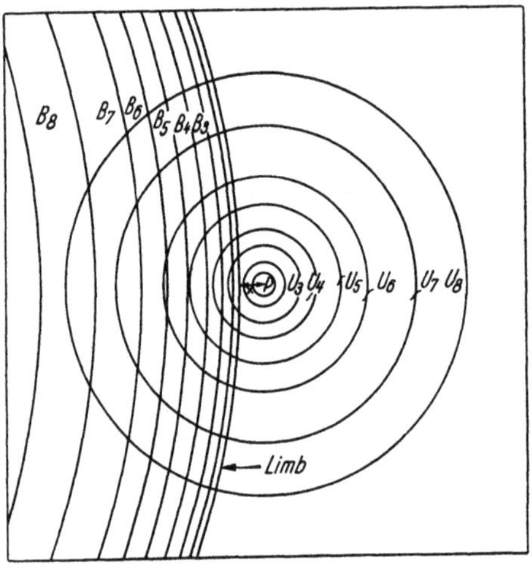

Fig. 3. The geometry of KINMAN's method of determining scattered light (after KINMAN).

If the radial distribution of scattered light beyond the limb can be observed, KINMAN's graphical method may be used to evaulate the constants of the scattering functions $s(\varrho) = \alpha\, e^{-\beta\varrho}$ and $s(\varrho) = A(B+\varrho^2)^{-1}$. The limb region of the Sun is drawn to scale and divided into curved strips, each having a width proportional to the mean surface brightness, B_n. A second system of rings, representing the scattering, is centered at point p at distance x from the limb, as in Fig. 3. The light which is scattered from the annulus U_i onto its center is

$$I(x)_{u_i} = s_i \sum_n \sigma_{in} B_n, \tag{8.12}$$

where σ_{in} is the area common to the annulus U_i and the strip B_n, and s_i is the mean value of the scattering function for zone i. The summation for all zones equals the measured surface brightness $I(x)$ at the point p. KINMAN selected 8 annuli about p, and 8 positions x, and solved for the coefficients s_i from the set of simultaneous equations. The central five zones from $\varrho = 0$ to $\varrho = 40''$ fitted the exponential form of $s(\varrho)$; the outer zones U_6, U_7 and U_8 were fitted to the inverse power function. For his observations, KINMAN found that the total surface brightness of the light scattered from an infinite area of unit surface brightness was 0.09. This figure corresponds closely to the scattered light at

[1] H. C. VAN DE HULST: Atmospheres of the Earth and Planets, ed. KUIPER, 2nd ed. Chicago, Ill.: Chicago University Press 1952.
[2] DE JAGER: Rech. Obs. Utrecht **13** (1952).
[3] E. PETTIT and S. B. NICHOLSON: Astrophys. Journ. **62**, 202 (1925).
[4] T. D. KINMAN: Monthly Notices Roy. Astronom. Soc. London **113**, 613 (1953).
[5] R. MICHARD: Ann. d'Astrophys. **16**, 217 (1953).
[6] A. K. PIERCE: Astrophys. Journ. **120**, 221 (1954).

the center of the Sun's disk and can be considered typical of most observations. The graphical method may be used to determine the scattered light at any point of the disk of the Sun. The correction to the observed limb darkening is

$$i(\varrho) = \frac{I(\varrho) + s(\varrho)}{1 + s(0)} \tag{8.13}$$

where $s(0)$ is the magnitude of the scattered light at the center of the disk.

γ) Correction for slit width and height. A finite scanning aperture introduces small errors into the limb darkening curves (see Fig. 4). A straight slit becomes a chord and a wide slit covers a large range of intensities near the limb. RAUDENBUSCH[1] assumes the limb darkening to be of the form given in Eq. (8.4), which can be written as:

$$I(x, y) = A + B\sqrt{1 - x^2 - y^2}. \tag{8.14}$$

Correction for slit length is made through the integral equation

$$i = \frac{1}{h} \int_0^h I(x, y)\, dy, \tag{8.15}$$

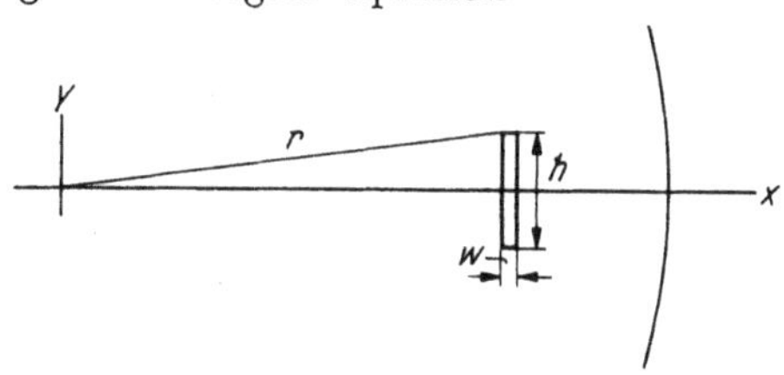

Fig. 4. Correction for slit width and height.

which relates the observed intensity i to the true distribution $I(x, y)$. The integration of Eq. (8.15) can be performed with the result that

$$i(x) = A + \frac{B}{h}\left\{\frac{h}{2}\sqrt{1 - x^2 - h^2} - \frac{1}{2}(1 - x^2) \arcsin\left(\frac{h}{\sqrt{1 - x^2}}\right)\right\}. \tag{8.16}$$

By expanding in powers of h, which is a small quantity, we obtain

$$i(x) = A + B\left[\sqrt{1 - x^2} - \frac{h^2}{4\sqrt{1 - x^2}} + \frac{h^2}{12\sqrt{1 - x^2}} + \cdots\right], \tag{8.17}$$

and hence in the first approximation

$$I(x) = \frac{B h^2}{6\sqrt{1 - x^2}} + i(x). \tag{8.18}$$

The correction for slit width may be included in the seeing error, as was done by RAUDENBUSCH, or may be treated separately. The integral equation to be solved is

$$i(x) = \frac{1}{2w} \int_{-w}^{+w} I(r + x)\, dx, \tag{8.19}$$

which is readily evaluated by a series expansion.

δ) Correction for time constant in drift curves. A direct photoelectric determination of limb darkening is much more accurate than photographic methods except possibly near the extreme limb. In the usual drift method, the ascending limb is lower and the descending or following limb is higher than the true intensity profile because of the time constant of the total recording system. ABEOT demonstrated that this error can be partially eliminated by selecting the center of symmetry of the drift curve and by averaging the galvanometer deflections for each limb.

[1] H. RAUDENBUSCH: Astronom. Nachr. **266**, 301 (1938).

If the response of the system to a step function can be found, the general problem can be solved[1]. Let the solar radiation incident upon the receiver, when converted to units of current, be $i(t^*)$, a function of time, measured from the incident limb. Further, let $K(t)$ express the output of the receiver and recorder in response to a unit step in i at the receiver. Then, the following equation holds:

$$\delta(t) = i(0)\, K(t) + \int\limits_0^t K(t - t^*)\, i'(t^*)\, dt^* \tag{8.20}$$

where $\delta(t)$ is the response of the system and i' is the time rate of change of the energy incident on the receiver. This integral equation can be solved for i' if $K(t)$ is known. The latter is easily obtained from observation by impressing step functions on the system. Finally $i(t)$ may be obtained by numerical integration.

9. Observations of limb darkening at eclipses. At the time of a solar eclipse the Moon, moving at the rate of 0".5 per second of time, acts as an occulting disk. It covers or uncovers successive arcs of the Sun's surface and changes the phase of the crescent. JULIUS[2] observed that the integrated intensity from the crescent depends not only on the phase but also on the intensity distribution over the Sun's disk. Furthermore, he was able to develop a theory and to secure observations which gave the intensity distribution. In principle, the method should be most satisfactory at the extreme limb, but the analysis of MINNAERT[3] and of EVANS[4] and the later observational results of others[5-8], clearly show that the irregular profile of the Moon's limb and the uncertainty introduced into the solution by the errors of observation make the method unsuitable for finding limb darkening.

ALLEN notes that, in the annular phase of an eclipse, the area of the exposed Sun is a constant; hence the total light variation is a function only of the limb darkening. The availability of annular eclipses for this program have been investigated by SAITO and HATA[9].

SCHWARZSCHILD[10] suggested the use of cinematography with slitless spectrographs. If sufficient image scale is used, a single point on the limb can be studied. Furthermore, results for different wavelengths are obtained. This procedure was employed by LINDBLAD and KRISTENSON[11] at the solar eclipse of July 9, 1945. They used an objective prism before a Zeiss film-camera having a focal length of 64 cm—dispersion 3.8 mm $H\gamma$ to $H\varepsilon$. In the continuous spectrum at λ 4100, LINDBLAD and KRISTENSON found that the intensity at the extreme limb could be expressed by

$$j(h) = c_1\, 10^{-2.28\,h}, \tag{9.1}$$

at λ 4800 by

$$j(h) = c_2\, 10^{-1.95\,h} \tag{9.2}$$

[1] A.K. PIERCE, R.R. MCMATH, L. GOLDBERG and O.C. MOHLER: Astrophys. Journ. **112**, 289 (1950).

[2] W.H. JULIUS: Astrophys. Journ. **23**, 312 (1906); **37**, 225 (1913).

[3] M. MINNAERT: Monthly Notices Roy. Astronom. Soc. London **89**, 197 (1928).

[4] D.S. EVANS: Observatory **67**, 138 (1947).

[5] J.G. FERWERDA, J. UITTERDIJK and A.J. WESSELINK: Bull. Astronom. Netherl. **9**, 81 (1940).

[6] R.A. RICHARDSON and E.O. HULBURT: Astrophys. Journ. **111**, 99 (1950).

[7] K. OSAWA: Ann. Tokyo Astr. Obs., Ser. II **3**, 53 (1951).

[8] F.J. HEYDEN, C.A. BECK, D.J. LOVELL, W.V. FUSSELL, J.H. HANCOCK and E.O. HULBURT: Astrophys. Journ. **118**, 412 (1953).

[9] K. SAITO and S. HATA: Tokyo Astronom. Bull., Ser. II, No. 80, 807 (1956).

[10] K. SCHWARZSCHILD: Nachr. Ges. Wiss. Göttingen **5**, No. 2 (1906).

[11] H. KRISTENSON: Stockh. Ann. **17**, No. 1 (1951); **18**, No. 5 (1955).

and at λ 3660 the coefficient of h is 1.29, where h is in seconds of arc. Unfortunately, the connection to measures on the disk cannot be made accurately.

The method employed by TEN BRUGGENCATE, GOLLNOW, and JAGER[1] does not depend on the total light received, but rather on direct photographs of a partial eclipse. Observations made at the eclipse of April 28, 1949, with the solar tower at Göttingen (Sun's diameter 285 mm) gave the intensity distribution near the limb as well as the intensity profile of the Moon's limb. The latter was used to correct the solar observations for the effect of seeing. Unfortunately, in spite of the obvious advantage of having a well-determined correction for diffusion, the observational times cannot be selected for moments of very best seeing. For example, in the above work the half-width of the diffusion profile was 2''6; in the 1954 eclipse work of BALLARIO and GODOLI[2] the half-width was about 4''; in the work of SCHEFFLER[3] σ equaled 2''5 to 3''2.

It seems likely that the majority of the future work on limb darkening will be done with photoelectric techniques because of their high precision and linearity. It will be necessary to use coronographic-type optics to reduce instrumental scattered light. Furthermore, observations must be made from high altitude balloons or rockets to eliminate skylight and seeing factors. The effects of granulation can perhaps most easily be determined by photographic means, but photoelectric techniques, using scanning apertures as small as 0''1 by 0''1 of arc, may also be feasible if the band pass of the monochromator or filter is not too narrow.

c) Observed solar energy distribution.

10. Observational techniques. A knowledge of $I_\lambda(\tau=0, \mu=1)$, the monochromatic intensity radiated by the photosphere into the normal direction, is required in order to solve Eq. (6.1) for the source function: $B_\lambda^*(\tau) \cdot I_\lambda(0, 1)$. If, as is generally assumed, local thermodynamic equilibrium prevails, then we may write

$$B_\lambda(\tau) \equiv I_\lambda(0, 1) \cdot B_\lambda^*(\tau) = \frac{2hc^2}{\lambda^5} \frac{1}{e^{hc/k\lambda T} - 1} \tag{10.1}$$

and from this expression derive the march of T with τ in the Sun's atmosphere.

Many difficulties stand in the way of an accurate determination of $I_\lambda(0, 1)$, for one must calibrate the spectral transmission of the telescope and spectrometer, evaluate the response of the receiver, and extrapolate the observations to zero air mass by LANGLEY's method. Finally, the spectral purity of the observations must be determined in order to make the correction for finite resolving power in the presence of absorption lines. In the past, two methods have been used to obtain the spectral characteristics of the equipment. The first requires that one build duplicate instruments and use them in series and separately. When used in series $T_{12} = T_1 \cdot T_2$, where T is the transmission at any specific wavelength; thus observations separately and in combination will yield either T_1 or T_2. It is further required in this method that one use a "black body" receiver or a receiver whose spectral response is known. In the second method a standard source, such as a black body, is used to calibrate the over-all response of the system. In this procedure, however, the great difference in brightness, and in spectral distribution, between the standard and the Sun is a major problem.

---

[1] P. TEN BRUGGENCATE, H. GOLLNOW and F.W. JAGER: Z. Astrophys. **27**, 223 (1950).

[2] M.C. BALLARIO and G. GODOLI: Mem. Obs. Astrophys. Arcetri **70**, 93 (1955).

[3] H. SCHEFFLER: Astronom. Nachr. **282**, 49 (1955). — Also E. LAMLA and H. SCHEFFLER: Z. Astrophys. **40**, 93 (1956).

11. Rocket ultraviolet region. Short wavelength regions of the Sun's spectrum, inaccessible to observers even on mountain tops, were first observed in October 1946 with instruments flown in V-2 rockets to elevations of 88 km[1]. Recently[2] first and second order spectra have been obtained with instruments employing concave diffraction gratings of 40 cm radius of curvature and mirror jaw slits to expand the field of view[3]. The intensities throughout the spectrum were obtained from calibrated exposures using as a standard a carbon arc whose spectral intensity distribution was known. The structure of the ultraviolet spectrum is determined by the great density of Fraunhofer lines: few windows exist. The windows at λ 2977 and λ 2913, when observed with a resolution of 0.3 Å, appear to be nearly free of line absorption. In Table 2 are listed, in the second column, the observed energy in microwatts/cm²/Å and in the third, the calculated values for $T = 6000°$ K, based on a solar constant of 2.00 cal cm^{-2} min^{-1}.

Table 2. *Observed and computed radiation through "windows" in the rocket ultra-violet solar spectrum.* MICROWATTS cm^{-2} Å^{-1}

λ	J (obs.)	J (calc. 6000° K)
2977	11.6	9.70
2913	10.4	9.04
2758	4.7	7.48
2700	5.2	6.91

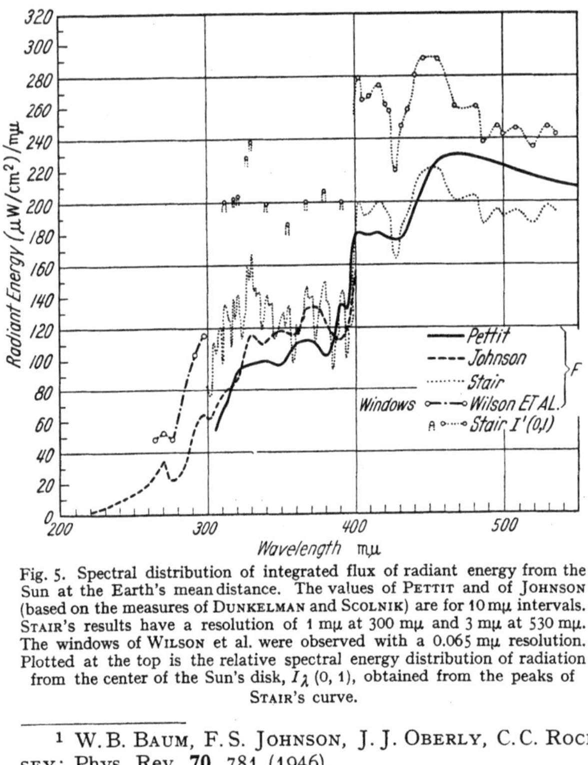

Fig. 5. Spectral distribution of integrated flux of radiant energy from the Sun at the Earth's mean distance. The values of PETTIT and of JOHNSON (based on the measures of DUNKELMAN and SCOLNIK) are for 10 mμ intervals. STAIR's results have a resolution of 1 mμ at 300 mμ and 3 mμ at 530 mμ. The windows of WILSON et al. were observed with a 0.065 mμ resolution. Plotted at the top is the relative spectral energy distribution of radiation from the center of the Sun's disk, I_λ (0, 1), obtained from the peaks of STAIR's curve.

A remarkable spectrum, Fig. 16b, showing 30 conspicuous emission lines in the region 977 to 1817 Å, was obtained February 21, 1955 with an Aerobee rocket, fired in New Mexico, USA[4]. On this photograph the continuous spectrum extends to about 1550 Å with Fraunhofer structure visible to about 1680 Å. It is stated by JOHNSON *et al.* that the emission lines show no limb darkening, while the effect is conspicuous in the region of Fraunhofer absorption.

12. Energy distribution in the ultraviolet. ABBOT[5], PETTIT[6], STAIR[7], STAIR, JOHNSON, and BAGG[8],

[1] W.B. BAUM, F.S. JOHNSON, J.J. OBERLY, C.C. ROCKWOOD, C.V. STRAIN and R. TOUSEY: Phys. Rev. **70**, 781 (1946).

[2] N.L. WILSON, R. TOUSEY, J.D. PURCELL, F.S. JOHNSON and C.E. MORRE: Astrophys. Journ. **119**, 590 (1954).

[3] R. TOUSEY: The Sun, ed. KUIPER, Chap. 9. Chicago, Ill.: Chicago University Press 1953.

[4] F.S. JOHNSON, H.H. MALITSON, J.D. PURCELL and R. TOUSEY: Astronom. J. **60**, 165 (1955). — J. Opt. Soc. Amer. **45**, 900 (1955).

[5] C.G. ABBOT and others: Ann. Astrophys. Obs. Smithsonian Inst. **2** to **7** (1908, 1913, 1922, 1932, 1942).

[6] E. PETTIT: Astrophys. Journ. **75**, 185 (1932); **91**, 159 (1940).

[7] R. STAIR: J. Res. Nat. Bur. Stand. **46**, 353 (1951); **49**, 227 (1952).

[8] R. STAIR, R.G. JOHNSON and T.C. BAGG: J. Res. Nat. Bur. Stand. **53**, 113 (1954).

CANAVAGGIA and CHALONGE[1] and DUNKELMAN and SCOLNIK[2] have determined spectral irradiance in the accessible ultraviolet. Like the still shorter wavelengths the region is dominated by absorption lines and very little of the continuum can be observed. (For example see the list of windows by CANAVAGGIA and CHALONGE[1] and by MICHARD[3].) Fig. 5 illustrates the low dispersion profile of the ultraviolet spectrum. The values of PETTIT, and of DUNKELMAN and SCOLNIK, as quoted by JOHNSON, are in microwatts/cm²/100 Å while the values of STAIR are for a resolving power 3 to 10 times greater. The effect of the absorption lines is clearly revealed. A new investigation of the intensity distribution in the interval $\lambda\lambda$ 3100–3800 Å has been made by THOMPSON at Dunsink using a high dispersion concave grating spectrograph [6].

Some observers have measured the integrated radiation from the whole disk, F_λ, while others have measured the intensity of the radiation from the center of the disk $I_\lambda(0, 0)$. The relationship between the flux and the radiation in the normal direction at the center of the disk is simply

$$\frac{I_\lambda(0,0)}{F_\lambda} = \frac{\int r\,dr}{\int \frac{I_\lambda(0,\vartheta)}{I_\lambda(0,0)}\,r\,dr}, \tag{12.1}$$

an expression first used by MINNAERT in his reduction of ABBOT'S results. The ultraviolet peaks of STAIR'S curve, and his curve redward of 4000 Å have been computed in accordance with Eqs. (12.1), for which values are listed in Table 3.

Table 3. *Ratio of the intensity in the normal direction to the mean intensity.*

λ Å	I_0/F	λ Å	I_0/F	λ Å	I_0/F	λ Å	I_0/F
3200	1.455	5500	1.245	11000	1.125	21000	1.070
3400	1.414	6000	1.225	12000	1.118	22000	1.068
3600	1.383	6500	1.260	13000	1.111	23000	1.067
3800	1.395	7000	1.188	14000	1.105	24000	1.066
4000	1.395	7500	1.172	15000	1.095	25000	1.065
4200	1.360	8000	1.160	16000	1.085	μ	
4400	1.330	8500	1.150	17000	1.079	3.5	1.045
4600	1.307	9000	1.145	18000	1.076	8.3	1.024
4800	1.289	9500	1.138	19000	1.074	10.2	1.020
5000	1.276	10000	1.124	20000	1.071		

13. Energy distribution $\lambda\lambda$ 4000–25000 Å. Absorption lines crowd the spectrum between $\lambda\lambda$ 4000 and 5000 Å, but to the redward of the Mg, *b* group at 5180 Å their density markedly decreases. By 6000 Å many intervals are relatively free of line absorption and here the continuum is well established even with low dispersion spectrometers. In the region $\lambda\lambda$ 4204–6678 Å a specific attempt was made by PLASKETT[4] to measure the monochromatic intensity at selected windows chosen to be as free as possible of Fraunhofer lines. Since his spectral resolution was high, the true continuum was closely defined. MULDERS[5], who applied corrections for Fraunhofer line absorption to ABBOT'S measures, obtained good agreement with PLASKETT'S curve except in the violet and ultraviolet where the corrections became large.

[1] R. CANAVAGGIA and CHALONGE: Ann. d'Astrophys. **9**, 143 (1946). — R. CANAVAGGIA, D. CHALONGE, M. EGGER-MOREAU and H. OZIOL: Ann. d'Astrophys. **13**, 355 (1950).

[2] L. DUNKELMAN and R. SCOLNIK: J. Opt. Soc. Amer. **42**, 876 (1952). — Results quoted by F.S. JOHNSON: J. Meteorology **11**, 431 (1954).

[3] R. MICHARD: Bull. Astronom. Netherl. **11**, 227 (1950).

[4] H.H. PLASKETT: Publ. Dom. Astrophys. Obs. **2**, 242 (1923).

[5] G.F.W. MULDERS: Z. Astrophys. **11**, 132 (1935).

CHALONGE and his coworkers approached the problem by assuming that in the two intervals $\lambda\lambda$ 3700—5000 and $\lambda\lambda$ 3200—3700, divided by the Balmer limit, separate blackbody color temperatures could be assigned. A plot of $\log I$ versus $1/\lambda$ by CANAVAGGIA *et al.*[1] (the method of gradients) yielded two straight lines defined by the windows, the highest maxima of their observations, and gave temperatures T_1 (λ 3700⁺) = 7130° and T_2 (λ 3700⁻) = 5900° with a Balmer discontinuity I_{3700^+}/I_{3700^-} = 0.125. MICHARD[2] corrected PETTIT's measures for the influence of Fraunhofer lines and found satisfactory agreement with CHALONGE's and CANAVAGGIA's spectrophotometric results. MICHARD's values are: color temperatures $T_1 = 7270°$, $T_2 = 5550°$, and a Balmer jump $D = 0.14$.

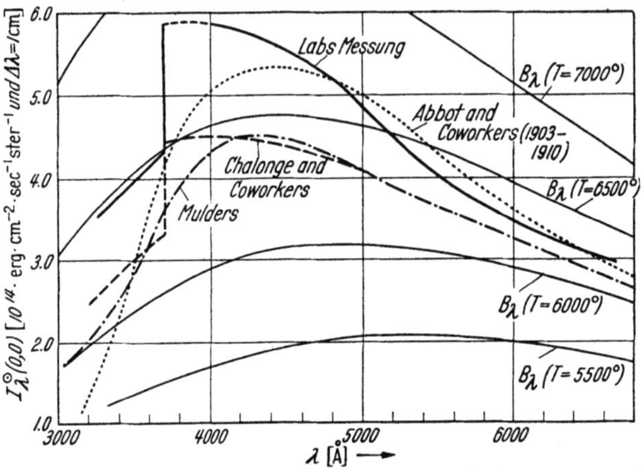

Fig. 6. LABS' measures of the continuum, obtained at the Pic-du-Midi, compared with the results of ABBOT, CHALONGE and MULDERS (after LABS).

Recently, at the Pic-du-Midi, LABS[3] has made a new series of measurements of the profile of the continuous spectrum from λ 3300 to λ 7000 Å and its absolute intensity at λ 5263. On either side of the Balmer limit Labs finds $T_1 = 7540°$ K, $T_2 = 6225°$ K and $D = 0.13$. These results, obtained with a glass/quartz prism double monochromator, rest on the known energy distribution and calibration of a standard carbon arc (temperature: 3949° ± 10° K) and on atmospheric extinction measures obtained independently with interference filters. A comparison of LABS' intensity values with the measures of other observers is shown in Fig. 6. The probable error in LABS' solar color temperature values is given as ± 250°; CHALONGE and CANAVAGGIA quote ± 200° for their work.

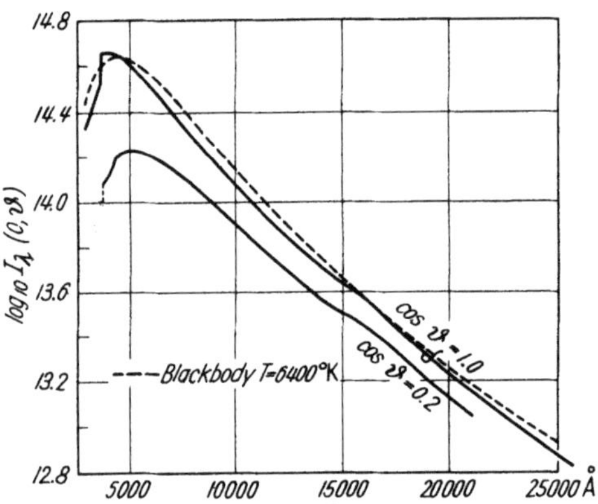

Fig. 7. Intensity of the continuum spectrum for the Sun's center and cos ϑ = 0.2.

MAKAROVA[4] has used the solar instalation at Kutchino to obtain the energy distribution in the continious spectrum. Measurements were obtained in absolute units by photographic photometry in the intervall $\lambda\lambda$ 3700—8000 Å and with a dispersion of 1.6 Å/mm. The resulting distribution, $I_\lambda(0,0)$, is very similar to

[1] R. CANAVAGGIA and D. CHALONGE: Ann. d'Astrophys. **9**, 143 (1946). — R. CANAVAGGIA, D. CHALONGE, M. EGGER-MOREAU and H. OZIOL-PELTEY: Ann. d'Astrophys. **13**, 355 (1950).

[2] R. MICHARD: Bull. Astronom. Netherl. **11**, 227 (1950).

[3] D. LABS: Sitzgsber. Heidelberg. Akad. Wiss. Symposium über Probleme der Spektralphotometrie. Heidelberg: Springer 1957.

[4] E. A. MAKAROVA: Astronomicheskij Zhurnal **34**, 539 (1957).

Table 4. *Solar energy distribution.*

Wavelength μ is in microns, the mean zero air mass spectral irradiance H_μ is in watts · cm^{-2} micron^{-1}, and P_μ is the percentage of the solar constant associated with wavelengths shorter than wavelength μ. These data are from F. S. Johnson: J. Meteorology **11**, 431 (1954). Column 4 is the decadic logarithm of the intensity of the *continuous spectrum* in the normal direction, after M. Minnaert [5] but modified beyond $\mu = 1.0$ (see text). Unit: erg · cm^{-2} sec^{-1} per $d\lambda = 1$ and solid angle $d\omega = 1.0$.

μ	H_μ	P_μ	$\log I_\lambda(0,1)$	μ	H_μ	P_μ	$\log I_\lambda(0,1)$	μ	H_μ	P_μ	$\log I_\lambda(0,1)$	μ	H_μ	P_μ	$\log I_\lambda(0,1)$	μ	H_μ	P_μ	$\log I_\lambda(0,1)$
0.22	0.0030	0.02		0.365	0.129	5.89		0.505	0.197	24.2		0.70	0.144	48.8	14.408	2.6	0.00445	96.90	
0.225	0.0042	0.03		0.37⁻	0.133	6.36	14.525	0.51	0.196	24.9		0.71	0.141	49.8		2.7	0.00390	97.21	
0.23	0.0052	0.05		0.37	0.133	6.36	14.655	0.515	0.189	25.6		0.72	0.137	50.8		2.8	0.00343	97.47	
0.235	0.0054	0.07		0.375	0.132	6.84		0.52	0.187	26.3	14.600	0.73	0.134	51.8		2.9	0.00303	97.72	
0.24	0.0058	0.09		0.38	0.123	7.29	14.659	0.525	0.192	26.9		0.74	0.130	52.7		3.0	0.00268	97.90	
0.245	0.0064	0.11		0.285	0.115	7.72		0.53	0.195	27.6		0.75	0.127	53.7	14.350	3.1	0.00230	98.08	
0.25	0.0064	0.13		0.39	0.112	8.13		0.535	0.197	28.3		0.80	0.1127	57.9	14.290	3.2	0.00214	98.24	
0.255	0.010	0.16		0.395	0.120	8.54		0.54	0.198	29.0	14.580	0.85	0.1003	61.7	14.236	3.3	0.00191	98.39	
0.26	0.013	0.20		0.40	0.154	9.03	14.663	0.545	0.198	29.8		0.90	0.0895	65.1	14.182	3.4	0.00171	98.52	
0.265	0.020	0.27		0.405	0.188	9.65		0.55	0.195	30.5		0.95	0.0803	68.1	14.132	3.5	0.00153	98.63	
0.27	0.025	0.34		0.41	0.194	10.3		0.555	0.192	31.2		1.00	0.0725	70.9	14.082	3.6	0.00139	98.74	
0.275	0.022	0.43		0.415	0.192	11.0		0.56	0.190	31.8	14.560	1.1	0.0606	75.7	13.982	3.7	0.00125	98.83	
0.28	0.024	0.51		0.42	0.192	11.7	14.662	0.565	0.189	32.5		1.2	0.0501	79.6	13.888	3.8	0.00114	98.91	
0.285	0.034	0.62		0.425	0.189	12.4		0.57	0.187	33.2		1.3	0.0406	82.9	13.799	3.9	0.00103	98.99	
0.29	0.052	0.77		0.43	0.178	13.0		0.575	0.187	33.9		1.4	0.0328	85.5	13.713	4.0	0.00095	99.05	
0.295	0.063	0.98		0.435	0.182	13.7		0.58	0.187	34.5	14.540	1.45			13.675	4.1	0.00087	99.13	
0.30	0.061	1.23		0.44	0.203	14.4	14.656	0.585	0.185	35.2		1.5	0.0267	87.6	13.640	4.2	0.00080	99.18	
0.305	0.067	1.43		0.445	0.215	15.1		0.59	0.184	35.9		1.55			13.606	4.3	0.00073	99.23	
0.31	0.076	1.69		0.45	0.220	15.9		0.595	0.183	36.5		1.6	0.0220	89.4	13.572	4.4	0.00067	99.29	
0.315	0.082	1.97		0.455	0.219	16.7		0.60	0.181	37.2	14.520	1.65			13.532	4.5	0.00061	99.33	
0.32	0.085	2.26	14.386	0.46	0.216	17.5	14.646	0.61	0.177	38.4		1.7	0.0182	90.83	13.492	4.6	0.00056	99.38	
0.325	0.102	2.60		0.465	0.215	18.2		0.62	0.174	39.7	14.500	1.8	0.0152	92.03	13.398	4.7	0.00051	99.41	
0.33	0.115	3.02		0.47	0.217	19.0		0.63	0.170	40.9		1.9	0.01274	93.02	13.310	4.8	0.00048	99.45	
0.335	0.111	3.40		0.475	0.220	19.8		0.64	0.166	42.1	14.477	2.0	0.01079	93.87	13.228	4.9	0.00044	99.48	
0.34	0.111	3.80	14.450	0.48	0.216	20.6	14.636	0.65	0.162	43.3		2.1	0.00917	94.58	13.151	5.0	0.00042	99.51	
0.345	0.117	4.21		0.485	0.203	21.3		0.66	0.159	44.5	14.455	2.2	0.00785	95.20	13.075	6.0	0.00021	99.74	
0.35	0.118	4.63		0.49	0.199	22.0		0.67	0.155	45.6		2.3	0.00676	95.71	13.004	7.0	0.00012	99.86	
0.355	0.116	5.04		0.495	0.204	22.8		0.68	0.151	46.7	14.408	2.4	0.00585	96.18	12.932				
0.36	0.116	5.47	14.502	0.50	0.198	23.5	14.619	0.69	0.148	47.8		2.5	0.00509	96.57	12.865				

LABS' curve (Fig. 6) but differs in that there is a dip at λ 4600 Å and the peak between $\lambda\lambda$ 6800—7400 Å is more pronounced. The absolute values of energy lie between those found by Abbot 1902—10 and 1920—22.

For a definitive curve MINNAERT [5] adopts 5780° K for the wavelengths short of 3700 Å and 7150° K at the longer wavelengths. In Table 4 and Fig. 7, we give his curve for the continuum radiation in the normal direction but modified in the infrared to fit the results of PEYTURAUX[1] and PIERCE[2]. It is difficult to estimate the uncertainty in these results; an error of 0.025 in Log $I_\lambda(0, 1)$ or 6% in $I_\lambda(0, 1)$ seems quite possible.

14. Far infrared. Beyond 2.5 μ very little work has been done. The infrared radiation in the spectral interval 0.8 to 3.5 μ has recently been measured by JAROSLAVCEV[3]. ADEL[4] found that his observations of the energy distribution profile, 8 to 14 μ, could be represented by temperatures from 6000 to 7000° K. However, as he recognized, for these temperatures and wavelengths, the shape of the curve is quite insensitive to the temperature. Rather, the problem must be attacked by determining the true intensity ratio between the Sun and a standard source at each wavelength. Until this is done, it seems likely that the best estimate of the course of the curve can be made from astrophysical theory. Because of the great absorption of the H^- ions in the far infrared, only the outermost boundary layers radiate to space. The boundary temperature is variously disputed to be between 3800 to 4500° K. If we adopt 4000° K, by the Rayleigh-Jeans law we expect the absolute intensity to be $\frac{4}{6}$ of the value that we would have extrapolated from shorter wavelengths using $T=6000°$.

A remarkable extension of the photospheric solar spectrum to 2.5 mm has been made by SINTON[5]. His detecting equipment consisted of a parabolic searchlight mirror used to focus the energy passing through a black paper filter onto a sensitive Golay infrared detector. The transparency of the 0.9 to 3 mm window is far from complete and as yet no absolute values for the spectral energy distribution are available.

d) Observational models of the photosphere.

15. Analysis of the limb-darkening observations. Limb darkening and absolute energy distribution, which together define the mean radiance in different directions, per square centimeter of the photosphere, are determined by the physical conditions throughout the atmosphere. To discover these conditions, one may fit the observed limb darkening and energy distribution to predicted values obtained from a sequence of theoretical models with assumed quantities, or one may attempt to deduce the parameters of an empirical model directly from the observations. The latter method has its limitations in the outermost layers, as we shall see in later paragraphs, but it best fits the true Sun. Polarization[6-9] and refraction[10] effects are very small; hence they have not been used in the derivation a model, though their observation may provide a valuable check.

[1] R. PEYTURAUX: Ann. d'Astrophys. **15**, 302 (1952).
[2] A.K. PIERCE: Astrophys. Journ. **119**, 312 (1954).
[3] I.N. JAROSLAVCEV: Isv. Akad. Nauk. USSR., Ser. Geofiz. No. 8, 984 (1956).
[4] A. ADEL: Astrophys. Journ. **89**, 1 (1939).
[5] W.M. SINTON: Phys. Rev. **86**, 424 (1952). — J. Opt. Soc. Amer. **45**, 975 (1955). — The energy has also been detected by THEISSING, CAPLAN and STELLE: J. Opt. Soc. Amer. **45**, 405 (1955). — H.H. THEISSING and P.W. CAPLAN: Inst. Radio Eng. Trans. **4**, 582 (1956)
[6] B. LYOT: C. R. Acad. Sci., Paris **226**, 25, 137 (1948).
[7] A.D. CODE: Astrophys. Journ. **112**, 22 (1950).
[8] J.C. PECKER: Ann. d'Astrophys. **13**, 294 and 319 (1950).
[9] H.H. VOIGT: Z. Astrophys. **28**, 176 (1951).
[10] P. PROISY: Ann. d'Astrophys. **12**, 123 (1949).

Solar granulation is positive evidence of inhomogeneities in temperature and density in the photosphere. After a very uncertain correction for seeing and scattered light is applied, the observed brightness difference between granule and intergranular area is about 25% [*1*], which by PLANCK's law corresponds to $\Delta T \approx 300°$.

The solution of Eq. (6.1) for the source function involves a minimum of postulates concerning the physical conditions. But the strict application of the integral equation to the solar problem has been questioned first by REDMAN[1] and later by EVANS[2], for it supposes that the emitting layers are plane parallel and homogeneous in extent. Because of turbulence, REDMAN in his analysis of the profiles of lines at the extreme limb considered that the roughness of the Sun's surface would reduce an apparent $\cos\vartheta = 0.05$ to an effective value of 0.20, and an apparent $\cos\vartheta = 0.30$ to an effective $\cos\vartheta = 0.39$. These figures are based on a model of the photosphere with a highly idealized wavy surface—a model for which the limitations were fully recognized by REDMAN. In a similar manner, PECKER[3] introduced rugosities to explain the center-to-limb variation of CH lines.

The validity of the application of Eq. (6.1) to the solar atmosphere depends in great measure on whether the photosphere is jet-like or wave-like in structure and also on the optical thickness of jet or wave. MOHLER[4] concludes, from his studies of the small chromospheric structures observed at the time of eclipse, that there is an approximate equality between the number of jets and the number of granules. Doppler shifts in Fraunhofer lines indicate a velocity range of 0 to 1.5 km/sec for granules and a r.m.s. velocity of about 0.4 km/sec. When these figures are combined with the granule life-times of 2 to 12 min (and a mean of 7 min) as observed by MACRIS[5], the result is a radial movement of 170 km, which is a sizeable fraction of the thickness of the photosphere. The above arguments in which we try to equate jet to granule are open to slight question for there is good evidence that the line cores, from which the Doppler shifts are determined, are formed in the low chromosphere, and not in the zone of granulation (see Sect. 36). Nevertheless, it seems reasonable to postulate that the jet structure of the chromosphere extends down into, if not through, the photosphere. *If* this is a close approximation, and if the jets are not optically thick in the continuum—though they may be in the absorption lines—then the equation for the emergent intensity I into direction $\mu = \cos\vartheta$ is

$$\overline{I_\lambda^*(0,\mu)} = \int_0^\infty \overline{B_\lambda^*(\tau_\lambda)}\, e^{-\frac{\tau_\lambda}{\mu}} \frac{d\tau_\lambda}{\mu}, \tag{15.1}$$

where $\overline{B^*(\tau_\lambda)}$, the monochromatic source function, is assumed to be a suitable mean of the elements of different physical conditions.

Eq. (15.1), which is derived for plane parallel layers, was extended by UNSÖLD to the spherical Sun. By means of the cosine law, and for h small (Fig. 8a), one obtains

$$l = h \sec\vartheta \left(1 - \frac{h}{2R} \tan^2\vartheta + \cdots\right), \tag{15.2}$$

[1] R. O. REDMAN: Monthly Notices Roy. Astronom. Soc. London **103**, 173 (1943).
[2] D. S. EVANS: Monthly Notices Roy. Astronom. Soc. London **107**, 433 (1947).
[3] J. C. PECKER: Ann. d'Astrophys. **12**, 9 (1949); **14**, 152 (1951).
[4] O. C. MOHLER: Monthly Notices Roy. Astronom. Soc. London **111**, 630 (1951).
[5] C. MACRIS: Ann. d'Astrophys. **16**, 19 (1953). — Observatory **75**, 122 (1955).

where h/R is of the order of 0.0005 for the solar photosphere. An error greater than 1% in l may arise when $\cos\vartheta < 0.15$. High speed machine calculation makes feasible the integration of Eq. (15.1) for the general case including curvature. PAGEL[1] has given the details for computing:

$$\left.\begin{aligned} I_\lambda(x) &= R\int\limits_{-\infty}^{+\infty} \eta\,\varrho\,k_\lambda(h)\,B_\lambda(h)\,\mathrm{e}^{-t_\lambda(y)}\,d y,\\ &= R\int\limits_{+y_0}^{-y_0} \eta\,\varrho\,k_\lambda(h)\,B_\lambda(h)\,\mathrm{e}^{-t_\lambda(y)}\,(-d y), \end{aligned}\right\} \tag{15.3}$$

where the geometrical factors are shown in Fig. 8b. In the above equation η is the ratio of equilibrium opacity to actual opacity, $k_0^{(t)}/k_0$. The arc of Fig. 8b represents the depth τ_0 above which emission is assumed to be zero and for which Pagel adopts a value of 10^{-5}. Auxiliary quantities are:

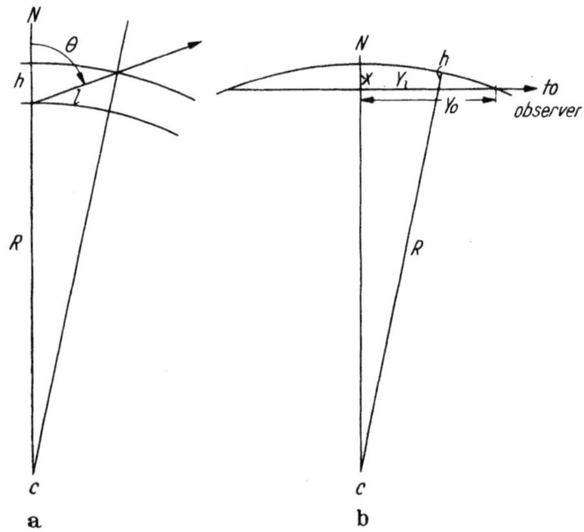

Fig. 8a and b. Geometry of the trajectory of a ray in the Sun's atmosphere.

$$t_\lambda(y_1) = R\int\limits_{y_0}^{y_1} \varrho\,k_\lambda(h)\,(-d y), \tag{15.4}$$

$$y_0 = (2x - x^2)^{\frac{1}{2}}, \tag{15.5}$$

and

$$(1-h)^2 = y^2 + (1-x)^2. \tag{15.6}$$

Having selected an x and thus y_0 one easily computes h. Then, for known tabular values of T and ϱk_λ as a function of h, one can perform the quadratures expressed by Eqs. (15.4) and (15.3).

The solutions of Eq. (15.1) have an international character, and many elegant results have been obtained. We state first and then establish some of the simple and approximate relations derived by BARBIER[2] from EDDINGTON's work and further developed by UNSÖLD[3] and KOURGANOFF [3].

The Eddington-Barbier relations are, for the intensity:

$$I_\lambda(0,\mu) \approx B_\lambda(\mu), \tag{15.7}$$

for the flux:

$$F_\lambda(0) \approx B_\lambda(\tfrac{2}{3}); \tag{15.8}$$

and, in particular, for the intensity at the center:

$$I_\lambda(0,1) \approx B_\lambda(1), \tag{15.9}$$

and at the limb:

$$I_\lambda(0,0) = B_\lambda(0). \tag{15.10}$$

---

[1] B. E. J. PAGEL: Monthly Notices Roy. Astronom. Soc. London 116, 608 (1956).
[2] D. BARBIER: Ann. d'Astrophys. 6, 113 (1943).
[3] A. UNSÖLD: Ann. d. Phys. 3, 124 (1948).

Again, we remind the reader of the limitation to plane parallel layers. BARBIER and UNSÖLD developed B_λ in a Taylor series about the point τ_λ^* (to be specified later):

$$\left.\begin{aligned} I_\lambda(0,\vartheta) = \int \Big[B_\lambda(\tau_\lambda^*) + (\tau_\lambda - \tau_\lambda^*)\left(\frac{dB_\lambda}{d\tau_\lambda}\right)_{\tau_\lambda^*} + \\ + \frac{(\tau_\lambda-\tau_\lambda^*)^2}{2}\left(\frac{d^2B_\lambda}{d\tau_\lambda^2}\right)_{\tau_\lambda^*} + \cdots\Big] e^{\tau_\lambda \sec\vartheta} \sec\vartheta\, d\tau, \end{aligned}\right\} \tag{15.11}$$

or

$$\left.\begin{aligned} I_\lambda(0,\vartheta) = B_\lambda(\tau_\lambda^*) + [\cos\vartheta - \tau_\lambda^*]\left(\frac{dB}{d\tau_\lambda}\right)_{\tau_\lambda^*} + \\ + \frac{1}{2}[\cos^2\vartheta + (\cos\vartheta - \tau_\lambda^*)^2]\left(\frac{d^2B_\lambda}{d\tau_\lambda^*}\right)_{\tau_\lambda^*} + \cdots. \end{aligned}\right\} \tag{15.12}$$

On the assumption that $\tau_\lambda^* = \mu = \cos\vartheta$, BARBIER noted that the second term of (15.12) becomes zero, and that the third has a minimum at this point, from which Eq. (15.7) follows. The Eddington-Barbier relations are rigorously true if the source function is linear; that is, if

$$B(\tau) = a + b\tau, \tag{15.13}$$

then

$$I(\mu) = B(\tau^* = \mu). \tag{15.14}$$

That they are approximately true in the general case has been demonstrated by KOURGANOFF [3]. Unfortunately this wide generality makes the solution of the integral equation somewhat unstable, particularly since the full range of $I(\mu)$ is not observed.

It is customary to graduate limb darkening observations, and extrapolate to the limb by a few terms of a series expansion or by other empirical formulae. LUNDBLAD[1] chose

$$\frac{B_\lambda(\tau_\lambda)}{I_\lambda(0,1)} = a_\lambda + b_\lambda\tau_\lambda + c_\lambda\tau_\lambda^2 + \cdots, \tag{15.15}$$

which, by substitution in Eq. (15.1), gives the following expression for the limb darkening:

$$\frac{I_\lambda(0,\mu)}{I_\lambda(0,1)} = a_\lambda + b_\lambda\mu + 2c_\lambda\mu^2 + \cdots; \qquad \cos\vartheta = \mu. \tag{15.16}$$

SYKES[2] pointed out that Eq. (15.15) is an expansion about $\cos\vartheta = \mu = 0$, in practice an unobservable point. He suggested substituting $\mu = e^\xi$, $\tau = e^\eta$, in Eq. (15.1) which then becomes

$$I(\xi) = \int_{-\infty}^{+\infty} \exp[-(\xi-\eta)\, e^{-(\xi-\eta)}]\, B(\eta)\, d\eta. \tag{15.17}$$

Now this is the form

$$g(x) = \int_{-\infty}^{+\infty} h(x-y)\, f(y)\, dy, \tag{15.18}$$

for which many solutions are known. VAN DE HULST'S[3] method of power series gives an acceptable solution in this case. If h has finite moments $M_n \equiv \int_{-\infty}^{+\infty} z^n h(z)\, dz$,

---

[1] R. LUNDBLAD: Astrophys. Journ. **58**, 113 (1923).

[2] J. B. SYKES: Monthly Notices Roy. Astronom. Soc. London **113**, 198 (1953).

[3] H. C. VAN DE HULST: Bull. Astronom. Netherl. **10**, 75 (1946).

for all n, and if $g(x)$ can be expanded in series $\sum c_n x^n$ then the solution of (15.18) is

$$f(y) = \sum_n c_n P_n(y), \tag{15.19}$$

where

$$P_u(y) = \sum_m {}^nC_m \lambda_{n-m} y^m. \tag{15.20}$$

The coefficients of λ are given by

$$\sum_k (-1)^k \lambda_{r-k} {}^rC_k M_k = \left.\begin{matrix} 1 \text{ if } r = 0, \\ 0 \text{ if } r \neq 0. \end{matrix}\right\} \tag{15.21}$$

The polynomials P_n and the values of λ may be found in SYKES' paper. Thus the limb darkening is represented by

$$I(\xi) = 1 + a\xi + b\xi^2 + c\xi^3 + \cdots, \tag{15.22}$$

in which the coefficients of ξ, over the range of observations, form a decreasing set of values and are effectively the deriviatives of $I_\lambda(\xi)$ at $\xi = 0$ or $\mu = 1.0$.

KOURGANOFF[1] selected

$$\frac{B_\lambda(\tau_\lambda)}{I_\lambda(0,1)} = a_\lambda + b_\lambda \tau_\lambda + c_\lambda E_2(\tau_\lambda) + \cdots, \tag{15.23}$$

where E_2 is the integral exponential function. By integration one gets

$$\frac{I_\lambda(\mu)}{I_\lambda(0,1)} = a_\lambda + b_\lambda \mu + c_\lambda \left[1 - \mu \ln\left(1 + \frac{1}{\mu}\right)\right] + \cdots. \tag{15.24}$$

At the present stage of accuracy of limb darkening observations there is no practical difference between expressions (15.16), (15.22) and (15.24); three terms fit the values of limb darkening within the errors of observations. KOURGANOFF'S representation, however, is derived from a source function that was found by him to give an excellent fit for integrated radiation. Later work by PIERCE and ALLER[2] showed that it is an equally good representation of limb darkening for monochromatic radiation. Furthermore, (15.23) gives the correct asymptotic behavior at $\tau \to 0$ and $\tau \to \infty$ (for an ideal Sun, i.e. no convection, etc.).

The flux at the surface of the Sun is given by

$$F = \int_0^{\pi/2} 2\pi I(0,\vartheta) \sin\vartheta \cos\vartheta \, d\vartheta. \tag{15.25}$$

Using the limb-darkening formula (15.24) we find that the flux integral has the value

$$\frac{F_\lambda}{I_\lambda(0,1)} = 2\pi \left[\frac{a_\lambda + c_\lambda}{2} + \frac{b_\lambda}{3} - c_\lambda\left(\frac{2}{3}\ln 2 - \frac{1}{6}\right)\right] + \cdots. \tag{15.26}$$

16. Temperature distribution versus depth. From the Eddington-Barbier relations we see that the source function $B_\lambda^*(\tau)$ is well established only in the interval $\tau = 0.1$ to 1.0, and it is the empirical form of the limb-darkening equations [Eq. (15.15) etc.] that determines the extrapolation of $B_\lambda^*(\tau)$ to the surface and to deeper layers. If $I_\lambda(0,1)$ is known and if local thermodynamic equilibrium prevails, then Eqs. (10.1) and (15.15) (or its equivalent) relate T and τ_λ. Typical curves of temperature T versus optical depth $\tau_{\lambda 5000\,\text{Å}}$ are given in Fig. 9. The instability of the solutions is clearly shown.

[1] V. KOURGANOFF: C. R. Acad. Sci., Paris **228**, 201 (1949).
[2] A.K. PIERCE and L.H. ALLER: Astrophys. Journ. **114**, 145 (1951).

From observations made in different wavelengths a (T, τ, λ)-surface may be constructed. Then, for a series of cuts corresponding to fixed temperatures, T_i, the $(\tau, \lambda)_{T_i}$ curves reveal the wavelength dependence of the integrated continuous absorption coefficient,

$$(\tau_\lambda)_T = \int_x^0 \varrho\, k_\lambda\, dx \tag{16.1}$$

as shown in Fig. 10. The wavelength dependence of the monochromatic absorption coefficient is found by differentiation of Eq. (16.1). This is the subject of the next section.

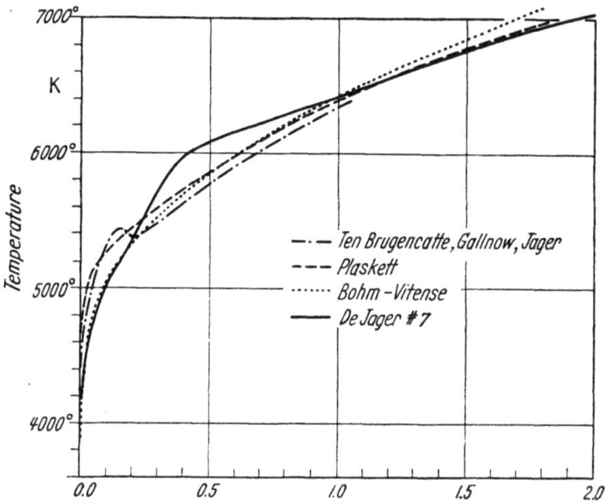

Fig. 9. Temperature distribution versus optical depth τ by different observers.

17. The continuous absorption coefficient. For a historical background to the work on the coefficient of continuous absorption the reader is referred to the discussion by STRÖMGREN [7]; the early theoretical position is covered by CHANDRASEKHAR'S paper[1]. The investigations of CHALONGE and KOURGANOFF[2], CHANDRASEKHAR and MUNCH[3], and BARBIER[4] have amplified and clearly proven WILDT'S[5] suggestion that in the wavelength interval $\lambda\lambda$ 4000—20000 Å, the negative hydrogen ion is the principal absorber of quanta in the solar photosphere. The solar atmosphere is composed of 1 metal atom to about 10^4 hydrogen atoms. At solar temperatures, in a first approximation, each metal atom releases one electron which may attach itself to one of the abundant hydrogen atoms to form a negative hydrogen ion, H^-. Thus the concentration of negative hydrogen ions is proportional to the electron pressure; in thermodynamic equilibrium their concentration is given explicitly by the Saha equation which tells us that for a typical photospheric layer the ratio of the number of negative hydrogen ions to hydrogen atoms is about $N_{H^-}/N_H \approx 10^{-7}$.

The most extensive calculations of the bound-to-free and free-to-free absorption coefficient, $\alpha(H^-)$, are those of CHANDRASEKHAR[6] and CHANDRASEKHAR and BREEN[7]. In more recent calculations by GELTMAN[8] of the bound-to-free continuous absorption coefficient, use was made of certain necessary conditions, other than energy minimization. These were: Equality of matrix elements of length, velocity, and acceleration, and the $\langle r^2\rangle_{Av}$ and the f-sum rules[9]. GELTMAN'S absorption coefficient is very similar in shape to that of CHANDRASEKHAR'S but

[1] S. CHANDRASEKHAR: Rev. Mod. Phys. **16**, 301 (1944).
[2] D. CHALONGE and V. KOURGANOFF: Ann. d'Astrophys. **9**, 69 (1946).
[3] S. CHANDRASEKHAR and G. MUNCH: Astrophys. Journ. **104**, 446 (1946).
[4] D. BARBIER: Ann. d'Astrophys. **9**, 173 (1946).
[5] R. WILDT: Astrophys. Journ. **89**, 295 (1939); **90**, 611 (1939).
[6] S. CHANDRASEKHAR: Astrophys. Journ. **102**, 223, 395 (1945).
[7] S. CHANDRASEKHAR and F.H. BREEN: Astrophys. Journ. **104**, 430 (1946).
[8] S. GELTMAN: Phys. Rev. **104**, 346 (1956).
[9] S. CHANDRASEKHAR and M.K. KROGDAHL: Astrophys. Journ. **98**, 205 (1943).

is 10% lower at the peak; below λ 4000 Å it exceeds CHANDRASEKHAR'S curve by approximately 10%. The computed integrated cross-section, corresponding to electron photodetachment currents, is in agreement with the laboratory results of BRANSCOMB and SMITH[1]. In their experiment electrons were liberated at the intersection of a beam of H^- ions and an intense beam of radiation from a tungsten filament. By modulation of the light beam, the collector picked up a small alternating number of electrons which was a measure of the probability of photodetachment for H^- irradiated by the incident spectral intensity distribution.

A second source of absorption is recognized in the strength of the hydrogen spectrum and the observation of a Balmer discontinuity. These indicate that there is a slight contribution by neutral hydrogen to the continuous absorption coefficient. The Pfund, Brackett, Paschen and Balmer continua, which start at 22793, 14578, 8204 and 3647 Å respectively, all absorb light short of these wavelengths in amounts proportional to $n(\lambda/\lambda_0)^3$, where n is the principal quantum number of the state of the atom. Because of the large excitation potential of these levels — 10.19 volts for the Balmer, 12.09 for the Paschen continuum, etc., up to 13.59 for ionization—there are very few atoms in the solar atmosphere capable of absorbing in these continua. For example, for the ratio of the number of atoms in the third quantum level capable of absorbing the Paschen continuum to the number of atoms in the ground state [4], the Boltzmann relation gives $N_{H'}/N_H = 6 \times 10^{-10}$, for $T = 6000°$. However, this number rises very quickly in the deeper layers and the H^- ion gradually loses out.

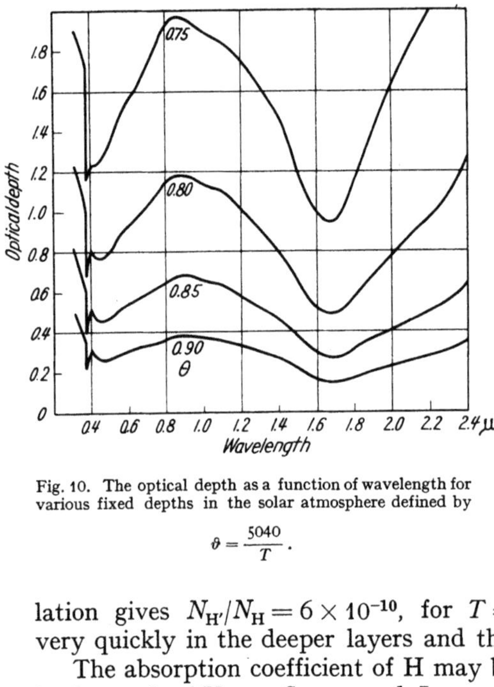

Fig. 10. The optical depth as a function of wavelength for various fixed depths in the solar atmosphere defined by

$$\vartheta = \frac{5040}{T}.$$

The absorption coefficient of H may be found in the tables of UNSÖLD [1] or in the work of UENO, SAITO, and JUGAKU[2].

If the variation of k_λ with λ is nearly the same throughout the surface layers, then LUNDBLAD[3] shows that, in a first approximation,

$$\tau_\lambda = \int_x^0 k_\lambda \varrho(x)\, dx \tag{17.1}$$

becomes

$$\frac{\tau_{\lambda_i}}{\tau_{\lambda_0}} = \frac{k_{\lambda_i}}{k_{\lambda_0}}. \tag{17.2}$$

The good agreement between the observed $\tau_\lambda(T)$ curve shown in Fig. 10 and the computed absorption coefficient for the H^- ion invites still closer inspection.

A modification of Eq. (17.1) is:

$$\frac{d\tau_\lambda}{dT} = k_\lambda \varrho(x) \frac{dx}{dT}, \tag{17.3}$$

[1] L. M. BRANSCOMB and S. J. SMITH: Phys. Rev. **98**, 1028 (1955).
[2] S. UENO, S. SAITO and J. JUGAKU: Contr. Kyoto Inst. Astrophys. No. 43 (1954).
[3] R. LUNDBLAD: Astrophys. Journ. **58**, 113 (1923).

in which $\varrho(x)$ and dx/dT are constant for a given temperature, T, corresponding to a fixed depth in the atmosphere. The left-hand member, $(d\tau/dT)_\lambda$, is derivable from observation, while k_λ of the right member may be obtained from quantum mechanical calculation, or from laboratory measurements. Hence one should have an excellent check on all phases of the problem. For the source function we select

$$\frac{B_\lambda(\tau_\lambda)}{I_\lambda(0,1)} = \varphi_\lambda(\tau_\lambda) = a_\lambda + b_\lambda \tau_\lambda + c_\lambda E_2(\tau_\lambda), \tag{17.4}$$

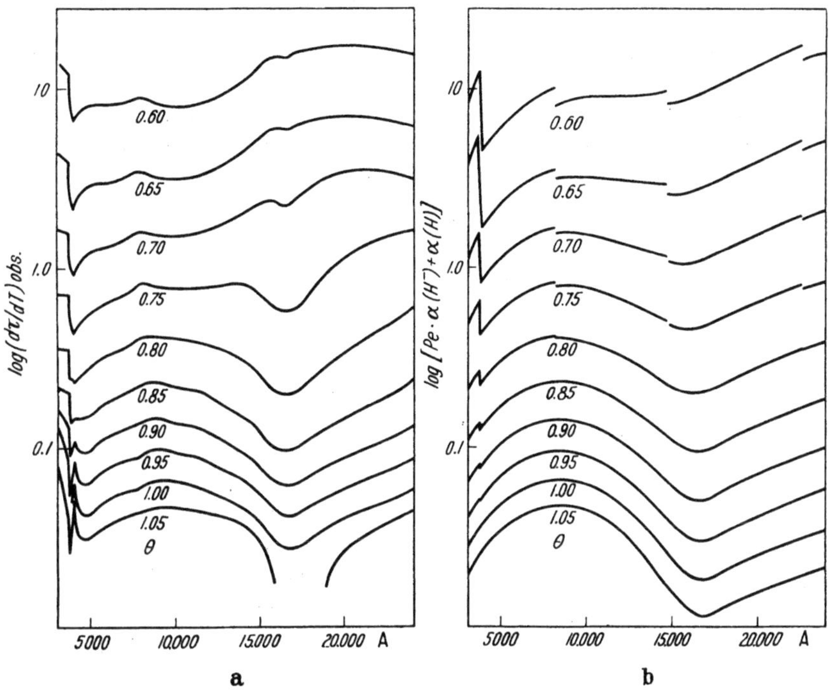

Fig. 11a and b. (a) Observed absorption coefficient (the individual curves have been shifted vertically an arbitrary amount). (b) Computed absorption coefficient. CHANDRASEKHAR's values for H^- and MINNAERT's [5] values of P_e were used.

which, together with the Planck equation

$$\varphi_\lambda(\tau_\lambda) \cdot I_\lambda(0,1) = \frac{c_1}{\lambda^5} \frac{1}{e^{c_2/\lambda T} - 1}, \tag{17.5}$$

gives, by differentiation, the desired gradient

$$\left(\frac{d\tau}{dT}\right)_\lambda = \frac{q}{T^2} \frac{\varphi(p+\varphi)}{p} \frac{1}{b_\lambda - c_\lambda E_1(\tau)} \tag{17.6}$$

where

$$q = \frac{c_2}{\lambda} \quad \text{and} \quad p = \frac{c_1}{\lambda^5 I_\lambda(0,1)}. \tag{17.7}$$

The observed $\text{Log}(d\tau/dT)_\lambda$ curves are illustrated in Fig. 11a for the temperatures shown ($\Theta = 5040/T$). An arbitrary vertical shift of the curves has been introduced.

If scattering is neglected, then the combined absorption coefficients of H and H^- for each wavelength is

$$k_\lambda = \frac{1-x}{m_H} \left[\alpha_\lambda(\mathrm{H}) + P_e \alpha_\lambda(\mathrm{H}^-)\right], \tag{17.8}$$

where x is the fraction of ionized H atoms. The logarithm of k_λ, illustrated in Fig. 11 b, is derived from Eq. (17.8), in which P_e was taken from the mean model of MINNAERT [5]. A moderately good agreement is found between the observed and computed absorption coefficients.

The observed absorption coefficient, when compared to the calculated one, strongly suggests the presence of several of the absorption continua of neutral hydrogen; however, the latter lack the sharp fronts predicted by KRAMERS' equation. But, sharp fronts are not to be expected, for in the deep layers of the solar atmosphere the confluence of the lines at the series limit will smooth any discontinuities. Furthermore, the abrupt break at the head of the Balmer series, shown in Fig. 11 a, is illusory.

A detailed calculation of the continuous absorption coefficient has been made by VITENSE[1]. In addition to the above absorbers, the absorption by Fe, Si, Mg and a few minor elements have been computed, in the absence of other information, by a hydrogen-like approximation:

$$\frac{\varkappa_\nu}{\text{atoms in the } n\text{-th state}} = \frac{1}{n^2} \, \frac{64\pi^4}{3\sqrt{3}} \, \frac{(Z+s)^4 m e^{10}}{c h^6 n^3} \cdot \frac{1}{\nu^3} . \tag{17.9}$$

The principal effect of the metals lies short of the Balmer limit. This, along with Rayleigh scattering, appears to be sufficient to explain the increased absorption in the violet, particularly in the higher solar layers, as was noted by MICHARD[2].

18. Model solar atmospheres [1] to [5], [7], [8]. An equilibrium model atmosphere is usually specified by a mean effective temperature T_e, surface gravity g, and by the relative abundances of the elements. From these quantities the depth variation of the temperature, total pressure, and the electron pressure may be computed. Furthermore, in principle, the absorption and emission coefficients of the continuous spectrum and line spectrum and hence the complete radiation field may be evaluated at each depth. More general and realistic models that include turbulence have recently been proposed, principally by the German school under UNSÖLD[3-6]. By assigning temperatures, areas, and velocities to rising, falling, and stationary elements an attempt is made to reproduce the mean model and the hypothetical picture of granulation. (See Sect. 38 and C. DE JAGER'S chapter in this volume.)

Hydrostatic equilibrium in the atmosphere requires a balance between the forces of gravity and outward pressure: per unit element,

$$dP = -g\varrho \, dh. \tag{18.1}$$

The total pressure P is made up of the gas pressure P_g, the radiation pressure P_r, and the turbulent pressure P_t. The latter arises from the large scale motions of groups of atoms in the same way that gas pressure results from the random motion of individual atoms, provided of course, that the turbulent elements are small with respect to the total system considered. MCCREA[7] developed the problem. Let ξ be the velocity component in the radial direction. Then if turbulence

[1] E. VITENSE: Z. Astrophys. **28**, 81 (1951).
[2] R. MICHARD: Ann. d'Astrophys. **16**, 217 (1953). See also G. MUNCH: Astrophys. Journ. **102**, 385 (1945).
[3] K.H. BÖHM: Z. Astrophys. **35**, 179 (1954).
[4] M. BRETZ: Z. Astrophys. **38**, 259 (1955).
[5] M. VOIGHT: Z. Astrophys. **40**, 157 (1956).
[6] E.H. SCHRÖTER: Z. Astrophys. **41**, 141 (1957).
[7] W.H. MCCREA: Monthly Notices Roy. Astronom. Soc. London **89**, 718 (1929).

is isotropic and Maxwellian we may write

$$\frac{1}{\varrho}\frac{dP_g}{dh}+\frac{1}{\varrho}\frac{d}{dh}\left(\frac{\varrho}{2}\xi_h^2\right)=g. \tag{18.2}$$

The turbulent factor becomes of importance only when the group velocities approach the velocity of sound.

The effect of radiation pressure is to reduce gravity by the amount g_r. Eq. (18.2) may be written

$$\frac{d}{d\tau}(P_g+\varrho\,\xi_h^2)=\frac{g-g_n}{\bar{k}} \tag{18.3}$$

where

$$g_r=\bar{k}\frac{\sigma}{c}T_e^4 \tag{18.4}$$

and $\bar{k}$ is the mean absorption obtained by weighting the absorption coefficient by the flux at each frequency:

$$\bar{k}=\frac{\int(k_\nu+s_\nu)F_\nu\,d\nu}{\int F_\nu\,d\nu}. \tag{18.5}$$

For the solar atmosphere $g_r/g=10^{-4}$, and is therefore negligible.

α) Strömgren's method of model atmospheres. In an important paper on the chemical composition of the solar atmosphere STRÖMGEN developed a very useful approximate theory of model atmospheres[1]. From the equations which define optical depth

$$\frac{d\tau}{dx}=-\bar{k}\,\varrho, \tag{18.6}$$

and hydrostatic equilibrium

$$\frac{dP}{dx}=-g\,\varrho, \tag{18.7}$$

we obtain

$$\frac{dP}{d\tau}=\frac{g}{\bar{k}}. \tag{18.8}$$

This equation cannot be integrated directly, because the opacity is a complex function of τ and P—usually given in tabular form. However, STRÖMGEN obtained an elegant approximate solution which clearly illustrates the interdependence of each physical parameter. The mean absorption coefficient per gram is considered to be due only to H^-; the absorption of the H^- atom, $\alpha(H^-)$, times the number of hydrogen atoms, $1/m_H$, multiplied by the ratio of the number of H^- ions to the number of neutral H atoms yields the opacity coefficient per gram—considered to be nearly all hydrogen:

$$\bar{k}=\frac{\alpha(H^-)}{m_H}\cdot\frac{N_{H^-}}{N_H}. \tag{18.9}$$

By SAHA'S equation, the ratio of the numbers of H^- ions to neutral H atoms is

$$\frac{N_{H^-}}{N_H}=P_e\cdot\Phi(T), \tag{18.10}$$

with

$$\mathrm{Log}\,\Phi(T)=-0.125+0.747\frac{5040}{T}-\frac{5}{2}\log T \tag{18.11}$$

in which the electron affinity of the H^- ion, namely 0.747 volts, has been substituted and the statistical weights of the ground states of H^- and H (1 and 2

[1] B. STRÖMGREN: Festschrift für E. STRÖMGREN — Publ. Copenhagen Obs. No. 127, 1940.

respectively) appear in the constant: $0.477 - \log \frac{2.2}{1}$. Table 5 gives Log $\Phi(T)$ as a function of T and Θ. In the photospheric layers the ratio N_{H^-}/N_H is always very small, i.e. for $\tau = 0.4$, $\vartheta = 0.9$, $P_e = 10$ dynes, $N_{H^-}/N_H = 10^{-7.8}$ When multiplied by $1/m_H$, however, the number of H^- per gram is large. The negative hydrogen ions are about 10^3 times more abundant than the neutral hydrogen atoms excited to the third quantum state under the same conditions and capable of absorbing in the Paschen continuum below 8200 Å.

Eqs. (18.8) and (18.9) combine to give

$$\frac{dP}{d\tau} = \frac{g}{\bar{k}} \cdot m_H \frac{1}{P_e \Phi(T)}, \tag{18.12}$$

an equation that can be solved if a relation between the total pressure and the electron pressure can be found.. STRÖMGEN introduced a simplifying assumption.

Table 5. Log $\Phi(T)$: H^-, Eq. (18.11); Log $\Phi_1(T)$: H, Eq. (18.22).

Θ	T	Log $\Phi(T)$	Log $\Phi_1(T)$	Θ	T	Log $\Phi(T)$	Log $\Phi_1(T)$
1.3	3877	−8.125	−9.173	0.85	5929	−8.923	−2.597
1.25	4032	−8.205	−8.451	0.8	6300	−9.026	−1.851
1.2	4200	−8.287	−7.727	0.75	6720	−9.133	−1.101
1.15	4383	−8.370	−7.001	0.7	7200	−9.245	−0.347
1.1	4582	−8.456	−6.273	0.65	7754	−9.363	+0.413
1.05	4800	−8.544	−5.543	0.6	8400	−9.488	+1.180
1.0	5040	−8.634	−4.811	0.55	9164	−9.619	+1.954
0.95	5305	−8.727	−4.076	0.5	10080	−9.761	+2.738
0.9	5600	−8.823	−3.338				

He let the number of free electrons per unit volume be equal to the number of metal atoms per unit volume. If the metals are singly ionized and the ionization of H is small (Table 5) as is nearly the case for the Sun, then

$$\frac{P_e}{P} = \frac{1}{A}, \tag{18.13}$$

where A is about 10^4. By substitution in Eq. (18.12) and by integration one obtains

$$P^2 = \frac{2g\, m_H A}{\bar{k}} \int_0^\tau \frac{d\tau}{\Phi(T)}, \tag{18.14}$$

from which P as a function of τ is determined, once T and hence $\Phi(T)$ is known as a function of τ. From Eq. (18.13) P_e is directly proportional to P.

β) Models with helium. Helium adds nothing to the absorption coefficient but adds to the total pressure and decreases the electron pressure by dilution. ALLER and PIERCE[1] computed P_g, P_e and $\bar{k}$ as a function of τ for a series of model atmospheres with the ratio of the number of helium atoms to the number of hydrogen atoms, $D = 0$, $\frac{1}{5}$, and $\frac{1}{20}$, and for different metal to hydrogen ratios. Recently, HUBENET[2] has computed differential corrections for values of $D = 0$ to 0.4 applicable to any model.

Let $D = \mathrm{He/H}$ by numbers of atoms, then

$$P_g \sim P_e A(1 + D), \tag{18.15}$$

[1] L.H. ALLER and A.K. PIERCE: Astrophys. Journ. **116**, 176 (1952).
[2] H. HUBENET: Proc. Kon. Nederl. Akad. Wetensch., Ser. B **59**, 211 (1956).

and the number of hydrogen atoms per gram is $\frac{1}{m_{\mathrm{H}}(1+4D)}$. Since $\bar{k}=\bar{k}(P_g, T)$ Eq. (18.12) may be multiplied by P_g and integrated to give

$$P_g^2 = 2g\int \frac{P_g}{\bar{k}(P_g, T)}\,d\tau. \tag{18.16}$$

By substitution of Eq. (18.15) for P_g, and the following expression for the absorption coefficient,

$$\bar{k} = \frac{1}{m_{\mathrm{H}}(1+4D)}\,\alpha(\mathrm{H}^-), \tag{18.17}$$

Eq. (18.16) becomes

$$P_g^2 = 2g\,m_{\mathrm{H}}\,A(1+D)\,(1+4D)\int\frac{d\tau}{\alpha(\mathrm{H}^-)}, \tag{18.18}$$

which can be solved by successive iterations. A reliable value for the abundance of helium is not available [5]. From observations of emission lines in the chromosphere and prominences UNSÖLD finds $D=0.20$. MICHARD[1] computed the Balmer discontinuity with two models, one with $D=0.0$ and the other with $D=0.2$. From a comparison between theory and observation he concludes that $D=0.0$ is more probable than $D=0.2$. For want of better information, D is usually chosen to be 0.1 or 0.2. Carbon, nitrogen, and oxygen are generally ignored, since tables of abundances indicate that the ratio $\Sigma(\mathrm{C, N, O})/\mathrm{H}$ is about 1/1000.

γ) Models which include the ionization of hydrogen. In the above models the ionization of H has been neglected. Although its ionization potential is almost twice that of the metals, it contributes an appreciable number of electrons because of its great abundance. In his analysis BARBIER[2] included the degree of ionization of hydrogen, x_{H}, and that of the metals, x_M.

Let E = number of electrons per atom of any kind $=\frac{\Sigma x_i \varepsilon_i}{\Sigma \varepsilon_i}$, where ε_i is the abundance and x_i is the degree of ionization, then

$$\frac{P_e}{P} = \frac{E}{1+E};\quad E = \frac{A\,x_{\mathrm{H}} + x_M}{A+1},\quad \text{where}\quad x_M \approx 1,\quad x_{\mathrm{H}} \ll 1,\quad A \approx 10^4. \tag{18.19}$$

Approximately

$$\frac{P_e}{P} = \frac{x_{\mathrm{H}}}{1+x_{\mathrm{H}}} + \frac{1}{A}\,\frac{x_M}{1+x_M}, \tag{18.20}$$

or

$$\frac{P_e}{P} = x_{\mathrm{H}} + \frac{1}{A} = \frac{N_{\mathrm{H}^+}}{N_{\mathrm{H}}} + \frac{1}{A}. \tag{18.21}$$

SAHA's equation for the degree of ionization of hydrogen,

$$\frac{N_{\mathrm{H}^+}}{N_{\mathrm{H}}} = \frac{\Phi_1(T)}{P_e};\qquad \mathrm{Log}\,\Phi_1(T) = -0.477 - 13.59\,\frac{5040}{T} + \frac{5}{2}\log T, \tag{18.22}$$

may be substituted in Eq. (18.21), which becomes

$$\frac{P_e}{P} = \frac{\Phi_1(T)}{P_e} + \frac{1}{A}. \tag{18.23}$$

---

[1] R. MICHARD: Ann. d'Astrophys. **16**, 217 (1953).
[2] D. BARBIER: Ann. d'Astrophys. **9**, 173 (1946).

Thus a better approximation can be made from the equations:

$$P^2 = \frac{2g\,m_{\mathrm{H}}}{\bar{k}} \int_0^{\tau_0} \frac{d\tau_0}{\Phi(\tau)\left[\frac{\Phi_1(T)}{P_e} + \frac{1}{A}\right]} \quad \text{and} \quad P_e = P\left[\frac{\Phi_1(T)}{P_e} + \frac{1}{A}\right]. \tag{18.24}$$

An exact analysis must be made numerically, however, by means of abundance tables and ionization tables for each element, as was first done by STRÖMGREN[1].

δ) Plaskett's model. PLASKETT[2] has used a method similar to BARBIER'S, together with SYKES'[3] representation of limb darkening (from the mean of five observers), to obtain evidence for an unstable layer lying below optical depths of 0.8. The condition for mechanical stability can be obtained from Table 6,

Table 6. *Physical conditions in the photosphere.*
(H. H. PLASKETT: Vistas in Astronomy, ed. by BEER. London: Pergamon Press 1955.)

τ	T (°K)	p (10^4 dyne cm^{-2})	ϱ (10^{-8} gm cm^{-3})	p_e (dyne cm^{-2})	x_M	x_{H} (10^{-4})	k_λ (cm^{-2} gm^{-1})	$z-z_0$ (km)
0.0	4649	0	0	0	—	—	0	
0.03	5033	1.7	4.2	1.7	0.89	0.1	0.09	318
0.06	5162	2.5	5.8	2.6	0.90	0.2	0.13	261
0.09	5246	3.1	7.1	3.2	0.91	0.1	0.15	227
0.12	5312	3.6	8.2	4.1	0.90	0.2	0.18	203
0.24	5505	5.1	11.2	6.9	0.93	0.4	0.26	145
0.36	5680	6.2	13.2	10.5	0.93	0.8	0.35	113
0.48	5838	6.9	14.4	17.7	0.93	1.6	0.53	93
0.72	6118	8.0	15.9	27.7	0.95	2.5	0.70	67
0.96	6356	8.8	16.8	46.5	0.96	4.3	1.01	49
1.20	6555	9.4	17.4	70	0.96	6.5	1.35	37
1.80	6959	10.3	17.9	153	0.98	13.9	2.42	19
3.00	7535	11.2	18.0	410	1.00	35.6	5.06	0

which reproduces his determination of the physical conditions in the photosphere. For no convection to occur, the following inequality holds ([*1*], p. 215):

$$\left(\frac{d\ln T}{d\ln P}\right)_{\text{photosphere}} < \left(\frac{d\ln T}{d\ln P}\right)_{\text{adiabatic}} = \frac{\gamma-1}{\mu}. \tag{18.25}$$

In a monatomic, unionized gas $\gamma = \frac{5}{3}$. Fig. 12 illustrates the gradients plotted as a function of P and of height z from an arbitrary zero. PLASKETT suggests that the excess of the photospheric $(d\ln T/d\ln P)$ over the adiabatic line at 0.4 is a consequence of the steep temperature gradient (dT/dz) in that region of the photosphere that radiates a sizable fraction of its thermal energy to outer space. Furthermore, this region is an unstable one, based on the evidence that the density levels off (ϱ may reach a maximum at height zero) in the region where the temperature gradient becomes very steep.

Unfortunately the observations of granulation are as yet too few in number to establish with certainty both the extent and the energy carried by the zone of granulation, but they suggest that this region and PLASKETT'S zone are identical. Photographs of granulation taken by RÖSCH[4] at Pic-du-Midi show granules to within 5 to 10″ from the limb. Thus the zone of disturbance is not limited

[1] B. STRÖMGREN: Publ. og mind. Medd. Kobenhavn Obs. No. 138 (1944).

[2] H. H. PLASKETT: Vistas in Astronomy, ed. by A. BEER, Vol. 1, p. 637. London and New York: Pergamon Press 1955.

[3] J. B. SYKES: Monthly Notices Roy. Astronom. Soc. London **113**, 198 (1953).

[4] J. RÖSCH: C. R. Acad. Sci., Paris **240**, 1630 (1955).

to the deeper layers but must extend to high levels, a result suggested by Doppler displacements and intensity variations in absorption lines[1]. It is strange perhaps that the motions in the granulation zone do not establish the adiabatic gradient. Apparently uncompensated radiation loss to space determines the temperature gradient and affects the lifetime and size of the granules in the observable part of the atmosphere.

ε) Schröter's inhomogeneous model. SCHRÖTER'S[2] two-stream model is typical of the more recent models that consider granulation and hence temperature inhomogeneities. These models are derived from the very difficult observations of the brightness ratio between granule and background and are thus uncertain to the extent that the observation are in error. SCHRÖTER adopts $(J_G - J_B)/J_B = K(\cos\vartheta = 1.0) = 0.45$ for the contrast ratio between granule and background and accepts the observation that the contrast increases from the center of the disk to about $\cos\vartheta = 0.25$ and then rapidly diminishes. The Eddington-Barbier relations are used to establish the temperature differences from a homogeneous model in two streams: for $\tau_G = \tau_B = 1$, $\Delta T = \pm 350°$; at $\tau = 0.3$ the difference increases to $\pm 400°$; and for $\tau < 0.1$ the temperature difference decreases to zero. As a proof of the model SCHRÖTER computes the average emission

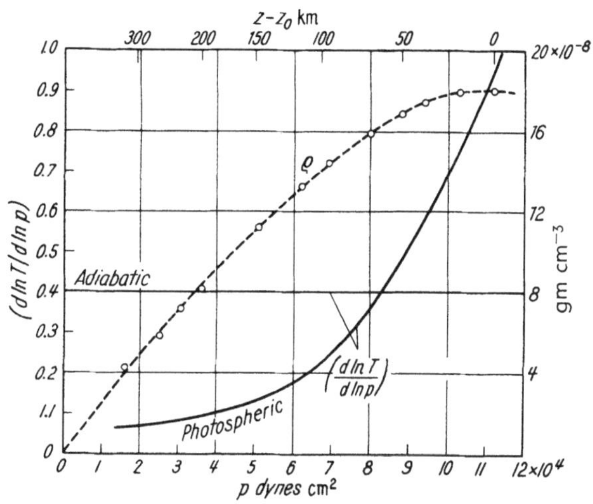

Fig. 12. $\frac{d\ln T}{d\ln p}$, photospheric and adiabatic models as a function of P (after PLASKETT). (Vistas in Astronomy Vol. 1, p. 644, ed. BEER. Pergamon Press: London 1955.)

$$\bar{I}(0,\mu) = \tfrac{1}{2}[J_G(0,\mu) + J_B(0,\mu)]$$

for the two streams and finds that the observed limb darkening is very well reproduced. He also finds that the ratio $J_G(0,\mu)/J_B(0,\mu)$ fits the observed center-to-limb variation of the contrast of the granules. Thus a simple two-stream model is a satisfactory second approximation for the solar atmosphere.

19. The high photosphere. Except at the extreme limb, it is evident that limb darkening is not influenced by the small amount of radiation from the outermost photospheric layers. Since, however, they are the very layers in which most of the Fraunhofer lines are formed, the line spectrum may serve as the best key for finding the physical conditions in the high photosphere. For example, DE JAGER[3] derived temperature corrections to a standard model from his observations of the limb darkening of the hydrogen lines. In a similar manner VITENSE[4] made use of the center-limb variation of the cores of strong lines of Fe and H. She found that the central intensities of these lines were best represented by a boundary temperature of 3800° K and by a very steep rise in temperature between $\tau = 0$ and $\tau = 0.1$. It has since been recognized that the cores of the first few

[1] R.R. McMATH, O.C. MOHLER, A.K. PIERCE and L. GOLDBERG: Astrophys. Journ. **124**, 1 (1956).

[2] E.H. SCHRÖTER: Z. Astrophys. **41**, 141 (1957).

[3] C. DE JAGER: Rech. Utrecht **13**, 1 (1952).

[4] E. VITENSE: Z. Astrophys. **34**, 209 (1954).

members of the Balmer series and the cores of the strongest iron lines are formed in the lower part of the chromosphere and are strongly affected by temperature inhomogeneities (Sect. 40). Furthermore, the temperature found from central intensities of strong Fraunhofer lines represent excitation temperatures of the upper levels of the transition and because of departure from thermodynamic equilibrium cannot be identified with the boundary temperature derived from the continuum.

PAGEL[1] combined the temperature distribution found by him, to best fit the center-to-limb behavior of infrared (1.5 μ) lines, with the result obtained by ATHAY, MENZEL, PECKER, and THOMAS[2], who analyzed observations of the con-

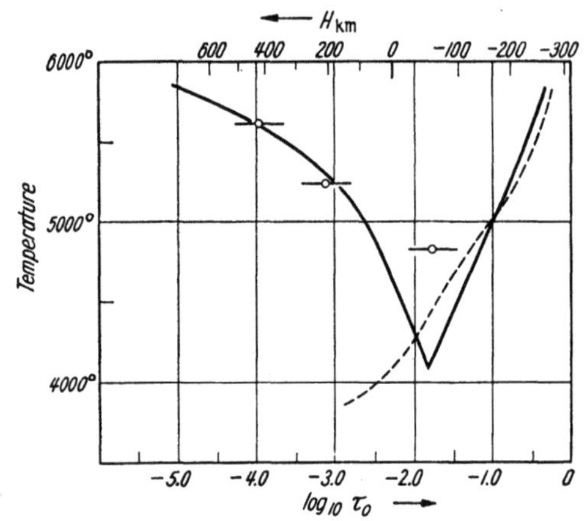

Fig. 13.

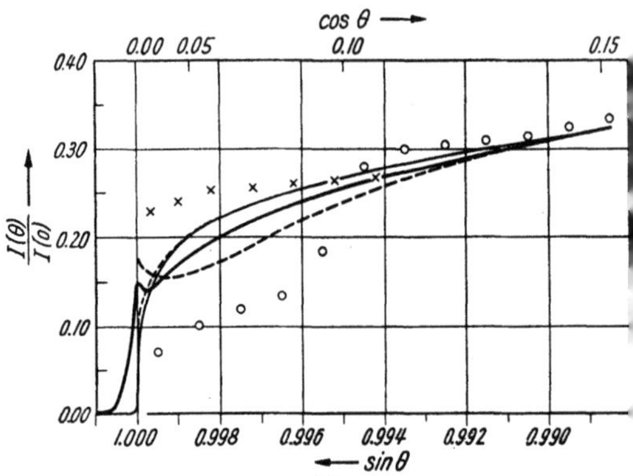

Fig. 14.

Fig. 13. Temperature distribution with depth in the Sun's atmosphere. Full line: PAGEL's model. Broken line: radiative equilibrium model of BÖHM. Circles: temperatures derived by ATHAY et al. (after PAGEL).

Fig. 14. Computed and observed darkening at extreme solar limb for λ 4700. The thick continuous line is PAGEL's computed limb darkening allowing for curvature; the broken line represents the plane-parallel approximation; BÖHM's[3] radiative-equilibrium model is represented by the thin continuous and broken lines. Filled circles: observations of HEYDEN et al.[4]. Crosses: WESSELINK et al.[5] (after PAGEL). [B. E. J. PAGEL: Monthly Notices R.A.S. 116, 608 (1956)].

tinuum photographed at the 1952 eclipse. From the T versus τ relation shown in Fig. 13, he computed a model atmosphere and also the darkening at the extreme limb, taking into account the curvature (Fig. 14). It is evident that the temperature inversion above $\tau = 0.015$ affects limb darkening in a region so narrow as to challenge the finest observational techniques.

A rediscussion of the eclipse observations of HEYDEN *et al.* has been presented by HUBENET and DE JAGER[6]. An attempt to find a satisfactory temperature distribution by making use of JULIUS' eclipse method failed. But, by using the raw observational data, they were able to derive a model (constant temperature between $\tau = 0.01$ to 0.05 and with a steep rise between $\tau = 0.07$ to 0.13 after which the slope is smaller) that reproduced the original observations.

20. Deviations from thermodynamic equilibrium. There is no evidence from observations of the spectrum of the continuum that there are any deviations

[1] B. E. J. PAGEL: Monthly Notices Roy. Astronom. Soc. London **116**, 608 (1956).
[2] R. G. ATHAY, D. H. MENZEL, J. C. PECKER and R. N. THOMAS: Astrophys. Journ. Suppl. **1**, 505 (1955).
[3] K. H. BÖHM: Z. Astrophys. **34**, 182 (1954).
[4] F. J. HEYDEN, C. A. BECK, D. J. LOVELL, W. B. FUSSELL, J. H. HANCOCK and E. O. HULBURT: Astrophys. Journ. **118**, 412 (1953).
[5] A. J. WESSELINK, J. G. FERWERDA and J. UITTERDIJK: Bull. Astronom Netherl. **9**, 81 (1940).
[6] H. HUBENET and C. DE JAGER: Bull. Astronom. Netherl. **13**, 43 (1956).

from thermodynamic equilibrium. PAGEL[1] has treated the form that the transfer equation must take when the Boltzmann and Saha equations fail to apply. The classical form of the transfer equation,

$$\cos\vartheta \frac{dI_\nu}{\varrho\, dx} = \varkappa_\nu (I_\nu - B_\nu), \tag{20.1}$$

can be written as

$$\cos\vartheta \frac{dI_\nu}{\varrho\, dx} = \varkappa_\nu I_\nu - \varkappa_\nu^{(T)} B_\nu, \tag{20.2}$$

where $\varkappa_\nu$ is the actual mass absorption coefficient and $\varkappa_\nu^{(T)}$ the corresponding coefficient with the assumption of thermodynamic equilibrium. For a gas in equilibrium, the absorption and emission coefficients are given by Kirchhoff-Planck equations

$$j_\nu\, d\nu = \varrho\, \varkappa_\nu^{(T)} \cdot B_\nu(T)\, d\nu \quad \text{and} \quad B_\nu(T) = \frac{2h\nu^3}{c^2} \frac{1}{e^{h\nu/kT} - 1}, \tag{20.3}$$

which, considered on an atomic scale, that of the principle of detailed balancing, implies also the validity of the Boltzmann distribution and the Saha ionization laws. WOOLLEY and STIBBS [4] compute that the average photoelectron, ejected from the H^- ion with an energy of 2 to 3 volts into a sea of electrons with a mean kinetic energy of 1 volt, will experience 10^4 electron-electron collisions before capture by an ion or by: $H + e \rightarrow H^- + h\nu$. Thus, the fast electron will rapidly lose energy and a nearly Maxwellian distribution of velocities will prevail. If this be so, the spontaneous emission of photons by electron capture is independent of the local radiation field.

By applying the principle of detailed balancing, PAGEL finds that the spontaneous emission per neutral hydrogen atom due to H^- is

$$\frac{j_\nu\, d\nu}{N_H} = P_e \left[a_{BF}(H^-) + a_{FF}(H^-)\right] B_\nu(T)\, d\nu, \tag{20.4}$$

where $a_{BF}(H^-)$ and $a_{FF}(H^-)$ are the effective bound-free and free-free cross sections per unit electron pressure (including the stimulated emission factor). The only deviation from equilibrium to be recognized is in P_e. For example, PAGEL finds in the high photosphere that radiative and collisional dissociation of H^- become of equal order of importance at $\tau = 0.01$ and that the ratio of the true number of H^- ions to the number computed in thermodynamic equilibrium is $N_{H^-}/N_{H^-}^{(T)} = 0.8$ for $T = 4100°$ K and $N_e = 10^{12}$. Thus, even in the low chromosphere, the deviations are not large. SWIHART[2] arrives at the same conclusion.

II. The Fraunhofer spectrum.

Introduction. We have seen in the preceding sections how precise observations of the continuous spectrum give information on the distribution of electron temperature, pressure and continuous absorption coefficient in the solar photosphere. There are, however, a variety of other important parameters of the photosphere which can only be deduced from the profiles and intensities of the absorption lines. The line spectrum reveals the chemical composition of the photosphere, its thermal and hydrodynamical properties, and the presence of electric and magnetic fields. Of special importance are the departures from thermodynamic

[1] B. E. J. PAGEL: Astrophys. Journ. **125**, 298 (1957).
[2] T. L. SWIHART: Astrophys. Journ. **123**, 143 (1956).

equilibrium, which are evidenced by inequalities among the electron temperatures deduced from the continuous spectrum, and the excitation and ionization temperatures inferred from the lines.

Ever since NEWTON'S discovery of the solar spectrum, spectroscopists have striven to increase the resolving power and to reduce or eliminate stray radiation in solar spectrographs. The resolving power of NEWTON'S arrangement was too low to reveal any absorption lines. WOLLASTON in 1802 observed four dark spaces in the spectrum, but the true beginning of research on the line spectrum dates from the visual mapping of nearly 600 lines by FRAUNHOFER in 1814. As a result of technical advances in the areas of optics, photography, and other types of energy detectors, increasing numbers of Fraunhofer lines have been brought into view in the ultraviolet and infrared, as well as in the visual regions of the spectrum, until today the number of observed Fraunhofer lines totals nearly 30000.

The significance of the Fraunhofer spectrum as arising from line absorption by chemical elements in the solar atmosphere was revealed by KIRCHHOFF'S laboratory observations of atomic spectra around 1860. During the remainder of the 19th century solar spectroscopy was devoted to the comparison of solar and laboratory spectra, chiefly for the purpose of identification of elements. But a certain amount of progress was made in deducing the physical conditions in the Sun by comparing its spectrum with those of different types of laboratory sources. The work of this period was climaxed by the publication by H. A. ROWLAND of his great table of wavelengths, identifications and eye-estimates of intensities covering the spectral region λ 2975—λ 7330. About 40 elements are identified in this table, which, in its revised form [9], is still the standard source of information on solar wavelengths and identifications in the visual and near ultraviolet spectral regions.

The turn of the present century marked the beginning of the modern era of quantitative interpretation of the Fraunhofer spectrum in which the application of physical theory to precise photometric measurements leads to the revelation of the structure, chemical composition and dynamics of the solar atmosphere. The required physical theory has four major components: (1) the quantum theory of radiation and atomic structure; (2) the theory of radiative transfer; (3) the theory of excitation and ionization; and (4) the theory of convection and turbulence.

The quantum theory of atomic structure has been developed in a long series of investigations following its initiation by PLANCK, BOHR and EINSTEIN in the early 1900's. Its application to the solar spectrum has been enormously successful, but progress in the numerical calculation of transition probabilities and collision cross sections for complex atoms is still disappointingly slow.

The theory of radiative transfer, as applied to the formation of absorption lines, is also well-established, following its initiation by SCHUSTER and SCHWARZSCHILD and its further development by STEWART, EDDINGTON, MILNE, UNSÖLD, MINNAERT, STRÖMGREN, CHANDRASEKHAR and many others. Given a model solar atmosphere in which the physical properties are specified in detail, the line profile can be calculated accurately by a variety of different methods, but only if the mechanism of re-emission is known. Entirely different profiles can result depending on whether the re-emission occurs according to coherent scattering, non-coherent scattering, or by KIRCHHOFF'S law. The difficulty is aggravated by the absence of an accurate quantitative theory of non-coherent scattering, by the probability that no one of the three mechanisms but a combination of them may be contributing to the line formation, and by the likelihood that different parts of the line may be formed by different mechanisms.

The ionization and dissociation equations are still employed in essentially the form first introduced by SAHA in 1922. There is no reason to question the validity of the theory when thermodynamic equilibrium exists, but this condition may not be satisfied in the highest layers of the photosphere. The question then arises whether the radiation temperature derived from limb darkening can be safely equated to the kinetic or electron temperature and also used in the calculation of the degree of ionization and excitation by the Saha and Boltzmann equations. It is to be hoped that further serious attention will be given to this important and difficult question.

In succeeding chapters of this volume, Dr. DE JAGER deals with convection and turbulence, which are revealed by the granulation of the solar surface and by the inhomogeneities in positions, widths and intensities of the Fraunhofer lines. The dynamical features of the solar atmosphere are difficult both to observe and to interpret, but they must be taken into account in all future studies of photospheric structure. It can be expected that the new instrumental advances which are being stimulated by recent improvements in the art of ruling gratings will greatly accelerate progress towards a satisfactory theory of solar gas dynamics.

Photometric measurements on the Fraunhofer spectrum are made in relation to the intensity of the neighboring continuous spectrum. The most valuable type of measurement is the intensity at each point of the line profile. Until recently, line profiles could easily be determined only for the outer parts of the strongest lines in the spectrum; the inner parts of the observed profiles have required elaborate corrections for scattered light and instrumental broadening. A review by PIERCE[1] in 1954 listed no more than about 75 lines of 9 elements for which profile measurements have been published. The difficulties associated with profile measurements have forced solar spectroscopists to resort to the measurement of the total absorption, or equivalent widths, of Fraunhofer lines. The equivalent width is a measure of the area inside the absorption line; since the intensity of the neighboring continuous spectrum is taken as unity, the equivalent width is expressed in wavelength or wave number units.

The great majority of physical investigations of the Fraunhofer spectrum since 1930 have dealt with the interpretation of equivalent widths in terms of the theory of the curve of growth. The curve-of-growth concept has been extremely useful in revealing certain gross features of the photosphere. For example, the discovery of large damping constants by MINNAERT and MULDERS made it clear that collisional rather than radiative damping was responsible for the wing broadening of strong lines. Other investigations have demonstrated that the r.m.s. velocity of photospheric atoms is too large to be attributed to purely thermal effects and must therefore be a consequence of turbulent mass motions. The inherent weakness of the curve-of-growth method is that it can give only average values of quantities like the turbulence and damping and virtually no information on their variation with height.

Many of the factors that control the equivalent widths of very strong and moderately strong lines, such as the Doppler velocities and damping constants, play no role in determining the intensities of very weak lines, which, from this point of view are better suited to the determination of abundances. But the intensities of weak lines are subject to large random errors of measurement when

[1] A.K. PIERCE: Proc. N.S.F. Conf. on Stellar Atmospheres, ed. M.H. WRUBEL. Bloomington, Ind., USA: Indiana Univ. 1954.

the resolving power is limited, and hence the abundances will be reasonably accurate only when a large number of lines can be averaged. The difficulty is that many elements, including most of the heavy ones that are strategic from the standpoint of element origin, are represented in the spectrum by no more than a half-dozen lines and consequently their abundances are poorly determined.

Very great strides have recently been made towards the perfection of the plane, reflection grating, which promise to initiate a new era in solar spectroscopy and in the physical investigation of the Sun. Striking success has been achieved with gratings ruled by Dr. H. W. BABCOCK at the Mt. Wilson and Palomar Observatories and installed at several solar observatories in the United States and in western Europe. Excellent spectral definition and resolving power have also been obtained at the Crimean Astrophysical Observatory in the U.S.S.R.[1] Some of the larger Babcock gratings are blazed in the fifth order in the green region of the spectrum, where the resolving power is about 600000 and the instrumental half-width is of the order of 0.01 Å. The special significance of these numbers is that the true half-widths of faint Fraunhofer lines are about three times greater than the instrumental half-widths and therefore the profiles of the weak lines may be directly observed. PIERCE[2] reports that the scattered light in the vacuum spectrograph[3] of the McMath-Hulbert Observatory amounts to only 0.7% of the continuum at λ 5896; the ghosts contribute an additional 1.8%. In the cores of the H and K lines the total contribution of scattering and ghosts is no more than about 2% according to a recent unpublished determination by MOHLER. Thus, the fact that both the profiles and the absolute central intensities of weak as well as strong lines may now be observed with relatively high precision presages a period of intensive re-analysis of the Fraunhofer spectrum in which heavy emphasis will be laid on the detailed elucidation of the photospheric structure from individual line profiles. Abundance investigations will continue to be based on the equivalent widths of weak lines, which can now be measured with much higher accuracy than heretofore. It should be remarked, however, that the problems of establishing the position of the continuum and of correcting for blending still deserve the most careful observational and theoretical attention.

In the succeeding sections, we shall first refer briefly to the observational data upon which investigations of the Fraunhofer spectrum are based, then sketch the theory of Fraunhofer-line formation, and finally review some of the more recent analytic studies of the spectrum. Refs. [*1*], [*4*], [*5*] and [*9*] contain excellent reviews of the subject, which have appeared during the past five years. Inasmuch as the earlier literature is well covered in these reviews, our own survey will emphasize the most recent developments.

a) Observational data.

21. Transmission of the Earth's atmosphere. Owing to absorption by the atomic and molecular constituents of the Earth's atmosphere, observation of the Fraunhofer spectrum from ground level is limited to a number of relatively narrow spectral regions, corresponding to the location of so-called "windows" in the Earth's atmosphere. No region of the spectrum is entirely free of atmospheric line absorption, but in the regions of the windows, the telluric absorption lines are relatively weak and widely separated. The absorption spectrum of the

[1] A. B. SEVERNY: Russ. Astronom. Journ. **33**, No. 1 (1956). — Observatory **76**, 241 (1956).
[2] A. K. PIERCE: J. Opt. Soc. Amer. **47**, 6 (1957).
[3] R. R. McMATH: Astrophys. Journ. **123**, 1 (1956).

Earth's atmosphere has been reviewed by GOLDBERG[1], and the location of the windows is shown in Fig. 15. The ultraviolet limit for ground observation is about λ 2935, but the radiation spectrum at shorter wavelengths has been observed from rockets to wavelengths as short as 1 to 2 Å. However, the photospheric or Fraunhofer spectrum has really been observed only to about λ 1600. The reason is that, at shorter wavelengths, the intensity of the continuous spectrum is too weak to be well observed in the exposure times possible with rockets, and hence only the line emission from the chromosphere and corona has been so recorded. The infrared limit of about 24 μ is fixed by water vapor absorption and an additional 9 or 10 octaves of the spectrum are passed through the atmosphere in the form of radio-radiation in the range from a few millimeters to about 20 meters. The radio-waves also have their origin in the chromosphere and corona.

Basic observational data on the Fraunhofer spectrum may be found in published catalogues of wavelengths, identifications and total intensities as well as in the

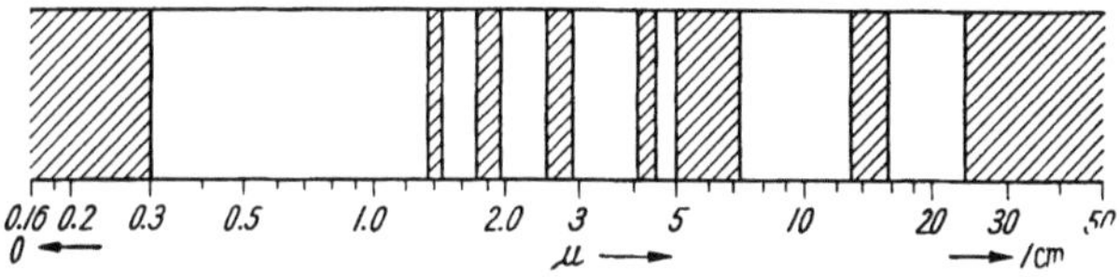

Fig. 15. Spectral windows in the earth's atmosphere. Absorption is essentially 100% complete in the crosshatched regions of the spectrum. The wavelength scale is in microns.

form of intensity tracings or atlases of the spectrum. The major sources of such data are listed in Refs. [*10*] to [*16*]. An excellent account of the techniques employed for the measurement and identification of Fraunhofer lines has been given in a recent article by C.E. MOORE in Ref. [*9*].

22. Wavelengths. The wavelengths of lines in the spectral region λ 2935 to λ 10604 have been measured in terms of the International Scale of 1928[2], with reference to a set of interferometrically determined standard wavelengths adopted by the International Astronomical Union. Infrared wavelengths longer than λ 10604 have been established by the method of coincidences, in which the positions of lines of unknown wavelengths are compared with those of known wavelengths in the overlapping higher orders of plane grating spectra[3]. The standard source of solar wavelengths and identifications in the visible and ultraviolet is still the Revision of ROWLAND'S Table [*10*] published in 1928, although the wavelengths given there are subject to small corrections[4] ranging from about −0.001 Å in the ultraviolet to about −0.008 Å in the red at Hα. The Revised Rowland Table has been further revised and extended in the ultraviolet region λ 2935—λ 3060 by BABCOCK, MOORE and COFFEEN and in the near infrared λ 6600—λ 13495 by BABCOCK and MOORE [*11*], the limits being set by the transparency of the Earth's atmosphere and by the infrared sensitivity of photographic plates. A new Revised Rowland Table covering the entire photographic region observable from the ground, i.e., λ 2935—λ 13495, is in preparation by C.E. MOORE. Over 200 new faint lines in the solar spectrum between λ 5000 and λ 6000 have been recorded and measured by ZALUBAS[5].

[1] L. GOLDBERG: The Solar System, Vol. II, Chap. 9, ed. G.P. KUIPER. Chicago, Ill.: Chicago University Press 1954.

[2] W. F. MEGGERS: Trans. Internat. Astronom. Union **3**, 81 (1928).

[3] O.C. MOHLER, A.K. PIERCE, R.R. MCMATH and L. GOLDBERG: Astrophys. Journ. **117**, 41 (1953).

[4] H. D. BABCOCK: Trans. Internat. Astronom. Union **7**, 151 (1950).

[5] R. ZALUBAS: Astrophys. Journ. **123**, 544 (1956).

The application of photoconductive cells to the observation of the Fraunhofer spectrum, beginning in 1947, has led to the compilation by MOHLER [*15*] of a new infrared table of solar lines, extending from λ 11984 to λ 25578, the resolution of the spectrum being comparable with that previously reserved only for the photographic region. This region of the spectrum is of particular interest both because it contains solar lines of relatively high excitation potential and because the continuous opacity of the photosphere passes through a minimum at about λ 16500. Both factors are favorable to the study of the deep layers of the atmosphere. A still further extension of the infrared spectrum is contained in the atlas and tables by MIGEOTTE and his co-workers [*16*] covering observations made at the Jungfraujoch in the region 2.8 to 23.7 μ. The Earth's atmosphere is completely opaque to solar radiation from 23.7 μ to the region of millimeter waves.

A new field of Fraunhofer line research has been initiated by the photography of the solar spectrum from high-altitude rockets above 100 km, which brings into view the resonance lines of atoms with high excitation potentials. The first observations were made in 1946 by a group of workers at the U.S. Naval Research Laboratory[1] and shortly thereafter by HOPFIELD and CLEARMAN[2]. Since 1948, the rocket spectroscopy of the solar ultraviolet has been carried out chiefly by the NRL group, and also by RENSE[3] and collaborators at the University of Colorado. The extensive bibliography that has resulted from these investigations is cited in the more recent papers referred to here. The solar spectrum has now been photographed[4] to an ultraviolet limit of 977 Å and observed at wavelengths shorter than 10 Å by means of photon counters with spectral sensitivities limited to narrow wavelength bands[5]. According to a report by TOUSEY[6], the solar continuum has been observed as far as λ 1550 and the Fraunhofer lines to λ 1650. Wavelengths and identifications have now been published for the region λ 2990 to λ 2085[7]. The resolving power in this region is about 0.5 Å and the scattered light rather large. Hence only relatively crude analyses are possible. However, the more stable and permanent observational platforms that will be provided by artificial satellites make it appear only a matter of time before the far ultraviolet spectrum is observed with the same precision now achieved for the visible spectrum. Figs. 16a and b are reproductions of the spectrum in the regions λ 2400—λ 1800 and λ 1800—λ 1000, respectively. The sharp drop in intensity below λ 2100 appears to be caused by a sudden increase in the opacity of the solar atmosphere, and may be attributed either to continuous absorption by metallic atoms or more probably to quasi-continuous absorption by abundant solar molecules such as NO, CO and H_2. The region at λ 1800 marks the transition between the Fraunhofer spectrum of the photosphere and the emission-line spectrum of the chromosphere at shorter wavelengths. The spectrum is dominated by the very strong Lyman-α emission at λ 1215.7, but also features several ionization stages of the abundant elements C, O and Si.

[1] E. DURAND, J. J. OBERLY and R. TOUSEY: Astrophys. Journ. **109**, 1 (1949).

[2] J. J. HOPFIELD and H. E. CLEARMAN: Phys. Rev. **73**, 877 (1949); H. E. CLEARMAN: Astrophys. Journ. **117**, 29 (1953).

[3] W. A. RENSE: Phys. Rev. **91**, 299 (1953).

[4] F. S. JOHNSON, J. D. PURCELL, R. TOUSEY and N. WILSON: Rocket Exploration of the Upper Atmosphere, p. 279. London: Pergamon Press 1954. — F. S. JOHNSON, H. H. MALITSON, J. D. PURCELL and R. TOUSEY: Astrophys. Journ. **1958** (in press).

[5] E. T. BYRAM, T. CHUBB and H. FRIEDMAN: Rocket Exploration of the Upper Atmosphere, p. 274, 276. London: Pergamon Press 1954.

[6] R. TOUSEY: Trans. Internat. Astronom. Union **9**, 179 (1955).

[7] N. L. WILSON, R. TOUSEY, J. D. PURCELL, F. S. JOHNSON and C. E. MOORE: Astrophys. Journ. **119**, 590 (1954).

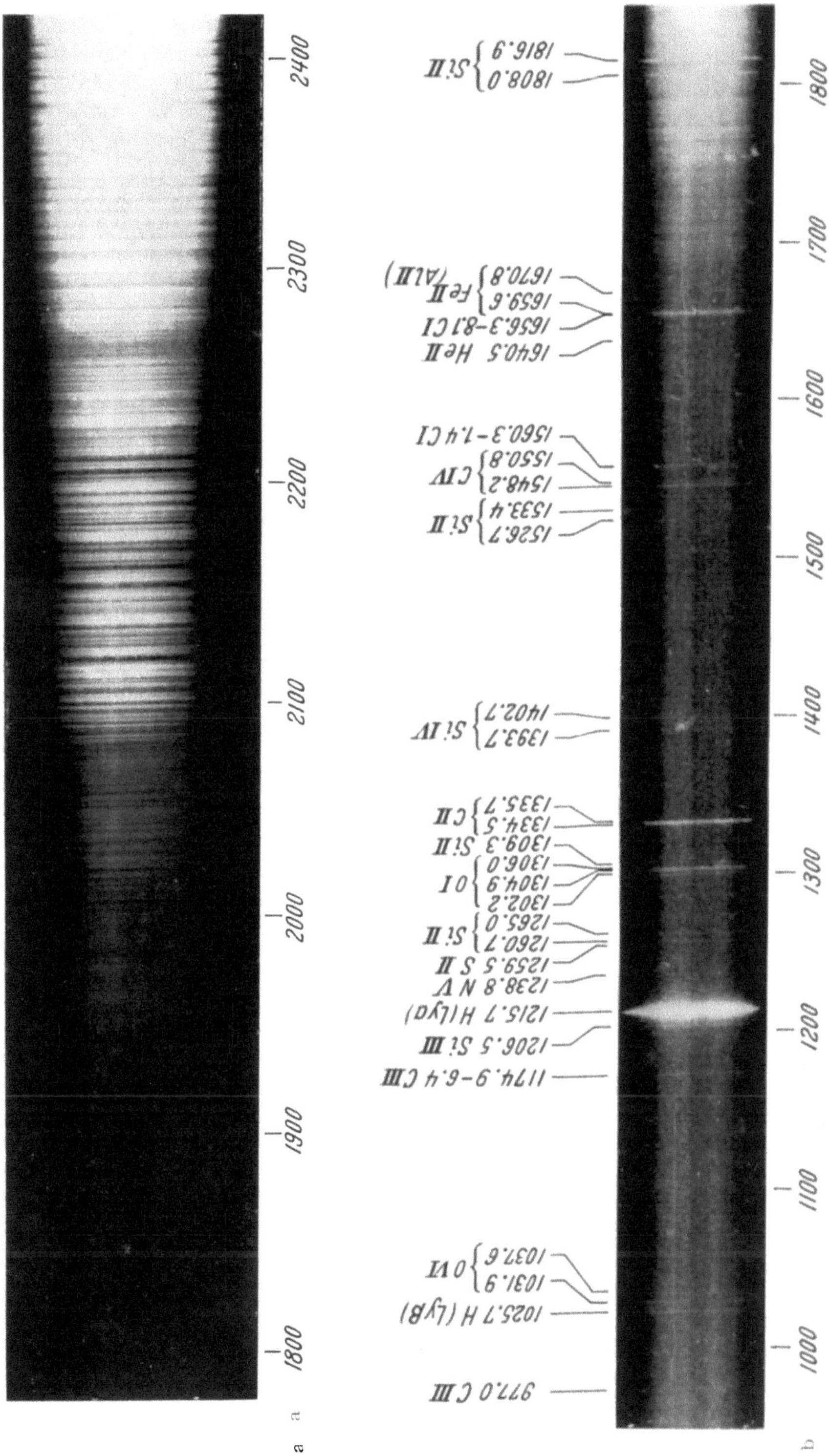

Fig. 16a and b. Photographs of the ultraviolet solar spectrum from Aerobee rockets at altitudes of over 100 km. Wavelength scales are in angstroms. (a) Wavelength region λ 2400 – λ 1800. (b) Wavelength region λ 1800 – λ 1000. Courtesy Dr. R. Tousey, Naval Research Laboratory, Washington, D.C.

23. Identifications. The first problem in the identification of Fraunhofer lines is to distinguish the lines of solar origin from those that are formed in the Earth's atmosphere (telluric lines). The latter are almost completely absent at wavelengths shorter than λ 5000, but become increasingly numerous as the wavelength increases. For example, MOHLER's infrared table contains a total of 7451 lines of which only 1786 are of solar origin. BABCOCK and MOORE [*11*] list four different methods of distinguishing telluric from solar lines:

1. Because the Sun rotates from east to west with a linear velocity of about 2 km/sec, solar lines observed at the east and west limbs show differential Doppler displacements amounting to about 0.08 Å at λ 6000, whereas the telluric lines are unshifted.
2. The strengths of telluric lines grow as the path length of sunlight through the Earth's atmosphere is increased, or, in other words, with increasing solar zenith distance. The intensities of water vapor lines also fluctuate with humidity.
3. Solar lines are usually changed in intensity and generally exhibit Zeeman splitting in the spectra of sunspots, whereas atmospheric lines are unaltered.
4. Solar lines are frequently wider and more diffuse than their telluric counterparts.

To these four methods must be added a fifth technique, which, however, can only be applied to photographic spectra of very high dispersion and resolving power. As pointed out in the introductory section, the Fraunhofer lines exhibit local fluctuations in position, width and intensity. On plates taken with the vacuum spectrograph of the McMath-Hulbert Observatory, these inhomogeneities in the solar lines are clearly visible to the eye, but the telluric lines are uniformly sharp (see Fig. 17). This method of segregating solar from telluric lines cannot be used for faint lines, for which the fluctuations are not readily apparent to the naked eye.

The straightforward and most widely used method of identifying solar lines with those of known atoms and molecules is by direct comparison with laboratory spectra. Identifications based on direct coincidences have also been supplemented by the prediction of numerous lines, which are not observed in the laboratory, but whose wavelengths are calculated from the differences between atomic energy levels derived from analyses of laboratory spectra. The prediction method has been particularly advantageous for the identification of faint lines, notably of Fe I, whose appearance in the Fraunhofer spectrum is favored by high temperature and long path in the photosphere, as well as for the identification of infrared lines, for which laboratory observations are very incomplete. The indispensable references for solar atomic identifications are the tables of atomic energy levels by MOORE[1] and their companion multiplet tables[2], and the infrared multiplet table of SWENSSON[3].

According to a recent compilation by MOORE [*9*], a total of sixty chemical elements have been fairly certainly identified in the Sun, with some evidence for the presence of seven others. Almost all of the elements are represented by their atomic spectra, with the exception of B and F, which are present in compounds only. The absence of the remaining elements may be explained on the grounds of low abundance, location of ultimate lines in the inaccessible ultraviolet, and the short mean lifetimes of certain radioactive atoms. The temperatures and pressures in the photosphere are such that the Fraunhofer spectrum is mainly

[1] C. E. MOORE: Nat. Bur. Stand. Circ. **1**, No. 467 (1949); **2** (1952).

[2] C. E. MOORE: Revised Multiplet Table. Princeton Obs. Contr. No. 20, 1945. — Ultraviolet Multiplet Table. Nat. Bur. Stand. Circ. Sec. I **1950**, No. 488; Sec. II **1952**.

[3] J. W. SWENSSON: Mém. Soc. Roy. Sci. Liège, Sér. IV **11**, Fasc. 1 (1951).

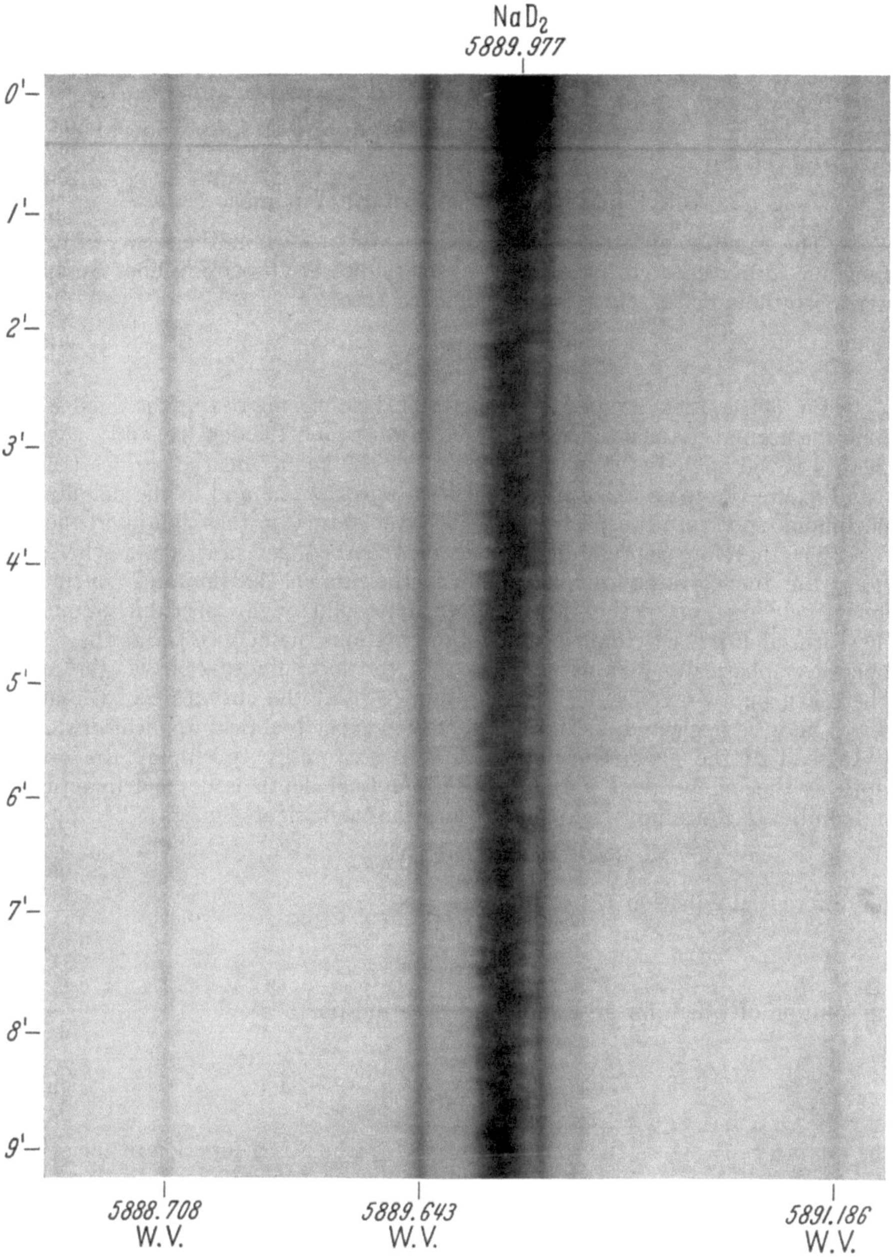

Fig. 17. The D_2 line of Na I in the Fraunhofer spectrum and neighboring lines of water vapor (W.V.). Photographed with the vacuum spectrograph of the McMath-Hulbert Observatory on Royal Pan emulsion at the center of the solar disk, 5th order of grating, slit width 125 μ, exposure 45 sec.

an atomic spectrum, but many molecular bands have also been identified. According to a recent survey by BROIDA and MOORE[1], 14 diatomic molecules are present in the Sun with reasonable certainty, whereas another 9 are possibly present.

[1] H. P. BROIDA and C. E. MOORE: Mém. Soc. Roy. Sci. Liège, Sér. IV **18** (1957).

Their lines are normally quite weak, with some exceptions, notably CN and C_2. Solar molecules are usually observed by means of their electronic spectra. The one exception is CO, for which the electronic transitions fall in the rocket ultraviolet below about λ 2000. Its solar abundance, however, is such that the fundamental[1] and first overtone vibrational bands[2] appear in absorption at 4.6 and 2.3 μ, respectively.

b) The formation of Fraunhofer lines.

24. The equation of transfer. The fundamental problem in the theory of Fraunhofer line formation is to predict the exact shapes of absorption lines by appropriate solutions of the equation of radiative transfer:

$$\mu \frac{d I_\nu}{\varrho\, d x} = (\varkappa_0 + \varkappa_\nu)\, I_\nu - j_0 - j_\nu . \tag{24.1}$$

In Eq. (24.1), I_ν is the specific intensity of the outgoing beam, x is the distance along the normal to the surface, measured inwards, ϱ is the density, and $\mu = \cos\vartheta$, where ϑ is the angle between the direction of the beam and the surface normal. $\varkappa_\nu$ and $\varkappa_0$ are the mass absorption coefficients in the line and in the neighboring continuous spectrum, respectively. The first term on the righthand side of Eq. (24.1) therefore represents the energy absorbed per unit mass, whereas j_ν and j_0 are the corresponding emission coefficients in the line and continuum, respectively, in ergs per unit mass, per unit solid angle, per unit frequency. The form of Eq. (24.1) implies two approximations, namely, (1) that the photosphere is spherically symmetric, and that the local fluctuations in the atmosphere can be averaged out (Sect. 35), and (2) that the curvature of the atmosphere may be neglected, as it usually may in most practical applications.

Instead of the geometric depth, x, it is customary to employ the optical depth as the independent variable. If the optical depth is defined in terms of the combined continuous and line absorption coefficients, viz.:

$$d\tau_\nu = (\varkappa_0 + \varkappa_\nu)\, \varrho\, d x , \tag{24.2}$$

the differential equation (23.1) becomes

$$\mu \frac{d I_\nu}{d \tau_\nu} = I_\nu - \frac{j_0 + j_\nu}{\varkappa_0 + \varkappa_\nu} , \tag{24.3}$$

the solution of which for an infinitely deep atmosphere is

$$I_\nu(\mu) = \int\limits_0^\infty \left(\frac{j_0 + j_\nu}{\varkappa_0 + \varkappa_\nu}\right) e^{-\tau_\nu/\mu}\, d\left(\frac{\tau_\nu}{\mu}\right). \tag{24.4}$$

The quantity $\frac{j_0 + j_\nu}{\varkappa_0 + \varkappa_\nu}$ is the source function S_ν, which is a function of the optical depth τ_ν. The emergent intensity I_ν is compounded of contributions S_ν from different levels of the atmosphere, each of which is absorbed exponentially by the overlying layers. In many practical applications, the optical depth is expressed in terms of the continuous absorption coefficient:

$$d\tau_0 = \varkappa_0\, \varrho\, d x , \tag{24.5}$$

and the parameter η_ν is introduced, where

$$\eta_\nu = \frac{\varkappa_\nu}{\varkappa_0} . \tag{24.6}$$

---

[1] J.L. LOCKE and G. HERZBERG: Canad. J. Phys. **31**, 504 (1953).
[2] L. GOLDBERG and E.A. MULLER: Astrophys. Journ. **118**, 397 (1953).

With this notation, the transfer equation takes the form

$$\mu \frac{dI_\nu}{d\tau_0} = (1+\eta_\nu)\, I_\nu - \left(\frac{j_0}{\varkappa_0} + \eta_\nu \frac{j_\nu}{\varkappa_\nu}\right), \tag{24.7}$$

whereupon the solution becomes

$$I_\nu(\mu) = \int\limits_0^\infty \left(\frac{j_0}{\varkappa_0} + \eta_\nu \frac{j_\nu}{\varkappa_\nu}\right) \exp\left(-\int\limits_0^{\tau_0} (1+\eta_\nu)\frac{dt}{\mu}\right) d\left(\frac{\tau_0}{\mu}\right). \tag{24.8}$$

The solution of the equation of transfer has engaged the attention of astrophysicists for over fifty years, beginning with the pioneering studies of SCHUSTER and SCHWARZSCHILD. However, despite many brilliant achievements, it is still not possible to reproduce in full detail the observed profiles of strong absorption lines, especially at their line centers, nor can the center-to-limb variations in the profiles be accurately predicted. Three separate but related problems enter into the solution of the radiative transfer equation. The first is the choice of the correct form of the emission coefficient j_ν, which requires exact knowledge of the mechanism by which the photospheric gases re-emit the energy they absorb. The second requires a knowledge of the structure of the photosphere, in particular the chemical composition and the variation with depth of the density and of the absorption and emission coefficients. Finally, given a mechanism of re-emission and a model photosphere, there remain the often formidable mathematical difficulties connected with the actual solution of the equation. Although a variety of exact and near-exact solutions of the equation of transfer have been carried out, some of them beautifully elegant, all have been based on relatively inexact approximations to the form of j_ν and to the structure of the photosphere. In the following sections we shall discuss these three aspects of the radiative transfer problem in order.

25. Re-emission mechanisms. *α) Coherent scattering.* Prior to about 1940, it was considered that the re-emission by excited atoms was governed by one or the other, or a combination, of the two mechanisms known as coherent scattering and pure absorption.

Consider a transition between a lower energy level 1 and an upper level 2. An atom in level 1 absorbs radiation of the proper frequency ν to be excited to level 2. It then promptly re-emits a photon of precisely the same frequency ν and returns to the ground level. The process is called *coherent scattering*. It is sometimes stated that an absorption line is formed by this process because the radiation absorbed from an outward moving beam along the line of sight is re-emitted or scattered in all directions. A more accurate statement [*5*] would be that the re-emitted photon is absorbed by an H^- ion and converted into continuous radiation before it can escape from the atmosphere. The coherent scattering mechanism, first introduced by SCHUSTER in 1905, is equivalent to MILNE's assumption of monochromatic radiative equilibrium as a means for solving the radiative transfer equation. If it is supposed that the total amount of energy absorbed in a given frequency ν is equal to that re-emitted in the same frequency, the emission coefficient, j_ν, takes the form

$$j_\nu = \varkappa_\nu \int I_\nu \frac{d\omega}{4\pi} = \varkappa_\nu J_\nu. \tag{25.1}$$

This equation implies that the populations of levels 1 and 2 are governed only by the intensity of ν-radiation and not by the temperature of the free electrons, since the latter do not interact in any way with the discrete levels. Furthermore, the source function J_ν varies across the line, since it has the same frequency dependence as the angularly-averaged value of I_ν.

β) Pure absorption. After an atom absorbs a photon of frequency ν, it may, before it can re-emit the same frequency, experience a collision of the second kind with a free electron or with a hydrogen atom, or it may suffer collisional or radiative ionization. The result is that the energy originally absorbed is converted into kinetic energy and eventually re-emitted in an entirely different frequency. The interaction between the free electrons and the excited states of the atom has the effect that the populations of the atomic levels are given by a Boltzmann distribution at the electron temperature, and further that the re-emission is governed by KIRCHHOFF'S law:

$$j_\nu = \varkappa_\nu B_\nu(T_e), \tag{25.2}$$

where $B_\nu(T_e)$ is the Planck function corresponding to the electron temperature. The re-emission is then said to occur in accordance with "local thermodynamic equilibrium", or "pure absorption". The source function is simply $B_\nu(T_e)$, which is of course constant over the entire line.

The assumption of pure absorption greatly simplifies the calculation of line profiles. It is generally considered valid when applied to transitions between highly-excited levels, on the basis that after excitation to the upper level, the atom is likely to re-radiate in a different frequency than that initially absorbed. Hence, a redistribution in frequency is effected. Furthermore, the lines of high excitation are formed principally in the deeper layers where the Boltzmann or excitation temperature is likely to be equal to the electron temperature.

When a very strong absorption line is formed by coherent scattering, its central intensity is zero, but the central intensity of a line formed by pure absorption is equal to the Planckian intensity at the temperature appropriate to the mean level of origin of the radiation. At the limb, the classical theory of coherent scattering predicts that the line radiation should be polarized[1] whereas lines formed by pure absorption should vanish at the limb, because the tangential line of sight for both the line and continuum radiation is essentially isothermal. The lines will not disappear, however, if the excitation temperature is lower than the electron temperature. Observations by REDMAN[2] show no evidence for polarization at the limb greater than about 10% of the amount predicted by the theory, but neither do most Fraunhofer lines disappear at the limb. Furthermore, the central intensities of even the strongest resonance lines are finite, although values as small as about 2% of the neighboring continuum have been measured by REDMAN[3] (see also Sect. 39).

γ) Combined absorption and coherent scattering. The failure of either the pure absorption or the coherent scattering mechanisms to account for observed line profiles suggested the possibility that a combination of the two processes might be operating. This hypothesis is expressed mathematically[4] by the introduction of a parameter ε, the fraction of excited atoms which are de-excited by super-elastic collisions or which undergo ionization. The line emission coefficient then becomes

$$j_\nu = \varepsilon \varkappa_\nu B_\nu + (1-\varepsilon) \varkappa_\nu J_\nu, \tag{25.3}$$

[1] H. ZANSTRA: Monthly Notices Roy. Astronom. Soc. London **101**, 250 (1941).
[2] R.O. REDMAN: Monthly Notices Roy. Astronom. Soc. London **101**, 266 (1941); **103**, 329 (1943).
[3] R.O. REDMAN: Monthly Notices Roy. Astronom. Soc. London **97**, 552 (1937).
[4] A.S. EDDINGTON: Monthly Notices Roy. Astronom. Soc. London **89**, 620 (1929).

where it is understood that both J_ν and B_ν are functions of τ_0. There are very good reasons for assuming[1] that the continuous emission coefficient j_0 is given by KIRCHHOFF'S law:

$$j_0 = \varkappa_0 B_\nu. \tag{25.4}$$

Hence, the equation of transfer (24.7) may be written in the well known Eddington-Strömgren form for combined absorption and coherent scattering:

$$\mu \frac{d I_\nu}{d \tau_0} = (1 + \eta_\nu) I_\nu - B_\nu (1 + \varepsilon \eta_\nu) - (1 - \varepsilon) \eta_\nu J_\nu. \tag{25.5}$$

Many interesting properties of the transfer equation and its solutions can be deduced from the integral form of the equation. Let the optical depth in the line plus continuum be τ_ν, so that

$$d \tau_\nu = (1 + \eta_\nu) d \tau_0. \tag{25.6}$$

If we set

$$L = \frac{1 + \varepsilon \eta_\nu}{1 + \eta_\nu}, \tag{25.7}$$

Eq. (25.5) becomes

$$\mu \frac{d I_\nu}{d \tau_\nu} = I_\nu - L B_\nu - (1 - L) J_\nu. \tag{25.8}$$

The source function is therefore

$$S_\nu = L B_\nu - (1 - L) J_\nu. \tag{25.9}$$

Making use of the integral equation for J_ν, which was first derived by SCHWARZSCHILD and MILNE by substituting the expression for I_ν in Eq. (24.8) into Eq. (25.1) we obtain (see [8], p. 260)

$$S_\nu = L B_\nu + \tfrac{1}{2} (1 - L) \int_0^\infty \mathrm{Ei}(|t_\nu - \tau_\nu|) S_\nu(t_\nu) d\tau_\nu, \tag{25.10}$$

where Ei is the exponential integral:

$$\mathrm{Ei}(x) = \int_0^\infty \frac{e^{-xy}}{y} dy, \tag{25.11}$$

and $y = 1/\mu$.

δ) Non-coherent scattering. Eqs. (25.5) and (25.10) have had considerable success in predicting the major features of solar line profiles on the basis of various kinds of assumed model photospheres. It fails, however, in many important details, particularly in not accounting for the wings of strong lines at the solar limb. These failures have led to the introduction of the concept of non-coherent scattering, which is based on the notion that, owing to the finite breadths of energy levels, radiation absorbed in one part of a line may be re-emitted elsewhere in the same line. The absorbed frequencies may thus be randomly re-distributed, even though each excited atom returns by a spontaneous radiative transition to the level from which it was excited. This concept was first suggested by EDDINGTON[2] and was later established by the calculations of SPITZER[3] and WOOLEY[4] and by the observational studies of HOUTGAST[5].

[1] H.H. PLASKETT: Vistas in Astronomy, Vol. I, p. 637. London: Pergamon Press 1955.
[2] A.S. EDDINGTON: Monthly Notices Roy. Astronom. Soc. London **89**, 620 (1929).
[3] L. SPITZER: Astrophys. Journ. **99**, 1 (1944).
[4] R. v. D. R. WOOLEY: Monthly Notices Roy. Astronom. Soc. London **98**, 624 (1938).
[5] J. HOUTGAST: Diss. Utrecht 1942.

In the solar and stellar atmospheres, the re-distribution of absorbed frequencies is brought about by radiative damping, by pressure broadening and by the random motions of the absorbing and emitting atoms. Radiative damping is important only for subordinate lines[1,2] since, in the absence of other broadening effects, the ground state has zero width and hence the absorbed quantum must be re-emitted with its frequency unchanged. The theory of pressure broadening in the solar atmosphere is very incomplete. No single theory has been developed that takes account simultaneously of the many kinds of interactions between the emitting and absorbing atoms and the surrounding charged particles. The theory has been well worked out, however, for the extreme cases of (1) collisional broadening, in which the atom is disturbed primarily by discrete encounters with neighboring atoms, ions and electrons, and (2) statistical broadening, in which the atom is subjected to a slowly-varying electric field produced by the totality of surrounding charged particles. For both extremes of collisional and statistical broadening, SPITZER has shown that the scattering will be completely non-coherent and hence it is fair to assume that non-coherency will prevail in the general case as well.

The importance of the Doppler effect as a re-distribution mechanism was pointed out by HENYEY[3] [1]. It is caused by the fact that a moving atom may absorb radiation from one direction and emit it in another. The absorbed and emitted frequencies will therefore not be identical. The influence of the Doppler effect on the scattering process has been investigated in considerable detail by MIYAMOTO[4], who has succeeded in this way in accounting semi-quantitatively for the shapes of the cores of very strong lines.

Whether the energy levels are broadened by the Doppler effect, by radiative damping, or by pressure broadening, the net result is that in most cases the frequency dependence of the absorption and emission coefficients is identical and hence the ratio $j_\nu/\varkappa_\nu$ is a constant for each line and independent of I_ν. Let us suppose, for example, that

$$j_\nu = \varkappa_\nu \bar{J}, \tag{25.12}$$

where $\bar{J}$ is independent of frequency. To determine the constant of proportionality $\bar{J}$, we use the condition that the total amount of energy absorbed and emitted in the line must be equal, regardless of whether the scattering is coherent or non-coherent. In other words,

$$\int j_\nu \, d\nu = \int J_\nu \varkappa_\nu \, d\nu = \bar{J} \int \varkappa_\nu \, d\nu \tag{25.13}$$

and therefore

$$\bar{J} = \frac{\int J_\nu \varkappa_\nu \, d\nu}{\int \varkappa_\nu \, d\nu}. \tag{25.14}$$

The equation of transfer for the case of combined continuous absorption, line absorption and completely non-coherent scattering then has the form

$$\mu \frac{dI_\nu}{d\tau_0} = (1+\eta_\nu) I_\nu - B_\nu (1+\varepsilon\eta_\nu) - (1-\varepsilon)\,\eta_\nu \frac{\int J_\nu \eta_\nu \, d\nu}{\int \eta_\nu \, d\nu}. \tag{25.15}$$

The more general case in which the scattering is only partially non-coherent has been discussed by ZANSTRA[5] and by WOOLEY and STIBBS [4]. ZANSTRA formulated

---

[1] See footnote 4, p. 45.

[2] L. SPITZER: Monthly Notices Roy. Astronom. Soc. London **96**, 794 (1936).

[3] L. G. HENYEY: Proc. Nat. Acad. Sci. U.S.A. **26**, 50 (1940).

[4] S. MIYAMOTO: Z. Astrophys. **35**, 145 (1954).

[5] H. ZANSTRA: Monthly Notices Roy. Astronom. Soc. London **101**, 273 (1941); **106**, 225 (1946).

the problem for a classical oscillator under the influence of phase-shift broadening (see Sect. 27), and concluded that the scattering would consist of coherent and non-coherent components, the intensity ratio being equal to the ratio of radiative to collisional damping. EDMONDS'[1] recent quantum-theory formulation shows that ZANSTRA's results are not applicable to non-coherency by statistical broadening and that the coherent component in fact amounts to no more than 5% of the total intensity.

26. Model atmospheres. No matter what mechanism of re-emission is assumed, the solution of the transfer equation requires a knowledge of the structure of the atmosphere, the most important parameters being the electron or kinetic temperature, the electron pressure, the total gas pressure, the turbulent velocity, and their dependence upon the optical depth τ_0. These parameters can then be employed to calculate the functions $B_\nu(\tau_0)$ and $\eta_\nu(\tau_0)$. It is commonly assumed that B_ν, the continuum source function, is given accurately by PLANCK's law, viz.,

$$B_\nu = \frac{2h\nu^3}{c^2} \frac{1}{e^{h\nu/kT_r} - 1}, \tag{26.1}$$

where T_r is the radiation temperature, which can be derived as a function of τ_0 from observations of limb darkening (see Sect. 17). Eq. (26.1) is valid only in the case of local thermodynamic equilibrium. The effects of line and continuous absorption cause T_r to vary with wavelength. PLASKETT[2] has given reasons for believing that under photospheric conditions, the kinetic and ionization temperatures will be equal to a mean value of the radiation temperature, but there is no indication as to how the mean value is to be determined. The existence of temperature inhomogeneities also leads to departures of the source function from the Planckian form[3].

The quantity η_ν is the ratio, $\varkappa_\nu/\varkappa_0$, of the line to the continuous absorption coefficient. The latter can be determined empirically from limb darkening, or calculated from the theory of continuous absorption by neutral hydrogen and negative hydrogen ions. Convenient tables for this purpose have been given by VITENSE[4] and by UENO, SAITO and JUGAKU[5]. The contribution by neutral hydrogen is slight in the upper and middle layers of the photosphere but increases with depth. The continuous absorption coefficient per gram of solar gas varies with both temperature and electron pressure, especially the latter, and also depends on the abundance of hydrogen relative to helium, since the latter does not contribute to the opacity. The use of empirical values of $\varkappa_0$ may perhaps be preferable to the theoretical values, since the latter assume that only neutral hydrogen and H^- are responsible for the opacity, whereas other unknown agents may also play a role.

The calculation of $\varkappa_\nu$, the line absorption coefficient per gram, usually involves a variety of uncertain assumptions. The line absorption coefficient is first of all proportional to $n_{r,s}$, the number of atoms per gram in stage of ionization r and level of excitation, s. It is related to the total number of atoms per gram through the Saha and Boltzmann equations, and therefore depends on the temperature and electron pressure, which are functions of the depth τ_0. It is usually assumed that the Boltzmann excitation temperature and the Saha ionization

---

[1] F. N. EDMONDS jr.: Astrophys. Journ. **121**, 418 (1955).
[2] H. H. PLASKETT: Vistas in Astronomy, Vol. I, p. 637. London: Pergamon Press 1955.
[3] K. H. BÖHM: Z. Astrophys. **35**, 179 (1954).
[4] E. VITENSE: Z. Astrophys. **28**, 81 (1951).
[5] S. UENO, S. SAITO and J. JUGAKU: Contrib. Kyoto Inst. Astrophys. **1954**, No. 43.

temperature are identical and equal to the electron and radiation temperatures. This assumption is valid for thermodynamic equilibrium but may be substantially in error for the outer layers of the photosphere.

27. Line broadening mechanisms[1]. *α) Doppler effect.* Next, it is necessary to know the frequency dependence of $\varkappa_\nu$ which depends on the mechanism of line broadening. The important mechanisms in the solar atmosphere are (1) thermal and turbulent broadening by the Doppler effect; (2) damping by radiation and collisions; and (3) pressure broadening. Certain metallic lines also show evidence of broadening by hyperfine structure[2,3].

The random Doppler effect leads to an absorption coefficient whose shape is determined entirely by the Gaussian distribution of atomic velocities, namely,

$$\varkappa_\nu = \varkappa_{\nu_0} \mathrm{e}^{-\left(\frac{\Delta\nu}{\Delta\nu_D}\right)^2}, \tag{27.1}$$

where $\varkappa_{\nu_0}$ is the absorption coefficient at the line center and $\Delta\nu_D$ is the so-called Doppler width, defined as the frequency displacement corresponding to the most probable *space* velocity. The central absorption coefficient is given by

$$\varkappa_{\nu_0} = n_{r,s} f \frac{\sqrt{\pi}\, e^2}{m c} \frac{1}{\Delta\nu_D}, \tag{27.2}$$

where f is the conventional f-value, and the other symbols have their usual meanings. A variety of observational evidence, both from the curve of growth and line profiles, (see the article by DE JAGER, in this volume) proves that the Doppler widths of metallic Fraunhofer lines are on the order of a factor of two greater than the thermal value for $T = 6000°$ K. The excess width is caused by the random turbulent motions of the solar gases. This requires that the Doppler width be given by the relation

$$\Delta\nu_D = \frac{\nu_0}{c}\left(\sqrt{\frac{2kT}{M}} + \xi_t\right), \tag{27.3}$$

where M is the mass of the atom and ξ_t is the most probable turbulent velocity.

It is not yet known with certainty how the turbulent velocity varies with depth in the photosphere. The depth dependence can be studied both by observing the center-limb variation of the half-widths of faint Fraunhofer lines, and also by determining the half-widths of lines that originate from different depths by virtue of their excitation characteristics. The analysis is complicated by the fact that the turbulence is anisotropic. The results obtained to date are contradictory. Some investigations suggest that the turbulence increases with depth, others that it decreases, and still others that it is independent of depth. Most investigators have adopted constant values of between one and two km/sec, but eventually the exact depth and angular dependence of the turbulent velocity will have to be taken into account (see Sects. 35 and 36).

β) Radiative damping. The absorption coefficient for radiative damping is given by the dispersion formula

$$\varkappa_\nu = n_{r,s} f \frac{\pi e^2}{m c} \frac{\gamma_r/4\pi}{(\nu - \nu_0)^2 + (\gamma_r/4\pi)^2}, \tag{27.4}$$

---

[1] For line broadening effects see the article by R. G. BREEN jr. in Vol. XXVII of this Encyclopedia.

[2] H. A. ABT: Astrophys. Journ. **115**, 199 (1952).

[3] W. R. HINDMARSH: Monthly Notices Roy. Astronom. Soc. London **115**, 270 (1955).

where γ_r is the sum of the reciprocal radiative mean lifetimes of the two energy states, in $\sec^{-1}$. It is therefore equal to the sum of the downward transition probabilities from both energy states. Values of γ_r may be calculated from tables of transition probabilities, as given for example, by LANDOLT-BÖRNSTEIN[1] and by ALLEN[2]. For most Fraunhofer lines its value is of the order of 10^8.

γ) Collisional damping. Collisional damping occurs in the solar photosphere as a result of discrete encounters with perturbing particles which produce large cumulative phase shifts in the radiated wave train during the course of the encounter. The instantaneous phase shift $\delta\nu$, is proportional to r^{-n}, where r is the distance of the disturbing particle and n is an integer which is determined by the nature of the disturbing particle. For an ion or an electron acting upon hydrogen, $n=2$ (linear Stark effect) while for most other radiating atoms $n=4$ (quadratic Stark effect). When the perturbing particle is a neutral atom of the same kind as the radiating atom, $n=3$; when it is a neutral atom of a different type, $n=6$ (van der Waals forces). It turns out that no matter what the value of n, the phase-shift theory of line broadening leads to an absorption coefficient which is identical with Eq. (27.4) for radiative damping, with the exception that γ_r is replaced by a collisional damping constant, γ_c, where

$$\gamma_c = \frac{2}{\tau}, \tag{27.5}$$

and τ is the so-called "mean free time" between collisions (the velocity divided by the mean free path). Formulae for γ_c have been given by LINDHOLM [5]. For the conditions that prevail in the photosphere, the collision damping for hydrogen lines is caused by the linear Stark effect of ions, while for metallic lines either the quadratic Stark effect of electrons or the van der Waals interactions with neutral hydrogen atoms predominates.

Observations show that in the moderate and very deep layers of the photosphere, the collisional damping, which is proportional to the number density of perturbing particles, exceeds radiative damping by factors of from 5 to 100. In the higher layers, however, radiative damping must be included and this can be done in general by replacing γ_r in Eq. (27.4) by $\gamma_r + \gamma_c$. Although the theory of collisional damping is not sufficiently refined for the accurate calculation of γ_c, it can be combined with laboratory experiments to give reasonably useful results[3].

δ) Statistical Stark broadening. The phase-shift theory of line broadening represents one of two extreme approaches to the general problem of pressure broadening, in which the energy levels are perturbed only by encounters with individual fast-moving particles. The phase-shift approximation is most nearly realized in the broadening of metallic lines by electrons and H atoms. In the other extreme, which seems to be more relevant to the hydrogen lines, the energy levels are considered to be broadened by the Stark effect of the slowly varying electric field generated by the surrounding configuration of charged particles. The theory of statistical broadening of the hydrogen lines was originally treated by HOLTSMARK for electrons and ions at rest with respect to the radiating H atom. In its application to stellar atmospheres, the Holtsmark theory implies that the electric field is steady and homogeneous during the time in which it perturbs

---

[1] LANDOLT-BÖRNSTEIN: Atom- and Molekularphysik, Part I: Atoms und Ionen, p. 260, 6th ed. Berlin: Springer 1950.

[2] C. W. ALLEN: Astrophysical Quantities. London: Athlone Press 1955.

[3] A. UNSÖLD and V. WEIDMANN: Vistas in Astronomy, Vol. I, p. 249. London: Pergamon Press 1955.

the individual H atom. It has been emphasized by ODGERS[1], however, that the effects of high-velocity electrons cannot be properly handled in the adiabatic type of approximation when the field of the electron fluctuates rapidly in a time comparable with the Bohr period of the atomic electron. The non-adiabatic case has recently been investigated quantum-mechanically by KIVEL, BLOOM and MARGENAU[2] and in the classical-path approximation by KOLB[3]. KOLB finds that the additional broadening due to high-velocity electrons is comparable to the static ion broadening at all densities and must be included in the theoretical treatment of hydrogen line profiles. This conclusion was confirmed by TURNER and DOHERTY[4], who showed that profiles of $H\beta$ formed in the shock tube could be better represented when electron broadening is taken into account. However, the problem of the hydrogen line broadening is not yet completely solved, partly because the approximations in the Kolb theory are not valid near the line center. This failure of the theory may not be too serious for the Sun, where the central parts of the hydrogen lines are probably widened by the random Doppler effect. Furthermore, the effect of the electrons on the wing broadening is allowed for only approximately by KOLB. A reliable theory for the line wings is urgently needed.

Attempts to test theories of Stark broadening against solar observations have been inconclusive. Thus, DE JAGER, MIGEOTTE and NEVEN[5] concluded that the observed profile of the Brackett-α line of hydrogen, $H_{4\text{-}5}$, could be represented by a profile computed on the basis of the Holtsmark theory. DOHERTY and ALLER[6] and SEVERNY[7] find, however, that the wings of the theoretical Holtsmark profiles for $H\alpha$ are shallower than those observed. SEVERNY concludes that Stark broadening is less effective for $H\alpha$ than is pressure broadening by neutral hydrogen, whereas DOHERTY and ALLER argue that the Holtsmark theory is incorrect and show that the Kolb theory gives better agreement with the solar observations.

ε) Combined Doppler and damping broadening. The absorption coefficient for metallic lines results from the combined broadening effects of random Doppler motions and damping by radiation and/or collisions. Each point of the damping-broadened profile of Eq. (27.4) is broadened into a Gaussian velocity distribution and the combined absorption coefficient is derived by summing the contributions of the separate error curves. The resulting integral must then be solved numerically. Rather complete tables of solutions for lines of astrophysical interest have been computed by HARRIS[8] and others. The absorption coefficient is written in the form

$$\varkappa_\nu = \varkappa_{\nu_0} \cdot H(a, v)\,, \tag{27.6}$$

where $\varkappa_{\nu_0}$ is the central absorption coefficient for pure Doppler broadening, given by Eq. (27.2) and $H(a, v)$ is tabulated as a function of the parameters a and v. Here

$$v = \frac{\Delta\nu}{\Delta\nu_D}\,, \tag{27.7}$$

[1] G. J. ODGERS: Astrophys. Journ. **116**, 444 (1952).
[2] B. KIVEL, S. BLOOM and H. MARGENAU: Phys. Rev. **98**, 495 (1955).
[3] A. C. KOLB: Thesis, University of Michigan 1956.
[4] E. B. TURNER and L. DOHERTY: Astronom. J. **60**, 158 (1955).
[5] C. DE JAGER, M. MIGEOTTE and L. NEVEN: Ann. d'Astrophys. **19**, 9 (1956).
[6] L. DOHERTY and L. H. ALLER: Proc. NSF Conf. Stellar Atmospheres, ed. by M. H. WRUBEL: Indiana University 1954.
[7] A. B. SEVERNY: Publ. Crimean Astrophys. Obs. **16**, 54 (1956).
[8] D. L. HARRIS: Astrophys. Journ. **108**, 112 (1948).

the distance from the line center in units of the Doppler width and

$$a = \frac{\gamma/4\pi}{\Delta\nu_D}, \tag{27.8}$$

the ratio of half-widths for damping and Doppler broadening, respectively.

In summary, the calculation of the line absorption coefficient requires first a knowledge of the structure of the atmosphere, the parameters of which are the density, temperature, electron pressure and velocity characteristics of the gases. The second requirement is for an accurate theory of line broadening, especially in the outer parts of the lines, where pressure broadening is operative. Finally, it is necessary to know certain atomic constants, the most important of which are the collisional damping constants, which vary from line to line and from element to element, and the f-values. The usual assumption of local thermodynamic equilibrium permits the equating of electron, excitation and ionization temperatures with the radiation temperature derived from limb darkening. This may be erroneous in the high layers, but there is at present no straightforward procedure for determining the extent of probable deviations.

With this brief assessment of the problem of Fraunhofer line formation, we shall now proceed to describe some of the approximate solutions that have been attempted.

28. Analytic solutions of the transfer equation. The calculation of Fraunhofer line profiles is always based on simplifying assumptions which chiefly concern (1) the structure of the atmosphere, and (2) the re-emission mechanism. The structure enters through the adoption of models of the atmosphere, by means of which B_ν is specified as a function of τ_0 and η_ν as a function of both τ_0 and ν. The simplest kind of model, which has come to be known as the Schuster-Schwarzschild (or S.S.) model, is that in which the continuous spectrum is radiated by PLANCK's law from a sharply-bounded photosphere and in which the absorption line is formed entirely in a gaseous layer above, the so-called reversing layer. This concept has been extremely useful in early explorations of the solar spectrum, but has now been virtually abandoned, since there is no sharp dividing line between the layers in which the continuous absorption and line formation take place. The exact solution of the S.S. problem with coherent scattering has been given by CHANDRASEKHAR [*2*].

Most analytic calculations of line profiles have in recent years been based upon the Milne-Eddington (or M.E.) model, in which both the continuum and line are formed in a single layer of infinite optical depth, with the further condition that η_ν is constant with depth. The continuum source function, B_ν, is always assumed to be a Planck function, and usually its dependence on optical depth is taken to be linear, i.e.,

$$B_\nu(\tau_0) = B_\nu^0 + B_\nu^1\,\tau_0, \tag{28.1}$$

the coefficients B_ν^0 and B_ν^1 being derived from theory or from limb-darkening observations. The observations show that this approximation is quite good except in the deepest layers, for which higher terms must be included. The constancy with depth of η_ν is a dubious assumption for most lines. Even for the K-line of Ca II, for which both the line and continuous absorption coefficients increase with optical depth, ALLER ([*8*], p. 271) finds that the ratio decreases by more than a factor of three between $\tau = 0.05$ and $\tau = 2.0$.

Inasmuch as there is no precise physical theory for the reemission, it has been necessary to assume one mechanism or another and to test the assumptions empirically against the observations. However, one cannot always be certain whether discrepancies should be attributed to errors in the model or to the assumed form of j_ν. The assumed mechanisms vary between the extremes of pure absorption ($\varepsilon = 1$) on the one hand and pure scattering ($\varepsilon = 0$) on the other. The scattering may also be assumed to be completely non-coherent or only partially so.

The frequency dependence of η_ν does not enter into the solution of the transfer equation for the M.E. model when the scattering is coherent. It must of course be taken into account whenever the scattering is non-coherent since it appears in the calculation of $\bar{J}$ in Eq. (25.15).

Accounts of the extensive literature on solutions of the transfer equation will be found in [2] to [5]. The classical EDDINGTON[1] approximation to the solution of the M.E. problem with coherent scattering [Eq. (25.5)] has been widely applied but has now been superseded by a variety of relatively simple exact solutions, the first of which was due to CHANDRASEKHAR[2]. CHANDRASEKHAR'S solution for the emergent intensity, with η_ν constant and B_ν linear with depth, was given in closed form in terms of the so-called $H(\mu)$ functions and their moments. More recent investigations have employed different methods with the aim of (1) simplifying numerical computations; (2) utilizing non-linear forms of B_ν; (3) obtaining the explicit form of $J_\nu(\tau)$ at each depth; and (4) developing techniques that could be applied to the non-coherent scattering problem. Thus, KOURGANOFF[3] showed that CHANDRASEKHAR'S method could be extended to yield the τ-dependence of J_ν. MENZEL and SEN[4] applied an operational method and HUANG[5] the variational method to obtain both the emergent intensity and $J_\nu(\tau)$. HUANG[6] has solved the M.E. problem by means of the Laplace transform method with the Planck function expressed as a polynomial in the optical depth. Similar solutions by Miss BUSBRIDGE[7] make use of the principle of invariance governing the law of diffuse reflection, first introduced by AMBARTSUMIAN[8] and generalized by CHANDRASEKHAR [2].

The principle of invariance appears to be an extremely powerful tool for the solution of the transfer problem with non-coherent scattering[9]. The problem has generally been attacked with the aid of two types of approximations, the first dealing with the structure of the atmosphere and usually involving the assumption of the Milne-Eddington model, and the second concerned with the manner in which radiation is interchanged between different parts of the line. The problem of the radiation interchange was treated by approximate methods[10-12] and also first solved exactly by the principle of invariance by SOBOLEV[13] for the simplified case in which both the Planck function and the continuous absorption coefficient are independent of depth. By the principle of invariance, a number of authors have now handled rigorously the problem of radiation interchange[14-17].

---

[1] A. S. EDDINGTON: Monthly Notices Roy. Astronom. Soc. London **89**, 620 (1929).
[2] S. CHANDRASEKHAR: Astrophys. Journ. **106**, 145 (1947).
[3] V. KOURGANOFF: Astrophys. Journ. **113**, 419 (1951). See also [3].
[4] D. H. MENZEL and H. K. SEN: Astrophys. Journ. **113**, 482 (1951).
[5] SU-SHU HUANG: Phys. Rev. **88**, 50 (1952); Astrophys. Journ. **117**, 215 (1953).
[6] SU-SHU HUANG: Ann. d'Astrophys. **15**, 352 (1952).
[7] I. W. BUSBRIDGE: Monthly Notices Roy. Astronom. Soc. London **113**, 52 (1953).
[8] V. A. AMBARTSUMIAN: Dokl. Acad. Sci. USSR. **38**, 229 (1943).
[9] I. W. BUSBRIDGE: Monthly Notics Roy. Astronom. Soc. London **115**, 661 (1955).
[10] G. MUNCH: Astrophys. Journ. **109**, 275 (1949).
[11] D. LABS: Z. Astrophys. **28**, 150 (1951).
[12] Z. SUEMOTO: Publ. Astronom. Soc. Japan **1**, 78 (1949).
[13] V. V. SOBOLEV: Russ. Astronom. J. **26**, 129 (1949).
[14] I. W. BUSBRIDGE: Monthly Notices Roy. Astronom. Soc. London **112**, 45 (1952); **113**, 52 (1953); **115**, 661 (1955). — Quart. J. (Oxford) (2) **6**, 218 (1955).
[15] V. V. SOBOLEV: Russ. Astronom. J. **31**, 231 (1954).
[16] D. N. STIBBS: Monthly Notices Roy. Astronom. Soc. London **113**, 493 (1953).
[17] S. UENO: Contrib. Inst. Astrophys. Kyoto **1955**, No. 58.

As a result of their work, the equation of transfer has been solved exactly with the following restrictions:

1. The atmospheric model is of the Milne-Eddington type, with a linear Planck function given by Eq. (28.1) and η_ν independent of depth.

2. The frequency dependence of the absorption coefficient is that of combined Doppler broadening and damping [Eq. (27.6)].

3. The scattering may be completely or only partially non-coherent[1].

4. Allowance may be made for a fractional amount of thermal re-emission by the introduction of the parameter ε in Eq. (25.15), ε being independent of both frequency and depth.

The effect on the non-coherent contours of departures from linearity of the Planck function has been discussed by LABS[2].

The question of small departures of η_ν from depth constancy has been treated analytically for coherent scattering by STRÖMGREN[3], and TUBERG[4] and extended to any value of μ by HITOTUYANAGI[5]. For non-coherent scattering, however, the arbitrary variation of B_ν, η_ν and ε with depth can only be handled by numerical integration and successive approximation. Calculations of this type have been made for the Balmer lines by DE JAGER[6] and for the K line by ALLER [8].

29. Observational tests of non-coherent scattering. We have already remarked that most Fraunhofer lines are not formed either by pure scattering or by pure absorption, but by a combination of the two. This is the reason for the introduction of the parameter ε into the equation of transfer. There are certain lines, however, for which one process or the other appears to be predominant. For example, resonance lines are most likely to be formed by scattering, and subordinate lines by pure absorption. Hence, the K-line of Ca^+ and the D-lines of Na have most frequently been employed in observational tests of theories of non-coherent scattering. The results have been somewhat conflicting. HOUTGAST[7] showed that the center-limb variations in the wings of several strong Fraunhofer lines could not be represented by either true absorption or coherent scattering and explained the discrepancies qualitatively by non-coherent scattering. More quantitative calculations on the K-line by MUNCH[8], SAVEDOFF[9], and ALLER [8] improve the fit with observation as compared with the pure coherent solution. SOBOLEV[10] also concludes that the theory of line formation by non-coherent scattering is in better agreement with observation than that of coherent scattering. UNSÖLD[11] drew attention to the importance of using an accurate model of the photosphere in testing for the existence of non-coherent scattering and criticized the use by HOUTGAST of a linear Planck function. However, ALLER'S calculation of the K-line profile was carried out by a numerical method in which the solar model was derived from limb darkening observations and the exact dependence of η_ν with depth was taken into account. The solution for

[1] See footnote 16, p. 52.
[2] D. LABS: Z. Astrophys. **34**, 173 (1954).
[3] B. STRÖMGREN: Astrophys. Journ. **86**, 1 (1937).
[4] M. TUBERG: Astrophys. Journ. **103**, 145 (1946).
[5] Z. HITOTUYANAGI: Publ. Astronom. Soc. Japan **5**, 96 (1953).
[6] C. DE JAGER: Rech. Obs. Utrecht **13**, 1 (1952).
[7] J. HOUTGAST: Diss. Utrecht 1942.
[8] G. MUNCH: Astrophys. Journ. **109**, 275 (1949).
[9] M. P. SAVEDOFF: Astrophys. Journ. **115**, 509 (1952).
[10] V. V. SOBOLEV: Russ. Astronom. J. **31**, 231 (1954).
[11] A. UNSÖLD: Vjschr. Astr. Ges. **78**, 201 (1943).

coherent scattering predicts emission in the line wings at the limb, but the introduction of non-coherent scattering brings the solution into very good agreement with the observations.

The difficulty of verifying experimentally non-coherent scattering is illustratet by several investigations of the D lines. HARRIS[1] and HITOTUYANAGI and INABA[2] have calculated the center-limb variation in the profiles on the assumption of coherent scattering and on the basis of reasonable physical models of the photosphere, including the variation of η_ν with depth. HARRIS finds that the observed line wings are slightly deeper and wider than the computed profiles and offers the possibility that the deviations may be due to the neglect of non-coherent scattering. On the other hand, HITOTUYANAGI and INABA obtain rather exact agreement with observation for a model atmosphere derived by ALLER and PIERCE[3], whereas HARRIS' calculations are based on a model by MUNCH. PRIESTER[4] uses a model by BÖHM-VITENSE[5], which is featured by a sharp drop in temperature in the outer layers, and also assumes pure absorption in reproducing his new observations of the D-line profiles at the center and limb of the Sun. But PLASKETT[6] has shown that only 0.46% of Na atoms in the excited 4^2P^0 level will be de-excited by collisions of the second kind with H atoms before re-radiating. It has also been found ([5], p. 133) that the probability of upward radiative excitation from 4^2P^0 is about 1% relative to that of spontaneous re-emission. It is therefore most unlikely that the D lines are formed by pure absorption. Although non-coherent scattering simulates pure absorption in the deep solar layers where the line wings (seen at the center of the disk) would be formed, appreciable departures are to be expected in the higher layers, which contribute to the wings when they are observed at the limb (see also Sect. 39).

The foregoing examples illustrate the difficulty of obtaining experimental proof of non-coherent scattering in the face of uncertainties in the solar model and in the collisional damping formulae which must be used to calculate the dependence of η_ν on the optical depth. Nevertheless, the theoretical evidence is so overwhelming that the reality of non-coherent scattering can hardly be doubted. It is therefore best to assume its validity and to look elsewhere for the sources of discrepancies with observation.

30. Numerical methods. In practical applications, the analytic solutions of the equation of transfer are best used when η_ν is constant with depth. When the variation of η_ν is relatively small, STRÖMGREN[7] has devised an approximate method in which the profile is represented in terms of suitably weighted mean values of the Planck function and of the quantity $(1+\varepsilon\eta_\nu)/(1+\eta_\nu)$. The mean value method, which may also be utilized in connection with CHANDRASEKHAR'S exact solution[8] has had many useful applications in the calculation of solar abundances and in the representation of the profiles of the D-lines and of the K-line. If the variation of η_ν with depth is too large to be handled by the mean-value method, the profile can be calculated precisely only by strictly numerical methods, which can be employed for any arbitrary dependence of η_ν upon the depth. One such method is based upon the direct numerical integration of the

---

1 D. L. HARRIS: Astrophys. Journ. **109**, 53 (1949).
2 Z. HITOTUYANAGI and H. INABA: Sci. Rep. Tôhoku Univ. Ser. I **38**, No. 3 (1954).
3 L. H. ALLER and A. K. PIERCE: Astrophys. Journ. **116**, 176 (1952).
4 W. PRIESTER: Z. Astrophys. **32**, 200 (1953).
5 E. BÖHM-VITENSE: Z. Astrophys. **34**, 209 (1954).
6 H. H. PLASKETT: Monthly Notices Roy. Astronom. Soc. London **115**, 256 (1955).
7 B. STRÖMGREN: Astrophys. Journ. **86**, 1 (1937).
8 M. TUBERG: Astrophys. Journ. **103**, 145 (1946).

transfer equation, after simplification by the Eddington approximation[1]. A second method, due to STRÖMGREN[2], attacks the solution of the corresponding integral equation (25.10) by an iteration process. This is the most accurate method that has so far been used for the calculation of profiles of strong lines, with non-coherent scattering included. Unfortunately, for the important case in which the line scattering coefficient is much greater than the line absorption coefficient, the Strömgren method converges only very slowly [8]. PLASKETT[3] has proposed a numerical method for the solution of the exact equation of transfer, Eq. (24.8), which involves no assumptions either as to the mechanism of re-emission or as to local thermodynamic equilibrium, and in which atomic transition probabilities for radiative excitation and ionization are used to derive values of the source functions and absorption coefficients as functions of the kinetic, excitation and ionization temperatures. The method seems to offer the possibility of yielding not only the abundance and the damping constants, but also the depth dependence of the excitation temperature, but no numerical applications have as yet been made.

A fourth numerical method, which is now being widely applied for abundance determinations, makes use of so-called weighting functions to represent the stratification of the atmosphere. The method is based on an idea of MILNE's that in an atmosphere with arbitrary stratification, the contribution of each layer to the line radiation can be represented by weighting factors which depend only on the model of the atmosphere. They are therefore functions of the optical depth and of the wavelength but are the same for all lines. The method is especially valuable when applied to weak lines and the wings of strong lines, for which it was first developed by UNSÖLD and by MINNAERT. The method was later extended to strong lines by PECKER and adapted to machine computation by ELSTE[4] and by TEN BRUGGENCATE, LÜST-KULKA and VOIGT[5].

MINNAERT[6] has emphasized the advantages inherent in the use of faint lines for analyses of the Fraunhofer spectrum. For example, the profiles and equivalent widths of faint lines are independent of the damping, which greatly simplifies the determination of abundances. Furthermore, no simplifying assumptions need be made with regard to the structure of the atmosphere. Finally, as we shall see later, the central intensities of strong lines are enhanced by a number of effects which are difficult to evaluate, such as fluorescence, non-coherent scattering and chromospheric emission. These effects play no role in the formation of weak lines. Until recently, the intensities of weak lines could be measured only with very limited accuracy, but the advent of spectrographs with very high resolving power has removed this limitation and hence the study of faint Fraunhofer lines is destined to receive an increasing share of attention in the immediate future. For this reason we shall outline the method of weighting functions in some detail.

The method can be equally well applied whether the re-emission occurs by scattering or absorption, or by both. However, since weak lines are most likely to arise from excited levels, the quanta are almost always re-emitted in a different frequency from that absorbed and hence it is usually assumed, e.g., in abundance determinations, that, for the faint lines, scattering may be neglected. We first

[1] A. PANNEKOEK: Publ. Astr. Inst. Amst. **4** (1935).
[2] See A. UNDERHILL: Publ. Dom. Astrophys. Obs. **8**, 357 (1951).
[3] See footnote 6, p. 54.
[4] G. ELSTE: Z. Astrophys. **37**, 201 (1955).
[5] P. TEN BRUGGENCATE, R. LÜST-KULKA and H. H. VOIGT: Nachr. Akad. Wiss. Göttingen **1955**, No. 3.
[6] M. MINNAERT: Trans. Internat. Astronom. Union **7**, 457 (1950).

define the line depth $r(\Delta\nu)$ as

$$r(\Delta\nu) = \frac{I_c(\mu) - I_\nu(\mu)}{I_c(\mu)}, \tag{30.1}$$

where I_ν and I_c are respectively the emergent intensities in the line and in the neighboring continuum. The exact solution of the equation of transfer may be written in the form

$$r^\mu(\Delta\lambda) = \int_0^\infty \frac{\varkappa_\nu}{\varkappa_{\lambda_0}} e^{-\tau_\nu/\mu} g_\lambda^\mu(\tau_{\lambda_0}) \frac{d\tau_{\lambda_0}}{\mu}, \tag{30.2}$$

where we use the subscripts ν and λ to refer to the absorption coefficients in the line and continuum, respectively. The wavelength λ_0 is arbitrary and is usually chosen to be λ 5000. For each region of the spectrum and position on the disk, the weight functions $g_\lambda^\mu(\tau_{\lambda_0})$ are functions of the optical depth at the standard wavelength λ_0. The line depth is the sum of the contributions of the absorption coefficient from each layer of the atmosphere, each weighted by its g-factor and decreased exponentially by absorption in the layers above.

The weight functions are given rigorously by

$$g_\lambda^\mu(\tau_{\lambda_0}) = \frac{\int_{\tau_{\lambda_0}}^\infty S(t_\lambda)\, e^{-t_\lambda/\mu} d\frac{t_\lambda}{\mu} - S(\tau_{\lambda_0})\, e^{-\tau_{\lambda_0}/\mu}}{\int_0^\infty S(t_\lambda)\, e^{-t_\lambda/\mu} d\frac{t_\lambda}{\mu}}, \tag{30.3}$$

where $S(t_\lambda)$ is the source function, and the denominator is $I_c(\mu)$, the intensity of the emergent radiation in the continuum at λ. Note that the usual assumption of local thermodynamic equilibrium (LTE), implies that the electron and excitation temperatures are equal. In this case, the source function may be set equal to B_ν, the Planck function appropriate to the electron temperature, which is derived as a function of optical depth from limb darkening observations. The calculation of the weighting functions is thus greatly simplified since they depend only on the properties of the model atmosphere and are the same for all lines.

Eq. (30.3) is not in the most suitable form for calculation because the numerator contains the difference between two terms of nearly equal magnitude. More useful is the differential form

$$I_c(\mu)\, g_\lambda^\mu(\tau_\lambda) = \int_0^\infty \frac{dS(t_\lambda)}{dt_\lambda} e^{-t_\lambda/\mu} dt_\lambda, \tag{30.4}$$

which shows that the weight function is determined by the gradient of the source function with depth. Obviously, an absorption line will not be formed unless the gradient is greater than zero. The numerical integration of Eq. (30.2) and (30.4) is most conveniently carried out with $\log\tau$ rather than τ as the independent variable[1], the chief reason being that the geometrical depth in the solar atmosphere is more nearly proportional to the logarithm of the optical depth.

The calculation of line profiles and equivalent widths by the weighting function method has been adapted to high-speed machine computation by Elste and applied both to the Fraunhofer spectrum and to the spectra of B stars[2].

[1] G. Elste: Z. Astrophys. **37**, 201 (1955).
[2] L. H. Aller, G. Elste and J. Jugaku: Astrophys. Journ. Suppl. **1957**, No. 25.

In ELSTE's notation, the substitution

$$x_0 = \text{Log}\ \tau_{\lambda_0}, \quad \tau_{\lambda_0} = 10^{x_0}, \quad d\tau_{\lambda_0} = \frac{10^{x_0}}{\text{Mod}}\, d x_0, \tag{30.5}$$

with Mod $=0.4343$ leads to the following form of Eq. (30.2):

$$r^{\mu}(\varDelta\lambda) = \int\limits_{-\infty}^{+\infty} \frac{1}{\mu}\,\frac{\varkappa_\nu}{\varkappa_{\lambda_0}}\,\frac{10^{x_0}}{\text{Mod}} \exp\left(-\int\limits_{-\infty}^{x_0} \frac{1}{\mu}\,\frac{\varkappa_\nu}{\varkappa_{\lambda_0}}\,\frac{10^{y}}{\text{Mod}}\, dy\right) g_\lambda^{\mu}(x_0)\, d x_0, \tag{30.6}$$

where λ_0 is usually taken to be 5000 Å.

The numerical calculation of the line intensity is made more convenient by the following substitution, which makes clear the functional dependence of the integrand:

$$\frac{1}{\mu}\,\frac{\varkappa_\nu}{\varkappa_{\lambda_0}}\,\frac{10^{x_0}}{\text{Mod}} = \frac{1}{\mu} M^* Z\,\frac{H(v,a)}{\xi}, \tag{30.7}$$

as a result of which Eq. (30.6) becomes

$$r^{\mu}(\varDelta\lambda) = \int\limits_{-\infty}^{+\infty} \frac{M^*}{\mu}\,\frac{Z}{\xi}\, H(v,a) \exp\left[-\int\limits_{-\infty}^{x_0} \frac{M^*}{\mu}\,\frac{Z}{\xi}\, H(v,a)\, dy\right] g_\lambda^{\mu}(x_0)\, d x_0. \tag{30.8}$$

In Eqs. (30.7) and (30.8), $H(v, a)$ gives the wavelength dependence of the absorption coefficient for combined Doppler broadening and damping, according to Eq. (27.6), while ξ is the most probable atomic velocity, compounded of thermal and turbulent components. Thus

$$\xi^2 = \frac{2kt}{M} + \xi_t^2. \tag{30.9}$$

The quantity M^* is independent of the depth and is given by

$$M^* = \frac{N_i}{N_{\text{H}}}\, g_{r,s} f_{r,s}\, \lambda\, \frac{\sqrt{\pi}\,\varepsilon^2}{m c}\, \frac{1}{\text{Mod}}\, \frac{1}{M_{\text{H}}\,\mu^0}\, 10^{\Theta_0 \varDelta\chi}, \tag{30.10}$$

where N_i/N_{H} is the number of atoms per gram of the given element in all stages of ionization, $g_{r,s}$ is the statistical weight of the lower level s, and μ^0 is the mean molecular weight of the un-ionized material relative to hydrogen. $\varDelta\chi$ is the energy in volts required to ionize the atom from the lower level of the transition and $\Theta_0 = 5040/T_0$, where T_0 is an arbitrary parameter approximately equal to the mean temperature of the atmosphere. For the Sun, Θ_0 may be taken as unity. The factor $\varDelta\chi$ is employed whenever the element in question is a neutral atom that is easily ionized under solar conditions. For neutral atoms that are difficult to ionize, and for ionized atoms, $\varDelta\chi$ is replaced by χ, the excitation potential of the lower level of the transition, measured from the ground state.

Both $H(v, a)/\xi$ and Z vary with depth, chiefly the latter, which has the form

$$Z = \frac{1}{\varkappa_{\lambda_0}}\,\frac{n_{r,s}}{\Sigma n_r}\,\frac{10^{x_0 - \Theta_0 \varDelta\chi}}{g_{r,s}} \tag{30.11}$$

in which $n_{r,s}/\sum\limits_r n_r$, the ratio of the number of atoms in level s and ionization stage r to the total number of all atoms of that element, is given by the Saha and Boltzmann equations. Many of the required partition functions have been tabulated by CLAAS[1] and by DE JAGER and NEVEN[2], but there are still important

[1] W. J. CLAAS: Rech. Obs. Utrecht **12**, 13 (1951).

[2] C. DE JAGER and L. NEVEN: Rech. Astr. Obs. Utrecht **13**, Part 4 (1957).

gaps for several of the heavy elements. Note that the factor $10^{\Theta_0 \Delta \chi}$ has been transferred from Z to M^* for convenience of calculation.

The equivalent width of the absorption line is obtained by integrating Eq. (30.8) over all frequencies. This leads to the expression

$$\frac{W_\lambda^\mu}{\lambda} = \int_{-\infty}^{+\infty} \frac{M^*}{\mu} \frac{\sqrt{\pi}}{c} Z(x_0)\, \Psi\left(\frac{Y}{\mu}, a\right) g_\lambda^\mu(x_0)\, d x_0, \tag{30.12}$$

where

$$\frac{Y}{\mu} = \frac{M^*}{\mu} \frac{1}{\xi} \int_{-\infty}^{x} Z\, d y, \tag{30.13}$$

and

$$\Psi\left(\frac{Y}{\mu}, a\right) = \frac{2}{\sqrt{\pi}} \int_0^\infty H(v, a)\, e^{-H(v,a)\, Y/\mu}\, d v \tag{30.14}$$

is the saturation function introduced by PECKER[1] in order to extend the theory of weighting functions to strong lines. The solution in terms of Eq. (30.12) is made possible by placing the quantity H/ξ in front of the integral in the exponent of Eq. (30.8). Note that Y is proportional to M^*, which is a measure of the total strength of the line and that as $Y \to 0$, $\Psi \to 1$. Hence for faint lines

$$\frac{W_\lambda^\mu}{\lambda} = \int_{-\infty}^{+\infty} \frac{M}{\mu} Z(x_0)\, g_\lambda^\mu(x_0)\, d x_0, \tag{30.15}$$

where

$$M = M^* \frac{\sqrt{\pi}}{c}. \tag{30.16}$$

We may introduce the symbols L_λ and $\mathscr{L}_\lambda$, such that for strong lines

$$W_\lambda^\mu = \frac{M}{\mu} \mathscr{L}_\lambda, \tag{30.17}$$

while for weak lines ($\Psi = 1$),

$$\frac{W_\lambda^\mu}{\lambda} = \frac{M}{\mu} L_\lambda. \tag{30.18}$$

We observe that L_λ is independent of M and hence, as is well known, the equivalent widths of faint lines vary linearly with the number of absorbing atoms. On the other hand $\mathscr{L}_\lambda$ depends on M in a complicated fashion through Eq. (30.12), (30.13), and (30.14). Both L_λ and $\mathscr{L}_\lambda$ are functions of the excitation and ionization potentials, the wavelength, and the atmospheric model.

c) The analysis of Fraunhofer lines.

31. Calculation of line profiles. We shall now explain how calculations based upon the above formulae may be applied to the analysis of Fraunhofer lines. The first step is the selection of a model atmosphere which specifies the electron pressure and temperature as a function of the optical depth at a standard wavelength λ_0. The numerous models that have been derived differ chiefly in the temperature distribution of the outermost layers, where $\operatorname{Log} \tau < -1.5$. The empirical model of ALLER and PIERCE[2] satisfies the observed limb darkening

---

[1] J. C. PECKER: Ann. d'Astrophys. **14**, 383 (1951).

[2] L. H. ALLER and A. K. PIERCE: Astrophys. Journ. **116**, 176 (1952).

and spectral energy distribution but is not physically meaningful for the very high layers from which little or no continuous radiation emanates. The high layers can be handled by means of theoretical models, the best of which appears to be that of BÖHM[1]. BÖHM's model agrees fairly well with the Aller-Pierce model in the intermediate and deep layers but has a boundary temperature of 3400° K as compared with 4500° K for the latter model. However, departures from thermodynamic equilibrium, which occur chiefly in the highest layers[2,3] generally tend to increase the state of ionization and hence the assumption of a higher boundary temperature simulates at least roughly the deviations from thermodynamical equilibrium.

In the present uncertain state of our knowledge concerning these departures, the Aller-Pierce temperature distribution is a reasonable one to adopt. These considerations have guided G. ELSTE in the adoption of the model given in Table 7 for use in connection with the Michigan solar abundance project[4]. The first column gives the logarithm of the optical depth at λ 5000, the second the quantity $\Theta = 5040/T$, and the third the logarithm of the electron pressure in dynes/cm². The electron pressure distribution was derived in the usual way for a He to H abundance ration of 1:5 and a metal abundance of 4×10^{-4} relative to hydrogen.

The next step is the calculation of the weighting functions, as a function of μ, λ and τ_{λ_0}, by numerical integration of Eq. (30.4). The results of ELSTE's calculations performed with the Michigan high-speed electronic computer (MIDAC) according to the model of Table 7 for a number of wavelengths, with $\mu = 1$, are collected in Table 8. Similar calculations were also performed for BÖHM's model, but the differences in the weighting functions are negligible for the deeper layers, Log $\tau > -1.2$, and not very important in the outer layers which make only a small contribution to the absorption of weak lines. HUBENET[5] has also reached the conclusion independently that the weighting functions are relatively insensitive to the choice of model.

Once the weighting functions are tabulated, the line profile may be computed by numerical integration according to Eq. (30.8). The only part of the integrand that is frequency dependent is $H(v, a)/\xi$, which for weak lines varies only slowly with depth in the atmosphere. If it is assumed initially to be constant with depth, the integral may be calculated for a number of different values of $H(v, a)/\xi$ and the shape of the absorption coefficient $H(v, a)$ derived by fitting the calculated values of $r(\Delta\lambda)$ to the observed profile. This procedure has recently been applied by SUEMOTO[6] to a number of weak lines with different effective depths of origin, observed at several values of μ, to investigate both the anisotropy of the turbulence and its dependence on optical depth. As a by-product of the computation, the parameter M^* can also be determined, since it is independent of both frequency and depth. The abundance of the element relative to hydrogen, $N_i/N_{\rm H}$, then follows by Eq. (30.10), if the absolute f-value of the line is known.

32. Contribution curves. When the integrand in Eq. (30.8) is plotted against x_0, the resulting curve is known as the contribution curve of the line depth, since it measures the contribution to the line from each atmospheric layer. Its physical significance has been discussed in detail by TEN BRUGGENCATE *et al.*[7].

[1] K.-H. BÖHM: Z. Astrophys. **34**, 182 (1954).
[2] K.-H. BÖHM: Z. Astrophys. **35**, 179 (1954).
[3] M. BRETZ: Z. Astrophys. **38**, 259 (1955).
[4] See L. H. ALLER: Vol. LI, this Encyclopedia.
[5] H. HUBENET: Proc. Kon. Nederl. Akad. Wetensch. Ser. B **59**, 480 (1956).
[6] Z. SUEMOTO: Monthly Notices Roy. Astronom. Soc. London **117**, 2 (1957).
[7] P. TEN BRUGGENCATE, R. LÜST-KULKA and H. H. VOIGT: Nachr. Akad. Wiss. Göttingen **1955**, No. 3.

Table 7. *Adopted model atmosphere.*

Log τ_0	Θ	Log P_e	Log τ_0	Θ	Log P_e	Log τ_0	Θ	Log P_e	Log τ_0	Θ	Log P_e
−3.0	1.118	−0.78	−2.0	1.090	−0.047	−1.0	0.990	0.826	0.0	0.783	1.837
−2.8	1.117	−0.64	−1.8	1.078	0.112	−0.8	0.957	0.998	+0.2	0.733	2.172
−2.6	1.113	−0.50	−1.6	1.064	0.282	−0.6	0.918	1.165	+0.4	0.682	2.54
−2.4	1.108	−0.36	−1.4	1.045	0.464	−0.4	0.877	1.335	+0.6	0.641	2.96
−2.2	1.100	−0.206	−1.2	1.021	0.645	−0.2	0.831	1.554	+0.8	0.613	3.30

Table 8. *Weighting functions, $g_\lambda(x_0)$, calculated with the Midac by G. Elste with $\mu = 1$.*

Log $\tau_{\lambda 5000}$	λ										
	4000	4500	5000	6000	7000	8000	10000	12000	14000	16500	22000
−3.0	0.840	0.845	0.834	0.770	0.698	0.626	0.548	0.525	0.503	0.432	0.369
−2.8	0.840	0.844	0.832	0.768	0.696	0.624	0.546	0.523	0.501	0.431	0.368
−2.6	0.840	0.842	0.830	0.765	0.693	0.620	0.543	0.520	0.498	0.430	0.366
−2.4	0.839	0.840	0.826	0.761	0.688	0.616	0.539	0.515	0.495	0.428	0.363
−2.2	0.839	0.836	0.821	0.754	0.681	0.609	0.532	0.509	0.490	0.426	0.359
−2.0	0.837	0.831	0.813	0.744	0.671	0.599	0.522	0.499	0.481	0.426	0.353
−1.8	0.835	0.824	0.802	0.731	0.656	0.585	0.509	0.485	0.470	0.420	0.344
−1.6	0.830	0.812	0.786	0.711	0.636	0.565	0.490	0.466	0.454	0.409	0.332
−1.4	0.820	0.795	0.763	0.684	0.608	0.538	0.465	0.440	0.432	0.398	0.315
−1.2	0.804	0.771	0.731	0.647	0.569	0.502	0.431	0.406	0.402	0.380	0.292
−1.0	0.777	0.735	0.686	0.597	0.519	0.455	0.388	0.363	0.363	0.357	0.261
−0.8	0.731	0.685	0.627	0.532	0.452	0.391	0.329	0.309	0.317	0.325	0.223
−0.6	0.666	0.615	0.550	0.450	0.371	0.316	0.261	0.246	0.257	0.282	0.174
−0.4	0.579	0.525	0.454	0.353	0.280	0.233	0.1913	0.1795	0.1936	0.2213	0.1194
−0.2	0.465	0.413	0.344	0.253	0.1894	0.1520	0.1202	0.1125	0.1256	0.1564	0.0669
0.0	0.353	0.287	0.229	0.1534	0.1057	0.0812	0.0614	0.0570	0.0639	0.0873	0.0245
+0.2	0.226	0.1672	0.1240	0.0765	0.0486	0.0346	0.0234	0.0208	0.0239	0.0336	0.00495
+0.4	0.1006	0.0693	0.0483	0.0265	0.0152	0.00968	0.00570	0.00426	0.00348	0.00284	0.000149
+0.6	0.0270	0.0174	0.01106	0.00526	0.00277	0.00160	0.000817	0.000585	0.000414	0.000234	0.0000175
+0.8	0.00302	0.00186	0.00112	0.00051	0.00025	0.000223	0.000105	0.000056	0.000031	0.000009	0.0000053

Three such curves are reproduced in Fig. 18 from the work of ELSTE[1]. They refer to Fe I lines, with $\Delta\chi = 3.5$ eV, and for which the quantity $\frac{1}{\mathrm{Mod}}\,\frac{N_i}{N_{\mathrm{H}}}\,g f \lambda = 10^{-10}$. The curves are drawn for the three parts of the line $\Delta\lambda = 0$, 0.04 and 0.08 Å and clearly illustrate how the outer parts of a line tend to be formed in relatively deep layers as compared with the line center.

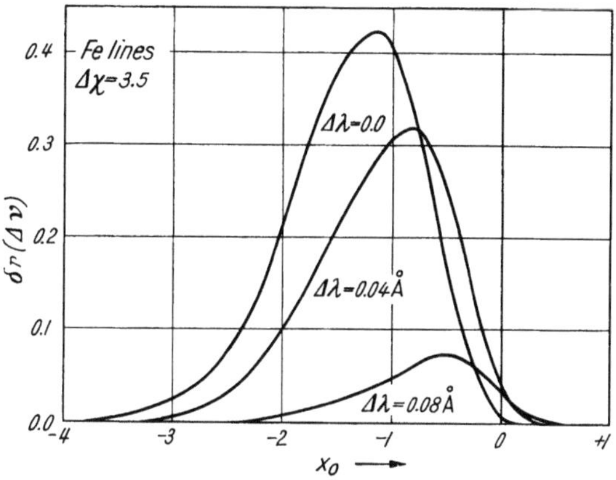

Fig. 18. Contribution curves for the depths of Fe I lines with $\Delta\chi = 3.5$ eV and $\lambda \sim 5500$ Å. The three curves give the contributions to the line depth, $\delta r(\Delta\nu)$, as a function of $x_0 = \mathrm{Log}\,\tau_0$, for three different parts of the line, $\Delta\lambda = 0$, 0.04 and 0.08 Å (after G. ELSTE).

In similar fashion, contribution curves appropriate to the equivalent widths are obtained by plotting the integrand of Eq. (30.12) against x_0. The resulting curves for the lines C I λ 5380 and Ni I λ 5436 are shown in Fig. 19 for the center of the disk ($\mu = 1.0$) and for the limb ($\mu = 0.1$). The curves, which are also due to ELSTE, give the contribution to the quantity $W_\lambda/2\Delta\lambda_D$ from each layer specified by x_0. They demonstrate strikingly that the effective-level of origin is raised from the center to the limb, and also that high excitation lines such as those of C I are formed in very deep layers as compared with low-excitation neutral metallic lines.

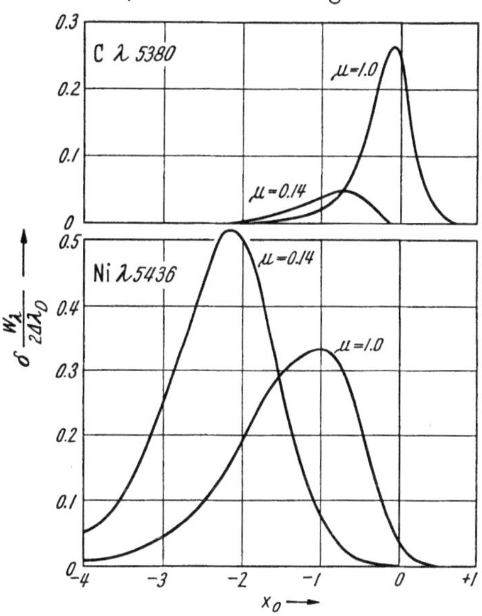

Fig. 19. Contribution curves for the equivalent widths of neutral nickel and carbon lines for the center of the disk and or $\mu = 0.14$ (after G. ELSTE).

33. The curve of growth[2]. The foregoing method of analysis can only be employed when accurate profiles of faint and medium-strong lines are available, and furthermore, its most useful application is in the investigation of atmospheric structure. For the determination of abundances, it is sufficient and usually preferable to rely on the equivalent widths of Fraunhofer lines. The analysis of equivalent widths is usually performed with the aid of the curve of growth, which relates the equivalent width of a line to the various atomic and atmospheric parameters that enter into its formation. We shall not attempt to summarize the many important investigations of the solar spectrum by the curve-of-growth method, beginning with the pioneering work of MINNAERT and MULDERS. Excellent reviews will be found in [*1*] and [*5*]. Until recently, these studies have been based upon schematic models of the photosphere, usually either of

[1] G. ELSTE: Z. Astrophys. **37**, 201 (1955).

[2] For a description of the curve-of-growth method cf. also D. BARBIER, Vol. L of this Encyclopedia, p. 360seq.

the Schuster or Milne-Eddington types. With the former, the equivalent width is calculated as a function of the optical depth in the line center: all lines are considered to arise from a layer above the region in which the continuous spectrum is formed. In the real photosphere, two lines having the same optical depth but occurring in different regions of the spectrum may have quite different intensities owing to the variation of the continuous absorption coefficient with wavelength. In the Milne-Eddington model, the influence of the continuous opacity is partially taken into account by expressing the equivalent width as a function of η_0, the ratio of the central line absorption coefficient to the continuous absorption coefficient outside the line. The use of curves of growth based upon these schematic models has yielded a great deal of valuable semi-quantitative information on the solar photosphere, including the recognition of collisions as the main source of line wing broadening, the existence of turbulent motions, and estimates of abundances.

It has become apparent, however, that the shape of the curve of growth, especially in the important transition region where small differences in equivalent width correspond to very large changes in the numbers of effective absorbing atoms, depends rather critically upon the choice of the solar model. Furthermore, the variation with height of the numbers of absorbing atoms can be quite different for lines of different excitation and ionization potentials. Finally, since the collisional damping constant and probably also the turbulent velocity vary with height, lines formed at different levels will be characterized by different values of these parameters. It is therefore not appropriate to employ a single theoretical curve of growth, based upon an idealized model, to interpret the equivalent widths of all lines of the same or different elements. The importance of the stratification effects has been emphasized by MINNAERT, who showed also how the method of weighting functions leads to an accurate expression for the equivalent widths of faint lines, which takes full account of the details of the model.

MINNAERT'S faint-line theory and its extension to stronger lines by PECKER have been adapted to machine computation by ELSTE, with particular reference to abundance determinations. By Eq. (30.18) the equivalent widths of faint lines are proportional to M, which is a constant for each line and contains the abundance, and the L_λ, which includes all factors dependent on the depth. We introduce the parameter C, such that

$$M = C \cdot \frac{N_i}{N_H}, \tag{33.1}$$

and hence

$$\operatorname{Log} C = \operatorname{Log} g f \lambda + \Theta_0 \Delta\chi + \operatorname{Log} L_\lambda - \operatorname{Log} \mu + \text{const}, \tag{33.2}$$

where

$$L_\lambda = \int_{-\infty}^{+\infty} Z(x_0)\, g_\lambda^\mu(x_0)\, dx_0. \tag{33.3}$$

For any given element, the function L_λ depends upon the wavelength, λ, and upon either $\Delta\chi$ or χ. Whether L_λ is a function of $\Delta\chi$ or of χ depends upon the ionization properties of the atom, which enter into the factor $n_{r,s}/\Sigma n_r$ in Eq. (30.11). Because of this factor, L_λ will also, in general, differ from element to element. However, following CLAAS[1], we may arrange the elements into groups with similar ionization properties. ELSTE has calculated values of L_λ for a number of elements and lines over wide ranges of χ, $\Delta\chi$ and λ, using the solar model

[1] W. J. CLAAS: Rech. Astr. Obs. Utrecht 12, 13 (1951).

given in Table 7. It turns out that there is one group of neutral elements (CLAAS' group I) for which L_λ has the same dependence on λ and $\Delta\chi$ for all elements. They are elements for which $I_0 \leqq 6.3$ eV and $I_1 > 14.4$ eV, where I_0 and I_1 are the ionization potentials of the neutral and singly-ionized atoms, respectively,

Table 9. *Coefficients a and b in* Eq. (33.4) *and* (33.5).

Element	*a*	*b*	Element	*a*	*b*	Element	*a*	*b*
C	− 6.35	− 3.685	V	+ 0.18	− 1.640	Zn	− 2.80	− 1.048
Mg	− 0.06	− 0.334	Cr	+ 0.37	− 0.878	Sr	+ 0.40	− 0.330
Si	− 1.00	− 1.033	Mn	+ 0.15	− 0.931	Y	+ 0.18	− 1.210
S	− 5.45	− 2.810	Fe	− 0.10	− 1.720	Zr	+ 0.18	− 1.673
Ca	+ 0.25	− 0.347	Co	− 0.10	− 1.596	Nb	+ 0.18	− 1.647
Sc	+ 0.18	− 1.365	Ni	− 0.26	− 1.205	Hg	− 6.00	− 2.100
Ti	+ 0.18	− 1.755	Cu	− 0.37	− 0.176			

and include Li, Na, Al, K, Ga, Rb and In. For this group of elements, a family of curves can be drawn, each of which gives Log L_λ as a function of $\Delta\chi$ for a specified wavelength. Inspection of the curves for other elements reveals that

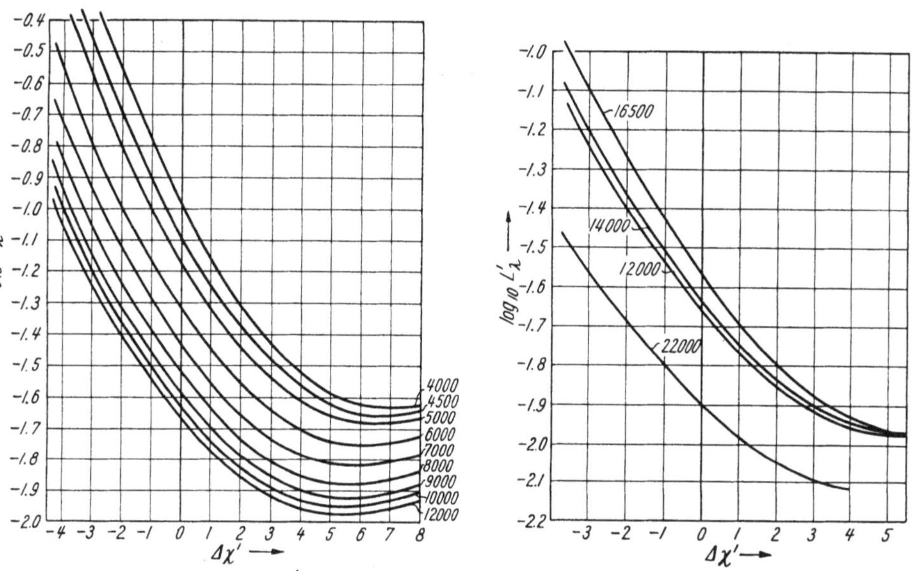

Fig. 20. The dependence of Log L'_λ upon $\Delta\chi'$ and λ, for neutral elements of relatively low ionization potential.

many of them can be derived merely by shifting the group I curves by appropriate amounts in the $\Delta\chi$ and Log L_λ scales. This makes possible the derivation of Log L_λ for all neutral elements for which L_λ is a function of $\Delta\chi$ from a single family of curves in which the quantity Log L'_λ is plotted against $\Delta\chi'$, where

$$\Delta\chi' = \Delta\chi + a, \tag{33.4}$$

and

$$\operatorname{Log} L_\lambda = \operatorname{Log} L'_\lambda + b. \tag{33.5}$$

Values of a and b are given in Table 9 for a number of elements of interest, whereas the L'-curves are shown in Fig. 20. For the elements in group I, a and b are zero, and hence the curves may be entered directly with $\Delta\chi_{\text{el}}$ to find Log L_λ.

For the elements in Table 9, the curves are entered with $\Delta\chi'$ to find $\text{Log}\, L'_\lambda$, after which $\text{Log}\, L_\lambda$ is obtained from Eq. (33.5).

The L-curves in Fig. 20 pertain to all neutral elements for which L is a function of $\Delta\chi$. For ionized elements and the neutral elements of Claas' group VII ($I_0 > 13.5$ eV, $I_1 > 14.4$ eV), for which L is a function of χ, the calculations were limited to selected values of λ and χ, according to Table 10.

34. Abundances of elements. Fig. 20 and Table 10 have provided the basis for a new determination of the abundances of the elements in the Sun by Goldberg, Müller and Aller[1]. The procedure for very weak lines ($\Psi = 1$) is to calculate Log C with Eq. (33.2), after which the abundance N_i/N_H follows directly from Eqs. (30.18) and (33.1). There are, however, many elements for which sufficiently faint lines are not available and hence use must be made of mediumstrong lines, which are partly saturated but are yet not so strong that damping plays a significant role in their broadening. For these stronger lines we must employ the Pecker theory embodied in Eqs. (30.12) to (30.14). By numerical integration of Eq. (30.12), a theoretical relation is derived between the equivalent width W_λ/λ and the parameter M, for pairs of values of λ and of $\Delta\chi$ or χ, whichever is appropriate. The graphical relationship between $\text{Log}\, W_\lambda/\lambda$ and $\text{Log}\, M$ is essentially the curve of growth for the line in question and for the assumed atmospheric model. An empirical curve of growth may also be derived by plotting observed values of $\text{Log}\, W_\lambda/\lambda$ against Log C for groups of lines of about the same wavelength and excitation characteristics. The sideways displacement between the empirical and theoretical curves then gives the abundance directly, by Eq. (33.1).

Table 10. *Values of* $\text{Log}\, L(\chi, \lambda)$ *for elements of group VII and ion lines.*

	λ	χ = 0	2	9	9.5
O I	5000		6.000	7.326	
	6000	5.808	5.895	7.117	7.243
	8000			6.798	

	λ	χ = 10.3	11.5
N I	7000	7.586	
	8000	7.441	
	10000	7.279	7.572

	λ	χ = 2.93
Sr II	4000	6.815

	λ	χ = 0.6	2.5
Ba II	4500		6.377
	6000	6.123	

	λ	χ = 8.83
Mg II	4500	8.059

	λ	χ = 8.08	9.8
Si II	4000		7.961
	6000	7.031	

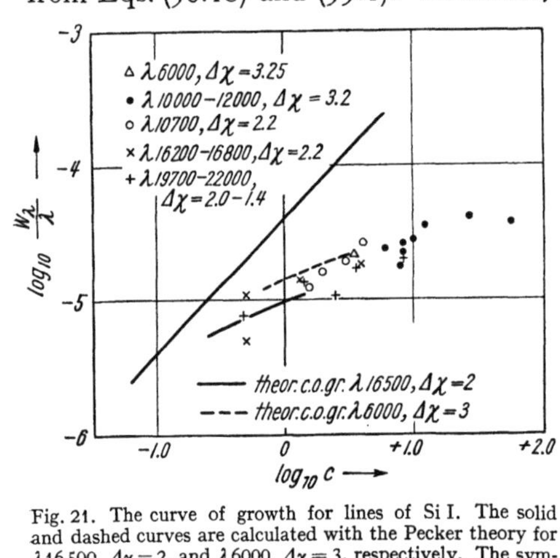

Fig. 21. The curve of growth for lines of Si I. The solid and dashed curves are calculated with the Pecker theory for λ 16500, $\Delta\chi = 2$, and λ 6000, $\Delta\chi = 3$, respectively. The symbols represent observational values of log W_λ/λ and values of Log C calculated by Eq. (33.2), with $\mu = 1$.

[1] See Aller: Vol. LI, this Encyclopedia.

The application of the Pecker theory by ELSTE to Si I is illustrated in Fig. 21, where the relation between Log W/λ and Log M has been calculated for ranges of values of M. For weak lines, the relationship is a straight line with slope 45°, but for stronger lines there is a separate curve of growth for each pair of values of λ and $\Delta\chi$. The value of W_λ/λ for which the line becomes saturated appears to depend on the rate at which the number of absorbing atoms increases with depth relative to the rate of increase of the continuous absorption. A line formed at a level below that of the continuum will be weaker than one formed in higher layers[1]. For comparison with the theory, the observed values of Log W_λ/λ are plotted against Log C for five different multiplets. There is some indication that the infrared lines do indeed fall below the visual lines in the diagram, as required by the theory, but the scatter of the points resulting from uncertainties in both the equivalent widths and the f-values precludes a more precise test.

35. Empirical determination of absorption coefficient. All model atmosphere calculations of line profiles require knowledge of the wavelength dependence of the absorption coefficient. It is usually assumed for lines other than hydrogen that the absorption coefficient is due to combined Doppler broadening and collisional damping, the thermal Doppler width and damping constant being dependent on the temperature and pressure. If the analysis is confined to weak and moderately-strong lines, the damping constant need not be known with high precision, but it is important to know the turbulent velocity and its dependence on both τ and μ. Most investigators have assumed that the turbulent velocity is independent of depth and have adopted values between 1 and 2 km/sec as suggested by curve of growth and line profile studies. There is now good evidence that the turbulence is anisotropic[2] and that the ratio of the tangential to the radial velocity is about 1.6[3,4]. However, the evidence for a depth dependence of ξ is conflicting. SUEMOTO finds that ξ increases with depth in apparent contradiction to the results of WADDELL and of ROGERSON[5] which suggest that ξ is constant with depth.

The discrepancies may be attributed partly to observational error and partly to differences in the methods of analysis. Of the five lines observed in common at the center of the disk by ROGERSON and SUEMOTO, the values of $\Delta\lambda_D$ for four lines are in reasonably good agreement, but for the fifth line, λ 6238 of Fe^+, ROGERSON's value of 0.064 Å is about 50% greater than SUEMOTO's. The results have been obtained by widely different methods of analysis. ROGERSON applied the method of Voigt profiles [*1*], whereas SUEMOTO used the more refined technique of calculating a theoretical profile by integration with an assumed model atmosphere and then determining the absorption coefficient by numerical fitting to the observed profile. Both methods contain serious disadvantages. The Voigt profile method is strictly applicable only to weak lines, whereas the method of numerical fitting necessitates extensive computations and the assumption of an atmospheric model.

An alternative method of extracting the wavelength dependence of the absorption coefficient retains the simplicity of the Voigt profile method and the accuracy of the numerical fitting method[6], but requires no advance knowledge of the model atmosphere. Similar in principle to those employed by ATHAY and

[1] ALLER, ELSTE and JUGAKU: Astrophys. Journn. Suppl. **1957**, No. 25.
[2] C. W. ALLEN: Monthly Notices Roy. Astronom. Soc. London **109**, 343 (1949).
[3] J. WADDELL: Thesis, University of Michigan 1956.
[4] Z. SUEMOTO: Monthly Notices Roy. Astronom. Soc. London **117**, 2 (1957).
[5] J. B. ROGERSON: Astrophys. Journ. **125**, 275 (1957).
[6] L. GOLDBERG: Astrophys. Journ. **127**, 308 (1958).

Thomas[1] and de Jager[2] for the derivation of the Doppler widths of the Balmer lines, the method requires the accurate observation of the profiles of two lines of known relative strengths in the same multiplet. It is assumed (1) that the source functions for the two lines are equal and independent of frequency; and (2) that the wavelength dependence of the absorption coefficient is independent of optical depth. It is then self-evident that at points of equal intensity in the two profiles the absorption coefficients are also equal. If $\Delta\lambda_a$ and $\Delta\lambda_b$ are the respective wavelength displacements for two points of equal intensity in profiles a and b, it follows that

$$\frac{\varphi(\Delta\lambda_b)}{\varphi(\Delta\lambda_a)} = \frac{(gf)_a}{(gf)_b} = k, \tag{35.1}$$

where $\varphi(\Delta\lambda)$ is the wavelength dependence of the absorption coefficient and gf is the relative line strength. The application of Eq. (35.1) to various points on the line profile yields the run of $\varphi(\Delta\lambda)$ with $\Delta\lambda$.

Table 11. *Values of $\Delta\lambda_D$ derived for* Fe^+ *lines* λ 6238 *and* λ 6248.

$\Delta\lambda_{6238}$	μ = 1.0	0.8	0.6	0.4	0.3
0.00	0.035	0.038	0.043	0.044	0.049
0.01	0.036	0.039	0.044	0.044	0.048
0.02	0.038	0.042	0.045	0.044	0.048
0.03	0.038	0.044	0.046	0.045	0.048
0.04	0.038	0.044	0.047	0.044	0.049
0.05	0.037	0.042	0.047	0.042	0.047
0.06	0.038	0.040	0.046	0.040	0.046
Suemoto:					
λ 6238	0.044	0.045	0.051	0.055	0.056
λ 6248	0.037	0.040	0.044	0.047	0.045

For the particular case in which the absorption coefficient in the line core is due to Doppler broadening only,

$$\varphi(\Delta\lambda) = e^{-\left(\frac{\Delta\lambda}{\Delta\lambda_D}\right)^2}, \tag{35.2}$$

and Eq. (35.1) permits a solution for $\Delta\lambda_D$, viz.:

$$(\Delta\lambda_D)^2 = \frac{0.4343}{\operatorname{Log} k}\left[(\Delta\lambda_a)^2 - (\Delta\lambda_b)^2\right]. \tag{35.3}$$

Although Eq. (35.3) is derived on the assumption that $\Delta\lambda_D$ is constant over the depth range of line formation, its application to different parts of the same line as well as to different lines can yield information on the depth dependence of $\Delta\lambda_D$. Similarly, the anisotropy of the turbulence may be inferred from center-limb observations. As an example, the method has been applied to Suemoto's observations of the lines $\lambda\lambda$ 6238 and 6248, which belong to the multiplet $b^4D - z^4P^0$ of Fe^+. These are the only two lines in Suemoto's list that fall in the same multiplet. Their relative strengths have not been measured in the laboratory, but have a theoretical LS-coupling ratio of 2. The results are given in Table 11, together with Suemoto's values, which were derived from the half-width of the absorption coefficient. The general agreement is good although there are interesting differences in detail. The essential constancy of the Doppler width throughout the line core suggests that it does not vary steeply with depth. The general increase of $\Delta\lambda_D$ towards the limb is apparently reversed at $\mu = 0.4$ but restored at $\mu = 0.3$, whereas Suemoto's values tend to drop off at $\mu = 0.3$.

The foregoing method seems susceptible to a wide variety of applications, since it may be applied not only to the investigation of turbulence, but also to the improvement of solar models.

36. Inhomogeneities in the photosphere. Thus far in this chapter, the Fraunhofer lines have been treated as though they are formed in a homogeneous atmosphere in which all physical parameters vary only with depth. The granulation

[1] R. G. Athay and R. N. Thomas: Astrophys. Journ. **127**, 96 (1958).
[2] C. de Jager: Rech. Astr. Obs. Utrecht **13**, 50 (1952).

alone is evidence that even in areas free of solar activity both the temperature and gas velocity fluctuate over regions as small as a few hundred kilometers in diameter. The mean fluctuations in temperature may amount to several hundred degrees and in velocity to several tenths of a kilometer per second.

The granulation is believed to originate in an unstable layer below $\tau = 1$. Columns of turbulent gases ascend from the unstable layer, while the cooler gases descend. RICHARDSON and SCHWARZSCHILD[1] reasoned that these motions should be revealed by local Doppler shifts in the Fraunhofer lines which could be observed under conditions of excellent seeing and very high spectroscopic resolving power. Measurements on plates taken with the 150-ft tower telescope and spectrograph at the Mt. Wilson Observatory revealed that the shifts were present, and an r.m.s. velocity of 0.37 km/sec was derived from a group of three neutral Fe lines at λ 6300. Good agreement with these results was later obtained by PLASKETT[2] from measurements on five lines of Fe I and one of Ni I near λ 5900.

Modern spectrographs, especially those employing Babcock gratings, plainly reveal the local Doppler shifts in the photospheric Fraunhofer lines. They can easily be observed visually everywhere on the solar disk in both normal and undisturbed regions[3] and are photographed routinely on every clear day with the vacuum spectrograph at the McMath-Hulbert Observatory[4]. Some typical photographs made at the McMath-Hulbert Observatory at times of good seeing are shown in Figs. 22 and 23. The Doppler shifts have been photographed even on days of less than average seeing, but when the definition of the image is better than about 3″ of arc, fluctuations in intensity as well as in velocity become clearly apparent. The line cores appear to be broken up into a series of elements with random Doppler displacements. The smallest elements observed have a thickness of about two seconds of arc, but this limit may be determined by atmospheric seeing.

Several qualitative conclusions may be drawn from Fig. 22. First, the velocity fluctuations about the mean position of each line are not correlated appreciably with changes of intensity in the continuum, as would be expected if the hot regions were systematically ascending and the cooler ones descending. The correlation may be masked by the smearing out of small elements by bad seeing. Second, if one intercompares different lines, it can be noted that both the intensity and velocity fluctuations in the different lines are in phase. In this respect, the behavior of the Fe and Fe^+ lines at λ 5169 is contrary to the notion that the intensity fluctuations in the lines are caused by temperature fluctuations in the photosphere, since it would then be expected that variations in the intensities of neutral and ionized lines would be out of phase. The behavior of the Fe and Fe^+ lines is further evidence for the absence of a simple connection between the granulation, as observed in the continuous spectrum, and the structures seen in the line centers. The reason is probably that the line cores are formed well above the continuous spectrum, perhaps even in the low chromosphere.

In principle, the measurement of small, local Doppler shifts is best carried out by bisecting the wavelength interval between points of equal intensity on opposite sides of the line. The measurements can be made manually[1] but a great

---

[1] R. S. RICHARDSON and M. SCHWARZSCHILD: Astrophys. Journ. **111**, 351 (1950).
[2] H. H. PLASKETT: Monthly Notices Roy. Astronom. Soc. London **114**, 251 (1954).
[3] H. A. VON KLÜBER: Observatory **76**, 68 (1956).
[4] R. R. MCMATH, O. C. MOHLER, A. K. PIERCE and L. GOLDBERG: Astrophys. Journ. **124** (1956).

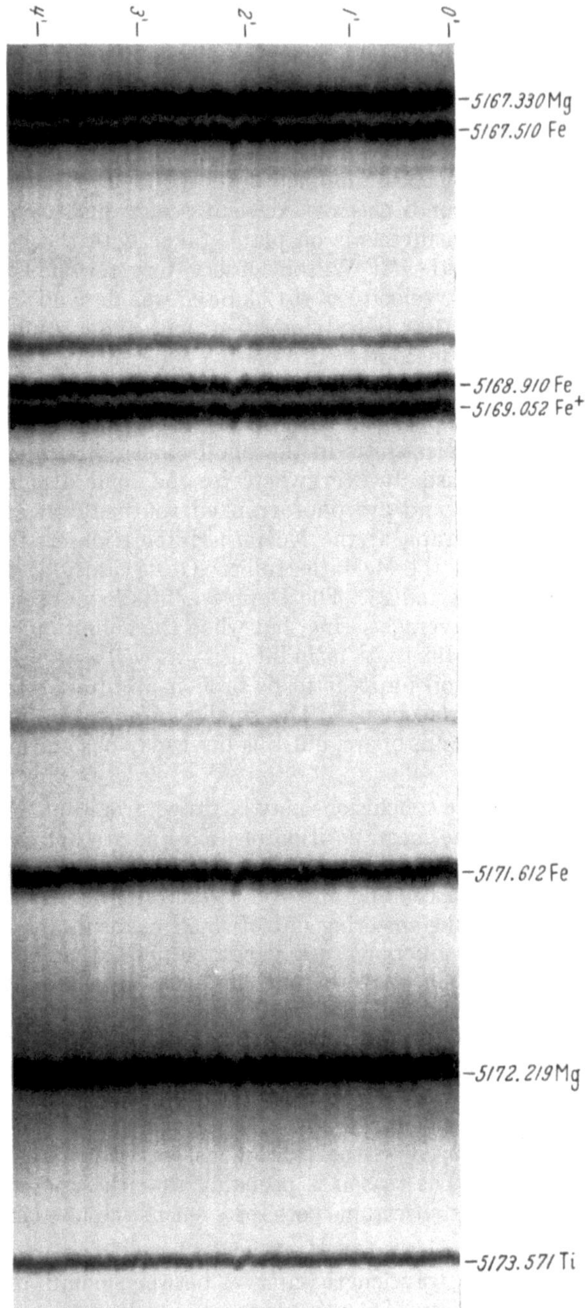

Fig. 22. Region of the Mg b lines in the Fraunhofer spectrum photographed with the vacuum spectrograph of the McMath-Hulbert Observatory.

saving in time and labor results from the use of automatic recording isophotometers[1]. Fig. 24 show typical isophotometer traces of the lines Ca I λ 5857.46 and Ni I λ 5857.76 made with the isophotometer of the McMath-Hulbert Ob-

[1] O. C. MOHLER and A. K. PIERCE: Astrophys. Journ. **125**, 285 (1957).

servatory. Measurements on records of this type have been made for a series of eleven lines by GOLDBERG, MOHLER, UNNO and BROWN[1] at various points on the solar disk. Table 12 gives the values of ξ, the random, turbulent velocity, derived by multiplying the root mean square of the radial velocity by $\sqrt{2}$. The third column of Table 12 gives the Rowland intensity and the fourth column the value of $\Delta\chi$ for neutral elements or χ for ionized elements. It would be

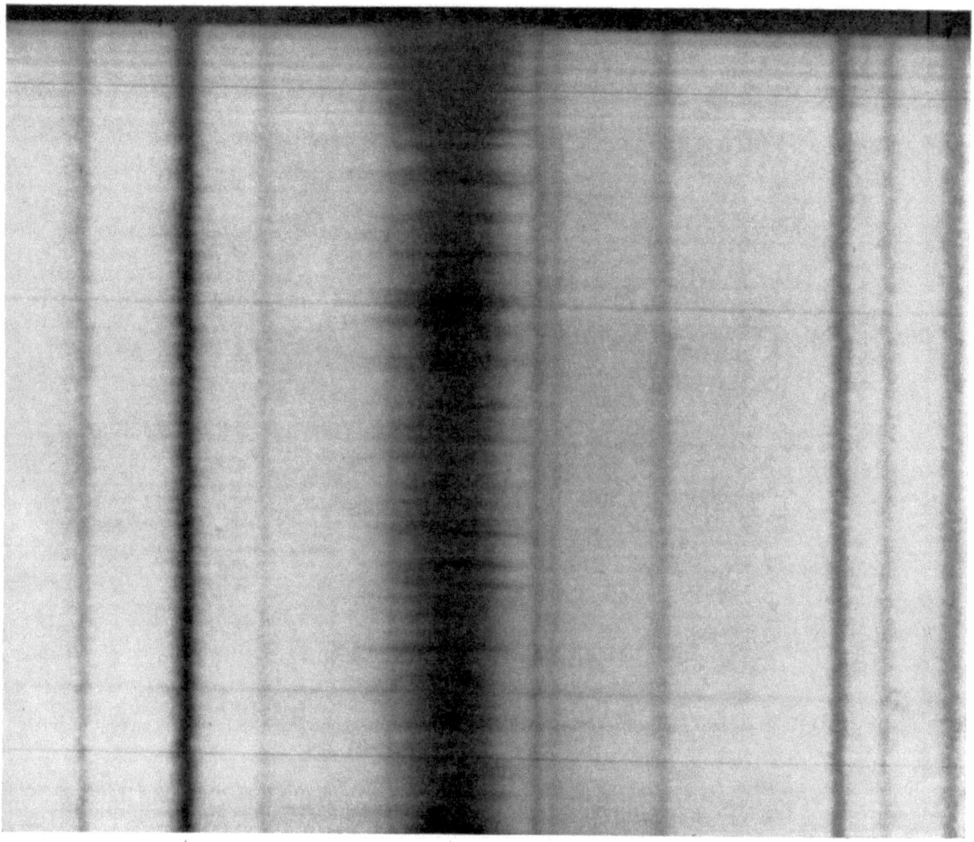

Fig. 23. Region of the Hβ line in the Fraunhofer spectrum photographed with the vacuum spectrograph of the McMath-Hulbert Observatory.

interesting to look for a dependence of ξ upon optical depth in the continuum. However, the scatter in the measurements, which seems chiefly due to variable seeing conditions, apparently masks the depth dependence, although there is some evidence that the shifts in the very strong MgI line are larger than those of the other lines. In general, the smearing effects of bad seeing are such that the absolute values of ξ in Table 12 are not too significant except as they represent minimum values. The most interesting feature of the table is the systematic dependence of ξ upon μ, as shown by the plot of the mean value of ξ, averaged

[1] L. GOLDBERG, O. C. MOHLER, W. UNNO and JACQUELINE BROWN: To be published in Astrophys. Journ. **1958**.

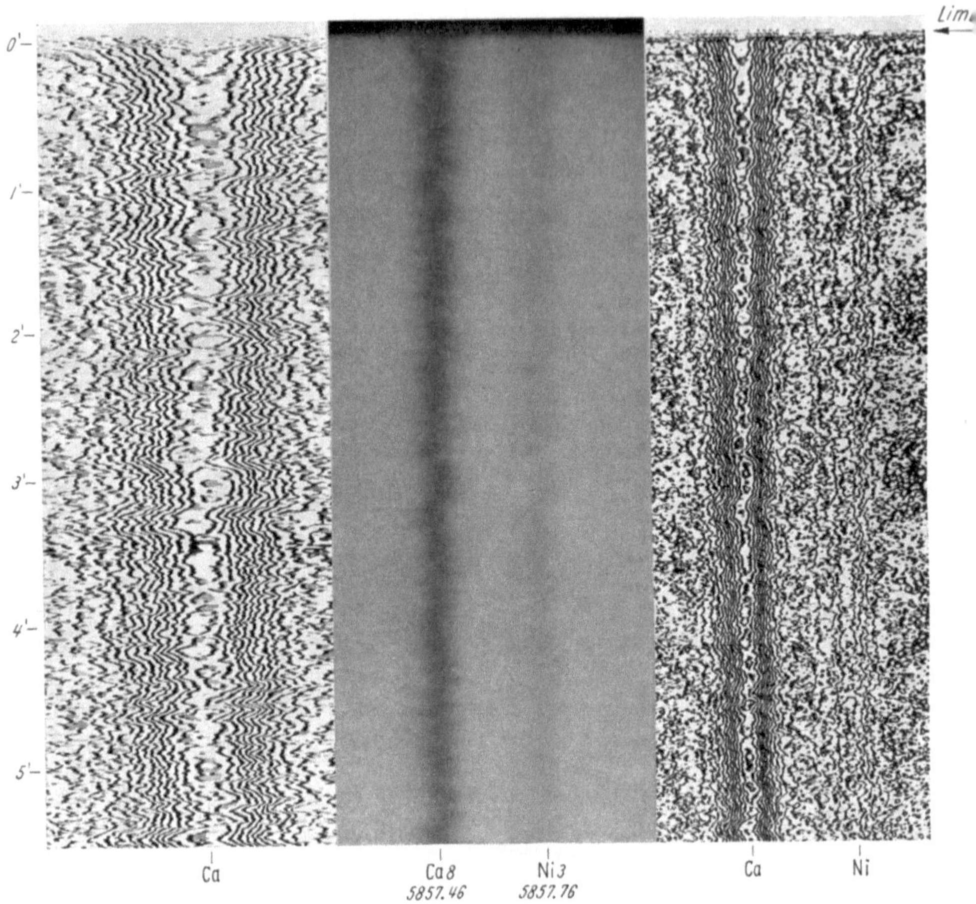

Fig. 24. The Fraunhofer lines Ca I λ 5857.46 and Ni I λ 5857.76, as they appear in photographs and on isophotometer tracings.

Table 12. *Values of ξ, the random, turbulent velocity, in km/sec, measured on iosphotometer tracings of selected lines at different points on the solar radius.* R.I. = Rowland intensity.

Element	λ	R. I.	$\Delta\chi$ or χ	ξ					
				$\mu=1.00$	$\mu=0.98$	$\mu=0.82$	$\mu=0.72$	$\mu=0.58$	$\mu=0.35$
Fe I	5165.4	2	3.66				0.31	0.44	0.29
Fe I	5655.5	2	3.62	0.24	0.21	0.19		0.32	
Ti I	5173.7	2	6.81	0.27	0.35		0.36	0.39	0.26
Sc II	5657.9	2	1.50	0.22	0.21	0.22		0.29	
Fe I	5662.5	4	3.70			0.24		0.31	
Fe I	5658.8	4	4.48			0.21		0.25	
Fe I	4625.1	5	4.63			0.33	0.34	0.38	
Cr I	4626.2	5	5.78	0.26	0.28	0.26	0.32	0.36	0.30
Ba II	5853.7	5	0.60	0.27	0.25	0.28	0.33	0.40	
Ca I	5857.5	8	3.17			0.27	0.35	0.40	0.32
Mg I	5172.7	20	4.91			0.44	0.40	0.47	0.27
			Mean:	0.25	0.26	0.27	0.34	0.36	0.29

over all the lines, against μ in Fig. 25. The increase of ξ to a maximum at about $\mu=0.6$ and its subsequent decrease is shown by all of the lines measured.

The physical significance of the center-limb variation in ξ is not yet clear. It must be noted that measurements of local Doppler shifts refer to the macro-turbulence of gas masses larger than the average thickness of the line forming region, whereas the Doppler widths derived from line profiles (see Sect. 35) give the sum of the squares of the micro- and macroturbulent velocities. The latter are always larger than the former by a factor of about five, but the ratio would probably be smaller under conditions of perfect seeing. Since the microturbulence is anisotropic, it is probable that the initial increase in the macroturbulence with decreasing μ may also be explained at least in part on the same basis. The decrease at $\mu = 0.35$ may result from the merging of the individual gas clouds when viewed tangentially at the limb. WADDELL[1] has also suggested that the interpretation of the center-limb variation must take account of the fact that the effective thickness of the line absorbing layer increases toward the limb, and hence that the smallest and presumably most energetic elements of the macroturbulence seen at the center of the disk would not be included in the limb measurements.

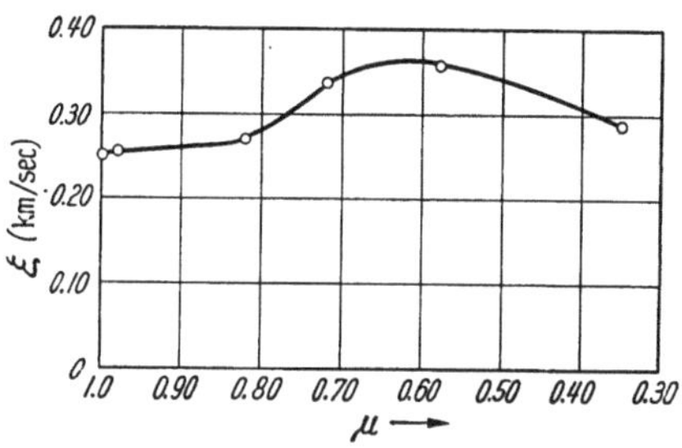

Fig. 25. Mean variation of ξ, the random, turbulent velocity in km/sec deduced from local Doppler shifts, as a function of μ.

Some portion of the center-limb variation in the local Doppler shifts probably reflects a variation in macroturbulence with height. The evidence for such a variation in the photosphere is not at all definite, but there seems to be no question but that the macroturbulence in the chromosphere is substantially greater than in the photosphere. Unpublished measurements of local Doppler shifts by GOLDBERG, MOHLER and BROWN give the following results for the lines Hβ, Hα and K, the centers of which originate in successively higher layers of the chromosphere:

Line	ξ (km/sec)
Hβ	0.64
Hα	1.2
K	1.5

Similar results for Hβ and Hα are quoted by DE JAGER in the following chapter of this volume.

37. Asymmetries in local Doppler shifts. As far as can be determined, positive and negative values of local Doppler shifts in weak and moderately-strong Fraunhofer lines, are about equal in number but some very strong lines, e.g., λ 5172 of Mg I, show a preponderance of violet shifts, which make the violet edge of the line core appear more "jagged" than the red. The resulting line asymmetries are clearly present in isophotes of the λ 5172 Mg I line. The same effect appears in the Hβ line[2] and can be inferred also from the increased contrast of spectroheliograms made on the violet edges of strong lines[3]. It has also been found[2] that about 70% of the structures in the Hβ line (see Fig. 23), which is formed in the low chromosphere, are associated with violet shifts in neighboring moderately strong photospheric metallic lines.

[1] JOHN H. WADDELL III: Thesis, University of Michigan 1956.
[2] L. GOLDBERG, O. C. MOHLER and J. D. BROWN: Nature, Lond. **179**, 369 (1957).
[3] L. D'AZAMBUJA: Ann. Obs. Paris, Meudon **8**, Fasc. II (1930).

The detailed investigation of local Doppler shifts in the Fraunhofer spectrum, which has hardly begun, offers a most promising area for the observational study of the hydrodynamics of the photosphere.

38. Center-limb studies of Fraunhofer lines. One of the important goals of solar physics is the derivation of a model of the atmosphere, which, when combined with a theory of Fraunhofer line formation, can reproduce accurately the line profiles and intensities observed all over the solar disk. Solar models derived from observations of the continuous spectrum are uncertain in the high layers both because of the blanketing effects of absorption lines and because of the seeing difficulties encountered in measurements at the extreme limb. Some investigators, notably DE JAGER[1] and BÖHM-VITENSE[2] have attempted to correct the continuum models by making use of the center-limb variations in the profiles of strong lines, e.g., the Balmer lines. This procedure has been criticized by ATHAY and THOMAS[3], on the grounds that the residual intensities of the hydrogen lines are determined by the excitation temperature, which, in the high layers, is undoubtedly different from the electron temperature. Since the profiles of strong lines are affected by uncertain knowledge of both the damping constants and the re-emission mechanism, much attention has been given in recent years to the center-limb variations in the profiles and intensities of weak and middle-strong lines. The result seems to be that no single homogeneous model of the photosphere can predict successfully the center-limb observations. In general, the discrepancies between theory and observation are greatest for lines formed at high atmospheric levels.

There appear to be three major reasons for the failure of the weak-line theory. First, it is necessary to take account of temperature and velocity inhomogeneities. Second, the dependence of the turbulence upon depth and direction is not well known. Third, departures from thermodynamic equilibrium in the high layers may result in ionization and excitation temperatures which differ from the electron temperature and which may differ both from element to element and from one line to another of the same element. These conclusions are derived from a number of recent investigations based on accurate observational data and analysis by the method of weighting functions. A theoretical study by BÖHM[4] has shown that temperature fluctuations on the order of $\pm 500°$ result in a wavelength dependence of the mean continuum intensity that deviates considerably from that of a Planck function in the wavelength region relevant to the ionization of metals. The temperature fluctuations are incorporated into an idealized "three-stream model" consisting of hot and cool gas columns, ascending and descending, respectively, and stationary columns of intermediate temperature. The actual temperature fluctuations deduced from brightness measurements of the granulation are much smaller—about $\pm 100°$. The difference is ascribed to the smearing of the small, very bright granules by bad seeing. BÖHM shows further that for a solar model in non-grey radiative equilibrium, the steep decrease in temperature in the layers above $\tau_0 = 0.1$ leads to departures from thermodynamic equilibrium, one result being that the ionization temperature for Fe I at $\tau_0 = 0.01$ is 250° higher than the excitation temperature for the low atomic levels. The inclusion of these two effects in the theory appears to give good agreement with observation for a number of strong and weak Fe I lines, but not for the Na *D*-lines. A more detailed application of the Böhm model by

[1] C. DE JAGER: Rech. Astr. Obs. Utrecht **13**, Part 1 (1952).
[2] E. BÖHM-VITENSE: Z. Astrophys. **34**, 209 (1954).
[3] R. G. ATHAY and R. N. THOMAS: Astrophys. Journ. **127**, 96 (1958).
[4] K.-H. BÖHM: Z. Astrophys. **35**, 179 (1954). — Publ. Astronom. Soc. Pacific **67**, 20 (1955).

Miss BRETZ[1] also indicates that the center-limb variation in the equivalent widths of metallic lines may be explained by inequalities between the ionization and excitation temperatures.

The Böhm three-stream model has been further refined by VOIGT[2], who supposes that the temperature inhomogeneities diminish with height and also that the stream velocities decrease with height while the microturbulence increases. These refinements are introduced to explain certain features of the center-limb changes in the profiles of the infrared O I triplet, λ 7774, including the violet asymmetry, which is greatest at the center of the disk and disappears at the limb. The center-limb variation of the central intensities seems to require that the excitation temperature exceed the electron temperature in the highest layers by about 200°. VOIGT comments that the apparent increase of the microturbulence with height to a value of about 3 km/sec at $\tau_0 = 0.02$ cannot wholly be distinguished from a possible anisotropy in the motions.

ELSTE'S[3] observations of the profiles of a number of weak and middle-strong metallic lines could also not be represented adequately by calculations based on local thermodynamic equilibrium. Satisfactory agreement between theory and observation was obtained for lines formed deep in the atmosphere, but the discrepancies increase with height. The results suggest that the ionization temperature in the high layers is greater than the local electron temperature. PAGEL[4] finds that his observations of the equivalent widths of nine Fe I lines near λ 15500 suggest a temperature inversion very close to the surface, at $\tau_0 = 0.015$. MITCHELL[5] attempted to correct the continuum model of the photosphere from center-limb measurements of the intensities of near infrared lines ($\lambda \sim 8000$ Å) carefully selected for sensitivity to pressure and temperature, again with the result that the corrections are different for different lines.

39. Deviations from local thermodynamic equilibrium (LTE). The foregoing examples, which are by no means complete, illustrate the importance of refining solar models to take account of velocity and temperature inhomogeneities and particularly of departures from thermodynamic equilibrium. The usual assumptions of thermodynamic equilibrium and pure absorption imply that the source function is equal to the Planck function appropriate to the local electron temperature and also that there is a complete re-distribution of absorbed frequencies as a consequence of electron collisions. The validity of the LTE assumption has never been proven. The second assumption is of course invalid in the case of coherent scattering, but even if the scattering is completely non-coherent the source function for all lines will not necessarily be described by a single Planck function at the local electron temperature.

An interesting empirical attack on the nature of the departures from LTE has recently been formulated by PECKER[6], who suggests that there may be very large inequalities between the electron and excitation temperatures, particularly in the high layers of the photosphere. The assumption of LTE implies that at each point in a given line profile the residual intensity is equal to the intensity of the Planck radiation appropriate to the value of T_{exc} in the layer at which the sum of the optical depths in the line and continuum are equal to unity:

$$\tau_c + \tau_l = 1 . \tag{39.1}$$

[1] M. BRETZ: Z. Astrophys. **38**, 259 (1955).
[2] H. VOIGT: Z. Astrophys. **40**, 157 (1956).
[3] G. ELSTE: Z. Astrophys. **37**, 184 (1955).
[4] B. E. J. PAGEL: Monthly Notices Roy. Astronom. Soc. London **115**, 493 (1955).
[5] W. E. MITCHELL jr.: Thesis, University of Michigan 1958.
[6] J. C. PECKER: C. R. Acad. Sci., Paris **245**, 499 (1957).

PECKER measures the central intensities of a large number of lines of Ti I over a wide range of excitation potential and uses the Planck formula to convert the intensities into T_{exc}. From a model of the photosphere he then calculates τ_l for the center of each line and from Eq. (39.1) obtains a relation between T_{exc} and τ_c. If the assumption of LTE is valid, this relation should be identical with that between T_e and τ_c which is found from measures of the continuous spectrum. He finds instead that although the results are compatible with LTE for the higher excitation potentials (1.73 to 1.88 eV), the deviations for the lower excitation potential lines are as great as 900° K. In other words, the departures from LTE are greatest in the upper layers of the photosphere.

If PECKER'S conclusions are correct, and their validity rests both on the accuracy of measurement of the central intensities and on the accuracy of the assumed solar model, the theory of the equivalent widths of lines given in the preceding section may have to be revised drastically[1]. According to the usual assumption of LTE, the weighting functions are given by Eq. (30.3) with $S = B_\lambda(T_e)$. In the Pecker theory, however, Eq. (30.3) would be written as

$$g_\lambda^\mu(\tau_{\lambda_0}) = \frac{\int\limits_{\tau_{\lambda_0}}^{\infty} B_\lambda(T_e)\,e^{-t_\lambda/\mu}\,d\,\frac{t_\lambda}{\mu} - B_\lambda(T_{\mathrm{exc}})\,e^{-\tau_{\lambda_0}/\mu}}{\int\limits_0^{\infty} B_\lambda(T_e)\,e^{-\tau_\lambda/\mu}\,d\,\frac{\tau_\lambda}{\mu}}, \tag{39.2}$$

reflecting the fact that the source functions for the line and continuum are now different. Thus, in the absence of LTE, the g-functions must be computed separately for each multiplet. The resulting modifications in the theoretical curves of growth might cause a considerable revision in existing determinations of abundances and turbulent velocities.

THOMAS[2] attacks the problem of deviations from thermodynamic equilibrium from theoretical considerations. He shows first that, owing to the re-distribution of frequencies by the Doppler effect, the mechanism of line formation in the Doppler core is formally equivalent to that of pure absorption or of completely non-coherent scattering. Hence, the source function for the line core is independent of frequency and is given by

$$S_\nu = \frac{2h\nu^3}{c^2}\left(\frac{n_L}{n_u}\,\frac{g_L}{g_u} - 1\right)^{-1}, \tag{39.3}$$

where n_L and n_u are the number densities of atoms in the lower and upper levels of the transition, respectively, and the g's are the corresponding statistical weights. In thermodynamic equilibrium, n_L/n_u is given by the Boltzmann equation at the electron temperature, and the source function reduces to the Planck function:

$$S_\nu = B_\nu(T_e)\,. \tag{39.4}$$

In the general non-equilibrium case, however, n_L/n_u is fixed by the equations of statistical equilibrium and depends on the mean intensity of the radiation field, on the local electron temperature and density, and on the radiative and collisional cross-sections. In a formal sense, the deviations from LTE may be described by replacing T_e in Eq. (39.4) by the excitation temperature T_{exc} which is defined by the number ratio n_L/n_u through the Boltzmann equation.

[1] J. C. PECKER: C. R. Acad. Sci., Paris **245**, 639 (1957).
[2] R. N. THOMAS: Astrophys. Journ. **125**, 260 (1957).

THOMAS finds that quite generally the value of the source function is in between that fixed by collisional excitation and radiative de-excitation and that fixed by photoelectric ionization and recapture. A rough analysis shows that the source function for resonance lines and certain strong subordinate lines departs significantly from that obtained for either LTE or coherent scattering.

A subsequent investigation by JEFFERIES and THOMAS[1] deals with the simple case of a two-level atom for which ionizations to the continuum are neglected. The depth-variation of the source function is solved by the Eddington approximation. It is found that (1) $T_{\text{exc}} < T_e$ for the upper atmosphere, whereas $T_{\text{exc}} > T_e$ for the lower regions; and as a consequence (2) the computed line profiles are deeper than those calculated for LTE. The authors have emphasized the crude nature of their preliminary numerical results, which must be verified by more extensive numerical computation by machine methods.

40. Central intensities of strong Fraunhofer lines. No theory of line formation has succeeded in accounting quantitatively for the observed relatively high central intensities of very strong Fraunhofer lines. The difficulties have been summarized by MINNAERT [5]. Coherent scattering results in lines with completely black centers, which are never observed. Furthermore, since the centers of strong lines originate in the high layers of the atmosphere, where collisions are relatively infrequent, pure absorption in the ordinary sense would not be expected to occur. Fluorescence effects can be expected to contribute appreciably to the central intensities, and non-coherent scattering also fills in the line centers[2,3]. DEMPSTER[4] showed that fluorescence would account for about half of the observed central intensity in the *D*-lines, the remainder being presumably due to non-coherent scattering. However, BRAY[5] concludes that the *D*-line cores are formed entirely by coherent scattering, although photoionization accounts for only 25% of the observed emission. The remainder is attributed to de-excitation of the upper levels by superelastic electron collisions, to the superposition of granular elements having different line-of-sight velocities, and to chromospheric emission.

The concept of chromospheric emission as a mechanism for filling in the centers of strong Fraunhofer lines may be erroneous, since it implies that the source function is greater in the chromosphere than in the photosphere, which may or may not be the case. Nevertheless, the detailed theory of the central intensity must allow for the formation of the cores of strong lines in the chromosphere.

Up until about 1930, it was commonly believed that all Fraunhofer lines were formed in a "reversing layer", which was synonymous with the lower part of the chromosphere. Later, this concept was almost completely discarded in favor of the notion that both the line and continuous absorption occur in the photosphere and that absorption by the chromosphere is negligible even at the centers of strong lines[6]. Quantitative studies of the flash spectrum photographed during eclipses now reveal that the radial optical thickness of the chromosphere is about 50 at the center of Hα[7], and nearly 2×10^4 at the center of the *K*-line[8].

---

1 J. T. JEFFERIES and R. N. THOMAS: Astrophys. Journ. **1958** (in press).
2 V. V. SOBOLEV: Russ. Astronom. J. **31**, 231 (1954).
3 I. BUSBRIDGE: Monthly Notices Roy. Astronom. Soc. Lond. **113**, 52 (1953).
4 R. R. DEMPSTER: Astrophys. Journ. **96**, 295 (1942).
5 R. J. BRAY: Monthly Notices Roy. Astronom. Soc. London **116**, 395 (1956).
6 H. C. VAN DE HULST: Handbook of the Solar System, Vol. 1, Chap. 5, ed. by G. P. KUIPER. Chicago, Ill.: University Chicago Press 1953.
7 R. G. ATHAY and R. N. THOMAS: Astrophys. Journ. Suppl. **1**, No. 12. — E. BÖHM-VITENSE: Z. Astrophys. **36**, 145 (1955).
8 S. MIYAMOTO: Z. Astrophys. **31**, 282 (1953).

Further, ZIRKER[1] finds that the optical depth at the center of the H-line of Ca^+ is already unity at a height of 2000 km, and that there is an appreciable chromospheric contribution to the centers of singly-ionized metallic lines as weak as Rowland intensity 10. Similar conclusions have been reached from disk studies of the shapes of the cores of strong lines and of the center-limb variation in the Fraunhofer line structure observed with the vacuum spectrograph[2]. The theory of the central intensities of strong Fraunhofer lines must therefore be investigated in terms of chromospheric rather than photospheric models.

One of the most interesting problems connected with the profiles of strong Fraunhofer lines concerns the peculiar shapes that are often found in the line cores. The most extreme case is that of the H and K lines of Ca II, which exhibit well-defined double reversals at their centers, as shown in Fig. 26 from photoelectric tracings made by O. C. MOHLER with the vacuum spectrograph at the McMath-Hulbert Observatory. The intensities of both the absorption and emission components show pronounced local fluctuations, but the profiles shown represent fairly accurately the average intensities in undisturbed regions at the center of the disk. The average central intensity of the K-line is 2.2%, relative to the highest maxima between the absorption lines at λ 4000, which is about one-third the value previously obtained by HOUTGAST[3]. The central emission reversal is asymmetric in the sense that the violet emission component is almost always stronger than the red component, but frequently the components are equal in intensity and occasionally the red component is the stronger of the two. Complete observational details, including the center-limb variation of the profiles, will be found in a forthcoming paper by GOLDBERG, MÜLLER and MOHLER[4].

The central reversals of the H and K lines are also observed in the spectra of a great many stars of spectral types, G, K and M. WILSON[5] has discovered a striking correlation between the absolute visual magnitudes of these stars and the total width of the K-line emission as measured visually on the spectrograms. The sun fits this relation very closely. WILSON and BAPPU[6] find that the correlation holds over a range of about 16 magnitudes.

It can be seen from Fig. 26 that the emission hump is more pronounced in the K-line than in the H-line, which suggests that the sharpness of the maximum is a function of optical depth. This is confirmed by the shapes of the cores of other strong Fraunhofer lines, such as the infrared triplet of Ca^+, the b-group of Mg I (see Fig. 27) and numerous strong lines of Fe I. Thus it is evident that a tendency toward emission reversal is a characteristic feature of the cores of nearly all strong Fraunhofer lines and the correlation found by Wilson suggests that the reversals are connected with rather fundamental properties which are common to all late-type stellar atmospheres.

The central reversals in the Fraunhofer lines pose two questions. The first relates to the mechanism by which the double reversal is formed and to the physical parameters of the stellar atmosphere that govern the emission widths, and the second to the manner in which these parameters are determined by the luminosity. The answer to the first question depends on the character of the source

[1] J. ZIRKER: Thesis, Harvard University 1956.

[2] R. R. McMATH, O. C. MOHLER, A. K. PIERCE and L. GOLDBERG: Astrophys. Journ. **124** (1956).

[3] J. HOUTGAST: Diss. Utrecht 1942.

[4] L. GOLDBERG, E. A. MÜLLER and O. C. MOHLER: Astrophys. Journ. to be published in **1958**.

[5] O. C. WILSON: Conf. on Stellar Atmospheres, ed. by M. H. WRUBEL: Indiana University 1954.

[6] O.C. WILSON and M. K. V. BAPPU: Astrophys. Journ. **125**, 661 (1957).

function in the lower part of the chromosphere, where the cores of strong Fraunhofer lines are formed. If the source function is independent of frequency across the line core, the double reversal suggests a maximum in the depth dependence of the source function. It has not yet been determined whether such a simple layer model can reproduce the observed line profiles at the limb as well as at the disk center. The physical significance of a maximum in the source function is also not clear.

The double reversal can also occur as a result of a maximum in the frequency dependence of the source function. This idea has been developed in an interesting series of papers by MIYAMOTO[1], who shows that the double reversal can be attributed to non-coherent scattering, the redistribution of frequencies being brought about by the random Doppler effect. If the scattering were purely coherent, a quantum with frequency at the line center would be prevented from escaping by virtue of the very high central opacity of the line. However, when redistribution occurs, a quantum created at the line center is eventually likely to be re-emitted just outside the Doppler core, where it has a higher probability of escaping. According to MIYAMOTO, this process gives rise to the emission maxima in the H and K lines and to the emission humps in other metallic lines. The width of the emission is essentially proportional to the Doppler width and depends only secondarily on chromospheric thickness. Similar conclusions have been reached on the basis of the layer model[2,3]. Since the Doppler widths are undoubtedly turbulent in origin, it would seem likely that turbulence is the principal atmospheric parameter that determines the emission line widths. The correlation between line width and absolute magnitude then implies that the degree of turbulence in stellar chromospheres is controlled by the rate of energy generation in the stellar interiors.

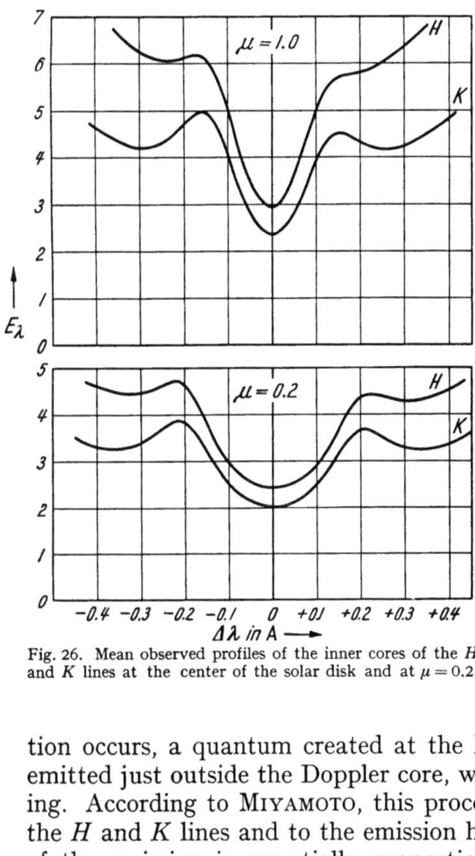

Fig. 26. Mean observed profiles of the inner cores of the H and K lines at the center of the solar disk and at $\mu = 0.2$.

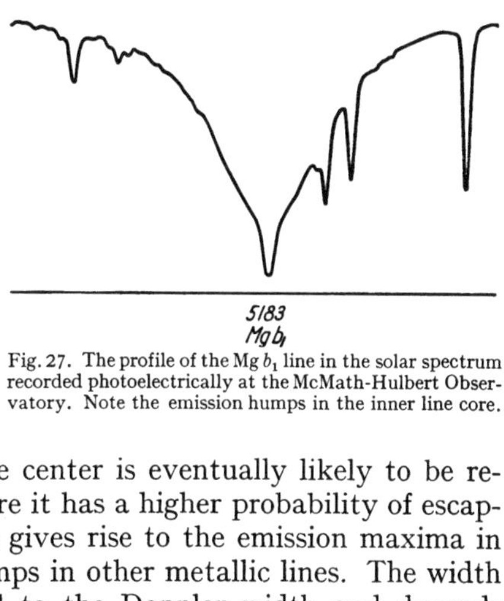

Fig. 27. The profile of the Mg b_1 line in the solar spectrum recorded photoelectrically at the McMath-Hulbert Observatory. Note the emission humps in the inner line core.

41. Wavelength displacements. Accurate absolute measurements on the Fraunhofer spectrum reveal that solar wavelengths are systematically larger than

[1] S. MIYAMOTO: Z. Astrophys. **31**, 282 (1953); **35**, 145 (1954); Publ. Astronom. Soc. Japan **5**, 142 (1954); **6**, 140, 150 (1954).

[2] O. C. WILSON: Astrophys. Journ. **126**, 525 (1957).

[3] L. GOLDBERG: Astrophys. Journ. **126**, 318 (1957).

vacuum laboratory wavelengths by amounts ranging from a few thousandths to a little more than one hundredth of an angstrom. Very early measurements, at the beginning of this century, showed that the red shift is greater at the limb than at the center of the disk, and that the displacements increase both with wavelength and line intensity.

An explanation for the displacements has long been sought in terms of the gravitational red shift predicted by the general theory of relativity, which amounts to 0.0127 Å at λ 6000. However, although the observed shifts have about the correct order of magnitude at the limb, the center-limb increase and the scatter shown by different lines proves that additional factors contribute to the displacements. ST. JOHN's detailed investigation in 1928 seemed to show that the observations are consistent with the Einstein effect provided that upward radial currents with velocities between 0 and 0.2 km/sec were also postulated.

It was later pointed out by Miss ADAM that phase-disturbing collisions between the absorbing atoms and neighboring atoms of hydrogen and helium could result in line shifts of the same order of magnitude as the Einstein shift. Furthermore, since the frequency of collisions decreases with height, and since, according to the Lindholm theory, collisional displacements are to the red, the observed small shifts at the center of the disk require improbably large upward motions, in the neighborhood of 0.6 km/sec. SPITZER attempted to remove this difficulty by theoretical considerations which indicated that most collisional shifts would be to the violet. Unfortunately, the collisional theory is not yet sufficiently precise to be conclusive as to the direction of the shift, nor can it account for the variations in the amount of the shift between different lines[1].

Interesting empirical relationships between the red shifts and the characteristics and disk positions of the lines have been derived by Miss ADAM[2] and by FREUNDLICH and FORBES[3]. Miss ADAM concludes that "the outstanding observational characteristic of the solar red shift is its dependence upon line strength, rather than on any other factor". She finds the red shifts to be correlated with the square of the wavelength multiplied by the logarithm of the total absorption in ergs. The physical significance of this relationship is not at all obvious. Measurements on six infrared lines by L. HERZBERG[4] suggest that the red shifts are term dependent.

Like other properties of Fraunhofer lines, the wavelength displacements are undoubtedly influenced by the temperature and velocity inhomogeneities which are associated with the granulation. A re-examination of the observational data from this point of view has recently been carried out by SCHRÖTER[5], on the basis of an inhomogeneous two-stream model of the photosphere. It is shown that each Fraunhofer line is a composite of contributions from hot regions moving upward and from cooler regions moving downward. The resulting profile is slightly asymmetric and is accompanied by a line shift, additional to the Einstein shift, whose direction and amount depend upon the particular excitation properties of the line, as well as upon the disk position. The theory of SCHRÖTER predicts a dependence of red shift upon line strength which is in good agreement with existing observations, and in general promises to supply the hitherto missing ingredient in the interpretation of the observed line displacements.

---

[1] M. G. ADAM: Monthly Notices Roy. Astronom. Soc. London **112**, 567 (1952).
[2] M. G. ADAM: Monthly Notices Roy. Astronom. Soc. London **115**, 405 (1955).
[3] E. FINLAY-FREUNDLICH and E. G. FORBES: Ann. d'Astrophys. **19**, 183, 215 (1956).
[4] L. HERZBERG: Canad. J. Phys. **35**, 766 (1957).
[5] E. H. SCHRÖTER: Z. Astrophys. **41**, 141 (1957).

General references.

[1] UNSÖLD, A.: Physik der Sternatmosphären, 2nd ed. Berlin: Springer 1955. — A comprehensive and detailed account of the physics of stellar atmospheres: Stellar atmospheres in thermal equilibrium. Continuous spectrum and formation of stellar atmospheres. Measurement of the intensity distribution in Fraunhofer lines. Physical foundations of the theory of Fraunhofer lines. The origin of the Fraunhofer lines. Physics of the Sun. UNSÖLD's 57 page bibliography covers the literature in the above fields to 1955.

[2] CHANDRASEKHAR, S.: Radiative transfer. Oxford: Clarendon Press 1950.

[3] KOURGANOFF, V.: Basic methods in transfer problems. Oxford: Clarendon Press 1952. — Ref. [2] and [3] present in detail all the known mathematical methods and techniques employed in the solution of the equation of radiative transfer for plane parallel atmospheres.

[4] WOOLLEY, R. V. D. R., and D. W. N. STIBBS: The outer layers of a star. Oxford: Clarendon Press 1953. — Especially valuable for its presentation of the physical concept of transfer problems both for the continuum and for absorption lines.

[5] MINNAERT, M.: The Solar System, Vol. I: The Sun. Chap. III: the Photosphere, ed. G. P. KUIPER. Chicago, Ill.: Chicago University Press. 1953. — An excellent review of the empirical models of the photosphere. Part one treats the continuum, part two the Fraunhofer spectrum. Contains an extensive bibliography up to 1953.

[6] RIGHINI, G.: Draft Report Commission 12 International Astronomical Union 1958. — A very brief summary of the most recent work on the solar photosphere.

[7] STRÖMGREN, B.: Astrophysics, ed. by HYNEK, Chap. V: The growth of our knowledge of the physics of the stars. New York: McGraw-Hill 1951. — A historical review of the work on the continuum and line spectrum.

[8] ALLER, L. H.: The atmospheres of the sun and stars. New York: Ronald Press Co. 1953. — A textbook covering the fundamentals and leading to the modern theories and observations of the sun and stars.

[9] MOORE, C. E.: Handbook of the Solar System, Vol. I, Chap. 4, The Identification of Solar Lines, ed. by G. P. KUIPER. Chicago, Ill.: Chicago University Press 1953. — Summarizes techniques and results connected with measurements of wavelengths and identifications of solar lines.

[10] ST. JOHN, C. E., C. E. MOORE, L. M. WARE, E. F. ADAMS and H. D. BABCOCK: A Revision of ROWLAND's Preliminary Table of Solar Spectrum Wavelengths. Published by Carnegie Institution of Washington. No. 396, 1928. — Contains wavelengths, identifications and eye estimates of intensities for Fraunhofer lines in the wavelength region λ 2975—λ 7331, with an infrared addendum λ 7333—λ 10218. A new Revised Rowland Table covering the entire photographic region observable from the ground, i.e., λ 2935 to λ 13495 is in preparation by C. E. MOORE.

[11] BABCOCK, H. D., and C. E. MOORE: The Solar Spectrum, λ 6600—λ 13495. Published by Carnegie Institution of Washington, No. 579, 1947. Wavelengths, identifications and eye estimates of intensities. An ultraviolet extension to the 1928 Revised Rowland, λ 2935—λ 3060, has also been published by H. D. BABCOCK, C. E. MOORE and M. F. COFFEEN: Astrophys. Journ. **107**, 287 (1948).

[12] MINNAERT, M., G. F. W. MULDERS and J. HOUTGAST: Photometric Atlas of the Solar Spectrum. Amsterdam: D. Schnabel 1940. — Complete intensity map of Fraunhofer spectrum in the region λ 3332—λ 8771.

[13] Photometric Catalogue of Fraunhofer Lines λ 6600—λ 8770. Rech. Astr. Obs. Utrecht **12**, Part 2 (1951). — Contains catalogue of equivalent widths measured from Utrecht Atlas. Measurements in the region λ 4500—λ 6600 have also been completed but not yet published.

[14] MOHLER, O. C., A. K. PIERCE, R. R. MCMATH and L. GOLDBERG: Photometric Atlas of the Near Infrared Solar Spectrum, λ 8465—λ 25242. Ann Arbor: University of Michigan Press 1950. — Extends the Utrecht Atlas to the near infrared by observations made with photoconductive cells.

[15] MOHLER, O. C.: A Table of Solar Spectrum Wavelengths 11984 Å to 25578 Å. Ann Arbor: University of Michigan Press 1955. — The companion table to the Michigan infrared atlas, containing wavelengths, identifications and equivalent widths of both solar and telluric lines.

[16] MIGEOTTE, M., L. NEVEN and J. SWENSSON: The Solar Spectrum from 2.8 to 23.7 Microns. Mém. Soc. Roy. Sci. Liège, Special Volume No. **1** (1956); No. **2** (1957). In three parts. Parts I and II contain an intensity atlas and a table of wavelengths, respectively. Part III, still in preparation, will consist of an atlas of laboratory spectra of molecules present in the earth's atmosphere.

Structure and Dynamics of the Solar Atmosphere.

By

C. DE JAGER.

With 136 Figures.

1. Introduction. In this article we deal only with the *photosphere* as far as its dynamics is concerned. So we treat successively convection and turbulence of the photosphere, granulation, temperature-inhomogeneities and photospheric micro- and macro-turbulence (Chap. A I). In Chap. B I the sunspots and faculae are discussed. The other parts of this article are concerned with the structure and dynamics of what might be called the outer solar atmosphere: the *chromosphere* (Chap. A II and B II) and the *corona* (Chap. C). In Chap. B III we deal with flares and, more generally, with centres of activity; in Chap. C II with the solar radio-emission. The sunspot cycle, the solar rotation and the general solar magnetic field are treated in Chap. D; E discusses the relations between solar and terrestrial phenomena. The prominences, although being coronal features, are treated immediately after the disturbed chromosphere (flares) because their temperatures and densities are more like those of the chromosphere than of the corona (Chap. B IV).

The general trend followed in the various chapters is to discuss first the normal Sun, then its disturbed parts. Furthermore we deal with the individual structures before treating their relations to other individuals. (E.g. we start the sunspot chapter with a discussion of the structure and physical properties of spots, neglecting the fact that they occur mainly in groups.) Then there follow the large scale relations of these phenomena, in space and in time. According to this lay-out the solar cycle is treated in Chap. D after the description of the spots, faculae, flares and prominences. Of course this lay-out cannot be followed quite rigorously, and reference to the Sun's variability is often made in the chapters preceding Chap. D.

The basical physical theories, like those on absorption and emission of radiation, the formation of Fraunhofer lines, the atomic processes, hydrodynamics and hydromagnetics will not be discussed extensively; reference is made to other volumes of the Encyclopedia of Physics, the books listed at the end of this article, and to GOLDBERG and PIERCE's paper on the normal photosphere.

Many thanks for their helpful remarks are due to Profs. MINNAERT and UNSÖLD who kindly read the whole manuscript. I am greatly indebted to many colleagues, too numerous to be mentioned here, for sending illustrations, or for their kindness to read parts of the manuscript and for their stimulating criticisms. It is my sincere hope to have done justice to their work in this review paper.

A. The undisturbed photosphere and chromosphere.

I. The undisturbed photosphere.

a) Granulation.

2. Size and life times; photometric data. The normal solar photosphere, outside the spots and faculae, is not homogeneously bright but shows a granular structure. According to the best observations the granulation seems to consist of bright, more or less roundish blotches on a dark background. THIESSEN[1] observing visually with a 60 cm refractor and using the full aperture, noted even their polygonal form, also suggested many years earlier by other visual observers; cf.[2]. Occasionally they are observed on very good photographs (Fig. 1).

Of the modern granulation photographs one might mention those of KRAT[3] (Poulkovo) and those obtained by RÖSCH[4] at the observatory at the Pic du Midi (altitude 2900 meters), where, due to the good atmospheric conditions and to a refined observational technique, photographs of the solar granulation could

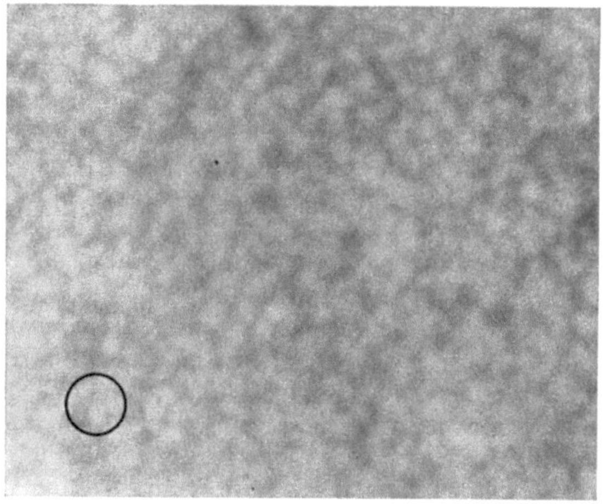

a

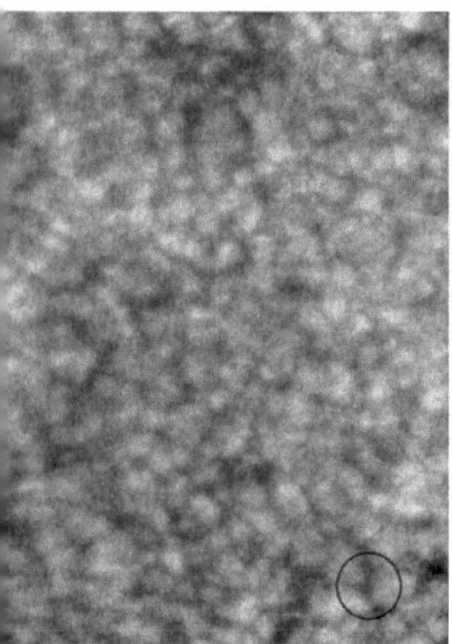

b

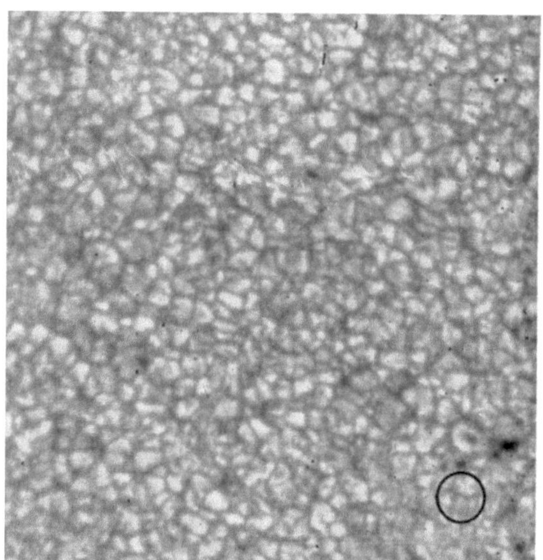

c

Fig. 1a–c. Modern granulation photographs. (a) Granulation photographed at the Pic du Midi Observatory, alt. 2900 metres; 28. II. 1957. Convection currents in the telescope habe been reduced by cooling the focal plane. Objective diameter 25 cm; exposure time 0.002 sec. The circle has a diameter of 5″. Courtesy J. RÖSCH. (b) Photograph obtained from a manned balloon, alt. 6.5 km; 1.IV. 1957. Objective diameter 28 cm; exposure time 0.002 sec. Circle diameter 5″. Courtesy D. E. BLACKWELL, D. W. DEWHIRST and A. DOLLFUS. (c) Photograph obtained from an unmanned balloon alt. 25 km. Objective diameter 30 cm; exposure time 0.001 sec. Circle diameter 5″. Courtesy M. SCHWARZSCHILD, J. B. ROGERSON and J. W. EVANS.

be obtained which equal or perhaps surpass the classical observations of JANSSEN, CHEVALIER, HANSKY, all made around 1900. In RÖSCH's observations the focal

[1] G. THIESSEN: Z. Astrophys. **35**, 237 (1955).

[2] C. DE JAGER: Trans. Internat. Astronom. Union **9**, 727 (1957). With references.

[3] V.A. KRAT: Isw. Gl. Astr. Obs. Poulkovo **152**, 1 (1954); **155**, 17 (1956).

[4] J. RÖSCH: C. R. Acad. Sci. Paris **240**, 1630 (1955); **243**, 478 (1956).

plane of the 23 cm telescope is cooled, to avoid convection currents in the instrument. Furthermore the photographs are taken in a rapid succession (20 frames per second with an exposure time of 0.005 sec per frame) which enables one to distinguish between the real granulation and the granules deformed by atmospheric agitation. More recently attempts to eliminate the influence of atmospheric turbulence were made by BLACKWELL, DEWHIRST and DOLLFUS[1] from observations from a manned balloon at 6.5 km altitude (exposure times 0.0002 sec) and by SCHWARZSCHILD, ROGERSON and EVANS[2] from an unmanned balloon, rising up to 25 km. In the latter case the telescope, objective diameter 30 cm, was pointed at the Sun by a photo-electric Sun-following device; the exposure time was 0.001 sec.

The diameters of the granules are 400 to 1000 km according to JANSSEN'S classical observations (Meudon, 1896). The bigger elements which also occur on his photographs are generally complexes of smaller ones. Most of the previous and modern observations confirm these results: HANSKY (Poulkovo, 1905) notes a mean diameter of 700 km; CHEVALIER (Zo-Ze, 1912) and KEENAN (Yerkes; 1938, 1939) measured mean diameters of 1000 km. More recently MACRIS (1953), reducing LYOT'S observations made at the Pic du Midi, found a mean diameter of 1000 km; these results are confirmed by MILLER (Rocky Point, N.Y.; unpublished). RÖSCH'S observations show a flat frequency maximum between 700 and 1000 km; the photographs contain granules down to 400 km, which is the theoretical resolving power of his objective (0".5). Naturally, smaller grains could be observed only with a reduced intensity ratio, e.g. grains with a diameter of 0".5 (360 km) would have a mean intensity fluctuation of about 0.2 times the true mean fluctuation; grains with a diameter of 250 km would have a top intensity reduced to 0.08[3]. The mean intensity fluctuation of granules with a mean diameter of 1" (725 km) would even be slightly reduced, to 0.8. Even if one would not be willing wholly to accept UBEROI'S conclusion[4] that the "measured intensity and scale of the granules are not even close to their actual value" it is obvious that the flat maximum between 0".5 and 1".5 of the observed distribution function of granule diameters is not real, but ought to be shifted towards smaller diameters. On the basis of such considerations, RICHARDSON and SCHWARZSCHILD[5] and KRAT[6] suggested that the smaller granules should be preponderant, and they advanced the hypothesis that the true sizes of the granules are of the order of 200 km or even 150 km.

The problem of the true granulation size was brought nearer to a solution by the magnificent photographs obtained by SCHWARZSCHILD *et al.* These pictures (Fig. 1c) show that the granules have an average size of the order of 1" (700 km). Greater granules occur, up to diameters of 2".5 (1800 km); the smallest elements have diameters down to 0".3 (200 km).

There seems to be a faint, perhaps negligible correlation of the granule number N_g per unit area with the daily sunspot number[7] R, the correlation coefficient between N_g and R is $+0.5$.

Brightness measurements in granules are very difficult because of the small sizes of the objects, and the great influence of scattered light, and image-deforma-

[1] D.E. BLACKWELL, D.W. DEWHIRST and A. DOLLFUS: Nature, Lond. **180**, 211 (1957).
[2] M. SCHWARZSCHILD, J.B. ROGERSON and J.W. EVANS: Astronom. J. **63**, 54 (1958).
[3] H.H. PLASKETT: Monthly Notices Roy. Astronom. Soc. London **114**, 266 (1954).
[4] M.S. UBEROI: Astrophys. Journ. **122**, 466 (1955).
[5] R.S. RICHARDSON and M. SCHWARZSCHILD: Astrophys. Journ. **111**, 351 (1950).
[6] Cf. footnote 3, p. 81.
[7] C. MACRIS and D. ELIAS: Ann. d'Astrophys. **18**, 143 (1955). — W.A. MILLER: Astronom. J. **61**, 187 (1956).

tions due to atmospheric agitation. WALDMEIER[1], THIESSEN[2] and KRAT[3] communicated values for $\Delta I/I$ (granules with respect to the dark surroundings) of 30 to 40% (5600 Å). Smaller values were found by SCHWARZSCHILD, who estimated that the r.m.s. intensity fluctuation, if corrected for instrumental effects, is between 0.05 and 0.1. Hence, the average relative intensity difference "mean of bright elements" *minus* "mean of dark surroundings" is between 0.1 and 0.2.

The above intensity difference should correspond to an average difference in the radiation temperature of between 120° and 230°. These temperature differences refer to the mean level from where the continuous radiation emerges, that means to photospheric layers with an optical depth of about unity.

The *life time* of the granulation elements is of the order of some minutes. It can be determined by statistical methods as was done by TEN BRUGGENCATE and GROTRIAN[4] who, treating the 5″ granulation, found that after 3.2 minutes $1/e$ of the original granulation elements still exist. RÖSCH[5], in a similar investigation, found life times of the order of 5 to 10 minutes, somewhat greater than the above result. KRAT[3] found life times of 1 to 2 min.

3. Centre-limb variation and the upper limit of the granulation zone. According to WALDMEIER[6] the contrast in the granulation diminishes towards the limb, and it is reduced to zero at $\cos\vartheta = 0.34$. This latter assertion is not verified by new observations; on RÖSCH's photographs the granulation is in any case still visible up to $r/R = 0.995$, this means down to $\cos\vartheta = 0.10$. From observations of PLASKETT[7] and SCHRÖTER[8] the contrast seems to increase slightly towards the limb at least till $\cos\vartheta = 0.4$. According to the former, the contrast is at least equal to the central disk value for $\cos\vartheta = 0.12$. It is true that very close to the limb the contrast seems to be smaller, but such an effect could perhaps be explained by perspectivic foreshortening which reduces the short axis of the elliptical images to a value smaller than the diameter of the tremor disk of the telescope. This effect has a great influence on the contrast (Fig. 2). RÖSCH's limb observations give the impression that granules are visible until 5″ from the limb so that the dynamic structures that manifest themselves as granules exist in the photosphere up to an optical depth of about 0.1. Quantitative decisions have to await a photometric investigation of the limb granulation.

4. Radial velocities; correlation curves. Radial velocity determinations suffer from the same observational difficulty as all granulation observations: because of the small apparent sizes of the granules the measured radial velocities may be much smaller than the true values. The communicated measurements refer to complexes of granules and are smaller than the mean radial velocities of the individual granules. It is with this in mind that one should interpret RICHARDSON and SCHWARZSCHILD's[9] mean radial velocity value, determined from a photograph of the solar spectrum ($\lambda = 6292 - 6307$ Å) taken at Mt. Wilson in 1949 (dispersion 0.5 Å/mm). The three measured Fe lines show a wavy form; the most probable radial velocity component is $\sqrt{\overline{\xi^2}} = 0.37$ km/sec; the "element

[1] M. WALDMEIER: Helv. phys. Acta **13**, 14 (1939).
[2] G. THIESSEN: Naturwiss. **37**, 427 (1950).
[3] Cf. footnote 3, p. 81.
[4] P. TEN BRUGGENCATE and W. GROTRIAN: Z. Astrophys. **12**, 323 (1936).
[5] J. RÖSCH: Private comm. (1958).
[6] M. WALDMEIER: Helv. phys. Acta **13**, 14 (1939).
[7] H.H. PLASKETT: Monthly Notices Roy. Astronom. Soc. London **96**, 402 (1936).
[8] E.H. SCHRÖTER: Z. Astrophys. **41**, 141 (1957).
[9] R.S. RICHARDSON and M. SCHWARZSCHILD: Astrophys. Journ. **111**, 351 (1950).

diameters" are uncertain (it is 2200 km = 3" according to STUART and RUSH[1] see however the criticism by WLERICK[2]). The real diameters of the granulation elements may be estimated from the data with a method indicated by SCHWARZSCHILD and RICHARDSON: Assume that one 3" element contains N real elements; assume further that the mass motion velocities of the individual, unresolved granulation elements have a mean value of 1.4 km/sec—this is the sum of micro- and macroturbulence derived from profiles of faint lines (ROGERSON[3]). Then $1.4 = 0.37\sqrt{N}$; from which $\sqrt{N} = 3.8$. Since $\sqrt{N}$ is the ratio between the diameters of the 3" elements and the "real" elements, we find for these latter a mean diameter of 600 km = 0".8, nearly equal to the granule diameters measured on direct photographs.

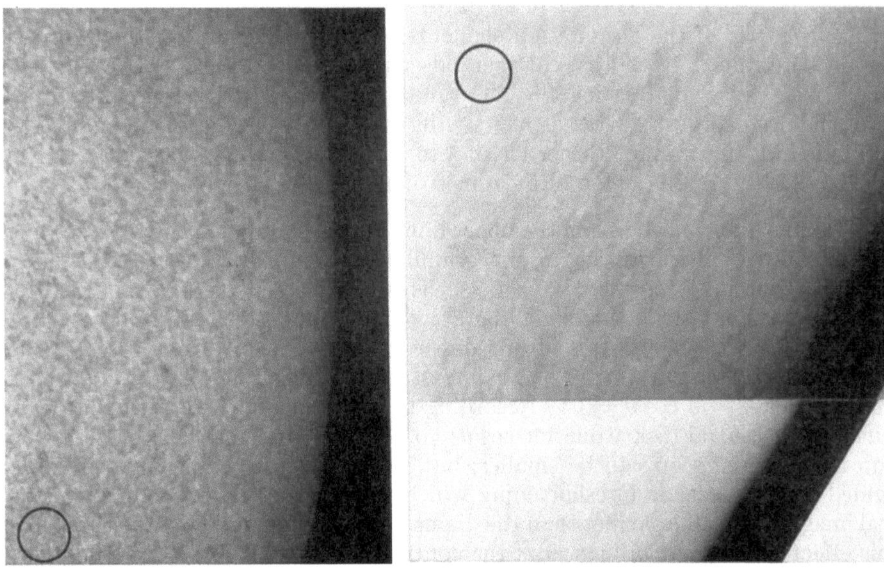

Fig. 2. Granulation near the Sun's limb. Right figure: Pic du Midi, 4. III. 1955. The lower part of the figure, reproduced at greater intensity shows the granulation up to some seconds of arc from the limb. The circle has a diameter of 5". Courtesy J. RÖSCH. Left figure: Potsdam, 11. VIII. 1948. The circle has a diameter of 30". In reproducing, the centre-limb variation of the continuous spectrum has been compensated artificially, thereby, however, reducing the apparent contrast near the limb. Courtesy E. H. SCHRÖTER.

It should be noted that RICHARDSON and SCHWARZSCHILD found by the same reasoning an average granule diameter of 150 km instead of 600; this is because they took 2" for the observed mean granule diameters and 2.8 km/sec for the microturbulent velocities. The first value may be too small and the second too great. A mean turbulent velocity component as high as 2.8 km/sec is only reached in the transition region between chromosphere and photosphere (cf. Sect. 11).

Next we give some r.m.s. velocities derived from later observations: (a) From the study of an Fe line at 6560 Å observed at Meudon DE JAGER[4] found $\sqrt{\overline{\xi^2}} = 0.50$ km/sec. (b) With the Michigan vacuum spectrograph[5] the "random velocities" (this notion is not further explained in the quoted paper) of the lines Cr

[1] F.E. STUART and J.H. RUSH: Astrophys. Journ. **120**, 245 (1954).
[2] G. WLERICK: Smithson. Contrib. Astrophys. **2** (3), 25 (1957).
[3] J.B. ROGERSON: Astrophys. Journ. **125**, 275 (1957).
[4] C. DE JAGER: Bull. astronom. Inst. Netherl. **13**, 133 (1957).
[5] R.R. MCMATH, O.C. MOHLER, A.K. PIERCE and L. GOLDBERG: Astrophys. Journ. **124**, 1 (1956).

4626 Å and Ba^+ 5853 Å measured in the disk centre are 0.70 and 0.44 km/sec; the velocities increase towards the limb (see Sect. 11). Occasionally the wavelength of the zig-zag pattern of the Fraunhofer lines is as small as 2″. SCHRÖTER[1], from middlestrong metal lines, found $\sqrt{\overline{\xi^2}} = 0.29$ km/sec.

Applying the same reasoning as above the Michigan observations, we find, assuming that the mean diameters of the observed granules are 2″ or 3″, that their true diameters are 1.″0 and 0.″6, or 1.″5 and 0.″9 respectively. This result is in accordance with that derived from RICHARDSON and SCHWARZSCHILD'S observations and with that found from direct photography.

The great uncertainty we meet in interpreting measurements of random radial velocities of Fraunhofer lines is the lack of resolving power: the observations always give a sort of indeterminate mean value referring to complexes of elements. Only for the big elements of the motion field might the observations be more certain. Therefore auto-correlation functions determined from the radial velocity measurements of the granulation are valuable. By means of auto-correlation curves the observations can be split into a significant part for long wavelengths and a less certain or wholly uncertain part for the small distances (cf. also UBEROI[2]). The significance of the auto-correlation curves, determined from the SCHWARZSCHILD-RICHARDSON observations[3,4] is that they show a secondary maximum near 15000 km. It was confirmed by WLERICK[5] who found a secondary maximum at 12000 km. This points to the occurrence of large scale motions in the photosphere, superposed on the small scale motions of the granules. It will be shown later that this large scale motion pattern is perhaps the photospheric part of the chromospheric large scale pattern which is so obvious in spectroheliograms, especially in those made in the core of the K line. Other indications for big scale motions in the photosphere are given by HART[6] who finds successive velocity maxima or minima separated by distances of the order of 26000 km. The r.m.s. velocity component is 0.16 km/sec.

When speaking about the "photosphere" one should not forget that the medium strong Fe lines from which SCHWARZSCHILD and RICHARDSON determined the *radial velocities*, originate in fairly high photospheric layers, perhaps near $\tau_0 = 0.1$ or 0.2 (cf. Sect. 11). (Here, as usual, τ_0 is the monochromatic optical depth for $\lambda = 5010$ Å.) On the other hand the *brightness measurements* refer mainly to layers with an optical depth of the order unity. Since a difference of one unit in $\log \tau$ is about equivalent to a difference of the geometrical height of 125 km it depends on the vertical extension of the granulation elements whether there will be any cross-correlation between brightness and velocity measurements.

Such a cross-correlation, with correlation coefficients -0.30 and -0.28 was observed by RUSH and STEWART and by PLASKETT[7] (the negative signs indicate that bright matter is moving upwards). In both cases the smallness of this value indicates the presence of a disturbing oscillatory velocity field which in PLASKETT'S case appears to have an average period of 5″ and an amplitude of about 0.5 km/sec.

Appreciable cross-correlation between the small scale fluctuations in brightness and velocity should not occur when the real granulation elements have

[1] E. H. SCHRÖTER: Z. Astrophys. **45**, 68 (1958).
[2] M. S. UBEROI: Astrophys. Journ. **122**, 466 (1955).
[3] F. N. FRENKIEL and M. SCHWARZSCHILD: Astrophys. Journ. **116**, 422 (1952).
[4] Cf. footnote 9, p. 83.
[5] Cf. footnote 2, p. 84.
[6] A. B. HART: Monthly Notices Roy. Astronom. Soc. London **116**, 38 (1956).
[7] H. H. PLASKETT: Monthly Notices Roy. Astronom. Soc. London **114**, 251 (1954).

vertical diameters smaller than 100 km. It would be interesting to examine the correlation between brightness observations and radial velocities deduced from lines formed in still higher layers: neutral lines of low excitation potential and molecular lines have a mean optical depth of formation $\tau_0 \approx 0.1$ to 0.01.

b) Temperature and density inhomogeneities in the photosphere.

5. The hydrogen lines; the inhomogeneous model. Any theory of Fraunhofer lines should take account of the fact that the observed spectrum is—at least roughly—a mean of two different spectra, one for the cold and one for the hot elements: elements which differ considerably in mean temperature and structure. Furthermore the hot elements move upward and the cold ones go down. Even with the best of the presently existing spectrographs one has not yet succeeded in obtaining spectra of the real individual granules. The smallest elements of which spectra have been obtained are still greater than 1″. This makes clear that a critical discussion of the Fraunhofer spectrum might reveal information on the true temperature and density fluctuations in the photosphere. So what is wanted is a theory on the formation of the Fraunhofer lines in which the inhomogeneities are incorporated, with such parameters as the values of the temperature and pressure differences, the up- and downward velocity-components of the elements, and also the relative sizes of the hot and cold "elements". The simplest model for the inhomogeneous photosphere is obtained by assuming that the vertical extension of the elements is great as compared with the scale height (thickness) of the photosphere. This hypothesis seems amply justified. On this assumption the photosphere may be represented by adjacent columns of matter, each of which having its own effective temperature and producing its own spectrum. The observed spectrum is the weighted mean spectrum, weighted according to the product of their sizes and the intensity of the emitted continuous radiation.

The wings of the hydrogen lines, especially of the Balmer and Paschen lines, are suitable to such a treatment because their depression in the spectrum depends only on T, not on the pressure, and furthermore, because it depends so strongly on the temperature that the mean value of the depressions for two atmospheres with different (T, τ) relations is not equal to the depression computed for the mean of both relations.

The ratio between the selective absorption coefficient $\varkappa_\nu$ in the line wings and the continuous absorption coefficient $\varkappa_\lambda$ is[1]

$$\frac{\varkappa_\nu}{\varkappa_\lambda} \approx \frac{F_0^{\frac{3}{2}} \cdot 10^{-13.59\,\Theta\left(1-\frac{1}{n^2}\right)}}{\varkappa_\lambda(T) \cdot P_e} \approx \frac{f(T) \cdot P_e}{\varkappa_\lambda(T) \cdot P_e} = \varphi(T) \tag{5.1}$$

(with $F_0 = 46.8\,(P_e/T)^{\frac{2}{3}}$, $\Theta = 5040/T$ and P_e being the electron pressure) if the continuous absorption coefficient is only due to H^- absorption. This is the case for the Balmer lines, and in good approximation also for the Paschen lines and for the high series members of the Brackett series.

So the discussion of the hydrogen lines in order to determine the ΔT values of the photosphere is simple and fairly straight-forward. First the line profile is computed on the basis of a mean model of the photosphere. It then appears that the computed line profile is narrower than the observed one. To explain this, let us assume that the photosphere is inhomogeneous and should be replaced by a multi-column model. Each column is defined by

$$\Delta\Theta = \Theta_{\text{column}}(\tau) - \Theta_{\text{mean model}}(\tau),$$

[1] C. DE JAGER: Rech. Obs. Utrecht **13** (1) (1952).

assumed to be constant for the region where the line profile is mainly formed ($\Theta = 5040/T$). The distribution function of $\Delta\Theta$ values was assumed rectangular for positive $\Delta\Theta$'s (colder columns) and exponential for the hotter columns (Fig. 3).

By this definition the areas of the Sun occupied by hot and cold elements are equal, and hence the continuous radiation of the multi-column model is equal to that of the mean model, since in good approximation (applying WIEN's law):

$$\frac{\Delta I}{I} = \frac{c_2}{\lambda}\frac{\Delta T}{T^2} = -\frac{c_2\,\Delta\Theta}{5040\cdot\lambda}. \tag{5.2}$$

The physical picture on which the adopted temperature distribution is based, is that of a photospheric stratum with nearly constant temperature, while hot elements are embedded in it. Their number decreases with increasing temperature differences.

Let $\delta(\Delta\Theta)$ and $I_\lambda(\Delta\Theta)$ be the depressions and the intensity of the radiation in the adjacent continuous spectrum, computed for each column, and let $p(\Delta\Theta)$ be the distribution functions of $\Delta\Theta$ values ($\int p(\Delta\Theta)\,d\Delta\Theta = 1$), then the mean depression is

$$\delta_{\text{obs}} = \frac{\int\limits_{-\infty}^{\infty} p(\Delta\Theta)\, I_\lambda(\Delta\Theta)\,\delta(\Delta\Theta)\, d(\Delta\Theta)}{\int\limits_{-\infty}^{\infty} p(\Delta\Theta)\cdot I_\lambda(\Delta\Theta)\, d(\Delta\Theta)}. \tag{5.3}$$

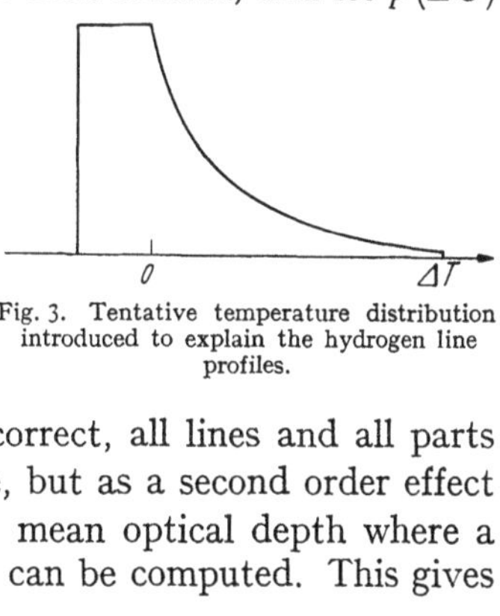

Fig. 3. Tentative temperature distribution introduced to explain the hydrogen line profiles.

A comparison with the observed $\delta(\Delta\lambda)$ functions gives an empirical determination of $\overline{\Delta\Theta}$.

In principle, when the basic assumptions are correct, all lines and all parts of the line profiles should give the same $\overline{\Delta\Theta}$-value, but as a second order effect $\overline{\Delta\Theta}$ might vary with depth. We define τ^* as the mean optical depth where a certain line depression $\delta(\Delta\lambda)$ is mainly formed; τ^* can be computed. This gives a relation between the observational $\overline{\Delta\Theta}$ and τ^*.

Computations[1] according to formula (5.3) for the Balmer lines $H_\alpha \ldots H_\delta$, for the Paschen line H_{3-5} and the Brackett line H_{4-5} have shown that the temperature differences (at the same optical depths) are small or zero at small optical depths ($\tau_0 \equiv \tau_{5010\,\text{A}} \approx 0.2$), but they increase to a maximum value $\Delta T \approx 500°$ near $\tau_0 = 1$, and decrease in deeper layers ($\Delta T = 150°$ at $\tau_0 \approx 2$). Here ΔT is the temperature difference between the mean of the hot elements and the mean of the cold ones.

By means of tentative photospheric models for the various columns these temperature differences at the same *optical* depths are converted into temperature differences at the same *geometrical* depths (hence at the same P_g-values). Fig. 5 contains the $(T, \text{Log}\, P_g)$ curves thus derived.

6. The line profiles according to the inhomogeneous model. The preceding section gave information on the inhomogeneous model in the medium and deep photospheric layers. Since most Fraunhofer lines are formed in the higher layers, the observed line profiles and their centre-limb variation might be used to construct an inhomogeneous model of the high and medium photosphere.

In such an attempt, by BÖHM[2], the distribution function of ΔT was assumed to be bell-shaped, approximated by a three-column model, in which a considerable

[1] C. DE JAGER: Bull. astronom. Inst. Netherl. **1959**.

[2] K.H. BÖHM: Z. Astrophys. **35**, 179 (1954).

part of the elements has the structure of the mean model. The model was modified later by VOIGT[1].

After having shown that modern homogeneous models are unable to explain at the same time the centre-limb variation of strong and weak Fe I lines and of the Na *D* lines BÖHM tried to improve the agreement (1) by introducing temperature inhomogeneities and (2) by taking into account the deviations from thermodynamical equilibrium. This latter effect is important since the lines discussed by BÖHM are all formed in the higher parts of the photosphere ($\tau_0 \approx 0.05$), where such deviations are likely to have influence.

In BÖHM's three-stream model one of the three columns, with an area $A = 0.5$, has a mean (T, τ) distribution and the two others, both with $A = 0.25$, have higher or lower temperatures. The "central model" is BÖHM's theoretical radiative-equilibrium model[2]. The two other models are defined in such a way that at one and the same geometrical depth the Kirchhoff-Planck functions for volume elements of the three columns have the ratios 0.4:1:1.6, so that the mean $B(h)$ value is equal to the value adopted for the mean model. The introduction of these B-ratios was based on theoretical considerations on the temperature variations in granules (VITENSE), which predict fluctuations of this order. Pressure equality at the same geometrical depth was assumed, an assumption demanded by hydrodynamical arguments. The centre-limb variations of Fraunhofer lines, computed for the three-stream model, are in better agreement with the observations than those computed for a homogeneous atmosphere. Still there remain some minor differences, even when deviations from thermodynamical equilibrium are taken into account.

The agreement is still somewhat better if the computations are made with SCHRÖTER's[3] two-element model. The temperature differences between the two kinds of elements were computed on the basis of the observed contrast in the granules and on their supposed centre-limb variation[4]. The assumptions are: (a) that the granular contrast in the disk centre is 0.45—a value of which we now know that it is perhaps a factor 2 or 3 too great and (b) that the contrast reaches a maximum near $\cos\vartheta = 0.3$ and decreases closer to the limb. This leads to a model in which $\Delta T = 0$ for small optical depths. The argument for making $\Delta T = 0$ for $\tau_0 < 0.1$, based here on assumptions on the granulation contrast at the extreme limb is uncertain. Other arguments are: (1) The theory (VITENSE[5]) shows that the granular energy is wholly radiated and interchanged with the surroundings near and above $\tau_0 = 0.3$. (2) Discussions of the centre-limb variation of the red-shifts of lines indicate that near the limb ΔT is small or zero (cf. Sect. 7). (3) The solar continuous spectrum near 2000 Å, which is emitted mainly by the limb regions of the Sun can fully be explained by a homogeneous photosphere[6]. (4) The narrow cores of the infrared O I lines near the Sun's limb are incompatible with an upper photospheric model consisting of two streaming columns of matter (VOIGT[1]). (5) PAGEL's study[7] of equivalent widths of infrared Fe I lines, absorption lines for which deviations from thermodynamic equilibrium are certainly small, does not indicate temperature inhomogeneities in the upper photosphere;

[1] H.H. VOIGT: Z. Astrophys. **40**, 157 (1956).
[2] K.H. BÖHM: Z. Astrophys. **34**, 182 (1954).
[3] E.H. SCHRÖTER: Z. Astrophys. **41**, 141 (1957).
[4] A first attempt to construct a two-column model on the basis of intensities of granules was made by V.A. KRAT: Isw. Gl. Astr. Obs. Poulkovo **152**, 1 (1954).
[5] E.V. VITENSE: Z. Astrophys. **32**, 135 (1953).
[6] C. DE JAGER: Bull. astronom. Inst. Netherl. **13**, 275 (1957).
[7] B.E. PAGEL: Monthly Notices Roy. Astronom. Soc. London **115**, 493 (1955).

(6) the same conclusion follows from an investigation of eclipse photometry of the Sun's limb[1]. However, it should be remarked that NEWKIRK[2] from a study of some infrared CO lines, found weak indications for $\Delta T \approx 500°$ hence $2\Delta T = 1000°$ near $\tau_0 \approx 0.05$.

SCHRÖTER'S ΔT values are computed with respect to the empirical homogeneous model Vitense II, which could correctly explain the centre-limb variation of the Na D lines. Let $I(\cos\vartheta)$ be the observed radiation and $E(\tau)$ the black-body radiation in the photosphere. With the Eddington-Barbier approximation $I_\lambda(\cos\vartheta) = B_\lambda(\tau_\lambda = \cos\vartheta)$, and with

$$(I_{\text{gran}} - I_{\text{intergr}})/I_{\text{intergr}} = 0.45 \equiv K, \tag{6.1}$$

and by applying WIEN'S law [cf. Eq. (5.2)] ΔT is found from:

$$\frac{\Delta T(\tau_0 = \cos\vartheta)}{T_{\text{Vitense II}}(\tau_0 = \cos\vartheta)} = \frac{\lambda\, K\, T_{\text{Vitense II}}(\tau_0 = \cos\vartheta)}{c_2}. \tag{6.2}$$

For $(\tau_0)_{\text{gran}} = (\tau_0)_{\text{intergr}} = 1$: $\Delta T = 700°$. It increases up to 800° near $\tau_0 = 0.3$, and is zero at $\tau_0 = 0.1$.

The further conditions are (a) pressure equality at the same geometrical heights, (b) the total energy flux summed up for both elements is equal to that for the homogeneous model and (c) at each geometrical height upflow = downflow:

$$m_{\text{gran}} \cdot v_{\text{gran}} = m_{\text{intergr}} \cdot v_{\text{intergr}}. \tag{6.3}$$

With these conditions the two-stream model can be constructed. It predicts a granular contrast increasing from 1.0 at $\cos\vartheta = 1.0$ to 1.25 at $\cos\vartheta = 0.3$ and decreasing steeply to 0.8 at $\cos\vartheta = 0.1$. It accounts well for the centre-limb variation of the continuous radiation. Also the intensities of some Fraunhofer lines can be explained equally well w th this model as with that of BÖHM; the model computations give a fair representation of the centre-limb variation of (1) faint metal lines, (2) the wings of strong Fe lines, and (3) the wings of the Na-D-lines. This latter point (3) is obvious since SCHRÖTER'S model is in the high layers identical with the model Vitense II which is known (PRIESTER[3]) to explain the Na-D-lines and their centre-limb variation. But the good agreement for the other lines is another justification for the assumption that the temperature differences are small in the outer photospheric layers.

This latter point is noteworthy in one respect: The strong Fraunhofer lines (e.g. those studied by BÖHM and SCHRÖTER) originate mainly in the high photosphere, near $\tau_0 = 0.05 \ldots 0.1$ and BÖHM introduced the temperature-inhomogeneities precisely in the high photosphere to explain some as yet unexplained properties of the Fraunhofer lines. However, the new trend is to reduce ΔT to zero in this critical region. The value of SCHRÖTER'S investigation lies in his demonstration that (1) the choice of a somewhat higher temperature in this uppermost region, connected with (2) temperature- inhomogeneities in deeper layers permits one to explain a number of Fraunhofer line features.

7. Widening and displacements of Fraunhofer lines in the inhomogeneous model. In the computations described above the main property of the inhomogeneous models is the difference in temperature between the columns; no use was made

[1] H. HUBENET and C. DE JAGER: Bull astronom. Inst. Netherl. **13**, 43 (1956).
[2] G. NEWKIRK: Astrophys. Journ. **125**, 571 (1957).
[3] W. PRIESTER: Z. Astrophys. **32**, 200 (1953).

of the fact that the matter is streaming. This aspect has been introduced in Fraunhofer line computations by VOIGT[1] and SCHRÖTER (l. c.).

The O I lines near 7775 Å are very sensitive to temperature variations because of their high excitation potential (9.1 eV). The lines are slightly asymmetric and this asymmetry depends on the distance from the centre of the disk. It can be explained by assuming (a) that the photosphere consists of hot and cold columns of matter and (b) that the hot matter is streaming upward and the cool matter goes down. The hot columns produce a strong absorption line shifted towards the violet, the cold ones produce a red-shifted weak line. The observed profile, which is the weighted sum of both, is asymmetric.

VOIGT's three-stream model, introduced to explain the O I line profiles, is a modification of BÖHM's; in the latter model the temperature fluctuations occur in the whole photosphere. In the new model ΔT and v are zero for $\tau_0 < 0.1$, as in SCHRÖTER's model (τ_0 is the monochromatic optical depth at 5010 Å).

Another new aspect introduced by VOIGT is that of a microturbulent velocity component varying with depth. One might wonder how so many parameters (temperature difference; convective motions, microturbulence) can all be derived in a semi-empirical way from one set of centre-limb observations of these three lines, but with a great amount of labour the various variables can be separated to a certain extent. Of course, the author, in his final choice, has also been guided by some other sources of information, e.g. in determining the variation of the microturbulent velocity component ξ_0 with depth. Nevertheless, in principle, the various unknowns can be found from the O I line profiles alone:

(a) The asymmetry gives information about the variation of the convective velocities with height. The violet halfwidth of the lines is greater than the red halfwidth, and the effect diminishes towards the limb. This is explained by convective motions increasing with depth. The best agreement with the observations is obtained for a model in which the upward velocity in the hot elements is $+3$ km/sec in deep layers; the corresponding velocity in the cold ones is -2 km/sec. Both values go to zero in the high solar layers.

(b) The centre-limb variation of the observed halfwidths is explained by a microturbulent velocity component, zero at $\tau_0 = 1.5$, and increasing with decreasing depth up to 3 km/sec at $\tau_0 = 0.015$.

(c) The temperature differences could, in principle, be derived from the centre-limb variation of the equivalent widths. In fact, BÖHM's ΔT values were kept unchanged in the regions belom $\tau_0 = 0.3$.

(d) The central depressions of the lines near the limb are greater than should be the case if computed with BÖHM's model. It is assumed that this is due to a "superexcitation" in the high layers. This suggestion, and the proposed explanation: influence of far ultraviolet coronal and chromospheric radiation do not seem sound. The chromosphere is opaque for this kind of radiation, but the additional depression near the limb might be due to chromospheric absorption. The contribution of the low chromosphere with its fairly high temperature in still rather dense layers, was neglected in VOIGT's computations.

In the investigation described above, the asymmetry and its variation between centre and limb were used to obtain information on the flow of matter in the photosphere. However, the asymmetry is a second order effect; it is certainly easier to compare absolute line shifts with theoretical computations, but the measurements are difficult to make.

---

[1] Cf. footnote 1, p. 88.

The most accurate set of absolute line shift observations is perhaps that of Miss ADAM[1]; Fig. 4 gives some of her measures (cf. FINLAY-FREUNDLICH and FORBES[2]). Three characteristic aspects are:

1. All measured shifts are red-shifts, but nearly all values are smaller than the predicted relativistic red-shift

$$\frac{\Delta\lambda}{\lambda} = \frac{GM}{Rc^2} = 2.12\times 10^{-6} \mathrel{\widehat{=}} 0.636\ \text{km/sec} \quad \text{or} \quad 12.9\ \text{mÅ} \quad \text{at} \quad 6100\ \text{Å}.$$

Only some values, for strong lines and for lines close to the limb, are slightly greater than the relativistic one. Hence there must be a violet-shifting mechanism which counteracts the relativistic red-shift.

2. There is a limb effect: the lines are more displaced towards the red the nearer they are to the limb.

3. The displacement curves, although having nearly the same form for all faint lines are shifted, one with respect to the other, in a vertical direction. The effect may slightly depend on the level of origin of the line: the strongest lines seem to show the greatest red-shifts. Lines with an equivalent width $W_\lambda \approx 170$ m Å have about the relativistic red-shift. For $W_\lambda > 200$ mÅ, we have $\Delta\lambda > \Delta\lambda_{\text{rel}}$.

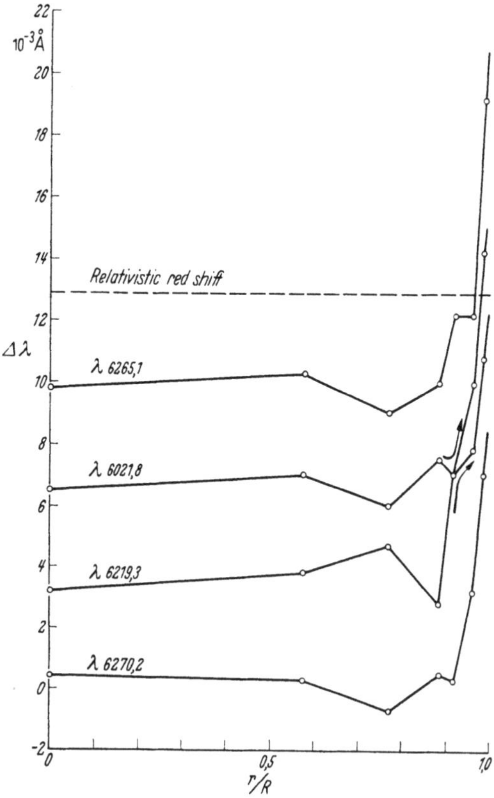

Fig. 4. Absolute shifts of Fraunhofer lines as a function of the distance to the disk centre [M. G. ADAM, cf. also E. FINLAY-FREUNDLICH and E. G. FORBES: Ann. d'Astrophys. **19**, 183 (1956)]. The lines have Rowland intensities 3 (6270), 5 (6265) and 6. The dashed line gives the predicted relativistic red-shift. Theory (SCHRÖTER) predicts a cosine curve, violet-shifted with respect to the dashed line, with $\Delta\lambda = 12.9$ Å for $r/R = 1.0$ and smaller violet-shifts for other values of r/R.

This relative violet-shift can be understood on the basis of the inhomogeneous solar model[3,4]. The observed spectrum is a weighted mean between that of the hot ascending masses and the cool downward moving ones; hot elements have the greatest weight. Also the dependence on the equivalent width is clear: stronger lines are formed in higher level, where the convective motions are small or negligible.

The limb effect was computed by SCHRÖTER[4] for middlestrong lines for which the theory is quite reliable, and for these a good agreement with observations is obtained. The effect is proportional to $\cos\vartheta$ and is zero at the limb; there we should deal with the pure relativistic effect.

---

[1] M. G. ADAM: Monthly Notices Roy. Astronom. Soc. London **112**, 548 (1952).

[2] E. F. FINDLAY-FREUNDLICH and E. G. FORBES: Ann. d'Astrophys. **19**, 183 and 215 (1956).

[3] G. THIESSEN: Z. Astrophys. **35**, 237 (1955).

[4] E. H. SCHRÖTER: Z. Astrophys. **41**, 141 (1957).

Whether the proposed explanation is correct depends on the accuracy of the limb measures. According to the theory no red-shift greater than 0.636 km/sec should occur; observations of lines of various strength should all converge towards the relativistic value for $\cos\vartheta \to 0$. This does not wholly agree with the observations.

It does not seem impossible that the unexplained red-shift component at the very limb is due to phase shifting collisions. In the case of VAN DER WAALS forces, LINDHOLM[1] predicted a red-shift of $\Delta\nu = 0.36\,\gamma$ with γ being the damping constant. This case applies only to lines originating from transitions in the lowest atomic levels, hence to the strongest lines. For the CaI line at 4227 Å (a resonance line) the empirical and the computed values[2] of Log γ are equal to 9.2. This would give an additional red-shift of 0.85×10^{-3} Å, insufficient to explain the observations, but greater γ-values occur for some other lines.

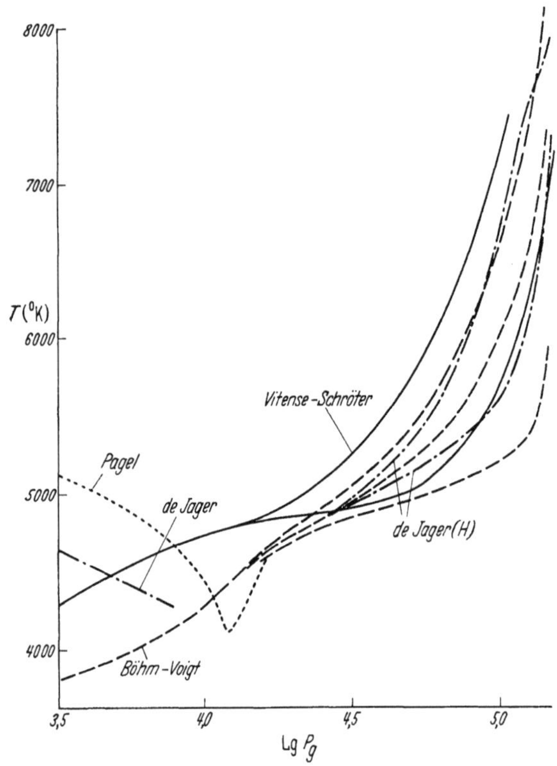

Fig. 5. Temperature distribution in various inhomogeneous models and in the transition region to the chromosphere, as a function of gas pressure P_g.

A comparison of the various inhomogeneous models is given in Fig. 5 (models Böhm-Voigt, Vitense II-Schröter and the hydrogen lines model). Notwithstanding all uncertainties it is gratifying that the values of the temperature fluctuations are all of the same order. Near $\tau_0 = 1$, at the same optical depth they are of the order of 800°. It is, however, obvious that the agreement is not very good yet and that we are only at the beginning. Much work has still to be done. We should further notice that the temperature differences *in*crease with the *geometrical* depth, but that they tend to *de*crease with increasing *optical* depth, which is due to the differences in transparency between the various columns. A very important problem still to be solved is that the temperature differences found observationally from Fraunhofer line studies are considerably greater than those found fron direct granulation photometry. This is one of the first points to be clarified before further progress can be made in this field.

The precise course of the temperature variation near the Sun's limb is still very uncertain as is shown in the figure from a comparison of the (T, Log P_g) curves of the models described above with the models of the transition region photosphere-chromosphere of PAGEL[3] and DE JAGER[4] (cf. Sect. 18).

[1] E. LINDHOLM: Ark. Math. Astronom. Phys. B **28** (3) (1941).
[2] A. UNSÖLD: Physik der Sternatmosphären, 2. Aufl., p. 467f. Berlin: Springer 1955.
[3] B.E.J. PAGEL: Monthly Notices Roy. Astronom. Soc. London **116**, 608 (1957).
[4] C. DE JAGER: Bull. astronom. Inst. Netherl. **13**, 275 (1957).

c) Micro- and macro-turbulence in the photosphere.

8. Micro- or macro-turbulence? The term "turbulence" has too often been used in a wrong sense. So let us start with a definition: a field of motion is called turbulent when the motions are so irregular that it does not seem appropriate or when it is virtually impossible to describe by mathematical formulae the motions of each of the individual volume elements, and when it is necessary to describe them by statistical methods. "Turbulence" always implies a macroscopic description of mass motions: In any field of irregular, "turbulent", motions there are "elements" which are so small that their internal motions are laminar, but when viewed on a larger scale the motion field is so irregular that it is permissible to speak about turbulent motions. As a rule the turbulent motions are provoked by a large-scale dynamic instability of the velocity field (convection, a stirring motions, etc.) which furnishes the energy of the field.

In this line of thought we cannot speak about "turbulence" or "turbulence elements" in the case of one whirl, a sunspot or a Bénard cell. A complex of whirls might under certain circumstances be called a turbulent field, but a complex of Bénard cells, in the classical sense, certainly not. (Bénard cells are formed by laminar motion.)

Confusion may arise when the term "turbulence element" is introduced. In a statistical turbulence theory there is no place for such a notion. It is advisable to avoid this term as much as possible and to introduce other well defined notions, such as the correlation distance, or the wavelengths or wavenumbers (cm^{-1}) of the Fourier-components of the velocity field obtained by decomposing the field, separately for each Cartesian velocity component. The only case in which the term "element" may be useful is in denoting the greatest wavelength L_0 occuring in the Fourier analysis of the field of motions. It is mainly in this range of wavelengths that the energy from the kinetic energy of mass motions is fed into the system. Through the action of the non-linear terms in the equation of motion the energy is transmitted to other wavelengths, mainly to shorter ones, to be finally destroyed by viscosity and dissipated into heat for small wavelengths. So another important quantity that occurs in a statistical turbulence theory is the wavelength L_1 for which viscous dissipation of energy is of the same order as the transfer of energy to Fourier-components with still smaller wavelengths.

It is clear that L_0 and L_1 are not exactly defined but that they are order-of-magnitude quantities. In physical experiments L_0 is supposed to be of the same order as the experimental volume diameter; in a stellar atmosphere or a compressible gas in a gravity field it is generally assumed (following a suggestion of M. SCHWARZSCHILD[1]) that $L_0 \approx H$, the scale height of the atmosphere, but there is no firm foundation for this assertion.

In general, the Fourier analysis of the velocity field will contain all L values between L_0 and L_1, and it can be shown that in isotropic incompressible turbulence, we have the relation:

$$\frac{\xi(L)}{\xi_0} = \left(\frac{L}{L_0}\right)^{\frac{1}{3}} \tag{8.1}$$

where $\xi(L)$ is the mean difference of the velocity component between any two points of the field at a distance L, and ξ_0 is the mean velocity component difference between points at a distance L_0. Often ξ_0 is identified with the mean amplitude of the velocity component of the field; this assumption is more or less justified by the fact that ξ_0 is the greatest velocity component occurring in the field.

[1] R. S. RICHARDSON and M. SCHWARZSCHILD: Astrophys. Journ. **111**, 359 (1950).

The expression (8.1) can be proved as follows: if ξ is the mean velocity component difference between two points at a distance L, then a volume element with a diameter L can be treated as moving homogeneously with velocity ξ with respect to the surroundings. Its kinetic energy is transformed to smaller wavelengths. The energy transformation dE/dt (ergs cm^{-3} sec^{-1}) is $-W\xi/L^3$ where W is the frictional resistance of the element = const. $\varrho L^2 \xi^2$. Hence $dE/dt = -\text{const}\, \varrho \xi^3/L$, and since dE/dt should be constant for all L values between L_0 and L_1, this means that $\xi \propto L^{\frac{1}{3}}$ for $L_1 < L < L_0$. This relation applies only to incompressible turbulence but it is a good approximation in general for subsonic velocities.

In the interpretation of Fraunhofer line observations we have to distinguish between two extreme cases:

(a) The values of L_0 and of the absorption coefficient $\varkappa$ (cm^{-1}) are such that $L_0 \varkappa \ll 1$. In that case the "line of sight passes through many elements" as is usually said, and there is no formal difference between the turbulent motions and the thermal motions of the gas; we may write

$$\xi^2_{\text{total}} = \frac{2RT}{\mu} + \xi_0^2 .$$

In this expression ξ_0 is a measure for the turbulent velocity component defined by representing the velocity distribution by $\exp[-(\xi/\xi_0)^2]$. Also in the following part of this paper ξ_0 (sometimes called ξ_{turb}, ξ_t or ξ) is assumed to be thus defined.

This is the case of fully developed *micro-turbulence*. In the Fraunhofer spectrum micro-turbulence produces a line widening, and an increase of the equivalent widths of strong or medium strong lines.

(b) $L_0 \varkappa \gg 1$. In this case we are dealing with fully developed *macro-turbulence plus* a part of the *micro-turbulent field*. Only those components of the velocity field for which $L\varkappa \ll 1$ contribute fully to the micro-turbulent line widening, as described above. The components for which $L\varkappa \gg 1$ only produce local Doppler displacements of spectral lines. A further complication is that the solar L_0 value is rather small. By lack of resolving power of telescopes and spectrographs the spectra of some individual elements are summed up in most spectrographs; this summation causes a line widening whereas the equivalent width remains constant at variance to the case of microturbulence.

Next one might ask whether we are dealing in the solar atmosphere with macro- or with micro-turbulence. The answer is that, for each observed part of the Fraunhofer spectrum (whether it is the continuous spectrum or any part of a line profile), one always observes mainly the radiation from that part of the atmosphere where the total optical depth $\tau \approx 1$. In rough approximation $\tau \approx 1$ at that geometrical depth where $\varkappa H \approx 1$. So if L_0 may be put equal to H, we are dealing mainly with the case of $\varkappa L_0 \approx 1$, a case, intermediate between the two limiting ones discussed above; the lines are displaced by macroturbulence and widened by microturbulence, but both on a reduced scale (the microturbulent widening as well as the macroturbulent displacements correspond both to velocities smaller than ξ_0).

A second way in which macro-turbulence enters into the picture is the following: the various points of a line profile may be formed at quite different depths in the solar atmosphere. For strong lines these depths may easily vary by more than an amount L_0. So each of the profile points is widened according to the local micro-turbulence and shifted according to the local macro-turbulence which may vary over distances $\gg L_0$.

9. Mean mass motions deduced from the profiles of Fraunhofer lines and from the curve of growth. Micro-turbulence produces a line widening and, consequently, an increase of the equivalent width: it can be studied from the form of the *curve of growth*. Macro-turbulence produces a shift of the whole line; unresolved macro-turbulence does not change the equivalent widths. It has no influence on the curve of growth. When studying *line profiles* we deal with micro-turbulence plus unresolved macro-turbulence. In each case one should ascertain to what optical depth the observed features refer.

α) The curve of growth. Data on micro-turbulence, deduced from the position of the flat part of the curve of growth, refer to other solar layers than data derived from profiles of faint lines. As VAN REGEMORTER[1] showed, the lines on the flat part of the curve of growth are formed in the highest photospheric layers, as a rule at mean optical depths five to ten times less than those of similar lines from the linear part of the curve of growth. So ξ_t-values[2] deduced from these two sources refer to layers at about one scale height distant and are therefore not comparable without further precaution; they can certainly not be used to derive a "photospheric spectrum of turbulence".

K. O. WRIGHT's value for Doppler widening as deduced from the curve of growth

$$\Delta\lambda = \frac{\lambda}{c}\left(\frac{2kT}{m} + \xi_t^2\right)^{\frac{1}{2}}$$

gives[3] $\xi_t = 1.6$ km/sec if $T = 0$. Here ξ_t is the micro-turbulent velocity component. With $T \sim 4700°$ and $m = 9.3 \times 10^{-23}$ g (Fe) we obtain $\xi_t = 1.0$ km/sec. This value is deduced from lines on the flat part of the curve of growth, which are perhaps formed mainly at optical depths of the order of $\tau_0 = 0.01$ to 0.05. This value agrees well with analogous determinations of SU-SHU-HUANG[4]: $\xi_t = 1.3$ km/sec, and of CLAAS[5]: 1.2 km/sec. As a mean value one might take 1.2 km/sec.

β) The study of line profiles gives information on micro-turbulence and (unresolved) macro-turbulence:

(i) From the widths of the middle strong lines in the Utrecht Solar Atlas VAN DE HULST[6] deduced $\xi_t = 1.7$ km/sec, valid for the centre of the disk. It was shown by ROGERSON[7] that this value is perhaps too great, VAN DE HULST's lines being influenced by self-absorption. For really weak lines (central depth < 0.2) $(\overline{\xi_t^2})^{\frac{1}{2}} = 1.4$ km/sec. There is no indication for a variation of this value with depth.

(ii) REICHEL[8] and VOIGT[9] found the following results from a study of the (Doppler) cores of Fraunhofer lines:

	Line	ξ_t (km/sec)
REICHEL	Sr II; 4078	2
	strong arc lines	1.1
	faint arc lines	1.0
VOIGT	O I; 7776	3.0

(iii) ALLEN's attempt to measure the centre-limb variation of the line widths of relatively faint lines was criticized by UNSÖLD[10]: the deduced increase of ξ_t

[1] H. VAN REGEMORTER: Thesis, Paris 1958.
[2] We speak about ξ_t, not about ξ_0, to denote that probably $\xi_t < \xi_0$ (see Sect. 8).
[3] K.O. WRIGHT: Astrophys. Journ. **99**, 249 (1949).
[4] SU-SHU-HUANG: Astrophys. Journ. **112**, 418 (1950).
[5] W. J. CLAAS: Rech. Obs. Utrecht **12** (1) (1951).
[6] H.C. VAN DE HULST: Bull. astronom. Inst. Netherl. **10**, 79 (1946).
[7] J.B. ROGERSON: Astrophys. Journ. **125**, 275 (1957).
[8] M. REICHEL: Z. Astrophys. **33**, 79 (1953).
[9] H.H. VOIGT: Z. Astrophys. **40**, 157 (1956).
[10] A. UNSÖLD: Physik der Sternatmosphären, 2. Aufl., p. 459. Berlin: Springer 1955.

from centre to limb (from 2.2 km/sec to 3.1 km/sec) is not correct, because the lines are rather strong so that their observed widths are greater than the widths of the absorption coefficient profiles. Only by correcting appropriately for self-absorption can this difficulty be removed. The alternative would be to use lines with equivalent widths smaller than about 30 mÅ. The importance of such data is that they refer to greater depths than the foregoing values.

(iv) ALLEN's investigation was repeated with higher resolving power (6×10^5) by SUEMOTO[1], who could resort to the above criticism by taking the detailed theory of line formation into account in interpreting the observations. For every point on the disk faint lines appear to have greater Doppler widths than stronger lines. This suggests a (micro- + macro-) turbulent velocity component *de*creasing with height in the photosphere. However, all lines are broader and shallower near the limb than in the disk centre, which suggests an *in*crease of the turbulent velocity component towards the limb, hence perhaps towards higher layers.

This difficulty was removed with ALLEN's suggestion that the convective motions are anisotropic in the mean photosphere: at $\tau_{6300}\approx 0.4$, the velocity should be greater in the horizontal direction than vertically. The value of the *micro*turbulent velocity component near $\tau_{6300}=0.4$ is 1.0 km/sec, in accordance with other determinations at that depth. In a similar kind of investigation (resolving power 3×10^5) WADDELL[2] found the motion field at $\tau=0.3$ anisotropic with $\xi_{\text{rad}}=1.8$ km/sec and $\xi_{\text{tang}}=3.0$ km/sec. There are no indications for a change of these values with depth.

(v) WEHLAU and WEHLAU[3], with the aid of line profiles measured in the Utrecht Atlas of the Solar Spectrum found $\xi=2.0$ km/sec for weak neutral metal lines, $\xi=1.6$ km/sec for Ti^+ and $\xi=2.1$ and 3.5 km/sec for molecular lines. The latter are known to be formed in the transition region between photosphere and chromosphere.

10. Mass motions at the Sun's limb and in the low chromosphere; conclusions. There is a gradual transition from the photosphere to the low chromosphere, and so this review of random mass motions in the photosphere is not complete without a summary of mass motion velocities in the transition region between photosphere and chromosphere.

First we summarize the *photospheric* determinations (all velocities in km/sec); as usual in this article ξ denotes the r.m.s. value of the velocity component perpendicular to the surface (cf. Sect. 4).

Table 1. *Macroturbulence from granular Doppler shifts.*

	ξ_t	Estimated mean τ_0
RICHARDSON-SCHWARZSCHILD	0.37	0.2
DE JAGER	0.50	0.2
MCMATH *et al.*; Ba^+ (centre to limb)	0.44 to 0.6	0.4 to 0.2
MCMATH *et al.*; Cr (centre to limb)	0.70 to 0.9	0.2 to 0.1
SCHRÖTER	0.29	0.2

The true values are perhaps greater because of lack of resolution. HART's[4] measurements indicating an increase of ξ_t towards the limb but at the same time a

[1] Z. SUEMOTO: Monthly Notices Roy. Astronom. Soc. London **117**, 2 (1957).
[2] J.H. WADDELL: Astrophys. Journ. **127**, 284 (1958).
[3] A.W. and W.H. WEHLAU: Publ. Astronom. Soc. Pacific **68**, 358 (1956).
[4] A.B. HART: Monthly Notices Roy. Astronom. Soc. London **116**, 489 (1956).

decrease with decreasing optical depth (0.16 and 0.12 km/sec at $\tau_0 \approx 0.4$—Ni and Ti—and $\tau_0 \approx 0.3$—Na—respectively) suffer perhaps too much from lack of

Table 2. *Micro- and macro-turbulence from line profiles.*

	ξ_t	Estimated mean τ_0
VAN DE HULST, medium strong metal lines	1.7	0.10
ROGERSON, faint metal lines	1.4	0.10
REICHEL, Sr II	2	0.03
REICHEL, strong arc lines	1.1	0.10
REICHEL, faint arc lines	1.0	0.15
VOIGT, O I	3.0	0.02
SUEMOTO	1.0	0.3
WEHLAU, faint metals	2.0	0.2
Ti⁺	1.6	0.3
CH	2.1	0.05
CN	3.5	0.05
WADDELL radial	1.8	0.3
WADDELL tangential	3.0	0.3

Table 3. *Micro-turbulence from curve of growth.*

	ξ_t	estimated mean τ_0
WRIGHT	1.0	0.02
HUANG	1.3	0.02
CLAAS	1.2	0.02

resolving power, while also the derivation of ξ-values from lines of different elements may easily introduce systematic errors. The latter remark applies also to the work of SUEMOTO and WADDELL.

These values are collected in Fig. 6.

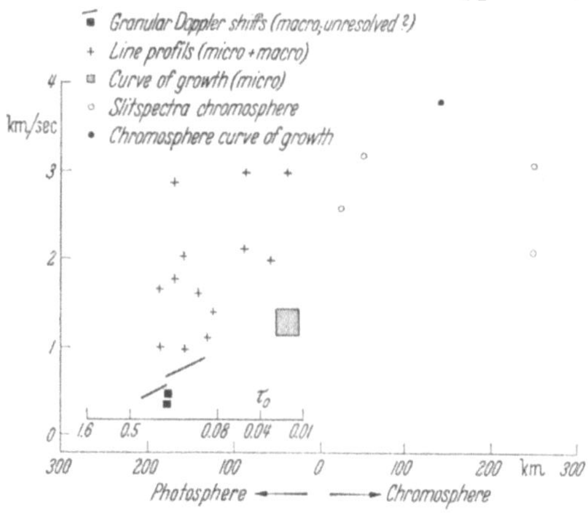

Fig. 6. The increase of the turbulent velocity component with height in the photosphere and low chromosphere.

In the absence of modern computations of the mean optical depth, the τ_0 values given here are roughly estimated with the aid of MINNAERT's[1] table of mean depths of line formation. From a comparison of MINNAERT's results (valid for a model atmosphere with a limb temperature of 4900°) with the new results of TEN BRUGGENCATE *et al.*[2] (for BÖHM's model atmosphere with a limb temperature of 3800°) it appears that MINNAERT's mean optical depths are about two to four times those of TEN BRUGGENCATE *et al.* Since we believe that the correct limb temperature of the Sun is about 4300° we take MINNAERT's values, and divide them by two. Furthermore it should be kept in mind that MINNAERT's results concern weak lines. The mean optical depths for lines on the flat portion of the curve of growth are about five to ten times smaller, those on the damping part are about one half (VAN REGEMORTER). The Doppler

[1] M. MINNAERT: Bull. astronom. Inst. Netherl. **10**, 399 (1948).

[2] P. TEN BRUGGENCATE, R. LÜST-KULKA and H.H. VOIGT: Veröff. Göttingen **6**, (110), 255 (1955).

cores of strong lines are formed still higher in the photosphere. The relation between τ_0 and h in Fig. 6 was taken from our model given in Table 4 (Sect. 14) of the present paper.

Turbulence determinations in the *low chromosphere* refer either to the component radial to the Sun's surface (ξ) or to the line-of-sight component (η). In the lowest 700 km there are two sources of information (cf. also Sect. 17), viz.
— line profiles (slit spectra; tangential component) (Redman and Suemoto[1], open circles in Fig. 6);
— curve of growth (flash spectra; tangential component) (Böhm-Vitense[2], filled circle in Fig. 6).

Fig. 6 shows the increase of the mean turbulent velocity component with height; there is a smooth transition between photosphere and chromosphere. There is no clear difference between the radial and tangential components at the same heights. This suggests that in the photosphere and the low chromosphere the microturbulent velocities are essentially isotropic. It appears further that the micro-turbulent velocity component is about one half of the sum of micro- and macro-turbulence. Since both effects add quadratically ξ_{macro} should be about $1.7\,\xi_{\text{micro}}$.

d) Theory of convection and turbulence in the solar atmosphere.

11. Theory of the solar convection zone. We adopt the following model of the convection zone. The observations show that the convection of the granulation zone is of the non-stationary type: matter ascends in hot bubbles, outside these bubbles cool matter moves downward (Fig. 7). Anticipating we notice that according to Vitense the convective elements may ascend up to $\tau_0 \approx 0.3$. Since the lower limit of the photosphere (virtually $\tau_0 = 2$ to 4) is only 100 km below the $\tau_0 = 0.3$ level, it is clear how extremely thin the visible region of the convection zone is.

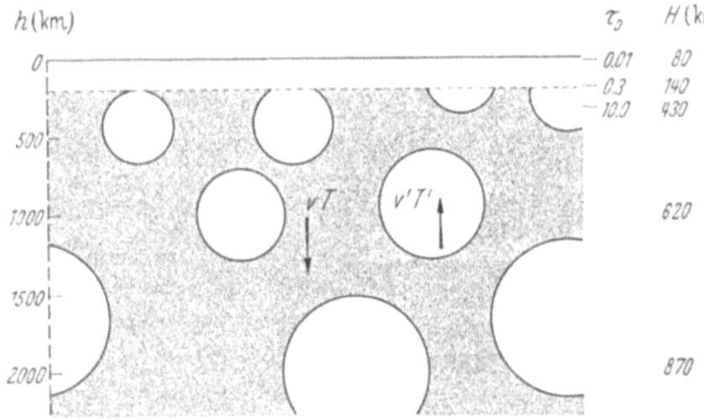

Fig. 7. Schematic picture of the upper part of the convection zone.

Granulation elements are structures which belong principally to the subphotospheric part of the Sun and which happen to penetrate with their very tops just into the photospheric region. So the structures and sizes of the elements are defined rather by the conditions in the deeper layers of the Sun than in the just visible upper parts. This makes it possible to predict the order-of-magnitude of the size of the granulation elements. It is generally assumed that in an atmosphere with a density gradient the convection elements have diameters L_0 of the same order as the scale height H of the atmosphere (but one should keep in mind that in such an approximated reasoning the ratio of L_0 and H may deviate from unity by factors π or so). When ascending or descending these elements dissolve again after having gone over a path equal to their own diameter (Prandtl's mixing length). The semi-qualitative argumentation for this reasoning goes as follows: elements which ascend adiabatically will expand. When

[1] R. O. Redman and Z. Suemoto: Monthly Notices Roy. Astronom. Soc. London **114**, 524 (1954).
[2] E. Böhm-Vitense: Z. Astrophys. **36**, 145 (1955).

they are about one scale height higher up, the elements mix because of their greater diameters and so they disappear. New elements originate, with diameters of the same order as the local scale height.

In the Sun the scale height H decreases very quickly, almost discontinuously, at a depth of about 250 km (see the column H at the right side of Fig. 7). In such a case the above reasoning can no longer be valid, and the observed granules will have diameters of the same order as the scale height near the 500 km level. Even if the upper part of an element, when arriving at the top of the granulation zone, would split up into smaller elements with mean diameters of about 150 km, the large-scale unequal heating from the lower levels remains and will produce hot and cool regions of the photosphere with a characteristic size of about 500 km. (In the figure the granulation elements have been given mean sizes equal to H. It seems as if the figure suggests that the diameter of an element decreases when ascending, but the reverse is true. The smaller elements originate after the dissolution of the big ones).

Thus we arrive, also from this point of view, at an average granular diameter of about 500 km. We recall that analogous values were suggested earlier (a) from the observations (Sect. 2), (b) from a discussion of radial velocity measurements (Sect. 4).

We next turn to a quantitative computation of the conditions in the convection zone. Since the present article deals with the visible, photospheric phenomena, our main concern will be with temperature, pressure, velocities and energy transport in that curious transition region where the granulation elements are on the point of disappearing and where the convection region merges into the non-convective upper part of the photosphere.

For that purpose we schematize the elements by cylinders—even if this is not completely true, it is a sufficient approximation in the very thin part of the convection zone discussed by us. For the mathematical description we define the following parameters, all functions of the depth h:

T and T' are the temperatures outside and inside the elements; it is assumed that in all elements the temperature is the same single-valued function of depth. We define $\Delta T = T' - T$.

P is the gas pressure. We assume pressure equality at the same height in- and outside the elements.

v and v' are the descending and ascending velocities of the surrounding matter and the granulation elements. Since the upward and downward mass transports must be equal, and if the areas of both elements are equal: $v'/v = T'/T$. For description purposes it is convenient to describe the motions of turbulence elements with respect to the surrounding medium. To that end we introduce $V = v' - v$.

To solve for these four unknowns, four relations between them should be found. A first attempt was made by DE JAGER[1], who, following earlier discussions of SIEDENTOPF and WOOLLEY, used three relations between these four unknowns. For the fourth relation, that between ΔT and T, some trial functions were introduced. That solution was finally chosen which gave the best agreement with the observed limb darkening of the Sun and the observed brightness differences between granulation elements and surroundings. The problem was rediscussed by E. VITENSE[2] who introduced a fourth relation—one between the relative loss of energy by a convective element and the logarithmic temperature-pressure gradients. In that way a more or less unambiguous treatment of the problem

[1] C. DE JAGER: Proc. Acad. Amsterd. **51**, 731 (1948).

[2] E. VITENSE: Z. Astrophys. **32**, 135 (1953).

could be given. Still there remains the problem how to choose the diameter L_0 of the granulation elements. VITENSE chose it consecutively equal to the local scale height H and to twice this value.

SCHATZMAN'S theory of the solar granulation[1], which appeared almost simultaneously with VITENSE'S, is essentially based on the same relations in the deep part of the convection zone. A difference is that SCHATZMAN assumed that the visible granules are manifestations of upward moving compression waves, initiated by the deeper situated convective motions. The observations (SUEMOTO, MCMATH *et al.*, WADELL), however, seem to support VITENSE'S image rather than SCHATZMAN'S: matter moving upward over the limit of the convection zone ($\tau_0 \approx 0.8$) and pushing further to $\tau_0 \approx 0.3$; the main dissipation of energy taking place for $\tau > 0.5$.

In the following part of this section VITENSE'S discussion will be reviewed. As indicated in the first part of this section the choice $L_0 = 2H$ will be more in agreement with reality than $L_0 = H$—at least in the upper part of the convection zone, where we adopt granular diameters of 500 km whereas the scale height is about half this value.

The *first relation* refers to the velocities of the elements, derived from the equality between the upward force

$$K = -\text{volume} \cdot g \cdot \Delta\varrho = F\varrho g L_0 \frac{\Delta\varrho}{\varrho} = -F\varrho g L_0 \frac{\Delta T}{T}$$

and the frictional resistance

$$W = F\varrho V^2.$$

Here, ϱ is the density of the surrounding matter; $\varrho - \Delta\varrho$ is the density in the element; FL_0 is the volume of the element.

This gives SIEDENTOPF'S relation:

$$V^2 = g L_0 \frac{\Delta T}{T}. \tag{11.1}$$

Formula (11.1) gives the velocity if convection is fully developed.

In order to elaborate this relation, a formula for $\Delta T/T$ will be developed. The mean temperature difference between element and surroundings is estimated as follows: The relative heating of an element in an interval Δh is

$$\Delta T = \Delta h \left(\frac{dT'}{dh} - \frac{dT}{dh}\right).$$

If we write: $dh = -H\,d\ln P = -RT/\mu g \cdot d\ln P$, we obtain over an interval L_0:

$$\frac{\overline{\Delta T}}{T} = \frac{L_0}{2H}(\nabla - \nabla') \tag{11.2}$$

with $\nabla' = \dfrac{d\log T}{d\log P}$ for the granulation elements, while ∇ refers to the surrounding matter.

Hence (11.1) becomes

$$\overline{V^2} = \frac{g L_0^2}{2H}(\nabla - \nabla'). \tag{11.3}$$

Relations for the logarithmic temperature-pressure gradients ∇ and ∇' as well as that of $\nabla_{\text{rad}} = \dfrac{d\log T}{d\log P}$ for an atmosphere in radiative equilibrium are found next. First of all, for an atmosphere in radiative equilibrium the total energy flux πF is:

$$\pi F = \sigma T_e^4 = \frac{16}{3} \frac{\sigma T^4}{\varkappa\varrho H} \nabla_{\text{rad}}. \tag{11.4}$$

[1] E. SCHATZMAN: Misc. Inst. Astr., Univ. Bruxelles (2) **90** (1953); **92** (1954). Cf. also C. WHITNEY: Smithson. Contrib. Astrophys. **2**, 365 (1958).

This formula defines the value of the radiative logarithmic temperature-density gradient. Naturally, in the convection zone the true gradients are smaller than the radiative gradient since a part of the energy transport is convective. In fact we have $\nabla_{\text{ad}} \leqq \nabla' \leqq \nabla \leqq \nabla_{\text{rad}}$, and the total energy flux

$$\pi F = \pi F_{\text{conv}} + \pi F_{\text{rad}}\,. \tag{11.5}$$

In this latter formula we insert Eq. (11.4) for πF. Further

$$\pi F_{\text{rad}} = \frac{16}{3}\,\frac{\sigma T^4}{\bar{\varkappa}\,\varrho H}\,\nabla. \tag{11.6}$$

The amount of energy transported convectively is generally written as (cf. BIERMANN[1]):

$$\pi F_{\text{conv}} = c_p\,\varrho\,\overline{\Delta T}\;\overline{V} \tag{11.7a}$$

wheres ZHEVAKIN[2] has argued that this expression should be:

$$\pi F_{\text{conv}} = \tfrac{1}{2}\,C\,\varrho\,\overline{\Delta T}\;\overline{V} \tag{11.7b}$$

with

$$C = c_p - P\left(\frac{\partial\, 1/\varrho}{\partial T}\right)_P; \qquad c_v < C < c_p.$$

c_p = specific heat at constant pressure; $\overline{\Delta T}$ and $\overline{V}$ are the mean temperature- and velocity differences between the element and its surroundings.

The differences between results computed with (11.7a) and (11.7b) are not fundamental.

If we substitute in (11.7): $\varrho = P/RT$ and eliminate $\Delta T/T$ from (11.7) and (11.1), the resulting expression is useful to estimate V in the deep layers where the energy transport is mainly convective. In the higher layers the energy exchange should be taken into account. From (11.2) and (11.7) we obtain:

$$\pi F_{\text{conv}} = c_p\,T\,\varrho\,V\,\frac{L_0}{2H}\,(\nabla - \nabla')\,. \tag{11.8}$$

This latter formula will be discussed below.

In the uppermost part of the convective zone the value of ∇' depends on the relative amount of energy radiated by the convective element. So, following UNSÖLD and VITENSE, we define

$$\Gamma = \frac{\text{energy surplus}}{\text{energy radiated during life time of element}}\,.$$

The energy surplus of the element with respect to an equal volume of the surroundings is equal to $c_p\varrho\Delta T\,L_0^3$ (L_0^3 is the volume of the element).

Since the elements are optically thick ($\bar{\varkappa}\varrho L_0 \gg 1$), the loss of energy by radiation from the elements (area $\approx 3\,L_0^2$) is written as a product of the three factors: emission per cm², area, and life-time:

$$\pi F = \frac{4\sigma}{3}\,\frac{\Delta T^4}{\Delta\bar{\tau}}\cdot 3\,L_0^2\cdot\frac{L_0}{V}\,.$$

With $d\bar{\tau} = -\bar{\varkappa}\varrho\,dh = \bar{\varkappa}\varrho H\,d\ln P = \bar{\varkappa}\varrho H\,\dfrac{dT}{T\nabla}$, the expression for Γ becomes

$$\Gamma = \frac{c_p\,\varrho\,T\,\overline{V}}{16\sigma\,T^4}\,\bar{\varkappa}\,\varrho\,L_0. \tag{11.9}$$

[1] L. BIERMANN: Z. Astrophys. **21**, 320 (1942).

[2] S.A. ZHEVAKIN: Astron. Zhurn. **33**, 137 (1956).

On the other hand the energy surplus is proportional to $(\nabla' - \nabla)$, and the radiation of energy is proportional to $(\nabla_{\rm ad} - \nabla')$ with the same proportionality constants. Hence

$$\Gamma = \frac{\nabla' - \nabla}{\nabla_{\rm ad} - \nabla'}. \tag{11.10}$$

Substitute (11.6), (11.8) and (11.9) in (11.5). This gives

$$\nabla = \frac{\nabla_{\rm rad} + \frac{3}{2}\Gamma\nabla'}{1 + \frac{3}{2}\Gamma}. \tag{11.11}$$

Then, after eliminating ∇ from (11.10) and (11.11), we obtain the fourth equation:

$$\nabla' = \frac{\nabla_{\rm rad} + \Gamma(1 + \frac{3}{2}\Gamma)\nabla_{\rm ad}}{1 + \Gamma(1 + \frac{3}{2}\Gamma)}. \tag{11.12}$$

The simultaneous solution in steps of Eqs. (11.3), (11.4), (11.11), (11.12) permits one to find the variation with depth of Log P_g and of T (for the convective elements as well as for the matter in between), and to find the relative velocity $V = v' - v$ of the ascending matter with respect to the downgoing matter. For each geometrical depth the velocities in both columns are then found with $v\varrho = -v'\varrho'$ or (because of the pressure equality) $vT' = -v'T$, which yields

$$v = -\frac{V\,T}{\Delta T} \quad \text{and} \quad v' = \frac{V\,T'}{\Delta T}. \tag{11.13}$$

(The upward velocity is taken positive.)

The quantities such as c_p, $\bar{\varkappa}$, which occur in the formulae, are functions of T and P. They can be computed and are tabulated, e.g. in UNSÖLD, Physik der Sternatmosphären [2]. Values of the coefficients of viscosity and thermal conductivity under conditions prevailing in the convection zone have been computed by EDMONDS[1] and OSTER[2].

12. The upper limit of the convection zone, and the development of turbulence in the photosphere. The upper limit of the convection zone, solved from the equation $\nabla = \nabla_{\rm ad}$ is $\tau_0 \approx 0.8$ in most photospheric models. But granulation observations close to the limb indicate the presence of the elements at greater heights. Also radial velocity observations of middle strong lines (generally originating at small optical depths, $\tau_0 = 0.1 \ldots 0.3$) clearly show the radial velocities of the elements, the upward motions corresponding to the hottest elements. So the problem is why the convective elements ascend to $\tau_0 = 0.1 \ldots 0.3$ and why they do not stop at $\tau_0 = 0.8$, or stated in another way: Why did the hot elements not yet wholly radiate their energy surplus when arriving at $\tau_0 = 0.8$ [Eq. (11.8) shows that $V = 0$ if $\Delta T = 0$].

To that end we consider the entropy diagram for the model atmosphere developed in Sect. 12 (Fig. 8). The entropy may be read for any given value of P_g and T from tables given by UNSÖLD. In deep layers the entropy is constant, higher up it decreases, to rise again at greater heights. This latter part is the radiative zone. Convection elements originating in the decreasing part of the entropy curve move along a horizontal line to the left in the entropy diagram when ascending into the photosphere. VITENSE showed in a detailed discussion of some peculiar cases that elements which originate at the upper level of the convection zone can move still to $\tau_0 \approx 0.3$ before the energy surplus is redistributed between hot and cold elements.

[1] F.N. EDMONDS: Astrophys. Journ. **125**, 535 (1957).

[2] L. OSTER: Z. Astrophys. **42**, 228 (1956).

Since the Reynolds number of the convection elements is great, they will be turbulent: the ascending and descending motions cannot be laminar. A hierarchy of turbulent motions will develop, the velocity spectrum of which can be computed with Eq. (8.1).

But also the non-convective, radiative part of the photosphere is turbulent, due to the formation of acoustic noise, originating in the granulation zone[1]. If the velocity field of the granulation zone is decomposed into its curl-free and its divergence-free parts, the former part produces waves which may pass through the non-convective upper region and enter into the chromosphere. They produce a turbulence field which may be described as a field of sound waves propagating in all directions. If analysed according to FOURIER the field presents velocity

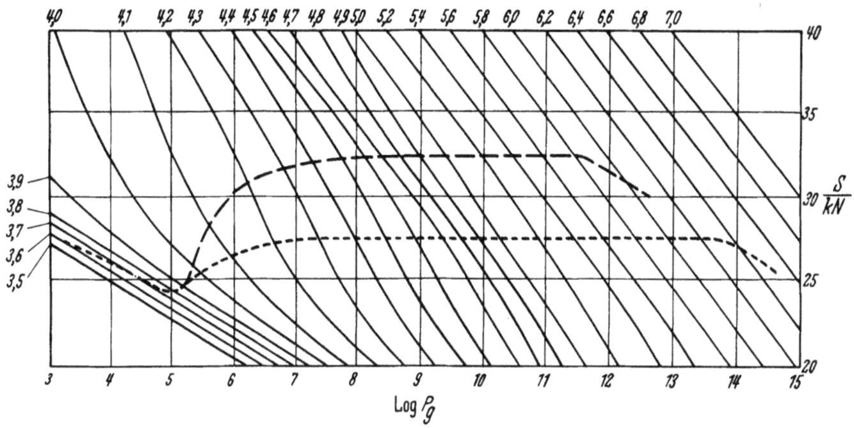

Fig. 8. The photosphere and the convection zone in the entropy diagram. Between the radiative part of the photosphere (left side), and the adiabatic part is the transition zone (Log $P = 5 \ldots 6$). The two curves are computed with the assumptions $L_0 = H$ (— — —) and $L_0 = 2H$ (----). The solid curves are labeled with the Log T-values. After E. BÖHM-VITENSE: Z. Astrophys. **32**, 135 (1953).

amplitudes $\approx \Delta v$ (UNSÖLD[2]). If c is the velocity of propagation of plane noise waves in the solar atmosphere, then the acoustical energy flux $\varrho\,\overline{(\Delta v)^2}\,c$ is constant, if dissipation of energy is unimportant. Hence the mean turbulent velocity should increase with height according to $(\varrho c)^{-\frac{1}{2}}$. A comparison of the observed values of ξ_t (Fig. 6 and 12) with those predicted according to this formula shows that the law is fullfilled qualitatively only, and that dissipation of energy already plays an important role throughout the whole upper photosphere.

13. The magnetic fields associated with the convection elements. In a turbulent plasma electric currents originate between regions of different electron densities and temperatures; these produce weak local magnetic fields. The fields may increase in strength if there is a way to transform kinetic energy $\frac{1}{2}\varrho\overline{v^2}$ into magnetic energy $\overline{H^2}/8\pi$. This transformation can be obtained by the "spaghetti process". Since the electrical conductivity is very great (cf. KOPECKY[3]) the lines of force can be considered as "frozen" into the gas: in matter at rest changes of the magnetic field are counteracted by induction currents and the field changes only when the lines of force are transported with the material. As a rule turbulent motions thus produce a stretching of the lines of force, increasing the lengths of the tubes of force, and because the amount of matter in a given tube remains

[1] Cf. L. BIERMANN: Trans. Internat. Astronom. Union (Dublin Meeting) **11**, 750 (1957).
[2] A. UNSÖLD: Z. Naturforsch. **7**a, 121 (1952) and Physik der Sternatmosphären, p. 667. Berlin: Springer 1955.
[3] M. KOPECKY: Bull. astronom. Inst. Czechoslov. **8**, 71 (1957).

constant the effect is that the lines of force are pressed together, thus increasing the local magnetic fields[1]. According to FERMI[2], BIERMANN[3], ELSASSER[4], CHANDRASEKHAR[5] and others this process will continue till equipartition between turbulent and magnetic energy is reached, i.e. when $\overline{H^2}/8\pi \approx \frac{\varrho \overline{v^2}}{2}$. With $\varrho = 3 \cdot 10^{-7}$ and $v = 1.5$ km/sec we obtain $H \approx 250$ Gauss, a field that is measurable, at least in principle. COWLING[6] has critizised the "spaghetti-process" arguing that the hypothesis of equipartition is based on analogy considerations that are perhaps not wholly sound. Only in the case of ALFVÉN's magnetohydrodynamic waves may there be equipartition between material kinetic energy and the energy of the disturbance field due to the wave. On the other hand analogously as, in a turbulent field, energy is transformed from long wavelength- to short wavelength-components, the influence of turbulence may be more and more to twist the existing lines of force rather than to generate large-scale fields from small-scale ones. Similar arguments were presented by BATCHELOR[7], but UNSÖLD[8], from a comparison of observed turbulent energies and magnetic fields, found empirical support for the first suggestion.

The measurement of the magnetic fields in tiny structures such as the granules presents the greatest difficulties.

An observation by KIEPENHEUER[9] might indicate the existence of such fields. With his photo-electric magnetometer he could measure considerable fluctuating fields when the solar image was displaced only slightly over the entrance slit of his magnetometer, provided that the solar image was in focus and that the seeing was good. With the given slit dimensions and assuming that one granule plus surroundings occupies an area of 5×10^5 km^2 one finds that an r.m.s. fluctuation of 10 Gauss corresponds to a granular field $H_g = 400$ Gauss, of the same order as the predicted fields.

With regard to the uncertainties of both data the agreement might be called good, but the investigation should necessarily be made anew with a small slit, covering only a few granules. It is noteworthy, in this respect, that BEGGS and VON KLÜBER[10], who made some preliminary measurements of this kind (slit length $\approx 4''$) did observe many strong local fields, up to 100 Gauss, but mainly in the sunspot belt. These measures seem to refer principally to "invisible spots" or to pores but not to the normal granulation.

14. A model of the photosphere incorporating convection and turbulence. A model is now given of the undisturbed photosphere and of the low chromosphere in which the temperature inhomogeneities, convection and turbulence are incorporated. We extend this model up into the low chromosphere. The cores of many Fraunhofer lines are formed in this region of the chromosphere so that knowledge of its structure is necessary for those who deal with the theory of Fraunhofer lines. The model is based on the following data:

1. *Temperatures:* (a) Photosphere: A two-column model being a mean of the models of Sect. 7. (b) Low chromosphere: Average of homogeneous models of PAGEL and DE JAGER (see also Fig. 15).

2. *Mass motions:* (a) Convective velocities: Voigt III. (b) Turbulent velocities in the photosphere and chromosphere: Fig. 6 (Sect. 11).

[1] H. ALFVEN: Monthly Notices Roy. Astronom. Soc. London **107**, 211 (1947).
[2] E. FERMI: Phys. Rev. **75**, 1169 (1949).
[3] L. BIERMANN: Z. Naturforsch. **5**a, 65 (1950). — Ann. Phys. **10**, 413 (1952).
[4] W.M. ELSASSER: Phys. Rev. **79**, 183 (1950).
[5] S. CHANDRASEKHAR: Observatory **75**, 161 (1955).
[6] T.G. COWLING: Vistas in Astronomy **1**, 313 (1955).
[7] G.K. BATCHELOR: Proc. Roy. Soc. Lond. **201**, 405 (1950).
[8] A. UNSÖLD: Physik der Sternatmosphären, 2. Aufl., p. 590. Berlin: Springer 1955.
[9] K.O. KIEPENHEUER: Astrophys. Journ. **117**, 447 (1953).
[10] D.W. BEGGS and H. VON KLÜBER: Nature, Lond. **178**, 1412 (1956).

3. The *chemical abundances* and the absorption coefficients are those adopted by VITENSE.

We assume pressure equality at the same geometrical height. With the given $(T; \operatorname{Log} P_g)$ values the electron pressures and the relation between height and optical depth can be computed. The equations are:

$$d\tau_0/dP_g = \varkappa_0/g_{\text{eff}}, \tag{14.1}$$

$$\varkappa_0 = \varkappa_0(P_e; T), \tag{14.2}$$

$$P_e = P_e(P_g; T). \tag{14.3}$$

In (14.1):

$$g_{\text{eff}} = \frac{1}{\varrho}\frac{d}{dh}\left(P_g + \frac{1}{2}\xi_t^2\right); \qquad g_{\text{eff}} \approx g_{\text{geom}}. \tag{14.4}$$

For ξ_t we take the quadratic sum of the micro-turbulent and macro-turbulent components. Tables giving the relation (14.2) are computed by E. VITENSE; for relation (14.3) by A. ROSA. The working model of the photosphere and the low chromosphere is given in the accompanying table. The subscripts *c*, *a* and *h* denote the cold, average and hot elements. The temperature fluctuations are chosen smaller than the empirical values shown in Fig. 5, which might be somewhat exaggerated. For the present model the granular contrast at 5000 Å is: $(I_h - I_c)/I_a = 0.2$. An average model has been included in the table since it seems interesting in some cases, to investigate the differences between an average model and an inhomogeneous one. The average model is essentially BÖHM's, but it has been adapted to the new measurements of the solar energy-wavelength curve[1,2]; a higher limb temperature has been assumed, and the model has been extended into the chromosphere.

Table 4. *An inhomogeneous model of the photosphere and low chromosphere.*

Log P_g	h (km)	T_{el} (°K)			Log P_e			$\tau_{5000\,\text{Å}}$			Vertical motion components (km/sec): convection components		
		c	a	h	c	a	h	c	a	h	up-ward	down-ward	ξ_{micro}
3.40	−97		4850			−0.35			0.0007				3.5
3.60	−42		4700			−0.19			0.0018				3.2
3.80	0		4480			−0.11			0.0045				2.9
4.00	41		4430			+0.03			0.010				2.6
4.20	83	4600	4600	4600		0.27			0.024				2.3
4.40	126	4840	4840	4850	0.58	0.58	0.58		0.056		0.15	0.1	2.0
4.50	149	4990	5000	5030	0.72	0.70	0.68		0.087		0.4	0.2	1.7
4.60	172	5140	5160	5220	0.76	0.77	0.73		0.12		0.5	0.35	1.5
4.70	197	5310	5370	5560	0.89	0.96	0.95		0.18		0.8	0.5	1.2
4.80	222	5530	5680	5770	1.16	1.21	1.27	0.28	0.30	0.30	1.2	0.75	0.9
4.90	250	5810	6140	6450	1.46	1.54	1.69	0.47	0.50	0.50	1.8	1.2	0.6
5.00	280	6280	6760	7460	1.93	2.09	2.36	0.90	1.02	1.04	2.6	1.8	0.4
5.05	296	6760	7380	8060	2.47	2.47	2.89	1.50	1.74	2.15	2.8	2.2	0.2
5.10	314	7570	8100	8570	2.86	2.83	3.06	2.0	3.2	3.6			
5.15	333	8050	8450	8780	3.34	3.11	3.29	5.1	6.8	6.7			
5.175	342	8150	8550	8880	3.43	3.20	3.4	8.1	10	9.1			

[1] D. LABS: Z. Astrophys. **44**, 37 (1957).

[2] G. F. SITNIK: Dokl. Akad. Nauk S.S.S.R. **110**, 193 (1956). — E. A. MAKAROVA: Astron. Zhurn. **34**, 539 (1957).

II. The chromosphere.

a) The chromosphere at the limb.

15. Problems of the chromosphere; transitions to photosphere and corona. The chromosphere has been discovered during total solar eclipses thanks to the circumstance that, generally, the Moon's apparent diameter is only slightly greater than that of the Sun. Before second contact the Sun's disk reduces increasingly fast to a thin and thinner crescent, in the last seconds only shining through the lunar valleys—forming the Baily beads. After that the solar Fraunhofer spectrum disappears completely and a faint emission spectrum becomes visible. The neutral emission lines have about the same relative intensities as the Fraunhofer absorption lines but the spark lines are enhanced. The greatest part of the emission lines is visible only during the first seconds after second contact and the last seconds before third contact. The short duration of visibility explains the name flash spectrum.

Let h be the distance between the Sun's limb and the projected Moon's limb. The Table 5 below[1] gives an impression of the duration of visibility of the lines; it lists the projected lunar limb distances h_0 where the total line emission of a slab of the chromosphere, 1 cm thick, perpendicular to the limb has decreased to 10^{11} erg sec^{-1} (solid angle)$^{-1}$. The last trace of the lines disappears generally when the intensity has decreased down to about 10^{10}, but the precise value depends on the properties of the spectrograph, the exposure times and the photographic plates.

Table 5.

	h_0 (km)	Average duration of visibility (sec)
Weak neutral metals (Fe, Ti)	800	2
Weak metal ions (Fe^+, Ti^+ etc.) . . .	800	2
Strong neutral metals (Fe, Ca, Cr, Al)	2500	6
Strong metal ions (Ti^+, Sr^+)	3000	7
He^+ 4686	3000	7 (small emission gradient)
Neutral He lines (except D_3)	5000	11
D_3 line of He	7500	17
Hα	12000	25
H and K lines of Ca^+	14000	30

Since the chromospheric crescent is very thin the flash spectrum can well be observed with a slitless spectrograph (camera equipped with an objective prism). The interpretation of the observed line intensities is difficult since the intensities are related to the emission per volume element by a double integration: The *surface brightness* F at each point of the chromosphere at a distance x to the limb is obtained by integrating the emission $\mathcal{E}$ per volume element over the line of sight passing the Sun's limb at a distance x. (A complication is that the observed surface brightness does not properly reflect the number of emitting atoms in the line of sight: self-absorption plays a role.) The *intensities in the flash spectrum* E are obtained by integrating the surface brightness F over a slice of the chromosphere, perpendicular to the Sun's limb, from $x=h$ to infinity; h is the projected distance of the Moon's limb to that of the Sun (Fig. 10).

To find F from E and $\mathcal{E}$ from F (elimination of the influence of self-absorption), the observed integrated emission curve $E(h)$ is generally expanded into a sum of exponentials. For if $\mathcal{E}(x)=\sum_{n=1}^{N}\mathcal{E}_n(x)=\sum_{n=1}^{N}\mathcal{E}_n(0)\,e^{-\alpha_n x}$, a simple computation

[1] The values in the table were extracted from the measurements of J. HOUTGAST: Rech. Obs. Utrecht **13**, Part 3 (1957); other observers give similar values.

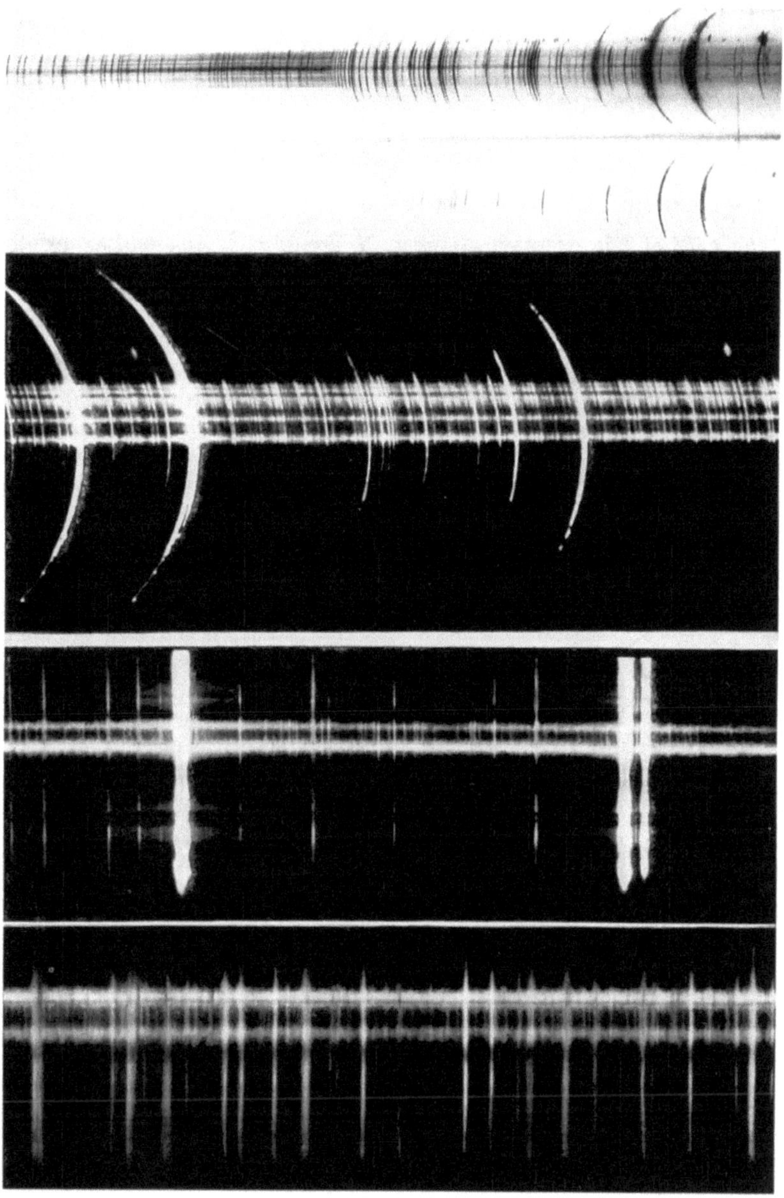

Fig. 9. The Khartoum eclipse (25. II. 1952) will be famous for a long time for the magnificent chromospheric spectra obtained that day. Upper two spectra: The flash spectrum from 3250 to 4150 Å obtained by the joint High Altitude Observatory—Naval Research Laboratory expedition (U.S.A.). The two overexposed strong lines at the right are *H* and *K* (Ca^+). In the lower spectrum (at 0.01 of the upper's intensity) the splitting of H_ε and *H* (Ca^+) is well visible. The upper spectrum clearly shows the Balmer continuum for $\lambda < 3680$ Å. Cf. also R. G. Athay, D. E. Billings, J. W. Evans and W. O. Roberts: Astrophys. Journ. **120**, 94 (1954). Courtesy J. W. Evans. Third spectrum: Enlargement of the flash spectrum taken by Houtgast and Zwaan (Utrecht, Holland). The spectrum contains the bright *K* line (3933 Å) the *H* line and Hδ (4101 Å), and shows moreover a wealth of faint metal lines. Cf. also J. Houtgast: Rech. Obs. Utrecht, **13**, 3 (1957). Courtesy J. Houtgast. Lowest two spectra: Portions of the slitspectra taken by Redman (Cambridge, G.B.). The upper spectrum gives the region around *K*, *H* and H_ε, the lower contains the region 3580 to 3614 Å. The just resolvable doublet at the left consist of 3580.93 (Sc II) and 3581.20 (Fe I); the strongest line at the right is 3613.84 (Sc II). Cf. also R. O. Redman: Vistas in Astronomy **1**, 713 (1956). Courtesy R. O. Redman.

involving two integrations and neglecting self-absorption shows that

$$F(x) = \sqrt{2\pi R} \sum_{n=1}^{N} \mathcal{E}_n(0)\, \alpha_n^{-\frac{1}{2}}\, e^{-\alpha_n x} \tag{15.1}$$

and

$$E(h) = \sqrt{2\pi R} \sum_{n=1}^{N} \mathcal{E}_n(0)\, \alpha_n^{-\frac{3}{2}}\, e^{-\alpha_n x}. \tag{15.2}$$

R is the solar radius.

In the case of one exponential term: $\mathcal{E} = \mathcal{E}(0)\, e^{-\alpha x}$ the emission gradient is equal to the density gradient—if self-absorption can be neglected. However, as a rule, things are more complicated.

In interpreting the deduced value of $\mathcal{E}$ in terms of number of atoms, or of physical parameters like the kinetic temperature, it should be realized that for many atoms the condition of local thermodynamic equilibrium is far from being fulfilled. Because of the great transparency of the chromosphere and because of dynamic phenomena connected with the dissipation of mechanical energy and the spicular motions, great differences can occur between the radiation temperature of the incident radiation and the local kinetic (electron) temperature of the gas, thus causing deviations from BOLTZMANN'S and SAHA'S laws. A major part of chromospheric studies is directed to the finding of the exact population of the atomic levels, both along observational and theoretical paths.

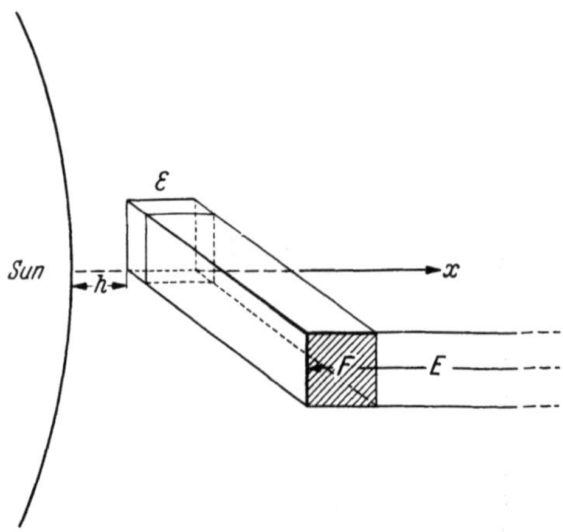

Fig. 10. The relation between the emission $\mathcal{E}$ per cm³ in the chromosphere, the chromospheric surface brightness F (cm⁻²), and the measured intensity E in the flash spectrum.

Some of the difficulties met in discussing slitless spectra do not occur in working out slit spectra; in principle, these give the surface brightness; moreover they yield line profiles, which might be used to find the kinetic properties of the gas. Difficulties are, however, that the surface brightness obtained on slit spectra is modified by scintillation (e.g. 10″ scintillation would give a smoothing of more than 7000 km—more than the "thickness" of the chromosphere) and that the exact distance of the slit to the Sun's limb is less well known than for flash spectra, for which the scale of height may be calibrated with an accuracy between 100 and 1000 km (varying from eclipse to eclipse) with the help of an exact timing of the observations. The uncertainty of this zeropoint is in the latter case due to the rather arbitrary definition of the moment of second or third contact. This instant is generally defined as the moment that the Fraunhofer spectrum changes into the emission spectrum or inversely; it is uncertain because of lunar limb mountains.

Limb observations of the chromosphere, using some selectively receiving optical instrument (spectrohelioscope; Lyot filter) show that the upper chromosphere has a very spiky appearance. Earlier observers compared it with a burning prairy. The chromospheric spicules, small, shortlived, needle-like structures rising and descending with velocities well over 10 km/sec are only observed above 5000 km. Below that height either the spicules do not exist, or their number is so great that they cannot longer be seen individually. Spectroheliographic studies on the disk show the spicules from above and suggest that they cluster together in larger structures. The smallest structures, which might be identified with spicules, may occur down to about 1000 km above the photosphere.

So attempts to represent the chromosphere by any model must be based on this two-element concept in which the dynamics of the spicular motion should be taken into account. The spicules of the chromosphere may penetrate into the corona up to heights of some ten thousands of kilometers. There is hardly any distinction between long spicules and short eruptive prominences.

Finally there is the problem how to explain the occurrence and the extension of the chromosphere: its heating, and its support—by turbulence or by magnetic fields—and the transitions to photosphere and corona. Visually the transition between photosphere and chromosphere is very abrupt, but the physical parameters (T, ϱ, etc.) merge continuously into each other and the difference is merely one of opacity. The difference between the chromosphere and the corona with its high temperature is more fundamental and takes place in an interval of space of the order of some thousands of kilometers. The problem of this latter transition is that of the heating of the low corona.

16. Considerations on the population of the H and He atomic levels in non-thermodynamic equilibrium. An essential feature of the chromosphere is the absence of local thermodynamic equilibrium (L.T.E.), which occurs because of the difference between the chromospheric electron temperature and the radiation temperature of the incident (diluted) photospheric radiation (to which the chromosphere is mainly transparent), further because of the different kinetic temperatures of the hot and cold elements and because of the occurrence of shock-wave phenomena in the upper part of the chromosphere. So it is not possible to compute the population of the atomic levels by BOLTZMANN'S and SAHA'S laws, thus to identify in each volume element the (kinetic) electron temperature T_{el} with the ionization and excitation temperatures T_{ion} and T_{ex}, but a more detailed discussion is necessary. (It has been verified that the differences between the kinetic temperatures T_{el} and T_{atom} are negligibly small in the astrophysical case[1].) The non-L.T.E. computations are mostly based on a reasoning like the following. Let N_n be the population of the n-th level of an atom or ion in thermodynamic equilibrium (characterized by the electron temperature T_{el} and the electron density N_e); N_n is expressed in cm^{-3}. Let the true population be $b_n N_n$; so that b_n is a factor indicating the deviation from thermodynamic equilibrium. The total number of outgoing transitions from the level n to all other levels j is $b_n N_n \sum_j P_{nj}$ where P_{nj} is for each transition a function depending on T_{el} and N_e. This number is equal to the total number of incoming transitions to the level n: $\sum_j b_j N_j P_{jn}$. Here the infinite number of atomic levels is approximated by a finite number N, which is in any case permissible since up from a certain n the value of $b_n N_n$ decreases quickly with increasing n, owing to the depopulating influence of collisions between the atom and surrounding particles. Further N may be reduced by grouping together a number of high levels. In this way

[1] P. L. BHATNAGAR, M. KROOK, D. H. MENZEL and R. N. THOMAS: Vistas in Astronomy **1**, 296 (1955).

one obtains N equations with N unknowns:

$$\sum_{j=1}^{N} b_j N_j P_{jn} - b_n N_n \sum_{j=1}^{N} P_{nj} = 0.$$

The solution of this system gives the b_n.

The practical computation for the conditions prevailing in the chromosphere has not yet been brought to a final solution in spite of hard work that has been involved, especially after 1945. One great difficulty is that even for atoms like hydrogen and helium many necessary data, especially those on collisional cross sections are not available. For other atoms the greatest part of the f-values and virtually all collisional cross sections are unknown.

Another difficulty is that the local radiation field is hard to compute. It consists of two parts: (a) the photospheric and coronal radiation fields, as modified by chromospheric absorption and (b) the own radiation (especially the lines) emitted by the chromosphere. However, the chromospheric absorption and emission can be computed only when the populations of the various atomic levels are known throughout the chromosphere, and by taking the equations of radiation transfer into account, both for the line radiation and the continuous radiation.

So, for a given chromospheric model the computation of b_n should proceed in steps: after a first solution of b_n at various heights in the chromosphere, the modified radiation field is recomputed (again for different heights) and a new solution for b_n is made, etc. Such a program has not yet been executed, but a number of approximate solutions has been published especially for hydrogen and also for helium.

The hydrogen case was first discussed by GIOVANELLI[1], who neglected inter-level transitions such as $2S \to 2P$ etc., and approximated the local radiation field by a Planck curve with a temperature of 5000°, neglecting the line radiation. The results are of course uncertain but they show that $b_n \gg 1$ for small n, approaching unity for $n > 10$. (MATSUSHIMA[2] showed that this is actually the case only for $n > 20$.)

Another conclusion, that the $2S$ state would be more densely populated than the $2P$ state, because of the former's metastability, was criticized by PURCELL[3] who showed that the collisional cross section $2S \to 2P$ is great, causing the population ratio $2P/2S$ to approach the thermodynamic equilibrium value 3.

In later theoretical computations the influence of the chromosphere's own radiation field was taken into account (JEFFERIES[4], ZIRIN[5], THOMAS *et al.*[6]) and at the same time it has been attempted to extract b_n from the observations (see e.g.[6]). The new results qualitatively confirm GIOVANELLI's first result ($b_n \gg 1$ for small n and approaching unity for increasing n). PECKER and PECKER[7] drew attention to the great differences that still occur between the various b_n determinations.

---

[1] R. G. GIOVANELLI: Austral. J. Sci. Res. A **1**, 275, 289 (1948).
[2] S. MATSUSHIMA: Astrophys. Journ. **115**, 544 (1952).
[3] E. M. PURCELL: Astrophys. Journ. **116**, 457 (1952).
[4] J. T. JEFFERIES: Austral. J. Phys. **6**, 22 (1953).
[5] H. ZIRIN: Astrophys. Journ. **123**, 536 (1956); **126**, 159 (1957).
[6] R. G. ATHAY, D. H. MENZEL, S. MATSUSHIMA, J. C. PECKER and R. N. THOMAS: Papers in Astrophys. Journ. Suppl. **1**, No. 12 (1956).
[7] CH. PECKER and J. C. PECKER: Contrib. Inst. Astrophys. Paris, Ser. A **213** (1956).

For the case of helium reference is made to the investigations on the non-equilibrium spectrum of helium in stellar atmospheres by WELLMANN[1]. Similar computations as those described above for H were made for He by JEFFERIES[2] and ZIRIN[3]. The last computations were made for four T_{el}-values ranging between 25000° and 2×10^5 °K. The photospheric radiation field was approximated by $T_{rad}=6000°$; the chromospheric contribution was neglected.

Another set of b_n computations has been made for two cases[4]: $T_{el}=10800°$ and 8600°, for $\text{Log}\, N_e=11.5$, conditions assumed to prevail near a height of 2000 km in the hot elements of the chromosphere. The incident photospheric radiation was assumed equal to the continuous spectrum for $\lambda>912$ Å integrated over the whole disk. The Lyman continuous radiation was approximated by a Planck curve with $T_c=6000°$. The level scheme was approximated by dividing it into ten "levels"; the tenth containing the number of He II ions—virtually equal to the number of groundstate He II ions. The ten equations obtained by equalizing the incoming and outgoing transitions were solved.

The principal result is the great overpopulation of the triplet states, due to the metastability of the 2^3S level. Also the metastable $2\,^1S$ level is overpopulated but to a less degree since the electrons may jump from $2\,^1S$ via $2\,^1P$ to the ground level. The overpopulation becomes smaller for increasing n-values. In practice accurate knowledge of b_n for He is rather unimportant; because of the strong dependence of N_n on T_{el} ($\chi\approx 20$ eV for He I and 50 eV for He II) the population of the He levels is primarily determined by T_{el}.

For the observational determination of the relative populations of the atomic levels a useful expression is next derived. It is based on the assumptions that collisional de-excitation is unimportant, that the gas is transparent and that the continuous radiation of the gas is negligible. The monochromatic radiation emitted in a line is according to KIRCHHOFF's law: $E=\varepsilon/\varkappa$, and introducing the Einstein transition probabilities one obtains:

$$E=\frac{A_{nm}N_m}{(B_{mn}N_n-B_{nm}N_m)}\,. \tag{16.1}$$

With the well-known relations

$$B_{nm}=\frac{\lambda^3 A_{nm}}{8\pi h}\,;\qquad B_{mn}=\frac{\lambda^3 g_m A_{nm}}{8\pi h g_n}\,,$$

where g_m/g_n is the ratio of statistical weights, the expression (16.1) becomes

$$E=\frac{8\pi h}{\lambda^3}\left(\frac{g_m}{g_n}\,\frac{N_n}{N_m}-1\right)^{-1}. \tag{16.2}$$

In thermodynamical equilibrium according to BOLTZMANN's law:

$$N_n/N_m=g_n\exp\left(h\,c/k\,\lambda\,T\right)/g_m; \tag{16.3}$$

and (16.2) reduces to PLANCK's law.

In the general case the ratio N_n/N_m may be derived from the observed emission of radiation E (cm^{-3}) by means of (16.2).

The observed selective emission may be described by determining the function T_{ex} that satisfies the Planck function (16.2) and (16.3) for a given measured

[1] P. WELLMANN: Cf. Vistas in Astronomy **1**, 303 (1956).
[2] J. T. JEFFERIES: Austral. J. Phys. **8**, 335 (1955).
[3] H. ZIRIN: Astrophys. Journ. **124**, 451 (1956).
[4] C. DE JAGER and B. J. M. DE GROOT: Bull. astronom. Inst. Netherl. **14**, 19 (1957).

radiation intensity. Then T_{ex} is defined as the excitation temperature for the transition $n \rightarrow m$; the source function for that specific radiation is given by PLANCK's law in which T is put equal to T_{ex}.

17. "Turbulence" in the chromosphere. When discussing chromospheric mass motions we have to distinguish between the random "turbulent" motions, which prevail in the lower part of the chromosphere and the systematic motions shown by the spicules of the high chromosphere. The motions of the individual spicules seem to be anisotropic, although in a big field of spicules many directions of motions occur—due to the inclinations of the spicules—so that in such a field the general velocity distribution may still look like isotropic turbulence. However, in the following the term "turbulence" will be used only for the random subsonic velocities and insofar as the term is applied in connection with individually observed spicules it will only be to denote microturbulence in the spicules.

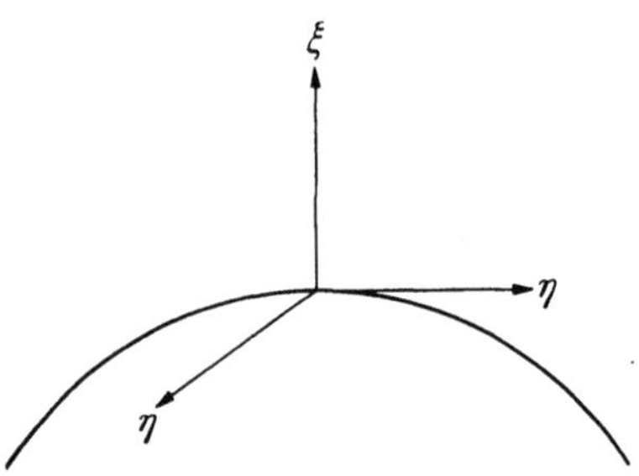

Fig. 11. Definition of the direction of the components ξ and η of the turbulent motion field. The mean turbulent velocity components ξ_0 and η_0 are defined in such a way that the distribution of the velocity components can be represented by $\exp(-(\xi/\xi_0)^2)$. In the text the subscript 0 is usually dropped.

There are two sources of information (Fig. 11). Widths of line-profiles measured in slit spectra of the chromosphere give information on $(\eta^2 + 2RT_{el}/\mu)$. If T_{el} is known, η, the r.m.s. sight-line turbulence component (tangential to the Sun's surface) can be computed. The values of ξ, denoting the component of turbulent motions perpendicular to the Sun's surface may be found from the density scale height of the chromosphere, after eliminating the temperature term. It may also be derived from the profiles of chromospheric lines observed in the Fraunhofer spectrum of the disk centre. The obvious difficulty in computng ξ or η is always that T_{el} must be known, its value might be uncertain.

α) Components tangential to the Sun's surface. (i) REDMAN's slitspectra[1] obtained at the eclipse of February 25, 1952 are the major source of information. Metal line spectra were obtained up to 1400 km from the limb (the Na D lines); Fe I and Ti II could be observed till 1200 km, the helium lines till 2600 km and the other lines only to about 400 km above the surface. Many lines are influenced by self-absorption, but this effect can be eliminated by plotting the observed widths of lines of the same element against line intensities. By extrapolating this relation to zero intensity the "true" width can be obtained. These half-widths are converted into η-values, after correcting for the temperature term. For the metal lines this T_{el} term is unimportant, but for He it is of importance. REDMAN and SUEMOTO assumed $T_{el} = 6000°$. We have recomputed η by assuming (a) that T_{el} for helium $= 10000°$, an assumption for which support will be given later, and (b) that T_{el} (He) $= 20000°$ another temperature mentioned in the literature.

In REDMAN and SUEMOTO's results the height-entry is the position of the spectrograph slit with respect to the Sun's limb. Then the chromospheric centre of gravity of the emitted radiation would be about half a scale height higher in the chromosphere, if seeing is neglected, if an exponential distribution of the source function is assumed, and if the lines are optically thin[2]. However, "seeing"

[1] R. O. REDMAN and Z. SUEMOTO: Monthly Notices Roy. Astronom. Soc. London **114**, 524 (1954).

[2] A. UNSÖLD: Physik der Sternatmosphären, 2. Aufl., p. 619. Berlin: Springer 1955.

will cause light from a projected height range of between 1000 and 7000 km to enter the slit; owing to the nonlinear $F(h)$ function (see Fig. 10) the contribution of the lower layers will enter with a greater weight so that the effective height of emission of observed light will be reduced and can even be lower than the projected slit distance.

In the table below h is the distance between the slit and the Sun's limb.

Rare earth's, faint metallic lines		Fe I, Ti II, NaD		Helium		
				h (km)	η (km/sec)	
h (km)	η (km/sec)	h (km)	η (km/sec)		10000°	20000°
< 50	2.7	600	3.9	600	8.9	6.2
< 100	3.2	600	3.9	1100	9.8	7.2
600	3.9	600	4.8	1100	9.8	7.2
< 500	2.1	1200	9	1400	8.6	5.8
< 500	3.1	1200	9	2000	11.6	9.4
		1400	9.4	2600	14.3	12.8

A comparison of the *helium* data with those for the other lines shows that the former deviate for $h < 1500$ km. This can be understood: the helium atomic levels with their great excitation potentials (≈ 20 eV) are not excited in the lowest chromospheric levels and helium radiation is only emitted in appreciable quantities above 1000 or 1500 km. Unpublished computations which I made for various chromospheric models show that these all predict the centre-of-gravity of the helium emission between 1500 and 3500 km. Also the observed variation of the helium flash-line intensities [$E(h)$ functions] indicates that $\mathcal{E}(h)$ reaches a maximum value somewhere near 2000 km. Hence, the η-values determined when the spectrograph slit was located below 1500 km refer virtually to greater heights in the chromosphere.

The widths of the *Balmer line profiles* ($H_\varepsilon - H_{31}$) which were also measured by Redman and Suemoto cannot directly be used for turbulence determinations since the sum of the thermal widening and of the Stark effect is much greater than the turbulent widening component; e.g. the measured total halfwidths are 0.3 to 0.5 Å, and the "turbulence width" $\lambda\xi/c = 0.15$ Å. Since the contributions add more or less quadratically, turbulence has hardly any influence on the resulting width.

The picture obtained by Redman's systematic investigation may be completed by some other turbulence determinations from chromospheric line widths.

(ii) At heights between 4000 and 13000 km, hence in the region where the spicules can be seen, either in groups or individually, Michard[1] (Arcetri and Meudon), E. v. P. Smith[2] and R. G. Athay[3] determined Hα profiles (and D_3 profiles[3]) of individual spicules. The slit of the spectrograph was placed tangentially to the Sun's limb. The crumbled aspect of the Hα emission line is due to its being composed of the individual spicular profiles, each widened according to microturbulence and thermal motions, and displaced according to the mass motions of the spicule as a whole. These latter motion components are distributed (Michard) by a bellshaped function with a r.m.s. value of 6.0 km/sec. The measured whole half-width of Hα ranged between 0.5 and 1.8 Å (Michard) and between 1.2 and 1.8 Å [2,3] The origin of this difference is not known. Ob-

[1] R. Michard: Ann. d'Astrophys. **19**, 1 (1956); **21** (1958).
[2] E. v. P. Smith: Astrophys. Journ. **126**, 529 (1957).
[3] R. G. Athay: Ann. d'Astrophys. **21**, 98 (1958).

viously one always risks to observe complexes of spicules instead of one single specimen; this tends to enlarge the observed profiles. From a study of the published spectra it does not become clear whether they really refer to single spicules or rather to coarse mottles (for the definition of coarse mottles see Sect. 23). For that reason the smallest value (0.5 Å) would seem to be the most reliable.

However, E. v. P. SMITH remarks that small prominences have narrower Hα profiles than spicules (their halfwidth is about 0.5 Å) so that a part of the difference between MICHARD and the others might originate from the incorporation of small prominences in the measurements. So we are inclined tentatively to adopt for the halfwidth of spicules between 8000 and 10000 km: $2\Delta\lambda=0.8$ to 1.2 Å. This would yield the following kinetic temperatures:

Assumed ξ_{micro}	0	15	20 km/sec
$2\Delta\lambda=0.8$	29000°	15000°	5000°
1.0	45000°	32000°	21000°
1.2	65000°	52000°	41000°

As an average value for $h=9000$ km $T_{\text{kin}}=35000°\pm15000°$ can be assumed. The observations do not yet permit to extract information about the microturbulence in the spicules, as long as the temperature is not yet precisely known. In principle, however, it is possible to separate macroscopic mass motions and microturbulence in the high chromosphere. In the medium chromosphere (2000 to 4000 km), where the spicular motion may be well developed but where the individual spicules cannot yet be seen, this procedure cannot be applied.

(iii) The half widths of the H and K lines of Ca^+ were measured by UNSÖLD[1] at Mt. Wilson in 1929:

h	H	K
2500 km	0.82 Å	0.87 Å
5600 km	0.73 Å	0.82 Å

The lines have a nearly rectangular profile, the equivalent width is practically equal to the observed halfwidth, indicating that the chromosphere is optically thick in the line centres. Since the (radiation) damping is small as compared with the Doppler width parameter (hence $a=0$) the curve of growth consists only of a linear and a flat part. Hence the lines are formed in the flat part of the curve of growth, a conclusion supported by the fact that the width of the K line is only 10% greater than that of the H line. At this part of the curve of growth $W/\Delta\lambda_D\approx4$ to 5 leading to $\eta=15$ to 12 km/sec at a mean height of 4000 km. Of course this η-value is very uncertain.

(iv) (a) Interferometric observations of the D_3 line at the limb outside eclipses[2] have shown that the line width is virtually independent of the height between 1500 and 4500 km. With $T_{\text{kin}}=10000$ and 15000° the resulting η-values are 16.2 and 15.4 km/sec. As a round value we take $\eta=16$ km/sec between 2000 and 4500 km.

(b) In the low chromosphere (2000 km) KRAT and KRAT[3] found the width of D_3 to correspond with $\eta=11.6$ km/sec (assuming $T=10000°$).

[1] A. UNSÖLD: Astrophys. Journ **69**, 209 (1929).

[2] P. J. TREANOR: Monthly Notices Roy. Astronom. Soc. London **117**, 22 (1957).

[3] W. A. KRAT and T. W. KRAT: Isw. Gl. Astr. Obs. Poulkovo **19** (152), 20 (1954); **19** (153), 1 (1955).

(c) At 2000 and 4000 km we computed from MICHARD'S[1] halfwidths of the D_3 line measured outside eclipses at Arcetri $\eta = 11.4$ and 14.6 km/sec (assumed $T = 15\,000°$).

(v) Other η-values were obtained by E. BÖHM-VITENSE[2] from the position of the flat part of the curve of growth of Ti II lines, observed by CILLIÉ and MENZEL[3]; the table gives the tangential microturbulent velocity components:

h	140	970	1800 km
η	3.8	5.5	8.6 km/sec

The scale of height in this table was derived by us[4] from that of CILLIÉ and MENZEL by adding 280 km.

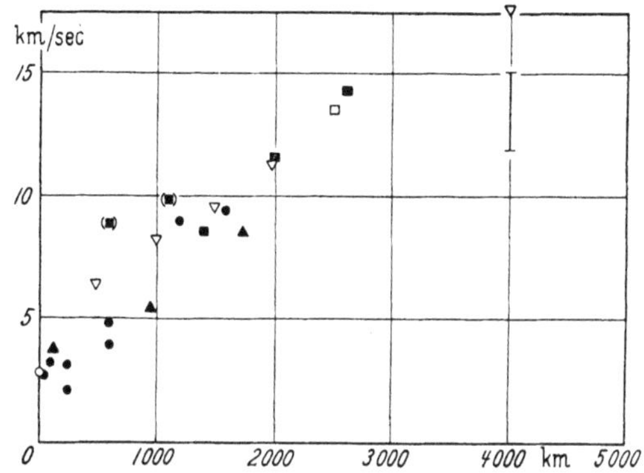

Fig. 12. Variation of the turbulence components ξ_0 and η_0 with height above the photosphere. Open and filled symbols denote vertical (ξ) and tangential (η) components. For further details see text.

β) Components radial to the surface. (i) The ξ-values, radial to the surface, may be obtained from widths of Fraunhofer line profiles in the centre of the disk. We recall that in Sect. 11 values between 1 and 3 km/sec were found for the photosphere. At $h = 0$, by interpolation from Fig. 6, $\xi = 2.75$ km/sec.

(ii) The profile of the infrared helium line[5] at 10832 Å averaged over the centre of the disk and the four limbs has a halfwidth of 0.91 Å. With $T_{\rm el} = 10000°$ this yields $\xi = 13.6$ km/sec. We assume that this line is formed mainly at $h = 2500$ km.

(iii) From the disk profiles of the K_3 and H_3 lines KHOKLOVA (footnote 5, p. 125) finds $\xi \approx 7$ km/sec at $h \approx 4500$ km.

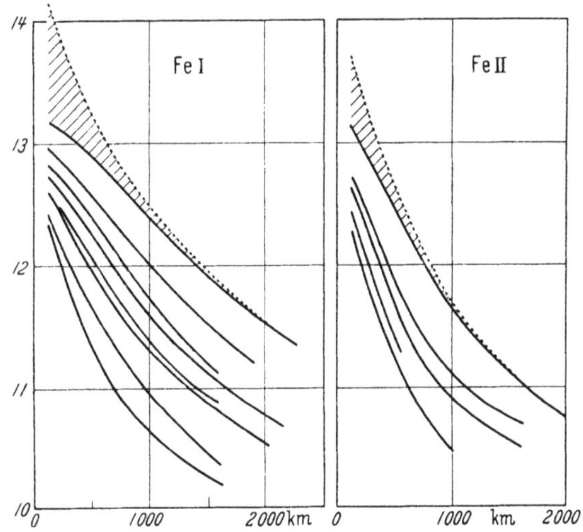

Fig. 13. Measured intensities in the flash spectrum; eclipse of 25 Febr. 1952. Left: Fe I; right Fe II lines. The ordinate is Log E [ergs sec^{-1} (solid angle)$^{-1}$ for a chromospheric slab, 1 cm thick]. The curves, if shifted horizontally upward fill a fanshaped region (shaded), the upper limit of which indicates the $E(h)$-curve for a hypothetical line without selfabsorption. These upper limits have roughly the same shape for Fe I and Fe II. (From J. HOUTGAST: Rech. Obs. Utrecht **13**, 3 (1957).)

(iv) Another way to derive ξ values is by means of the observed intensity gradients of emission lines in the chromosphere. If self-absorption can be neglected, if the excitation and/or ionization conditions do not vary with height and if the source function $\mathcal{E}(h)$ is exponential, also $E(h)$ is exponential. In that

[1] R. MICHARD: Ann. d'Astrophys. **20**, 1 (1957).
[2] E. BÖHM-VITENSE: Z. Astrophys. **36**, 145 (1955).
[3] G. G. CILLIÉ and D. H. MENZEL: Harvard Circ. **410** (1935).
[4] C. DE JAGER: Bull. astronom. Inst. Netherl. **13**, 275 (1957).
[5] O. C. MOHLER and L. GOLDBERG: Astrophys. Journ **124**, 13 (1956).

case the emission scale height is equal to the density scale height. These stipulations are serious enough, but they can be solved with the aid of HOUTGAST'S observations[1] of the flash intensities of Fe I and Fe II lines made at Khartoum at the eclipse of 25 February 1952. The problem to determine the amount of self-absorption was solved by shifting all Log $E(h)$ curves for Fe I into a vertical direction. For great heights all curves appear to coincide, having the same gradients. This proves that in that region self-absorption is unimportant. In lower heights the complex of curves is fan-shaped; the upper limiting curve of the fan indicates the emission curve free from self-absorption. An analogous treatment applies to the Fe II lines (Fig. 13).

It appears that these self-absorption-free Log $E(h)$ curves for Fe I and Fe II are nearly parallel. This means that also the second condition is fulfilled:

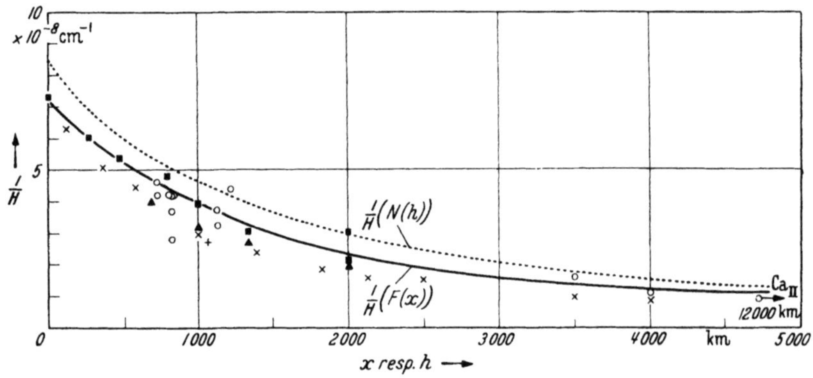

Fig. 14. Chromospheric gradients. Solid curve: observed gradient $d \ln F(x)/dx$ as a function of the distance x to the limb; F corrected for selfabsorption (cf. Fig. 9b). Dotted: derived gradient of the atomic numbers $d \ln N(h)/dh$ as a function of the height h above the photosphere. After VAN DE HULST; cf. A. UNSÖLD: Physik der Sternatmosphären, 2. Aufl., p. 620. 1955. ■ D. H. MENZEL; ▲ S. A. MITCHELL and E. T. R. WILLIAMS; ● R. WILDT; + G. G. CILLIE and D. H. MENZEL. Added (crosses): 1952 gradients of J. HOUTGAST: Rech. Obs. Utrecht **13**, 3 (1957).

for, if e.g. the ionization and excitation increased with height, the Fe I curve would be steeper than the Fe II curve. In the present case it is permissible to identify the Fe I—Fe II gradient with the true particle gradient for Fe. Between $h = 500$ and 2200 km the observations are represented by the same scale heights as usually found by other authors (see Fig. 14). So we derive:

h	500	1000	1500	2000	4000 km
H	160	210	270	330	670 km
ξ	6.5	8.2	9.6	11.3	17.7 km/sec.

Also in this case the temperature term in H is unimportant.

The η- and ξ-values derived in this section (Fig. 12) are in good agreement.

So the conclusions are: (a) the mean (micro- and macro-) turbulent velocity component increases from 2.7 km/sec at the top of the photosphere up to 10 km/sec at 1800 km and to about 14 km/sec at 3000 km. Above that height it remains perhaps constant; its value above 5000 km is unknown. (b) At least up to 3000 km there is no indication for anisotropy in the turbulence field. (c) It is not yet certain that the amount of turbulence differs for the hot and cold elements of the chromosphere, although KHOKLOVA'S result (part iii) seems to point into that direction.

[1] J. HOUTGAST: Recherches Obs. Utrecht **13**, part 3 (1957).

18. The intensity profile of the Sun's limb. The limb region is the transition between photosphere and chromosphere where the total optical depth of the photosphere, measured in the continuous spectrum, and tangentially to the limb, becomes of the order unity or smaller. It is the region where the Sun becomes transparent for light of the visual wavelength region. We assume that this transition region ends where the continuous absorption coefficient is small as compared with the scattering coefficient, i.e. where the emitted blackbody radiation is negligible as compared with the scattered photospheric light (again: for the visual wavelength region; in the far ultraviolet the situation is quite different).

Observations of the relative intensity variation near the Sun's limb are due to LINDBLAD and KRISTENSON[1] and to HENRIKSEN[2] at the eclipse of 1945. These intensities have not been calibrated on an absolute scale. Observations of the continuous spectrum from the very limb up into the corona were made at the Khartoum eclipse (25 February 1952) by ATHAY, BILLINGS, EVANS and ROBERTS[3]. The latter observations have been calibrated on an absolute intensity scale. In the corresponding region the *gradients* agree with those derived by LINDBLAD and KRISTENSON and HENRIKSEN. The scale of *absolute intensities* was initially in error, being too high by a factor two. After correcting[4], the agreement with other observers of the same eclipse was quite satisfactory (estimated probable error of absolute calibration ≈ 0.2 in logarithm). The observations in the visual region refer to 4700 Å. Besides measurements were made at two wavelengths at either sides of the Balmer jump.

The interpretation of observations of the continuous spectrum, made in the flash spectrum with a slitless spectrograph is still of a higher degree of difficulty than the interpretation of line strength measurements (cf. Sect. 15, Fig. 10) because the observations do not even yield $E(h)$ but are related to this function by still another integration and one summation: The intensity I measured in one point in the continuous spectrum is the integral over the surface brightness $E(h, \lambda)$ for other wavelengths and h-values (weighted by the plate sensitivity). Moreover at each point in the observed spectrum the radiation is the sum of the contributions from the two sides of the Sun's limb; the latter effect is only important at great distances to the limb. These effects have been eliminated by ATHAY, MENZEL, PECKER and THOMAS[5], principally by making use of the profile of the atmospheric A_{00} band (O_2) at 7621 Å. This band is doubled in the observed spectrum, which shows the contribution of the E and W limb.

In this way the authors obtain the crescent intensities E in the continuous spectrum for a chromospheric slab, 1 cm thick, and by further numerical differentiation the surface brightness F in the continuous spectrum is found. The curve, thus obtained shows a limb brightening, but it is uncertain whether this brightening has any reality after two differentiations of the original observations. It is known that differentiation of an observed function tends to enormously increase small existing deviations from a smooth curve. Furthermore we recall that photographic observations—on which the discussion is based—are not sufficiently accurate for a discussion of limb observations of the Sun[6].

[1] B. LINDBLAD and H. KRISTENSON: Convegno Volta **11**, 61 (1953).
[2] S. W. HENRIKSEN: Astrophys. Journ. **120**, 521 (1954).
[3] R. G. ATHAY, D. E. BILLINGS, J. W. EVANS and W. O. ROBERTS: Astrophys. Journ. **120**, 94 (1954).
[4] R. G. ATHAY, D. H. MENZEL and F. Q. ORRALL: Smithson. Contr. Astrophys. **2**, No. 4, 35 (1957).
[5] R. G. ATHAY, D. H. MENZEL, J. C. PECKER and R. N. THOMAS: Astrophys. Journ. Suppl. **1**, 505 (1955).
[6] H. HUBENET and C. DE JAGER: Bull. astronom. Inst. Netherl. **13**, 43 (1956).

The emergent radiation may be computed for any model of the transition region photosphere-chromosphere with the expression (see Fig. 10)

$$\left.\begin{aligned} I_\lambda(x) &= \int_0^\infty S(\tau'_\lambda)\, e^{-\tau'_\lambda}\, d\tau'_\lambda, \\ \tau'_\lambda &= \int_{-\infty}^{y} \varkappa_\lambda\, dy. \end{aligned}\right\} \tag{18.1}$$

τ'_λ is the monochromatic optical depth measured along a line passing the Sun at a distance x; $\varkappa_\lambda$ is the monochromatic absorption coefficient per cm; $S(\tau'_\lambda)$ is the source function.

In computing S one should take into account the influence of scattering (for an example see DE JAGER[1]) and that of deviations from local thermodynamic equilibrium. For a treatment of this latter case cf. PAGEL[2]. These two effects have the character of corrections; more important is the choice of the temperature-height function.

Computations of the Sun's limb profile[1], with the aid of some tentative low chromospheric models varying according to the assumed turbulence functions and to the temperature and helium abundance, indicated that the best agreement would perhaps occur for $T(0) = 4300°$ and T increasing upward, $N(H)/N(He) \approx 5$, and for the $\xi(h)$ function of Fig. 12.

	$h=$		
	0 km	250 km	500 km
PAGEL	4700°	5400°	5650°
DE JAGER . . .	4300°	4500°	4800°

A simultaneous investigation of the Sun's limb profile, by PAGEL[3], was based on a lower helium abundance (10%) and on the assumption that $\xi_t(h) = 0$ throughout the region, where the limb radiation is emitted. The non-validity of the Bolzmann-Saha approximation was taken into account by discussing the ionization equilibrium of hydrogen and H^-. The resulting temperature distribution (see the Table above) fully explains the limb profile at 4700 and at 3640 Å.

The radiation emitted in the visual spectral region, at heights between 500 and 1000 km, seems to be for a great part scattered radiation. In the low chromosphere scattering to bound electrons (Rayleigh scattering) has usually been thought negligible as compared with Thomson scattering (by free electrons), but this assumption may not be wholly correct[1].

19. The spectrum, temperatures and densities of the low chromosphere. In this section we sum up temperature determinations in the low chromosphere.

α) *In the far ultraviolet photospheric continuous spectrum* the radial optical depth is of the order of unity in the low chromosphere. So, if scattering may be neglected, the observed radiation temperature yields the local electron temperature. A reasonable approximation is that the emitted radiation emerges from $\tau_\lambda \approx \frac{2}{3}$, thus from a layer somewhere in the low chromosphere. If scattering plays a rôle, the source function $S(\tau_\lambda)$ is not equal to the black-body radiation $B(\tau)$, generally it is smaller, down to $B/2$. The radiation temperatures given in the table on p. 119 are based on rocket observations of the ultraviolet solar spectrum: photographic spectral observations above 1800 Å and photon counter observations below 1500 Å.

[1] C. DE JAGER: Bull. astronom. Inst. Netherl. **13**, 275 (1957).
[2] B. E. J. PAGEL: Astrophys. Journ **125**, 298 (1957).
[3] B. E. J. PAGEL: Monthly Notices Roy. Astronom. Soc. London **116**, 608 (1956).

λ (Å)	T_{rad} (°)	Ref.	Remarks	Estimated height above photosphere (km)
2250	5000	1		— 100
2000	4500	1		0
1450	4050	2		450
1250	3900	2		600
1215	4050	3	upper limit	650
1050	5700	4	very uncertain	850

β) *The molecule bands in the Fraunhofer and chromospheric spectrum* were investigated by D.V. THOMAS[5] and PARKER[6] (chromosphere), GOLDBERG and MÜLLER[7], NEWKIRK[8] and PECKER[9] (Fraunhofer). The bands originate in the transition region photosphere-chromosphere. In the Fraunhofer spectrum the mean optical depth (centre of disk) is 0.07 to 0.13, with an appreciable chromospheric contribution[9]. In the flash spectrum the bands are observable up to $h = 600$ km.

The photospheric spectra yield excitation temperatures near the limb of 4300°[7] and of 4600°[8]; PECKER's investigation, based on some model computations yields after some trials the following provisional values:

h (km)	0	250	500
T (°)	4600	3900	4100
$\log P_g$	4.1	3.2	2.4

The flash-spectrum observations of the CN band at 3850 Å lead to $T_{ex} = 7000°$ at $h \approx 0$ and 4500° at $h < 600$ km[5]. For the C_2 band[6] at $h \approx 400$ km $T_{rot} = 4600° \pm 400°$.

γ) *The continuous emission close to the Sun's limb* measured at the 1952 eclipse (footnote 5, p. 117) for 4700 and 3640 Å at $h = 1000$ km can be derived from a representation of the observations by a series of exponentials. Assuming that the Balmer continuum is due to free-bound transitions, we have for the source function (cf. [10]):

$$\mathcal{E}_{3640} = 2.63\ 10^{-33}\, N_p\, N_e\, T_{el}^{-\frac{3}{2}}. \tag{19.1}$$

We assume that the continuous radiation at 4700 Å is due to scattering of photospheric radiation by free electrons. This assumption is justified for the particle concentrations at $h = 1000$ km and for $T_{el} = 5000°$; it is no longer valid for a temperature near 20000° (cf. [10]). With this assumption the source function is:

$$\mathcal{E}_{4700} = \frac{4\pi\,\sigma_e\, F_{ph}\, N_e}{2} = 9.4\ 10^{-29}\, N_e \tag{19.2}$$

with

$$\sigma_e = \frac{8}{3}\left(\frac{e^2}{m\,c^2}\right)^2.$$

F_{ph} is the photospheric radiation flux at 4700 Å.

If we assume that $N_e = N_p$, then N_e can be solved from (19.2) and T_{el} from (19.1). It is a difficulty that the H^- emission is not wholly negligible so that the solution is obtained

[1] F. S. JOHNSON, J. D. PURCELL and R. TOUSEY: Bull. Amer. Phys. Soc. **29**, (4) (1954).
[2] H. FRIEDMAN: Ann. Géophys. **11** (1955).
[3] E. J. BYRAM, T. CHUBB, H. FRIEDMAN and N. GAILAR: Phys. Rev. **91**, 1278 (1953).
[4] R. TOUSEY, J. D. PURCELL and K. WATENABE: Phys. Rev. **83**, 792 (1951).
[5] D. V. THOMAS: Monthly Notices Roy. Astronom. Soc. London **118**, 458 (1958).
[6] J. M. PARKER: Astrophys. Journ. **121**, 731 (1955).
[7] L. GOLDBERG and E. A. MÜLLER: Astrophys. Journ **118**, 397 (1953).
[8] G. NEWKIRK: Astrophys. Journ. **125**, 571 (1957).
[9] J. C. PECKER: in: Les molécules dans les astres. Liège colloquium, 1956, p. 332, 1957.
[10] H. ZANSTRA: Circ. Amsterd. **1** (1950).

by iteration. Two possible solutions are[1]

$\log N_e$	11.43	11.54
T_{el}	6100°	6400°

for $h = 1000$ km.

δ) The width of the Balmer line profiles is another source of information. On spectra, obtained by REDMAN at the eclipse of 1952, the halfwidths of the lines $H_{2-7} \ldots H_{2-31}$ were measured by REDMAN and SUEMOTO[2]. The width, measured on three spectra taken between 600 and 1200 km decreases with increasing n; for $n > 14$ it increases again. The first effect is due to the decreasing influence of self-absorption, the latter to the increasing influence of the Stark effect. The Doppler widening is negligible for these lines. Adopting some N_e values (it was assumed that $\log N_e = 11.0$, 11.5 and 12.0 respectively) the width of the emission coefficient profiles can be computed. To obtain from this the width of the actual line profile necessitates an evaluation of the effect of self-absorption. The ratio of optical depths τ_0 in the centres of the various lines can be computed. The absolute values of τ_0 can be derived with the aid of an estimate of τ_0 for H_{2-30}, a line virtually free from self-absorption. It appears that self-absorption becomes important ($\tau_0 > 0.3$) for $n < 18$. For a line like H_{2-7}, $\tau_0 \approx 10$! This result is in agreement with a finding of ATHAY and THOMAS[3] that at $h = 1000$ km the optical depth of H_{11} is 2.2. For the case treated by REDMAN and SUEMOTO the optical depth of H_{11} would be 2.2 if $T_{el} = 8000°$ and $\log N_e = 11.5$.

A comparison of the halfwidths computed under these assumptions with the observed values yields the conclusion that the observations can be explained nearly as well with $T_{el} \approx 10000°$ or with 6000° (for $\log N_e = 11.5$, in accordance with the N_e value found in the preceding paragraph). The higher temperature may still be the better of the two. This temperature refers to heights between 600 and 1200 km. The importance of this determination is that use is made of emission line observations without necessarily having to treat the difficult problem of deviations from the equilibrium which occurs nearly always in discussions of the chromospheric line emission.

ε) Conclusions from the high members of the Balmer series. Now that we know (i) that the high members of the Balmer series are free of self-absorption, and since (ii) it seems permissible to assume that for these high series members the departures from thermodynamic equilibrium are small[4] (thus assuming $b_n \approx 1$ for n greater than—say—20), we may use the observed intensities of the high Balmer series members to find N_e and T.

The observations[5] give the total intensity in the lines for a slice of the chromosphere 1 cm thick, bordered at one side by the Moon's limb. Assuming an exponential distribution of $E_n(h)$, $\mathcal{E}_n(h)$ may be found (the subscripts refer to the number of the line in the Balmer series). Further

$$\mathcal{E}_n = N_n A_{2n} h\nu$$

where N_n is determined by BOLTZMANN's and SAHA's law, hence by T_{el} and N_e. Applying this to the lines H_{20}, H_{24} and H_{28} we find the following results, expressed as a relation between T_{el} and $\log N_e$. The table contains $\log N_e$ for three heights and four values of T_{el}.

	T_{el}			
	7500°	10000°	12500°	15000°
$h = 1070$ km	11.54	11.66	11.70	11.65
1750	11.24	11.35	11.38	11.29
2400	10.94	11.05	11.08	10.95

[1] R. G. ATHAY, D. H. MENZEL, J. C. PECKER and R. N. THOMAS: Astrophys. Journ. Suppl. series **1**, 505 (1955).

[2] R. O. REDMAN and Z. SUEMOTO: Monthly Notices Roy. Astronom. Soc. London **114**, 524 (1954).

[3] R. G. ATHAY and R. N. THOMAS: Astrophys. Journ. Suppl. series **1**, 491 (1955).

[4] S. MATSUSHIMA: Astrophys. Journ. **115**, 544 (1952).

[5] R. G. ATHAY, D. E. BILLINGS, J. W. EVANS and W. O. ROBERTS: Astrophys. Journ. **120**, 94 (1954).

This result shows that the observed Balmer line intensities do hardly permit a temperature determination, but that they indicate log N_e values of 11.6, 11.3 and 11.0 at $h=1070$, 1750 and 2400 km, in agreement with the value found in subsection γ above.

ζ) *The low Balmer lines* with their great self-absorption offer greater difficulties, but the problem of self-absorption can be solved by constructing curves of growth for the Balmer lines (BÖHM-VITENSE[1]). The amount of self-absorption gives N_2, the number of second quantum state hydrogen atoms. On the other hand the total number of hydrogen atoms (N_H) at the same height can be found from the observed number of Ti^+ ions. This latter number (N_{Ti^+}) can be determined observationally from the strengths of the Ti II lines, using the absolute f-values. Since virtually all Ti is Ti^+, N_H is found from N_{Ti^+} with the known relative abundance of Ti.

It is at this point that this otherwise very elegant treatment may be criticized, since it is not certain that the Ti^+ and H lines are emitted by the same kind of elements. The H lines may be emitted by the hot elements and the metal lines by the cold ones.

A comparison of N_2 with N_H gives the excitation temperature of the second level of hydrogen (values near 6000°) and this value may be converted into electron temperatures with the aid of b_n computations[2]. So we obtain[3]:

	$h=$	
	950	1780
T_{el} from T_{ex}	8500°	8500°
T_{el} from T_{ion}	9300°	9200°

Here also the Cillié-Menzel scale of heights has been corrected by adding 280 km.

The problem has been discussed thoroughly by ATHAY and THOMAS[4]. The parts of the chromosphere where the cores of the Balmer lines are emitted are opaque to the Lyman lines. Under these circumstances the observed excitation temperatures of the Balmer lines (3700 to 4500°) are explained by electron temperatures between 10000 and 20000°.

η) *Combination of observations of Ti II and Fe I lines* may yield the ionization temperature of Fe (l. c.). BÖHM-VITENSE's computation is based on the complete ionization of Ti and Fe. From N_{Ti^+} we obtain N_{Fe^+} with the known abundance ratio, whereas observations of Fe I lines yield N_{Fe}. With the N_e values assumed to be known from other sources the ionization temperature T_{ion}(Fe) is obtained. It is not possible to find this value using only the Fe I and Fe II lines, since absolute f values are not known for Fe II. The ionization temperature of Fe turns out to be virtually constant with height, equal to 4900° between 0 and 2000 km. There is as yet no way to transform this value into electron temperatures, but for the time being we are inclined to assume that the differences between T_{ion}(Fe) and T_{el} are not great for this complex atom with its low ionization potential.

ϑ) *The other metal lines* have been subject of an extensive quantitative discussion by ZIRKER[5] of the HAO slitless spectra and REDMAN's slit spectra, both obtained at the eclipse of 1952. It appeared that the homogeneous chromospheric model of ATHAY, MENZEL, PECKER and THOMAS fails to predict the metal lines satisfactorily, on the basis of a local thermodynamic equilibrium (L.T.E.) computation. The predicted emission is too low. Similar difficulties appear when the line intensity is computed with the two-element model derived by ATHAY and MENZEL from the helium lines (with, in the low chromosphere, cold elements at $6000° < T_{el} < 11000°$ and hot elements at $15000° < T_{el} < 40000°$). These negative results were thought to be an indication that the assumption of L.T.E. was not correct.

Next, in an observational approach, the effect of self-absorption was eliminated by a curve-of-growth method. The ionization- and excitation-temperatures derived from the various metal lines may differ but are generally in the range 5000 to 6000° thus confirming the results of BÖHM-VITENSE. The corresponding electron temperatures do not seem to deviate much from these values.

---

[1] E. BÖHM-VITENSE: Z. Astrophys. **36**, 145 (1955).

[2] It is true that GIOVANELLI's early b_n-values, which were used for this conversion, are unreliable. But the error in his *excitation temperatures*, which depend on the *ratio* of successive b_n-values is not important for the actual degree of accuracy of chromospheric studies.

[3] C. DE JAGER: Bull. astronom. Inst. Netherl. **13**, 275 (1957).

[4] R. G. ATHAY and R. N. THOMAS: Astrophys. Journ. **127**, 96 (1958).

[5] J. B. ZIRKER: Technical Report High Alt. Obs. Boulder 1956.

ι) The double inversion in the central part of the H and K lines of Ca^+ in the disk centre (Fig. 17, Sect. 23) forms a special case that has often been studied. The K_3 absorption has been attributed to non-coherent scattering (MIYAMOTO[1]), but this cannot explain the observed asymmetry of the K_3 line. Since also the central absorption found by MUSTEL and TSAP[2] in the infrared Ca^+ triplet is displaced relatively to the bright reversal, these central line profiles should perhaps be explained by absorption by moving gases in the upper chromosphere.

The excitation mechanism of the bright (K_2) component must be electronic impact rather than recombination (MUSTEL[3]): the observed intensity is much too high for recombination whereas the impact mechanism fairly describes the observed profiles, if one takes $N_e = 5 \times 10^{11}\ \text{cm}^{-3}$ and $T_{el} = 5500°$.

Summary (see also Fig. 15). The data on the continuous radiation and the metal lines supply the arguments for the existence of cold elements in the low

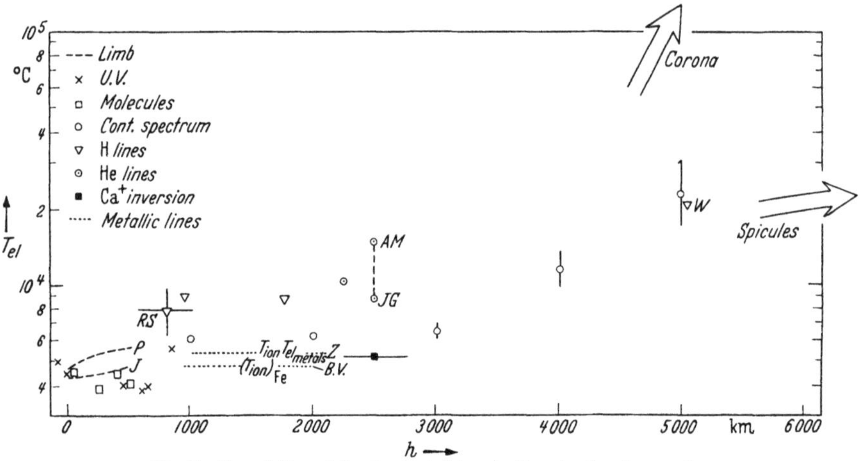

Fig. 15. Compilation of temperature determinations in the chromosphere.

chromosphere, with T_{el} increasing from 4000 to 4500° at $h=0$ till about 6000° at 3000 km. The data on the hydrogen emission indicate the presence of hotter elements with T_{el} being about 10000° at 1500 km.

20. The medium chromosphere (1500 to 6000 km). There are only few determinations of temperature and density in the medium part of the chromosphere.

α) From the continuous radiation at 4700 and 3640 Å ATHAY, MENZEL, PECKER and THOMAS[4] found, assuming that the radiation at 4700° is due to electron scattering ($\propto N_e$) and that at 3640 Å to free-bound transitions ($\propto N_e^2 T_e^{\frac{3}{2}}$):

h (km)	2000	3000	4000	5000
$\log N_e$	10.94	10.53	10.28	10.08
T_e	6400	6400	10000	17800
		7000	13600	31600

Near 6000 km the temperatures are uncertain since the observations of ATHAY, BILLINGS, EVANS and ROBERTS at Khartoum show that the coronal lines have their maximum below 10000 km, perhaps at 8000 km, and that the corona might even exist as low as 6000 km. Apart from this problem there is the question how to interpret the above data in terms of a two-element model, e.g. the 4700 Å radiation may come from one kind of elements and the

[1] S. MIYAMOTO: Z. Astrophys. **31**, 282 (1954).
[2] E. R. MUSTEL and T. T. TSAP: Isw. Krymsk. Obs. **16**, 67 (1956).
[3] E. R. MUSTEL: Isw. Krymsk. Astrophys. Obs. **11**, 165 (1954).
[4] Cf. footnote 1, p. 120.

free-bound Balmer emission comes from the other element. This problem of the splitting of the chromosphere into two kinds of elements is discussed in Sect. 26.

β) In Sect. 17 we referred to MICHARD's, SMITH's and ATHAY's measurements of the widths of Hα at the base of the visible spicule region; there we assumed at $h = 9000$ km a kinetic temperature of $35000° \pm 15000°$ for single spicules.

γ) On the other hand WOLTJER[1] determined T_{el} from outside eclipse photometry of Hα. The value of $N_e^2 T_{el}^{-\frac{3}{2}}$ was determined from the high Balmer lines and the number N_2 of second quantum state hydrogen atoms was derived from a study of the observed Hα profiles. The observations show that N_2 is about 50 times larger in the spicules than between them. From $N_e^2 T_{el}^{-\frac{3}{2}}$ and N_2 the values of N_H, N_e and T_{el} for the spicules were derived using GIOVANELLI's tables. The mean temperature at 5000 km is 21000°, and Log $N_e^{sp} = 10.6$. WOLTJER's spicular temperature near 5000 km appears to agree well with the value found at and above the same height level by ATHAY and MENZEL[2] (see subsection δ below).

In view of the great uncertainties of the above described reduction the agreement with the measurements referred to in subsection β is not bad.

δ) *The helium data* observed at the flash spectrum at the Khartoum eclipse were used by ATHAY and MENZEL[2]. A comparison of the numbers of particles in the various levels, with their excitation potentials leads to a well-known fact: The triplet levels are overpopulated with respect to the singlets, and the *relative* distribution of atoms over the triplet states can be described by a lower excitation temperature (5000°) than that of the singlets (15000°). This is perhaps due to the metastability of the $2\,^3S$ level. A two-element model of the chromosphere was derived from a combination of the helium data with the continuum data at 3640 and 4700 Å. Between 1000 and 3000 km the temperature of the cold elements rises only slowly from about 6000 to 7000°, whereas the hot elements should have a temperature of about 19000° between 1500 and 4000 km. Above 4000 km the temperature of the inter-spicular matter rises steeply to emerge into the corona; the spicular matter keeps $T \approx 2 \times 10^4$ °K.

In a rediscussion of the helium problem based on a detailed treatment of the helium equilibrium DE JAGER and DE GROOT[3] could show that the intensities of all helium lines, could be accounted for with $T_{el} \approx 9000°$ and 20000° at $h = 2000$ and 4000 km respectively. These temperatures agree with other T_{el} determinations, based on the hydrogen lines. A similar result (12000° and 6000° for the hot and the cold elements of the medium chromosphere) was found by SHKLOVSKI and KONONOVITCH[4] from a rediscussion of the helium problem. The chromospheric ultra violet radiation is important for the helium excitation. So we are inclined to adopt a hot-element temperature of about 10^4 °K near $h = 2000$ km, and $2 \cdot 10^4$ °K near 4000 km, and to regard the hot elements as the sources of the helium and the hydrogen emission (cf. Sect. 19).

ε) *From the infrared helium line*[5,6], observed as an absorption line in the infrared Fraunhofer spectrum at 10832 Å, further information on the temperature and density of the medium chromosphere may be gathered. The equivalent width varies irregularly over the disk; it is strong at some places and disappears at others. Nevertheless the mean profile, derived for the undisturbed Sun is fairly constant. The equivalent width in the disk centre is 68 mÅ. The mean central depression is 10% and the width, rather constant over the disk is 0.19 Å, leading to a turbulent velocity component of 13.5 km/sec (assuming $T_{el} = 10000°$). This line cannot be formed in the cold regions of the chromosphere. An integration performed with VAN DE HULST's model (in which the temperature is about 5000° in the region where the line is presumably formed—2000 km above the photosphere) yields a central depression of 7×10^{-5}. So the line must be emitted by the hot regions. Knowing this, the line profile can be used without any theoretical supposition for estimating the area occupied by the hot elements. The observed profile, which has a central depression of 0.10 is a mean over hot and cold regions. Since the depression of a hot-element profile cannot be greater than 100%, at least 0.1 of the Sun must be covered by hot elements (in that case the line profiles of the hot elements would have a zero residual intensity). It is even possible to fix this value to within narrower limits: The profile is fairly well Dopplerian and the derived turbulent velocity component matches well with other determinations of this quantity which proves that the line profile of a single hot element cannot be very saturated and has a depression smaller than 0.4 or 0.5. So the *lower* limit of A, the area occupied by the hot elements, can be put at 0.20 or 0.25 at

[1] L. WOLTJER: Bull. astronom. Inst. Netherl. **12**, 165 (1954).
[2] R. G. ATHAY and D. H. MENZEL: Astrophys. Journ. **123**, 285 (1956).
[3] C. DE JAGER and B. H. M. DE GROOT: Bull. astronom. Inst. Netherl. **14**, 21 (1957).
[4] I. S. SHKLOVSKI and E. B. KONONOVITCH: Astron. Zhurn. **35**, 37 (1958).
[5] L. D'AZAMBUJA: Bull. Astronomique **11**, 358 (1939).
[6] O. C. MOHLER and L. GOLDBERG: Astrophys. Journ. **124**, 13 (1956).

$h = 2000$ km. Anticipating Sects. 26 and 27, we notice that radio data fix the *upper* limit of A at about the same values.

Next the observations will be used to estimate the temperature. We take the densities adopted by WOLTJER; the density has a much smaller influence on the line strength than the temperature, because of the extremely high excitation potential of the 2^3S level (≈ 20 eV). Assuming the b_n factors of Sect. 16 and the temperatures of WOLTJER's model a central depression is computed of 22%. This value refers to *one* element; in order to obtain the mean value it should be multiplied by A. By taking $A = 0.5$ we obtain agreement between the observed and the computed central intensities. It was shown above that $A \approx 0.25$; so, higher temperatures would be needed to predict the observed line depression. This inference is, however, counteracted by the fact that perhaps not all helium absorption takes place in the hot elements near 2000 km, but that a small part of it is produced in the cold elements at higher levels, where the cold-element temperature increases steeply to coronal values. Since a factor 2 in the depression has a very small influence on T this leads to the conclusions (a) that WOLTJER's temperatures are substantially correct at 2500 km where this line is mainly formed: $T = 10800°$, and (b) that $A \approx \frac{1}{4}$ at that height.

Summary (see Fig. 15). In the medium chromosphere the helium and hydrogen lines point to temperatures of 10^4 to 2×10^4 °K; the latter value is probably reached near a height of 4000 km. The data on the continuous emission indicate $T < 10^4$ °K for $h < 4000$ km; above 4000 km T rises steeply to coronal values. These data agree with those found in the preceding section and show that also in the medium chromosphere, at least below 4000 km, two elements occur: the hot ones emit H and He lines and the cold ones ($T \approx 6000$ to $7000°$) emit the continuous radiation and are responsible for the emission of the central parts of the Ca^+ lines.

Near 5000 km the situation is very complicated: the chromosphere merges into the corona near that height. Above 6000 km the hydrogen emitting elements —the spicules—are cold elements in a hot corona.

A further discussion of the temperature determinations described in this section and in Sect. 19 will be presented in Sect. 26.

b) The chromosphere on the disk.

21. Height of emission of the central parts of some Fraunhofer lines. Spectroheliograms made in the central parts of strong Fraunhofer lines offer a possibility to study the solar chromosphere projected on the disk: at these wavelengths the selective absorption coefficient is so great that the chromosphere is opaque for radiation of that wavelength. For an interpretation one should know the heights of the chromospheric layers that contribute mainly to the structures observed in spectroheliograms. These heights are of the same order as the height h^* in the chromosphere where the optical depth at the wavelength considered is unity. Here h^* is found from the equation:

$$\int_{h^*}^{\infty} k_{\Delta\lambda}(h)\, N(h)\, dh = 1$$

where $k_{\Delta\lambda}$ (cm²) is the absorption coefficient per particle at a wavelength distance $\Delta\lambda$ to the line centre, and N is the number of particles per cm³.

The accompanying table gives this value of h^* for the first four Balmer lines, for the K line of Ca^+, and for some other lines in which spectroheliograms have been made.

The "height of emission" h^*, given in the table, is computed in two different ways. For some lines it was found by a direct integration over the chromosphere, using a chromospheric model; for the others it was derived only for $\Delta\lambda = 0$ from the measured top heights z of the chromospheric emission lines above the Sun's limb, assuming that this top-height is the point where the optical depth, measured along a tangent of the Sun, is unity. We used the observed heights measured on *spectroheliograms*, since in these latter observations the top height z is the highest point where the central intensity of the line is still of the same order as the disk value. The tangential optical depth is about unity at the height where the intensity shows a deflection point, hence where the chromosphere changes from opacity into transparency. The "heights" of the lines measured by *eclipse observations* cannot be used for our purpose since they refer to points where the last trace of the line can be seen. There the line is optically thin.

Next we want to find from this given height z (when the *tangential* optical depth is unity) the height h^* in the chromosphere where the *radial* optical depth is unity. The following procedure was followed. Let us assume that for one particular line the scale height H of the chromosphere is constant for the considered line. In reasonable approximation h^* is the height where $N(h^*)\, k(h^*)\, H = 1$. When a line has a top height z, this means that $N(z)\, k(z)\, H_t = 1$, where H_t is the scale height in tangential direction: $H_t = \sqrt{2RH}$. Then h^* is the height where $N(h^*) = N(z)\, H_t/H$ and since $N(h) \propto \exp(-h/H)$ this yields:

$$h^* = z - 1.15\, H \,\mathrm{Log}\, \frac{2R}{H}\,.$$

For z we have mainly used the heights of chromospheric emission lines measured by D'Azambuja[1] on spectroheliograms. The scale heights of the chromospheric

Table 6. *The mean chromospheric heights of emission (h^*) of Fraunhofer lines.*

Line	$\Delta\lambda$	h^* ($\cdot 10^{-3}$ km) from model computations	h^* ($\cdot 10^{-3}$ km) from observed z
Hα	0.0	5.0[2]; 3.5[3]	5.0
	0.2	5.0	
	0.4	4.0	
	0.6	3.0	
	0.8	1.0	
Hβ	0.0	3.0[2]; 2.5[3]	
	0.2	2.5	
	0.4	1.0	
Hγ	0.0	2.0[2,3]	
	0.2	1.0	
Hδ	0.0	1.5[2]; 1.0[3]	
	0.2	0.8	
K(Ca$^+$)	0.0	3.9[4]; 4.5[5]	3.0
	0.2	3.2	
	0.3 (K2)	2.0	1.6
	0.6 (K1)	0.6	
	1.0	0.2	
	1.5	−0.1	

Element	λ	h^* ($\cdot 10^{-3}$ km) from z
Mg	3838	1.2
	5184	0.3
Fe	4202	0.1
	4384	0.4
Ca	4227	0.8
Ca$^+$	8542	1.4
	8498	
Na	5890	0.3
Sr$^+$	4078	0.9
		From model computations
He	10832	2.0
	5876	2.0

[1] L. D'Azambuja: Ann. Obs. Meudon **8**, II (1930).
[2] C. de Jager: Bull. astronom. Inst. Netherl. **13**, 133 (1957).
[3] R. G. Athay and R. N. Thomas: Astrophys. Journ. **127**, 96 (1958).
[4] Kindly communicated by Mr. L. D. de Feiter: NERA Obs. Holland.
[5] V. L. Khokhlova: Astron. Zhurn. **35** (1958).

lines are known from many eclipse observations; in our computations we have used VAN DE HULST's compilation (see Fig. 14).

Auxiliary table:

z ($\times 10^{-3}$ km)	0.5	1.0	1.5	2.5	5.0	8.0
h ($\times 10^{-3}$ km)	−0.1	0.3	0.7	1.4	3.0	5.0

Table 6 will serve in the interpretation of the spectroheliograms. It shows that most metal-line spectroheliograms refer to the low chromosphere, below 1500 km, and that only spectroheliograms obtained with the central parts of the Balmer lines, with the H and K lines and the infrared helium line refer to higher levels. For Hα, Hβ and the Ca$^+$ K line it is the whole central part that originates from the high chromosphere.

It should be kept in mind that spectroheliograms taken in two different lines cannot be compared directly, even if the radiation of both is emitted by the same layers. The aspects of the spectroheliograms can be affected considerably by Doppler motions of the lines. These motions hardly affect the line centre, but they do have a great influence on the intensities in the steep parts of the profiles.

Anticipating Sect. 23—cf. especially Fig. 19—we note that in spectroheliograms of the mean chromosphere taken with a high image resolving power *three types of elements* can be distinguished.

(a) The *small granules* (the "*fine mottling*"), diameters of 700 to 1600 km, are only observable with the highest resolving power.

(b) These granules may cluster together to form a *coarse mottle* with diameters between 2000 and 8000 km.

(c) These coarse mottles often are arranged in a *coarse network* with a characteristic length ("meshwidth") of 25000 to 50000 km.

These three structures occur in all kinds of spectroheliograms but with different relative importances. Furthermore an element can be dark in one kind of spectroheliogram and bright in another and these factors greatly determine the different aspects of spectroheliograms.

The following description of spectroheliograms will centre around the problem of the visibility of the three types of elements in the different parts of the chromosphere.

22. Spectroheliograms referring to the low chromosphere. In this section we discuss spectroheliograms concerning the lowest 1000 km of the chromosphere, viz. all metal line spectroheliograms with the exception of those of Ca$^+$. The most systematic series of observations is that of D'AZAMBUJA[1]; the observations are discussed by REICHEL[2], on suggestions advanced already earlier by UNSÖLD.

D'AZAMBUJA made a distinction between three classes of metal lines, those with well developed wings (class 1), those with faintly developed wings (class 3) and the intermediate class (2). The lines consist generally of a central core with a rather steep intensity gradient at $\Delta\lambda = \pm 3\Delta\lambda_D$ ($\Delta\lambda_D$ is the Doppler width), and a wing part. The "width" of the lines, measured in the spectrum, generally refers to the width of the central part.

There appear to be no fundamental differences between the undisturbed parts of spectroheliograms made in the cores of these lines. All these images show dark mottles on a brighter field; the diameters are 2000 to 6000 km. Their mutual distances are of the order of 10000 to 20000 km. Clearly we are dealing here

[1] L. D'AZAMBUJA: Ann. Obs. Meudon **8**, II (1930).

[2] M. REICHEL: Z. Astrophys. **33**, 79 (1954).

with the coarse mottling. That the diameters are smaller than those given at the end of section 21 proves that the contrast of these mottles is small.

The fine (granular) mottling has never been observed well on metal line spectroheliograms. However, the network is well visible.

So, when speaking of the visibility of the granulation on metal spectroheliograms, it is the visibility of the network and the coarse mottling that is principally meant. It is rather precarious to speak in this connection of "granulation" if it is not clearly stated that these granules are objects other than the photospheric ones.

The visibility of the coarse mottling as a function of the distance $\Delta\lambda$ to the line centre was estimated by D'AZAMBUJA in an arbitrary scale (0 ... 5) (Fig. 16). The general picture is that the visibility maxima occur at the two boundaries of the core, hence at the steep parts of the line. The visibility is a minimum in the centre (however, in nearly all cases, not zero!) and it decreases to zero in the wing part, at $\Delta\lambda \approx 5\Delta\lambda_D \dots 10\Delta\lambda_D$. An important detail is that the visibility curve is generally asymmetric: the violet maximum is more pronounced than the red one, the asymmetry is greatest for the strongest lines; it is faint or zero for the weaker ones.

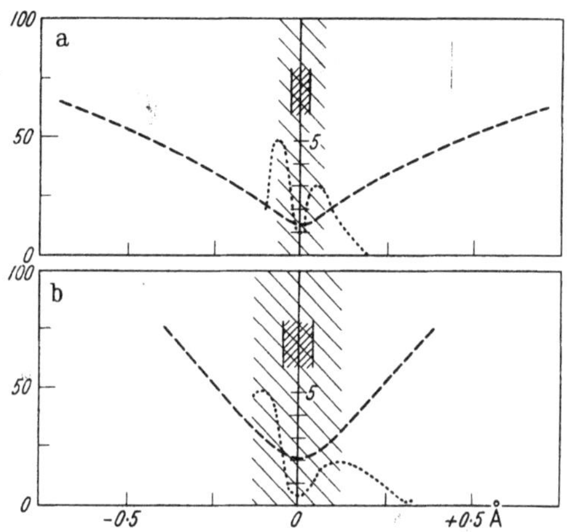

Fig. 16a and b. The visibility of chromospheric mottling on metal line spectroheliograms as a function of the wavelength distance to the line centre. (a) Ca 4227. (b) Na 5890. Dotted line: visibility curve in an arbitrary scale 1 ... 5. Dashed line: line profile. Shaded regions: central part of the line and width of the spectrograph slit.

That the maxima of the visibility curves occur in the steep parts of the line profiles suggests as a possible explanation that they are produced by Doppler effects in the mottling. Analogously one explains the reduced visibility for greater distances to the line centre. The *position* of the maximum depends nearly entirely on the form of the line profile; the *intensity* of the fluctuations depends also on the amount of the shift. If we assume, with REICHEL, that the distribution of the velocity components in the direction of vision is Maxwellian, $f(\xi) \propto \exp(-\xi/\xi_0)^2$, then the value of ξ_0 can be determined if the photometric value of the intensity fluctuations is known. An estimate of these latter values was made by D'AZAMBUJA from which REICHEL finds $\xi_0 = 0.4$ km/sec.

On the basis of the above described mechanism it would be expected that the intensity fluctuations are zero in the line centre, contrary to observations. This difference may have been caused by the fact that the selecting slit had a finite width in D'AZAMBUJA'S observations, and/or by the possibility that real intensity fluctuations do also occur in the line centre. Bright mottles in the line centre might be due to a local increase in the radiation temperature or to a local increase in the microturbulent velocity component. The latter effect makes the line core broader and shallower.

Also the asymmetry in the visibility curve of the mottling on metal line spectroheliograms can be accounted for. The phenomenon is explained by assuming that the ascending mottles (violet displaced lines) have a greater microturbulence than the descending ones: in that case the contrast is obviously greater

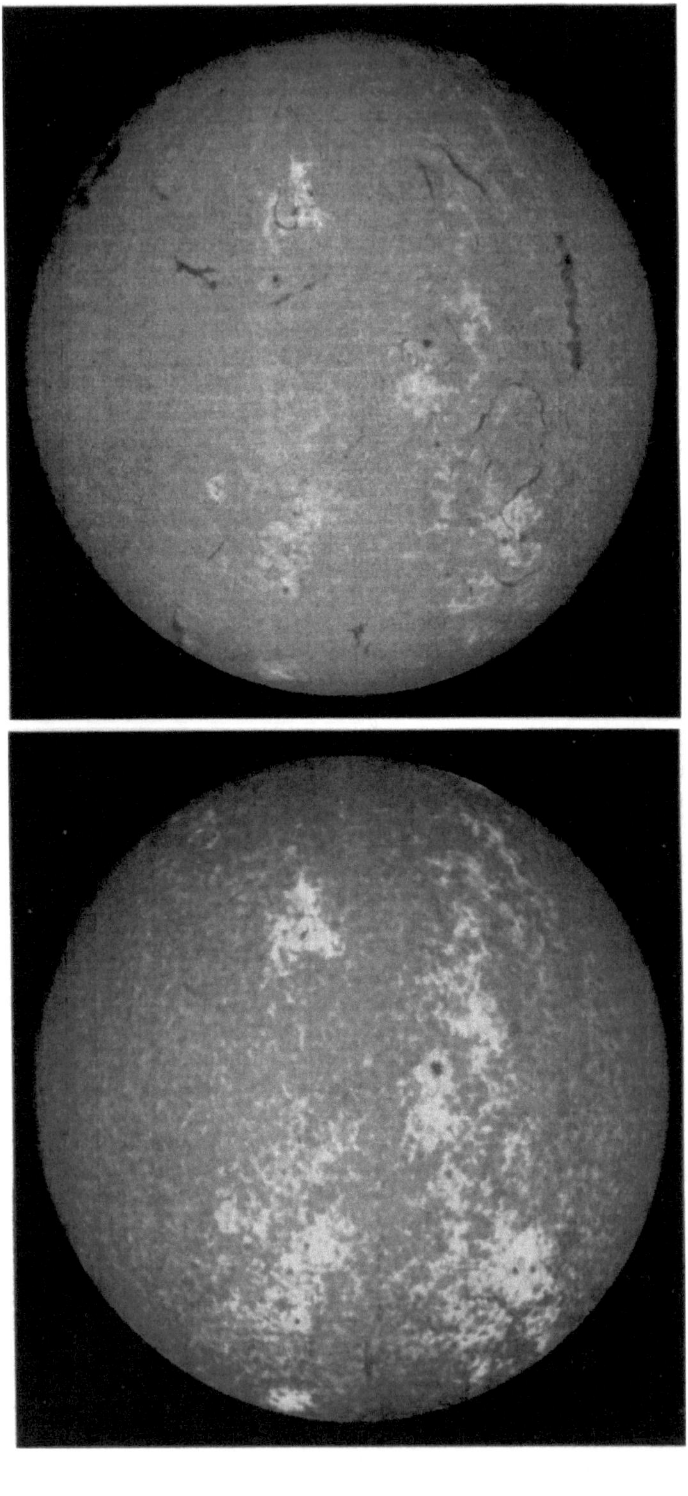

Fig. 17. The chromosphere on the disk. Left: K_{23} spectroheliogram of July, 2, 1957; 09.23 to 09.27 UT. Right: Hα_{23} spectroheliogram of the same day; 08.48 to 08.54 UT. North is at the top; East on the left side. Courtesy M. C. BALLARIO, Arcetri.

in the violet side of the lines. That the effect is less clearly visible in faint line spectroheliograms is due to the finite width of the selecting slit: the core profile of these narrow lines is partly washed out.

23. Spectroheliograms referring to the medium chromosphere. (Ca^+). It is principally the He and Ca^+ spectroheliograms and those made in $H\gamma$, $H\beta$ and the wings of $H\alpha$ ($\Delta\lambda > 0.5$ Å) that refer to the medium chromosphere (1000 km $< h < 4000$ km). The infrared helium spectroheliograms at 10832 Å made by D'Azambuja are a striking example of observational mastership, but the long exposure times (6^h!) have wiped out nearly all finer details of the images. Since furthermore the He 5876 spectroheliograms show only the excited regions of the Sun the present discussion, which refers only to the normal Sun, is largely restricted to a description of the Ca^+ spectroheliograms and in particular to those made in the K line. The $H\alpha$ wing spectroheliograms, which are also emitted by the medium chromosphere, are described mainly in the following Sect. 24 together with the spectroheliograms of the centre of $H\alpha$. However, a short comparison of Ca^+ and $H\alpha_{1,2}$ spectroheliograms is already given at the end of this section.

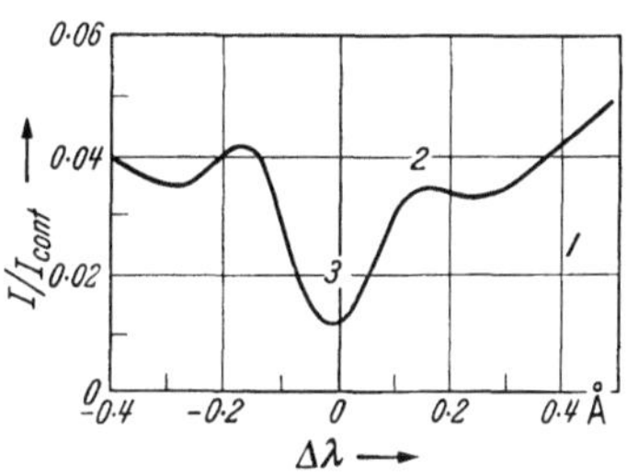

Fig. 18. Core profile of the K line of Ca^+. The figures *1*, *2* and *3* denote the position of the second slit in taking K line spectroheliograms. Courtesy L. Mohler, McMath Hulbert Obs.

The standard wavelengths at which daily observations are made at a number of observatories are designated by the subscripts 1, 2 and 3 (Fig. 17, and the table in Sect. 21). Fig. 17 shows the characteristic aspect of the H and K lines: the central depression (down to 1.5 ± 0.3% of the adjacent continuum), surrounded by two emission components; the intensity of the violet component being nearly always stronger than that of the red component. According to Goldberg and Müller[1] the absolute intensities of the emission peaks fluctuate from point to point; occasionally the red component may even be the strongest.

The K_1 spectroheliograms, which refer to the very basis of the chromosphere, do not show details other than the normal metal line spectroheliograms (Sect. 24). The coarse mottling is well visible as bright elements of 5000 km diameter, as well as the coarse network, formed by these mottles. In K_2 this mottling and its network are very pronounced; there the halfwidth of the coarse mottling is perhaps 8000 km. Photospheric details such as small sunspots are already covered by the higher absorbing layers of the chromosphere. Still higher up, in spectroheliograms made in $K_{2,3}$ and K_3 it appears that the 8000 km mottling in turn consists of many smaller ones with diameters of the order of 1″ (700 km); they are visible only on spectroheliograms made in the best atmospheric (and instrumental) conditions (Hale and Elleman, 1903; D'Azambuja and Deslandres,

	Estimated distance between two maxima (km)	Estimated diameters of elements (km)	Visible in
Fine mottling	1000 (in coarse mottles) up to 6000 (outside c.m.)	700 to 1600	$K_{2,3}$
Coarse mottling	10000 to 20000	2000 to 8000	} metals
Network	50000		} Ca^+

[1] L. Goldberg and E. Müller: Astrophys. Journ. **1958**.

1910). So the K_3 spectroheliograms (see Fig. 19) are the best suitable to show at the same time the three characteristic chromospheric elements to which we referred at the end of Sect. 21 (the values in the table are rough estimates; no quantitative analysis being as yet performed). The coarse mottling is identical with the "bright Ca^+ flocculi" often found in the older literature.

The mean brightness fluctuation on K_{232} spectroheliograms has been determined by ROGERSON[1] on Mt. Wilson spectroheliograms: $\overline{(\Delta I^2)}^{\frac{1}{2}}=0.107$ and 0.124

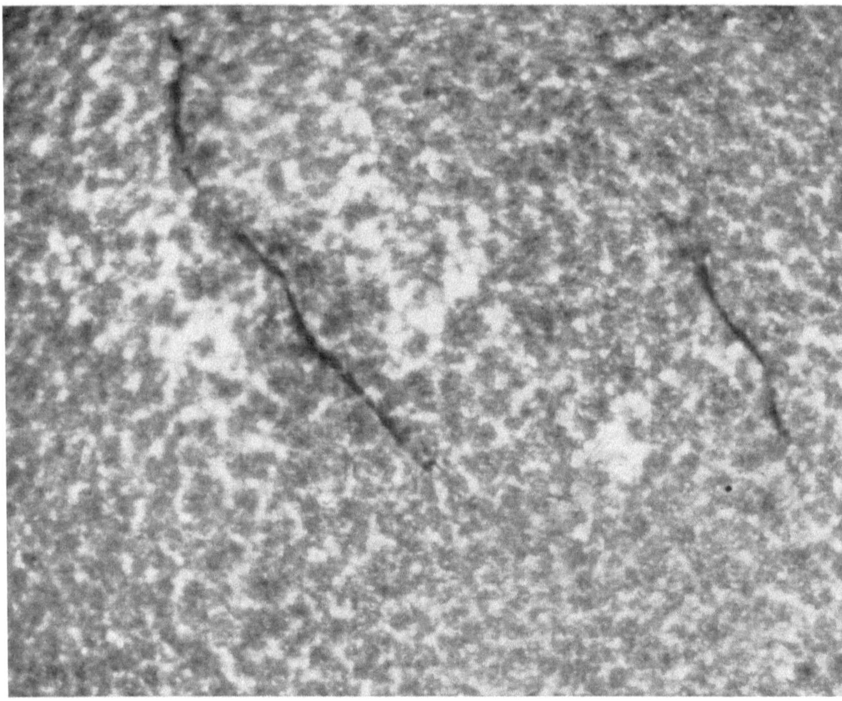

Fig. 19. Part of a K_3 spectroheliogram obtained May 25, 1922 with high geometric resolution. The image shows the chromospheric granulae (fine mottles), the coarse mottling and the network, both in the quiet chromosphere and in a faint facular region. Slitwidth 0.14 Å. Scale: 1 mm = 6700 km. North is at the top, West is on the right side. Courtesy L. D'AZAMBUJA, Meudon.

in two studied cases. The intensity fluctuations are largely confined to elements in the range 25000 to 70000 km, a conclusion that is also obvious from a quick glance at Ca^+ spectroheliograms.

The coarse mottling and the network, being also observed in K_1 and in metal line spectroheliograms and in the photospheric granulation spectrum occur at the basis of the chromosphere and in the high photosphere. The fine mottling, although having the same sizes as the photospheric granules, should not be identified with the latter, since the fine mottling seems not to be observed in spectroheliograms made in lines emitted below 1000 km. We infer from this observation that the fine mottles originate above 1000 or 2000 km; an observation that should perhaps suggest that a splitting of the chromosphere into two kinds of elements occurs somewhere near or above 1000 km. This suggestion is supported by the lack of fine structure in the Hδ line[2] (centre of disk).

[1] J. B. ROGERSON: Astrophys. Journ. **121**, 204 (1955).
[2] E. H. SCHRÖTER: Z. Astrophys. **45**, 68 (1958).

The fine K_3 mottles cluster together in the coarse mottling, but they occur also as more isolated bright points in the space between. We estimate that about 30% of the disk is covered with the fine mottles, but such an estimate, subjective as it is, should necessarily be replaced by a quantitative photometric analysis. It is curious enough that more than sixty years after the invention of the spectroheliograph such an analysis has not yet been made.

The life times of the coarse Ca^+ mottles of the undisturbed Sun are between 6 and 14 hours[1]; the network has a greater life time.

The explanation of the Ca^+ spectroheliograms has not very much advanced. The difficulties of a quantitative explanation of the central part of these resonance line profiles are enormous. The complicated chromospheric structure is the proper part of the problem. Furthermore, deviations from thermodynamic equilibrium and problems of redistribution in the line core are certainly important. The problem can perhaps be separated into some sub-problems. First of all there is the question whether the Ca^+ line profile with its double inversion is the real profile of *one* kind of chromospheric elements. It seems (MUSTEL, footnote 2, p. 122) that the K_3 absorption line is due to absorption in other gases than those emitting K_2. Next, the question arises what a fine bright mottle is. Is it a spicule seen from above? The following difficulty is to explain why these fine mottles cluster and show density fluctuations with the observed characteristic lengths.

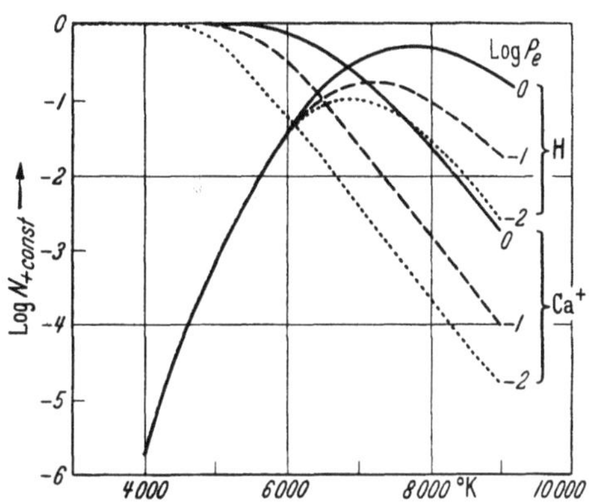

Fig. 20. The relative population of the second level of hydrogen and that of the ground level of Ca^+, as functions of the temperature and electron-pressure, computed for local thermodynamical equilibrium.

Anticipating the description of Hα spectroheliograms in Sect. 24 we notice an important difference between spectroheliograms in Ca^+ and those in H$\alpha_{1,2}$ (emitted by the same layers). The mottles are bright in K_3 whereas they are dark in Hα_2 and Hα_1 (cf. Fig. 18). This can be explained by assuming that the Ca^+ and Hα_{12} structures are formed mainly in the region where the ionization temperatures is near 6000°, (this temperature was derived for chromospheric metals and for H by Mrs. BÖHM-VITENSE) and that the structures are due to local temperature variations. For $T \approx 6000°$ the values of dN/dT have inverse signs (Fig. 20). In both cases (Hα and Ca^+) the mottles should have higher ionization temperatures than their surroundings.

A similar conclusion (higher T in bright Ca^+ mottles) is obtained by MUSTEL[2] who further ascribes the difference in aspect of Ca^+ and Hα spectroheliograms—the first being mottled, the latter being more fibrous—to differences in the excitation conditions. The central emission in Hα is due to recombinations ($\propto N_e^2$), that in Ca^+ to electron impacts (depending mainly on T_{el}) so that fibrils and vortices on Hα pictures might be regions of greater density (cf. Fig. 37) and the mottles on Hα and K_2 (Fig. 18) are regions of higher temperatures.

24. Spectroheliograms of the chromosphere above 3000 km. Spectroheliograms made in the core of Hα ($\Delta\lambda < 0.6$ Å) are emitted by layers above 3000 km, those made at $\Delta\lambda > 0.6$ Å are formed in the lower regions and show in broad

[1] C. MACRIS: Rend. Accad. naz. Lincei Roma, Ser. VIII **21**, 303 (1956).
[2] E. R. MUSTEL: Isw. Krymsk. Astrophys. Obs. **9**, 25 (1952); **11**, 102 (1954).

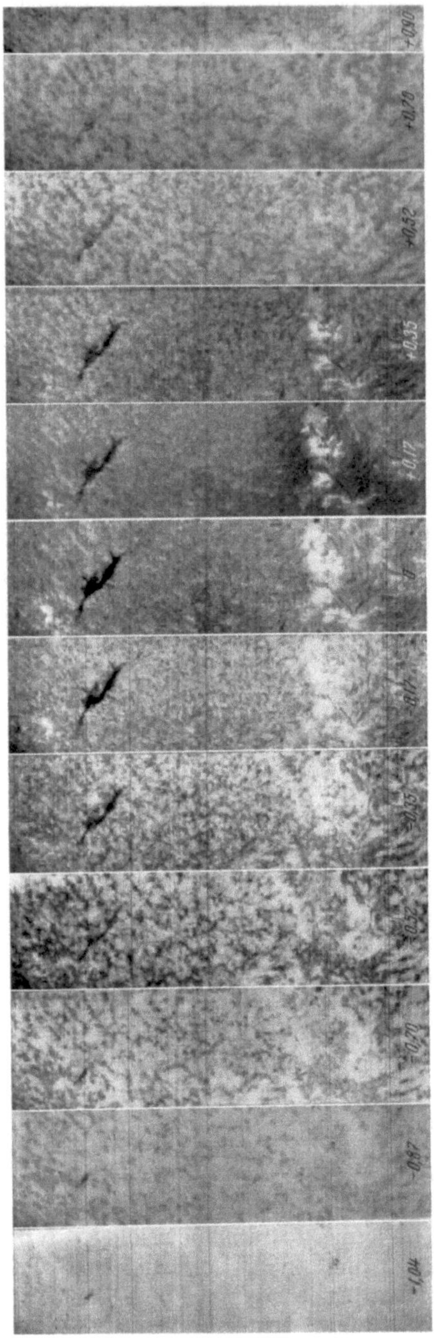

Fig. 21a. Parts of Hα spectroheliograms obtained July 26, 1949 in different parts of the line profile; the wavelength differences with respect to the line centre (Å) are written on the lower parts of each strip. The Sun's limb is in the upper left corner, the centre is below. Note the variation with wavelength and the different aspects of (a) the quiet Sun's mottling, (b) a filament with one ascending part, $v \approx 40$ km (upper parts of the strip), (c) a new centre of activity (at about ¼ of the height from below), and, faintly visible, somewhat below it (d) an older centre of activity. Courtesy M. D'AZAMBUJA, Meudon Observatory.

outline the same phenomena as the spectroheliograms described in the foregoing section.

An inspection of Hα spectroheliograms, made in various parts of the line profile, shows that in the central part of the line the fine mottling dominates, whereas in the steep parts of the profile a coarse mottling is the most spectacular structure visible. Both kinds of mottles can, however, be traced throughout the whole line profile (Fig. 21). For a detailed description cf. DESLANDRES[1] and DE JAGER[2].

The intensity fluctuations in the line centre (centre of disk) have been measured by ROGERSON[3], DE JAGER[2] and DE JAGER and SERVAGEAN[4]. The measurements are difficult and easily lead to too small values because of the smoothing effects of the atmosphere and the instrument which tends to enlarge the structures and to reduce their contrast; the table below gives $\overline{(\Delta I^2)}^{\frac{1}{2}}$ in the disk centre and line centre:

ROGERSON[3]		DE JAGER[2], 1953		DE JAGER and SERVAGEAN[4]
solar image diameter: 7 cm	167 mm	uncorrected	corrected	uncorrected
0.026	0.039	0.050	0.09	0.047

DE JAGER'S 1953 results obtained from a selected high resolution plate are perhaps the best. The fluctuation of $\overline{(\Delta I^2)}^{\frac{1}{2}}$ through the line profile, both for disk centre and for the mean of the north and south limbs ($r/R = 0.95$) is shown in Fig. 21c (cf.[4]);

[1] H. DESLANDRES: Ann. Obs. Meudon **4**, (1) (1910).
[2] C. DE JAGER: Bull. astronom. Inst. Netherl. **13**, 133 (1957).
[3] J. B. ROGERSON: Astrophys. Journ. **121**, 204 (1955).
[4] C. DE JAGER and R. SERVAGEAN: Bull. astronom. Inst. Netherl. **1959**.

the maxima at $\Delta\lambda = \pm 0.5$ Å are clearly visible especially on the limb-spectroheliograms.

The fine mottles, visible mainly in the core of the Hα line appear as a rule dark against a brighter background, but a smaller number of bright mottles do also occur. The properties of the dark fine mottles may be summarized as follows (DE JAGER[1]): in the line centre we observe mainly the fine mottling, but the diameters are somewhat greater than for the fine mottling in K_3 (1700 km or greater, perhaps partly caused by lack of resolving power) and their number is smaller. About 20% or less of the solar image is occupied by the fine mottles, if a diameter of 2000 km is assumed; if their diameters are taken equal to 1000 km, those for the K_3-fine mottles, then only 5% of the disk would be covered by them. The dark fine mottles are discernable up to $\Delta\lambda = 1.0$ Å and perhaps further, with a higher contrast at the red side of the line. Their life times are of the order of 5 min (also MACRIS[2] and SCHRÖTER (footnote 2, p. 130) found life times of 2 to 10 min, whereas KIEPENHEUER[3] reports longer life times); it is tempting to identify them with the spicules seen from above. The distribution of the number of fine mottles as a function of $\Delta\lambda$ depends on the spicule absorption line profile and the wavelength distribution of the centres of these profiles due to Doppler shifts. Each spicule produces an absorption line with a rather low radiation temperature in its centre. This line profile is superposed on the profile produced by the interspicular matter. Because the interspicular matter is greatly transparent in Hα this latter profile is the photospheric or low chromospheric one. Assuming that the distribution of the vertical velocity components ξ is Gaussian and symmetrical around $\Delta\lambda = 0$, the observed distribution of counted spicule numbers with $\Delta\lambda$ can be explained by taking $\sqrt{\overline{\xi_0^2}} = 18$ km/sec, which is only slightly smaller than the outward velocity of the spicules of 20 to 30 km/sec, measured, however, some 2000 to 5000 km higher.

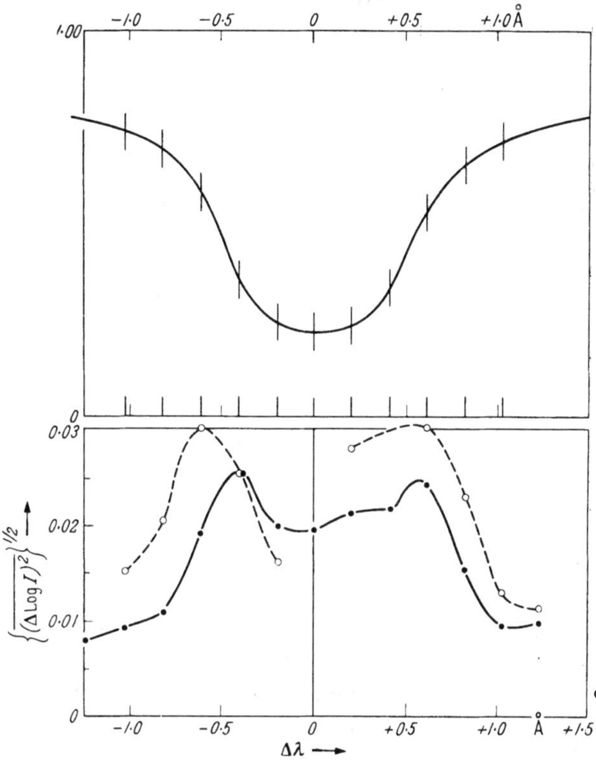

Fig. 21 b and c. (b) Line profile of Hα; vertical lines refer to the points in Fig. 21 c. (c) Root mean square of the intensity fluctuations, on Hα spectroheliograms as a function of $\Delta\lambda$. Solid line: centre of disk. Dashed: $r/R = 0.95$. [C. DE JAGER and R. SERVAGEAN: Bull. astronom. Inst. Netherl. **1959**.]

Also *the coarse mottling* is visible even in the line centre where it appears as a bright network of fine mottles (life time $\approx 1^d$ or sometimes even longer).

[1] See footnote 2, p. 132.
[2] C. MACRIS: Rend. Accad. naz. Lincei Roma, Ser. VIII **21**, 419 (1956).
[3] K. O. KIEPENHEUER: Z. Astrophys. **42**, 209 (1957).

In the steep parts of the line profile the coarse mottling appears pronounced. The two kinds of coarse elements found on the Ca^+ spectroheliograms are also visible on Hα pictures, but, the general aspect of Hα spectroheliograms is less coarse than the K pictures owing to the fact that Hα spectroheliograms show mainly the 8000 km elements; the network is less developed, but it is visible.

KIEPENHEUER[1] has described the features observable on Hα filtergrams (=pictures taken through a birefringent filter) of the quiet Sun. His description agrees with that given above with the exception that the life time of the smallest details is estimated to be of the order of one day, contrary to the above description where the fine mottles have life times of 5 min whereas the coarse mottling (being composed of many fine mottles) has life times of about one day.

It is clear that the greater passband of a filter (0.5—0.7 Å and a number of "sidelobes") may produce some confusion, especially when comparing centre and limb: the fine mottles still visible in the disk centre are near the limb somewhat surpassed by the greater contrast of the coarse mottles. Also the sign-inversion of the coarse mottling in the Hα-centre is invisible on a filtergram. On the other hand a filtergram has the advantage of a short exposure time and hence, in principle, a better image resolving power than a spectroheliograph.

The maximum of visibility of the coarse mottling in Hα occurs near $\Delta\lambda =$ 0.5 Å, the red maximum being more pronounced than the violet one (Fig. 21). This can be explained by another observation, viz. that there is a correlation between a local line widening and a local line displacement. When the line is displaced towards the red it is widened; it is narrowed when it is violet-displaced[2]. So if the displacement is interpreted in terms of ascending and descending motions, the descending gas masses produce wider lines than the ascending ones. The r.m.s. value of the line widening is 0.065 Å, the r.m.s. of the line displacement corresponds with a r.m.s. velocity component of 1.14 km/sec (DE JAGER); SCHRÖTER[2] finds 0.85 km/sec.

This result should be compared with other observations of the line shifts in the coarse mottling.

1. Observations of the displacements of Hβ by GOLDBERG, MOHLER and BROWN[3] refer to elements which are 2000 km wide, and yield $(\overline{\xi^2})^{\frac{1}{2}} = 0.64$ km/sec; SCHRÖTER[2] finds for Hβ and Hγ: 0.77 and 0.49 km/sec respectively.

2. In Sect. 22 reference was made to REICHEL's result derived from D'AZAMBUJA's metal line spectroheliograms.

3. We further mention HART's result (Sect. 4). From RICHARDSON and SCHWARZSCHILD's granular radial velocity observations (cf. Sect. 4) one derives for the r.m.s. velocity component of the 15000 km elements: $(\xi_0)_{15\,000} = \sqrt{0.18} \times$ 0.37 km/sec = 0.16 km/sec. (Here 0.18 is the relative autocorrelation for the 15000 km elements and 0.37 is the measured r.m.s. velocity component.)

Height		$(\overline{\xi^2})^{\frac{1}{2}}$
−100 km	SCHWARZSCHILD and RICHARDSON; HART	0.16 km/sec
+500	metal line spectra; REICHEL	0.40
1000	$H\gamma$ spectra; SCHRÖTER	0.49
2000	Hβ spectra; GOLDBERG, *et al.*; SCHRÖTER	0.64; 0.77
4000	Hα spectra; DE JAGER; SCHRÖTER	1.14; 0.85

The table shows the increase with height of the mean upward velocity component of the coarse mottling. We are therefore inclined to interpret the coarse mottling

[1] K. O. KIEPENHEUER: Z. Astrophys. **42**, 209 (1957).
[2] Footnote 2, p. 132 and footnote 2, p. 130.
[3] L. GOLDBERG, O. C. MOHLER and J. D. BROWN: Nature, Lond. **179**, 369 (1957).

and its network as enormous ascending and descending elements in the solar chromosphere, composed of many fine mottles ("spicules"). The ascending motions occur already in the high photospheric levels and they persist up to a height of at least 5000 km. with a mean velocity component continually increasing with height.

The characteristic profiles of the coarse mottling in Hα (broader profiles of ascending elements) might be explained either by a local increase of the microturbulence in the individual spicules (the mean value supposed to be 15 km/sec might change by ± 5 km/sec), or by a local variation of the excitation temperature; the mean value, if supposed to be 5000° might change by $\pm 100°$ (cf. footnote 2, p. 132). According to SCHRÖTER (footnote 2, p. 132) the second possibility is more probable.

c) Structure and dynamics of the chromosphere; transition to the corona.

25. The spicules; diameters, numbers, motions and lifetimes. The spicules of the upper chromosphere were for the first time described by SECCHI. Their mean heights are 7500 km at the equator and 7800 km at the pole, according to the daily Arcetri measures. It is 7300 and 9500 km respectively according to counts made on Sacramento Peak films (LIPPINCOTT[1]). Their number is 30% higher at the poles than at the equator[2], a frequency minimum occurs at $b=35°$, where the number is 10% less than at the equator.

There are definition differences between various authors: ROBERTS, who introduced the word, referred only to the fairly large active spikes (lengths 4500 to 8000 km above the "normal" chromosphere) that appear most frequently in the polar regions. MOHLER used it in the same way. MENZEL and EVANS[3] describe them as cores of luminous gas travelling out at high speed to a limb distance of perhaps 15000 km, where they fade into invisibility with no trace of slowing down. At any one moment 50 or more of these spicules are visible along the solar limb. This description refers obviously to the longest specimens, much ressembling small quasi-eruptive prominences. In exceptional cases they may be very long and then they can hardly be distinguished from prominences or surges. At the eclipse of 25 February 1952 WALDMEIER observed a thin polar "spicule" of 80000 km. In the three minutes of totality the change of height was well visible. It had the same inclination as the neighbouring polar rays. As a rule such long spicules are only observed within 30° of the poles (LIPPINCOTT[1]).

Other definitions refer also to the smaller ones of which a much greater number is visible all around the limb. In this review we will use the word "spicule" both for the long specimens as well as for the small spike-like general structure of the chromosphere, and in theoretical discussions the definition will even be extended to regions below 5000 km, tentatively identifying one kind of the low-chromospheric elements with spicules.

The spicules often show strong inclinations up to an angle of 30% with the Sun's limb. Fig. 23, after LIPPINCOTT[1] is in substantial agreement with an earlier finding by BUGOSLAVSKAYA[2] that polar spicules tend to slope towards the equator whereas those in active zones (20 to 40°) tend to slope towards the poles.

α) The diameters are often difficult to estimate because of their smallness. Some authors[4] give diameters of 2000 km, but the images may be widened by atmospheric agitation. DUNN[5] estimates that the best values are 500 to 600 km.

[1] S. L. LIPPINCOTT: Smithson. Contrib. Astrophys. **2**, no. 2 (1957).
[2] E. J. BUGOSLAVSKAYA: Astron. Zhurn. **23**, 225 (1946).
[3] D. H. MENZEL and J. W. EVANS: Convegno Volta **1953**, 119.
[4] Cf. J. H. RUSH and W. O. ROBERTS: Austral. J. Phys. **7**, 230 (1954).
[5] R. B. DUNN: Astronom. J. **62**, 141 (1957).

β) *The numbers of spicules* visible at the Sun's limb can be counted fairly easily if one limits the counts to the few long ones. For the shorter ones the counts are influenced by mutual obscuration (causing the counted number at a certain distance to the limb to be smaller than the true number) as well as by partial obscuration by the Sun's limb (falsifying the height distribution: not all

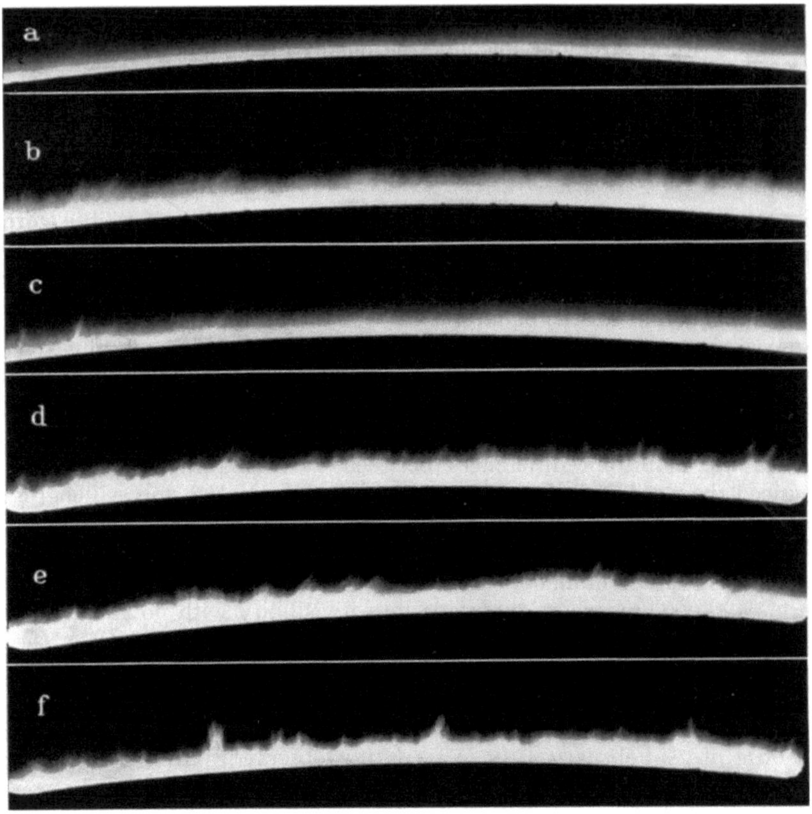

Fig. 22. The Sun's limb through an Hα filter (pass band 3.5 Å)

simultaneous exposures { a 3 Å from centre of Hα, b In Hα } intensity of light 0.1 of images c–f; c 3 Å from centre of Hα; d In Hα }
e, f exposures in Hα, like d, but at other instants.

Courtesy R. B. DUNN, Sacramento Peak.

spicules occur exactly at the limb). Both effects can be eliminated (WOLTJER[1], RUSH and ROBERTS[2]); the theory can only be applied to the highest region with few spicules. Let

$O(d)\,dh$ be the total number of spicules observed at the whole solar limb with observed top heights between $h-\frac{1}{2}\,dh$ and $h+\frac{1}{2}\,dh$ from the limb;

$Q(h)\,dh$ be the number after correcting for superposition in the line of sight;

$F(l)\,dl$ be the number of spicules with *true* top heights between $l-\frac{1}{2}\,dl$ and $l+\frac{1}{2}\,dl$ from the photosphere. (l is not the length but its component in the vertical direction.)

[1] L. WOLTJER: Bull. astronom. Inst. Netherl. **12**, 165 (1954).

[2] Cf. footnote 4, p. 135.

The latter function should be known. The counts $O(h)$, plotted in Fig. 24 (logarithmically) were taken by several investigators[1-4]. The apparent difficulty in comparing the various values is that the gradients and the densities of the photographic material differ from one observer to the other. LYOT e.g. was known to use rather soft photographic material. Also the pass band of the filter is wide in some cases (LYOT, 3 Å), narrow in others. MOHLER[3] studied eclipse plates, taken without birefringent filter. The various aspects of the chromosphere observed with different exposure times can best be judged from Fig. 22 which shows some of the photographs figuring in a spicule film taken by DUNN.

Because of the strong decrease of intensity with height, each set of counts is useful only in a limited range of heights. We have tried to

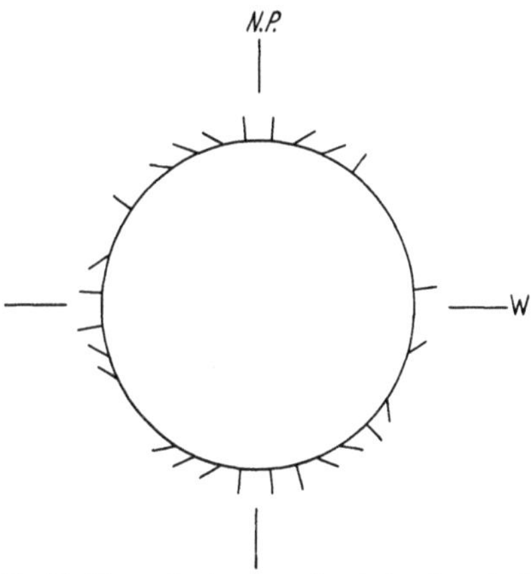

Fig. 23. The predominant quadrants for the spicule slopes. No slopes are shown for those arcs on which the slopes were indeterminant as a rule, or where there was sufficient disagreement between the slopes on different surveys to render the estimates meaningless. Courtesy S. L. LIPPINCOTT.

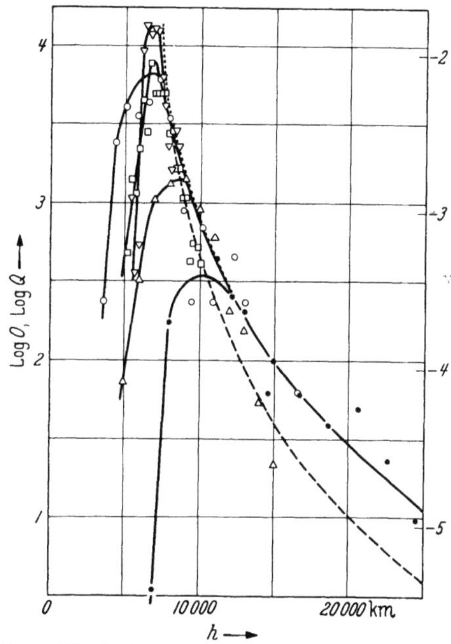

Fig. 24. Distribution of topheights of spicules as a function of the distance to the limb. For the definition of $O(h)$ (uppe. envelope of the solid curves) and $Q(h)$ (dotted curve) see textr The dashed line gives the relative area $A(h)$ occupied by spif cules (adopted diameters 600 km); the corresponding scale o- ordinates is on the right side ot the diagram. ● RUSH and ROBERTS; △ LIPPINCOTT; ○ MOHLER; ▽ WOLTJER; □ DIZER.

combine the various sets of spicule counts into one figure (24). Some authors give counted numbers of top heights per unit arc length, and thus derive absolute $O(h)$ values; in other cases the values of the function $O(h)$ are only relatively known. These latter curves were brought into agreement with the others by a vertical shift. In general there were no difficulties in adjusting the various counts, except for MOHLER's eclipse observations to which a height correction of 3000 km has been applied (MOHLER's heights were given with respect to the Moon's limb). Also the observations of RUSH and ROBERTS were slightly corrected (the base of their spicule counts is reduced to 6000 km above the Sun's limb instead of their assumed value 7500 km). The Sun's limb was taken to be best visible on LYOT's films taken with a broad filter (counts of DIZER and WOLTJER). The counted histograms are uncertain for great heights (few counted spicules) and for small heights (mutual overlapping).

[1] Cf. footnotes 1 and 4, p. 135.
[2] Cf. footnote 1, p. 136.
[3] O. C. MOHLER: Monthly Notices Roy. Astronom. Soc. London **111**, 630 (1951).
[4] M. DIZER: C. R. Acad. Sci., Paris **235**, 1016 (1952).

For the further discussion let two or more spicules be counted as one if their axis distance measured along the limb is smaller than y radians. The mean number of spicules in one interval y is:

$$m = \frac{y\,Q}{2\pi}. \tag{25.1}$$

On the basis of a Poisson distribution of spicules along the limb the number of counted spicules is

$$O(h)\,dh = \frac{2\pi}{y}\sum_{1}^{\infty}\frac{m^n}{n!}\,e^{-m} \equiv \frac{2\pi}{y}\,m'. \tag{25.2}$$

Hence

$$Q/O = m/m'. \tag{25.3}$$

Obviously $m' < 1$. The observed $O(h)$ curve shows a maximum near 6800 km, presumably because at that height $m' \approx 1$ and because the spicules cannot be seen at lower heights. At that height $O\,\Delta h = 10^4$ for $\Delta h = 1000$ km. If $m' = 1$ at 6800 km, we find with (25.2), that $y = 440$ km, in agreement with DUNN's estimate of the spicule diameters. As a next step, $Q(h)$ is found with (25.2) and (25.3); cf. the dotted curve in Fig. 24.

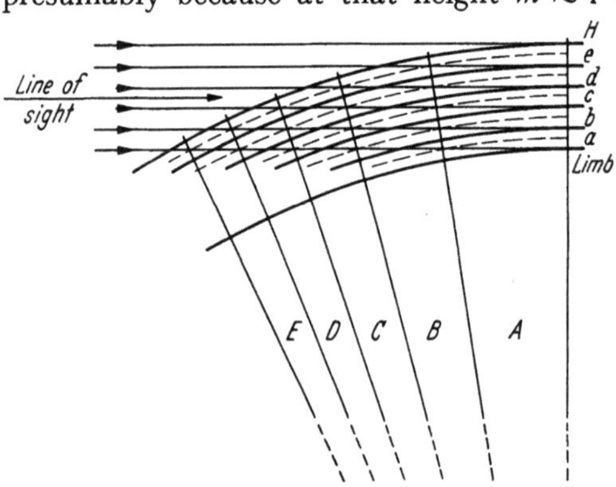

Fig. 25. Effect of obscuration by the limb on the apparent heights of spicules distributed along the line of sight. After J. H. RUSH and W. O. ROBERTS: Austral. J. Phys. 7, 243 (1954).

The computation of $F(l)\,dl$ was done after a method given by RUSH and ROBERTS[1] (Fig. 25). Let the spicules attain a maximum height H. The arbitrary height ranges $a \ldots e$ define the angles $A \ldots E$. Since the maximum range e cannot contain spicules from lower heights $F_e = Q_e$. Then, further

$$F_d = Q_d - \frac{B}{A}\,Q_e;$$

$$F_c = Q_c - \frac{B}{A}\,Q_d - \frac{C}{A}\,Q_e, \quad \text{and so on}.$$

The resulting distribution of top lengths of spicules (dashed line in Fig. 24) is only reliable for $h \gtrsim 7000$ km. With an average diameter of 1200 km we thus obtain that at 8000 and 10000 km respectively 1.3 and 0.2% of the solar surface is occupied by spicules (right ordinate scale). If the spicule diameters were 600 km, these values should be four times smaller.

The statistics of Fig. 24 refers to heights above 7000 km. It is difficult to obtain data on the numbers of spicules attaining smaller heights because of the influence of mutual obscuration. An attempt to determine $F(h)$ in the range of small heights, by ATHAY and THOMAS[2], based on LIPPINCOTT's counts, yielded the result that $F(h)$ seems to *de*crease below $h = 9000$ km.

[1] Cf. footnote 4, p. 135.
[2] R. G. ATHAY and R. N. THOMAS: Astrophys. Journ. 125, 804 (1957).

We recall that the following tentative values of $A(h)$, now interpreted as the relative area of the hot elements of the chromosphere below 5000 km, were derived from spectral data:
(Sect. 24): $h = 5000$ km $A(h) \approx 2\%$ (fine mottles in Hα, assumed diameter 600 km).
(Sect. 20): $h = 2500$ km $20\% < A(h) < 30\%$ (He 10832).
(Sect. 23): $h = 2000$ km $A(h) \approx 30\%$ (fine Ca^+ mottles; A rather uncertain).

γ) The average life time, defined as the time during which a spicule is higher than the mean level, is 2 min as determined by DIZER[1], it is 3.5 min according to RUSH and ROBERTS[2] and 5 min according to LIPPINCOTT[3]. The values do not wholly agree, presumably because of differences in the exposure times. The various investigations refer to different mean levels: 6500 km for DIZER, 9000 km for LIPPINCOTT, 14000 km for RUSH and ROBERTS.

δ) The velocities and motions. Most observers report that the spicules ascend with approximately constant velocity, then stop abruptly and fade away (about 40% of the studied cases), or descend with about the same velocity. About 25% of the descending ones first stay at maximum height for a minute or so. At 6500 km DIZER found a mean velocity of 20 km/sec (40 km/sec in disturbed regions), at 9000 km LIPPINCOTT found rates of growth and of recession of 19 km per sec respectively, at 14000 km RUSH and ROBERTS found 30 km/sec. There are many deviating cases, some reaching speeds up to 200 km/sec and others presenting irregular or surge-like motions. There is an approximate linear relation between the mean ascending velocity component and the maximum height[2]; the fastest spicules thus reaching the greatest heights: $\bar{\xi}\,(\text{km/sec}) = 40\,\dfrac{(h - 8000\,\text{km})}{10000}$, a relation valid up from $h = 12500$ km, where $\bar{\xi} \approx 20$ km/sec.

The assumption of constant motion was investigated in detail by ATHAY and THOMAS[4] who found indications that the statistics can as well or perhaps better be described by (parabolic) motions in a gravitation field than by that of constant motion followed by abrupt standstill.

Of course the observed velocities are partly apparent: one does not measure velocities but the rate at which the gas becomes visible and invisible in Hα. Observations of Doppler shifts are useful to eliminate this difficulty, but when the component normal to the surface is observed, one only obtains information about the deep layers ($h \approx 3000$ to 4000 km). At that height the r.m.s. *vertical velocity component* might be $(\overline{\xi^2})^{\frac{1}{2}} = 18$ km/sec as was found[5] from a statistical discussion of fine mottles in Hα spectroheliograms made at various $\Delta\lambda$-values. At $h = 6000$ km MICHARD[6] found that the r.m.s. *tangential (sightline) velocity component* is $(\overline{\xi^2})^{\frac{1}{2}} = 19.5$ km/sec, about equal to the vertical component at that height. This indicates perhaps that many spicules are greatly inclined, as can also be inferred from spicular photographs.

26. An empirical two-element model of the chromosphere. The results obtained in the preceding sections will be assembled to construct a working model of the chromosphere. It should contain at least the following functions of the height h (an asterisk denotes spicular values or values concerning the hot elements of the chromosphere below 4000 km): T; T^*; N_e; N_e^*; ξ; ξ^*; A. In principle, knowledge of T, N_e and the abundances is sufficient to find the numbers of atoms in various

[1] Cf. footnote 4, p. 137.
[2] Cf. footnote 4, p. 135.
[3] Cf. footnote 1, p. 135.
[4] R. G. ATHAY and R. N. THOMAS: Astrophys. Journ. **125**, 804 (1957).
[5] C. DE JAGER: Bull. astronom. Inst. Netherl. **13**, 133 (1957).
[6] Cf. footnote 1, p. 113.

states of excitation and ionization—if the ionization and excitation relations are known. We are still far from that goal. Further the microturbulent velocity component function is badly known in the low chromosphere and not at all above 5000 km. Neither do we know whether ξ_{micro} differs for the hot and cold elements. This makes clear that the model described here cannot be more than a tentative working model.

α) The temperatures (see Figs. 15 and 26). We assume that the dark fine mottles in Hα spectroheliograms and the bright fine mottles in Ca^{+}_{23} are the hot elements of the low chromosphere.

They emit and absorb the hydrogen and helium lines. Near 2000 km their electron-temperature is about 10000°. Higher up, above 5000 km, the fine mottles are invisible at the disk owing to the transparency of the upper chromosphere for Hα (they could perhaps be discovered on a Lyα filtergram[1]), but above 6000 km one observes at the limb the spicules, which emit Hα light, hence they are colder than the surrounding corona. In that region their electron temperature ranges perhaps between 2×10^4 degrees and 3 to 6×10^4 degrees.

If we accept that the fine mottles are the spicules, then T^* rises slowly from 10000° near 1000 km to about 40000° near 10000 km. Hence $T<T^*$ in the low chromosphere and $T>T^*$ in the higher chromosphere; at an intermediate point $T=T^*$. Below 3500 km the temperature T of the cold elements may increase slowly from about 5000° at 500 km to 6500° at 3000 km. Very uncertain is the location of the height where T begins to increase to coronal values, but this cannot be much above 5000 km, for close to that height the coronal lines begin to appear. They have their maximum at or below 8000 km[2]. The data on the continuous chromospheric spectrum (footnote 5, p. 117) show that the steep increase of $T(h)$ begins between 4000 and 5000 km. Above 5000 km certainly $T>25000°$, and according to ATHAY and MENZEL $T\approx10^5$ degrees.

As a result of this discussion we provisionally assume the T and T^* curves indicated by the solid lines in Fig. 26 and in the table at the end of this section.

β) The densities. It is not correct to determine the ratio of densities in the two kinds of elements e.g. by assuming pressure equality in and between the spicules, for it is not probable that this assumption is realized in these dynamic features in which supersonic velocities and magnetic effects play a rôle. It should be tried to determine the densities separately for each of the elements. There are only few such independent determinations, cf. the table. (For comparison the last column gives Log N_e derived from radio-observations.)

h	log N_e	log N_e^*	ref.	Log N_e from radio-observations
1000 km	11.5		Sect. 19, γ	10.8
1070		11.6	Sect. 19, ε	
1750		11.3	Sect. 19, ε	10.7
2400		11.0	Sect. 19, ε	10.5
5000		10.6	Sect. 20, γ	9.4 to 9.8
Corona at 10000 km	8.3		VAN DE HULST[3]	8.2 to 9.0

The "coronal" density at 10000 km is based on a slight extrapolation of the coronal values, measured somewhat higher. In the low chromosphere the densities

[1] Cf. R. MERCURE, S. C. MILLER, W. A. RENSE and F. STUART: J. Geophys. Res. **61**, 571 (1956).

[2] R. G. ATHAY and W. O. ROBERTS: Astrophys. Journ. Suppl. **1**, 491 (1955).

[3] H. C. VAN DE HULST: Bull. astronom. Inst. Netherl. **11**, 135 (1950).

are about equal in both kinds of elements, but at 5000 km the spicules seem to be somewhat denser than the interspicular (coronal) matter. The table indicates that both at $h = 1000$ and 5000 to 8000 km there is roughly pressure equality between the two kinds of elements: $N_e \times T_{\text{eff}} \approx N_e^* \times T_{\text{eff}}^*$, where, in defining T_{eff} and T_{eff}^*, the contribution of turbulent pressure should of course be added to the thermal pressure.

γ) The turbulent velocities. For these functions we accept the $\xi(h)$ function of Fig. 12, assuming that $\xi(h) = \xi^*(h)$. In the region above 5000 km where only the spicules are relevant no assumption can be made as yet.

δ) The $A(h)$ function giving the relative area of the Sun covered with spicules (or fine mottles—in the low chromosphere) is given in Fig. 24 and Sect. 25. In the low chromosphere ($h < 5000$ km) we tentatively adopt for the area of the hot elements: $A = 0.15$ at 2000 km and 0.1 at 3000 km. That the hot elements cannot occupy a greater relative area than about 0.25 near 3000 km is also proved by radiodata: otherwise the computed intensity of the radiospectrum at cm-waves would be much too high as compared with the observations.

Table 7. *A working model of the chromosphere based on empirical data.*

h (km)	$T_{\text{ex}}(N_2)$	T_{ion} (H)	T_{ion} (Fe)	Log N	Log N_e	T_{el}	Log N_e^*	T_{el}^*	ξ	A
				Cold elements			Hot elements			
0	(6000°)	(5600°)	(4900°)	15.6		4400°		4400°	3 km/sec	
1000	(6300)	(5800)		13.5	11.3	5000		9000	7	
2000	6300	5700	4900	12.8	10.8	5500	11.3	12000	12	0.15
3000	6250	5400		12.3	10.5	6000	10.9	15000	15	0.1
4000	6250	5400		11.9	10.0		10.6		16	0.05
				Interspicular matter			Spicules			
6000	(5500)			9.4	9.4		10.9	25000		0.02
8000				9.0	9.0	$> 10^5$				0.01
10000				8.8	8.8			40000		0.001

$T_{\text{ex}}(N_2)$ = excitation temperature for the second level of hydrogen
T_{ion} (H) = ionization temperature for hydrogen
T_{ion} (Fe) = ionization temperature for Fe.
N = total number of heavy particles per cm³.
(after E. Böhm-Vitense: Z. Astrophys. **36**, 145 (1955); scale of heights decreased by 550 km.)
N_e = total number of electrons per cm³.
T_{el} = kinetic temperatures of the electron gas.
ξ = micro- and macro-turbulent vertical velocity component.
A = relative area occupied by the hot elements of the chromosphere (below 5000 km) and by the spicules ($h > 5000$ km).

The hot elements of the low chromosphere and the spicules are placed in the same column, but their identity has not yet been wholly proved, and in the Athay-Menzel-Thomas model it is even suggested that the cold elements are at the basis of the spicules.

ε) The working model is given in the accompanying table. The electron densities are mainly taken from the table in subsection β. The excitation and ionization temperatures are from Böhm-Vitense's model of the chromosphere[1]. The spicular temperature at 5000 to 10000 km is from Michard, E. Smith and Athay. The model of the (uppermost) transition region to the corona is based on Athay and Menzel's nonspicular temperature distribution, on Piddington's discussion of the radiospectrum of the quiet Sun (Sect. 28α) and on Oster's[2] discussion of the transition region chromosphere-corona.

[1] E. Böhm-Vitense: Z. Astrophys. **36**, 145 (1955).
[2] L. Oster: Z. Astrophys. **40**, 28 (1956).

When the given model is used to compute the pressure-variation, it appears that already at $h \approx 3500$ km the pressure becomes smaller than the coronal pressure at $h \approx 10000$ km. At this latter height we have $\text{Log}\, N_e = 8.5$ and $T = 10^6$ degrees, leading to $\text{Log}\, P_g = -1.1$. It can easily be seen that any of the actually existing models presents this difficulty, which indicates the presence of an additional pressure term. A large scale solar magnetic field of 1.0 Gauss (about equal to the observed general magnetic field) would be required to make the magnetic stress equal to the low coronal gas pressure. In the presence of such a field the acoustic energy coming from below will flow mainly along the lines of force up into the transition region[1], supporting and heating the low corona.

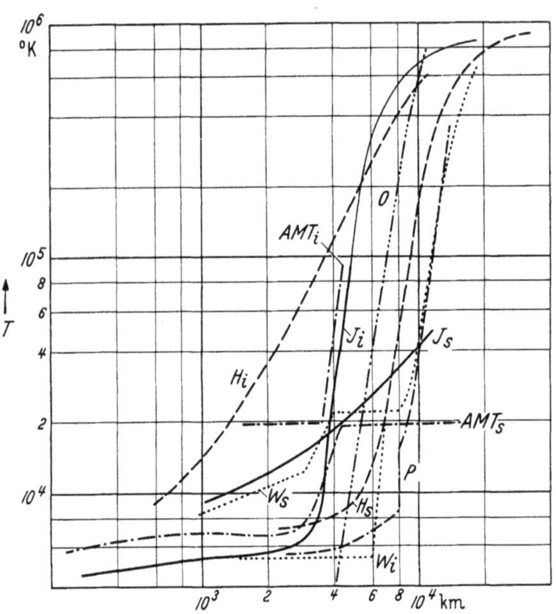

Fig. 26. Comparison of chromospheric models. Subscripts *s* and *i* denote spicules and interspicular matter. H: HAGEN; P: PIDDINGTON; W: WOLTJER; AMT: ATHAY, MENZEL, THOMAS; O: OSTER; J: DE JAGER.

ζ) *Comparison with other two-element models.* The temperature distribution of the present model is shown in Fig. 26 together with the models of WOLTJER[2] (mainly based on Hα intensities), and those of ATHAY-MENZEL-THOMAS[3,4] (based on H and He emission intensities in continuum and lines), PIDDINGTON[5], and HAGEN[6] (based on cm and dm-wave radio data). There seem to be great differences between the four models but this is only apparently, so. One of the differences between our adopted model (J in Fig. 26) and those of HAGEN, PIDDINGTON and WOLTJER is the position of the steep temperature gradient; it is somewhat lower in our model, in accordance with the ATHAY-MENZEL-THOMAS (A.M.T.) model. In the radiomodels (H and P in Fig. 26) it is near 10000 km but the scale of heights is indeterminate in any radiomodel, whereas the position of this steep part of the curve seems fairly well fixed below 8000 km after the eclipse observations of 1952. Our adopted spicular temperature of 2×10^4 °K near 5000 km agrees with WOLTJER and A.M.T. In broad outline our model has the best resemblance with the A.M.T. model, but the hot elements of the low chromosphere have a somewhat higher temperature (20000°) in the A.M.T. model, a temperature mainly based on He-data. That this latter temperature is too high or that the assumed A-value in the A.M.T.

[1] R. LÜST: Z. Astrophys. **37**, 67 (1955).

[2] L. WOLTJER: Bull. astronom. Inst. Netherl. **12**, 165 (1954).

[3] R. G. ATHAY and D. H. MENZEL: **123**, 285 (1956).

[4] R. N. THOMAS and R. G. ATHAY: Radioastronomy (ed. H. C. VAN DE HULST). I.A.U. symposium No 4, p. 279, 1957.

[5] J. H. PIDDINGTON: Proc. Roy. Soc. Lond., Ser. A **203**, 417 (1950). — Astrophys. Journ **119**, 531 (1954).

[6] J. P. HAGEN: Radioastronomy (ed. H. C. VAN DE HULST). I.A.U. symposium No. 4, p. 263, Cambridge 1957. Cf. also footnote 2, p. 144.

model (0.9 at $h \approx 3000$ km) is too great, is shown by a comparison of the A.M.T. model with radio data: the central disk brightness temperature, computed according to the A.M.T. model, is much higher in cm-wavelengths than the observed brightness temperature[1].

The model derived here much ressembles the one derived by SHKLOVSKI and KONONOVITCH[1] on the basis of optical and radio data.

Very uncertain are the conditions in the transition region between chromosphere and corona. Sources of information are the far ultraviolet spectrum of the chromosphere and the radio spectrum of the Sun in the wavelength region below about 50 cm; this problem is discussed in the following sections.

27. The radio spectrum of the undisturbed chromosphere[2]. Only the radio radiation of wavelengths shorter than about 3 cm comes mainly from the chromosphere around and below 5000 km; since the continuous absorption coefficient increases with increasing wavelength, a great part of the radiation of longer wavelengths (but <1 meter) is emitted by the higher chromospheric regions and by the transition region to the corona.

We will discuss successively: the continuous radiation of the disk centre; the centre-limb variation of the continuous radiation (including eclipse observations) and the line spectrum of the chromosphere.

α) *The radiation temperature* $T_r(0)$ of the *disk centre* has been measured up from a wavelength of 0.4 cm (cf. Fig. 96 of Sect. 83). The increase of T_r with increasing λ clearly shows the increase in temperature from 6000 to 7000° in the low chromosphere, up to more than 5×10^5 degrees in the corona. This curve should be interpreted in terms of a two-element model of the chromosphere.

The radiation temperature in the disk centre is

$$T_r = \int_0^\infty T_{\mathrm{el}}(\tau)\, e^{-\tau}\, d\tau \tag{27.1}$$

with

$$\tau = 0.10\, f^{-2} \int_z^\infty N_e^2\, T_{\mathrm{el}}^{-\frac{3}{2}}\, dz. \tag{27.2}$$

In rough approximation $T_r \approx T_{\mathrm{el}}$ at the height where $\tau = 1$, hence, where approximately

$$0.10\, f^{-2} N_e^2\, T_{\mathrm{el}}^{-\frac{3}{2}}\, H = 1. \tag{27.3}$$

Here H is the "scale height" of the function $N_e^2\, T_{\mathrm{el}}^{-\frac{3}{2}}$; in the hot elements of the low chromosphere, where $T \approx$ constant, H is half the scale height of N_e, i.e. ≈ 1000 km (cf. the table in Sect. 27b). In the transition region to the corona it may be of the order of 200 km, in the lower chromosphere somewhat greater.

The observed radiation temperature in the millimetre region is 6000 .. 7000°. With formula (27.3) we find, with $T_{\mathrm{el}} = 6500°$, $f = 5 \times 10^{10}$ Hz ($\lambda = 6$ mm) and with $H = 1000$ and 200 km: Log $N_e = 10.1$ and 10.5, values prevailing between $h =$ 2000 to 4000 km according to various chromospheric models. In that region the influence of the hot elements on the emitted radio radiation is not yet wholly negligible: with $T_{\mathrm{hot}} = 10000°$ and $A = 0.25$, the value of T_{cold} becomes 5000° at $h \approx 3000$ km. If the temperatures in hot and cold elements are constant below that height, T_r should increase for decreasing wavelength smaller than 6 mm, if the relative number of hot elements increases downward.

---

[1] I. S. SHKLOVSKI and E. B. KONONOVITCH: Astron. Zhurn. **35**, 37 (1958).

[2] For the *theory* of the solar radio radiation reference is made to part D, II, a and to Sects. 82 to 84.

At greater heights (longer wavelengths) the spicular contribution is negligible and the observations might be used, as was done by PIDDINGTON[1], to determine the temperature-density relation in the transition layer chromosphere-corona:

h (km):	3000	5000	10000	15000
Log N_e:	10.5	10.0	9.3	8.9
T (°K):	5200	6250	33000	555000

As noted before, the zeropoint in the h-scale is indeterminate in radio observations.

β) *The centre-limb variation* at 8.5 mm, found from eclipse observations (Fig. 97, Sect. 84), was interpreted by COATES[2] in terms of a two-element model of the chromosphere, but it is not certain whether such a refined model may quantitatively be deduced from one set of eclipse data.

γ) *The only spectral line observed* as yet in radio-astrophysics is the 21 cm line of ground-state neutral hydrogen. This line does not occur in the solar spectrum because the number of neutral hydrogen atoms is too small above the level where the optical depth of the continuous radiation at 21 cm is of the order unity. The most important lines that might be produced by neutral hydrogen have been summarized by WILD[3], who computed their transition probability:

λ (cm)	Transition	Transition probability (sec^{-1})
21.11	$1S-1S$	2.9×10^{-15}
3.03	$2P-2S$	6.5×10^{-7}
10.2	$3P-3S$	1.0×10^{-7}

There is a greater probability of finding the line at 3.03 cm than to find the 21 cm line because the smaller number of atoms in the second quantum state of hydrogen ($10^{-10.19\Theta}\approx 10^{-5}\dots 10^{-10}$) is compensated by the greater value of the transition probability (ratio of A-values: 4×10^8) and because at $\lambda=3$ cm the number of neutral H atoms above the layer with $\tau=1$ is much greater than at 21 cm.

Whether the line will appear in emission or absorption depends on the relative populations of the $2\,{}^2S_{\frac{1}{2}}$ and the $2\,{}^2P_{\frac{3}{2}}$ levels, as compared to the populations predicted for the radiation temperature of the emergent radiation [cf. Eq. (16.2)]. Since the relative populations of the two levels will have the thermodynamic equilibrium value (PURCELL[4], see also Sect. 16), the aspect of the line (emission or absorption) depends on the temperature gradient of the chromosphere. The line has not yet been observed and theoretical computations[5] show that even for favourable chromospheric models the emission peak will exceed the continuum only by 10^{-3} to 10^{-6}.

28. The far ultraviolet emission line spectrum. Rocket observations[6] have revealed about thirty emission lines in the far ultra violet solar spectrum between 950 and 1850 Å. The spectrum shown in Fig. 27 was obtained with a normal incidence concave grating spectrograph (dispersion 41 Å/mm); the solar image

[1] Footnote 5, p. 142.
[2] R. J. COATES: Astrophys. Journ. **128**, 83 (1958).
[3] J. P. WILD: Astrophys. Journ. **115**, 206 (1952).
[4] E. M. PURCELL: Astrophys. Journ. **116**, 457 (1952).
[5] C. DE JAGER: in 1958 (Paris) Symposium on Radio Astronomy. (1959.)
[6] F. S. JOHNSON, H. H. MALITSON, J. D. PURCELL and R. TOUSEY: Astronom. J. **60**, 165 (1955). W. E. BEHRING, H. MCALLISTER and W. A. RENSE: Astrophys. Journ. **127**, 676 (1958).

was obtained by a F/10 telescope of 20 cm focal length, pointed at the Sun by means of an automatic following device. Of the emission lines some are due to highly ionized atoms such as O VI but there is also one due to O I; the strongest line is Ly α; further there is Ly β, and there occur lines of Si I, II, III, IV; C I, II, III, IV; N V, Fe II, He II, Al II. The intensity of the continuous spectrum increases with increasing wavelengths, and up from about 1550 Å the true continuum dominates above the stray light intensity. A discontinuity, tentatively ascribed[1] to Al I, occurs near 2090 Å. At somewhat shorter wavelengths the Fraunhofer lines disappear, and emission lines appear at about the same wavelengths. The emission lines refer to rather high temperatures and may yield information about the highest parts of the chromosphere, presumably the part where the temperature of the interspicular matter and of the top part of the spicules increases steeply towards the corona. An estimate of the temperatures necessary to produce line emissions of these ions may be obtained with an ionization formula derived by WOOLLEY and ALLEN[2]. Let x be the fraction of ions in the higher state of ionization. We have $x/(1-x) = S/\alpha$; where S is the collisional ionization coefficient and α is the recombination coefficient. The neglect of radiative excitation is allowed, since (a) the photospheric and chromospheric radiation flux of short wavelengths is negligible, and (b) the coronal radiation is so diluted that it has no influence [cf. ELWERT[3] and Eq. (72.5)]. According to WOOLLEY and ALLEN[4]:

$$S = 3 \times 10^{-8}\, T^{\frac{1}{2}} \chi^{-2} \exp(-11600\,\chi/T)\,,$$
$$\alpha = 1.5 \times 10^{-11} Z^2\, T^{-\frac{1}{2}}\,,$$
$$Z = \text{degree of ionization}\,,$$
$$\chi = \text{ionization energy in electron volts.}$$

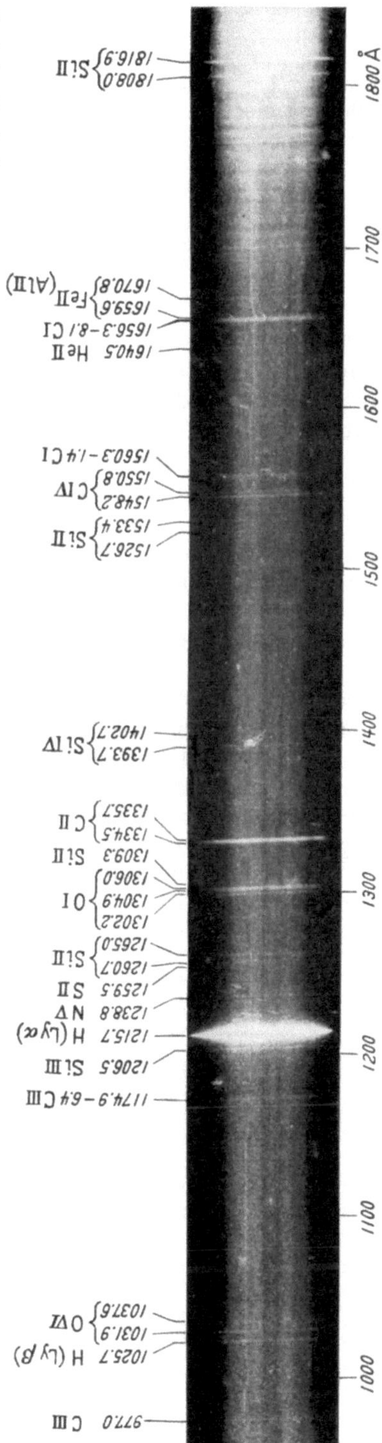

Fig. 27. The solar spectrum photographed at an altitude of 114 km on February 21, 1955 by U.S. Naval Research Laboratory. By means of a sun-following device the solar image was kept fixed on the slit thus enabling to obtain the centre limb variation of the lines. (Official United States Navy Photograph.) Courtesy R. TOUSEY.

[1] Cf. footnote 6, p. 144.

[2] R. v. d. R. WOOLLEY and C. W. ALLEN: Monthly Notices Roy. Astronom. Soc. London **110**, 358 (1950).

[3] G. ELWERT: Z. Naturforsch. **9**a, 637 (1954).

[4] R. v. d. R. WOOLLEY and C. W. ALLEN: Monthly Notices Roy. Astronom. Soc. London **108**, 292 (1948).

Hence the electron temperature of the dividing level is

$$T = 5 \times 10^{-4} Z^2 \chi^2 \exp(11600\,\chi/T).$$

Suppose $x = 1 - x$;—for a rough discussion this treatment is sufficient—then the following electron temperatures are found

He II	68000°	C I	15000°	N V	172000°	O I	19500°
		C IV	103000°			O VI	260000°
Al II	20000°	Si I	8000°	S II	26000°	Fe II	19000°
		Si IV	2000°				

This list clearly shows the wide range of temperatures that occur in the transition region chromosphere-corona.

In order to find the heights that are related to these empirically determined temperatures, it would be necessary to know (a) the absolute intensities, (b) the oscillator strengths of the emitted line radiations, (c) the electron densities and (d) the ion densities. The absolute intensities have been given for a number of lines[1,3]; the oscillator strengths are lacking for all lines except for those of hydrogen.

The Ly α line. In the ultra violet spectrum Ly α is the most prominent feature. The extended wings shown on the photograph are due to the great overexposure and are instrumental. The true width of the line has been determined by making use of the fact that the line just falls in a narrow transmission "window" of the Earth's absorption spectrum. From a comparison of the increase of line intensity with decreasing air mass it was concluded[1] that the total halfwidth must be equal to or less than 0.3 Å. Other measurements[2] suggest a greater width (0.8 Å).

The integrated radiation of Ly α has been measured from 1949 onwards[3], initially with photon counters and thermoluminescent phosphors, later with spectrographs (photographically) and ion chambers. The ion chamber measurements have the greatest reliability. The chambers have LiF windows and are filled with NO gas. The transmission region is 1100 to 1350 Å.

The table below gives the total Ly α emission as measured[3] by NO ion chambers.

Time of firing (UT)	Sun's condition	Incident Ly α flux ergs cm^{-2} sec^{-1}
18. 10. 1955, 22.50	quiet	5.7 (−1; +3)
21. 10. 1955, 00.15	quiet	4.0 (±0.8)
4. 11. 1955, 15.30	quiet	9.2 (±3)
20. 7. 1956, 19.15	late in imp. 1 flare	6.1 (±0.5)
25. 7. 1956, 21.13	quiet	6.7 (±0.3)
29. 7. 1957, 22.00	quiet	6.1 (±0.3)
23. 3. 1958, 18.08	quiet	6.3 (±0.3)

The great intensity fluctuations obtained earlier with less reliable instruments are not confirmed by the new observations. It seems that the total Ly α radiation of the Sun is fairly constant (also during the occurrence of weak flares) and equal

[1] E. T. BYRAM, T. A. CHUBB, H. FRIEDMAN and J. E. KUPPERIAN jr.: Astrophys. Journ. **124**, 480 (1956). — IGY Rocket Report Series **1**, 190 (1958).

[2] S. C. MILLER, R. MERCURE and W. A. RENSE: Astrophys. Journ. **124**, 580 (1956).

[3] T. A. CHUBB, R. W. KREPLIN, J. E. KUPPERIAN jr., H. FRIEDMAN and R. W. KREPLIN: Astrophys. Journ. **128**, 738 (1958).

to (6.3 ± 0.4) erg cm^{-2} sec^{-1} on top of the Earth's atmosphere. Rocket observations made in 1957 even showed no evidence for large increases in Ly α during *strong* flares[1].

Since the chromosphere is optically thick for Ly α, the emission comes mainly from regions where the optical depth for Ly α absorption is of the order unity. These regions are to be found in the transition chromosphere-corona or in the very top parts of the spicules.

From a detailed discussion of the radiation conditions in the region where this line is formed, ATHAY and THOMAS[2] could find the electron temperature of

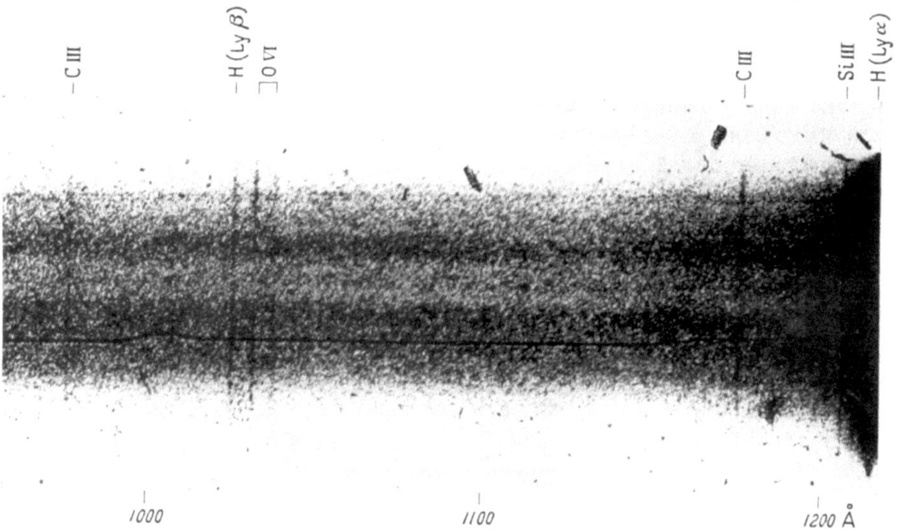

Fig. 28. High contrast enlargement of a part of Fig. 27 showing the limb brightening of the O VI lines. Ly β and the C III line have nearly constant intensity over the disk. (Official United States Navy Photograph.)

this region, this being $2 \times 10^4 < T_{\text{el}} < 4.5 \times 10^4$. A similar, slightly greater value was estimated by DE JAGER[3] on the basis of GIOVANELLI's b_n computations.

These values agree well with the upper limit of T_{el} in the region where the line is formed, deduced from the line-width: The observations yield

$$\Delta\lambda_D \leq 0.30\,\text{Å}/2\sqrt{\ln 2} = 0.18\,\text{Å},$$

where

$$\Delta\lambda_D = \frac{\lambda}{c}\sqrt{2RT + \xi^2}. \tag{28.1}$$

Assuming $\xi = 30$ km/sec, this being the mean mass velocity component for the spicules at the top of the chromosphere, we find from (28.1): $T_{\text{el}} \leqq 65\,000°$.

An extremely impressive achievement is the observation of the centre-limb variation of the far ultra-violet emission lines, by making use of a photoelectric Sun-following device. Fig. 28 shows a spectral region where the difference in centre-limb variation of Ly β and that of the O VI lines is visible; O VI is strength-

[1] H. FRIEDMAN, T. A. CHUBB, J. E. KUPPERIAN jr., R. W. KREPLIN and J. C. LINDSAY: Comm. at 1957 (Toronto) meeting of the U.G.G.I.

[2] R. G. ATHAY and R. N. THOMAS: Astrophys. Journ **124**, 586 (1956).

[3] C. DE JAGER: in Solar Eclipses and the Ionosphere, p. 174. London: Pergamon press 1956.

ened towards the limb, Ly β hardly so[1]. Theoretically a slight limb-brightening of Ly α would be expected (DE JAGER[2]).

The continuous and line emission spectrum of the chromosphere for 100 Å $< \lambda <$ 1100 Å have been computed by OSTER[3], on the basis of an adopted model (see Fig. 26) of the transition region chromosphere-corona. The results are shown in Fig. 83 (Sect. 74).

29. Evidences for variations along the Sun's limb. There are evidences that the chromosphere has no circular symmetry. We summarize the various indications, starting in the photosphere and the low chromosphere and going up into the higher regions.

α) From observations of the centre-limb variation of the g and K lines of Ca in the polar and equatorial solar regions, DAS and ABHYANKAR[4] found indications that during sunspot minimum the equator is cooler than the poles by about 90°. From a study of weak Fe II lines at sunspot maximum a similar result was found by BECKERS[5], who found moreover indications for a low secondary maximum ($\Delta T \approx 20°$ at $b = \pm 25°$). This shows that nonspherical symmetry not only occurs in the chromosphere but even in the photosphere.

β) The infrared absorption line of helium at 10832 Å ($2^3S - 3^3P$) has a greater equivalent width near the equator than near the poles[6]. (The distribution of the equivalent widths of this line on the disk follows closely the radio isophotes at $\lambda = 150$ cm[7].) The mean values of the equvalent widths (mÅ) observed between 1951 and 1954 are:

Centre	Limb	
	N and S	E and W
67	86	132

That the line appears as an absorption line demonstrates that its exitation temperature T_{ex} is smaller than the radiation temperature T_{rad} for 10830 Å (cf. last part of Sect. 16). OSTER[3] found $T_{ex} = 1200°$. The strengthening towards the limb is mainly caused by the greater number of absorbing particles (proportional to sec ϑ). The weakening towards the poles might be caused either by a higher polar T_{ex} or by a greater number of spicules in the equatorial region. Since the spicules are likely to be more preponderant in the polar regions, the first conclusion is correct: $(T_{ex})_{poles} > (T_{ex})_{eq}$.

γ) Hα_{12} spectroheliograms show the dark mottles at the limb clearer near the poles than near the equator. That the local wing depressions are greater near the polar limbs is, as in the helium case, due to a greater polar excitation temperature; we estimate the temperature difference at 50 to 100°.

δ) The daily measures of the height of Hα, made at Arcetri since 1925, show that during sunspot minimum the Hα chromosphere is higher at the poles than at the equator (indicating a greater polar excitation temperature), whereas the chromosphere is nearly symmetrical at sunspot maximum[8].

[1] Cf. footnote 2, p. 146.
[2] Cf. footnote 3, p. 147.
[3] L. OSTER: Z. Astrophys. **40**, 28 (1956).
[4] A. K. DAS and K. D. ABHYANKAR: Nature, Lond. **172**, 496 (1953). — Vistas in Astronomy **1**, 658 (1956).
[5] J. M. BECKERS: Bull. astronom. Inst. Netherl. **1959**, with further references.
[6] O. C. MOHLER and L. GOLDBERG: Astrophys. Journ. **124**, 13 (1956).
[7] Personal remark by Dr. O. C. MOHLER.
[8] M. G. FRACASTORO: Oss. e Mem. Arcetri **64**, 33 (1947).

ε) Outside-eclipse observations of the chromospheric Hα and D_3 lines give some slight indications[1] that Hα is stronger near the poles than near the equator (intensity ratio ≈ 1.2) whereas the reverse might be the case for D_3 (intensity ratio ≈ 0.8).

ζ) Also the radio observations of the intensity distribution in the Sun at decimeter and meter waves show decidedly a polar asymmetry (cf. Fig. 97). Since the continuous absorption coefficient in the radio-range decreases with increasing temperature, the observed polar flattening might indicate that the polar corona is slightly hotter than the equatorial corona. The density differences between poles and equatorial regions seem insufficient to explain the observations.

Summarizing we note that most chromospheric observations indicate that the poles have higher excitation temperatures than the equatorial regions. This temperature excess, already perceptible as low as in the photosphere, persists in higher layers and is observable in any case up to the transition layer chromosphere-corona.

30. Support and heating of the chromosphere. To the old question how the chromosphere is supported, in other words, how it can have the observed vertical extension, the answer given in 1928 by McCrea is still valid: the great scale height is due to the intense turbulent motions, having between 2000 and 4000 km an average value of 15 km/sec. In the upper parts also the high temperature and the magnetic stress contributes to the observed chromospheric extension.

A much more intricate problem is that of the heating of the chromosphere. The low chromosphere has a more or less constant temperature of 5000 to 6000°, rising rapidly to coronal values at heights between 4000 and 6000 km. Embedded in this main part are the fine mottles; they seem to have higher temperatures than the main part. So dissipation of energy occurs mainly in the mottling, and in the transition layer chromosphere-corona.

The mechanism, usually adopted is that of heating from below, the source of heating being the turbulence of the photosphere, which in turn originates from the lower convective motions. So, ultimately the upper chromosphere and the corona are heated by energy coming from the convection zone. The main part of the convective energy (the solar flux) is of course dissipated in the convection zone but a small fraction of it moves upward in the form of travelling sound waves. It is only the curl-free part of the velocity field that is able to do so, and of it only those components with frequencies above a critical one[2], of the order of the reciprocal of the atmospheric scale height H; $\lambda \leqq 4\pi H$; accoustic waves with longer wavelengths are reflected. Since the coronal temperature is much higher than that of the photosphere and chromosphere the corona can only be heated by directed motions. The energy of the sound waves can be transformed into heat by means of hydromagnetic waves, which are perhaps generated by shock waves.

The development of a theory of the aerodynamic *generation of sound* was started by Lighthill[3]; on the basis of this theory Proudman[4] (cf. also [5]) estimated the amount of the energy flux transformed into accoustic noise to be $P_a = 38\,\varrho M^5 \varepsilon\,(\text{erg cm}^{-3}\,\text{sec}^{-1})$, where ϱ is the density, M the Mach number for the turbulent motions ($= \xi_{\text{turb}}/v_s$) and ε is the mean rate of energy dissipation per unit

[1] R. O. Bishop: Monthly Notices Roy. Astronom. Soc. London **116**, 593 (1957).
[2] E. Schrödinger: Phys. Z. **18**, 443 (1917).
[3] M. J. Lighthill: Proc. Roy. Soc. Lond., Ser. A **211**, 564 (1952).
[4] I. Proudman: Proc. Roy. Soc. Lond., Ser. A **214**, 119 (1952).
[5] W. Unno and K. Kawabata: Publ. Astr. Soc. Japan **7**, 21 (1955).

mass in the convection zone. BIERMANN[1] estimated that the exponent of M might be somewhat lower, about 3 to 4. In the presence of a magnetic field the generation of sound is more efficient[2].

Since the turbulent energy flux is proportional to $\xi^2 v_s$ and since the fraction transformed into noise is proportional to $\xi^3 - \xi^5$, this shows the strong dependence on ξ of the energy available for heating the chromosphere and corona. With BIERMANN[1] it may be asked whether this is perhaps a way to explain the variations of chromosphere and corona during the solar cycle.

The next problem is how the generated energy is dissipated. As noticed in Sect. 12 the average microturbulent velocity component (or, what is about the same: the velocity amplitude of the accoustic noise) increases with height as long as $\xi/v_s < 1$. The accoustic waves transform into *shock waves* when ξ/v_s becomes of the order unity. Then the dissipation of energy becomes appreciable (SCHATZMAN[3], following suggestion of BIERMANN and SCHWARZSCHILD), and this may explain why the chromospheric mottling becomes well visible only above $h \approx$ 1000 km: at that height $\xi \approx v_s$, cf. Fig. 12. Nevertheless, in the chromosphere below 400 km a considerable part of the dissipated energy is still lost by radiation and hence does not contribute to the heating. The dissipated energy is used for heating the gas in a more efficient way at heights where $\log N_e < 9.5$; there the dissipation, being $\frac{1}{2}\,\varrho\, v_s^3/H$, ergs cm^{-3} sec^{-1} where H is the scale height, exceeds the emission of the gas[4] ($= 5.5 \times 10^{-22}\, T^{-\frac{1}{2}} N_e^2$; mainly Lyman free-bound hydrogen emission). This explains why at heights above about 4000 km the temperature suddendly rises to coronal values (see the chromospheric model; Table 7). By backward conduction of the absorbed energy this steep rise is smoothed (for computations of the thermal conductivity of the chromospheric matter, cf. OSTER[5]).

The energy of these shockwaves may be transformed into heat with a mechanism suggested by PIDDINGTON[6]: in magnetic field regions the heating may be caused by *hydromagnetic waves*. The heating due to Joule loss of hydromagnetic waves is negligible[7,8], but more heat is dissipated by frictional loss of energy between an oscillating ion plasma and the neutral atoms which tend to move independently but which are forced to oscillate with the ions under the impact of the ions. The intensity of the collisions, hence the efficiency of the effect, increases when the hydromagnetic waves reach regions of smaller density where the binding between the neutral and the ionized gases is weaker. The greatest heating occurs in the region where the ion and neutral atom densities are about equal; that is above 4000 or 5000 km. Particles with energies of the order of 100 eV are emitted in sufficient numbers to explain the quiet corona. This mechanism explains the overnormal brightness of magnetic field regions; reversely the brightness of bright mottles and flocculi may indicate the presence of a magnetic field.

The precise distribution of temperature with height in the chromosphere does not only depend on the amount of energy absorbed at various heights but also on the amounts reradiated and lost by thermal conduction, and further on the exchange of radiation between the hot and the cold elements, hence on the detailed

[1] L. BIERMANN: Trans. Internat. Astronom. Union **9**, 750 (1957). With list of references.

[2] R. M. KULSRUD: Astrophys. Journ. **121**, 461 (1955).

[3] E. SCHATZMAN: Ann. d'Astrophys. **12**, 203 (1949).

[4] Cf. [*2*], p. 666 and C. DE JAGER: in "Plasma Physics and Astrophys. Application", 1958 Summer School, Varenna 1959.

[5] L. OSTER: Z. Astrophys. **42**, 228 (1957).

[6] J. H. PIDDINGTON: Monthly Notices Roy. Astronom. Soc. London **116**, 314 (1956) Observatory **76**, 21 (1956).

[7] T. G. COWLING: in: The Sun (ed. G. P. KUIPER) Chicago 1953.

[8] J. H. PIDDINGTON: Monthly Notices Roy. Astronom. Soc. London **114**, 638 (1954).

energy balance. With ATHAY and THOMAS[1] one might ask for a $T(h)$ distribution such, that at each height there is thermal stability: an increase (or decrease) of T_{el} at a certain height should correspond to such an increase (or decrease) of energy radiation, that the temperature comes back to its earlier value. Two such regions of stability may occur in the chromosphere:

1. A region of neutral hydrogen with a temperature of the order of 6000°. The emission is principally due to bound-bound and free-bound emissions of electrons in the field of hydrogen atoms.

2. In the second region helium becomes appreciably ionized. The emission is mainly due to bound-bound and free-bound emissions of He II. For this stability region the thermodynamic equilibrium computations predict temperatures between 20000° and 25000°.

A third stability region apparently occurs in the corona where there is equilibrium between the inflow of mechanical heat and the loss by radiation, conduction and evaporation.

ATHAY and THOMAS suggested that the two chromospheric stability regions are identical with a cool low chromospheric region and a hotter upper chromospheric region. Alternatively one might, however, identify them with the spicular and the non-spicular matter.

B. The disturbed parts of the photosphere and chromosphere.

I. Sunspots.

a) The individual spots.

31. The appearance of sunspots. The appearance of a characteristic sunspot is clear from the accompanying illustration (Fig. 29). A typical spot consists of a dark umbra surrounded by a brighter penumbra. The intensity jumps almost discontinuously between umbra, penumbra and photosphere, whereas it seems to be nearly constant in the penumbra, less so in the umbra. For roundish spots, not too small, there exists a definite relation between the diameter of the umbra (U) and penumbra (P). According to NICHOLSON (1933) and WALDMEIER (1939): $P/U = 3.2 - 36P$ (P in solar diameters), with a mean relative scatter of $\pm 10\%$. The mean umbral area of large spots (area $> 5 \times 10^{-4}$ visible hemisphere) is 0.16 of that of the whole spot[2]. The penumbra always shows a bright filamentary structure consisting of greatly elongated granules. They are more or less radial with respect to the umbra. The life time of these penumbral elongated granules is 30 min (MACRIS[3]). Very rarely a bright ring is observed at the inner border of the penumbra[4, 5], about 2 to 3% brighter than the average penumbra.

The umbra proper is not homogeneous. CHEVALIER (1913) and later THIESSEN (1952), from visual observations, found that the umbrae of spots may show small bright points like granules. This was confirmed photographically by RÖSCH[6] (Fig. 30). In a number of spots a part of the umbra is filled with small bright points, with diameters smaller than those of the granules, perhaps smaller than 350 km. They do not form the mosaic structure exhibited by the photospheric granules but appear more as isolated points. Their life time is not yet known.

[1] R. G. ATHAY and R. N. THOMAS: Astrophys. Journ. **123**, 299 (1956).
[2] A. W. F. EDWARDS: Observatory **77**, 69 (1957).
[3] C. MACRIS: Ann. d'Astrophys. **16**, 19 (1953).
[4] Cf. A. K. DAS and A. S. RAMANATHAN: Z. Astrophys. **32**, 91 (1953).
[5] R. ANANTHAKRISHNAN: Nature, Lond. **168**, 291 (1951).
[6] J. RÖSCH: l'Astronomie **71**, 129 (1957). — C. R. Acad. Sci., Paris **243**, 478 (1957).

Medium and large spots may show a bright irregular structure round the penumbra. It is called the "bright ring". In the disk centre the "ring" is about 2 to 3% brighter than the photospheric background[1]. At $r/R = 0.87$, $\Delta I/I$ is

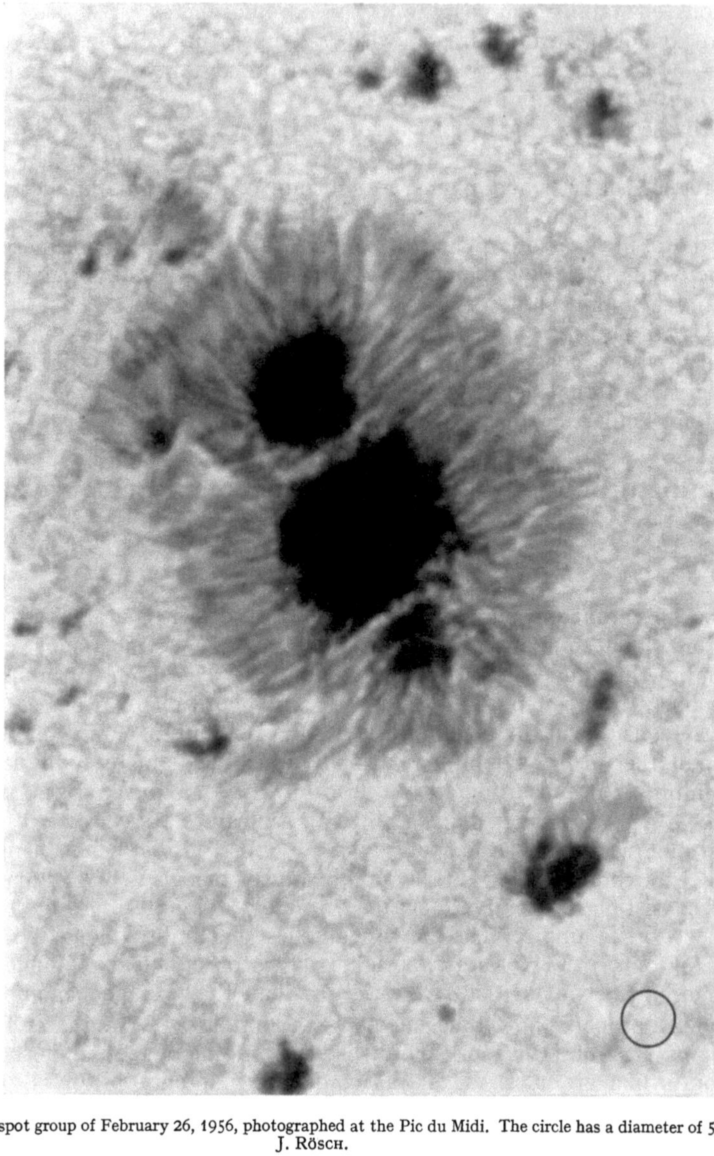

Fig. 29. Sunspot group of February 26, 1956, photographed at the Pic du Midi. The circle has a diameter of 5″. Courtesy J. RÖSCH.

reduced to half this value and the ring is invisible at the Sun's limb, indicating that the ring is produced by a radiation excess in the deep photosphere. The mean relation between the diameters of penumbra (P) and ring (R) is $P/R = 0.62 + 3.6\,P$ (P in solar diameters).

32. The magnetic fields of spots. The magnetic field of spots is mostly perpendicular to the solar surface. Generally the field is of the order of some hundreds

[1] Cf. footnote 3, p. 151.

to some thousands of Gauss. Measurements of the maximum field strengths of the most important spots are made daily at the observatories of Mt. Wilson, Potsdam, Simeis and Poulkovo.

α) Principle of measurement (Fig. 31 a). For the measurement of strong fields in which the lines of force are in the direction of vision, use is made of the classical Zeeman splitting of lines in a magnetic field; in the central part of the disk (longitudinal field) the line splits into its two circularly polarised σ components with distances $\pm \Delta\lambda$;

$$\Delta\lambda = g \frac{e\,\lambda^2}{4\pi m c^2} H = 4.7 \times 10^{-5} g\,\lambda^2 H$$

(λ in cm; H in Gauss).

So, e.g. for $\lambda = 6000$ Å, $g = 1$, and $H = 3000$ Gauss, the distance $2\Delta\lambda$ between the

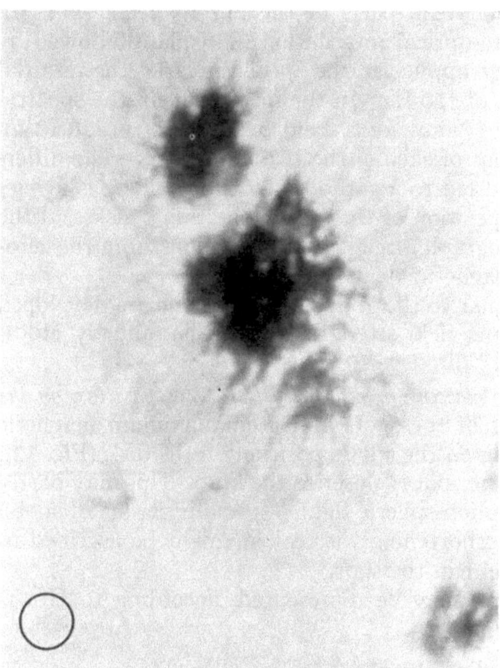

Fig. 30.

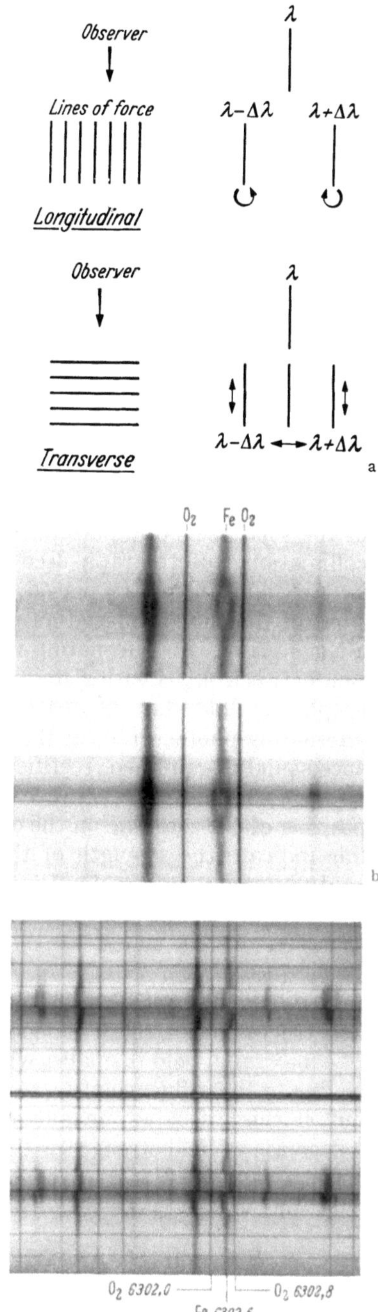

Fig. 31 a—c.

Fig. 30. Umbral granules. Reproduction from the same photograph as the one shown in Fig. 29.

Fig. 31 a—c. Zeeman splitting of Fraunhofer lines. (a) Schematic representation of Zeeman splitting. [Drawing after H. von Klüber: Vistas in Astronomy **1**, 751 (1956).] (b) Spectra of the Fe line 6302.5 Å in and around a spot. Spot near disk centre. (c) Spectrum of a sunspot. Small quarter wave strips are placed before the spectrograph slit, and the image of the Sun in front of the slit was split by a double image prism into identical images, each polarized perpendicular to the other. The spectra, thus obtained are identical but for the opposite sense of polarization in corresponding strips. Longitudinal magnetic field component ≈ 3000 Gauss. The telluric O_2 lines are not influenced. Courtesy H. von Klüber.

two line-components is 0.1 Å. For the "normal" splitting of *singulet* lines, due to the orbital magnetic moments, the gyromagnetic ratio (Landé g-factor) $=1$ Mostly $g \neq 1$ due to the contribution of the electronic spin.

Some lines which are often used are:

λ 5250.4	(Fe)	$g=2$	$g\lambda^2 \times 10^8 = 0.83$
6173.3	(Fe)	$2\frac{1}{2}$	0.95
6302.5	(Fe)	$2\frac{1}{2}$	1.00
6733.2	(Fe)	$2\frac{1}{2}$	1.14

Further we have

10288.9	(Si)	2	2.12

The accuracy of measurement is roughly ± 50 Gauss; the smallest fields which can be detected in this way have a strength of the order of 100 Gauss.

For the detection of smaller fields (0.1 to 10 Gauss) H.W. BABCOCK[1] has developed an apparatus which permits in an average case to make a complete magnetogram of the whole disk in about 90 minutes. (The time for a whole scan of the Sun may vary between 15 min and 3 hours depending on the complexity of the solar magnetic pattern.) In the focal plane of a 25 meter Littrow (autocollimation) spectrograph used in the fifth order (11 mm/Å), two slits followed by two multipliers are placed in the steep parts of the line λ 5250.2 (Fe). In front of the slit are placed an electro-optical retardation ($\lambda/4$) plate followed by a Nicol prism. An alternating voltage applied to the $\lambda/4$-plate varies the retardation between $\pm\lambda/4$ with a frequency of 120 Hz. In the focal plane of the spectrograph the line shift (8×10^{-5} Å $=10^{-3}$ mm for a field of 1 Gauss) produces an alternating photocurrent at the output of each of the two multipliers. The difference signal is amplified, rectified and fed to a cathode-ray tube. There is a one-to-one correspondence between the position of the slit on the Sun's disk and the position of the zero-line on the oscillograph screen. The deflections from the zero-line indicate the strength of the magnetic field (Fig. 44).

Instruments in broad outline similar to the above described one are developed by BEGGS and VON KLÜBER[2] (limiting field strength ≈ 1 Gauss), and by MOGILEVSKIJ, VELLER and VALD-PERLOV[3] (limit ≈ 10 Gauss).

β) The relation between diameter, maximum magnetic field and position on the Sun's disk. Between the diameters of the spots and their maximum magnetic field, there holds a relation depending on the spot's position on the disk (Fig. 32). The maximum field is smaller when the spot is closer to the limb. This may partly be due to the influence of scattered photospheric light (especially for small spots) but, if this effect of "magnetic foreshortening" is real, it might be ascribed to a decrease of field strength with height in the spot.

The relation $H_M(A)_{\vartheta=0}$ (Fig. 32b) may be represented, according to HOUTGAST and VAN SLUITERS[4] by the formula:

$$H = 3700 \frac{A}{A+66} \text{ (Gauss)} \qquad (A \text{ in } 10^{-6} \text{ visible hemisphere}).$$

For comparison we also show the relation derived by NICHOLSON[5], who gives somewhat lower H_M values. Both relations point to maximum mean field strengths

[1] H. W. BABCOCK: Astrophys. Journ. **118**, 387 (1953).
[2] D. W. BEGGS and H. VON KLÜBER: Nature, Lond. **178**, 1412 (1956).
[3] M. A. MOGILEVSKIJ, A. E. VELLER and V. M. VALD-PERLOV: Dokl. Akad. Nauk., SSSR. **95**, 957 (1955).
[4] J. HOUTGAST and A. VAN SLUITERS: Bull. astronom. Inst. Netherl. **10**, 325 (1948).
[5] S. B. NICHOLSON: Publ. Astronom. Soc. Pacific **45**, 51 (1933).

of about 3500 Gauss although individual cases may yield greater values (VON KLÜBER once measured a field of 4500 Gauss).

MATTIG[1] has questioned the value of the $H_M(\vartheta)$ curves of Fig. 32a because of the great influence of scintillation on the H_M values: scintillation increases the contribution of scattered photospheric light in the spot-centre, thus apparently reducing H_M near the limb. Even for a great spot (diameter 50″; $A = 660 \times 10^{-6}$ visible hemisphere) a fairly small scintillation (2″) suffices to change a *true* limb-*in*crease of H_M into an *observed de*crease of H_M towards the limb, as is shown by the table:

	Disk centre	$\cos\vartheta = 0.2$
H_M (obs)	3100 Gauss	2330 Gauss
H_M (true)	3170 Gauss	3940 Gauss

In any case the conclusion must be that the "observed" decrease of H_M towards the limb is an upper limit. Likewise, the increase of H with geometrical depth z

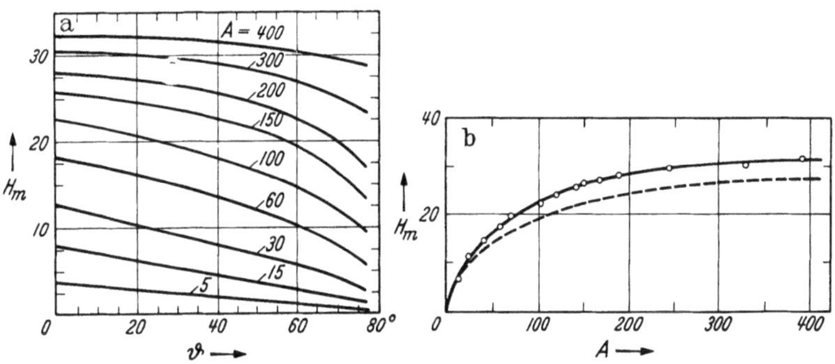

Fig. 32a and b. (a) Centre-limb variation of the maximum fieldstrength H_M for spots of different areas A. (b) Variation of H_M with A in the disk centre. Solid line after HOUTGAST and VAN SLUITERS; dashed line after NICHOLSON. Unit of H_M: 100 Gauss, of A: 10^{-6} visible hemisphere. J. HOUTGAST and A. VAN SLUITERS: Bull. astronom. Inst. Netherl. **10**, 325 (1948). S. B. NICHOLSON: Publ. Astronom. Soc. Pacific **45**, 51 (1933).

in the spot, deduced by various authors from the $H_M(\vartheta)$ curve, is an upper limit. Anticipating Sect. 33, we notice that the function $H(z)$ is also greatly influenced by the uncertainties of the umbral model. With WALDMEIER's umbral model and one of the curves of Fig. 32a HOUTGAST and VAN SLUITERS[2] found $dH/dz = 5.8$ Gauss/km. With MICHARD's model[3] the same $H_M(\vartheta)$ curve yields 1.5 Gauss/km, and with MATTIG's[1] 18 Gauss/km!

For comparison we note (anticipating subsection δ below) that HOUTGAST and VAN SLUITERS[2] found from the divergence of the lines of force in a spot $dH/dz = 0.5$ Gauss/km; and from the measurement of lines of different equivalent widths (corresponding with different levels) they found 2.5 Gauss/km. With MICHARD's and MATTIG's models this latter value would be 0.5 and 6.5 Gauss/km.

γ) Magnetic field distribution in and around the spots. The magnetic field distribution in and around the spots as determined by MATTIG[4] is

$$H = H_M (1 - \varrho^4)\, e^{-2\varrho^2}, \tag{32.1}$$

where ϱ is the distance to the spot centre expressed in the penumbral radius. The variation of field strength over the umbra and penumbra is quite regular

[1] W. MATTIG: Z. Astrophys. **44**, 280 (1958).
[2] Cf. footnote 4, p. 154.
[3] R. MICHARD: Ann. d'Astrophys. **16**, 217 (1953).
[4] W. MATTIG: Z. Astrophys. **31**, 273 (1953).

and does not show the discontinuities of the visual image (Fig. 33). In one case, spot Mt. Wilson Nr. 10793, a spot with a symmetrical penumbra but with a rather asymmetric umbra, the field was determined in various directions; it appears that it is fairly circularly symmetric.

The formula (32.1) is an interpolation formula, and does not take account of the weak fields that may occur in the facular field surrounding the spots.

δ) Divergence of the lines of force. With increasing distance to the centre of the spot the angle Θ between the magnetic lines of force and the normal to the solar surface increases, from 0° in the centre of the spot to nearly 90° near the border of the penumbra. According to HALE and NICHOLSON[1], who determined

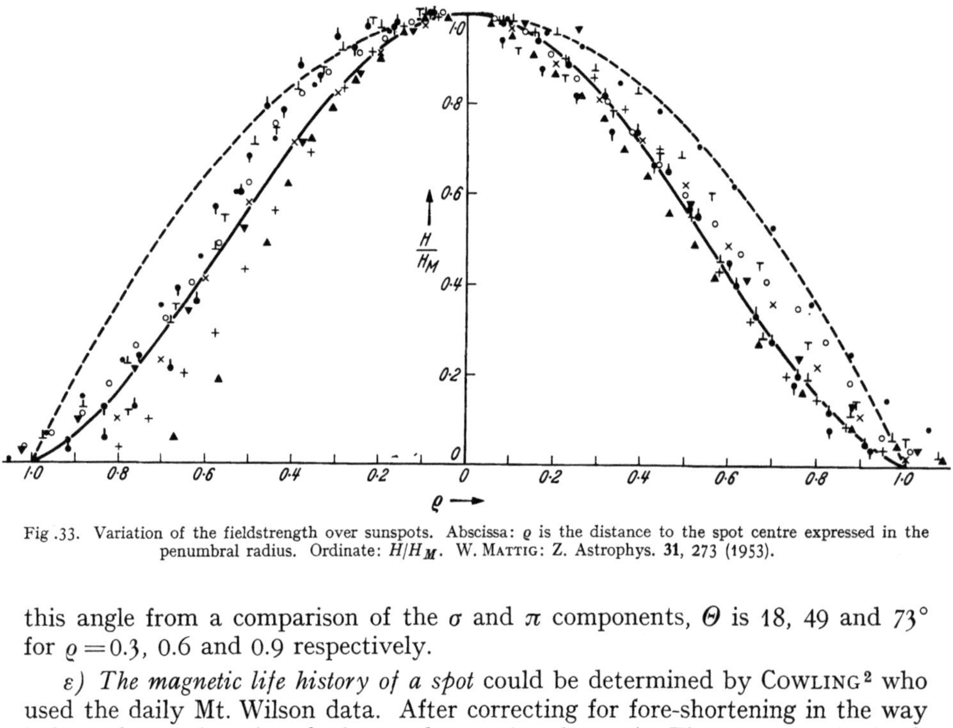

Fig. 33. Variation of the fieldstrength over sunspots. Abscissa: ϱ is the distance to the spot centre expressed in the penumbral radius. Ordinate: H/H_M. W. MATTIG: Z. Astrophys. **31**, 273 (1953).

this angle from a comparison of the σ and π components, Θ is 18, 49 and 73° for $\varrho = 0.3$, 0.6 and 0.9 respectively.

ε) The magnetic life history of a spot could be determined by COWLING[2] who used the daily Mt. Wilson data. After correcting for fore-shortening in the way indicated in subsection β above, the results shown in Fig. 34 were obtained.

The maximum magnetic field of an average well developed spot changes quickly in the beginning and end stages of a spot's life, but it remains nearly constant during the main life history of the spots. Maximum field and area reach a maximum in about 12 days, then the area decreases and the field remains nearly constant until the disappearance of the spot. COWLING remarks that the field seems to decay by the successive disappearing of the outer lines of force below the photosphere, not by a general diffusion. According to GROTRIAN and BRUNNCKOW[3] short periodic fluctuations with daily amplitudes of 100 to 200 Gauss occur during the stage of maximum development of the spot. A statistical discussion of the daily Mt. Wilson measurements by GROTRIAN and KÜNZEL[4] yields values of 250 and 500 Gauss for the daily fluctuations of field strength.

[1] G. E. HALE and S. B. NICHOLSON: Papers Mt. Wilson Obs. V (I, II) (1938).
[2] T. G. COWLING: Monthly Notices Roy. Astronom. Soc. London **106**, 218 (1946).
[3] K. BRUNNCKOW and W. GROTRIAN: Z. Astrophys. **26**, 313 (1949).
[4] W. GROTRIAN and H. KÜNZEL: Z. Astrophys. **26**, 325 (1949).

ζ) Magnetic flux. By means of the known relations between the spot diameters and their maximum magnetic fields and with the known $H(\varrho)$ relations the magnetic flux Φ can be computed from $\Phi = \int H\, df$. With $\Theta = \frac{\pi}{2}\varrho$, MATTIG obtained $0.25\, H_M\, \pi b^2$ where, as usual, b is the radius of the penumbra. With this formula the variation of the magnetic flux with time is easily computed for the case of Fig. 34. It appears that Φ first increases quickly. After reaching a rather sharp maximum together with the maximum area, it decreases proportionally with the area (A), when H_M remains constant. Finally, when both A and H_M decrease, Φ decreases very quickly.

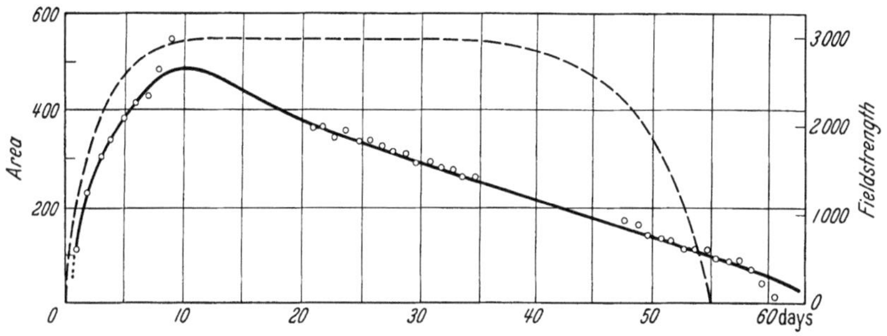

Fig. 34. Variation of the area and the field of the leading members of a spotgroup. Solid line: area (left ordinate scale unit: 10^{-6} visible hemisphere. Dashed: H_M (right scale), unit: Gauss. T. G. COWLING: Monthly Notices Roy. Astronom Soc. London **106**, 218 (1946).

33. Radiation and spectrum of the umbra; model of the umbra. *α) Photometry of spots.* The intensity profile of the spots seems to consist of a rather flat penumbral part ($I/I_{\text{phot}} \approx 0.75$) and an umbral part ($I/I_{\text{phot}} \approx 0.2$ to 0.3)[1]. The transitions between photosphere, penumbra and umbra are steep but are smoothed by scattered light and lack of resolving power. The biggest spots are also the darkest. The umbra has not a flat central intensity profile (the umbral granules have been mentioned), but the intensity may increase slowly towards the penumbra. This phenomenon, however, may also be due to scattering of photospheric light. The occurrence of bright "rings" in the photosphere around the spot and (rarely) in the penumbra has been mentioned (Sect. 31). There is no doubt that the profiles become more discontinuous with increasing quality of the images (KORN[2]). Fig. 35a contains a typical intensity profile in a big spot after DAS and RAMANATHAN[3], and Fig. 35b shows an extraordinary case of an extremely bright ring segment around the umbra, once photographed by RÖSCH.

β) Continuous radiation. The mean value of the integrated radiation of the umbral centre, measured with a thermocouple on a large spot is 0.4 ± 0.10 times the integrated radiation of the photosphere[4]. Of course, great variations occur from spot to spot. The most reliable observations for spots near the disk centre are close to 0.3. With $T_{\text{eff}}(\text{Sun}) = 5785°$ K, we obtain $T_{\text{eff}}(\text{spot}) = (0.4)^{\frac{1}{4}} \times 5785°\,\text{K} = 4600° \pm 300°$. Many investigators measured the monochromatic intensity ratios umbra/photosphere, as functions of wavelength and distance to the disk centre. The results show a considerable scatter, especially for small spots, mainly owing to measuring errors (the elimination of the influence of atmospheric agitation and also of instrumental and atmospheric scattering is

[1] R. S. RICHARDSON: Astrophys. Journ. **78**, 359 (1933).
[2] J. KORN: Astronom. Nachr. **270**, 105 (1940).
[3] A. K. DAS and A. S. RAMANATHAN: Z. Astrophys. **32**, 91 (1953).
[4] T. W. WORMELL: Monthly Notices Roy. Astronom. Soc. London **96**, 736 (1936).

difficult), partly perhaps a consequence of real differences between one spot and another.

An important observation made by WANDERS is that the ratio of the radiation umbra/photosphere does not depend on the distance to the disk centre. This was confirmed by measurements on about 30 spots by MICHARD[1] who also noted that the ratio decreases somewhat in the ultraviolet (but MICHARD's values at $\lambda \leqq 0.4\,\mu$ are certainly influenced by the numerous Fraunhofer lines in that region and might be liable to corrections). From the data of MICHARD and (for the infrared) those of PETTIT and NICHOLSON, we collect (cf. also ALLEN[2]) the following results:

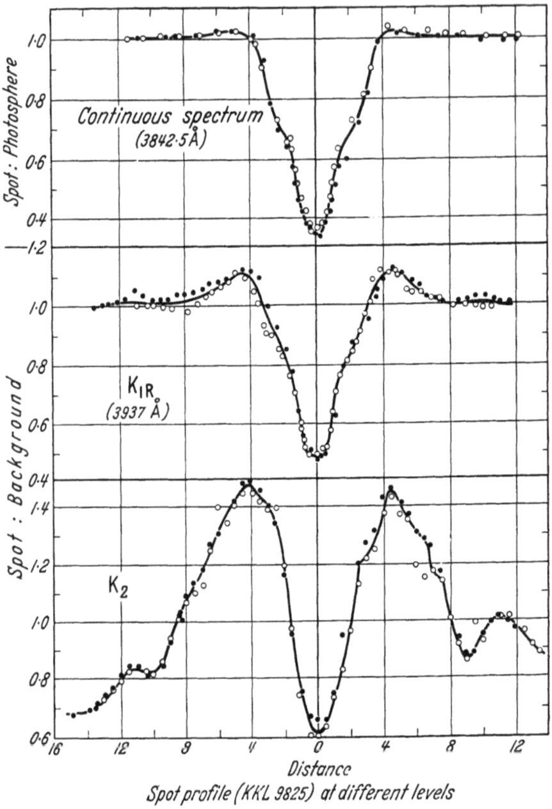

a

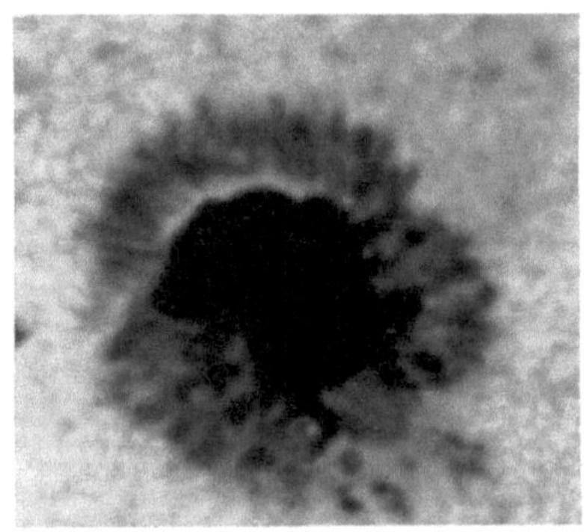

b

Fig. 35a and b. (a) Spot profiles at different levels. The figure clearly shows the increasing strength of the bright ring in higher chromospheric levels. A. K. DAS and A. S. RAMANATHAN: Z. Astrophys. **32**, 91 (1953). (b) Sunspot photographed at the Pic du Midi, 25 February, 1957, showing extraordinary brightening of a part of the umbral limb. Courtesy J. RÖSCH.

for the whole disk (except $\lambda = 3300$ Å)

$\lambda\,(\mu)$	0.4	0.5	0.6	0.8	1.0	1.5	2.0
$I(U)/I(\mathrm{ph})$	0.17	0.22	0.28	0.37	0.46	0.64	0.68

for $\lambda = 3300$ Å

$\cos\vartheta$	0.96	0.87	0.80	0.63	0.48	0.27
$I(U)/I(\mathrm{ph})$	0.12	0.06	0.12	0.06	0.06	0.03

RAMANATHAN's data[3], yielding some 10% greater values are not included in this summary.

[1] Footnote 3, p. 155.
[2] C. W. ALLEN: Astrophysical Quantities, p. 149. 1955.
[3] A. S. RAMANATHAN: Z. Astrophys. **34**, 169 (1954).

The model of the photosphere in the umbra of a "typical spot" (with according to MICHARD's assumption, an umbral area of 80×10^{-6} of the visual hemisphere) can now easily be derived. Since $I(U)/I(\mathrm{ph})$ does not depend on r/R, the ratio of the source functions $S_u(\tau)/S_{\mathrm{ph}}(\tau)$ is independent of τ, and equal to the above ratios. It is permissible to take the source function equal to the black-body function. With WIEN's law we may put (following UNSÖLD):

$$\mathrm{Log_e}\,[I_u(\lambda)/I_{\mathrm{ph}}(\lambda)] = \frac{c_2}{5040\,\lambda}\,\Delta\Theta .$$

The data from the above table lead to $\Delta\Theta = 0.25$ for all wavelengths above 4000 Å. Hence the spot model can be obtained from the photospheric model by adding 0.25 to all Θ-values. Because of this low temperature the relative abundance of H^- ions in sunspots is much smaller than in the photosphere, as was found by T. V. KRAT[1] from a detailed discussion of the continuous spectra of large spots. The continuous absorption is chiefly due to free-free and free-bound transitions in the field of H^+ ions. The model is given in Table 8, p. 161.

γ) Line spectra. A comparison of line strengths in the umbrae of spots and in the photosphere permits to determine the ratios of temperatures, and pressures in the umbra and photosphere. A preliminary analysis of CH. E. MOORE, based on the calibrated Rowland intensities of the lines yielded $\Delta\Theta = 0.19$ and Δ Log $P_e = -0.22$ (umbra minus photosphere). A more refined attack, from TEN BRUGGENCATE and VON KLÜBER[2], based on a careful photometry and on curve-of-growth studies yielded $\Delta\Theta = 0.29$ and Δ Log $P_e = -1.43$. The excitation temperature in the spot is 3900°. Similarly excitation temperatures of 3800 to 4200° were found by ABHYANKAR and RAMANATHAN[3] from a study of 226 metal lines in spectra of four sunspots. The relatively low electron pressure in the spots was also found by T. V. KRAT[1] from the equivalent widths of the resonance lines 4227 of Ca I and K of Ca II. By comparing the ionization formulae for the two lines and by assuming ionization temperatures of 3800 and 5040° for spot and photosphere he found $(P_e)_{\mathrm{phot}}/(P_e)_{\mathrm{spot}} \approx 40$.

It appeared[2] that the horizontal part of the curve of growth was higher for the spot than for the photosphere. This might of course be ascribed to a greater microturbulence but, as an alternative solution, magnetic intensification was suggested[2]. In a longitudinal field a "normal" line will be split into a doublet of opposite circular polarisation. This will not yield any intensification since each doublet component absorbs only continuous light of its own polarization (BABCOCK[4], WARWICK[5], HUBENET[6]). But if each of the doublet components consists of many subcomponents (anomalous Zeeman splitting) an intensification may be obtained, similar to the normal fine structure intensification of spectral lines. (This case has been treated in detail by HUBENET[6].) In the asymptotic case of very strong fields a line consisting of n components and situated on the flat part of the curve of growth will be intensified $n/2$ times. Generally the intensification factor will be much less than that. A similar reasoning applies to the triplet splitting in a transverse field; it was discussed by UNNO[7].

---

[1] T. V. KRAT: Isw. Gl. Astr. Obs. Poulkovo **137**, 1 (1948).
[2] P. TEN BRUGGENCATE and H. VON KLÜBER: Veröff. Göttingen **78** (1944).
[3] K. D. ABHYANKAR and A. S. RAMANATHAN: Astrophys. Journ. **121**, 739 (1955).
[4] H. W. BABCOCK: Astrophys. Journ. **110**, 126 (1949).
[5] J. W. WARWICK: Harvard Reprint **348** (1951). — Z. Astrophys. **35**, 245 (1955).
[6] H. HUBENET: Z. Astrophys. **34**, 110 (1954).
[7] W. UNNO: Publ. Astr. Soc. Japan **8**, 108 (1957).

A further, presumably small intensification will be obtained in an inhomogeneous field: if the field strength H and, hence, the splitting distance vary with depth the self-absorption in the line wings will be reduced (HUBENET[1]); furthermore each line component becomes asymmetric.

Both aspects (intensification by anomalous Zeeman splitting and in an inhomogeneous field) have been investigated observationally. MATTIG[2] computed a curve of growth on the basis of an umbral model of a typical spot, assuming an inhomogeneous field with $dH/dh = 5$ Gauss. Line profiles were computed after PECKER'S exact method. The computed intensification was negligibly small. WARWICK[3], assuming a constant magnetic field, predicted that the intensification of lines with anomalous splitting patterns is too faint to explain the rise of the curve of growth. ZWAAN[4] also investigated the intensification due to abnormal Zeeman splitting, using TEN BRUGGENCATE and VON KLÜBER'S curve of growth. The linear part of this curve could be determined with more certainty by means of high dispersion Mt. Wilson spot-spectra taken by MULDERS. On the flat part there appear no systematic deviations for lines with few or with many Zeeman components.

So the differences between curves of growth for spots and the photosphere must be due to *microturbulence*. If for the photosphere $(\overline{\xi_t^2})^{\frac{1}{2}} = 1.2$ km/sec and $T_{\text{exc}} = 5000°$, and if in the spot $T_{\text{exc}} = 3800°$, we obtain with the aid of data published by MATTIG: $(\overline{\xi_t^2})^{\frac{1}{2}}_{\text{spot}} = 1.9$ and 3.0 km/sec, for ionized and neutral lines respectively. These values apply to a mean spot. There are indications that ξ_t increases with height (MATTIG) and that it is smaller for bigger spots (ZWAAN). In the spectrum of spots close to the limb the Hα line is displaced towards the limb[5], showing that Hα is emitted by higher levels than the photospheric spectrum. The core of Hα in spots appears to come from $h^* \approx 1\,700$ km (cf. Table 6 for the normal chromosphere).

δ) The band spectra. The molecular C_2 lines are enhanced in the spots with respect to the photosphere[6] (ratio of equivalent widths $= 3.2$), whereas the CH lines have nearly equal strengths in both[7]. These data, compared with dissociation computations[8] indicate that $P_{\text{phot}}/P_{\text{spot}} < 1$, where the pressures, of course, refer to the levels in spots and the photosphere where the molecular lines are formed.

From a study of line intensities in the Swan band of C_2 at 5165 Å RICHARDSON[6] found the excitation temperature of the sunspots, with respect to that of the photosphere: $\Delta\Theta_{\text{exc}} = 0.17$.

We summarize the $\Delta\Theta$ values for umbra and photosphere:

continuous radiation	$\tau_0 \approx 1.0$;	$\Delta\Theta = 0.25$
line spectrum	0.3;	$= 0.29$
intensities in Swan bands	0.2	$= 0.17$.

From these data a model of a typical umbra may be derived, following MICHARD. Assuming an average value $\Delta\Theta = 0.25$, the (Θ, τ_λ)-relation for the average spot

---

[1] Cf. footnote 6, p. 159.
[2] Cf. footnote 1, p. 155.
[3] J. W. WARWICK: Z. Astrophys. **35**, 245 (1955).
[4] C. ZWAAN: Bull. astronom. Inst. Netherl. 1959.
[5] W. MATTIG: Naturwiss. **45**, 104 (1958).
[6] R. S. RICHARDSON: Astrophys. Journ. **73**, 216 (1931).
[7] R. S. RICHARDSON: Astrophys. Journ. **77**, 195 (1933).
[8] C. DE JAGER and L. NEVEN: in: Molécules dans les astres, Liège colloquium 1956, p. 357, 1957.

may be derived from the photospheric model by adding 0.25 to all photospheric Θ values. MICHARD'S (Θ, τ)-relation is somewhat more complicated, but it is questionable whether his refinements are really vindicated by the observations. His model was extrapolated and recomputed by MATTIG, who assumed at $\text{Log}\,\tau_0 = -3$: $T = 3330°$.

The values of P_e and P_g were derived by MATTIG in several approximations according to BARBIER'S method. The contribution of the magnetic pressure to the gas pressure has rightly not been taken into account: The fundamental equation of magnetohydrodynamics is (with $v = 0$):

$$\frac{1}{4\pi}(\boldsymbol{H}\,\text{curl}\,\boldsymbol{H}) + \text{grad}\,p = \varrho\,\boldsymbol{g},$$

the z-component (assuming radial symmetry) being:

$$\frac{1}{4\pi} H_r \left(\frac{\delta H_r}{\delta z} - \frac{\delta H_z}{\delta r}\right) = \varrho\,g - \frac{\delta p}{\delta z}.$$

In the spot's centre $H_r = 0$, so that the equations reduce to those used for the normal photosphere. Also the contribution of the turbulent pressure has not been added; it is small but it does not seem wholly negligible.

Table 8. *Model of an umbra after Michard and Mattig.*

τ_0	T	Log P_g	Log P_e	h (km)
0.001	3330°	3.91	− 1.43	0
0.005	3520	4.28	− 1.00	57
0.02	3720	4.59	− 0.59	110
0.10	4060	4.94	− 0.05	171
0.30	4400	5.17	+ 0.38	224
0.70	4720	5.35	0.70	261
1.00	4870	5.42	0.83	278
1.50	5080	5.52	0.99	299
2.00	5250	5.58	1.11	314

In the model thus obtained (see the Table 4) P_e is smaller than at the same optical depths in the photosphere, as is required by the observations (MOORE, TEN BRUGGENCATE and VON KLÜBER, MICHARD, KRAT). So we find: at $\tau_0 = 0.01$; 0.1 and 1: $\Delta\,\text{Log}\,P_e$ (umbra *minus* photosphere) $= -0.7$; -0.8 and -1.2. Further, the pressure gradient at the same temperature is greater in the spot than in the photosphere. Also BERDITSHEVSKAIA[1] predicts, on the basis of a radiative spot model, that

$$(dP/dh)_{\text{spot}} > (dP/dh)_{\text{phot}}.$$

The answer to the very fundamental problem at what height pressure equality occurs between spot and photosphere has not yet been given, but it might be found in investigations like those of MUSTEL[2] who draw attention to the observation that the H_2 and K_2 emission peaks are considerably narrower above sunspots than above the adjacent disk parts. He suggested this to be caused by a reduced density above the spot. This hypothesis is in agreement with direct visual and photographic observations that the chromosphere is lowered above spots, and this might be related to the Evershed effect (outflow of gases in the low chromosphere produce a locally reduced density). Theory (Sect. 39) predicts

[1] V. S. BERDITSHEVSKAIA: Astron. Zhurn. **31**, 51 (1954).
[2] E. R. MUSTEL: Isw. Krysmk. Astrophys. Obs. **13**, 96 (1955).

a lower pressure in the spot. So pressure inequality occurs throughout the chromosphere and photosphere.

The comparison of the geometrical depth scales of umbra and photosphere shows that a spot is perhaps a bit less transparent than the photosphere. This new result disagrees with MICHARD'S who found that the spots are *much more* transparent than the photosphere. This important difference is due to the fact that MICHARD'S pressures are about ten times smaller than those of MATTIG, and this latter difference is for a factor 2 caused by a difference in P_e (MATTIG'S P_e-values were determined theoretically from $T(\tau)$, MICHARD'S were derived empirically); for the greater part—a factor 5—it is caused by differences in the adopted chemical abundances [MATTIG used WEIDEMANN'S modern $N(\mathrm{H})/N$ (metals) ratio of 2×10^4]. For the time being, we are inclined to attach some more weight to MATTIG'S result: a spot is a photospheric region cooler than its surroundings and perhaps slightly less transparent.

34. Motions in sunspots; the Evershed effect. The Evershed effect is a radial mass motion with velocity v_E, mainly in the penumbrae of the spots; it can best be studied near the Sun's limb. The effect was discovered in 1909, later investigators examined its variation with height above the spot, with the distance to the spot centre, with the spot area, with the magnetic field, and with the distance to the centre of the disk. Although it should be borne in mind (ABETTI[1]) that the individual v_E values may vary greatly in amount and direction (the weighted mean value being about 2 km/sec but values as great as 6 km/sec being observed), a fairly complete picture may be built up; it is summarized below:

The value and direction of motion vary with the intensity and the excitation potentials of the lines, this signifies perhaps: with height. Faint metal lines show an outstreaming motion up to 2 km/sec; stronger lines show no motion at all, and the strongest lines (Ca^+, $\mathrm{H}_\alpha \ldots \mathrm{H}_\gamma$) exhibit an instreaming motion up to 3 km/sec. Since the strongest lines originate from the highest (chromospheric) layers this means that in photospheric layers matter is outstreaming. At heights of about 500 to 1000 km above the surface, v_E is zero and, at heights greater than 2000 km, it is instreaming. The effect, discovered by ST. JOHN, was confirmed by the Arcetri observers. The picture is completed and further confirmed by the observation that for a number of faint lines in a spot with $A = 350\times 10^{-6}$ visible hemisphere, examined by ABETTI and MICHARD[2], v_E decreases towards the limb: for $\cos\vartheta = 1.0$, $v_E = 2.6$ km/sec, and for $\cos\vartheta = 0.5$, $v_E \approx 1$ km/sec. This also indicates a decrease of v_E with height.

The value of v_E varies in the spot as a function of the distance to the spot centre. It is virtually zero near the border of the umbra, reaches a maximum value near the border of the penumbra and then decays outward into the photosphere (KINMAN[3]; see Fig. 36). It is nearly zero at a distance to the centre of the spot of about 1.5 times the penumbral radius.

The maximum value of v_E depends linearly on the umbral radius ($U/2$). KINMAN derives the relation

$$(v_E)_{\max} = (1.7 \pm 0.2)\,\frac{U}{2} + 0.56 \pm 0.17 \text{ km/sec},$$

where U is expressed in 10^4 km. Consequently $(v_E)_{\max}$ depends also on the area of the spots; this latter relation, being derived by MICHARD from ABETTI'S

[1] G. ABETTI: Publ. R. Obs. Arcetri **50**, 47 (1932).

[2] R. MICHARD: Ann. d'Astrophys. **14**, 101 (1951).

[3] T. D. KINMAN: Monthly Notices Roy. Astronom. Soc. London **112**, 425 (1952); **113**, 613 (1953).

observations, is represented in the following table; the velocities have been reduced to the disk centre.

A ($\times 10^{-6}$ hemisphere)	34	70	104	230	340
$(v_E)_{max}$ (km/sec) (reduced to disk centre)	1.0	1.8	2.7	3.7	2.8
Number of cases	1	5	6	9	2

There is some divergence between ABETTI's and KINMAN's results: for a spot with $A = 350 \times 10^{-6}$ visible hemisphere, $U/2 = 8800$ km, one finds $(v_E)_{max} = 3.0$

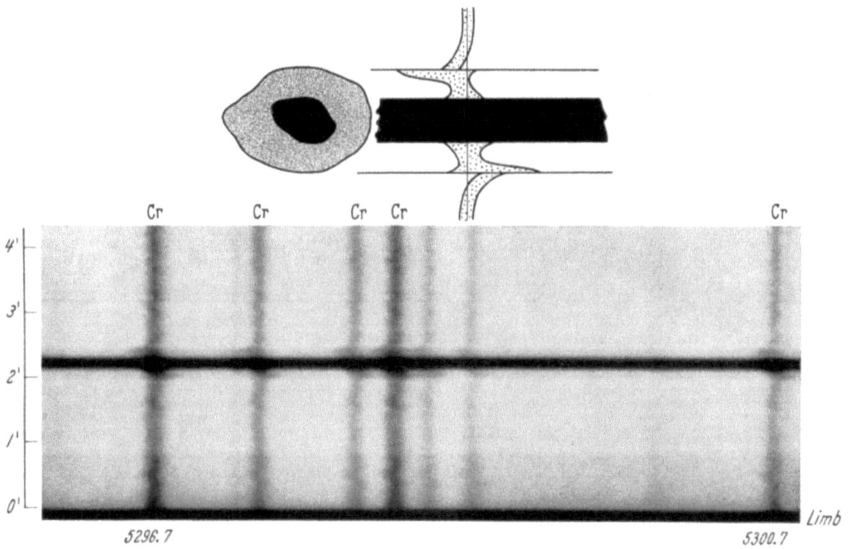

Fig. 36. The Evershed effect. Spot spectrum obtained 2. Sept., 1955, with the McMath-Hulbert Observatory vacuum spectrograph. Cf. also R. R. McMATH, O. C. MOHLER, A. K. PIERCE and L. GOLDBERG: Astrophys. Journ. **124**, 1 (1956). Courtesy McMath-Hulbert observatory.

and 2.1 km/sec, respectively. The time for the streaming matter to reach maximum velocity can be computed by a simple integration. For five spots investigated by KINMAN this value is independent of the umbral diameter and equal to 1.3×10^4 sec.

Finally, there is a relation between $(v_E)_{max}$ and the maximum magnetic fields of the spots (MICHARD[1]):

$H_m \times 10^{-3}$ (Gauss)	1.3	2.2	2.9	3.2	3.5
$(v_E)_m$ (km/sec)	1.3	2.0	2.5	3.7	4.0
n	4	2	5	9	9

The above picture of outstreaming and instreaming motions in and around the spots makes it clear that we should expect cyclonic and anticyclonic motions in the gas masses surrounding the spots. The Coriolis force, which produces a deflection to the right on the norhtern hemisphere is $C = 2m v_E u \sin\varphi$ where m is the mass of a particle with a speed v_E, and u is the angular velocity of the Sun.

[1] Cf. footnote 2, p. 162.

If the Evershed effect is active over a distance l, which is traversed by the particles in a time $t=l/v_E$, the relative deflection is approximately

$$\frac{s}{l}=v_E u \frac{l^2}{v_E^2 l}\sin\varphi = u\frac{l}{v_E}\sin\varphi,$$

and with $\varphi=30°$, $v=1$ km/sec, $l=6\times10^4$ km, $u=2.7\times10^{-6}\,\text{sec}^{-1}$, one finds $s/l\approx0.08$. These cyclons are actually observed in hydrogen and calcium spectroheliograms (Fig. 37) where they are produced by the instreaming motion and, hence, have the same character as the terrestrial cyclons with a counterclock motion on the northern hemisphere.

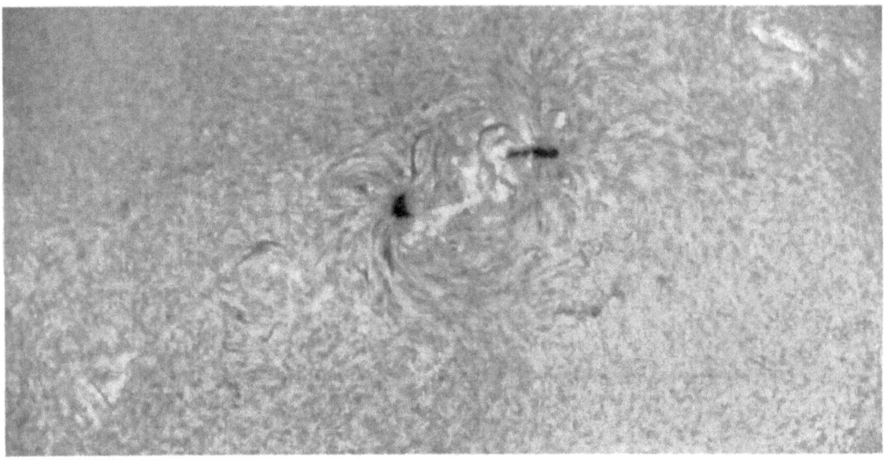

Fig. 37. A part of an $H\alpha_{2,3}$ spectroheliogram of 20. IX. 1926 showing vortices and fibrils (cf. also Sect. 43). Courtesy L. D'AZAMBUJA, Meudon.

The Coriolis force is the cause that v_E is not purely radial but that it has a small tangential component. For faint metal lines this tangential component has been observed (ABETTI and CALAMAI[1]). As can be expected, the sign of rotation is anticyclonic, that is inverse as compared with the hydrogen and calcium vortices because for the metal lines the motions are outstreaming, producing a clockwise motion on the northern hemisphere.

35. Structure, motions and magnetic fields in a typical spot. With the data collected in the preceding sections we will try to construct a model of a "typical large spot". This attempt of course suffers to the same evil as all discussions of sunspots, that the various properties, measured for different spots are combined to construct a model of an "average" spot. It has not yet been tried to measure all the parameters listed below for one and the same spot, and to construct a model of a "real" spot.

In Sect. 33 the temperature, pressures, and optical depth have already been given for a spot with an umbral area of 80×10^{-6} visible hemisphere (the area of the visible hemisphere is $2\pi R_\odot^2$). For such a spot we have, according to the relations already given:

Total area $A=350\times10^{-6}$ visible hemisphere.
Umbral diameter $U=17500$ km; relative intensity of umbra: 0.3.
Penumbral diameter $P=37000$ km; relative intensity of penumbra: 0.8.
Diameter of bright ring $R=50000$ km; relative intensity: 1.03.

[1] G. CALAMAI: Publ. R. Obs. Arcetri **52**, 39 (1934).

Maximum field strength $H_M = 3100$ Gauss.
Total magnetic flux $\Phi = 8 \times 10^{21}$ Gauss cm^2.
Maximum Evershed effect in photosphere 3 km/sec.

If we call r the distance to the centre, in km, and i the inclination of the lines of force to the Sun's surface, we have

r (km)	H (Gauss)	i	v_E (km/sec) in photosphere	r (km)	H (Gauss)	i	v_E (km/sec) in photosphere
0	3100	0°		14000	600	70°	2.4
2000	3000	10	0	18000			1.5
6000	2500	30	1.8	24000	(10)		0.3
10000	1500	50	2.7				

The variation of $(v_E)_{\max}$ with height above the photosphere:

$$h = 0\text{ km}: \quad (v_E)_{\max} = +3\text{ km/sec}$$
$$500\text{ km} \qquad 0\text{ km/sec}$$
$$2000\text{ km} \qquad -4\text{ km/sec}.$$

With the models of the spot of Sect. 33 and of this section, it is possible to understand the heat balance of the spot, and the occurrence of the bright ring.

We assume (cf. KINMAN[1]) that matter moves sideward through a thin superficial layer under absorption and emission of radiation. In the penumbral part it is gradually heated by the hotter layers beneath it. If the equation of motion is known, the heat balance may be computed.

For an ideal gas in the steady state (for which $\partial/\partial t$-terms are zero) we have according to the first law of thermodynamics (written in terms of the Lagrangian time derivative):

$$\frac{Dq}{Dt} = \frac{De}{Dt} + P\frac{D(1/\varrho)}{Dt} = \frac{v}{\gamma-1}\left(\frac{1}{\varrho}\frac{dP}{ds} - \gamma P\frac{d(1/\varrho)}{ds}\right) \tag{35.1}$$

where q and e are the heat absorption and the internal energy per unit mass; ϱ is the density, P is the gas pressure.

EULER's equation of motion in the steady state form is:

$$v\left(v\frac{dv}{ds} + \frac{1}{\varrho}\frac{dP}{ds}\right) = 0, \tag{35.2}$$

and the equation of continuity for radial horizontal flow is

$$\frac{1}{\varrho}\frac{\partial\varrho}{\partial s} + \frac{1}{v}\frac{\partial v}{\partial s} + \frac{1}{s} = 0. \tag{35.3}$$

From Eqs. (35.1) to (35.3) one obtaines after eliminating P and ϱ:

$$\frac{Dq}{Dt} = \left(c_p T - \frac{v^2}{\gamma-1}\right)\frac{dv}{ds} + \frac{c_p T v}{s}. \tag{35.4}$$

This expression gives the heat absorbed per second per unit mass as functions of the horizontal velocity component v, the temperature T and the distance s from the centre of the spot. In this expression the pressure does not occur, which is favourable in view of the uncertain value of the gas pressure and of the magnetic pressure contribution.

The computation was made by KINMAN[1] for a spot with a penumbral diameter of 2.2×10^4 km for which the intensity profile was measured by RICHARDSON. The temperature variation in the spot is found by assuming that the measured intensities are proportional to T^4. For the (Evershed) velocities v_E a mean curve

[1] T. D. KINMAN: Monthly Notices Roy. Astronom. Soc. London **113**, 613 (1953).

was derived from the observations. It appears that Dq/Dt is of the order of 10^8 erg g^{-1} sec^{-1}; it is >0 for $s<2\times10^4$ km, negative for $s>2\times10^4$ km and it approaches zero for $s>4\times10^4$ km. Hence, energy is absorbed in the umbra and in about half of the penumbral area, and it is emitted outside this region. Although the total amount of emitted energy is large, yet the bright ring will have a relatively slight intensity, as the energy is emitted in a large area.

b) Group of spots.

Of basic importance for the understanding of the sunspots is: (a) that they occur mainly in groups, (b) that spots are transient phenomena. In this subdivision we first summarize our present knowledge of spot groups. In Sect. 39 theories of sunspots will be discussed.

36. Magnetic classification of spot groups; polarity law; N-S asymmetry. Magnetic fields of spots are regularly measured at Mt. Wilson (by a visual method), Potsdam (photographic), Simeis and Poulkovo. The Mt. Wilson results are published in the Publications of the Astronomical Society of the Pacific, those of Potsdam in the Astronomische Nachrichten.

The bipolar spot group, consisting of two neighbouring spots of opposite polarity, is considered as the basic phenomenon and the following generally adopted classification scheme is based on this concept. In reading this scheme it should be kept in mind that a spot may not always be visible, even when the field is still measurable. For that reason we prefer to speak of magnetic field strength maxima (centres), rather than about spots. The spots always occur in chromospheric faculae ("plages"). The position of the magnetic field strength maxim(a)(um) with respect to the faculae is contained in the following classification scheme. The preceding (p) part of a facula is the western part.

Unipolar groups

α = faculae symmetrically distributed around the field strength centre.
αp = centre in preceding part of elongated field.
αf = centre in following part of field.

Bipolar groups; two neighbouring field strength maxima of opposite polarity.

β = area of preceding and following centres about equal.
βp = preceding centre is the main member.
βf = following centre is the main member.
$\beta\gamma$ = no clear borderline between North and South polarity regions.

γ = *Multipolar groups;* no regularity discernable.

A $\left\{\begin{matrix}\text{preceding}\\ \text{following}\end{matrix}\right\}$ l, added to the classification symbols, means that on the day of observation the group $\left\{\begin{matrix}\text{appeared}\\ \text{disappeared}\end{matrix}\right\}$ at the $\left\{\begin{matrix}\text{East}\\ \text{West}\end{matrix}\right\}$ limb.

A $\left\{\begin{matrix}\text{preceding}\\ \text{following}\end{matrix}\right\}$ d means that the group $\left\{\begin{matrix}\text{originated}\\ \text{died}\end{matrix}\right\}$ at the day of observation.

Of all 6384 groups classified between 1919 and 1946, 8.6% were α; 91% were β and 0.4% were γ[1]. This shows that normal spot groups are generally bipolar, the other groups are exceptions. Moreover, many of the single spots have lost their companion or they are accompanied by an "invisible spot": a facular region with a magnetic field strength maximum without the corresponding spot.

Generally the two members have opposite polarity. This is one aspect of the important *polarity law*. The other aspect is that, generally, all preceding spots in the northern hemisphere have the same polarity, opposite to that of the preceding spots of the southern hemisphere. Finally the distribution of polarity

[1] R. S. RICHARDSON: Astrophys. Journ. **107**, 78 (1948).

among the two members of a group changes sign with each new solar cycle. In the past cycles the polarities were distributed as follows:

	Hemisphere			
	northern		southern	
	p	*f*	*p*	*f*
Cycle 1933 to 1945	*N*	*S*	*S*	*N*
1945 to 1954	*S*	*N*	*N*	*S*
after 1954	*N*	*S*	*S*	*N*

The mean magnetic flux of the *p* spots is about three times greater than that of the *f* spots. Hence the total magnetic flux changes sign between the sunspot cycles, producing a small field variation with a period of 22 years. The amplitude of this variation is of the order of 10^{22} Gauss cm², of the same order as the total flux of the poloidal solar magnetic field (cf. Sect. 104).

37. Development of spot groups: the E-W asymmetry. A spot starts its life as a small pore in the general granulation field, diameter ≈2000 to 3000 km. Most of these pores soon disappear, some remain as long as one day or still longer and may develop further. If the spots are stable, spot groups are formed which within a week may cover a solar area with a diameter up to 10° and more. The growth of a spot group that reaches full development can best be illustrated by Fig. 38, in which the various stages of the spot group are indicated by WALDMEIER's[1] designations *A* ... *H*. The table below gives a short description.

A.	Single spot or group; pores without penumbra, no bipolar structure	1st day
B.	Bipolar group	
C.	Bipolar group; major spot shows penumbra	2nd day
D.	Both major spots have penumbrae; length of group < 10°	
E.	Great bipolar group; many smal spots; length ≳ 10°	5th–10th day
F.	Maximum type; length > 15°	
G.	Great bipolar group, without smaller companions, length ≳ 10°	10th–15th day
H.	Unipolar spot with penumbra; > 2°.5	
J.	Unipolar spot with penumbra; *P*<2°.5.	

The description refers to the complete development; in most cases only stage *A* is reached, or the development is *A*-*B*-*A* or *A*-*B*-*C*-*B*-*A* ... etc. The postmaximum phase in most cases takes much more time than the ascending part: sometimes the descending part takes two or even more solar rotations. Only rarely does the reverse occur. In general the spot group has a short life time and within ten of fifteen days it is frequently reduced to a single unipolar spot. The mean decay time of spot groups is ten days; out of ten thousand spot groups studied by GNEVISHEV[2] only 3700 ($=e^{-1}\times 10000$) had a life time as long as 10 days. Only 40 (=0.4%) stayed 50 days, three (=0.03%) stayed 100 days and one 150 days. The life time depends also on the maximum group area. Roughly, it is 10 days for a group with area 10×10^{-5} visible hemisphere; it is 40^d for a group with area 40×10^{-5}. The life times are greatest at the lowest latitudes; spots at high latitudes are always short-lived.

By means of auto-correlation analysis[3] of sunspot *numbers* *R* (cf. Sect. 96) and total spot *areas* *A* the mean life times appeared to be 1.7 month for *R* and 1.4 month for *A*. The difference, 0.3 month, is probably real, and is due to the

[1] M. WALDMEIER: Z. Astrophys. **16**, 285 (1938).
[2] M. N. GNEVISHEV: Poulkowo Obs. Circ. **24** (1938).
[3] C. W. ALLEN: Monthly Notices Roy. Astronom. Soc. London **117**, 174 (1957).

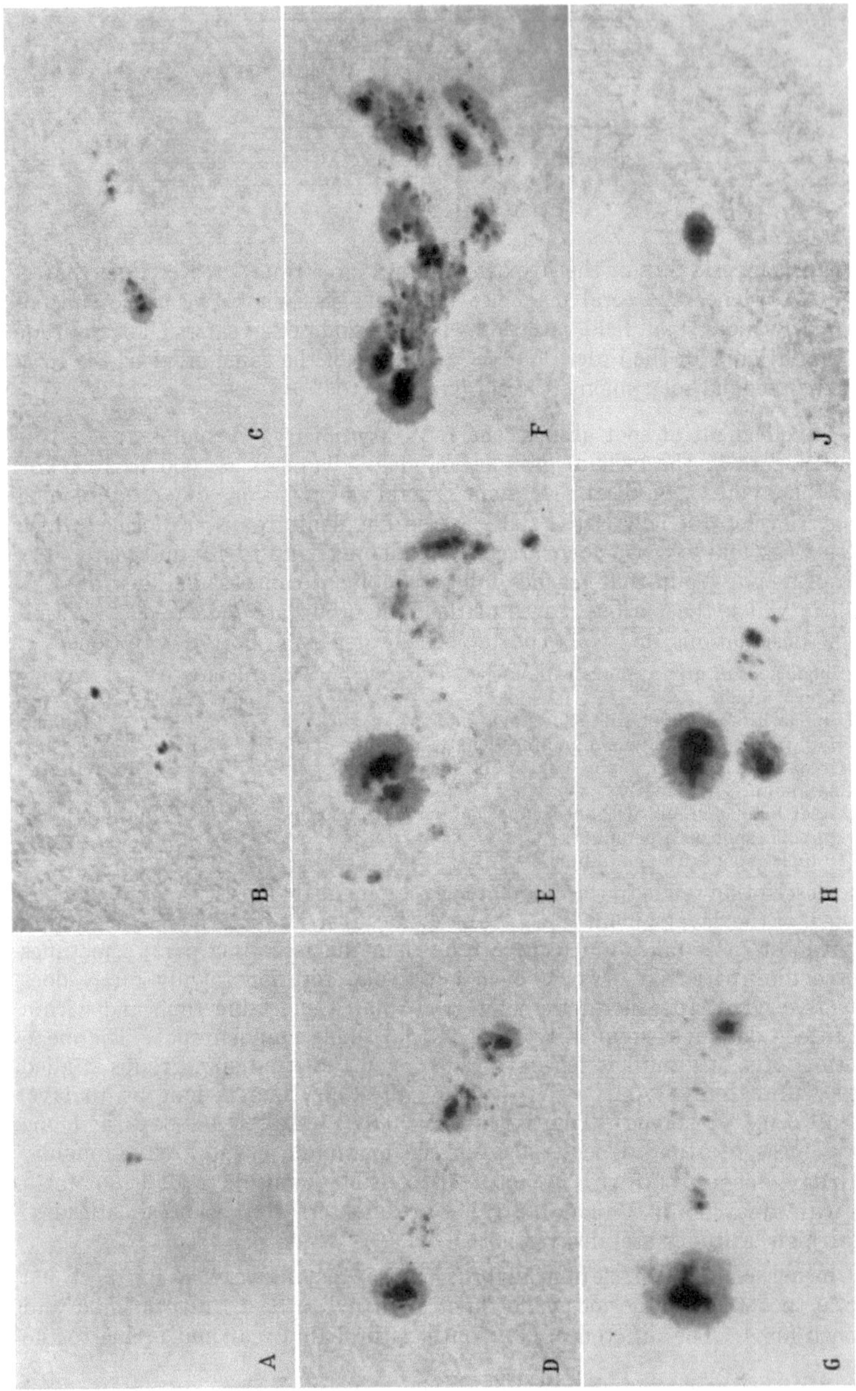

Fig. 38 A—J. Sunspots at various stages of their development.
A: 33. 5. 1956 B: 11. 8. 1948 C: 8. 7. 1956 D: 21. 8. 1956 E: 12. 8. 1956 F: 9. 9. 1956 G: 30. 5. 1956 H: 21. 8. 1956 J: 31 8. 1949
Courtesy W. MATTIG, Potsdam.

dissolution of large spots in many smaller ones in the second phase of their lifes. (These mean life times for R and A are greater than the above quoted ones since in taking total R and A values more weight is given to the large groups.)

When the group is in phase F, it contains a great number of spots, amounting to 40 or more. At such a moment the total area of the group may be of the order of 1000×10^{-6} visible hemisphere.

The future development of a group is often already reflected in the primary part of the development curve (Fig. 39). The initial increase in the rate of growth is well marked for large spots; for smaller ones it is nearly absent; these latter spots reach their maxima earlier than the larger ones.

The spot group covers a more or less elliptical region, the great axis only slightly inclined towards the Sun's equator in such a way that the p spot has a smaller

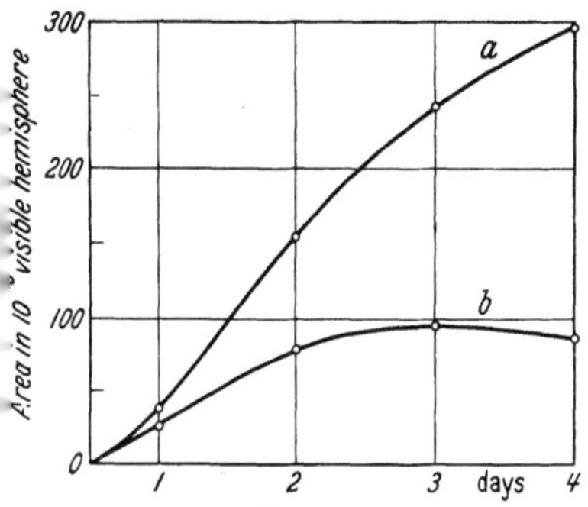

Fig. 39.

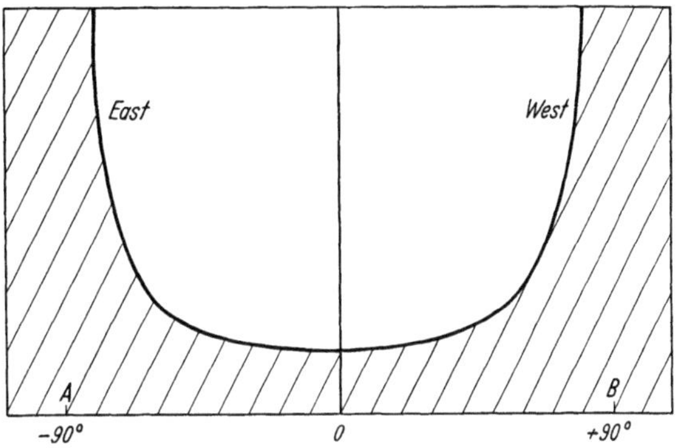

Fig. 40.

Fig. 39. Initial rate of growth of spots with $\binom{a}{b}$ areas $\gtrless 150 \times 10^{-6}$ visible hemisphere on the fourth day. J. C. DOBBIE: Observatory **62**, 289 (1939).

Fig. 40. East-West asymmetry in the origin of sunspots. After M. MINNAERT.

latitude than the f spot. The inclination i_A of the axis increases with increasing heliographic latitude, cf. the following table, summarizing investigations of BRUNNER and JOY. The individual values from which the means have been taken scatter greatly, but the general trend is well visible.

Heliographic latitude b (°)	0–4	5–9	10–14	15–19	20–24	25–29	30–34
i_A	2.°2	3.°0	5.°5	6.°5	9.°4	11.°9	14.°9

Further i_A depends on the life history of the spot: for a mean latitude it is

7.°8 for class B groups
6.°5 for class C-E groups
4.°5 for class F groups.

The *E-W asymmetry* was discovered by MAUNDER[1] in 1907: at the eastern half more spotgroups seem to *originate* than at the western half of the visible disk; more disappear at the western half. This is obviously due to an effect of statistical selection, because it seems more than doubtful that the Earth has any influence on the Sun. The effect was explained by SCHUSTER[2]. Let there be a mean time difference ΔT between the moment of birth and that of first visibility

[1] A. S. D. MAUNDER: Monthly Notices Roy. Astronom. Soc. London **67**, 451 (1907).
[2] A. SCHUSTER: Proc. Roy. Soc. Lond., Ser. A **85**, 309 (1911).

of the spots. Further, the visibility of all spots decreases towards the limb, due to perspectivic foreshortening ($\propto \cos\vartheta$). Combined with the Sun's rotation this gives a preponderance of first discoveries at the eastern half, as is clear from a diagram given by MINNAERT[1] (Fig. 40): the abscissa gives the heliographic longitude (supposing for simplicity that the Sun's axis is perpendicular to the ecliptical plane, the spot being at the equator). The vertical axis is the time axis. Spots originate at equidistant points at all longitudes between the points A and B. Because of the solar rotation they drift towards the right. The locus-time curves form a system of parallel lines, inclination $(14°/\text{day})^{-1}$. If the visibility were equal at each point of the disk the number of first appearances would be given by the number of intersections, between -90 and $+90°$, of these lines with a horizontal one at height ΔT. There would be no asymmetry in the visibility. Actually, for a spot at heliocentric angle ϑ the delay time becomes $\Delta T \sec\vartheta$. In that case more spots attain visibility at the eastern half than at the western half of the disk.

A detailed discussion by WALDMEIER[2] and FERRARI D'OCCHIEPPO[3] based on extensive observational material yielded the best agreement with the observations for a time ΔT of growth from birth to visibility $\Delta T = 0\overset{d}{.}8$; the observations on the growth of spots suggest $\Delta T \approx 0\overset{d}{.}5$.

It has been claimed that the E-W asymmetry is also valid for the visibility proper: at any time there should be more spots visible at the eastern half than at the western half. However, new statistical investigations[4] have made this asymmetry doubtful.

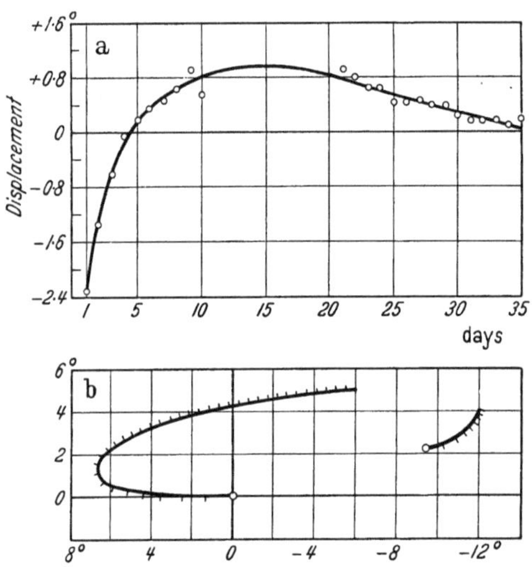

Fig. 41 a and b. (a) Mean displacement in longitude of 30 days lived leading spots. Monthly Notices Roy. Astronom. Soc. London **85**, 553 (1925). (b) Relative motions of the spots of bipolar groups. M. WALDMEIER: Astr. Mitt. Zürich **138**, 451 (1939).

38. Internal motions of the spots in groups. The spot groups exhibit systematic internal motions, especially in the first days after the origin of the spots. The main trend is a general divergence: initially the p- and f-spots move in opposite directions, but the preceding spots move westward more rapidly than the f-spots move eastward. Fig. 41a shows the mean displacement for some 30-day p-spots (the f-spots generally last shorter than the p-spots). In determining the relative displacement in longitude, the influence of the differential solar rotation given by Eq. (102.1) has been eliminated. What remains, are the differential spot motions in longitude.

The westward motion of the p-spot is most rapid immediately after the origin of the spot. The motion slows down and the spot comes to rest when the group has obtained its greatest development. At that moment the spot has shifted over a distance of the order of 10° towards the west. Then the dissolution of the group sets in and the p-spot moves back eastward, but at a much lesser

[1] M. MINNAERT: Astronom. Nachr. **269**, 48 (1939).
[2] M. WALDMEIER: Astr. Mitt. Zürich **138**, 460 (1939).
[3] K. FERRARI D'OCCHIEPPO: Mitt. Sternwarte Wien **5**, Nr. 1 (1951).
[4] A. BRUZEK: Z. Astrophys. **33**, 267 (1954).

speed (0°.08 to 0°.1 per day) and at the end of its development the spot has often come back to near its birthplace. The f-spot has then already disappeared.

The f-spot moves eastward with a mean speed of 0°.2 per day immediately after its birth; after about 5 days it comes to a standstill. For a typical example (which, however, shows a more than normal motion in latitude) see Fig. 41b.

In this connection it should be remarked that the backward motion of the p-spot reduces to a standstill if for the solar rotational velocity the values are adopted suggested by the observation of long-lived facular regions (M. D'AZAMBUJA, private communication, cf. also Sect. 102).

The latitude motions are generally small. The maximum mean values are of the order of 0°.02 to 0°.04 per day. The general latitude drift of sunspots will be discussed in more detail in Sect. 104 in connection with the problem of the general internal motions.

39. What is a sunspot? The answer to this question can not yet been given in all details, but there are indications in what direction to look for it.

The life time of the magnetic fields of spots must be a very long one (COWLING[1]). The conductivity of matter is great in the Sun's interior, so that any change in the magnetic field will be counteracted by induction currents: the matter can move but very slowly with respect to the magnetic fields, or in the common term: the fields are "frozen" into the conducting material.

We estimate the decay time of the fields with the following computation based on MAXWELL's equations in which the magnetic permeability μ and the dielectric constant ε are taken as unity, and in which the displacement current is assumed negligibly small as compared with the current density. The latter is $\boldsymbol{j} = \sigma \boldsymbol{E}$, in which σ is the conductivity, $\boldsymbol{E}$ is the electric vector connected with $\boldsymbol{H}$ by the equation of electromagnetic induction:

$$\operatorname{curl} \boldsymbol{E} = -\partial \boldsymbol{H}/\partial t.$$

Hence

$$\partial \boldsymbol{H}/\partial t = -\operatorname{curl} \boldsymbol{j}/\sigma$$

and with

$$\operatorname{curl} \boldsymbol{H} = 4\pi \boldsymbol{j},$$

we obtain

$$\partial \boldsymbol{H}/\partial t = \frac{1}{4\pi\sigma} \nabla^2 \boldsymbol{H}. \tag{39.1}$$

This equation is of the diffusion type. Apparently the field decays by diffusion and by subsequent neutralization. The characteristic time t_0 for this diffusion process is formed by approximating (39.1) as $4\pi\sigma H/t_0 = H/l^2$, where l is the scale length of the field variations; t_0 appears to be independent of H. With $l = 3000$ km, $\sigma = 3 \times 10^{-8}$ e.m.u. we obtain $t_0 = 10^3$ years, which is very long as compared with the life time of a typical spot, 10 or 20 days. So it is obvious that the magnetic fields of spots exist a long time before and after the visibility of the spot. These fields must somehow be brought to visibility, perhaps by violent convection in the granulation zone, in the lower part of which they originate, maybe by the "spaghetti process" (cf. Sect. 13). According to this process, turbulent energy is changed into magnetic energy till equality is reached between the mean turbulent energy $\frac{1}{2} m v^2$ and the magnetic energy $H^2/8\pi$. With VITENSE's values: $\varrho = 10^{-4}$, $v = 0.5$ km/sec, valid for the mean region of the convection zone, we obtain $H = 1700$ Gauss, which is of the correct order of magnitude. Greater field strengths should be produced in deeper layers of this zone.

The image of the formation of sunspots, thus obtained, is that of magnetic tubes, initially slowly twisting in the deep parts of the convection zone. In

[1] T. G. COWLING: in: The Sun (ed. G. P. KUIPER), [*1*], p. 545. 1953.

this respect, it perhaps is essential that the characteristic diameter of a well developed spot group, 10^5 km, is of the same order as the assumed thickness of the convection zone, and that the time for a convection element to rise through the zone is of the same order as the life time of great spot groups[1]. Somehow these fields are brought to the surface and produce there the associated phenomena. That spots are not observed near the solar poles is sometimes suggested to be due to the inhibition of convection by the poloidal magnetic field, which has its greatest vertical component near the poles. In the equatorial regions the poloidal field does not appear, whereas the large, but weak toroidal fields occurring in low latitudes might be caused by the generating effects described above.

Any theory of sunspots should explain why spots are found in *pairs* with *opposite polarities*. Two fairly early suggestions are (a) those of HALE who thought of a horseshoe-like field reaching with both ends at the surface, and (b) that of BJERKNESS who thought of long toroidal fields closed around the Sun's axis and lying below the surface. As PARKER[2] showed, such a horizontal magnetic flux tube in an electrically conducting atmosphere is dynamically unstable; it is buoyant and thus tends to rise, to become visible as sunspots. This applies equally well to toroids as to twisted magnetic fields[3]; hence supporting the two early suggestions, but modified in such a way that the fields exist long before, in the form of twisted or toroidal tubes, in the lower invisible part of the convection zone.

Another paper discussing the toroidal theory in a modern form is by BULLARD[4]. ALFVÉN[5], on the other hand, assuming that the internal solar field is mainly orientated in meridional planes, suggests that sunspots are associated with magnetohydrodynamic disturbances propagating along the lines of force of this field, and this view is supported by arguments of FERRARO and KENDALL[6]: the magneto-mechanical forces arising from the convective motions which should bring the spots upward, seem too great to be compatible with the observed photospheric large-scale velocity fields.

It is difficult to decide on observational grounds between the various suggestions, although the observations of the evolution of magnetic regions by the BABCOCKS perhaps favors the concept of a submerged field that occasionally rises in loops to the surface.

If next we turn to the *individual spots*, the question arises why the spots are cooler than the surrounding photosphere. The low surface temperature of the spots cannot be due to a cooling at birth after which the spots are heated by the surrounding photosphere: the opacity is too small and the observed sharpness of the spot intensity profile indicates the existence of another, conservative, cooling mechanism. It is probable that the low temperature is kept by the strong magnetic fields inhibiting convective transport of energy, thus causing a steep temperature gradient in the spot's outer part (BIERMANN[7]). A strong magnetic field suppresses convective motions by the effect of Foucault currents. No convection will occur above the depth where the magnetic energy density is of the same order as the energy of turbulent motions. For $H = 2000$ Gauss and

[1] C. DE JAGER: Vistas in Astronomy **3** (1959).
[2] E. N. PARKER: Astrophys. Journ. **121**, 491 (1955).
[3] P. H. ROBERTS: Astrophys. Journ **124**, 430 (1956).
[4] E. C. BULLARD: Vistas in Astronomy **1**, 685 (1956).
[5] H. ALFVÉN: Cosmical Electrodynamics. 1950.
[6] V. C. A. FERRARO and P. C. KENDALL: Astrophys. Journ. **124**, 443 (1956).
[7] L. BIERMANN: Vjschr. Astr. Ges. **76**, 194 (1941).

for Böhm-Vitense's model of the granulation zone this equality occurs for Log $P_g = 7.5$, hence near the upper part of the convection zone, but still well below the photosphere. In the deep layers convection is not prevented. This might perhaps be the reason why Rösch's observations show but faintly developed umbral granulation elements, which do not even occur in all spots.

A second cause for the low temperatures of sunspots might be the magnetic evaporation of hot ions and electrons in a divergent magnetic field[1]. A hydrogen ion and electron pair with energy W traversing the sunspot's magnetic field undergoes an upward force $(W/H)\,\text{grad}\,\boldsymbol{H}$. This force is for an average spot greater than the gravitation force on the ion if $W > 3$ eV. So the hot ion pairs will be evaporated from the spot. The mechanism is effective in regions where the Larmor frequency of the ions exceeds the collision frequency, and also effective in the upper less dense parts of the atmosphere.

So a spot proper is rather shallow, as far as cooling is concerned, but the magnetic field, to which the cooling is due, may penetrate deep into the granulation zone. On the basis of this image Sweet[2] suggested to explain the Evershed effect and the bright ring: the convective energy flux, coming from below, is prevented to ascend in the spot region and escapes sideward. An estimate shows that the computed velocities are of the right order of magnitude, but it is difficult to explain with this model why the Evershed effect is also observed above the umbra.

According to Schlüter and Temesvary[3] the pressure in the spots should be less than in the neighbouring regions. It is this pressure difference that prevents the magnetic field from expanding sideward.

The above picture of sunspots is just the opposite from the one generally assumed before 1940 when it was thought that sunspots were big convective regions, where the low temperatures follow from the adiabatic expansion of the ascending gas. We now know that in the uppermost part of the spots convection is inhibited by the magnetic field.

II. Photospheric and chromospheric faculae.

The faculae are bright structures on the disk, mainly visible in spectroheliograms of the strong Fraunhofer lines (Figs. 18, 19 and 37) and also, faintly, visible in the continuous spectrum, but only close to the Sun's limb. The first are called the *chromospheric faculae*, the last are the *photospheric faculae*, but both names apply to the same structures viewed in high and low parts of the solar atmosphere. There is a strong relationship between faculae and sunspots: spots, even the smallest, always occur in faculae, although the converse is not true: often small facular areas occur without spots.

a) The faculae proper.

40. Photospheric faculae; structures and relation to sunspots; development. The photospheric faculae appear as bright structures around sunspots, with areas about four times greater than those of the spot groups. Their latitude distribution is nearly the same as that of the spots, but they may occur in a somewhat (15°) broader belt, to about 10° higher and 5° lower than the spots.

[1] W. H. Bostick: Discussion remark in I.A.U. symposium **6**, 273 (1958).
[2] P. A. Sweet: Vistas in Astronomy **1**, 675 (1956).
[3] A. Schlüter and S. Temesvary: I.A.U. symposium **6**, 263 (1958).

The mean latitude is somewhat higher than that of the spots. Small rather short-lived faculae appear in latitudes around $b = 70°$. These *polar faculae* will be treated in Sect. 42.

At the minimum of sunspot activity the mean percentage of the facular *areas* not connected with spots is 30%; this value decreases to 6% at the maximum of solar activity. For the total number of facular *centres* these values are 60 and 25% respectively. This shows that the faculae which contain spots have a greater mean area than those which do not.

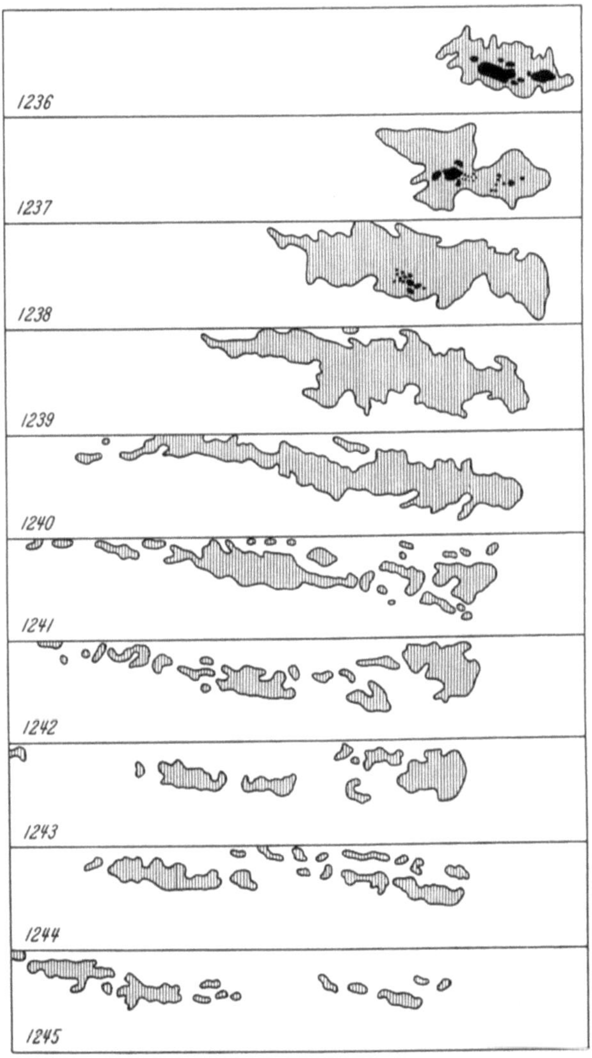

Fig. 42. Development of a big spotgroup and of the corresponding facular field during solar rotations No 1236—1245, February to October 1946. M. WALDMEIER: Ergebnisse und Probleme der Sonnenphysik, p. 203. Leipzig 1955.

Upon close examination (TEN BRUGGENCATE[1], WALDMEIER[2]) the faculae appear as a network of aligned mottles. The alignements have widths equal to the diameters of the knots: 5000 to 10000 km and lengths of the order of 50000 km; figures that greatly resemble the coarse mottling observed in spectroheliograms. There is more correspondence: the mottles have diameters of the order of 4000 to 6000 km, and they are composed of facular granules with diameters of the order of 1000 km or somewhat greater. So we find again the three kinds of elements already observed in spectroheliograms. In the photospheric faculae these characteristic chromospheric structures thus seem to descend down to the upper photospheric layers. A photometric investigation of TEN BRUGGENCATE[1] has shown that at least part of the background of the facular region is about 10% brighter than that of the surrounding photosphere. The granules, knots and the alignements are superposed on this background.

[1] P. TEN BRUGGENCATE: Z. Astrophys. **19**, 59 (1939).
[2] M. WALDMEIER: Helv. phys. Acta **13**, 14 (1939).

The life times of the facular granules are about 6 to 7 minutes according to MACRIS[1] and 1 to 2 minutes according to KRAT[2,3] and GOLDBERG-ROGOSINSKAJA[3]. Earlier estimates gave life times of 1^h and more, but these probably refer to complexes of granules, perhaps to the coarse mottles.

The development of the spot faculae is closely related to that of the spots. Sometimes they originate together with the spots, often some hours to some days earlier. Initially the faculae show the bipolar character of the spot group; they cluster on the two main members and in some cases these two centres remain visible even after the disappearance of the *f*-spot. The facular centres follow the divergent motions of the spots.

The greatest facular brightness is reached after the first few days; later the facular fields become larger but the brightness diminishes. During their development they tend to extend in latitude direction, especially towards greater latitudes. The differential solar rotation causes the initially more or less circular field to elongate in longitude and to become oval. This makes clear why the axis of the oval facular field is not parallel to the Sun's equator but shows an inclination, analogous to that of the bipolar spot groups. After many solar rotations the long-lived facular fields may reach considerable lengths, as is illustrated by a characteristic example in Fig. 42. The faculae occurring in this figure were followed during at least ten solar rotations. The corresponding spots appeared only during the first three rotations. Observations show that in general the mean life time of the faculae is about three times greater than that of the corresponding spots. The end stage of the facular field is its resolution into many smaller centres, which gradually dissolve. Especially during sunspot maximum and some years thereafter the Sun is covered by many facular fields, which may be so numerous and so extended that they combine and form long alignments of facular fields in the sunspot belts.

The facular areas are largest near sunspot maximum. However, in the first few years after maximum, the zone occupied by faculae still extends in latitude, indicating that the *number* of facular centres may become greater, while their mean areas become smaller.

The faculae not associated with spots are smaller than the associated ones and their life times are smaller than those of the spot faculae, viz. two months or less. They never reach the great brightness of the spot faculae. As a rule, if a facular area is bright, a spot has occurred or will occur in it.

An E-W asymmetry of facular distribution has been claimed[4]; the faculae should be preponderant on the east side. However this assertion is not confirmed[5].

41. Photometry of photospheric faculae; model. As has already been mentioned in Sect. 40 the detailed photometry shows that the facular granules are superposed on a background which, itself, seems to be brighter than the neighbouring photosphere. The following table summarizes the values of the mean intensity fluctuations, as determined by TEN BRUGGENCATE:

Excess of granular background over photosphere 10%
Mean $\Delta I/I$ in granule region $\pm 10\%$

---

[1] C. MACRIS: Ann. d'Astrophys. **16**, 19 (1953).
[2] V. A. KRAT: Isw. Gl. Astr. Obs. Poulkovo **152**, (1954).
[3] V. A. KRAT and N. M. GOLDBERG-ROGOSINSKAJA: Isw. Gl. Astr. Obs. Poulkovo **155** (1956).
[4] E. W. MAUNDER: Monthly Notices Roy. Astronom. Soc. London **80**, 724 (1920).
[5] G. H. A. ARCHENHOLD: Observatory **64**, 165 (1941).

The brightest faculae are 40 to 45% brighter than the photosphere; MUSTEL[1] estimates that this latter value may even go up to 150%.

The centre-limb variation of the brightness of the unresolved faculae has been measured both for the integrated (WORMELL[2]) and for the "monochromatic" radiation; the values in the Table 9 give the brightness of the *unresolved* faculae expressed in that of the neighbouring undisturbed photosphere.

Table 9.

cos ϑ	3900 Å	4330 Å [3]	5000 Å	5780 Å [3]	Total radiation [2]
0.52	1.15 [4]		1.08 [4]		
0.6		1.04		1.02	
0.7		1.06		1.03	1.020
0.8		1.10		1.04	1.025
0.9	1.28 [4]	1.14	1.16 [4]	1.10	1.040
0.94	1.15 [5]		1.10 [5]		
0.95		1.17		1.15	1.060

It is not impossible that the values at 3900 and 4300 Å are not really photospheric and that there is a small chromospheric contribution from the numerous Fraunhofer lines in that region. So the true values might be smaller. On the other hand, all values given above do not refer to the true facular elements but to the unresolved mean facular areas, and for *that* reason the values for the true facular granules should be *greater* than the tabulated ones. According to WALDMEIER[6] maximum brightness is obtained at $r/R = 0.945$. Very close to the limb, the faculae are invisible, but this decrease of brightness near the very limb may be due to the effect of perspective; in reality the brightness may continue to increase towards the limb. The existence of the chromospheric faculae on top of the photospheric ones indicates, in any case, that the photospheric local rise in temperature continues at least some thousands of kilometers into the chromosphere.

An important fact is that the ratio of *total intensities* (last column of the table) depends on r/R; hence the faculae are not in radiative equilibrium. Obviously in the upper regions the facular temperature is higher than the photospheric temperature at the same monochromatic optical depth. Since the faculae disappear in the centre of the disk, the temperature increase in the high layers must be compensated by a relative decrease of temperature in the lowest.

τ_{4100}	T_{phot} °	T_{fac} °
0.01	4750	4830
0.10	5160	5410
0.3	5680	5930
0.6	6140	6200
1.0	6600	6560
3.0	8030	7670

Assuming $\Delta T \approx 100°$ at $\tau \approx 0$, a (T, τ) model of the facular region was derived by REICHEL[7] from each of the given sets of the photometric data. In his model ΔT reaches its maximum value at $\tau_{4100} = 0.2$ ($\Delta T = +350°$), it is zero at $\tau_{4100} = 0.9$; the extrapolated and uncertain value at $\tau_{4100} = 2.0$ is about $-300°$. The

[1] E. R. MUSTEL: in: Theoretische Astrophysik by V. A. AMBARTSUMIAN a. o., p. 320. Berlin 1957.

[2] T. W. WORMELL: Monthly Notices Roy. Astronom. Soc. London **96**, 736 (1936).

[3] R. S. RICHARDSON: Astrophys. Journ. **78**, 359 (1933).

[4] T. V. KRAT: Astron. Zhurn. **24**, 329 (1947).

[5] V. A. AMBARTSOUMIAN and N. KOSIREV: Poulkovo Obs. Circ. **2** (1932).

[6] M. WALDMEIER: Z. Astrophys. **26**, 147 (1949).

[7] M. REICHEL: Z. Astrophys. **33**, 79 (1953).

gas and electron pressures in this model of the mean faculae are nearly equal to those for the photosphere. It should be kept in mind that this model applies to the unresolved faculae ($\Delta I/I \approx 10\%$), and perhaps not to any physically existing part of the Sun.

42. The polar faculae. The polar faculae are small, roundish, and occur only in the polar regions. In contrast to the faculae of the main zone they are isolated mottles with a mean diameter of 2300 km[1]; the observed values ranging largely between 1800 and 3000 km. They seem evenly distributed over the polar regions up from latitudes of about 68°; in lower latitudes their surface density decreases with decreasing latitude. The (perspectivic) reduction of visibility towards the limb causes that maximum visibility occurs at a distance to the disk centre of 0.945 (the same distance as for the equatorial faculae), and this effect simulates the existence of two polar facular *belts* instead of polar *caps*.

The life times of the polar faculae vary from a few minutes to two or three days, with a mean life time of $0^{\mathrm{h}}.5$. Considerable intensity fluctuations may occur with an average period of $\frac{1}{4}$ hour.

Waldmeier[1] suggested that the polar faculae occur mainly in the years around the sunspot minimum and especially at the descending branch of the sunspot curve; their frequency seems to be less at the ascending branch. Another suggestion of Waldmeier, that the polar faculae have some connection with the polar rays of the corona and perhaps with the general solar magnetic field, is supported by the apparent agreement between the two following observations which, isolated as they are, still might be an indication: in 1955 the polar faculae were still well visible near the S-pole, whereas their number was greatly decreased near the N-pole. At the eclipse of 20 June 1955 the south polar rays had the same intensity as at the minimum, whereas the north rays were hardly visible.

43. Structure and development of chromospheric faculae. The chromospheric faculae are principally studied on H and on K (Ca II) spectroheliograms; they were investigated on metal line spectroheliograms by D'Azambuja. They have many properties in common with the photospheric faculae of which they form the chromospheric extension, but they are well visible on the whole disk. In the cores of strong lines they cover even the big sunspots. The most brilliant faculae are those observed in K_3 (Ca II).

The interpretation of the observations will give insight into, and must be based on the model of the disturbed chromosphere and of the profiles of the Fraunhofer lines. The theory should explain why faculae are bright in the centre of a certain line and dark in its wings, or why they are bright in the red wing of a line and dark in the violet one etc., cases that are actually observed.

Another question is what is the essential difference between the structure of the normal, undisturbed spectroheliograms and the chromospheric faculae. The chief difference seems simply to be the areal density, of hot elements, i.e. high resolution spectroheliograms (cf. Fig. 17) show faculae to be solar regions more densely beset with the same kind of bright fine mottles that occur on the normal chromosphere.

Much of what has been said of the structures and development of photospheric faculae (*p.f.*) remains true of the chromospheric faculae (*c.f.*). The apparent sizes of the *c.f.* may be somewhat greater than those of the *p.f.*, but this mainly reflects the greater intensity of the chromospheric structures. Also in the chromospheric faculae differences in size are found, e.g. between K_3 and Hα. A

[1] M. Waldmeier: Z. Astrophys. **38**, 37 (1955).

difference between *c.f.* and *p.f.* is that the relation of the *c.f.* with the general solar background is much clearer. A glance at the K_3 images shows e.g. how the *c.f.* quite naturally fit into the structure of the normal Sun: the normal network of coarse mottles continues into the facular area and the coarse network has the same mean size. On high resolution spectroheliograms we also see the 5000 km elements and the fine granular mottling, with a mean size of 1000 or 2000 km. The only difference with the undisturbed Sun seems to be that the number of fine granular mottles is greater in the faculae than outside. This number is greater in the 5000 km elements (which are composed of some tens of fine mottles) and it is also greater in the large scale alignements of the 5000 km elements, and in the space between the lines of this "network". On Hα filtergrams KIEPENHEUER[1] found two kinds of faculae. The "normal" Hα faculae consist of bright facular mottles, diameters <2000 km. In the spot groups the faculae show a more amorphous structure, not resolvable in single mottles.

Quantitative photometric measurements of faculae are not available, nor is it known whether the "background" between the granules is brighter as compared with that of the normal Sun. Precise photometric measurements of facular areas on spectroheliograms are highly desirable.

The size of the *c.f.* depends on the lines and the wavelengths in which the images are made. In many cases they are more or less elliptic, in other cases irregular. The ratio of the number of elongated to irregular groups increases before sunspot maximum and decreases after maximum[2], so that after maximum the irregular groups are dominant. The axes of the elliptical faculae show an inclination towards the solar equator, the west point being closest to it; the inclination ranges between 0 and 40°, in a few cases it reaches 90° (BUTLER[2]). It is not clear whether this inclination can be completely explained by the influence of the differential solar rotation: long-lived facular regions do not seem to change their axis appreciably in the course of their life.

The life time of the *photospheric* faculae, determined by autocorrelation analysis of the curve giving their total visible area versus time, is 80 days[3], but the *chromospheric* faculae may have life times of 200 to 300 days or more. So they exist much longer than the accompanying spots. The spots are generally formed shortly after the facular area, which is first condensed and brilliant, and which afterwards increase in area to become more irregular, dispersed and more patchy. The origin of spots can often be traced in the chromosphere some hours before the spots become visible[4]: then in Hα spectroheliograms a very bright *c.f.* appears at the place, where some hours later, the spots will be born. When, after some weeks or months, the spots have disappeared the facular region does still exist. Finally the facular area merges into the chromospheric network. Never has an abrupt disappearance of a facular region been observed.

Only in few cases has a life time shorter than half a rotation (14^d) been observed. Such a case is e.g. the remarkable onset of the new solar cycle, on August 13, 1953. A very small facula appeared at $+52°$ latitude, it remained visible till the next day. Two little spots appeared in this area, the highest spot group ever observed. Magnetic observations showed the inversion of magnetic polarity, as compared with other spot-group observations, indicating the start

[1] K. O. KIEPENHEUER: Z. Astrophys. **42**, 209 (1957).
[2] C. P. BUTLER: Monthly Notices Roy. Astronom. Soc. London **82**, 334 (1922); **84**, 134 (1924).
[3] C. W. ALLEN: Monthly Notices Roy. Astronom. Soc. London **117**, 174 (1957).
[4] M. WALDMEIER: Z. Astrophys. **14**, 91 (1937).

of the new cycle[1]. Also in other cases the start of a new cycle was accompanied by the occurrence of small, short-lived but bright faculae; in 1923 such a facula was observed at $+64°$ but without spots[2].

The fine structure of a facular field has been studied by LIPPINCOTT[3] on Hα filtergrams taken by LYOT, and by KIEPENHEUER[4] on Hα filtergrams taken at the Freiburg-Capri station. The chromosphere around sunspot groups is characterized by a complex of fine dark filaments (thickness ≈ 2000 km, average length 12000 km) that may extend to fairly great distances from the spots (see Fig. 37). Contrary to the solar vortices, they do not show the cyclonic form but their forms rather suggest a magnetic explanation. They are very thin, and only under good observing conditions are they observable up to great distances. According to LIPPINCOTT these "fibrils" show fluctuations in times of the order of $0^h\!.5$ (range: some minutes to more than one hour) but they persist for a longer time, of the order of one day (KIEPENHEUER[4]). In the same region of a centre of activity they are all orientated in the same way; their apparent velocities are ≈ 7 km per hour.

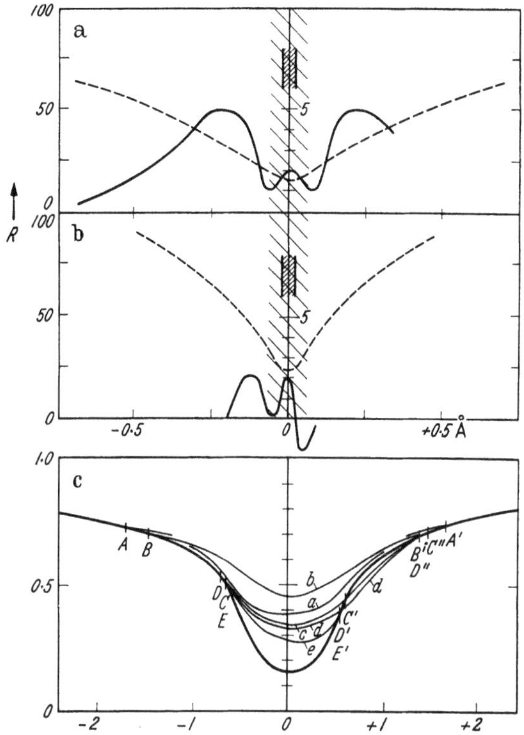

Fig. 43a–c. Visibility of chromospheric faculae, in an arbitrary scale 1 ... 5. Dashed: line profile. Shaded: line core, and slit width. (a) Fe 4384. (b) Fe 4202. Positive visibility denotes that faculae are brighter than the surroundings. L. D'AZAMBUJA: Ann. Obs. Meudon 8, fasc. 2 (1930). (c) Hα profiles in faculae. The intensity is expressed in the local continuum outside the line. Abscissa: wavelength distance in Å. M. A. ELLISON: Publ. Obs. Edinburgh 1, No. 5 (1952).

44. The "visibility", spectra and temperatures of chromospheric faculae. According to D'AZAMBUJA[5] the "visibility" of the chromospheric faculae generally shows a maximum value in the line centre, where the faculae are observed as bright structures. A visibility minimum occurs at some distance to the line centre, but further outward the visibility increases again to reach two pronounced maxima, generally higher than the central visibility maximum, at symmetrical points in the line wings. Also in the line wings the faculae are still bright (Fig. 43). With the known profiles of the Fraunhofer lines in the undisturbed photosphere, and with such visibility curves, the facular line profiles may be constructed, at least qualitatively. D'AZAMBUJA estimated that one unit of this intensity scale corresponds with an increase of intensity of 10%. The communicated curves (Fig. 43) are mean values and do not take account of the facular fine structure.

[1] H. W. DODSON: Publ. Astronom. Soc. Pacific **65**, 256 (1953).
[2] M. D'AZAMBUJA: L'Astronomie **67**, 430 (1953).
[3] S. L. LIPPINCOTT: Ann. d'Astrophys. **18**, 113 (1955).
[4] K. O. KIEPENHEUER: Z. Astrophys. **42**, 209 (1957).
[5] L. D'AZAMBUJA: Ann. Obs. Meudon **8**, II (1930).

Noteworthy is the Sr II line 4078 Å, where the faculae are not or hardly visible in the line centre, and where they are dark in the wings. Also the Fe I line 4202 Å is interesting; it shows bright faculae at the short wavelength wing and dark ones at the long wavelength side. The lines Fe I 4104 Å and 4132 Å do only reveal the faculae in the line centre.

The line profiles of Hα in facular areas of various intensity were studied by ELLISON[1] (Fig. 43c). The faint faculae are bright in the line centre, dark in the red wing and nearly invisible in the violet wing. Brilliant faculae are bright everywhere. In the higher members of the Balmer series up from Hγ, the faculae are dark.

The Ca^+ faculae are by far the brightest and the most extended. Especially in K_3 they are very bright and diffuse, whereas they are smaller and show sharp boundaries in K_1, as they do in the photospheric faculae. According to various authors the transition from diffuse to sharply bordered structures takes place between 2 and 6 Å. At $\Delta\lambda \gtrsim 10$ Å (line depression $\lesssim 20$%) the structures do not differ from the photospheric faculae.

The *facular spectra* have been investigated in several ways: A comparative study of *curves of growth* of faculae, of the adjacent photosphere and of solar type main sequence stars by MITROPOLSKAJA[2] has shown that the temperature difference: facula *minus* photosphere is 65 to 120°, if lines of neutral elements are used, and 190 to 300° for ionized atoms. If we take account of the facular fine structure and assume that $\frac{1}{3}$ of the light comes from the faculae, and $\frac{2}{3}$ from the interfacular photosphere, these values become 2 to 3 times greater. The analysis of these results suggests that the faculae and the photosphere differ mainly in their ionization temperatures.

A study of the *line profiles* shows that the central intensity of Hα and Hβ (and also of Hγ, for very bright faculae) exceeds that of the photospheric profile[3] and that the higher members of the Balmer series show dark faculae. This again shows that the excitation temperature of the faculae increases with height. For $h < 4000$ km the number of second-level H atoms is greater in the faculae than in the quiet chromosphere[4] while the reverse holds for N_{Ca^+} (cf. also Fig. 20).

Flash spectra of the chromosphere in active regions give information on the structure of the uppermost parts of the facular fields. The difference between the quiet and the active chromosphere was explained by ATHAY and THOMAS[5] by assuming in the active regions a larger proportion of the hotter chromospheric elements, thus confirming the viewpoint deduced from spectroheliographic studies. A similar viewpoint was expressed earlier by MUSTEL[6]; the bright hydrogen flocculi were assumed to be produced by a gas with $10^4 < T_{el} < 2 \times 10^4$ °K.

As to the *detailed structure of the facular line profiles*, REICHEL finds that the visibility minimum at the borders of the line core can be explained by a small increase of the microturbulent velocity component. The increase computed for various types of lines is in most cases of the order of 0.15 km/sec; for the Sr II line it should be 0.35 km/sec. The central brightening should be due to an increase in the radiation temperature for the layers where the line centres are formed; the increase in the wings is caused by this effect and by a higher excita-

[1] M. A. ELLISON: Proc. Roy. Obs. Edinburgh **1** (5) (1952).
[2] O. N. MITROPOLSKAJA: Isw. Krymsk. Astrophys. Obs. **8**, 93 (1952); **11**, 152 (1954); **15**, 130 (1955).
[3] E. R. MUSTEL: Isw. Krymsk. Astrophys. Obs. **7**, 3 (1951).
[4] V. A. KRAT, T. V. KRAT and L. M. PRAVDOUK: Isw. Gl. Astr. Obs. Poulkovo **158**, 1 (1958).
[5] R. G. ATHAY and R. N. THOMAS: Astrophys. Journ. **125**, 788 (1957).
[6] E. R. MUSTEL: Isw. Krymsk. Astrophys. Obs. **17**, 162 (1957).

tion and ionization temperature ("number effect"). KIEPENHEUER[1] and GIOVANELLI[2] have suggested that the faculae are heated by electric currents, induced by changes of the magnetic fields in the chromosphere and photosphere (cf. Sect. 73). In Sect. 30 reference is made to PIDDINGTON'S criticism to this mechanism and to the mechanism of heating by hydromagnetic oscillations of a partially ionized plasma in a magnetic field.

45. The magnetic fields of faculae. With the refined observational technique described in Sect. 32, BABCOCK and BABCOCK[3] in 1952 started a program on a routine basis of daily magnetic mapping of the Sun. Apart from the poloidal magnetic field of about 1 Gauss, only observable at high latitudes, there are two other kinds of fields observed in patches at lower latitudes. The stronger appear as areas of opposite magnetic polarity, the *bipolar magnetic* (*BM*) regions. A few of the strong magnetic fields are more complex and may be classed as *multipolar magnetic* (*MM*) regions. In the *BM* regions there is, generally, about equality of positive and negative magnetic fluxes. A very few areas of moderate or small magnetic intensities are *unipolar magnetic* (*UM*) regions.

In the *BM* regions the distribution of magnetic polarity among the preceding and following parts is the same as that in the preceding and following sunspot groups of the same cycle and the same hemisphere. So the *BM* regions, strong or weak, follow the polarity law of sunspots (cf. Sect. 36). Generally the following part of a facular region has the greatest latitude.

The *BM* regions have a similar but longer *life history* than the faculae. Initially the faculae are small (some minutes of arc in extension) and have a strong magnetic field of the order of several Gauss. The magnetic field always appears first. The typical strong *BM* region soon develops spots and faculae. As the region ages it expands; the spots disappear first, then the faculae, leaving the weak extended and patchy *BM* field. The magnetic field seems to be the fundamental physical entity, while the spots and the faculae are secondary. Many weak *BM* regions do not develop spots or faculae. The largest *BM* regions may extend over one tenth of the hemisphere, but the field, then, is generally weak, about a fraction of one Gauss. Some weak *BM* regions may persist for several days only; the main *BM* regions may sometimes last even up to eight or ten solar rotations as the faculae may do. Rapid fluctuations, of the order of one Gauss may occur in about half an hour.

Spots occur only in the *BM* regions when they are young. *Faculae* are always associated with *BM* regions when the field intensity is greater than 2 Gauss. Only in a few cases can the brighter Ca^+ faculae not be matched with *BM* regions. The positions of *filaments* near the faculae seem to be determined by the magnetic fields: often filaments mark the borderline between regions of opposite polarity in a *BM* region, or they occur around the region, thus forming a kind of general borderline. They may delimit *BM* regions on the high latitude side. The poleward motions of filaments observed by D'AZAMBUJA[4] (1°/rotation) generally results from an expansion of the *BM* regions (cf. Sect. 63).

There is a correlation between *BM* regions and regions of enhanced *coronal activity*. The bright coronal arches, which extend above the faculae, may prove that the vertical extent of the magnetic regions is of the same order as their horizontal extent.

---

[1] K. O. KIEPENHEUER: Ann. d'Astrophys. **9**, 42 (1946).
[2] R. G. GIOVANELLI: Monthly Notices Roy. Astronom. Soc. London **108**, 163 (1948).
[3] H. W. and H. D. BABCOCK: Astrophys. Journ **121**, 349 (1955).
[4] See footnote 1, p. 231.

The *UM* regions, which are relatively rare, occur in the same latitudes as the *BM* regions. They have rather small magnetic fields, ranging from 0.1 to 3 Gauss, and cover a rather large area, about 0.1 to 0.2 of the disk, with badly defined borderlines.

There is no trace of coronal or chromospheric brightening above *UM* regions. It is suggested (BABCOCK) that the *UM* regions are the remnants of old *BM* regions, and that they may be identified with the so far hypothetical *M* regions (cf. Sect. 47).

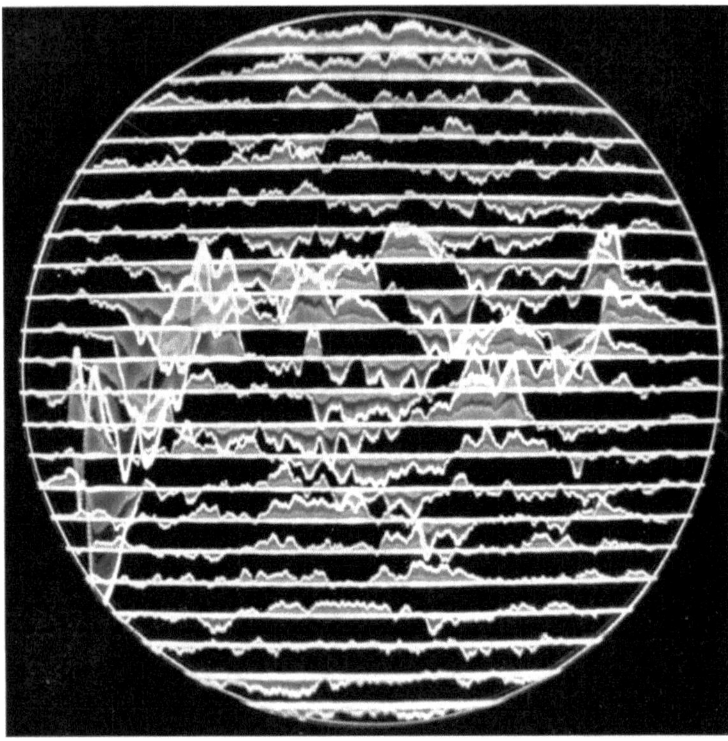

Fig. 44. Magnetoheliogram of July 19, 1953. The Sun's axis is vertical with North at the top, East on the right. A deflection equal to the interval between two straight lines indicates that $H \cos b$ is about 1 Gauss at the disk centre, 2 Gauss at the limb. For upward deflections H is directed towards the observer. The slit-height was about equal to the distances between the lines. One spot group only was visible, at 54° W, 8° S. Courtesy H. W. BABCOCK, Pasadena.

Also very short lived magnetic regions occur; they are described by KIEPENHEUER[1] and by BEGGS and VON KLÜBER[2]. The former, working near sunspot minimum, found local field patches which might persist for some hours. Their fields are of the order of a few Gauss in regions of a size corresponding to those of the big sunspot groups. They do not contain visible sunspots. The magnetograph shows them to be short-lived *BM* regions. BEGGS and VON KLÜBER, observing near sunspot maximum (slit length 4″) detected a large number of very small areas with fields up to 100 Gauss, in the sunspot belts, also in undisturbed parts of the Sun. Small visible spots sometimes developed in such areas one or two days later, indicating that the occurrence of a visual spot is preceded by the birth of a local magnetic field.

---

[1] K. O. KIEPENHEUER: Astrophys. Journ. **117**, 447 (1953).

[2] See footnote 10, p. 104.

b) The "centre of activity".

46. The development of a typical centre of activity[1]. A *centre of activity* (CA) is the totality of all visible phenomena accompanying the birth of sunspots (D'AZAMBUJA[2]. A similar definition was introduced by GNEVISHEV[3]). The complexity of the many optical phenomena which form together a centre of activity arises from the fact that none of them can apparently be considered as the primary one. The arrival of a magnetic disturbance at the Sun's surface may prove to be the essential and primary event.

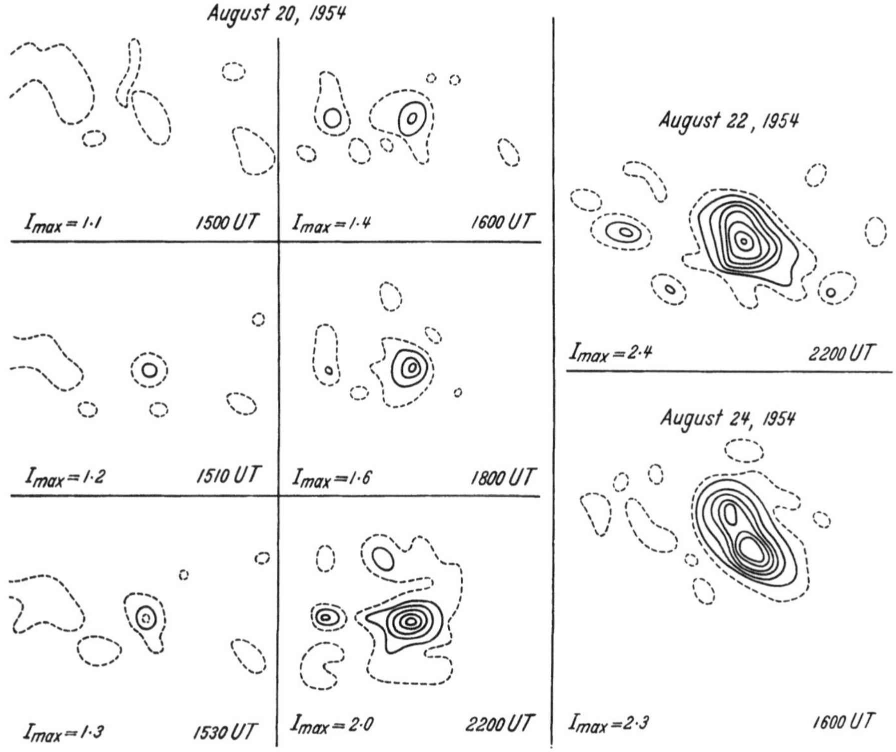

Fig. 45. Photometric development of a K_{232} facular field, 1954, August 20 to 24. Dotted lines indicate 1.1 isophotes. Full lines give 1.2; 1.4 ... etc. H. W. DODSON and E. R. HEDEMAN: Monthly Notices Roy. Astronom. Soc. London **116**, 428 (1956).

The observations of a *CA* refer to the photosphere, chromosphere and corona and are made by different kinds of observational techniques. It is a great difficulty that the corona can optically only be observed at the Sun's limb and this is also partly true of the prominences. Radio observations show the corona before the disk but these suffer from lack of resolving power.

The accompanying table gives the development of a typical, *completely developed CA* as described by KIEPENHEUER[4] (cf. also Fig. 47). This is of course an average example; numerous deviating cases occur. In many cases a centre of activity does not pass through the complete development as described here; its evolution may last shorter. For the description of the development of one

[1] Cf. also C. DE JAGER: Vistas in Astronomy **3** (1959).
[2] See footnote 2, p. 179.
[3] M. N. GNEVISHEV: Isw. Gl. Astr. Obs. Poulkovo **131**, 36 (1938).
[4] K. O. KIEPENHEUER: Convegno Volta **11**, 105 (1953).

Table 10. *Life history of a fully*

Day	Faculae	Spots
0		
+1	Medium size Hα mottle elongates and becomes bright facular point (Hα intensity ≈ 0.5 × continuous spectrum), that assumes EW elongation. Sometimes two bright faculae form (preceding facula first) connected by a dark flocculus. After the formation of spots the faculae as a rule decrease in brightness to "normal value"	In some cases spots originate within some hours after formation of facula, but in general spots do not yet occur; pores may exist; one of them may develop into leading spot in some hours to one day
2	Faculae increase in brightness and area	First (preceding) spot appears at W end of facular region
5		Following spot originated at E end of faculae; many smaller spots originate between *p* and *f* spots. Spots have opposite polarities
11	Brigthness and size continue to increase; linear dimensions up to 150000 km	Spots near maximum development
27	Size of faculae has still increased; brightness constant	Nearly all spots, except the *p* spot have disappeared
54	Decrease of facular brightness; facular area sometimes splits in two by filament ("dark lane")	As a rule all spots have disappeared
81	Faculae dissolve into faint network; still brighter than normal chromosphere; size has remained constant	
108	Faculae dissolved; only some orientation effect of mottles near filament	
135	Faint orientation of mottles near filament	
162–270		

particular short-lived *CA* we refer to a paper by DODSON and HEDEMAN[1] (the development of an active solar region, 1954, August 20 to 27).

A *CA* always originates in the spot zone. First a magnetic field is observed, without any visible phenomena. After one or two days it is followed by a small, bright facula which originates rather suddenly (Fig. 45), in which the first sunspots originate some hours to one day later. (BILLINGS[2], however, suggested that

[1] H. W. DODSON and E. R. HEDEMAN: Monthly Notices Roy. Astronom. Soc. London **116**, 428 (1956).

[2] D. E. BILLINGS: Astrophys. Journ. **120**, 184 (1954).

developed centre of activity.

Prominences; filaments; corona	Flares; surges	Area and magnetic field of *BM* region
		First observations of magnetic field
Coronal radiation at 5303 Å increases simultaneously with faculae, immediately after formation of spots. Bright coronal area of same extent as facular field		Field obtains bipolar character
Small unstable filaments orientate around faculae; faint prominence activity; coronal radiation at 5303 Å increases	First signs of flare activity after 2–5 days	Area of magnetic region between 40000–80000 km
Short-lived filaments occur; spot prominences	Flares, generally between main spots, accompanied by flare surges	
Close correlation between brightenings of yellow coronal line (5694 Å) and flare activity; coronal condensations	Top of flare activity	Irregularly variable field distribution
Stable filament forms at polar side of faculae; directed to *p*-spot; angle with meridian roughly 38°, with many deviating cases Maximum coronal brightness at 5303 Å	Only few flares	Maximum magnetic flux
Filament has increased in length to about 10^5 km; angle with meridian increases; decrease of coronal brightness at 5303 Å		
Prominence increases in length; angle with equator decreases		Field-strength decreases
Prominence shows maximum length; nearly parallel to equator		
Prominence decreases in length; centre-of-gravity of prominence seems to migrate into polar direction		Area has increased to $0.05 \times$ disk; field scattered
Structural changes of filament; dissolution. Development of coronal streamers, correlated with *UM* region		Area increases further up to $0.2 \times$ disk. *BM* region changes into *UM* region

the fields may sometimes precede the visible spots by one rotation.) These develop in the way indicated in Sect. 37; their relative motions are described in Sect. 38. Simultaneously the facular field increases in area and often also in brightness, as described in Sect. 43. After some days to one week the first flares occur and in a well developed *CA* the flare activity is greatest after 10 to 15 days, being correlated not only with the extent of the spot group but also with the change of this area with time—hence depending on the activity of the group. After one solar rotation the flare activity has virtually ceased and then a second phase starts, in which the facular field decreases in brightness and becomes more

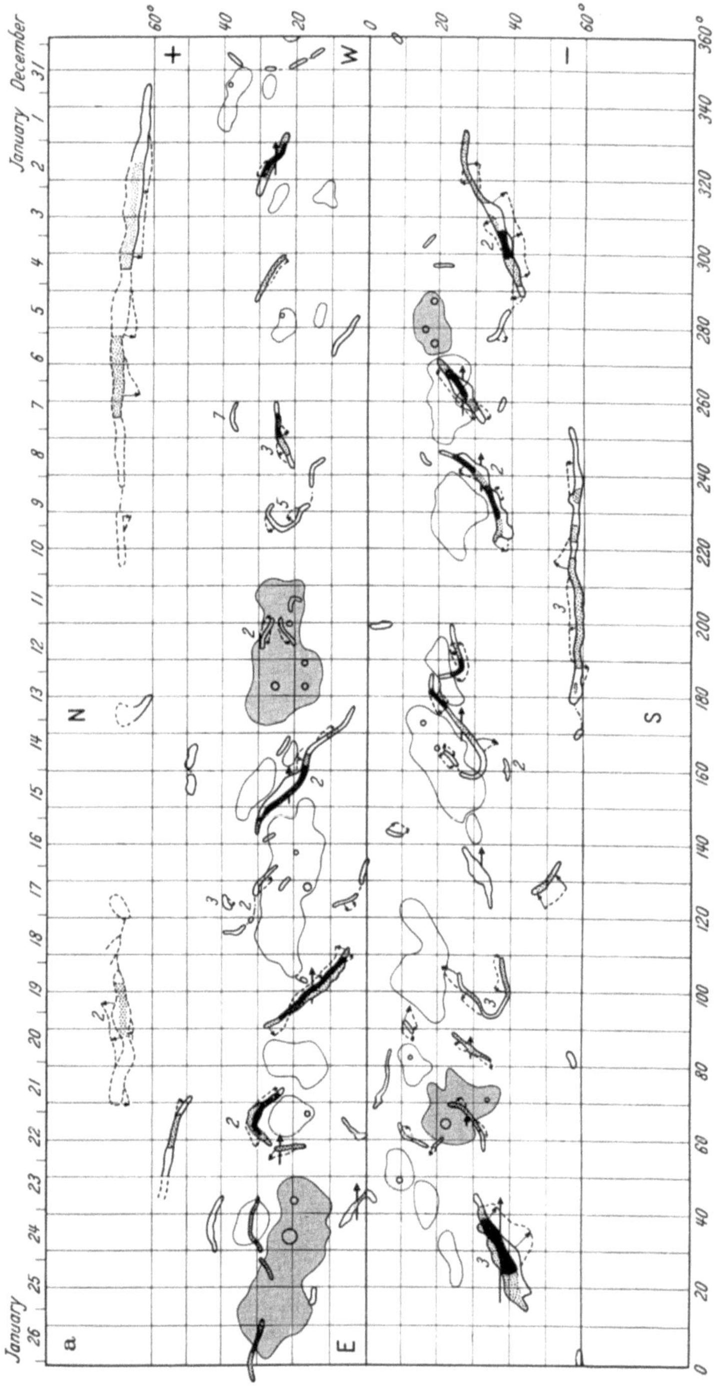

Fig. 46a and b. Two consecutive "cartes synoptiques". (a) Rotation 967; December 1925 to January 1926.

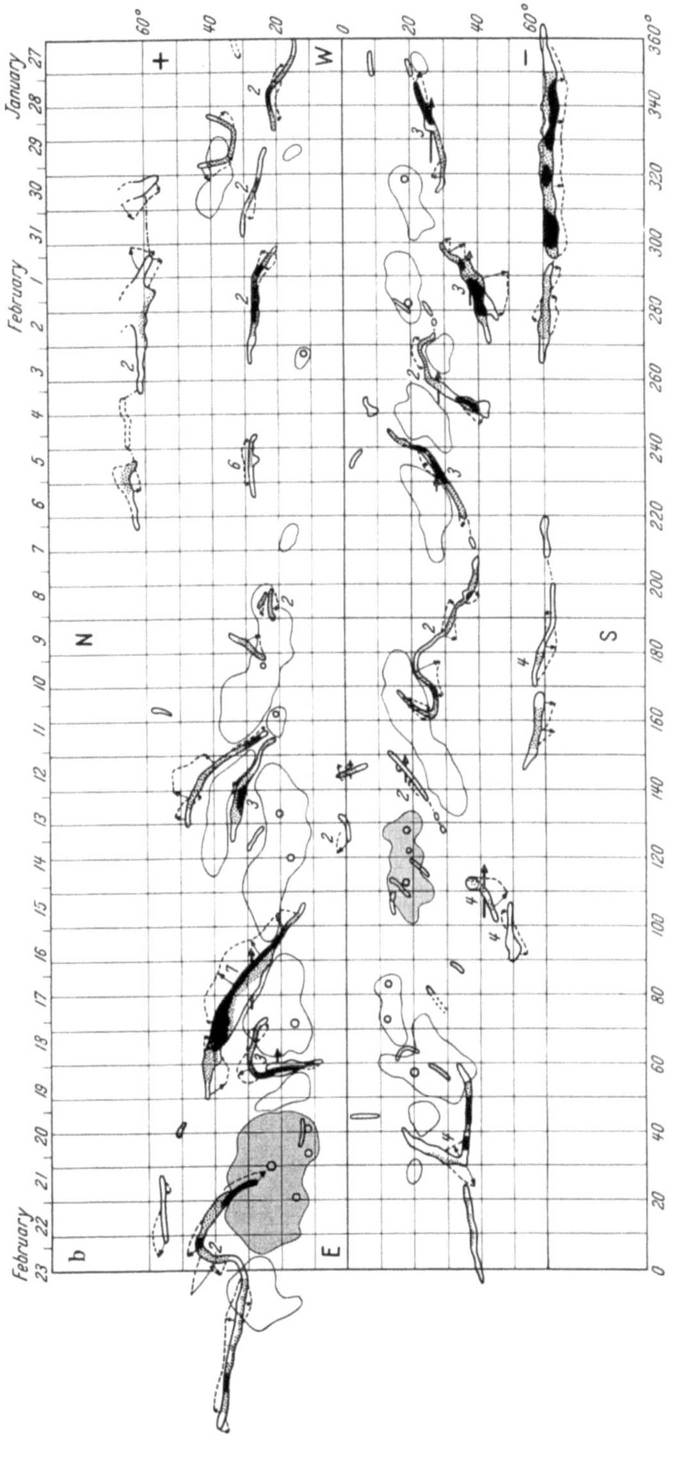

Fig. 46b. Rotation 968, January to February 1926; one year before sunspot maximum. The figures show the development of the faculae (shaded or open regions), of the spots (circles) and of the filaments (motions are indicated by arrows). L. D'AZAMBUJA: Ann. Obs. Meudon **6**, fasc. 5 (1932).

and more patchy and in which the long quiescent filaments are formed. After 5 to 10 rotations the optical *CA* has disappeared, but magnetically a *UM* region still remains visible for some time. It contains quiescent filaments and it seems to be connected with the long streamers of the corona.

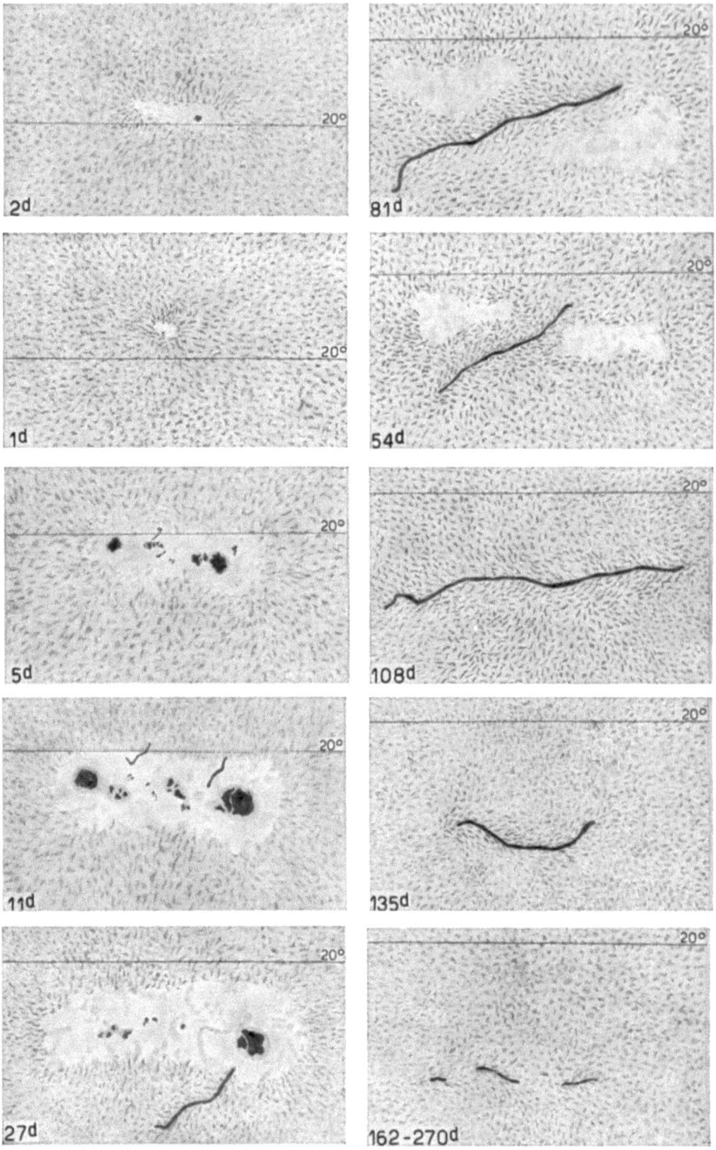

Fig. 47. Schematic development of a fully developed centre of activity, after K. O. KIEPENHEUER: Convegno Volta **11**, 107 (1953).

In the range of radio wavelengths there is a good correlation between the extent of the Ca^+ plage and the intensified solar background radiation at decimeter waves. The great outbursts may occur on all wavelengths, sometimes even on the centimeter and decimeter waves; they are correlated with solar flares. In the meter waves the pattern of radio events is more complicated, showing outbursts, noise storms, enhancements of the base level and type III bursts. The correlation with optical phenomena is less clear than in the decimeter region, but the first three events show some correlation with flare activity, whereas many type III bursts are correlated with the first phase of flares. So one might say that at decimeter wavelengths the increased radio radiation is principally correlated with faculae; the intensified radiation at meter wavelengths is rather correlated with flare activity and spots.

After the birth of the stable filaments, that is about 20 days after the birth of the *CA*, the filament makes an angle of about 38° with the meridian. Many non-facular filaments also seem to start their life with this angle. D'AZAMBUJA remarked that a line initially parallel to the meridian will after 21 days assume an angle of 38° under the influence of the differential solar rotation. This suggests that the spot and the quiescent filament might be formed at the same time but that the filament comes to visibility only after about 20 days, when the spot activity has virtually ended. Their development may be initially hampered by the disturbing influence of the magnetic fields. These are rather erratic, especially in the first three weeks when the spots are in full development. That many disturbing forces are active in this interval is also shown by the fact that flares occur in the *CA* mainly during this time.

Fig. 46 shows two consecutive "Cartes Synoptiques" prepared by D'AZAMBUJA; the map shows the many forms which the centres may obtain during their development.

47. Relations of centres of activity to regions of enhanced coronal activity and to *C* and *M* regions (see also Sect. 79 and 110). The active coronal regions, above the photospheric and chromospheric centres of activity are characterized by an increased intensity of the green coronal line (5303 Å, Fe XIV). Its brightness increases more or less parallel with that of the faculae, but it seems to decrease faster. These active coronal regions, the most active of which are also called *permanent coronal condensations*[1] have a greater density and a higher electron temperature than the undisturbed corona. They present short coronal rays and arches. In periods of great flare and spot-activity the yellow line (5694 Å; Ca XV) and a continuous spectrum become visible indicating a high electron density and high temperatures (excitation potential of the Ca XV line is 814 eV). They should be distinguished well from the *sporadic coronal condensations*, which have lower temperatures; they emit the coronal lines at 5303 and at 6379 Å and they emit Hα light (excitation potential 10.19 eV)[2]. Cf. also Sect. 79.

The active coronal regions should be distinguished sharply from the *C* regions of the corona, discovered by WALDMEIER. In the *C* regions the 5694 line is invisible, but the green line (5303 Å) is more strengthened than would be expected from the visible optical phenomena. Erroneously it is thought sometimes that a *C* region is any region where the 5303 line is bright, but when so defined the coronal centres of activity also would fall under this definition.

There is a correlation between the occurrence of a *C* region at the disk centre and the observation, some days later, of small or moderate geomagnetic storms on the Earth. These latter occur on the average about three days after the central meridian passage of a *C* region, indicating a travelling time of this order for the disturbing agent (presumably a corpuscular current) which leads to a mean velocity of about 500 km/sec. They are not related to any photospheric phenomena; although the *C* regions occur in the sunspot belt they seem to avoid the

[1] M. WALDMEIER and H. MÜLLER: Z. Astrophys. **27**, 58 (1950).
[2] M. WALDMEIER: Z. Astrophys. **40**, 221 (1956).

great sunspot groups, and occur principally above the fairly calm and spot-free areas of the photosphere. It has been suggested that the *C* regions mostly occur above the remnants of old centres of activity. At their place there seems to remain a region which is still active during a number of solar rotations.

Already before the discovery of the *C* regions, geophysicists had introduced the *M* regions of the Sun (BARTELS 1932), hypothetical regions, that were put responsible for the recurrent magnetic disturbances on the Earth. Statistics of geomagnetic disturbances based on the international magnetic character figures (which distinguish between magnetically quiet, and disturbed days), have shown that the weak or moderately disturbed days have a recurrence tendency with a mean period of 27 days, the rotation period of the Sun. The *M* regions are the "magnetically effective" regions; they may influence the Earth when they are situated such that the emitted stream of corpuscular or electromagnetic radiation may strike it. Note that it is not the big magnetic storms that are produced by the *M* regions. These latter storms are sporadic and are mainly due to discrete clouds of particles emitted in connection with great solar flares, with greater speed than the *M* particles. They do not show the 27 days recurrence.

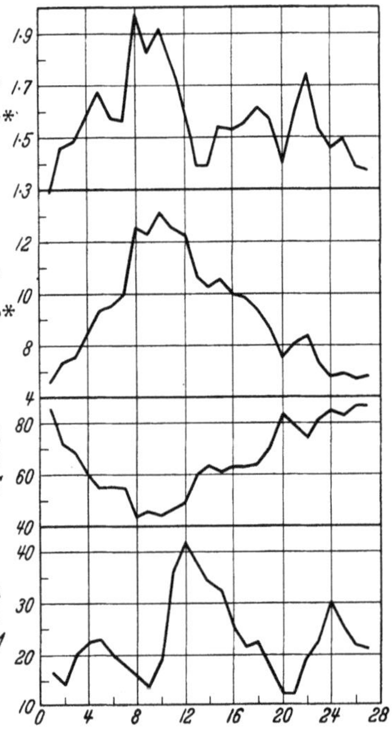

Fig. 48. Relation between geomagnetic activity M, sunspot numbers S and the "reduced" prominence and coronal activity P^* and C^*. For further explanation see text. M. WALDMEIER: Vistas in Astronomy 2, 808 (1957).

Like the *C* regions, the *M* regions occur in the sunspot belts, but they avoid the areas within 40° of the spotgroups, and the emission has a direction effect: the greatest intensity of the geomagnetic storms is reported when the Sun's axis has its greatest inclination towards the Earth so that the subterrestrial point on the Sun has its greatest solar latitude.

It is probable from the foregoing that the *C* and *M* regions are virtually identical. The only optical phenomena associated with *C* and *M* regions are long-lived filaments (these are known to occur in old centres of activity) and the coronal streamers which are so manifest on many eclipse photographs (ALLEN[1]). WALDMEIER[2] has confirmed that the long coronal streamers follow the same statistical distribution law as the quiescent prominences. Similar results were obtained by ROBERTS *et al.*[3]. We should localize the *M* regions at the basis of the streamers, in a region which, apart from the quiescent prominences, is optically quiet.

In the photospheric level the *M* regions were suggested to be identified with the Unipolar Magnetic (*UM*) regions on the solar magnetograms (BABCOCK and BABCOCK[4]), though the number of observations is actually still insufficient to draw a pertinent conclusion[5,6]. In magnetographic observations of the Sun

[1] C. W. ALLEN: Monthly Notices Roy. Astronom. Soc. London **104**, 13 (1944).
[2] M. WALDMEIER: Z. Astrophys. **42**, 34 (1957).
[3] W. O. ROBERTS and J. C. PECKER: J. Geophys. Res. **60**, 33 (1955).
[4] H. W. and H. D. BABCOCK: Astrophys. Journ. **121**, 349 (1955).
[5] M. B. WOOD: Astrophys. Journ. **124**, 447 (1956).
[6] H. W. BABCOCK: Astrophys. Journ. **126**, 224 (1957).

there was good correlation between a series of twelve consecutive central meridian passages of a big *UM* region and the periodical occurrence of a series of magnetic storms (27 days period). A delay of one to four days was observed between both events. The recurrence of cosmic ray maxima, in time associated with the *UM* region, has also been suggested[1].

The relation between the *M* regions and the various solar phenomena can best be illustrated by Fig. 48 (WALDMEIER[2]): It shows for one long series of geomagnetic storms, all due to the same *M* region: (a) the "reduced coronal activity" C^*, (b) the "reduced prominence activity" P^*, (c) the number of sunspots *S* and (d) the geomagnetic activity *M*. The "reduced coronal activity" is obtained by taking first the intensities of the coronal 5303 line as observed at the limb but ascribed to the central meridian passage by adding or subtracting seven days, and by dividing them next by the number *S* of sunspots. In this way a measure is obtained for coronal activity not correlated with spot activity. The same treatment applies mutatis mutandis to the "reduced prominence activity" *P*. The figure clearly shows the relations between the various phenomena as described above. Also the time delay of 3 and 2 days for the *M* peaks at days no. 12 and 24 is clear from the figure.

Thus the *M* (or *C*, or *UM*) regions are suggested to be the remnants of the centres of activity. When all solar activity, generally displayed by a centre of activity, has finished, there apparently remains a large region that is not yet wholly quiet but is able to emit a more or less continuous stream of particles into a narrow cone with a mean velocity of the order of 500 km/sec.

So the *C* (*M*, *UM*) regions will mostly occur between one half to two years after the origin of a centre of activity. Often an *M* region is destroyed when a new *CA* originates in its place. This explains why the *M* regions have their longest life times in the years after a sunspot maximum, when the disturbing influence of newly formed *CA*'s is reduced to a minimum.

Finally, mention should be made of an alternative possibility suggested by MUSTEL[3], to correlate geomagnetic disturbances with facular fields, hence with *CA*'s. With this assumption the travel time of the particles becomes much longer than with the first, leading to interplanetary velocities of 200 to 300 km/sec at the minimum of the solar cycle, and higher average speeds during periods of greater activity.

III. Flares and associated phenomena[4].

a) Monochromatic observations.

48. Description and classification. A flare is a sudden and short-lived brightening of a local region of the Sun's disk. Flares always occur in centres of activity, and especially in the central regions of the sunspot groups, often between the *p* and the *f* spots, or at a spot's penumbra, practically never on top of a spot. Rarely have flares been observed at distances greater than 10^5 km from the centre of a spot group. Only occasionally (5%) are they observed in faculae without spots, but these were nearly always cases in which the spots had disappeared shortly before or appeared shortly afterwards. Flarelike faculae, associated with the disparition of filaments, occur outside spot groups; they are discussed in Sect. 55.

[1] J. S. SIMPSON, H. W. BABCOCK and H. D. BABCOCK: Phys. Rev. **98**, 1402 (1955).
[2] M. WALDMEIER: Vistas in Astronomy **2**, 808 (1956).
[3] E. R. MUSTEL and O. N. MITROPOLSKAIA: Astron. Zhurn. **35**, 351 (1958) and Isw. Krymsk. Astrophys. Obs. **18**, 162 (1950).
[4] General reference: Joint discussion on solar flares; Trans. Internat. Astronom. Union **9**, 648 (1957).

Rarely are flares observed in white light, but in Hα or Ca^+ light where their intensity sometimes increases to tenfold its normal value, they may offer an impressive view.

Flares should be classified primarily according to their area, corrected for perspectivic forshortening. The classification is somewhat rough, but still satisfactory in three groups of increasing importance: 1, 2, 3, to which are added cases like 1^- (subflares, ELLISON); 3^+ (NEWTON; initially defined as flares showing outstanding terrestrial effects), and finer subdivisions 1^+ and 2^+.

It seems that in practice the correction for forseshortening is not always applied since flares of "importance 1" are observed even close to the limb where they should be invisible. This fact also explains the apparent increase of the Hα line width towards the limb[1].

The following table, largely extracted from a report of a working group on flare classification[2] (chairman: M. A. ELLISON) presented at the Dublin meeting of the IAU, gives mean and recommended values of areas and intensities of flares of the various "importances":

Table 11.

Imp.	Duration (min)		Recommended area limits	Mean area	Average Hα line width at max.	Approximate central intensity in Hα in units of adjacent continuous spectrum.		Relative frequency of 1, 2, 3 flares
	average	range	(10^{-6} visual hemisphere)		Å	mean	during flash stage	
1^-			< 100	72	1.5	0.6		
1	20	4– 43	100– 250	160	3.0	0.8	1.5	0.72
2	30	10– 90	250– 600	349	4.5	1.2	2.0	0.25
3	60	20–155	600–1200	973	8	1.4	2.5	0.03
3^+	180	50–430	> 1200		15	2	3	

Besides these types one also knows the classes 1^+ (=1.5; area of 1, central intensity of 2 or 3), and 2^+ (area of 2, central intensity of 3).

The statistical data on the flares of importance 1 are unreliable, because of observational reasons (atmospheric agitation; resolution of optics). The subflares exceed in number all other flares, but a reliable statistics has not yet been made.

Big flares (Fig. 49) are generally very irregular and filamentary (order of size of a filament $10^5 \times 10^4$ km²), whereas the small ones tend to circular shapes. That the correlation between the area and the Hα top intensity, expressed in the local continuous spectrum near Hα is rather poor is shown by Fig. 50 (cf also [3]).

Data on the time of appearance and disappearance, the area, the width and central intensity in Hα are published in the Quarterly Bulletin on Solar Activity of the International Astronomical Union. Provisional data on the importance and the time of occurrence may also be found in the daily maps of the Sun published by the Meudon Observatory (L. D'AZAMBUJA) and by the Fraunhofer Institut (K. O. KIEPENHEUER).

From a statistical discussion of the observed areas of 329 flares occurring at various distances from the disk centre J. WARWICK[4] could derive data on the distribution of *top heights*. The observed area is a combination of the area of the side of the flare (as it would be seen at the limb) and the top area (as it would

[1] Z. SVESTKA and L. FRITZOVA: Bull. Astr. Inst. Czech. **7**, 30 (1956).
[2] M. A. ELLISON: Trans. Internat. Astronom. Union **9**, 146 (1957)., with further references.
[3] H. W. DODSON, E. R. HEDEMAN and R. R. MCMATH: Astrophys. Journ. Suppl. **2**, 241 (1956).
[4] J. WARWICK: Astrophys. Journ. **121**, 376 (1955).

be seen in the centre of the disk). Due to perspective the observed flare area decreases statistically with increasing ϑ but not according to a cos ϑ-law; the limb areas are not zero. The derived height distribution $F(h)$ appears to be a function of the flare area, but in the average, for the whole of the observational material, $F(h) \propto \exp(-\beta h)$ with $1/\beta = 20300$ km. This result is roughly confirmed by an analysis of limb flares by C. S. WARWICK[1] and by GIOVANELLI and M. K. McCABE[2], who found average flare heights of 16000 km and 7300 km, respectively. That

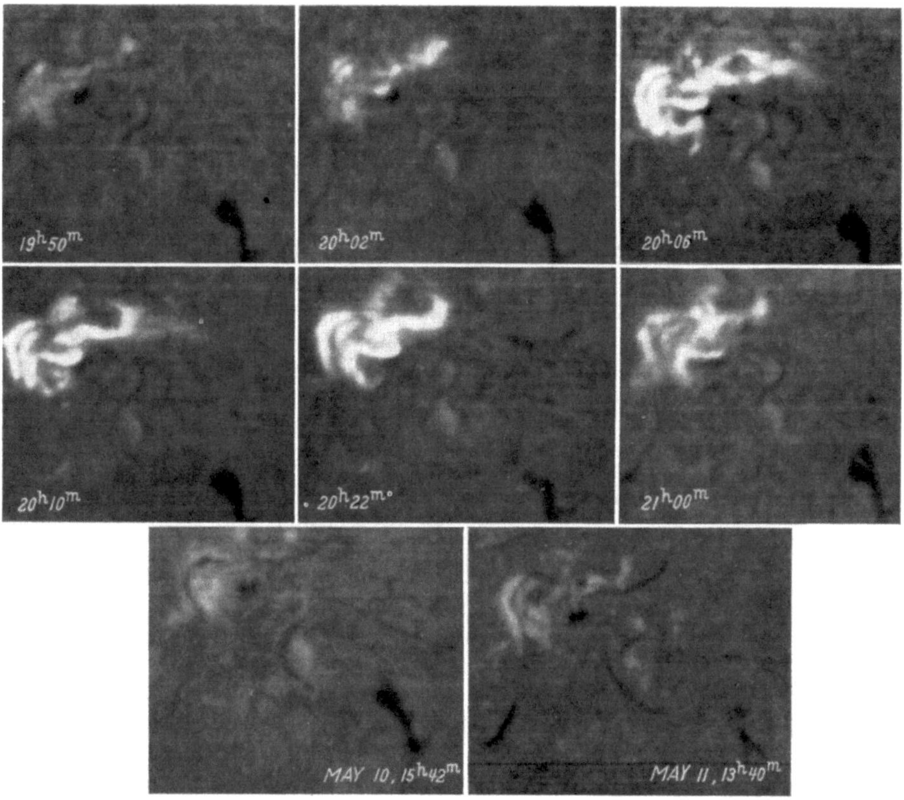

Fig. 49. Hα spectroheliograms, showing the development of the imp. 3 flare of May 10, 1949, and comparison with the same region before and after the flare. Cf. also Astrophys. Journ. **110**, 382 (1949).] Courtesy McMath-Hulbert Observatory.

limb flares sometimes may be fairly elevated and rise upward without being a surge was shown by D'AZAMBUJA[3] (cf. also NEWTON[4]) and later also by DODSON and McMATH[5]. Three very intense limb flares were described by ELLISON[6]; all three flares were detached, from the photosphere and only connected to it with bright fringes. While the flare may perhaps originate within the chromosphere much of its subsequent development appears to take place in coronal space.

So, in the majority of cases (70% according to SEVERNY and SHAPOSHNIKOVA[7]) flares are rather flat and extended but mainly stationary structures, occurring in

[1] C. S. WARWICK: Astrophys. Journ. **121**, 295 (1955).
[2] R. G. GIOVANELLI and M. K. McCABE: Austr. J. Phys. **11**, 130 (1958).
[3] L. D'AZAMBUJA: L'Astronomie **56**, 97 (1942).
[4] H. W. NEWTON: J. Brit. Astronom. Ass. **57**, 54 (1947).
[5] H. W. DODSON and R. R. McMATH: Astrophys. Journ. **115**, 78 (1952).
[6] M. A. ELLISON: Transact. I. A. U. **9**, 651 (1957).
[7] A. SEVERNY and E. SHAPOSHNIKOVA: Isw. Krymsk. Astrophys. Obs. **12**, 3 (1954).

or above the chromosphere, and enter with their upper part into the corona. The greatest heights seem to be reached for the smallest flares (WARWICK[1]). The stationary flares increase in height (speed <10 km/sec) during the time of brightness increase; maximum height is reached at the moment of maximum brightness. Sometimes great velocities are reported. Although some of the communicated observations of "limb flares" with appreciable outward motions may perhaps be due to confusion with bright surges or eruptive prominences, it is without doubt that some flares are not stationary and show considerable motions. (On 10. 2. 56 a limb flare was reported to have an initial expanding motion of 100 km/sec; it soon evolved into a surge and even into an ejected prominence; similar cases occurred on 24. 6. 56 and on 12. 11. 56[2]. A very remarkable flare on top of a prominence occurred on September 6, 1956[3].) Whether such objects should still be called flares or should be classified as "flare-associated" phenomena (Subdivision III c) or spot prominences is primarily an—uninteresting—matter of definition. Surges showing Hα line widths and intensities comparable to that in flares are familiar to routine observers of the Sun and the fact that designations like "eruptive prominences of importance 2" etc., are sometimes met in the literature reflects the difficulty.

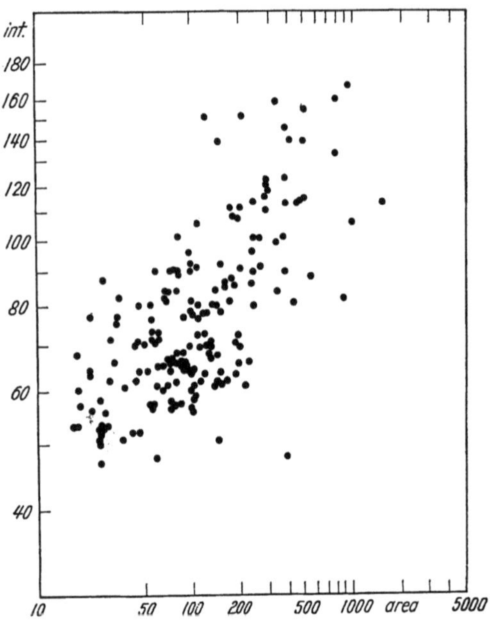

Fig. 50. Relation between the area, corrected for perspectivic foreshortening (abscissa; in 10^{-6} visible hemisphere) and central $H\alpha$ intensity (ordinate; in percent of continuous spectrum) of 184 flares. M. A. ELLISON: Trans. Internat. Astronom. Union 9, 146 (1957).

About 30% of the flares observed by SEVERNY and SHAPOSHNIKOVA[4] showed moving luminous formations with the same brightness as the stationary part of the flare and with velocities up to several hundreds of kilometers. Very often they look like moving luminous fronts and sometimes such a front is stopped near a spot or it bends around the spot[5]. Cf. also[6].

In this paper such moving phenomena will be called *flares* if (1) they last shortly and (2) originate suddenly (see the definition at the beginning of this section) and if (3) their Hα brightness and (4) their line width are comparable to the flare values (see Table 11 on p. 192 above).

49. Relation to faculae and spots; recurrence tendency, development and internal motions. Detailed photometric investigations of flares on high resolution photographs are hardly available (for an example cf. [7]), but visual and photographic observations show that most flares are not amorphous but have a fine-structure,

[1] Cf. footnote 4, p. 192.
[2] R. J. BRAY, R. E. LOUGHHEAD, V. R. BURGESS and M. K. MCCABE: Austral. J. Phys. **10**, 319 (1957).
[3] R. G. GIOVANELLI and M. K. MCCABE: Observatory **77**, 238 (1957).
[4] Cf. footnote 7, p. 193.
[5] A. SEVERNY: Isw. Krymsk. Astrophys. Obs. **17**, 129 (1957); **19**, 22 (1958).
[6] A. BRUZEK: Z. Astrophys. **44**, 183 (1958).
[7] D. E. BILLINGS and W. O. ROBERTS: Astrophys. Journ **118**, 429 (1953).

which follows the shape of the facular field. So, most flares can be considered as local brightenings of previously existing faculae. It is the bright mottles of the facular field that brighten up. After the flare the facular field may remain brighter and larger than it was before the flare. As a rule a flare does not accomplish a lasting change of the chromospheric structures[1]. Sometimes not the

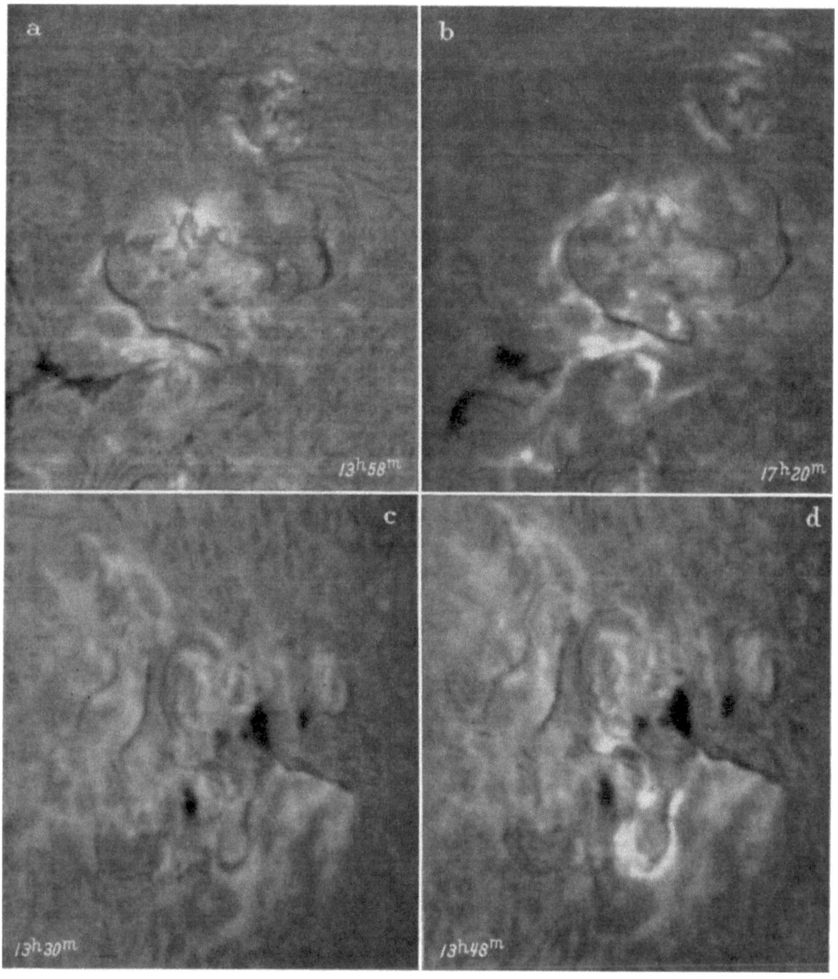

Fig. 51 a—d. Relations between flares and the structure of the facular field. (a) and (b) Hα spectroheliograms of Aug. 16, 1949, showing that certain flares are primarily the brightening of previously existing structures. (At the time of the flare the large dark filament with complex shape in the lower left contained parts with radial velocities of —70 km/sec: flare activation of filaments—cf. Sect. 55.) (c) and (d) The flare of May 19, 1951, closely bordered and followed the track of a filament. Courtesy H. W. Dodson-Prince, McMath-Hulbert Observatory.

facula but a dark filament is brightened; in other cases there is a close relationship in location and motion between certain flares and dark filaments (Fig. 51). Especially in weak flares the normal facular structure can be traced within the flare, but the mottles are brighter than normally. Bright flares often show a very complex pattern.

A flare may consist either of one coherent region or of a number of individual regions of enhanced luminosity, all in the same spot region. The parts may

[1] Cf. footnote 6, p. 194.

brighten up simultaneously or shortly after each other. A flare area seems to increase by a successive activation of the various parts of the faculae, not by a real "creeping" of the flare over the solar surface.

If E is the mean number of flares per day and R the sunspot number, then according to SVESTKA[1]: $E = 0.047R$ and $0.034R$ for the 17th and 18th cycle, while WALDMEIER[2] found $E = 0.044R$, $0.015R$, and $0.002R$ for flares of importance 1, 2, and 3 respectively. So, if $R = 150$ one may expect on the average nine flares per day, among which two of importance 2. SEVERNY and SHAPOSNIKOVA[3] noted that on the average one flare appears during each seven hours of a spot's life. BEHR and SIEDENTOPF[4] found for the whole solar cycles 1933 to 1944 and 1944 to 1954 the relation $N_f = a(R-10)$, which N_f = total number of flares per cycle, R = mean spot number, while $a = 1.98$ for 1933 to 1944, and $a = 1.47$ for 1944 to 1954.

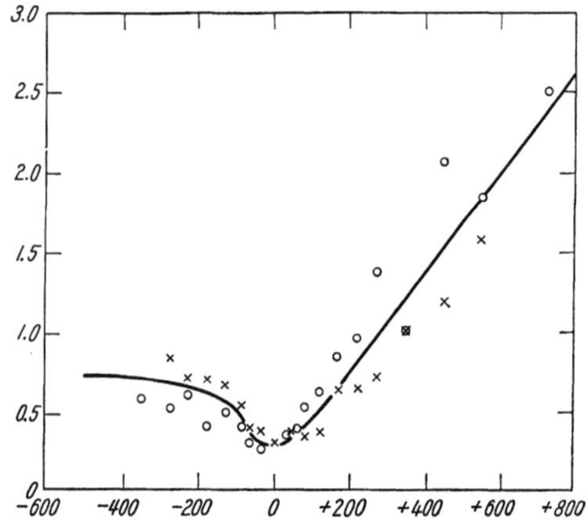

Fig. 52. The probability for a flare to occur (ordinate), as a function of the changes of sunspot area during the day preceding (circles) and following the eruption (crosses). Areas in 10^{-6} visible hemisphere. R. G. GIOVANELLI: Astrophys. Journ. **89**, 555 (1939).

However, statistics of this kind are not complete since the probability that a facular field will produce flares depends greatly on its stage of development (cf. Sect. 46). GIOVANELLI[5] found that the frequency E of flares per spot group is

$$E = A \cdot f\left(\frac{dA}{dt}\right),$$

(Fig. 52), where A is the area of the spotgroup, and t is the time. The rate of flare production is greatest during the ascending part of the development of the group. E seems to be independent of the maximum field-strength of the group, but this viewpoint is not accepted by all investigators[6]. The probability of flares increases if spots divide themselves or if spots of opposite polarity originate[7].

Most flares are confined to the first third part of the spot's life and the greatest number of flares is produced between the 8th and the 15th day of a well-developed spot-group's life. (As a matter of fact shorter lived groups produce flares during a shorter time interval.) The most important flares occur in spot-groups of classes C, D and E. The F stage, being at the maximum of the spots development curve, produces fewer flares.

Flares rarely occur near the unipolar α spots, relatively more near β groups and especially near the magnetically complex γ groups. When a spot is flare-active, it may produce 30 to 50 flares during one passage over the disk. Some very active centres may reach 80 flares in one disk passage. The great sunspot group of June 1955 which arose on the invisible hemisphere and became visible at the eastern limb on June 10 produced flares between June 12 and June 21[8].

[1] Z. SVESTKA: Bull. Astr. Inst. Czech. **7**, 9 (1956).
[2] M. WALDMEIER: Z. Astrophys. **16**, 276 (1938).
[3] A. B. SEVERNY and E. F. SHAPOSNIKOVA: Isw. Krymsk. Astrophys. Obs. **12**, 3 (1954).
[4] A. BEHR and H. SIEDENTOPF: Z. Astrophys. **30**, 177 (1952).
[5] R. G. GIOVANELLI: Astrophys. Journ. **89**, 555 (1939).
[6] B. WITTE: Solar research memorandum (23. X. 1951). High Altitude Observatory, Boulder.
[7] Cf. footnote 6, p. 194.
[8] K. LODÉN: Ark. Astr. **2**, 27 (1956).

Only in the Stockholm-Capri station (8^h observation per day) 32 flares were observed. The total number of reported flares was 53 among which four of importance 3. Besides there were many microflares.

Observations under very good conditions (Lyot, Pic du Midi) have shown that an active spot group may produce an overwhelming number of *microflares* (not to be confused with subflares; the latter are larger). The microflares have very small areas, but have central Hα intensities comparable to those of the

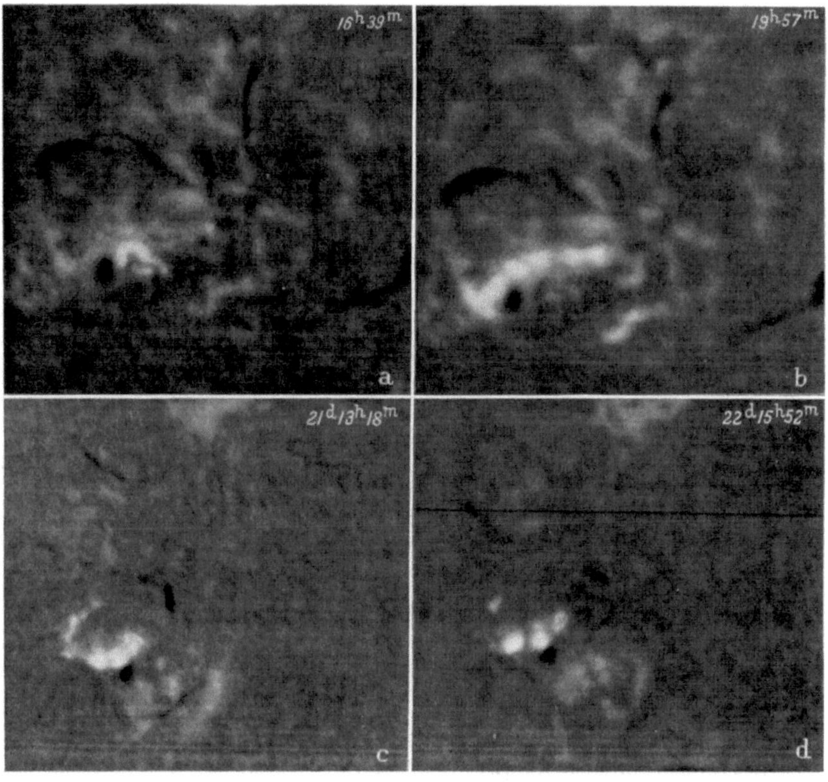

Fig. 53a–d. Recurrence tendency of flares. (a) and (b) Two flares developing in different parts of the same active area on the same day (Nov. 1, 1949). (c) and (d) Two flares which developed in identical positions on successive days (1950, July 21 and 22). Courtesy McMath-Hulbert Observatory.

other flares. Lyot once observed seventeen of them in one hour. Of some of them the diameters were smaller than 1″, the life times were 5 to 15 minutes, occasionally greater.

There might be a tendency for the flares to repeat themselves: in some cases it has been observed that some flares do recur in seemingly identical locations of a facular region within an interval of time of several hours or several days (Fig. 53a). But there are other cases that two consecutive flares, even when being rather close in position and time, have no parts in common (Fig. 53c–d).

A slight asymmetry in the disk distribution of flares has been reported[1–3]; more flares seem to occur at the eastern half than at the western half of the Sun.

[1] F. Link and J. Kleczek: C. R. Acad. Sci., Paris **227**, 827 (1948).

[2] A. Behr and H. Siedentopf: Z. Astrophys. **30**, 177 (1952).

[3] J. S. Hey and V. A. Hughes: Monthly Notices Roy. Astronom. Soc. London **115**, 605 (1956).

50. Light curves of flares. Since most observations are made in Hα the light curve is defined as the curve giving the relative brightness for the centre of Hα ($\Delta\lambda = 0$). Care should be taken when comparing observations obtained with birefringent filters (because of the large passband width) with spectroheliograms or spectra. The latter kind of observations are sufficiently monochromatic, the first should still be corrected for the influence of the finite bandwidth.

Light curves of flares have been obtained first by WALDMEIER (Zürich) and later especially by SEVERNY (Crimea), BALLARIO (Arcetri), and by McMATH and DODSON and associates. Most light curves are assymetric. The ascending

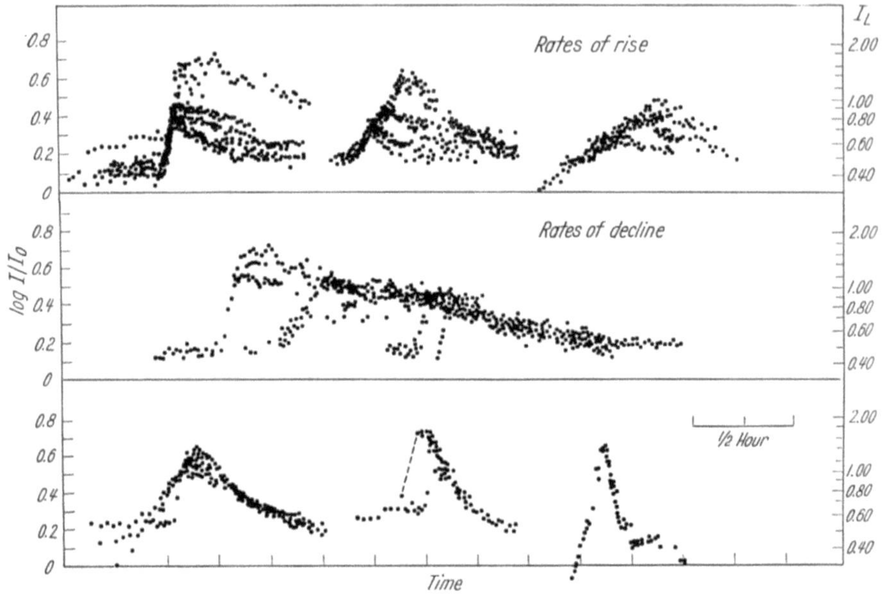

Fig. 54. Comparison of light curves for different flares. H. W. DODSON, E. R. HEDEMAN and R. R. McMATH: Astrophys. Journ. Suppl. 2, 241 (1956).

part lasts as a rule not longer than 5 or 10 minutes; the descending part lasts longer. In connection with the first phase with its rapid rise to maximum brightness ELLISON speaks of the "flash stage".

If A is the duration of the ascending part and D is that of the descending part then, averaging roughly, $A/D = 0.4$, 0.5, and 0.6 for flares of importance 1, 2, and 3 respectively.

According to DODSON[1] the flares prefer three rates of rise (Fig. 54). All are roughly exponential, with respectively: $\text{Log}\, I = 0.1$ per minute; 0.1 per 3 minutes and 0.1 per 7.5 minutes, but this latter case seems to be relatively rare.

For the decline there is one preferred rate, this being the slowest: $\text{Log}\, I = -0.1$ in 20 minutes. Other flares may decline faster, but the rate of decline is nearly always slower than the rate of increase. Once a flare was observed, of which the rates of rise and decline were nearly equal[2]. There is no clear relation between the rate of increase of brightness and the maximum brightness, but there is some evidence that very bright maxima correlate with a fast rate of rise. The slowest rates of rise and decline were observed for flares within 55° of the central meridian;

[1] H. W. DODSON: in: The Sun (ed. G. P. KUIPER), p. 692. Chicago 1953.
[2] H. W. DODSON and R. R. McMATH: Astrophys. Journ. 115, 78 (1952).

the fastest rates of decline occurred for flares within 40° of the limb. If the brightness variation is described by the interpolation formula

$$\operatorname{Log} I_t = \operatorname{Log}\{I_{\max} 10^{a(t-t_{\max})}\},$$

where t is measured in minutes, then the values of a_a (for the ascending branch) range between 0.40 and 0.031, the average is 0.10. For the descending branch a_d is negative and further always $|a_d|<|a_a|$; $|a_d| \leqq 0.025$ for 87% of all cases[1].

The light curves of Fig. 54 refer to the brightest regions of the flare. The different parts of a flare may show quite different light curves. Sometimes the less bright parts of a flare have a light curve which may be obtained from the main curve by a single multiplication by a constant factor. However, it occurs also fairly often that different parts of flares brighten up successively; there are

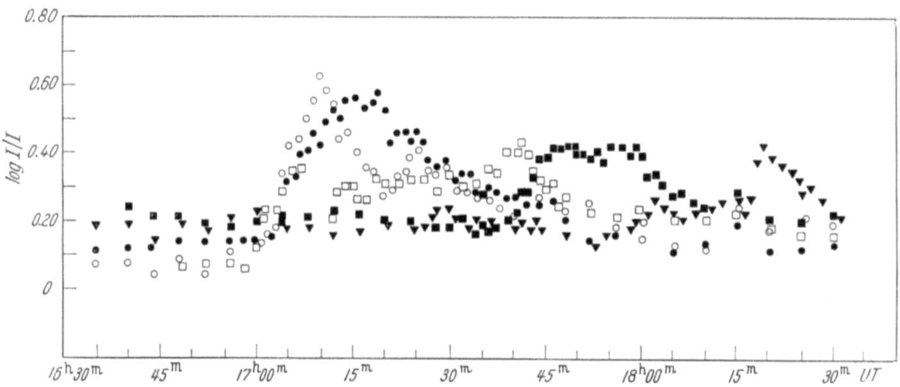

Fig. 55. Light curves of several flares developing at the same place; successive maxima show diminishing intensity (June 22, 1949). H. W. DODSON: in: The Sun (ed. G. P. KUIPER), p. 697. 1953.

various possibilities: the various light curves may be identical but for a shift along the time axis, but it is also possible that the parts show quite different light curves (Fig. 55).

There is a correlation between the position of the flare on the disk and the observed maximum intensity I_c, expressed in units of the local continuous spectrum near Hα (DODSON[2]).

For flares $I_c = 0.63 + 0.34 \cos\vartheta$,
for subflares $I_c = 0.36 - 0.26 \cos\vartheta$.

b) Spectrographic observations.

51. Hα spectra. Most flare spectra are obtained in Hα. So we start with a discussion of this line. The profiles of these relatively small objects have to be corrected for the influences of (a) scattered heterochromatic light; (b) grating ghosts and (c) the apparatus curve of the spectrograph; (a) is the most serious source of errors, (b) and (c) are unimportant for modern spectrographs.

The two parameters which are mostly used for describing the profile are the central intensity (expressed in units of the adjacent continuous spectrum $= AB/JB$ in Fig. 56) and the effective line width of the emission line spectrum XY; XY is the visibility range i.e. the wavelength region where the flare can be distinguished

[1] H. W. DODSON, E. R. HEDEMAN and R. R. MCMATH: Astrophys. Journ. Suppl. 2, 241 (1956).
[2] H. W. DODSON: Astronom. J. 60, 158 (1955).

in spectroheliograms, or in birefringent filters—with some care because of the large passband width. Fig. 56 (ELLISON[1]) shows some schematized Hα profiles of flares of types 1 and 2 and their relation to the normal solar Hα profile, to the profile in faculae, and in filaments and prominences.

Fig. 57 shows the development of the Hα spectrum[2] of the large (imp. 3^+) flare of August 31, 1956. Especially during the flash stage (second spectrum) the wings extend far (in this particular case a small continuous emission was moreover superimposed over the Hα line). As a rule the wings as measured in spectra extend further than is often reported from spectroheliographic observations. In a spectroheliograph or in a spectrograph used visually, the slight excess in brightness of the flare wing with respect to the wing of the normal solar Hα profile cannot be discovered by visual means. Its intensity excess is at about

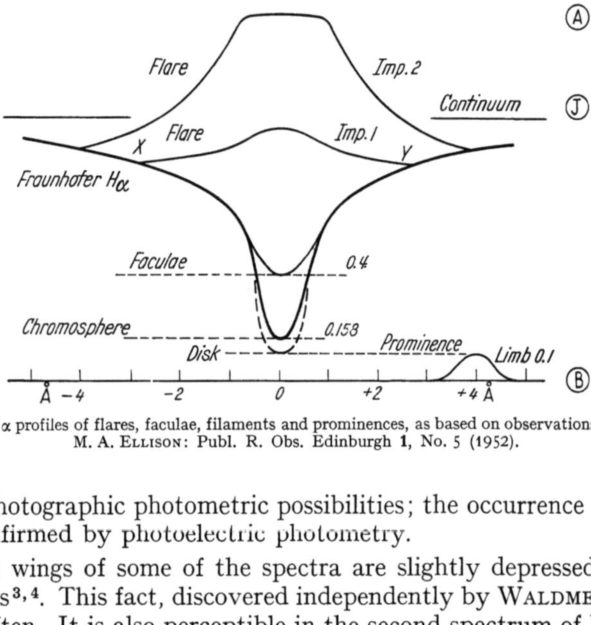

Fig. 56. Sketch of Hα profiles of flares, faculae, filaments and prominences, as based on observations made at Edinburgh. M. A. ELLISON: Publ. R. Obs. Edinburgh 1, No. 5 (1952).

the limit of photographic photometric possibilities; the occurrence of these wings should be confirmed by photoelectric photometry.

The violet wings of some of the spectra are slightly depressed as compared to the red ones[3,4]. This fact, discovered independently by WALDMEIER and ELLISON, occurs often. It is also perceptible in the second spectrum of Fig. 57. ELLISON ascribed it to the absorption of ascending gases expelled in all directions by the flare and hence producing thereby a broad shallow absorption "line" in the violet part of the line profile. The number N_2 of absorbing second quantum state hydrogen atoms in the line of sight is estimated[5] at 5×10^{12} cm^{-2} and the average upward velocity component is 100 km/sec, hence less than the velocities of the slow particles that are emitted by M regions, and considerably less than that of the particles which produce the big magnetic storms occurring after a solar flare (cf. Chap. E of this paper). These latter particles would produce a depression at $\Delta\lambda \approx -30$ Å, perhaps observable immediately after the start of the flare or perhaps at the moment of the flare's beginning. Radio observations suggest that these fast particles are emitted mainly at that very moment, leading to type III radiobursts.

[1] M. A. ELLISON: Publ. Roy. Obs. Edinburgh **1**, 75 (1952).
[2] C. DE JAGER: Bull. astronom. Inst. Netherl. **1959**.
[3] M. A. ELLISON: Vistas in Astronomy **2**, 799 (1957).
[4] M. A. ELLISON: Monthly Notices Roy. Astronom. Soc. London **109**, 3 (1949).
[5] T. V. KAZACHEVSKAJA and A. B. SEVERNY: Isw. Krymsk. Astrophys. Obs. **19**, 45 (1958).

A slight disk asymmetry was suggested[1]: the average line width of flares, east of the central meridian is by about 5% greater than that at the western half. BRUZEK[2] could not confirm the effect.

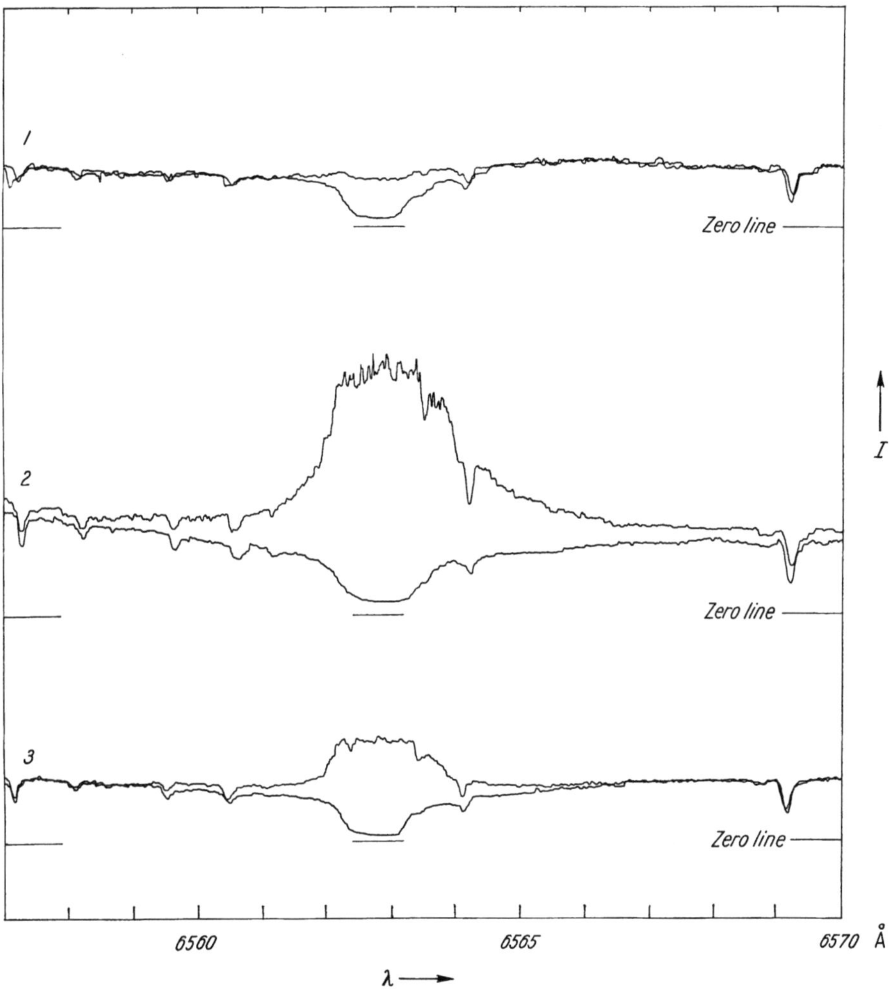

Fig. 57. Development of the spectrum of the great flare of August 31, 1956. The flare was of importance 3^+ and lasted from $12^h\,33^m$ UT till about 15^h UT. The tracings give direct intensities; each lower one gives the undisturbed Hα profile, the upper ones show the flare profile. 1. $12^h\,36^m\,22^s$ UT; shortly after the start. 2. $12^h\,40^m\,31^s$ UT; flash stage. 3. $12^h\,44^m\,10^s$ UT; on descending branch. C. de JAGER, Utrecht Observatory.

ELLISON[3] found an approximately linear relation between the width of Hα in flares and the central intensities of this line. For $2\Delta\lambda = 2$ Å the mean central intensity is about 0.65, for $2\Delta\lambda = 8$ or 9 Å it is about 1.5. For greater values of $2\Delta\lambda$ the central intensity increases but slowly. The relation was confirmed by SVESTKA[4].

---

[1] J. S. HEY and V. A. HUGHES: Monthly Notices Roy. Astronom. Soc. London **115**, 605 (1956).

[2] A. BRUZEK: Z. Astrophys. **38**, 1 (1955).

[3] Cf. footnote 4, p. 200.

[4] Z. SVESTKA: Bull. Astr. Inst. Czech. **2**, 153 (1951).

According to GOLDBERG *et al.*[1] the mean widths of the Hα line in flares increases from the centre of the disk towards the limb[1,2]; but SVESTKA and FRITZOVA[2] have suggested that this phenomenon is caused by an insufficient correction for perspectivic foreshortening, especially for limb flares (so that type 2 flares are called type 1 at the limb).

SEVERNY[3] has described the detailed fine structure (0''.4) observed in spectra of flares (Fig. 58), as well as in faculae. The emission granules last some minutes and sometimes longer; at their place the Fraunhofer lines are somewhat bent towards the violet (cf. also Sect. 53 for a description of the so called "moustaches").

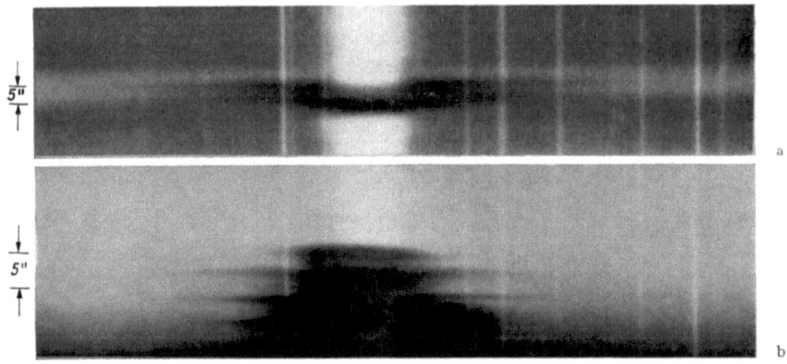

Fig. 58a and b. Hα spectra of flares. (a) Flare of July 16, 1957. The spectrum shows that the central Hα depression so often observed in less resolved spectra, is in this case due to some "moustaches" occurring simultaneously with the flare (cf. also Sect. 53), rather than to self-reversal. (b) Limb flare spectrum, showing fine structure. Courtesy A. B. SEVERNY, Crimean Observatory.

52. The interpretation of the Hα and other Balmer line profiles. *α) Central intensities.* During the flash stage the average central intensities for flares of importances 1, 2 and 3 are respectively: 1.5; 2.1 and 2.4 (cf. Table 11 on p. 192). Assuming that at that moment the flares are optically thick in Hα, the corresponding radiation temperature is often identified with the local kinetic temperature, but, as will be shown, this is a rough approximation.

For the type 3^+ flares the radiation temperature thus found is near 8000°.

β) The central reversal observed in many type 2 flares could perhaps for a part or even wholly be ascribed to scattered photospheric light in the telescope and spectrograph: it is not certain at all that this influence has always been correctly eliminated, especially since recent high resolution pictures[3] have shown the mottled and/or filamentary character of large flares: undisturbed chromospheric or facular matter with Hα in absorption may occur between the luminous flare points. Also „moustaches" may play a rôle in producing this effect (see Fig. 58 and 61, and Sect. 53).

In view of this uncertainty the proposed explanations for the central reversal, either by means of a cool mass of gas above or besides the flare[4], or by nonthermodynamical equilibrium conditions[4,5] will not be treated in detail.

[1] L. GOLDBERG, H. W. DODSON and E. A. MULLER: Astrophys. Journ. **120**, 83 (1954).
[2] A. BRUZEK: Z. Astrophys. **38**, 1 (1955).
[3] A. B. SEVERNY: Astr. Zhurn. **33**, 74 (1956). — Isw. Krymsk. Astrophys. Obs. **17**, 129 (1957); **19**, 46 (1958).
[4] Z. SVESTKA and L. FRITZOVA: Bull. Astr. Inst. Czech. **7**, 30 (1956).
[5] Z. SVESTKA: Publ. Astr. Inst. Czech. Acad. Sci. **32** (1957).

γ) Line broadening. An essential problem is whether the emission line profiles are widened by radiation damping or by Stark effect. If the broad line wings are ascribed to the Stark effect, by analogy with the photospheric spectrum (ELLISON and HOYLE[1]), a number of 10^{13} free electrons per cm^3 is necessary to explain the observed widths. In that case the higher members of the Balmer series should *a fortiori* be widened by the Stark effect. But SUEMOTO[2], MUSTEL and SEVERNY[3], SVESTKA[4] and GOLDBERG *et al.*[5] argued that the electron density was less than ELLISON and HOYLE initially assumed; in that case Hα is widened by radiation damping; whether some of the higher Balmer lines are also widened by radiation damping remained an open problem. In a later paper, based on high resolving power and large-dispersion spectra obtained with the Crimean tower telescope KAZACHEVSKAJA and SEVERNY[6] could state with more certainty that all Balmer lines, including Hα, are widened by the Stark effect.

In order to show the contributions of the two effects in the line wings, we should know the excitation and ionization temperatures in the flare. Often thermodynamic equilibrium is assumed, hence $T_{\text{el}} = T_{\text{ion}} = T_{\text{ex}}$, an assumption that may be grossly erroneous. This means that some of the conclusions which will be summarized below are preliminary. We write for the intensity of the emission line profile:

$$I(\Delta\lambda) = I_A(\Delta\lambda)\, e^{-\tau \sec\vartheta} + E\,(1 - e^{-\tau\sec\vartheta}). \tag{52.1}$$

I_A is the residual intensity in the undisturbed Hα line profile.

E is the source function of the flare, $E/I_{\text{cont.}}$ may be identified with the central intensity of the flare, if the flare is optically thick.

τ is the total optical depth of the flare at the wavelength distance $\Delta\lambda$.

When $\tau \ll 1$, as is the case in the far wings, (52.1) becomes

$$I - I_A = (E - I_A)\,\tau \sec\vartheta.$$

If H is the atmospheric scale height in the flare and $\varkappa$ is the absorption coefficient per atom in the second quantum state of the H atom (total number per cm^3: N_2), τ may be put equal to $\varkappa H N_2$. In the far wings

$$\varkappa_S = 1.4\times 10^{-29} N_e \Delta\lambda^{-\frac{5}{2}} \quad \text{(case of Stark effect)}$$

$$\varkappa_R = 5.0\times 10^{-17} \Delta\lambda^{-2} \quad \text{(case of radiation damping)}.$$

In this formula for $\varkappa_R$ the influence of the Doppler broadening on the core profile has been neglected. For a discussion of this case cf. SVESTKA (footnote 5, p. 202).

At $\Delta\lambda = 4$ Å we have approximately equality for $N_e = 7\times 10^{12}$. If $N_e > 7\times 10^{12}$ the Stark effect predominates; in the other case it is the radiation damping. So the next problem is to find the value of N_e:

We assume that the width is defined as the distance between the two points where the flare intensity exceeds that of the line by 5%, hence: $I - I_A = 1.05\, I_A$. In the two cases mentioned, this line width is respectively:

$$2\,\Delta\lambda_S = 2.9\; 10^{-12} \left[\frac{1.05\, I_A}{E - I_A} N_e N_2 H \sec\vartheta\right]^{\frac{2}{5}},$$

$$2\,\Delta\lambda_R = 7.1\; 10^{-9} \left[\frac{1.05\, I_A}{E - I_A} N_2 H \sec\vartheta\right]^{\frac{1}{2}}.$$

[1] M.A. ELLISON and F. HOYLE: Observatory **67**, 181 (1947).
[2] Z. SUEMOTO: Publ. Astr. Soc. Japan **3**, 110 (1951).
[3] E. R. MUSTEL and A. B. SEVERNY: Isw. Krymsk. Astrophys. Obs. **8**, 19 (1952).
[4] Cf. footnote 4, p. 201.
[5] L. GOLDBERG, H. W. DODSON and E. A. MÜLLER: Astrophys. Journ. **120**, 83 (1954).
[6] T. V. KAZACHEVSKAJA and A. B. SEVERNY: Isw. Krymsk. Astrophys. **19**, 46 (1958).

The further procedure involves four uncertain points. For a given flare we observe $E/I_{\text{cont.}}$; (1) this gives T_{rad} which we assume equal to T_{ex} and T_{ion}. (2) For any value of N_e we then find N_H with the known ionization formulae or tables, and (3) N_2 follows with BOLTZMANN'S law. (4) We assume H, the scale height of the flare, equal to the scale height in the upper part of the chromosphere: $\approx 10^8$ cm. Knowledge of the ionization and excitation conditions, and of the scale height of flares is actually lacking, but accepting the above four assumptions, it is possible to compute the values of $2\Delta\lambda_S$ and $2\Delta\lambda_R$ for $\sec\vartheta = 1$, for the various types of flares and different N_e values. The comparison with the observed $2\Delta\lambda$-values yields N_e and makes it possible to decide at the same time which of the two widening mechanisms is active. The table below gives the basic data for the Hα line in two types of flares and Fig. 59 shows the computed values of $2\Delta\lambda_S$ and $2\Delta\lambda_R$ for these two types of flares as functions of the electron density. The comparison with the observations (short horizontal bars) shows that in both types of flares the "best" value of Log N_e seems to be about 13.2, in agreement with modern values (footnote 6, p. 203). In the range of Log N_e values between about 12 and 13.5 the contribution of the two mechanisms are nearly equally important. Above Log $N_e = 12.7$ the Stark effect dominates slightly and below 12.0 the reverse is the case.

Imp.	$2\Delta\lambda$ Å	$E/I_{\text{cont.}}$	T
3^+	15.0	3	7650°
2	4.5	2	6900°

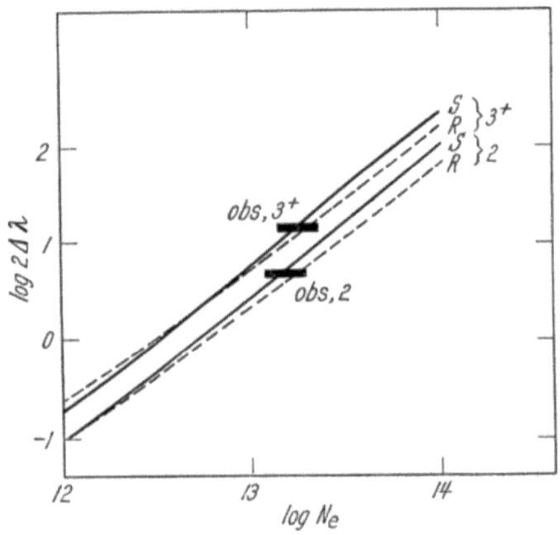

Fig. 59. Stark broadening and radiation damping in imp. 2 and 3^+ flares. For further details see text.

The four uncertainties listed above, of course, throw some doubt on the final result, and it is obvious that the above gives only an outline for a theory of flare spectra. However, for a more detailed discussion it is not sufficient to use only Hα observations. The material should be extended with profiles of other lines. A detailed analysis was published by MUSTEL and SEVERNY[1] on the basis of observations of Hα and Hγ ... Hζ in one flare. ŠVESTKA[2] and FRITZOVA[3] have discussed ELLISON'S observations[4] as well as those of MUSTEL and SEVERNY[1], of SUEMOTO[5] (H_{10}) and of RAO, DAS *et al.*[6] (Hα and Hβ), and found Log N_e slighlty smaller than 13.

Later investigations yielded higher N_e values, mainly because the new observations have been made with a greater image resolving power, which enables better to distinguish between the real brilliant flare knots and the background:

With the flare spectrograph of the Tokyo Observatory, which permits to photograph in one single exposure the whole spectrum from Hα to the Balmer

[1] E. R. MUSTEL and A. B. SEVERNY: Isw. Krymsk. Astrophys. Obs. **8**, 19 (1952).
[2] Z. ŠVESTKA: Publ. Astr. Inst. Czech. Acad. **32** (1957).
[3] Z. ŠVESTKA and L. FRITZOVA: Bull. Astr. Inst. Czech. **8**, 61 (1957).
[4] Cf. footnote 4, p. 200.
[5] Z. SUEMOTO: Publ. Astr. Soc. Japan **3**, 110 (1951).
[6] N. RAO, A. DAS *et al.*: Nature, Lond. **164**, 964 (1949).

continuum (dispersion 3.2 Å/mm) SUEMOTO[1] obtained spectra of a great number of flares (mainly imp. 2), showing the Balmer lines up to H_{14}. The flare emission line profile corrected for the contribution of the photosphere below is essentially found from our expression (52.1) by taking the profile equal to

$$1 - \exp\left(-\tau(\Delta\lambda)\sec\vartheta\right) = I_{\text{flare}}(\Delta\lambda)/(E - I_A). \tag{52.2}$$

(SUEMOTO's treatment was somewhat more refined since he first corrected the "photospheric" Hα profile for the influence of chromospheric line radiation.) The half-widths of the profiles of Hα to H_{14}, determined from (52.1) reach a minimum value for $n=8$. The observed widths can be explained by assuming that all the lines are widened by the Stark effect with an adopted flare excitation temperature of 5600° and $\text{Log}\, N_e \approx 13.3$. Simultaneously KAZACHEVSKAJA and SEVERNY[2] derived similar results ($T_{\text{el}} = 10^4$ °K and $\text{Log}\, N_e = 13.4$) from a study of high resolving power spectra obtained at the Crimean Observatory.

At this value of N_e all Balmer lines, including Hα must necessarily be widened by the Stark effect (cf. Fig. 59). The population of the low quantum levels ($n<3$ or 4) is determined by the own radiation field of the flare; the higher levels are excited by collisional excitation and ionization[3].

Reference is further made to the non-thermodynamic equilibrium discussions of ZIRIN[4] and JEFFERIES[5], leading to $T_{\text{el}} > 10^4$ °K. Since the He II line at 4686 Å is usually absent, even in spectra of many strong flares[6], T_{el} cannot greatly surpass this value.

δ) The results of this section may be summarized in the following *working model of a flare*. In strong flares the optical depth in the centre of Hα may be as great as 10^4, corresponding with $N_2 = 2\times 10^{16}$ above 1 cm^2 of the photosphere[7]. For medium type flares this value is $4\times 10^{15} \ldots 10^{16}$. For the smallest flares the optical thickness in the centre of Hα is of the order of 200.

Further $T_{\text{kin}} \approx 10000$ to 15000°; $T_{\text{ex}} \approx 6000$ to 8000°; $N_e \approx 2\times 10^{13}$. A slight overpopulation of the lower levels (up to a factor 1.4 for $n=3$) was ascribed[7] to photo-excitation by the flare's own radiation field. The temperatures are higher during the flash stage.

53. Other spectrographic observations; the continuous spectra and the "moustaches". Flare spectra do not only show the Balmer emission lines but also other strong lines, among which there are the He line 10832 Å, the Ca^+ lines, the Paschen 3—6 and 3—7 lines. ALLEN[8] listed 116 emission lines observed between 3900 and 6700 Å in the spectra of strong flares. In the strong limb flare of 21 February 1942 WALDMEIER[9] observed 100 emission lines between 5000 and 6700 Å, among which many lines of neutral and ionized metals. The fact that the Balmer series is only observed up to H_{15} is not due to their merging into each other but to their weakness, so that they only partially fill up the Fraunhofer lines without being visible as emission lines. The Balmer continuous emission spectrum generally is weak also, so that the same might be true of the high series members. The relation between the central intensity of Hα and that of the Na D_2 line and the He D_3 line[10] is reproduced in Fig. 60. Apparently weak

[1] Z. SUEMOTO: Publ. Astr. Soc. Japan **10** (1958).
[2] Cf. footnote 5, p. 200.
[3] A. B. SEVERNY: Isw. Krymsk. Astrophys. Obs. **19**, 72 (1958).
[4] H. ZIRIN: Astrophys. Journ. **124**, 451 (1956).
[5] J. T. JEFFERIES: Monthly Notices Roy. Astronom. Soc. London **117**, 493 (1957).
[6] V.A. KRAT and V.M. SOBOLEV: Isw. Gl. Astron. Obs. Poulkovo **162** (1958).
[7] Cf. footnote 5, p. 202.
[8] C. W. ALLEN: Monthly Notices Roy. Astronom. Soc. London **100**, 635 (1940).
[9] M. WALDMEIER: Z. Astrophys. **26**, 305 (1949).
[10] M. WALDMEIER: Z. Astrophys. **20**, 62 (1940).

flares produce an absorption in D_3; strong ones show emission lines. New observations show the relation to be perhaps less simple than in the figure; the D_3 absorption can also occur without a flare[1], and may persist after a flare.

It is often assumed that *continuous spectra* occur only in the brightest flares, but this is not according to fact. Nearly all flares of importance 2 and more seem to have continuous spectra[2], but this spectrum is generally very weak and can hardly be discovered. In most cases its intensity is of the order of 1% of the continuous radiation and only in some cases it amounts to higher values. In the very intense flare of July 25, 1946 it amounted to 10%[3]. With the data of the preceding section the optical depth in the continuous spectrum may be computed. With VITENSE's tables of $\varkappa_\lambda$, and assuming $\operatorname{Log} N_e = 13.3$; $\Theta = 0.7$, $E_{\text{flare}}/E_{\text{photosphere}} = 2.5$, and taking the scale height H as small as 100 km, one finds $\operatorname{Log} \tau_\lambda = \operatorname{Log}(\varkappa_\lambda N_H H) = 2.2$ and $I_{\text{flare}}/I_{\text{photosphere}} = 0.04$ which is in the range of the "observed" values. This again indicates that the emitting elements of flares are small and may have sizes of the order of 10^2 rather than of 10^3 km. MUSTEL's[4] suggestion to explain the continuous spectrum by a slight increase of the *photospheric* temperature, thus assuming that flares extend down into the photosphere does not seem well founded as follows from a detailed discussion by ŠVESTKA (footnote 5, p. 202). Sometimes the continuum is slightly polarized[5].

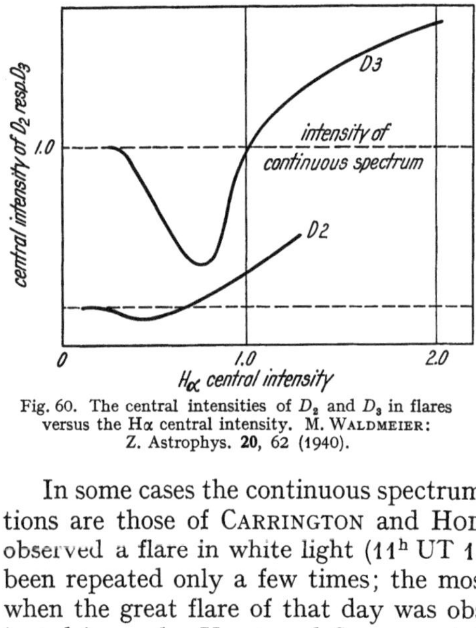

Fig. 60. The central intensities of D_2 and D_3 in flares versus the Hα central intensity. M. WALDMEIER: Z. Astrophys. **20**, 62 (1940).

In some cases the continuous spectrum is more striking. The classical observations are those of CARRINGTON and HODGSON (1859) who both, independently, observed a flare in white light (11^h UT 1. 9. 1859). This kind of observation has been repeated only a few times; the most impressing case is that of 23. 2. 1956 when the great flare of that day was observed visually in white light on a projected image by UNNO and SHIMIZU at the Tokyo Observatory[6]. The duration of visibility was about 5 min, the diameter was 3500 km. Also for the flares of July 7, 1937[7], March 5 and July 25, 1946[8], August 31, 1956 and March 23, 1958[9] there was an observable enhancement of the continuum.

Another classical observation is that of ELLERMAN's[10] "bombs" in which the continuous spectrum is brightened over a broad region (30 Å) around Hα, but where Hα and the other Fraunhofer lines are practically unchanged. The interest in this latter kind of observations has been revived in recent years by SEVERNY's high resolution- and high dispersion-observations of active faculae and of flares[11].

[1] V.A. KRAT and V.M. SOBOLEV: Isw. Gl. Astron. Obs. Poulkovo **162** (1958).
[2] Cf. footnote 10, p. 205.
[3] M. A. ELLISON: Observatory **66**, 358 (1946).
[4] E. R. MUSTEL: Trans. Internat. Astronom. Union **9**, 671 (1957).
[5] A.B. SEVERNY and W.L. KHOKHLOVA: Isw. Krymsk. Astrophys. Obs. **20**, 67 (1958).
[6] M. NOTUKI, T. HATANAKA and W. UNNO: Publ. Astr. Soc. Japan **8**, 52 (1956).
[7] M. WALDMEIER: Z. Astrophys. **20**, 46 (1940).
[8] M. D'AZAMBUJA: L'Astronomie **61**, 114 (1947). Cf. also Observatory **67**, 156 (1947).
[9] M. WALDMEIER: Z. Astrophys. **46**, 92 (1958).
[10] F. ELLERMAN: Astrophys. Journ. **46**, 298 (1917).
[11] A. B. SEVERNY: Astron. Zhurn. **33**, 74 (1956). — Isw. Krymsk. Astrophys. Obs. **17**, 129 (1957); **19**, 72 (1958). Cf. also Observatory **76**, 241 (1956).

These observations show that tiny structures, in some respects quite similar to the ELLERMAN bombs can be seen fairly often when only the image resolution is great enough. This latter kind of emission appears near growing spots and in the first development stage of flares, especially in the flash stage, and is concentrated in short-lived granules with diameters of the order of 400 km.

The difference between these so-called "moustaches" and the Ellerman bombs is the size (the bombs are larger) and perhaps the life time: DODSON-PRINCE and MOHLER (private communication) have observed a number of bombs with rather long life times.

Fig. 61 a—d. Moustaches. From above to below: (a) Moustaches in the K-line. (b) In H and Hε, during a flare. (c) Symmetrical Hα moustaches, and (d) asymmetrical moustaches in Hα. Courtesy A. B. SEVERNY, Crimean Observatory.

The difference between moustaches and flares is that the flare emission fills mainly the central part of the line and also the wings, whereas the moustaches are mainly brilliant in the line wings, leaving the line core practically undisturbed. This shows that they occur mainly at photospheric or chromospheric heights below the region where the Hα core is emitted, that is below about 2000 km, which is confirmed by a study of H and K moustaches by KHOKHLOVA[1] ($\tau = 0$ to 1.0). In limb observations they are also observed *above* the limb (Fig. 58).

Moustaches are always accompanied by a continuous emission which appears as a continuation of the moustaches in the form of thin threads through the spectrum. This continuous spectrum resembles that of an $A\,0$ star. At the place of the continuous emission the Fraunhofer lines are somewhat violetward displaced (≈ 2 km/sec).

SEVERNY remarked that the very broad wings (up to 10 to 15 Å) of flares in the flash stage often consist of a cluster of moustaches (cf. also Fig. 58).

[1] E. KHOKHLOVA: Isw. Krymsk. Astrophys. Obs. **19**, 115 (1958).

A typical Hα-moustache may extend up to 15 Å, the violet wing generally being brighter and broader than the red one, independent of the position on the disk. Sometimes only the violet wing is visible, indicating a great violet shift of the emission as a whole (up to 1000 km/sec; neither limbshift nor darkening). Often also the dark absorption line is shifted violetward, suggesting slow large scale motions, at a speed not exceeding 8 km/sec. The life time of these structures is proportional to the area, it is of the order of some minutes. The area increases with the line intensities. The amount of energy liberated in these granules is estimated to be of the order of 5×10^2 ergs cm^{-3} sec^{-1}; i.e. a thousand times more than the thermal photospheric energy density. Since the energy exchange with the surroundings goes at a fairly rapid rate, a source of energy should be required to maintain the observed emission rate during the life time of these formations.

The amount of energy emitted seems comparable to the energy produced in thermonuclear reactions. Another indication that thermonuclear reactions might be active in the generation of this radiation is found in the observation that the depression at 6560.8 Å, tentatively ascribed to deuterium (abundance of $D/H=4\times 10^{-5}$ according to SEVERNY[1]), increases in the moustaches. Also the amount of energy then generated is of the same order as the observed quantity.

It is not certain whether this identification is correct. The D line profile, as published by SEVERNY seems too narrow; it should have a width comparable to the Hα-core[2].

It is clear that the moustaches and "Ellerman bombs", although related to flares are different classes of objects. Flares are typical upper chromospheric phenomena; the moustaches are low chromospheric. They are explosion-like phenomena, a conclusion drawn from the observation that the expansion is independent of ϑ, hence is isotropic.

54. The ultra violet and X ray spectrum associated with flares. The first measurement of the X ray radiation of the Sun during (the last moments of) a flare was made in 1956.

Up to then the only source of information was the terrestrial ionosphere. Flares seem to have no perceptible influence on the E and F layers, which means that the flare spectrum does not show an excess of radiation in the wavelength interval 100—1000 Å. However, the D layer is appreciably influenced. Ionospheric observations point to a lowering of the reflecting ceiling by 15 km during the strongest flares. It is known that the *normal* D layer is produced by Ly α radiation, ionizing NO molecules which occur in a rather small abundance in the upper atmosphere, but this radiation seems insufficient to produce the lowering of the D layer. FRIEDMAN and CHUBB[3] have computed the following relation between the height of the D layer base and the required increase of Lyα radiation:

Height of D layer (km):	74.6	71.8	67.2	64.4	62.2	60.5
Increase in Lyα:	1	2.5	25	2.5×10^2	2.5×10^3	2.5×10^4.

For a large flare the Lyα radiation should increase by about 10^5. It seems improbable that the Lyα radiation can increase by so large a factor: the necessary radiation temperature in Lyα should increase from 6000 to 80000°[4]. But since the terrestrial atmosphere near $h=70$ km is also transparent for X-rays ($\lambda<5$ Å),

[1] A. B. SEVERNY: Astron. Zhurn. **34**, 328 (1957).
[2] C. DE JAGER: Rech. Obs. Utrecht **13**, No. 1 (1952).
[3] H. FRIEDMAN and T. A. CHUBB: Physics of the Ionosphere, p. 58. London 1955.
[4] C. DE JAGER: in: Solar Eclipses and the Ionosphere, p. 174. London 1955.

which moreover ionize all constituents—not only NO, FRIEDMAN and CHUBB predicted that the flares would prove to emit X-rays of wavelengths shorter than 6 Å. Their prediction found a beautiful confirmation[1] in the first rocket observations of a flare, made on July 20, 1956 by means of a rocket (15 cm diameter) brought by a balloon to a height of 21 km and fired at floating altitude by radio command after the detection of a flare of importance between 1 and 1^-. Observations were made in the last few minutes of the flare's activity. The rocket was equipped with an ionization chambre with a LiF window, filled with NO gas to measure Lyα radiation, and a Geiger counter with Be window for the measurement of X-rays between 0.05 and 10 Å.

No increase of Lyα radiation over normal was observed, but excess X-ray radiation proved to be present in the spectral region below 8 Å (though these X-rays were not strong enough to produce a radio fade out).

In the summer of 1957 new data could be obtained[1] on solar X-rays emitted during flares of importance 1^+, 2 and 3. The measurements were made with two-stage rockets fired from a ground station at an island in the Pacific. The total X-ray flux varied from about 10^{-3} ergs cm^{-2} sec^{-1} for the imp. 1 flare to about 10^{-2} ergs cm^{-2} sec^{-1} for the imp. 2 flare. For the imp. 3 flare the increase of hardness of the radiation seems more important than the increase of intensity. The largest flares produce the shortest wavelength emission: about 2 Å. The radiation was perceptible at a height as low as 63.5 km.

The Table 12 below surveys the most reliable data, obtained during and after the sunspot minimum of 1953/54 as compared with later observations.

Table 12.

Rocket	Time of firing (UT)	Energy below 8 Å (erg · cm^{-2} sec^{-1})	Energy 8–20 Å (erg · cm^{-2} sec^{-1})	Flare
A 14	15. 11. 1953, 22.40	$<6.7\times10^{-6}$	0.0015	
A 15	25. 11. 1953, 15.46	2.9×10^{-6}	0.0013	
A 16	1. 12. 1953, 15.29		0.0004	
A 34	18. 10. 1955, 22.50	0.065	0.0012	
D 8	20. 7. 1956, 19.17	5×10^{-3}		1^-
42 *F*	20. 8. 1957, 16.49	$>3\times10^{-3}$		1^+
45 *F*	29. 8. 1957, 21.12	$>2\times10^{-3}$		2
49 *F*	18. 9. 1957, 17.54	$>1.2\times10^{-4}$		3

The theoretical problem whether X-rays of so short a wavelength can be emitted by the Sun has been investigated by ELWERT[2], who computed the X-ray emission of coronal condensations, characterized by temperatures of 6×10^6 and 3×10^6 °K and a central electron density of 6×10^9 cm^{-3}. It appeared that the emitted intensity of radiation considerably increases as compared with the undisturbed corona ($T_{el}=10^6$ °K). The short wavelength end of the condensation spectrum is found at 6 to 7 Å, shorter than for the normal Sun, but still greater than the wavelength where radiation is actually observed. It is perhaps not impossible that small coronal knots have a still higher temperature. The yellow coronal line (Ca XV, 841 eV) which is closely connected with strong flares points to a temperature in coronal "hot spots" that may reach 10^7 °K. Such a temperature might be sufficient to emit X-rays near 3 Å (cf. also Sect. 74).

[1] H. FRIEDMAN, T. A. CHUBB, J. E. KUPPERIAN, R. W. KREPLIN and J. C. LINDSAY: IGY Rocket Report Series 1, 183 (1958).

[2] G. ELWERT: Z. Astrophys. 41, 67 (1956).

That radiation shorter than 1.5 Å has not been observed is explained by the fact that this wavelength limit corresponds to the ionization energy of the K-shell of Fe (ionization temperature 4×10^7 °K). Radiation of still shorter wavelengths can only be emitted if the K-shells of still heavier atoms are ionized, but these atoms have so low an abundance that no appreciable radiation intensity can be expected. Cf. further Sect. 74.

c) Dynamical phenomena associated with flares.

55. Influences on filaments and prominences: activation. As a rule stable filaments do not occur in an active centre, the filaments occurring are mostly short-lived and their life times may still be shortened by the occurrence of flares.

The influences of flares on filaments and prominence are various: (1) Together with a flare a high speed flare-filament (active dark flocculus) may appear. These objects are disk-projected flare surges (see next section). (2) A flare may activate an already existing filament, which as a matter of fact is at a considerable distance from the flare, because stable filaments do as a rule not occur closely to spot-groups. Often (3) such an activation results in a sudden disappearance of the filament. Finally, (4) flare-like faculae, occurring outside spot-groups may be connected with ascending filaments. In this section cases (2), (3) and (4) are discussed; case (1) is treated in Sect. 56. Reference is further made to Sect. 66 (on the activation and dissolution of filaments).

α) The activation of a filament is reflected by the increase of darkness, size and shape and of internal motions. The activation may last for a considerable time and it may in rare cases even start before the visual appearance of a flare. It may end with the complete dissolution of the filament. Some regularities in flare activation have been summarized by BRUZEK[1], who also reviews earlier descriptions by others. In all of his 23 observed cases the filaments obtained high radial velocities up to several hundreds of km, together with an increase of area and darkness. In 15 cases the filament returned to its original shape and position, in the other cases it disappeared. On the other hand 39 cases were observed in which closely neighbouring filaments were not influenced by flares. There is neither any dependence on brightness or area of the flares nor on the distance between flare and filament. The average duration of the activation process for non-disappearing filaments is about one hour. The motions are much more complicated than in the case of surges. It seems that filaments situated on the prolongation of threads of filamentary flares are more influenced than others. Assuming that the—unknown—activating disturbance starts together with the flare and at the same place, one obtains (from the time difference between start of flare and start of activation) a velocity of progression for this disturbance of 20 to 200 km/sec, average value 60 km/sec.

β) Disappearance of filaments. In the cases that filaments were observed to *dissolve* wholly under the influence of flares, they *were* already more or less unstable. In other cases it happens that disappeared filaments come back after one or two hours sometimes after a longer interval (days).

In the majority of the cases (60%, BRUZEK[2]) the sudden disappearances of filaments are not due to flares but are rather correlated with the origin of new centres of activity. It is known that the appearance of a new centre has a dissolving influence on all remnants, if still existing, of old centres, like filaments. The

[1] A. BRUZEK: Z. Astrophys. **28**, 277 (1951).
[2] A. BRUZEK: Z. Astrophys. **31**, 99 (1952).

disturbances originating from the new centre progress much slower than those due to flares, viz. with an average speed of 1 km/sec; their influence may extend to a distance of 25°. Filaments at this distance dissolve 3 to 5 days after the origin of its centre. BRUZEK[1] has suggested to associate this expansion with the general expansion velocity of young centres of activity as detected in the expansion of the emission line corona (5303 Å) of the *BM* regions, and of the Hα and Ca^+ faculae (speeds $\approx 0{,}3$ to 0,6 km/sec).

γ) Flare-like faculae. BRUZEK[2,3] and BECKER[4] have drawn attention to the fact that of the ascending filaments (sudden disappearances by ascension) that are not associated with the origin of new centres of activity, 70% are associated with *flare-like* Hα *faculae.* These faculae generally form a string of bright mottles or alignments of equal length as the filament at the basis of the disappeared filament or parallel to it. Sometimes two successions are formed, parallel to each other at some 5° apart. These flare-like faculae resemble the flares in some respects, they may attain a brightness equal to those of imp. 2 flares, but (1) they occur outside centres of activity (because they occur at the same place of initially quiescent filaments) and (2) their development is much slower. (The suddenness of commencement is one of the characteristics of flares—see definition, p. 191.) The brightening begins 20 to 40 minutes after the filament starts ascending; maximum brightness is attained some 30 to 100 minutes later; the decay is very slow (mean life time 3 hours). Sometimes the bright mottles move away from the basis of the filament with a velocity of some km/sec.

The observations that certain filaments are activated or dissolved whereas others, often closer to a flare, remain unchanged might suggest a magnetic or mechanical perturbation rather than the influence of radiation. However ÖHMAN[5] has suggested to explain this behaviour on the basis of a radiational disturbance by assuming that filaments that do not disappear are cold ones consisting for a considerable part of H_2 molecules. These are not affected by the ionizing influence of the Lyman continuous radiation of flares.

56. Flare-surges, -puffs and -filaments[6]. One should well distinguish between *flares* and the *flare-surges* that accompany flares. When seen at the limb the flare surges are bright; viewed against the disk they may as well be brighter or darker than the surroundings; often they are bright in their first phase and become dark in the next phase.

In general a *limb-surge*, according to the description of MENZEL and EVANS[7], consists of "streamers or knots of luminous gas ejected from active areas of the solar surface. The ejected filaments or knots exhibit a tendency to return to the Sun along or near the initial paths. The actual trajectory may be highly curved and run frequently parallel to paths of neighbouring moving material. The surges, as a class, possess the highest velocities of all the prominences, often attaining several hundreds of kilometers per second. The filaments are often quasi-oscillatory. As PETTIT has pointed out, surges may recur again and again in the same place. The similarity of successive apparitions of the filament, as

[1] A. BRUZEK: Z. Astrophys. **31**, 111 (1952).
[2] A. BRUZEK: Z. Astrophys. **42**, 76 (1957).
[3] A. BRUZEK: Observatory **72**, 154 (1952).
[4] U. BECKER: Z. Astrophys. **42**, 85 (1957).
[5] Y. ÖHMAN: Ark. Astr. **2**, 1 (1956).
[6] Cf. W. O. ROBERTS: Trans. Internat. Astronom. Union **9**, 658 (1957).
[7] D. H. MENZEL and J. S. EVANS: Convegno Volta **11**, 130 (1953).

to outline and curvature, strongly suggests that the same material is involved. At least the gas must be subject to nearly identical force fields.

Some surges fade to invisibility or near invisibility, which event may occur either during the rise or descent of the material. The indications are, however, that much of the invisible matter falls back to the solar surface although an appreciable fraction of it may well expand into the corona."

A typical flare-surge, described by MCMATH and MOHLER[1] appeared 12 min after a limb flare. It reached maximum extension (height of 60000 km) 13 min later. Then the top shrunk back and it subsided 24 min after that moment. So the mean upward velocity was about 100 km/sec, the downward velocity half of that.

The average time difference between the beginnings of a flare and an accompanying surge is of the order of 4 min, most of the surges starting after the flare (up to 30 min); only about 10% start before the flare[2].

The lifetime of a surge varies from 2 min for the smallest to about $2\frac{1}{2}$ hours for the largest (mean duration: 25^m [2-4]). They have a tendency to be recurrent. PETTIT once observed a centre of activity in which eight surges occurred in $4\frac{1}{2}$ hours. Flare surges may reach heights of 10^5 km; they are never to be identified with flares; they do not occur precisely over a flare but they may form part of it, and always occur in the neighbourhood of a flare. The average distance is about 15000 km[2], but the distance may amount up to 170000 km, once observed by BRUZEK[4]. The direction of the surges, projected on the disk are initially nearly radially away from the nearest major spot[5].

The frequency of flare-surges is given below (ELLISON[3]):

Importance	Number of flares	% of surges per flare
1	142	31
2	45	51
3 and 3+	9	78

BRUZEK[4], on the basis of a somewhat different flare-classification, consisting of four classes defined by area *and* line width, finds:

Class	I	II	III	IV
Number . .	118	113	68	17
Percentage .	51	52	62	82

On Hα spectra of important flares surges are common accompanying features. They present absorption spectra which may either be red- or violet-displaced. As seen at the limb, the matter may appear to have outward velocities up to 500 km/sec and more; if projected on the disk and observed as Doppler displacements the velocities seem generally to be smaller, which might be due to lack of absorbing power of the hydrogen atoms.

When observed on the disk in Hα (Fig. 62a) a surge starts its life as a bright or dark mottle, of about 10000 to 20000 km diameter ("high speed dark fila-

[1] R. R. MCMATH and O. C. MOHLER: Observatory 68, 110 (1948).
[2] K. LODÉN: Ark. Astr. 2, 153 (1958).
[3] M. A. ELLISON: Monthly Notices Roy. Astronom. Soc. London 109, 3 (1949).
[4] A. BRUZEK: Z. Astrophys. 28, 277 (1951).
[5] R. G. GIOVANELLI: Austral. J. Phys. 11, 191, 350 (1958). — E. F. SHAPOSHNIKOVA: Isw. Krymsk. Astrophys. Obs. 18, 151 (1958).

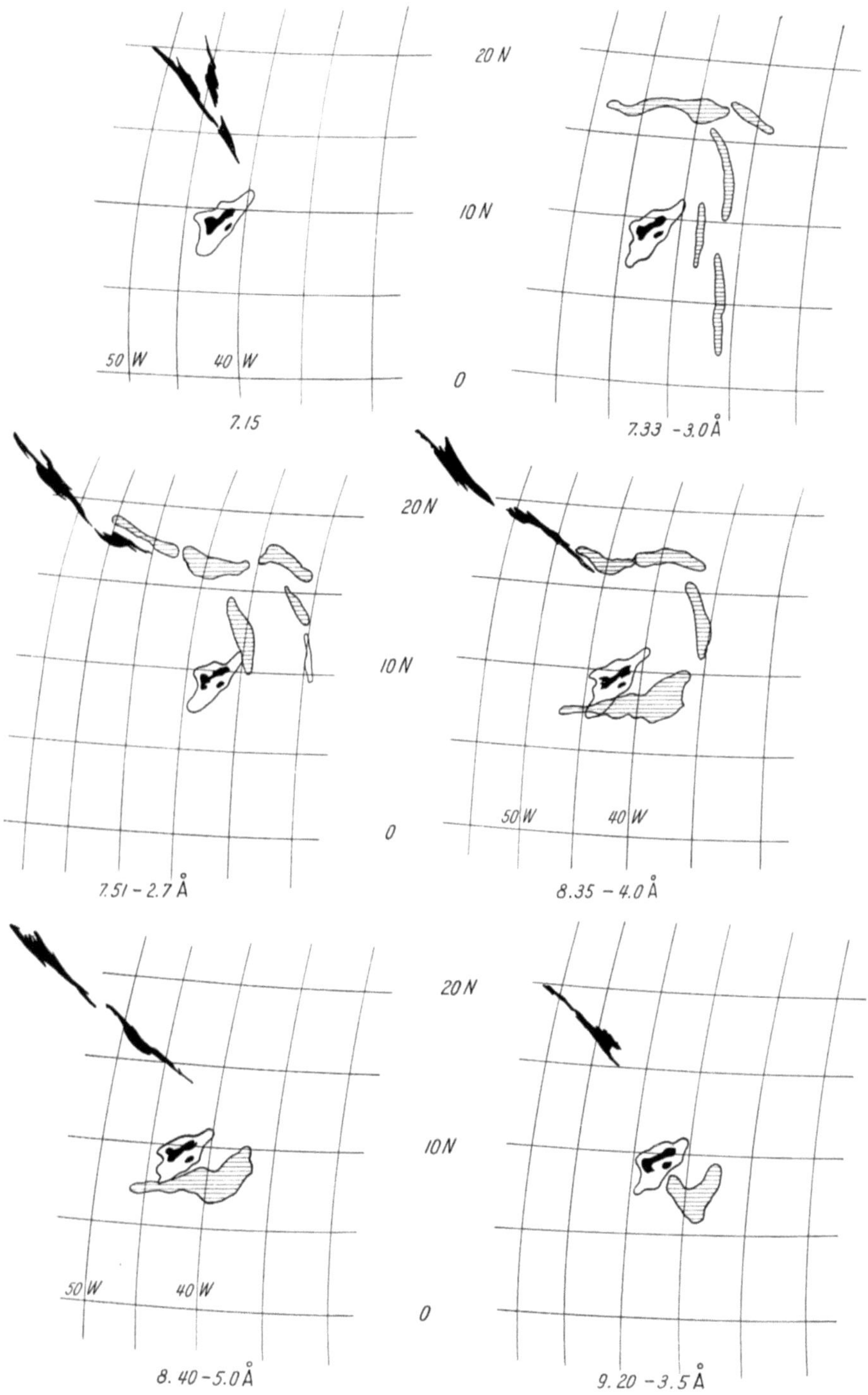

Fig. 62a. Flare surge of July 3, 1957, observed in Hα, in projection on the disk. The flare is shaded, the filament is black. Below the drawings are indicated the observation time (UT) and the effective line width of Hα on the flare spectrum, measured in a spectroheliograph. Courtesy H. Haupt, Kanzelhöhe.

ment"). In 90% of the cases the velocity of ascent increases first, the maximum, up to 200 km/sec and sometimes more, is reached in a few minutes; then the velocity begins to decrease. Simultaneously the filament becomes less visible, till the matter comes to rest. Shortly after that moment a red displaced absorption streak appears, nearly or precisely at the same place, presumably the same matter going down. There seems to be a discontinuity in the velocity time diagram (NEWTON[1]) near $v = 0$, but this might be due to the decrease of visibility of the filaments near $v = 0$ (BRUZEK[2]).

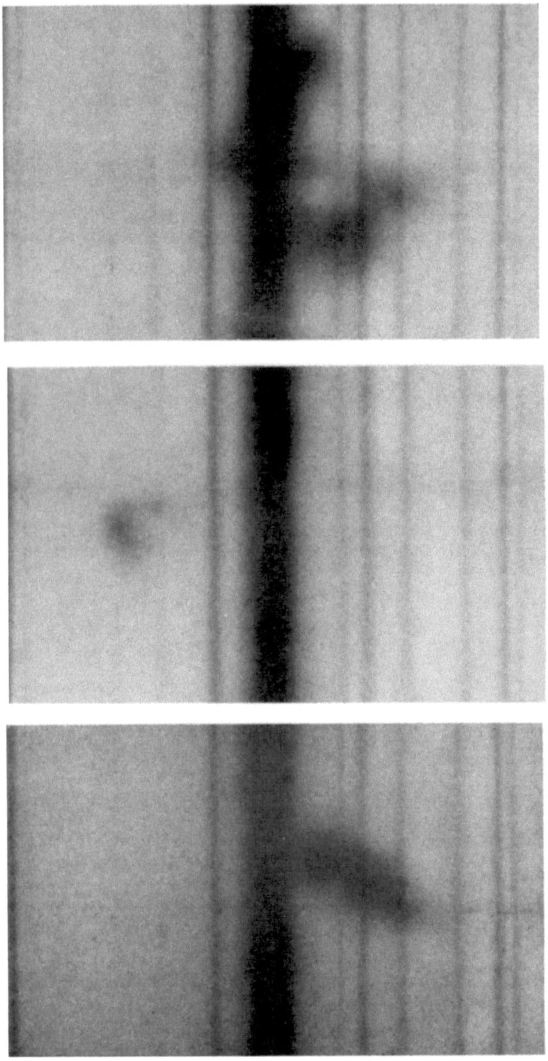

Fig. 62 b. Hα spectra of surges. For an estimate of the velocities: Hα is about 1 Å wide; at 6562 Å: 1 Å = 45.7 km/sec. From above to below: 1948, Sept. 19; 1951, June 7, 19.^{h}01; 1951, Sept. 15, 15.^{h}43 UT. Courtesy M. A. ELLISON, Edinburgh.

The reduced visibility is caused by the fact that the central intensity of Hα in surges is about equal to that of the quiet Sun spectrum so that surges are only well visible when presenting Doppler motions: Since the central intensity of Hα in filaments is about half that of the quiet Sun spectrum quiescent filaments are visible in the centre of Hα, and stationary surges hardly.

Sometimes surges are not only observed in monochromatic light but also, as dark structures, in white light[3]. These white light surges are only visible for some minutes ($\leqq 10^m$), during the first phase of a surge. As a rule they coincide more or less with the Hα structure. Often, details become earlier visible in white light than in Hα, suggesting that the white light picture refers to the fastest parts of the surge.

Particles with higher speeds than normal surges can be emitted during a flare's very early stages, at the time of onset[4-6]. The outward velocities observed in these cases amount to 750 km/sec and in one case[5] even reached 1330 km/sec.

[1] H. A. NEWTON: Monthly Notices Roy. Astronom. Soc. London **102**, 2 (1942).
[2] Cf. footnote 4, p. 212.
[3] U. BECKER: Naturwiss. **21**, 554 (1957).
[4] H. W. DODSON, E. R. HEDEMAN and J. CHAMBERLAIN: Astrophys. Journ. **107**, 66 (1953).
[5] J. D. WARWICK: Astrophys. Journ. **125**, 811 (1957).
[6] Cf. footnote 5, p. 212.

The great Doppler shifts make disk observations of these phenomena exceedingly difficult; they were discovered in limb observations both in Hα and Ca^+ spectroheliograms. These ejections differ from the flare surges in degree (greater velocities, sometimes exceeding the velocity of escape), and perhaps also in character: the surges may occur during the whole life of the flare whereas these fast ejections occur close to the time of onset. GIOVANELLI[1] found that such high-speed ejection are related to a rapid expansion of the flare in its initial stage: the *flare-puff*.

In the radio-range $^2/_3$ of the flare-puffs were accompanied by outbursts perhaps related to the emission of matter. The ejection may take place from a place close to the observed flare or at a greater distance, up to 100000 km. The apparent trajectories are nearly linear. These ejections may be connected with an appreciable motion in a lateral direction (cf. also Sects. 92 and 93).

57. Cosmic ray emission by the Sun[2,3]. The intensity of cosmic radiation on the Earth may increase temporarily after the occurrence of intense solar flares. So it seems appropriate to discuss the solar cosmic ray emission in this chapter, devoted to solar flares, although it looks very uncertain that the cosmic radiation is emitted by the flare proper. It seems more probable that a higher, hot coronal region, is responsible for this emission. The instrumental problems will not be treated in this section; the reader is referred to Vol. XLVI of this Encyclopedia.

Although the primary cosmic radiation can be well detected by means of balloons and even on high mountain tops it is recorded in its most "pure" form by means of rockets. What is observed at the Earth's surface is the secondary radiation which originates when a primary high energy nucleus, generally a proton, collides with oxygen or nitrogen nuclei in the upper terrestrial atmosphere. The disintegration products of this collision are unstable charged and neutral π-mesons (the meson component) as well as stable high energy nucleons (protons and neutrons: the nucleonic component). The neutral π^0-mesons, with life times of the order of 10^{-8} sec commonly decay whilst in flight into a pair of high energy γ-rays (the "soft component") and these γ-rays produce cascades of electrons and quanta. The π^+- and π^--mesons partly decay, yielding μ-mesons, partly they are captured. The μ-mesons penetrate down to sea-level (the "hard" component). The high energy nucleons, being the third component, lose their energy by other collisions with air atoms; at each collision lower energy nucleons (p; n) are produced with energies of 1 to 20 MeV. These collisions are called nuclear bursts or stars after their appearance in Wilson chambers and in nuclear emulsion plates. The disintegration neutrons of the low energy primaries have rather low energies so that they may be slowed down and be recorded by a neutron detector.

For the detection of charged particles principally two instruments are used. The *ionization chamber* is a large vessel, often spherical, with a volume of several hundreds of liters, with small concentric spherical electrodes. It records all charged particles, especially mesons; the instrument is equally sensitive to mesons coming from all directions; if not highly shielded, it may be sensitive to local low energy radiations, and nuclear bursts. The *Geiger-Müller counters* may be combined to form a telescope which permits the detection of particles from a given direction; also this instrument records mainly the high energy particles.

For the detection of the secondaries of the low energy primary radiation SIMPSON *et al.* have developed a neutron intensity monitor consisting of a local neutron producer, paraffin to slow down the neutrons to low velocity, and special proportional counters to detect the rate of capture of neutrons in the boron nuclei of the counter gas ($B^{10} + n \rightarrow Li^6 + He^4$). This type of instrument extends the observation of changes in cosmic ray intensity of the low energies and, hence, increases the possibility for detecting cosmic rays produced by solar flares.

Since the yield of secondary particles depends on the atmospheric pressure (causing greater absorption) and on the temperature (causing a change in the

---

[1] See footnote 5, p. 212.

[2] Cf. J. A. SIMPSON: Ann. Géophys. **11** (1955).

[3] J. A. SIMPSON: Proc. Nat. Acad. Sci. U.S.A. **43**, 42 (1957).

number of decaying μ-mesons), and also on the terrestrial magnetic field, correction terms must be known for these three effects. The temperature and pressure variations produce a daily component of considerable amplitude when measuring the meson radiation; the nucleonic component is sensitive only to changes in pressure. Another part of the daily variation is due to a modulation of the primary cosmic ray intensity in the Earth's magnetic field[1]. This latter effect is especially important for solar physics, since the field changes caused by enhanced solar ultraviolet or slow particle radiation may erroneously suggest a solar influence on the primary cosmic ray intensity. So care is needed. Phenomena which are thought to be due to modulation effects are:

1. The 27-day recurrences, with variations, generally not larger than 5% (still depending on the detecting instrument and also varying during the solar cycle[2-5]—see also Sect. 61). Correlogram analysis shows that more than one 27-day period may exist at the same time. Although they may occasionally persist up to nine solar rotations, the recurrence tendency is less for cosmic rays than for the so-called magnetic activity[6].

2. The Forbush type decrease[7], which starts rather suddenly, lasts for five to ten days, and may be as large as thirty to forty percent.

Whether or not the Forbush type decreases and the so-called cosmic ray storms, occurring together with geomagnetic storms, are essentially the same is still a matter of controversy[5,8].

Experience since 1942 has shown that the cosmic radiation from the Sun has relatively low energies. Since the relative yield of secondary nucleons, as compared with the unstable charged particle component increases with decreasing energy, an increase of low energy primary radiation may hardly or not at all be detectable with charged particle detectors but quite well with neutron detectors. For the flare of 19 November 1949 the increase of received radiation with a meson detecting telescope was 12%, whereas the increase of recorded neutrons was more than 400%. The neutron component appears to show a latitude effect greater than 200% at low altitudes and greater than 300% at high altitudes. This is also a consequence of the above statement: high energy particles hardly show any latitude effect since they are not deviated in the Earth's magnetic field.

The above makes clear that the Earth's magnetic field acts as a spectrum analyser for cosmic ray particles. So, by making simultaneously observations at different geomagnetic latitudes and with different kinds of instrument, it is possible to determine the relative portion of particles of different incident energies, or as it is usually said: of different rigidities. The *rigidity* of a particle with charge Ze and momentum P in a magnetic field is given by Pc/Ze, generally measured in units of 10^9 eV (GeV): all particles of the same rigidity will experience the same deflection in a magnetic field. An ion chamber or meson telescope permit to make measurements from 7 to 10 GeV to the highest rigidities. The neutron monitor extends for 10^9 eV to the higher rigidities. The vertical cut-off rigidity at the Earth's equator is 15 GeV; only particles with energies >60 GeV can approach the Earth from any direction.

---

[1] H. ELLIOT and P. ROTHWELL: I. A. U. symposium, Stockholm 1956.
[2] J. A. SIMPSON: Ann. Géophys. **11**, 305 (1955).
[3] J. A. SIMPSON: Phys. Rev. **94**, 426 (1954).
[4] S. E. FORBUSH: J. Geophys. Res. **59**, 525 (1954).
[5] R. R. BROWN: J. Geophys. Res. **61**, 639 (1956); **62**, 147 (1957).
[6] S. E. FORBUSH: I. A. U. symposium, Stockholm 1956.
[7] S. E. FORBUSH: Terr. Mag. Atm. Electr. **43**, 207 (1938).
[8] K. KAWABATE, I. KONDO, K. MURAKAMI and M. WADA: Rep. Ionosph. Res. Japan **11**, 229 (1957).

The first clear indication of cosmic rays produced together with a solar flare was on 28 February 1942, when a 3^+ flare occurred in the large spot group which passed the central meridian at the same day. (It was the radio noise of this same flare which led HEY to the discovery of the solar radio-radiation[1].) The increase of cosmic radiation which was of the order of 20%, was observed about one hour after the maximum of the flare; after one day it was followed by a decrease and a slow recovery afterwards, due to a disturbance of the terrestrial magnetic field by the slow particles, that also produce a magnetic storm. Cosmic rays, due to strong flares, are observed when the source occupies virtually any place on the visible solar disk; there is perhaps a slight E-W asymmetry[2].

The problem of the impact of cosmic rays on the Earth has been considered by various authors. FIROR[3] computed the Störmer orbits, assuming a dipole field for the Earth. A beam of particles from the Sun, with magnetic rigidities of 1 to 10 GeV will produce an increase in the cosmic ray flux in four well defined "impact regions", at the west of the sub-solar point, at average longitudes 4^h and 9^h and at average latitudes of $50° \pm 20°$. Besides, some cosmic radiation will also be received in two background zones encircling the Earth between latitudes 25 and 60°, with a faint maximum at about 20^h local longitude. The relative intensities in the three zones are 7:3:1. So if a detector is in one of the impact zones at the moment of a cosmic ray producing flare, an increase of intensity should be observed. Criticism to the theory refers to the fact that interplanetary space may not be empty ($N_e \approx 10^3$ cm^{-3} near the Earth) and it is doubtful whether in such a case the elementary Störmer theory may still be applied. The average number of particles in a cosmic ray storm is of the order of 1 cm^{-2} sec^{-1}.

Some data on cosmic ray bursts associated with great solar flares are given in the Table 13; it contains the biggest four flares observed. The flare of 7. 3. 1942 was not observed optically.

Table 13.

Date	Flare imp.	Solar coordinates	T_{max} (UT)	Line width of Hα (Å)	Time lag C.R.-flare max	Time lag C.R.-geomagn. storm
28. 2. 1942	3^+	04 E 07 N	12.00	4.1 (12.42)	1^h	$19^h.5$
7. 3. 1942		90 W 07 N?	04.42		1^h	
25. 7. 1946	3^+	15 E 22 N	16.27	15.9 (max)	$2^h.5$	
19. 11. 1949	3^+	70 W 02 S	10.32	22.9 (max)	$0^h.5$	$26^h.4$
23. 2. 1956	3^+	80 W 20 N	03.42	> 18 (max)	$0^h.2$	48^h

Of the 1949 and 1956 flares the neutron component was also measured. The general trend is that the neutron yield increases very steeply to a maximum, followed by a slow decrease which may last for many hours after the end of the visual flare.

The flare of 23. 2. 1956 is the one best studied. It started at $03^h 31^m \pm 1^m$ UT and was located rather close to the limb. Its maximum area (uncorrected for foreshortening) was 1.3×10^{-3} of the visible hemisphere. The last observations were made at $04^h 14^m$ UT (not end of flare). On the daylight side of the Earth a sudden ionospheric disturbance started at $03^h 30^m$ to $03^h 32^m$ UT. Solar radio-burst emissions were observed to start between $03^h 33^m$ and $03^h 35^m$ UT on all frequencies between 10000 and 19 MHz. Those at centimeter and decimeter

[1] Cf. Vol. LIII of this Encyclopedia.
[2] Cf. J. A. SIMPSON: Proc. Nat. Acad. Sci. U.S.A. **43**, 42 (1957).
[3] F. FIROR: Phys. Rev. **94**, 1017 (1954).

waves were of the strongest ever observed[1]. The first cosmic ray particles arrived at the Earth's nightside at $03^h 45^m \pm 1^m$ UT. The cosmic ray flux exceeded in intensity all previously observed cosmic ray bursts. At one of the more than 40 observing stations (Berkeley) the following relative increases were measured; they are characteristic for observations at medium geomagnetic latitudes.

Neutrons	maximum at $03^h 49^m$ UT	increase: 600%
Total mesons	$03^h 52^m$ UT	58%
Hard mesons	$03^h 53^m$ UT	38%.

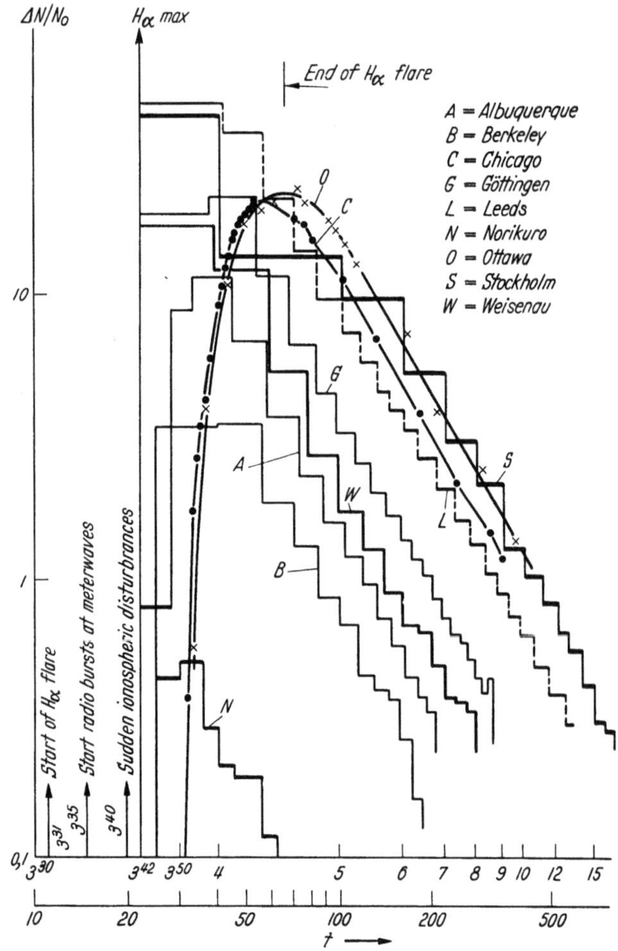

Fig. 63. Onset of cosmic ray burst of 23 February, 1956, at different observing stations. A. EHMERT and G. PFOTZER: Mitt. Inst. Phys. Stratosphäre, Weissenau **6** (1956).

The contribution of high energy particles was so great, this time, that there was even a substantial increase of meson activity near the geomagnetic equator: in India the average increase was 6%.

Some characteristic properties of this and other cosmic ray increases are listed below (cf. also [2-4]):

[1] H. TANAKA and T. KAKINUMA: Proc. Res. Inst. Atm. **4**, 74 (1956).
[2] R. R. BROWN: J. Geophys. Res. **61**, 639 (1956).
[3] P. MEYER, E. N. PARKER and J. A. SIMPSON: Phys. Rev. **104**, 768 (1956).
[4] A. EHMERT and G. PFOTZER: Mitt. Max Planck Inst. Phys. Stratosphäre **6** (1956).

(a) The onset of cosmic radiation was very sharp (Fig. 63). This eliminates the possible occurrence of interplanetary diffusion by magnetic fields. It is estimated[1] that the field between the Sun and the Earth must in any case be smaller than 3×10^{-6} Gauss to produce the observed sharpness of the onset.

(b) Of all intense cosmic ray bursts observed up to now the onset occurs *after* that of the flare (10 min to $2\frac{1}{2}$ hours). Since, however, protons of the observed rigidity must have velocities close to that of light, the solar "cosmic" rays must have been excited *after* the flare (other solutions to explain this delay, such as acceleration near the Earth or interplanetary diffusion, seem rather improbable for the explanation of character and time of the *onset*). The cosmic rays are perhaps emitted in the very hot coronal regions overlying a flare, where according to Boischot and Denisse also the type IV radiobursts are emitted, some time after a flare (see Chap. C II, Sect. 95). Further we refer to Waldmeier, who noticed that also the coronal emission (5694 and 5445 Å) in coronal condensations reaches a maximum about half an hour after the flare's beginning (see Chap. C I, Sect. 79).

(c) The onset was observed nearly simultaneously at the light and the dark side of the Earth, while it had the same intensity with no maximum near 12^h local time. Hence the stream did not contain neutral particles in appreciable numbers.

(d) The greatest intensity increase was observed by the (European) observers at sites between 23^h30^m and 02^h30^m hour local time; so the impact zone seems to be shifted westward by 3^h with respect to the predicted longitude. [This difference can be explained (Pfotzer[2]) by assuming a slight asymmetry of the terrestrial magnetic dipole field.] The time to reach maximum intensity was about 8 min in the main impact zone, whereas outside this zone (Chicago) it was more than twice this time interval. The onsets in non-impact zone stations like Chicago and Wellington (New Zealand) were about 5 min later than those in the impact zone. This delay was suggested to arise from scattering of the particles against some reflecting region outside the Earth's orbit. The radius R of this scattering region can be estimated from the difference in onset times, assuming that the impact zones are hit directly; $R\approx 1.1$ to 1.4 astronomical units.

(e) The slowly decaying tail of the intensity curve, lasting up to 15 hours after the end of the flare, had the same hyperbolic form in all terrestrial stations at the same geomagnetic latitude (not depending on the local time of the observing station) although the Earth turned around over more than 180 degrees in this time. So these tails must have been due to an isotropic radiation, coming from all directions. Since this radiation was originally of solar origin it must have been diffused in interplanetary space (Schlüter[3]). Hence the field-free inner region of the solar system must be surrounded by a spherical barrier shell which scatters the solar particles. This shell should contain disordered magnetic fields ($\approx 10^{-5}$ Gauss) extending from a solar distance of 1.4 astronomical units to perhaps double the distance or more. Others (footnote 8, p. 216) think that the particles are scattered by magnetized clouds around the earth, emitted by the same solar CA as the one where the cosmic ray storm originated. At the same time these clouds may be held responsible for the Forbush type decrease, preceding and following the cosmic ray storm.

---

[1] A. Ehmert: I.A.U. symposium, Stockholm 1956.
[2] G. Pfotzer: Mitt. Max Planck Inst. Phys. Stratosphäre **9** (1957).
[3] A. Schlüter: Z. Naturforsch. **6a**, 592 (1951).

(f) The marked latitude dependence of the intensity of radiation indicates the usual low energy spectrum. The decline of radiation was faster at low latitude

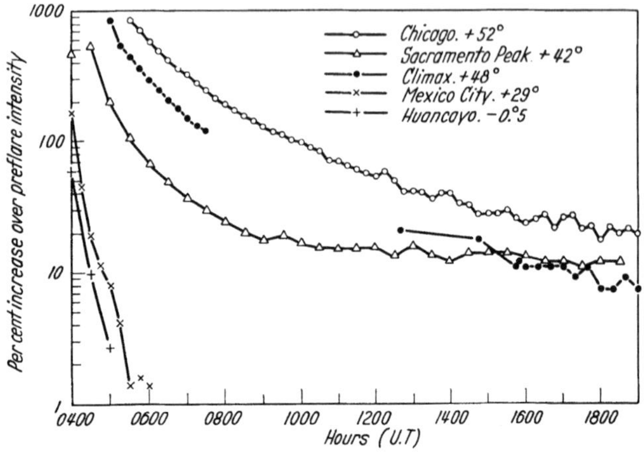

Fig. 64. The decline of the cosmic ray intensity on 23 February, 1956, measured with neutron detectors at stations of different geomagnetic latitudes. J. A. SIMPSON: Proc. Nat. Acad. Sci. U.S.A. **43**, 42 (1957).

stations than at high latitudes, showing that the spectrum after several hours contained only low energy particles (Fig. 64). From a mutual comparison of the observations at six stations at different geomagnetic latitudes and one balloon-borne neutron detector the spectrum could be derived[1] at three different moments (Fig. 65). The solar cosmic ray spectrum is much steeper than the general spectrum.

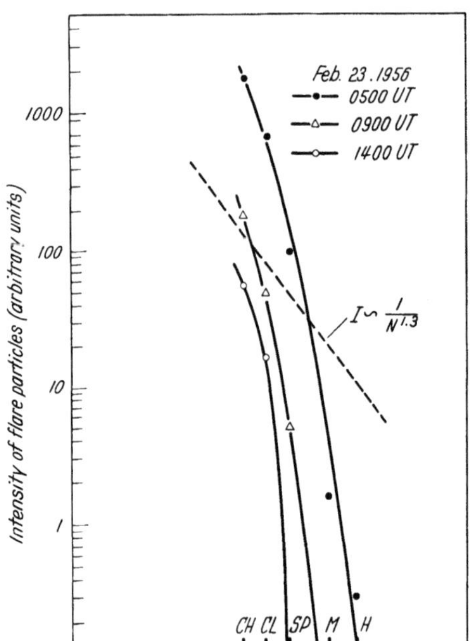

Fig. 65. The spectrum of the primary cosmic ray particles at three different instants as deduced from the data given in Fig. 64. The normal cosmic ray spectrum is shown by a dashed line with arbitrary intensity. J. A. SIMPSON: Proc. Nat. Acad. Sci. U.S.A. **43**, 42 (1957).

(g) The total kinetic energy of the cosmic ray particles emitted by the flare of 23. 2. 1956 may be estimated by assuming that the particles are protons which are all still within the above described interplanetary cavity at the time of maximum. The total energy of all particles in excess of 3 GeV is 1.4×10^{30} ergs. The low energy end of the spectrum might double this amount, but its precise amount is difficult to estimate (atmospheric absorption). This value is much smaller than the energy output of a big flare at flash stage in optical regions[2]: 10^{30} ergs/sec.

(h) Small flares, down to importance 1^+ also produce measurable effects. A statistical investigation (footnote 2, p. 216) showed that in-

[1] P. MEYER, E. N. PARKER and J. A. SIMPSON: Phys. Rev. **104**, 768 (1956).
[2] M. A. ELLISON: Trans. Internat. Astronom. Union **9**, 651 (1957).

struments in the impact zones record an increase of about 0.5% during the occurrence of flares; outside these zones nothing is recorded.

58. Theories on the origin of flares[1]. There is a tendency to ascribe the very origin of the flare phenomenon to nuclear processes near the solar surface. Various authors (AMBARTSOUMIAN[2], FOWLER, BURBIDGE and BURBIDGE[3]) have stressed the possibility that primary energy release might take place not only in the stars' centres but also in "hot spots" closely below the surface. These assumptions might have obtained some observational confirmation by SEVERNY's observations of the "moustaches" (cf. Sect. 53).

However, whether or not these assumptions are right, a flare process itself is not a photospheric phenomenon; it is chromospheric or it occurs in the corona. A flare is certainly not the primary phenomenon but is probably activated by more fundamental processes. Theory should explain how under certain circumstances a part of the chromosphere and corona obtains a very high density, which is sustained during some tens of minutes. (The density change in a flare seems to be more fundamental than the change in temperature; the latter does not much deviate from that of the chromospheric hot elements.)

The observations show that flares that occur in a bipolar group, mostly originate between the two spots, and that they are most common in the magnetically complex γ-groups. Their occurrence is correlated with changes in the group structure. SEVERNY's[4] magnetographic observations, obtained with a high resolution magnetograph have the flare shown to be related to an instability or even a collapse of the field. Their rise-time is usually short, of the order of 5 min.

These observations lead to the *neutral region theory*, which assumes that electric fields, originating by induction might give the high energies to the particles. These electrical fields are assumed to be produced by fluctuations or variations in the magnetic field; they are greater if dH/dt is greater. On the other hand the electrons should be free to be accelerated over great distances, not braked by collisions and by strong magnetic fields. So one should look for a magnetically neutral region where the field is weak but where dH/dt is great[5]. Could this be the reason why flares often originate between the main spots of a group? In a *complex* spot group such neutral regions are likely to occur. It is assumed that a flare originates in the neutral region between two spots, if at the same time dH/dt is so great that in one free path an electron gains an energy $\approx \frac{1}{2} kT$. The effective cross section for collisions of electrons with ions decreases with increasing velocity, so there will be a lower limit to the velocity above which an electron gains kT in one free path. Since the collisional cross section with neutral hydrogen is large, the mechanism only works in the parts of the chromosphere where hydrogen is mainly ionized. There the number of free electrons acquiring great velocities might be sufficient, leading to a local heating of the gas. This is the flare.

The neutral point theory of solar flares was critized by COWLING[6], who finds a discharge theory unacceptable on the basis of LENZ' law: "the effect of induced currents is to oppose the changes to which they are due" and because computations show that the accelerating layer should be only very thin (5 meters) according

[1] Cf. also A. UNSÖLD: Transact. Internat. Astronom. Union **9**, 667 (1957).

[2] V. A. AMBARTSOUMIAN: Coobčs. Bjurakan Obs. **13**, (1954); cf. V. KOURGANOFF: Nuovo Cim. Ser. X, **1** (4) (1955).

[3] W. A. FOWLER, G. R. BURBIDGE and E. M. BURBIDGE: Astrophys. Journ. Suppl. **2**, 167 (1955).

[4] A. B. SEVERNY: Astr. Zhurn. **35**, 335 (1958). — Isw. Krymsk. Astrophys. Obs. **20**, 22 (1958). Cf. also V. BUMBA: Isw. Krymsk. Astrophys. Obs. **19**, 105 (1958).

[5] R. G. GIOVANELLI: Monthly Notices Roy. Astronom. Soc. London **107**, 338 (1947).

[6] T. G. COWLING: The Sun (ed. G. P. KUIPER), p. 587. Chicago 1953.

to this theory. Replies by DUNGEY[1] and SWEET[2] show that these objections probably do not hold near a neutral region. Quantitative estimates by the latter show that the theory is able to predict flares of the observed duration and intensities.

The existing theories attempt to explain a flare as a local heating of the chromosphere. OSTER[3] computed the relaxation times for thermal cooling and found for chromospheric conditions decay times similar to the observed life times of flares. However, if it is true that flares have average heights of 16000 to 20000 km, thus if they are coronal for a considerable part, it is not a *heating* but a *cooling* mechanism that is wanted: the theory should explain how coronal matter can cool down suddenly to $T = 10^4$ °K. Perhaps this temperature decrease is related to the magnetic instability observed by SEVERNY, and the subsequent contraction of the plasma (footnote 4, p. 221). Cf. also Sect. 69.

59. The velocities of particles ejected by the Sun. In this section we summarize the various evidences for the emission of solar particles. There is a wide range of velocities ranging from less than tens of kilometers, directly measured with the Doppler technique up to the 3×10^5 km/sec assumed for the cosmic radiation emitted after the flares.

The optically observed velocities may perhaps be divided into two classes: first those emitted at the onset of a flare—some moustaches, some flare surges—with observed velocities up to and sometimes above the velocity of escape (618 km per sec), similar to the velocities sometimes observed for eruptive prominences. The other class consists of the slower flare surges with average velocities between 50 and 100 km/sec.

Further there are the depressions observed in the violet wing of Hα spectra of flares, indicating the occurrence of outstreaming matter with average velocities of 100 to 200 km/sec. The asymmetry is mainly observed in the flash and the second phase of the flares; whether it does also occur in the first phase is not known. In this connection we should refer to the various unsuccessful attempts to discover the corpuscular streams on their way to the Earth by the observation of Doppler displaced Fraunhofer lines, especially of the K line of Ca^+. These attempts failed because the observations generally were made too long a time after the appearance of flares, when the stream was already too rarified. As the Hα observations show, the best time to look for the effect is at, or very shortly after the beginning of the flare.

Other solar evidence for the emission of two kinds of solar particles is gained from the radio observations (cf. Chap. C II). Spectral observations, mainly by WILD and associates show the existence of two kinds of moving disturbances. During their passage through the corona they produce bursts of type II and type III. Both are correlated with solar flares. The type III bursts which occur during the onset of a flare have average velocities of 100000 km/sec (range: 10000 to 150000); the former, occurring mainly during the second phase of a flare, have velocities of 200 to 1500 km/sec (cf. also GIOVANELLI, footnote 5, p. 212).

Terrestrial evidence for the occurrence of particles emitted by the Sun is given by the magnetic storms (cf. Sect. 109). The big magnetic storms occur after flares; the particles have a mean travelling time of 1.7 day yielding a mean interplanetary velocity of 1000 km/sec. The recurrent storms (Sect. 110), pro-

[1] J. W. DUNGEY: I.A.U. symposium **6**, 135 (1958).
[2] P. A. SWEET: I.A.U. symposium **6**, 123 (1958).
[3] L. OSTER: Z. Astrophys. **44**, 26 (1958).

duced by the M regions have mean velocities of 500 km/sec, with a considerable scatter.

The following table tentatively assembles the above results; it gives the average velocities in km/sec.

Medium	Photosphere	Chromosphere and corona	Corona	Interplanetary space	
Observation	Moustaches	Sprays and Surges	Radio observations (spectral type between brackets)	Cosmic rays	Magnetic storms
Onset of flare . . .	up to 1000	100–800	100000 (III)	300000	1000–1600
2nd phase		100	1000 (II)		
Later effects . . .					

An essential result of this review is that the photospheric and low-chromospheric velocities given in the first two columns are smaller than those of the latter three columns. This suggests that the main acceleration process occurs in the corona. For the theoretical explanation of the acceleration mechanism two facts are perhaps essential: (1) The acceleration is produced in magnetically disturbed regions, and the highest velocities tend to be obtained in the strongest disturbed regions. (2) The relatively low velocities of the first two columns are observed on low temperature and high density objects producing emission and absorption in normal Fraunhofer lines like Hα and K (Ca^+), whereas the higher velocities are observed in the more teneous high temperature coronal plasma. Thus the acceleration mechanism must be such that it increases the *low* velocities of the cool *dense* matter to *higher* velocities of the hot less dense plasma, in the presence of a magnetic field. So, small velocities of many particles have to be transformed to high velocities of fewer ones, by means of a very effective energy transforming mechanism.

The existing theories attempt to explain this acceleration by the transformation of magnetic energy into mechanical energy. Assuming equality between turbulent and magnetic energy in a "knot" of chromospheric matter, it can be computed that the electric field induced by the dissolution of such a magnetic surge may be sufficient to produce particles of the observed energies. However, the efficiency of this mechanism is practically reduced to zero by the induction effect which tends to oppose and to reduce the field-changes. A way to overcome this difficulty seems to assume acceleration in a neutral or demagnetized medium.

With SCHLÜTER's "melon seed" theory[1] the initial velocities can be increased up to a factor of the order 5, under circumstances, depending on the local magnetic field and the local velocity of escape. In this theory it is supposed that a knot of highly conductive matter is contained in less conductive matter. Because of the great conductivity the lines of force of the surrounding field cannot penetrate into the knot, so that this knot acts as a diamagnetic body, surrounded by the lines of force. So the knot is not magnetized. By the action of the Maxwell tension of the adjacent magnetic field the lines of force have the tendency to stretch, thereby producing an accelerating force and causing the diamagnetic knot to be blown out as a melon seed. Further KIEPENHEUER's[2] theory might be mentioned in

[1] A. SCHLÜTER: Radioastronomy (ed. H. C. VAN DE HULST). I.A.U. symposium No. 4, p. 356, 1957.

[2] K. O. KIEPENHEUER: Radioastronomy (ed. H. C. VAN DE HULST). I.A.U. symposium, No. 4, p. 345, 1957.

which the particles are assumed to be set into motion by the effect of a solar magnetic storm. Also KOLPAKOV[1] suggested that the particles are accelerated by the component of an induced electric field perpendicular to the magnetic field of the spots. SWEET[2] examined the possibility of acceleration in a neutral region, originating in a spot group consisting of four members of opposite magnetic polarity. Let the spots move, leading to a field-change, then the voltage drop V over the neutral region is about equal to $1/c$ times the rate of flux decrease:

$$V = 1/c \frac{d}{dt} (\int H \, d\Sigma).$$

If the velocity is v, the length of the region l, the flux change $\approx v l H_0$. With $l = 10^9$ cm, $H_0 = 10^3$ Gauss; $V = 10^{10}$ eV if $v = 10$ km/sec, thus largely explaining the observed velocities. For another mechanism, based on shock waves generated after the collapse of the pre-flare field, cf. SEVERNY (footnote 4, p. 221).

MUSTEL[3] and WARWICK[4] have suggested to explain the high velocities of surges occurring in the first stage of a flare (the so-called "flare-sprays") by the radiation pressure, the first by pressure on Ca^+, exerting a drag on hydrogen atoms; the other by the radiation pressure of Ly α, thus making revive a rather old suggestion (MILNE 1924). In order that the mechanism is efficient the ultraviolet emission or Lyα should be greatly enhanced during the flare. For a critical discussion of this effect, in which it is shown that the radiation pressure of Lyα and of the Lyman continuous radiation of the *quiet* Sun is too small by many powers of ten, the reader is referred to UNSÖLD, Physik der Sternatmosphären, 2. Aufl., p. 695 ff.

IV. Filaments and prominences.

60. Classification of prominences and filaments. The prominences are visible as bright structures beyond the Sun's limb, in the corona where they appear in various forms, usually below 100000 km, the greater part below 30000 km. Their spectrum consists mainly of the Balmer and neutral helium lines and the H and K lines. The strength of Hα, Hβ and Hγ explains their pink-bluish colour, when seen during an eclipse. On spectroheliograms made in the cores of the corresponding Fraunhofer lines, they appear as dark ribbon-like absorption structures, generally called filaments, and sometimes dark (or: absorption-) markings or dark flocculi. Filaments and prominences are the same structures, viewed on and beyond the disk, but their aspects differ greatly depending on the brightness of the various parts of the prominence and many filaments may even disappear more or less when coming outside the limb. The great variety of their forms and structures have led various authors to the construction of classification schemes, some of which are fairly elaborated. The following scheme, the first column of which is mainly due to PETTIT[5], might be useful in most cases.

An alternative classification scheme was proposed by MENZEL and EVANS[6], who divided the prominences into two main classes according to the dynamically important viewpoint whether they descend (from *a*bove: A) or rise (from *b*elow: B). Since it is, moreover, reasonable to assume that prominence forms, support and

[1] P. E. KOLPAKOV: Astron. Zhurn. **34**, 222 (1957).
[2] P. A. SWEET: I.U.P.A.P. Conference on Cosmic Radiation, Varenna 1957.
[3] E. R. MUSTEL: in: Les processus nucléaires dans les astres, Liège coll. 1953, 223 (1954). — Astr. Zhurn. **32**, 177 (1955).
[4] J. W. WARWICK: Astrophys. Journ. **125**, 811 (1957).
[5] E. PETTIT: Astrophys. Journ. **98**, 6 (1943).
[6] D.H. MENZEL and D. S. EVANS: Convegno Volta **11**, 119 (1953)

	Main disk region	Pettit classification	Menzel-Evans classification
I. Quiescent prominences			
a) Normal quiescent prominences	undisturbed part of sunspot belt (outside centres of activity)	V	*N* *AN b, c, d, m*
b) Polar prominences	polar caps		
II. Moving prominences			
a) Active prominences	nonspot prominences; mainly in sunspot belt, but some eruptives occur in the polar regions or near spots	I	*AN*
b) Eruptive prominences		IIa, b	*BN*
c) Spot prominences			
c_1) Spot prominences proper	centres of activity	IIIa, b, c; VI	*AS*
c_2) Knot prominences ("condensations"), loops, and funnels	near flares		
d) Surges	as a rule in centres of activity; often accompanying flares	IIId	*BSs*
e) Spicules	whole disk with slight preponderance near poles		*BNs*

The very rare "tornado prominences"[1] (PETTIT's type IV) are not included in this scheme since too little is known of their relations (both morphological and evolutionary) to the solar activity regions. Tornadoes seem to show a helical form; once helical *motions* (≈ 4 km/sec) were suggested by Doppler shift observations (cf. [1]).

motions are regulated to a certain extent by electromagnetic forces, each of these groups is divided into spot prominences (*S*) and non-spot prominences (*N*). The subdivision is pictorial and distinguishes between *ASf:* funnels, *ASL:* loops; and the non-spot *A* prominences: *ANa:* coronal rain, *ANb:* tree trunk, *ANc:* tree, *ANd:* hedgerow, *ANm:* mound. The *B* prominences are divided into *BSs:* surges, *BSp:* puffs and *BNs:* spicules.

The non-spot prominences may be broadly identified with the quiescent prominences (truly quiescent or activated) and the spicules. The spot prominences of the Menzel-Evans classification belong to the second group. The active and eruptive prominences are often normal quiescent ones which happen to be disturbed by some reason, often by an existing or newly formed centre of activity.

The class designations like treetrunk, tree, hedgerow, mound do not necessarily represent a correct structural idea in three dimensions: they are usually sectional projections of the general ribbon-like structure of the quiescent prominences. "Mounds" do not show as circular disks on the disk spectroheliograms.

The Menzel-Evans classifications show the great variety of the prominence *shapes* that may occur. A classification scheme based on the prominence *motions* was developed by SEVERNY. For this classification, which applies mainly to the prominences of our type II, reference is made to Sect. 66.

The prominences show a particular *latitude distribution*, which, moreover, depends on the solar cycle (see Sect. 99). They are most abundant in two equatorial zones, between 10 and 40°, where upon close examination each zone appears to consist of two others: one confined to the strict neighbourhood of the spots (prominences of groups II, c, d) and one, at about 5 to 10° higher latitude (Ia;

[1] E. PETTIT: Publ. Astronom. Soc. Pacific **62**, 144 (1950).

IIa, b). The high latitude (polar) prominences appear for the first time some years before sunspot minimum, and with increasing activity in the following years the zone moves poleward; the migration continues till the time of next maximum. The polar prominences disappear shortly afterwards. So there are six prominence zones, three at the northern and three at the southern hemisphere.

The prominences are divided according to their *spectra* into three groups: the "*normal spectra*" contain emission lines of H, He (D_3) and Ca^+, whereas the *metallic line spectra* contain these lines and besides weaker lines, those of Fe II, the Mg *b* and the Na *D* lines. The difference is one of intensity rather than of excitation conditions. Also the faint "normal" spectra contain metal lines but these are too faint to be well visible. Precise absolute photometry is not available, but in Hα are the quiescent prominences generally by a factor 10 to 100 brighter than the prominences in the active regions, and in turn are the quiescent prominences exceeded in brightness (by a factor 2) by the bright knot-like spot prominences which are characteristic for coronal condensations, and which occur during periods of high flare activity. The *spectra* of these *active spot-prominences* seem to differ essentially from the above described kinds of spectra: they present lines of higher excitation (e.g. He II), indicating higher temperatures. In order to obtain a finer subdivision of the first two groups of spectra WALDMEIER[1] has introduced a spectral classification (I to V) according to the intensity ratio Fe II—Mg *b*.

The classifications described above refer for the greater part to the forms, motions and spectra of prominences; they are *descriptive*. An *evolutionary* classification, based on what happens to filaments in the course of their evolution and that of the Centre of Activity (CA) may give more insight into the nature of prominences. The treatment of prominences in this article is partly based on such a classification. Partly, because it is not possible to classify all prominences into a unique development scheme; there are too many deviating cases. The outline of the scheme followed here is:

1. First (spot) phase of CA: formation of short-lived spot-prominences, surges and coronal condensations.
2. Later (non-spot) phase of CA: formation of stable filaments.
3. Disintegration and "poleward motion".
4. The polar filaments are a disputed class, they may or may not result from the poleward motion of the quiescent prominences.

During the phases 2 and 3 activation and dissolution of the filaments may occur; this hardly happens to the polar filaments. Since the quiescent filaments have the simplest structures it is these objects that will be discussed first. The surges have already been discussed in Sect. 56, the spicules in Sect. 26.

a) The quiescent prominences.

61. Structures, dimensions and internal motions of quiescent prominences. At the time of their first appearance the quiescent filaments have an average length not exceeding 50000 km. However, they grow and after some solar rotations a filament may be fully developed. When projected on the disk it appears as a thin, dark, somewhat curved line, especially in Hα, less well visible in Ca^+ spectroheliograms (see Fig. 18). Viewed at the limb it appears to have a considerable area; it often consists of a number of arches each connected with the chromosphere. So a quiescent prominence appears to be a long, flat, sheet-like structure, nearly perpendicular to the surface (Fig. 66). The following

[1] M. WALDMEIER: Z. Astrophys. **28**, 208 (1951).

Table 14 gives the average length, height and thickness as well as the ranges of these quantities, as measured on spectroheliograms by PETTIT[1] and by L. and M. D'AZAMBUJA[2]. All values are given in kilometers. The lengths of individual arches are about twice their heights[3].

Table 14.

	L. and M. D'AZAMBUJA[2]		PETTIT[1]	
	average	range	average	range
Thickness . . .	6600	4000— 15000	10000	6000— 12000
Height	42000	15000—120000	50000	30000—100000
Length	200000	up to 1100000	200000	60000—600000

An extraordinary case was the long filament, generated by the great sunspot group of February 1946 (the second largest spot-group which occurred after 1874). It reached its greatest length, (1900000 km) in May 1946, when its thickness and height were 10000 and 100000 km. It was visible till the end of July.

The quiescent prominences reach greater heights than the average kind of prominence. The histogram of all prominence heights shows a steep decrease near 30000 km, so that this height seems to form a sort of natural platform for prominences[4]. About 10% of all prominences reach heights greater than 30000 km.

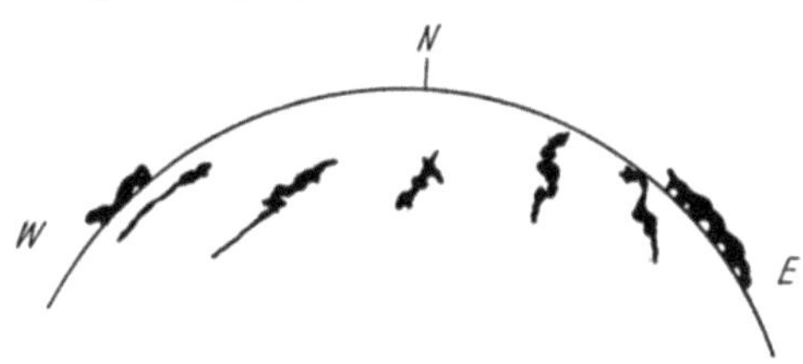

Fig. 66. Development of a quiescent prominence, viewed during half a solar rotation. From E to W: 1956, February 12, 14, 16 18, 20, 22 and 25. (Drawings after observations made at Meudon.)

The planes of the filaments are not exactly perpendicular to the Sun's surface: in the average case the apparent thickness appears to be minimum before central meridian passage (c.m.p.). Of 171 prominences[5] 134 were inclined westward, 36 eastward, one was perpendicular. A precise investigation of the effect[6] shows that the filaments are as a rule tilted away from the corresponding activity centre. The average tilt towards the west ($\approx 8°$) results from the fact that filaments occur preferently at the west side of the activity centre (ratio of W to E filaments $\approx 3:1$). This inclination may find a magnetic explanation: the variation of tilt with heliocentric distance to the centre depends on the magnetic field of the main spot of the centre.

On Hα pictures the lower part of many prominences often shows a bright rim with a mean brightness of 25% of the continuous photospheric spectrum. In this average case this rim is first visible at the west side, after c.m.p. at the east side, and only during a short interval between it is visible on both sides of the prominence. This shows that many filaments are opaque in Hα.

The filaments show a microstructure consisting of many filamentary threads with diameters of the order of 1000 km (Fig. 67); these have relative motions of 5 to 10 km/sec[1,7], without appreciably changing the forms of the prominences

[1] E. PETTIT: Astrophys. Journ. **76**, 9 (1932).
[2] L. and M. D'AZAMBUJA: Ann. Obs. Meudon **6**, fasc. VII (1948).
[3] A. K. DAS and K. SETHUMADHAVAN: Kodaikanal Obs. Bull. **127** (1949).
[4] D. E. BILLINGS and C. KOBER: Astronom. J. **62**, 242 (1957).
[5] M. ROUMENS: C. R. Acad. Sci., Paris **201**, 12 (1935).
[6] U. BECKER: Z. Astrophys. **40**, 65 (1956).
[7] H. W. NEWTON: Monthly Notices Roy. Astronom. Soc. London **94**, 472 (1932).

as a whole. The intensity variation perpendicular to the axes of these threads is consistent with the assumption of a Gaussian distribution of the emitting and absorbing atoms with respect to the threads' centre[1]. The kinetic and excita-

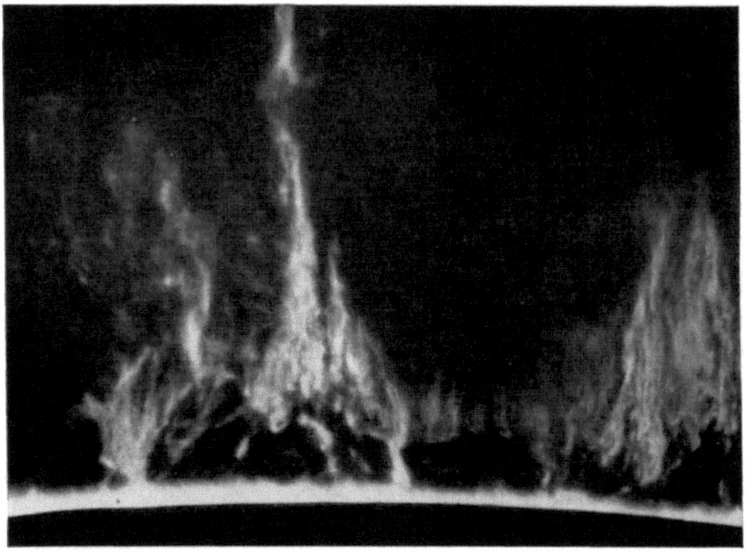

a

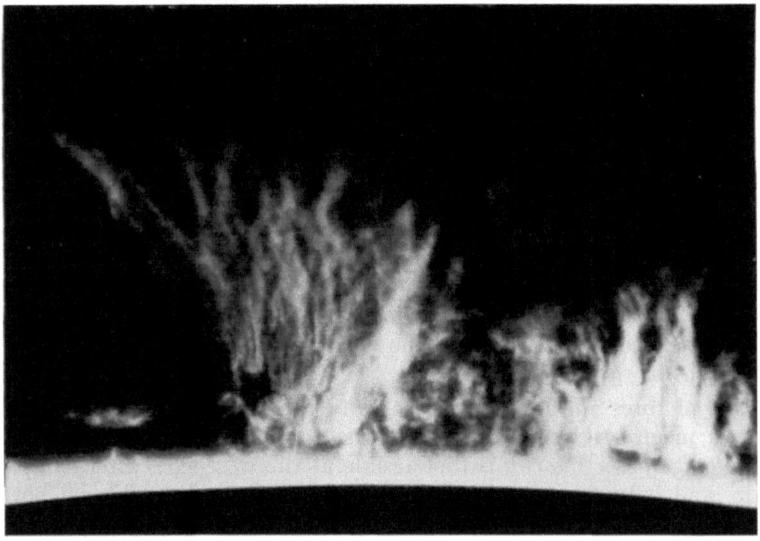

b

Fig. 67 a and b. Quiescent prominences (hedgerow type in the Menzel-Evans classification). Photographs obtained with an Hα filter constructed by R. B. DUNN; 3.4 Å wide. (a) 23. VIII. 1956; (b) 20. IX. 1957. Courtesy R. B. DUNN, Sacramento Peak Observatory.

tion temperatures do not seem to change across the threads. Most of them are optically thin for Hα.

62. Origin and evolution of quiescent prominences. Quiescent prominences show no substantial changes over periods of several hours to more than a day

[1] F. Q. ORRALL: Astronom. J. **61**, 10 (1956).

(PETTIT). The life histories of 363 well-developed quiescent filaments were studied by L. and M. D'AZAMBUJA in a fundamental investigation[1,3]. All these filaments *originated* in centres of activity, one third near spots, the others in spot-free facular fields. According to M. D'AZAMBUJA their mean lifetime is three rotations, distributed as shown in the second line of the Table 15 (giving percentages):

Table 15.

Lifetime (rotations)	0—1	1—2	2—3	3—4	4—5	5—6	6—7	7—8
Quiescent filaments	8	26	26	19	8	5	4	2
All filaments . . .	80	11	3	2	2	2	0.2	0.1

The values in the second line refer to a selected material: to the quiescent prominences of the spot belts living one cycle or longer, not to the active and the polar prominences. The lifetimes of the filaments in activity centres are shorter; the preponderance of short lifetimes may be estimated from the third line of the Table 15 giving the histogram of lifetimes of *all* filaments observed between 1928 and 1930 (GNEVISHEV[2]).

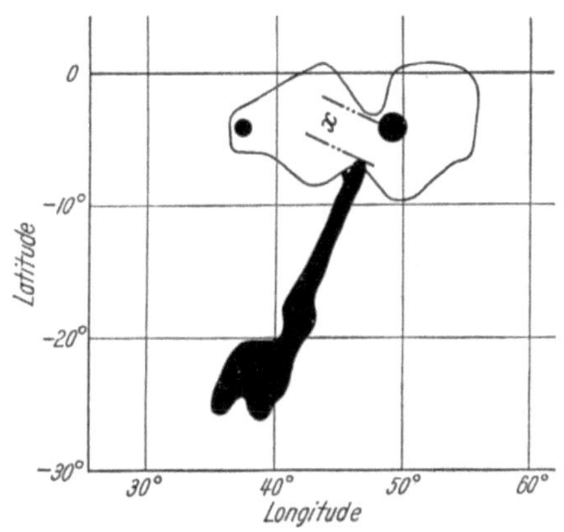

Fig. 68. Relative positions of filament, faculae and spots. L. and M. D'AZAMBUJA: Ann. Meudon 6, fasc. VII, Fig. 15 (1948).

α) *Filaments originating* ***in*** *spot groups* (about $\frac{1}{3}$ of the cases) start their life at the high latitude border of the group. In 80% of the cases the spot filaments point towards the leading spot[3]; in the other 20% they point towards the following spot (Fig. 68). The distance between the spot and the elongation of the filament is in the great majority of cases smaller than 10000 km; the average distance x between the end of a filament and the spot is between 10000 and 20000 km. On Hα spectroheliograms taken with a high resolving power the space between the filament and the spot contains a number of small dark mottles; they are bright and may form a continuous chain of some 10000 km on K_3 spectroheliograms.

During the first ten days after their origin the filaments are still more or less unstable, they may suddenly disappear and come back, but in the meantime they increase in length. A week after their origin the length may have doubled.

At the moment of occurrence of the filaments the spot-group has a certain age and has surpassed the stage of greatest activity. A calmer period has set in and in these fairly quiet surroundings the filament comes to being. Nearly right from the beginning it shows the characteristic form of a series of arches or trees described in the preceding section. In the average case the spot-groups have an age of about 0.9 rotations at the moment of birth of the prominences. At that moment the filaments have on the average an angle of about 38° with the meridian. It is curious to realise that this angle would be obtained under the influence of the differential rotation by a line initially directed North-South, in 0.77 rotations.

[1] M. D'AZAMBUJA: L'Astronomie **61**, 58 (1947).
[2] M.N. GNEVISHEV: Poulkovo Obs. Circ. **30**, 115 (1940).
[3] L. and M. D'AZAMBUJA: Ann. Obs. Meudon **6**, fasc. VII (1948).

This latter value agrees so well with the 0.9 rotation found above that it is tempting to assume that the unknown pre-filamentary field was orientated into such a North-South direction at the moment of birth of the centre of activity (Fig. 69).

β) *Filaments which originate* **outside** *spotgroups* show no clear differences as compared with those originating near the spots. The point of lowest latitude has as a rule the same latitudes as the spots. It is not yet clear why these filaments originate precisely at this place and not at another. Often such filaments originate as several independent parts which merge later. Their orientation is less regular than that of the filaments originating near spots, and they are as a rule smaller and less developed than the spot filaments.

γ) *During the further development* the filament increases in length at a rate $\Delta l/\Delta t \approx 10^5$ km/rotation. It obtaines its greatest extension after about three

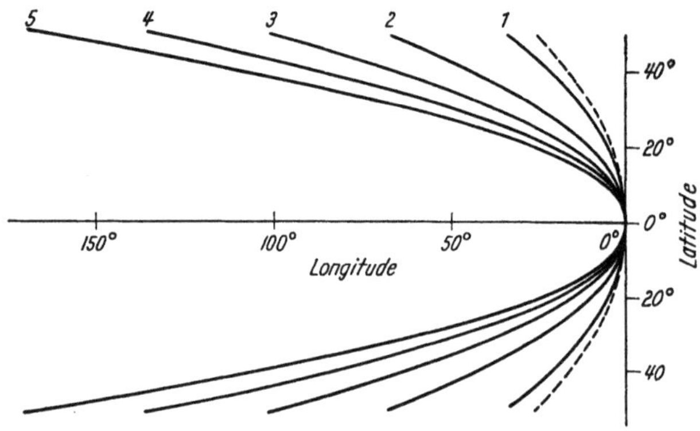

Fig. 69. Deformation of a meridian on the Sun by the solar rotation, after 1, 2, ... rotations. The broken line gives the deformation after 21 days. L. and M. D'AZAMBUJA: Ann. Meudon 6, fasc. VII, Fig. 26 (1948).

solar rotations. Statistically the time lag between the moments on which maximum length is attained, and that on which the spots have their maximum total area is 50 days for filaments with latitudes between 0 and 10°; it is 100 days for filaments between 10 and 30°, and 110 days for filaments at about 40° latitude.

After three rotations the main structure is as described in the preceding section. The filament has obtained its greatest length. Under the influence of the differential rotation the angle with the meridian has further decreased. At that phase it may occur that the filament divides the centre of activity into two bright parts and is situated in a dark lane in between. The poleward motion, to which it was subjected right from its origin, continues and after four or five rotations the filament, which is now nearly directed along the parallel, forms a kind of a high-latitude border line of the centre of activity, and it looks as if the filament is pushed upward by the expansion of the *BM* regions. The decrease of inclination of filaments due to the differential solar rotation is shown in Fig. 69.

At the end of its life the filament fragmentates and gradually dissolves, this process beginning with the parts closest to the equator. However, before that time many things may happen: During these latter phases of its life the filament may become associated with another one, it may become active or even eruptive. About half (46%) of the visible equatorial filaments undergo one temporary sudden disappearance (probably eruptive). Taking into account their time of invisibility, temporary sudden disappearance must be normal for virtually all quiescent filaments (see further Sect. 66).

63. The poleward migration and the polar prominences. The evolution described above refers mainly to the low latitude prominences, these forming the major part of all prominences. The polar prominences are oriented nearly parallel to the equator and sometimes form a nearly uninterrupted crown around the polar caps. Also for that reason their evolution is much more difficult to follow than that of the equatorial filaments. It sometimes happens that the remainder of a low latitude prominence, on its way poleward, joins the polar cap of prominences and finally becomes unrecognizable in the realm of the polar filaments.

The above description suggests that the polar caps are an asylum for old prominences, but one should be cautious not to confuse the latitude motion of *individual* prominences with that of a prominence *zone*. Both phenomena occur; the first seems due to the general increase of area of activity centres, which pushes their quiescent prominences upward. The other phenomenon is treated in Sect. 99. There is some disagreement on the hypothesis that equational prominences may reach the polar zone:

(i) Before reaching the polar caps the majority of the filaments already dissolve. However the D'AZAMBUJA's remark[1] that even after a disappearance lasting for two or three rotations a filament of about the same form sometimes reappears at the place where it would have been located if it had been drifting poleward undisturbedly. A number of such cases occur on the Meudon Synoptic Charts of the Sun.

(ii) The speed of poleward motion decreases with increasing latitude, so that it is doubtful whether a filament may reach a sufficiently high latitude. The answer is in fact the same as that given to the first consideration: if the filaments live long enough they will reach the polar zone, and D'AZAMBUJA estimates that a filament formed at $b = 30°$ needs about 3 years to reach the polar cap.

The poleward migration has been determined by L. and M. D'AZAMBUJA with the aid of the positions of 268 stable filaments, observed between 1919 and 1930. Their lifetimes ranged between 1 and 9 rotations (cf. Sect. 62). The resulting latitude drift is given in the table below:

					polar filaments	
b (°)	0—10	11—20	21—30	31—40	38—50	51—69
Δb(°/rot.)	+2.3 (15)	+1.6 (76)	+1.3 (112)	+1.2 (59)	+0.9 (21)	+0.8 (27)

The numbers between brackets give the number of prominences from which the means have been obtained. The scatter is considerable, 30% of the cases gave negative Δb values. Nevertheless the general tendency is clear, there is an average poleward movement, and Δb diminishes with increasing latitude. The above values are confirmed by the Kodaikanal observations[2] which moreover give

$b = 45° - 70°$	$70° - 90°$
Δb (°/rot.) $= 0.^\circ 6$	$1.^\circ 5$

A general remark that should be made in this connection is that the poleward migration should not primarily be interpreted as a mass motion of the prominences but rather as a displacement of the excitation region. A filament is displaced poleward by a gradual disintegration at the low latitude end and a corresponding growing at the high latitude part (cf. U. BECKER[3]).

[1] L. and M. D'AZAMBUJA: Ann. Obs. Meudon **6**, fasc. VII (1948).
[2] R. ANANTHAKRISHNAN and P. MADHAVAN NAYAR: Kodaikanal Obs. Bull. **137** (1954).
[3] U. BECKER: Z. Astrophys. **40**, 65 (1956).

64. The line spectra of prominences. Outside eclipses the prominences show the strongest lines only, generally of Rowland intensity greater than 50, but during eclipses, after long exposure times, the bright prominences show the greater part of the chromospheric lines: 40 members of the Balmer series, 26 of the Paschen series up to $P31$ and many lines of other neutral and ionized atoms.

The unusual excitation conditions in prominences become clear if one realises that the prominences are relatively low temperature objects ($T \approx 10^4$ °K) embedded in a hot medium with a kinetic temperature, about a hundred times higher. Their outer parts are subject to the intense ultraviolet and X-ray radiation of the corona, the inner parts are shielded against this radiation.

The observed solar spectrum above 800 or 900 Å does not differ much from a Planck curve with a temperature of about 5000°. In very rough approximation the energy-λ curve might be considered to be more or less constant for $\lambda < 800$ Å (cf. Fig. 83 and Sect. 73). So it is a priori clear that the excitation in the prominences will be regulated by a radiation function consisting of two parts: T_{rad} has a value between 4000 to 6000° for $\lambda > 800$ Å and T_{rad} has no definite value for $\lambda < 800$ Å. This introduction serves principally as a warning to take deduced prominence "temperatures" not too literary. As in the study of the chromosphere and corona one of the problems of prominence spectroscopy is that of finding the relations between the kinetic temperatures of the gas, their densities and the observed spectral emission (cf. also Sect. 16). A straightforward method to find these relations does not exist; one should proceed by successive approximations, starting with the most unambiguous methods.

α) *Kinetic temperatures* may be derived from widths of line-profiles. The deduced T_{kin} values are upper limits unless they have been corrected for self-absorption and micro-turbulence. (In this respect reference is made to the remark in Sect. 61 on the fine structure of filaments showing relative motions of 5 to 10 km/sec. These motions will partly contribute to the deduced "micro-turbulence" values). For *quiescent prominences* M.T. CONWAY[1], after correcting for self-absorption, found $T_{\text{kin}} = 14000°$. TEN BRUGGENCATE[2] found for the ratio of half-widths $D_3/\text{H}\alpha = 0.66$; in the case of thermal widening one would expect the ratio $\frac{5876}{4861} \times \frac{1}{2} = 0.6$. The good agreement argues for thermal widening without any contribution of turbulent motions. The deduced kinetic temperature for the heavy particles is 13000°. For a bright quiescent prominence SEVERNY[3] found $T_{\text{kin}} = 14000°$. ELLISON and REID[4], from Hβ and Hγ profiles for which self-absorption was unimportant, as could be shown, found $T_{\text{el}} \approx 10000$ to 12000°. These values agree well with each other.

For *active prominences* ZIRIN[5] found from the line-widths of Hα, D_3 and Hβ, after eliminating the influence of turbulence, $7000° < T_{\text{kin}} < 50000°$ with a mean value of 25000°. For the *eruptive prominence*[6] of June 24, 1956 ZIRIN[7] found a temperature of 3×10^4 °K from line widths and from a comparison of the relative intensities of He II 4686 and He I 4471.

---

[1] M.T. CONWAY: Monthly Notices Roy. Astronom. Soc. London **112**, 55 (1952). — Contr. Dunsink Obs. **3** (1952).

[2] P. TEN BRUGGENCATE: Convegno Volta **11**, 163 (1953). — Veröff. Göttingen **104** (1953).

[3] A.B. SEVERNY: Isw. Krymsk. Astrophys. Obs. **12**, 33 (1954).

[4] M.A. ELLISON and J.H. REID: Publ. Obs. Edinburgh **2**, No. 2 (1957).

[5] H. ZIRIN: Astrophys. Journ. **124**, 451 (1956).

[6] In the paper quoted the object is called a flare, but the published photographs as well as the communicated line widths and intensities rather seem to refer to an eruptive prominence. See further our remark on p. 194.

[7] H. ZIRIN: Astrophys. Journ. **126**, 159 (1957).

These results suggest that active prominences generally have higher kinetic temperatures and perhaps more turbulence than quiescent ones.

β) *The radiation temperature* in the centre of Hα in an optically thick prominence can be derived from the observed central intensity in Hα expressed in that of the continuous spectrum of the disk centre.

The observations show that the central intensity in the core of the emission line profile at the limb and that of the absorption line profile on the disk are about equal for strong prominences; the small differences need not be real:

	Absorption lines	Emission lines
M. T. CONWAY[1]	9.9% (4.2 to 14.8%)	10.4% (2.9 to 27.7%)
ELLISON[2]	8.2%	6.2% (0.8 to 13.0%)
TEN BRUGGENCATE[3]		11.5%

The values are expressed in the intensity of the continuum of the disk centre. Self-absorption in Hα emission lines appears to become important for a central intensity $\gtrsim 0.1$ of the continuum at the disk centre (ELLISON and REID, Fig. 5; cf. footnote 4, p. 232). The mean value, being 9%, may be used to find the radiation temperature in the centre of Hα, supposing that the radiation temperature in the continuous spectrum near Hα for the Sun's centre is 6150°. One finds $T_{\text{rad}} = 3700°$, considerably lower than T_{kin} for H and He found above. This radiation temperature may also be interpreted as an excitation temperature for the relative populations of the 2nd and 3rd levels of hydrogen (cf. Sect. 16). It is worth while to refer in this connection to the chromospheric spicules where radiation temperatures for Hα and electron temperatures were found which are of the same order as those found here. This proves that the excitation conditions for hydrogen in the spicules are not very different from those in prominences.

γ) The T_{el} and T_{ex} values found under the headings α and β are useful starting data, but they should be compared with *more refined computations*. ZIRIN[4] discussed the hydrogen equilibrium under the conditions prevailing in prominences. The populations of the lowest five levels of H and the continuum were considered. The fine structure of the atomic level scheme was neglected, and also the ultraviolet coronal radiation was assumed to be negligible. The populations of the hydrogen levels and the Hα emission were *computed* for various values of N_p (number of protons per cm³) and T_{el}. On the other hand the equivalent widths of Hα and Hβ of some, presumably quiet, prominences have been *measured* by TEN BRUGGENCATE[3] (the equivalent width of an emission line is expressed in the continuous radiation of a wavelength interval 1 Å wide in the spectrum of the Sun's centre). After an empirical correction for self-absorption (curve of growth) these values yield the numbers of excited H atoms (N_3 and N_4) in the line of sight. The equivalent width of Hα yields the ratio N_3/N_2. These three populations can be fairly described by $T_{\text{ex}} = 3700°$. Assuming $T_{\text{ion}} = T_{\text{ex}}$ and taking a prominence thickness of 6000 km (cf. Sect. 65) this yields $\text{Log}\, N_e$ ($= \text{Log}\, N_p$) $= 10.0$. A comparison of this deduced value of N_p and the population ratio for the levels 2 and 3 yields, with the aid of ZIRIN's population tables: $T_{\text{el}} \approx 25000°$. This is in good agreement with the average kinetic temperatures

[1] M. T. CONWAY: Monthly Notices Roy. Astronom. Soc. London **112**, 55 (1952).
[2] M. A. ELLISON: Publ. Edinburgh **1**, 5 (1952).
[3] P. TEN BRUGGENCATE: Convegno Volta **11**, 163 (1953). — Veröff. Göttingen **104** (1953).
[4] Cf. footnote 5, p. 232.

derived from line widths of active prominences, but it is higher than the value found for quiet ones.

A similar analysis by JEFFERIES[1] was based on detailed considerations of the H and He level populations. For the quiescent prominences studied by him, the best agreement with the observed intensities of Hα and D_3 was generally obtained for electron temperatures between 1.1×10^4 and 2.0×10^4 °K and $10^{10} < N_e < 5\times 10^{10}$.

δ) The use of equivalent widths of emission lines, as applied in the foregoing paragraph is permissible only when the line intensities can be corrected for self-absorption. A *curve of growth* for quiescent prominences was constructed by BRÜCK and MOSS[2] and again by BÖHM[3], assuming excitation and ionization temperatures of 3700°. Turbulent velocities, determined with the aid of the curve of growth range for various prominences from 5 to 30 km/sec with an average value of 15 km/sec.

Once such a curve of growth has been established, it may be used to correct the observed equivalent widths for the influence of self-absorption, and next to use these corrected data to find the relative numbers of emitting atoms. Such an analysis of an *eruptive prominence* spectrum taken by GROTRIAN was carried out by UNSÖLD[4]. The prominence showed lines of neutral metal atoms (Ca), metal ions (Sr^+, Cr^+ etc.) H, He and He^+, covering a wide range of excitation conditions. Formally the excitation conditions may be described by replacing in BOLTZMANN's and SAHA's law $\chi\Theta$ by the arbitrary function $\varphi(\chi)$, which should be determined empirically. The empirical $\varphi(\chi)$ function determined by UNSÖLD appears to be a straight line up to $\chi\approx 10$ eV—thus indicating no appreciable deviations from the equilibrium excitation conditions, with $T_{ex}\approx 3700°$. For greater χ-values up to 50 eV it is nearly horizontal. This result reflects in a sense the character of the solar energy wavelength as described in the first part of this section, having $T\approx 4000°$ for $\lambda > 800°$ Å and being approximately constant for shorter wavelengths. An attempt to explain this curve on the basis of the computed ultraviolet radiation field of the transition layer chromosphere-corona was made by OSTER[5].

ε) The foregoing discussion has shown how some regularity can be detected in the complicated radiation conditions in prominences. TEN BRUGGENCATE[6] made an interesting, though perhaps too formal attempt, to find the excitation and ionization conditions, from the given Hα and D_3 intensities, taking the deviations from equilibrium into account. For details reference is made to the original papers; the results are: $T_{el}=14400°$, and $n_{H^+}/(n_H+n_{H^+})=0.998$; $n_{He^+}/(n_{He}+n_{He^+})=0.29$, so that hydrogen is completely, helium half ionized. These results are in substantial agreement with those found in the first part of this section. They also agree with those found by SEVERNY[7] from the analysis of the H, He and Ca^+ lines of a bright prominences: $T_{kin}=14000°$, $\xi_t=7.5$ km/sec and Log $N_e=10.3$.

ζ) Finally we remark that the light of the Hα and D_3 lines is slightly polarized (LYOT 1937). For Hα the polarization is about 1%. The direction of maximum

[1] J.T. JEFFERIES: Monthly Notices Roy. Astronom. Soc. London **116**, 629 (1956).

[2] H.A. BRÜCK and W. MOSS: Monthly Notices Roy. Astronom. Soc. London **103**, 258 (1943); **105**, 17 (1945).

[3] K.H. BÖHM: Quoted in UNSÖLD, Physik der Sternatmosphären, 2. Aufl., p. 687. Berlin: Springer 1955.

[4] A. UNSÖLD: Physik der Sternatmosphären, 2. Aufl., S. 690. 1955.

[5] L. OSTER: Z. Astrophys. **40**, 28 (1956).

[6] P. TEN BRUGGENCATE: Convegno Volta **11**, 163 (1953). — Veröff. Göttingen **104** (1953).

[7] Cf. footnote 3, p. 232.

polarization is not perpendicular to the limb but inclined towards north and south on the northern and southern hemispheres respectively. For D_3 the degree of polarization is 0.009 to 0.015, inclinations up to 40° have been observed. THIESSEN[1] showed that the observations can be explained by resonance fluorescence in a magnetic field; an effect of some percents has been computed.

η) Further we note that the helium abundance of prominences was determined by various authors. UNSÖLD[2], from the observations of BRÜCK and MOSS (footnote 2, p. 234), found $N(\mathrm{He})/N(\mathrm{H}) = 0.15$. SEVERNY[3], using H$\alpha$ and the 10832 line of helium, found $N(\mathrm{He})/N(\mathrm{H}) = 0.2$. JEFFERIES[4], from H$\alpha$ and D_3 found 0.2.

We *summarize* the results of this section:

$$T_{\mathrm{kin}} \text{ for } \begin{pmatrix}\text{quiescent}\\ \text{active}\end{pmatrix} \text{ prominences: } \begin{pmatrix}15000°\\ 25000°\end{pmatrix};$$

$$T_{\mathrm{ex}} \text{ for } \chi \approx 3 \text{ eV} = 4000°;$$

$$T_{\mathrm{ex}} \text{ for } \chi \gtrsim 10 \text{ eV} = 15000°;$$

$$\xi_t \approx 4 \text{ to } 10 \text{ km/sec};$$

$$\mathrm{Log}\, N_e = 10.3;$$

fraction of neutral hydrogen: ≈ 0.002; of neutral helium: 0.5;

abundance ratio H/He = 5.

65. The continuous spectra of prominences. Observations show a distinct continuous spectrum only in bright prominences. Even these prominences are optically thin for continuous radiation since in white light they are invisible on the disk. The most important feature in continuous light is the Balmer discontinuity at 3647 Å. The Paschen continuous spectrum at 8206 Å was photographed by LYOT outside eclipses. In the region in between, a faint continuum is sometimes discernable; it is polarized perpendicularly to the limb. The degree of polarization is 15%. This radiation is most probably due to Thomson scattering.

For a quantitative analysis absolute intensity measurements are necessary. The Balmer and Paschen continua are primarily due to recombination of protons with electrons, so their intensities depend on the values of $N_e N_p (\approx N_e^2)$ and on the electron temperature T_{el}, according to[5]

$$\mathcal{E}(\nu)\, d\nu = \mathrm{const}\, \frac{N_e N_p}{T_{\mathrm{el}}^{\frac{3}{2}}}\, e^{-\frac{h\nu - \chi}{kT}}\, d\nu$$

where $h\nu$ is the excitation potential of the upper level involved and χ is the excitation potential of the second hydrogen level. On the other hand the intensity of the continuous radiation in the visible region for $\lambda > 3650$ Å is mainly due to scattering of photospheric light by free electrons, depending on N_e according to

$$\mathcal{E}_\lambda = I_\lambda \sigma N_e$$

where I_λ is the radiation flux of the photosphere at wavelength λ; σ is the scattering coefficient per free electron $= 6.67 \times 10^{-25}$ cm². From the knowledge of the

[1] G. THIESSEN: Z. Astrophys. **30**, 8 (1951).
[2] A. UNSÖLD: Physik der Sternatmosphären, 2. Aufl., p. 690. Berlin: Springer 1955.
[3] A.B. SEVERNY: Cf. Vistas in Astronomy **1**, 703 (1956).
[4] J.T. JEFFERIES: Monthly Notices Roy. Astronom. Soc. London **116**, 629 (1956).
[5] G.G. CILLIÉ: Monthly Notices Roy. Astronom. Soc. London **92**, 820 (1932); **96**, 771 (1936).

two *absolute* intensities, N_e and T_{el} can be found independently if the thickness of the prominence is known; the *ratio* of the intensities depends on both N_e and T_{el} so that, for assumed thickness and N_e values, T_{el} can be found and reversely. This latter method may be used if the spectra are not calibrated to absolute intensities.

For a prominence observed during the 1952 eclipse ZANSTRA and REDMAN[1] thus found the following relation between T_{el} and N_e; for each adopted value of T_{el} the thickness L is determined from the absolute intensities of the observed radiation:

T_{el}	N_e	L (km)
5000°	0.7×10^{10}	38000
10000°	2.2×10^{10}	12000
15000°	4.7×10^{10}	5900
20000°	1.0×10^{11}	2700

A value of T_{el} between 13000 and 15000° is perhaps closest to reality, as shown in the previous section; in this case the thickness (7000 km) is only somewhat greater than the observational value. (These observational values are greater than the actual thickness of the prominences which need not stand perpendicular to the line of sight[2].) From the *line spectrum* of the same prominence $T_{el}=13000°$ is found with thermodynamic equilibrium computations[3]. With non-equilibrium computations T_{el} becomes 20000° and accordingly $L=2700$ km and Log $N_e=11$. Values for N_e and T_{el} similar to ZANSTRA'S were derived by KOELBLOED and VELTMAN[4] and by HAUG[5] from a rediscussion of earlier eclipse observations of prominences, obtained in 1926 (DAVIDSON and STRATTON) and 1929 (DOORN). From the ratio of the Balmer free-bound emission to the remaining components of the continuous light at 3647 Å ORRALL and ATHAY[6] found $5200°<T_{el}<20000°$, and correspondingly $10^{10}<N_e<2.7\times10^{11}$.

The polarization of the light in the visual wavelength region was computed by BAUMBACH[7] in connection with a general discussion of the polarization of the corona. The computations were made for two wavelengths (6200 and 4330 Å), three scattering angles, and five heights above the Sun's limb, assuming that all the light is scattered by free electrons. The light emitted at a point M of the Sun's disk, scattered by an electron E into the direction of the observer O (angle $MEO=\alpha$) has the intensity components

$$\sigma_2(\alpha)=2\pi\left(\frac{e^2}{m\,c^2}\right)^2 N_e\cos\alpha \quad \text{in the plane } MEO,\ \perp EO;$$

$$\sigma_1(\alpha)=2\pi\left(\frac{e^2}{m\,c^2}\right)^2 N_e \quad \text{perpendicular to that plane, } \perp EO.$$

The total intensities are obtained after a double integration, over the line of vision and over the Sun's disk.

For normal values of height (10^4 to 10^5 km) a polarization degree of 15 to 40% is computed. It is greater at 4000 Å than at 6000 Å. Comparison with the

[1] H. ZANSTRA and R. O. REDMAN: Circ. Amsterd. **6** (1952).
[2] S. G. M. HAUG: Circ. Amsterd. **8** (1953).
[3] R. G. ATHAY and F. Q. ORRALL: Astrophys. Journ. **126**, 167 (1957).
[4] D. KOELBLOED and W. VELTMAN: Circ. Amsterd. **3** (1951).
[5] S. G. M. HAUG: Circ. Amsterd. **5a** (1952).
[6] F. Q. ORRALL and R. G. ATHAY: Astronom. J. **62**, 28 (1957).
[7] S. BAUMBACH: Astronom. Nachr. **267**, 273 (1939).

observed values of about 15% (LYOT) shows that the greater part of the emission in the visual wavelength region must be due to scattering by free electrons.

b) Moving prominences.

The group of the moving prominences comprises two kinds of objects. (1) A *quiescent prominence* may (a) be activated temporarily or the activation may result into (b) a dissolution or (c) a blow-off of the prominence. (2) The *sunspot prominences* are moving all their life.

SEVERNY has classified the moving prominences according to their types of motions[1]. There are three types: (a) eruptive prominences, rather rare; (b) electromagnetic (about 50%) and (c) irregular or turbulent prominences. The eruptive prominences are treated in Sect. 67. The electro-magnetic prominences are characterized by regular motions along paths resembling lines of force. These are mainly the sunspot filaments and the activated quiescent prominences. The irregular motions may occur among sunspot prominences but they occur mainly in the activated prominences.

66. Activation and dissolution of quiescent prominences; sudden disappearances. The *activation* is often only a temporary interruption in the life of a quiescent prominence. The velocities increase, the form may change in a complicated way. At such a moment, according to a description given by ELLISON[2] "the higher parts of the prominence develop into a complex pattern of streamers which show sightline components of velocity or the order of ± 30 km/sec. Hunting backwards and forwards over a range of 1 to 2 Å with the lineshifter, one observes rapid changes both of shape and luminosity between adjacent wavelengths. The prominence 'wriggles' as the line-shifter is moved". The total activation time may be as short as ten minutes but it may also last for hours, sometimes for days. A temporary or definite disappearance may be due to or connected with the activation. The general form is that of bright matter originating in the upper part of the prominence and streaming down, often along beautiful curves towards one or more "centres of attraction", chromospheric areas which may often be identified with sunspots. The loop structures are one of the most common features in active regions. Three types of motion are preponderant: (1) material becomes visible on top of the loop and streams down on either side; (2) material coming from the chromosphere streams upward along one side of the loop and down along the other; (3) no material is streaming but the whole loop, apparently attached to the chromosphere, sways back and forth. There may be similarity in the motions of the prominence knots. Often they appear to move in groups along similar paths, apparently under the influence of the same field of force[3].

A characteristic case of an activated prominence was discussed by ROTSCHILD, PECKER and ROBERTS[4]. The filament was initially of the quiescent type; it was visible at the limb during three days (23, 24, 25 July 1951) and suddenly disappeared on the last. During the whole period of visibility it was restricted to a sharply bounded volume of the corona. It displayed strong internal motions; bright knots, living for 15 to 20 minutes, showed rather regular uniform motions (of the electromagnetic type in SEVERNY's classification) of the order of 10 to 100 km/sec. The accelerations of the velocities seemed to be independent of

---

[1] A. B. SEVERNY: Dokl. Akad. Nauk SSSR. **82**, 25 (1952).

[2] M. A. ELLISON: Monthly Notices Roy. Astronom. Soc. London **116**, 624 (1957).

[3] H. W. DODSON: Monthly Notices Roy. Astronom. Soc. London **108**, 383 (1948).

[4] K. ROTSCHILD, J. C. PECKER and W. O. ROBERTS: Astrophys. Journ. **121**, 224 (1955).

that of gravity, moreover in the majority of the cases there was a tendency for the acceleration to reverse at a certain point in the trajectory.

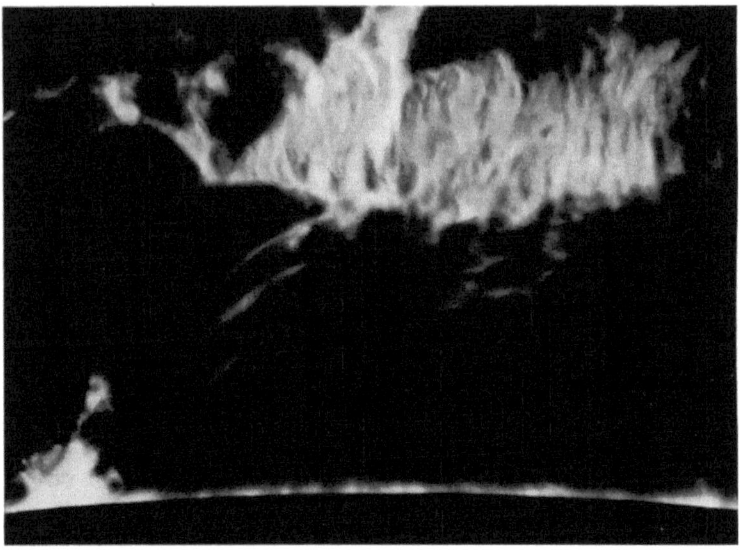

a

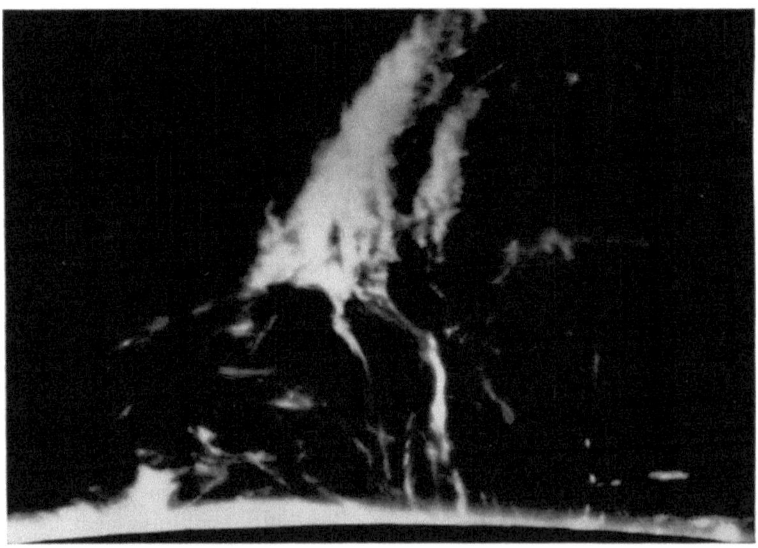

b

Fig. 70a and b. Active prominences, photographed 1956, September 18 (a) and October 4 (b), through an Hα filter constructed by DUNN, 3.4 Å wide. Courtesy R. B. DUNN, Sacramento Peak Observatory.

BILLINGS[1] has drawn attention to small unstable rather short-lived (≈ 15 min) closed loops that occur in active prominences. The diameters are of the order of 10^4 km. They occur frequently in the most active regions, characterized by emission of the yellow coronal line; there is a correlation between the velocity of the internal motions and the activity of the region.

[1] D.E. BILLINGS: Publ. Astronom. Soc. Pacific **69**, 162 (1957).

An interesting type of prominence is the *interactive prominence* (PETTIT's type Ia), consisting of two neighbouring prominences exchanging streamers. According to PETTIT[1] there is no evidence that the movement represents a complete circuit within the prominences.

The (sudden) *disappearance* that often follows the active stage may occur in various ways:

(i) In many cases material disappears into the chromosphere, while at the same time other matter seems to condense from the corona, supplying at least partially the loss of luminous matter. This process may continue even till the very moment that the whole prominence has disappeared. This is one of the possible causes of a sudden disappearance.

(ii) In other cases a prominence may dissolve suddenly without any clear reason, or

(iii) it may fly into the high corona, sometimes reaching velocities greater than the velocity of escape. This case, the eruptive or "ejected" prominences, will be discussed in the next section.

When studied on daily routine spectroheliograms or filtergrams it is often impossible to find out whether a filament, after its activation, has disappeared by gradual dissolution or whether the "sudden disappearance" is due to a blow off effect. This uncertainty is not wholly removed in the existing literature on sudden disappearances.

Activation and sudden disappearances often occur when a new centre of activity originates at the place of a quiescent prominence. The new centre tends to destruct the existing prominences. The violent destructing influence of flares has been mentioned (Sect. 55). In accordance with the above D'AZAMBUJA found that sudden disappearances happen mainly to equatorial filaments; in only 5% of the 206 cases studied by him, polar filaments were involved. Of all equatorial filaments 46% disappear at least once in their life, so, taking the invisible hemisphere into account, it may be that the greater part of all filaments disappear at least once. In 31 of the 206 observed cases the filament disappeared twice, in 4 cases it disappeared and reappeared three times. After the disappearance generally a chain of black mottles remains visible which after some time may combine into a new filament.

67. Eruptive prominences. Eruptive prominences do not form a special class but a special phase in the life of other types. The eruptive, upward motion of an initially quiescent prominence is a particular case of activation. These phenomena have been studied mainly by PETTIT[2], WALDMEIER[3], L. and M. D'AZAMBUJA[4], ELLISON[5] and others. Often the central part of the filament starts ascending with a velocity increasing with time. One or both ends of the filament may or may not remain attached to the chromosphere. When in the end phase the greatest part of the prominence has been blown out, a part may be seen streaming back near the legs. Usually the underside of the prominence becomes brighter shortly before the eruption. According to the D'AZAMBUJAS only one eruption in three leads to a final dissolution, in the other two thirds of the cases a filament is rebuilt at the same place and in the same form after one or some days. Obviously

[1] Cf. E. PETTIT: Chap. 6 in HYNEK, Astrophysics. New York: McGraw-Hill 1951.
[2] E. PETTIT: Publ. Yerkes Obs. **3**, 205 (1925). — Astrophys. Journ. **84**, 319 (1936).
[3] M. WALDMEIER: Z. Astrophys. **15**, 299 (1938).
[4] L. and M. D'AZAMBUJA: Ann. Obs. Meudon **6**, 104 (1948).
[5] M. A. ELLISON: Publ. Roy. Obs. Edinburgh **1**, 75 (1952). — Monthly Notices Roy. Astronom. Soc. London **116**, 624 (1957).

the field conditions favouring the growth of filaments may often survive the eruption. Eruptions occur everywhere on the disk, also in the zone of the polar prominences; mostly in the sunspot zone, but as a rule in regions without spots (since quiescent prominences do not occur near spot-groups).

Among the eruptive disappearances not associated with the origin of a new centre of activity, a great part (70%, according to BRUZEK[1]) are followed by the origin of bright *flare-like faculae*, that may obtain a top brightness sometimes equal to that of importance-2 flares, however, with a much slower development curve. This phenomenon has been described first by WALDMEIER[2], it has been discussed extensively by BRUZEK[1] and has been mentioned further by U. BECKER[3] and MARTRES[4]. The narrow relationship between the filament and the location of these bright faculae, often forming strings of bright mottles at the basis of or parallel to the disappeared filament, are, according to BRUZEK, a support for the hypothesis that the faculae and flares are heated by currents, induced in the chromosphere by a changing magnetic field. Since the general form and orientation of parts of the filaments seems to be related to the field in their surroundings, this might indicate that the ascending motion of the filament is due to a change of the magnetic field. Another indication that the motions are dictated primarily by the magnetic field is found in the observation that the force-field seems to change only for those prominences that change form rapidly, thus indicating a relation between the form changes and the force-field changes[5].

The *upward motions* of eruptive prominences have been investigated by PETTIT[6] and WALDMEIER[3], who found some regularities:

(a) Initially the upward velocities are sometimes small but may also be large, and increase to great values, which occasionally may exceed the velocity of escape (618 km/sec at the surface). The greatest observed velocity, observed on two occasions, is 1130 km/sec, the acceleration may reach $50\,g_\odot$[7,8]. Distances up to 1.5×10^6 km have been reached[9]. For time-velocity diagrams cf. PETTIT[10].

(b) In a number of cases it was observed that the increase of the velocity is abrupt, taking place in less than two minutes of time. PETTIT assumed this to be a general case and suggested even that at each discontinuity the ratio between both velocities is a small integer. This latter statement is certainly not always true and PAN PUH[11] has shown moreover that in certain cases the same height—*vs.*—time diagram may be represented by a smooth curve as well as by a discontinuous one. The smooth curves are nearly exponential and may, after WALDMEIER, be characterized by one parameter $p = d\,\mathrm{Log}\,h/dt$. If h and t are expressed in 10^3 km and hours, $\bar{p} = 0.33$; p ranges between 0.10 and 0.68. This result was confirmed by LARMORE[5] in a study of four prominences, including an active, an eruptive, a sunspot and an active-eruptive prominence. In no case sudden increases of the velocity were observed; all time—*vs.*—distance graphs could well be represented by parabolic curves. Nevertheless it is un-

[1] A. BRUZEK: Z. Astrophys. **42**, 76 (1957).
[2] M. WALDMEIER: Z. Astrophys. **15**, 299 (1938).
[3] U. BECKER: Z. Astrophys. **42**, 76 (1957).
[4] M. J. MARTRES: L'Astronomie **70**, 401 (1956).
[5] L. LARMORE: Astrophys. Journ. **118**, 436 (1953).
[6] E. PETTIT: Publ. Yerkes Obs. **3**, 205 (1925). — Astrophys. Journ. **84**, 319 (1936).
[7] D. H. MENZEL, E. v. P. SMITH, H. DE MASTUS, H. RAMSAY, G. SCHNABLE and R. LAWRENCE: Astronom. J. **61**, 186 (1956).
[8] K. BJERKE: Ark. Astr. **2**, No. 13, 145 (1957).
[9] R. R. MCMATH and E. PETTIT: Astrophys. Journ. **88**, 244 (1938).
[10] E. PETTIT: Publ. Astronom. Soc. Pacific **63**, 237 (1951).
[11] PAN PUH: Ann. Obs. Meudon **8**, fasc. IV (1939).

mistakable that there *are* cases in which the height—*vs.*—time diagram is clearly discontinuous so that it certainly may happen that the upward motion of filaments is accelerated discontinuously; the acceleration time being of the order of one to two minutes.

Although the ascent is as a rule accompanied by a general fading of the prominences, the cases of sudden accelerations seem to be accompanied by a flash of light emission[1]. For examples of light curves with such a flash, cf. [1, 2].

In some cases it happens that the various parts of the prominences move at different speed, e.g. in one case[3] of an eruption introduced by a flare the velocities ranged between 110 and 690 km/sec; still all parts were accelerated at the same moment.

In a few cases a sudden decrease of the upward velocity was reported[4] but this may be apparent and due to a local decrease of visibility in the highest knots.

(c) The trajectories are usually nearly straight lines; there is a tendency to be nearly radial, but trajectories at large angles to the radius do also occur. (On February 28, 1946 angles of 62 and 90° with the radial were observed; the velocities were 138 and 268 km/sec.) A case was reported[5] in which the prominence matter seemed to radiate from, or to be attracted by, a common centre in the corona.

(d) When ascending the forms do generally change only slowly. Details of the quiescent prominence may often be found back in the eruptive one. The ascension is accompanied by a faint expansion; in one well studied case (March 20, 1938 WALDMEIER) the linear dimension increased to the fourfold value over a distance of 10^6 km. This expansion is much less than would be expected on the basis of the decreasing coronal density (factor 10^2). In other cases[3, 6] the size did hardly increase twofold during an ascension over 4×10^5 km.

(e) In nearly all cases the ascending motions are accompanied by downward streamings towards an attraction centre, presumably a spot at some distance to the starting point of the eruption: the eruptions occur to prominences not belonging to centres of activity, in undisturbed regions of the Sun's disk. Sometimes the centres may excercise their attractive power even over distances of 10^5 km. In some cases it occurs that the attraction is so violent that the ascending prominence ends by disappearing into the low chromospheric matter *(quasi-eruptive stage)*. Often the eruption starts from near a spot (SHAPOSHNIKOVA[1]).

68. Sunspot prominences. In contrast to the objects discussed in the sections 61 to 66, the sunspot prominences occur in centres of activity. On spectroheliograms they are discernible as compact, short and dense structures, perhaps denser than prominences in general. In many cases they point towards the p spot of the group (see Fig. 68). It is clear that the inflowing streamers discussed in Sects. 66 and 67, which condense into visibility in the corona and flow towards an attraction centre, may be classified as sunspot prominences, if the attraction centre is a sunspot.

The more specific sunspot prominences consist mainly of two groups with many intermediate forms. At the limb they are shown as *loops* and *arches*

[1] A. B. SEVERNY: Dokl. Akad. Nauk SSSR. **53**, 475 (1950). — Isw. Krymsk. Astrophys. Obs. **10**, 9 (1953). — E. F. SHAPOSHNIKOVA: Isw. Krymsk. Astrophys. Obs. **18**, 151 (1958). Cf. [*10*], p. 333—334.

[2] TH. A. CRAGG: Publ. Astronom. Soc. Pacific **69**, 268 (1957).

[3] W. COMPER and R. KERN: Z. Astrophys. **43**, 20 (1957).

[4] Cf. footnote 5, p. 240.

[5] M. WALDMEIER: Z. Astrophys. **44**, 213 (1958).

[6] G. OLIVIERI: L'Astronomie **70**, 120 (1956).

(EVANS and MENZEL's *AC*1; PETTIT's IIIa) and as *funnel-type* prominences (*AS*1) (certain prominences of PETTIT's IIIb) related to the *coronal clouds* or *knots* (WALDMEIER). In most of these cases the visible matter is certain to move mainly downward. The condensation centre may last for a considerable time, so that there must be a constant replenishment from the corona.

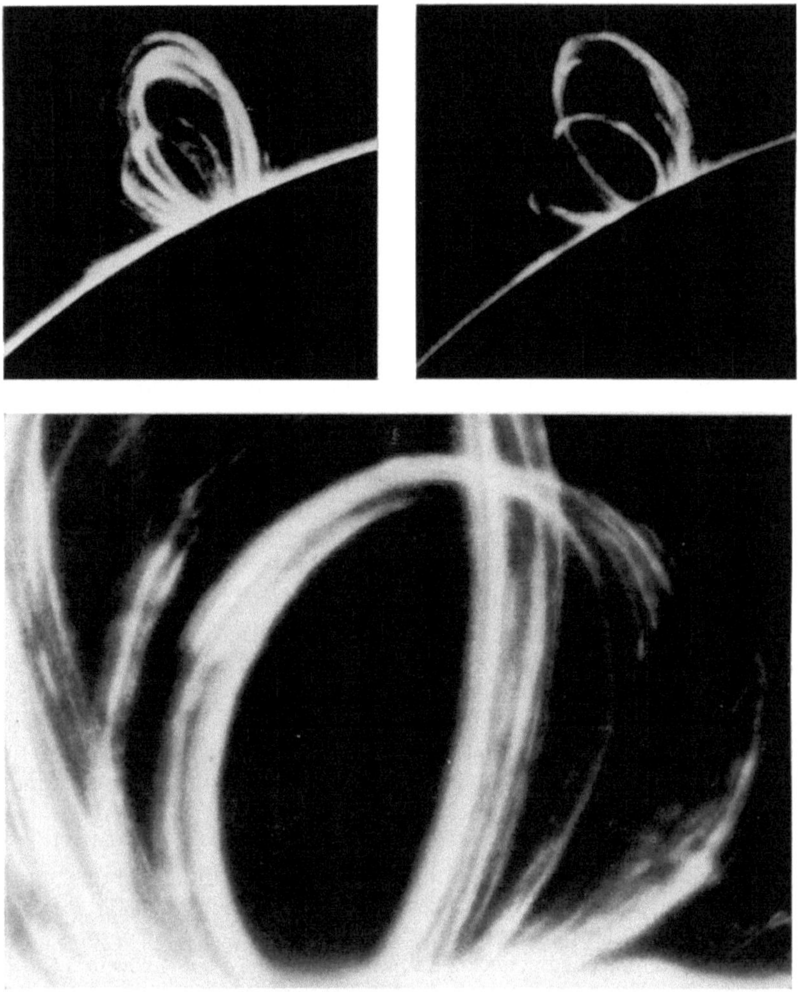

Fig. 71. Loop prominence of 28 June, 1957. Upper left: 10. 58 UT; Right: 13.44 UT. Photographs obtained at the Ondrejov Observatory with a Šolc Hα filter, centred at Hα; width 8 Å. Diameter of the original solar image 19 mm. For a description of the filter, see: Die Sterne 33, 151 (1957). Courtesy K. HERMANN-OTAVSKY, Praha. Lower picture: Detail of the same loop photographed at the Sacramento Peak Observatory with an Hα filter constructed by R. B. DUNN, 3.4 Å wide. The picture is a mirror image of the above ones. Diameter of the original solar image 27 cm. Courtesy R. B. DUNN Sacramento Peak Observatory.

DOLDER, BILLINGS and ROBERTS[1] have noticed that the active coronal regions (showing 5694 Å emission) are characterized by the following types of "active region prominences" (here called sunspot prominences): (a) surges, (b) scattered knots, (c) downward motion of knots and streamers towards an active centrum, (d) formation of continuous streamers and (e) sharp curvatures and focussing

[1] F. P. DOLDER, W. O. ROBERTS and D. E. BILLINGS: Astrophys. Journ. **120**, 112 (1954).

effects in trajectories. In this section (b) and (c) are classified as condensations, knots or funnels; (d) and (e) as loops.

The *loops* show specific structures, which strongly suggest their correlation with the magnetic field of the spot area to which they are related. Their field of motions is often homogeneous over distances up to 10^5 km; the motions are of the electromagnetic type (SEVERNY) and they show a focussing effect: matter originating from or converging towards one point (Fig. 71). The loops are "inflowing filaments moving along curved trajectories, which originate in the surrounding chromosphere or descend from coronal clouds formed above the spot area. They terminate at the boundary of the penumbra with a mean velocity of inflow of 48 km/sec. Their movements seem to be independent of the magnetic polarity of the attracting sunspot. The mean projected length is 61000 km.

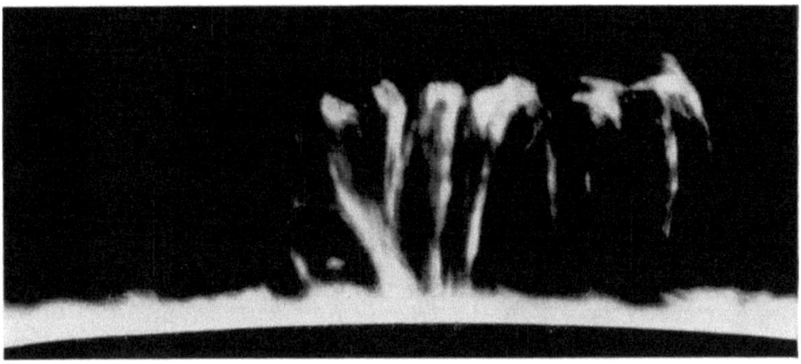

Fig. 72. Coronal prominence knots, approaching the funnel-type (Menzel-Evans classification). Picture obtained 11 September, 1956, with an Hα filter constructed by DUNN; 3.4 Å wide. Courtesy R. B. DUNN, Sacramento Peak Observatory.

(ELLISON[1]). The trajectories of the prominence knots of the outstandingly active region of February 26, 1946 suggest a magnetic field like that of a dipole buried at 0.2 R below the photosphere, with an approximately radial dipole axis[2]. In another case[3] this "dipole" be situated at 0.03 R below the surface.

In the loop prominences motions are observed going to and fro, from one attraction centre to another, often progressing into opposite directions along the same curved path. The presence of moving knots is often interpreted as proving that material motion is involved, not motion of the excitation mechanism alone, but this interpretation is not fully justified (cf. Sect. 69ε).

The *condensations, knots* or *funnels* consist in their idealized form of a number of bright knots formed at some distance above the Sun's limb, generally 50000 to 100000 km. All occur near the same level and are bright; they have nearly the same brightness as the chromosphere at the disk viewed in Hα (15% of the central disk continuous spectrum near Hα). They are connected with the chromosphere by thin, less bright threads of luminous matter. The threads may increase in thickness, so that the envelop of the downward moving matter takes a funnel-shape. The threads are often curved. The knots are strongly connected with flare activity; their occurrence in an activity centre is an indication that the centre will, or has exhibited flare activity.

The knots are the centres of the coronal condensations. They are in continuous agitation, and the brightness may change in a time interval of the order of 10 minutes. Their lifetime may be as short as that and sometimes the whole system

[1] M. A. ELLISON: Monthly Notices Roy. Astronom. Soc. London **104**, 22 (1944).
[2] M. CORRELL, M. HAZEN and J. BAHNG: Astrophys. Journ. **124**, 597 (1956).
[3] M. CORRELL and W. O. ROBERTS: Astrophys. Journ. **127**, 726 (1958).

of knots does not live longer than a quarter of an hour, while the rate of fading still seems to depend upon the size of the knots[1]. Usually the system is of longer life; the knots are continuously replaced by others, which form anew after and during the disappearance of the old knots. WALDMEIER noticed the Doppler motions of the threads; these motions are great at great distances to the limb, near the surface they are smaller and may even approach zero. This indicates a considerable curvature of the orbits. Often the motion is decelerated discontinuously like in eruptive prominences.

The system may extend, finally to form a coronal cloud from which streamers rain down into the spot area. Such a cloud may last for several days. Regions near these knots must have abnormally high temperatures. Coronal studies show that the knots occur in regions where the corona is greatly excited, often showing the yellow line of Ca XV, the line with the highest ionization potential of all solar lines.

The phenomena exhibited by the prominences, here described for an average case, may differ from case to case. They greatly depend on the amount of activity of the prominence; and the simple downward streaming of matter as described here may take rather complicated forms, going over into loop-like prominences; so there is no sharp distinction between the knot prominences and the loops.

A special class of sunspot prominences are the surges. These have been treated in Sect. 56 (flare-surges), but surges may also occur without flares. When projected on the disk they appear as "dark surges", also called "active dark flocculi". Sometimes, especially in the first stage, they may be bright, and in very few cases they were seen dark in white light; cf. also Sect. 56.

69. Considerations on the origin and motions of prominences. We are still far from a consistent prominence theory, but some aspects of the problem will be touched in the following. (For many problems concerning highly ionized gases, such as the prominences and the corona, reference is made to an important paper by OSTER[2], containing fundamental computational data on the viscosity, the electrical and thermal conductivity in stellar matter.)

α) Pressure equality between corona and prominences. It is perhaps essential that there is approximately pressure equality between the corona on the one hand, and the prominences (ZANSTRA[3]) and the spicules (for the latter see Sect. 26) on the other. Pressure equality $P_{\text{cor}} = P_{\text{prom}}$ would lead to

$$k N_{\text{cor}} (T_{\text{el}})_{\text{cor}} = k N_{\text{prot}} (T_{\text{el}})_{\text{prot}} \quad \text{or} \quad (T_{\text{el}})_{\text{prot}} / (T_{\text{el}})_{\text{cor}} = N_{\text{cor}} / N_{\text{prot}} .$$

At $h = 50000$ km in the corona $N_e = 2 \times 10^8$, $T_{\text{el}} = 7 \times 10^5$ and in an average prominence $N_e = 2 \times 10^{10}$ and $T_{\text{el}} = 15000°$. Hence $(N_e\, T_{\text{el}})_{\text{cor}} = 1.4 \times 10^{14}$ and $(N_e\, T_{\text{el}})_{\text{prot}} = 3.0 \times 10^{14}$. Since N_e is virtually proportional to the total number of particles, there is approximately equal pressure at the same horizontal level. In a vertical direction the corona and the prominences are not in hydrodynamical equilibrium, since the prominence has a density about a hundred times greater than that of the corona. So there must be a mechanism to support the prominences.

β) Relations to magnetic fields. It is tempting to suppose that many or all sunspot prominences with their characteristic forms and motions are related to the fields of the spots. But also the quiescent prominences have some connection with magnetic fields: they originate at the borders of *BM* regions or appear in the neutral part of such fields between the regions of opposite polarities, but

[1] A. B. SEVERNY: Isw. Krymsk. Astrophys. Obs. **10**, 9 (1953).
[2] L. OSTER: Z. Astrophys. **42**, 228 (1957).
[3] H. ZANSTRA: Observatory **67**, 10 (1947). — Vistas in Astronomy **1**, 256 (1955).

only at the moment that the spot and flare activity has essentially ceased and when the magnetic field has reduced to small values. They attain their greatest development at the borders of the *UM* (or *M*) regions. The polar drift of big prominences was suggested to be connected with the gradual expansion of ageing *UM* regions.

The long-lived *quiescent prominences* might perhaps represent an equilibrium configuration between the magnetic and turbulent (or thermal) energies in regions of weak fields: With $N_p \approx N_e = 2 \times 10^{10}$ cm^{-3} we have $\varrho = 2 \times 10^{-14}$ gram cm^{-3} and with $v_{\text{turb}} \approx v_{\text{th}} \approx 5$ km/sec: $\frac{1}{2} \varrho v_t^2 = 0.002$ erg cm^{-3}. Equalizing this quantity to the magnetic energy $H^2/8\pi$ yields $H = 0.2$ Gauss, a quite acceptable value for the borders of *UM* regions. This explains why the fields are essential for prominences and their evolution, a suggestion also supported by the observation that after a sudden disappearance a prominence in many cases comes back in essentially the same form.

The *sunspot prominences* obviously occur in regions of greater magnetic fields. Marshall[1] computed that for a pair of large sunspots, with $H_0 = 3500$ Gauss, with a separation of $l = 50000$ km, the dipole magnetic field H' at $R = 50000$ km above the spots is

$$H' = \frac{H_0\, l\, d^2}{4\pi R^3} = 3 \text{ Gauss}$$

leading to a magnetic energy density of 0.5 erg cm^{-3}, 250 times greater than the average kinetic energy density calculated above. Hence in this case the lines of force move freely, carrying the completely conducting matter with them. The characteristic difference between quiescent and sunspot prominences therefore seems to be one of balance between kinetic and magnetic energy.

γ) The condensation mechanism. Observations show that many prominences condense in the corona, initially forming a small luminous cloud that gradually grows, taking more complicated forms, often indicating field patterns. There seems to be a continuous streaming of coronal matter from the condensations through the prominences towards the chromosphere. The observations (a) that the corona is less dense around many prominences and (b) that downward streaming occurs much more often than upward motions, support the condensation hypothesis.

The formation of a prominence is related to the coronal thermal instability. A knot of coronal matter loses energy by free-free and free-bound transitions, and further by line radiation, and conduction. These losses are counteracted by an in-flow of mechanical energy. Equilibrium is apparently reached at $T \approx 10^6$ °K. The rates of the first mechanisms are

free-bound (Lyman)	$5.45 \times 10^{-22}\, T^{-\frac{1}{2}}\, N_e^2$ erg cm^{-3} sec^{-1},
free-free	$1.44 \times 10^{-27}\, T^{\frac{1}{2}}\, N_e^2$ erg cm^{-3} sec^{-1}.

They are equal for $W = 3.8 \times 10^5$ °K. If locally the coronal temperature decreases below that value the emission of radiation increases, producing a further cooling[2], etc. Stability is reached when the temperature becomes so low that a further decrease would reduce the ionization, thus reducing the number of electrons, yielding a decreasing rate of recombinations. As has been shown in Sect. 30, there are in the transparent regions of the solar atmosphere three layers

[1] L. Marshall: Astrophys. Journ. **126**, 177 (1957).
[2] K.O. Kiepenheuer, in: The Sun (ed. G.P. Kuiper), p. 430. Chicago 1953.

of stability, characterized by temperatures of 6000, 20000 and $\approx 10^6$ °K (ATHAY and THOMAS[1]). For a detailed discussion of the coronal radiation equilibrium see KLECZEK[2]. KIEPENHEUER'S mechanism, here described, accounts for the tendency, first observed by WALDMEIER[3], of loop prominences to follow the same pattern as the (monochromatic) coronal loop structures.

The next problem is why prominences do not originate everywhere in the corona. The answer consists of two parts: (a) there must be a mechanism to produce the *initial* cooling of the corona; (b) this initial cooling must be so great, and must extend over such a region, that the amount of radiation is greater than the absorption of heat from the corona, so that cooling can *continue.*

(a) *Cooling* may be produced by a change of the magnetic field (MARSHALL[4]). Consider the coronal electrons spiraling around the magnetic lines of force. The work done on an electron in one cycle is $\Delta W = -eE\,2\pi R = \frac{e}{c}\,\frac{\partial H}{\partial t}\,\pi R^2$. The radius of the electronic orbit is $R = cp/eH$ (p is the momentum of the electron). The fractional increase of electron energy per second is $f_H\,\Delta W/W$, where $f_H = eH/2\pi mc$ is the gyrofrequency, and $W = p^2/2m$. Hence, substituting,

$$f_H\,\frac{\Delta W}{W} = \frac{1}{H}\,\frac{\partial H}{\partial t}\,.$$

Thus, if the magnetic field changes from 4 to 2 Gauss, the kinetic temperature of the electrons changes by 50%.

(b) The *existence of a prominence* depends on its thickness. The emission of radiation should exceed the absorption of coronal energy[5]. A prominence is continuously bombarded from all sides by coronal matter with average velocities $v = \left(\frac{8}{\pi}\,\frac{RT}{\mu}\right)^{\frac{1}{2}}$. For $T = 10^6$ °K and $\mu = 1$, $v \approx 150$ km/sec, giving a gas current of 1.5×10^{15} electrons per $\mathrm{cm^{-2}\,sec^{-1}}$. The free path is $\lambda = (NQ)^{-1}$, where N is $4 \times 10^8\ \mathrm{cm^{-3}}$ and $2 \times 10^{10}\ \mathrm{cm^{-3}}$ for corona and prominence respectively; Q is the collision cross section for a charged particle, computed on the basis of a Coulomb interaction

$$Q = \frac{4\pi}{9}\,Z^2\,e^4\,k^{-2}\,T^{-2},$$

giving for the corona ($T = 10^6$ °K) $\lambda = 6000$ km, and for a prominence ($T = 15\,000°$) $\lambda = 10$ cm. So the free path in the corona is of the order of the thickness of a prominence. The small free path in the prominence produces a cooling of the incident coronal electron gas to prominence temperatures. The *absorbed* heat per particle is $\approx 2\,kT_{\mathrm{cor}}$, and this quantity multiplied by the gas current: 1.5×10^{15} electrons $\mathrm{cm^{-3}\,sec^{-1}}$ yields an absorbed energy $4 \times 10^5\ \mathrm{erg\,cm^{-2}\,sec^{-1}}$.

The prominence *emission* in the Lyman continuum is $4.5 \times 10^{-22}\,T^{-\frac{1}{2}}\,N_e^2\ \mathrm{cm^{-3}}$ $\mathrm{sec^{-1}}$. The emission of the prominence per $\mathrm{cm^2}$ is obtained by multiplying this quantity by H, the thickness, or the effective thickness if the prominence is optically thick.

With $T = 15\,000°$ and $N_e = 10^{10}\ \mathrm{cm^{-3}}$, the emission is greater than the absorption when $H > 10000$ km. This value is of the same order of magnitude as the observed thickness of prominences.

---

[1] R. G. ATHAY and R. N. THOMAS: Astrophys. Journ. **123**, 299 (1956).

[2] J. KLECZEK: Bull. Astr. Inst. Czech. **8**, 120 (1957).

[3] M. WALDMEIER: Astr. Mitt. Zürich **146** (1945).

[4] Cf. footnote 1, p. 245.

[5] Cf. A. UNSÖLD: Physik der Sternatmosphären, 2. Aufl., p. 698. Berlin: Springer 1955.

δ) *The visibility of prominences* depends on whether they emit a sufficient quantity of $H\alpha$ radiation to be seen through a birefringent filter through which most modern prominence observations are made. On the basis of ZIRIN's detailed computations[1] it appears that for a particle density of 10^{10} cm^{-3} the $H\alpha$ emission decreases by a factor 4 if T_{el} increases from 25000 to 50000° and by a factor 10 if it increases from 10^5 to 2×10^5 °K. The visibility also depends on the density: when the material is dense enough or cool enough to be optically thick in $Ly\alpha$, the $H\alpha$ intensity increases sharply. So, one kind of the sudden disappearances, namely that in which the prominence gradually fades into invisibility, might be explained either by an increase of T_{el}, perhaps due to an increase of the magnetic field (cf. subsection β, above), or by a decrease of the thickness. In this latter case the emission becomes smaller than the absorption of energy, eventually also leading to an increase in temperature etc.

In this connection ÖHMAN's[2] two temperature prominence model should be mentioned. In order to explain why certain prominences are not influenced by neighbouring flares (assuming that the disturbance is due to the flare's *radiation*— which seems in disagreement with the observation that a certain activation time is necessary) he assumes that these prominences consist partly of H_2 molecules. The low temperature necessary to form these molecules ($\approx 1700°$) could be produced by adiabatic expansion of hydrogen gas into the corona. Computations by ERICSSON[3] show that at a pressure of 0.06 dynes cm^{-2} the interface between the cooled expanding gas (1700°) and the shock wave in front of the expanding gas (6000°) has a velocity of 40 km/sec. These considerations might find a support in U. BECKER's observations of dark surges, visible in integrated light[4].

ε) *The streaming of prominence knots and the support of prominences.* The observed streaming velocities easily reach values of some tens of km, amounting in surges up to 700 km/sec. The *great* values are difficult to bring into agreement with UNSÖLD's[5] finding that the corona is an extremely viscous medium and show again that the *high* velocities are not produced by gas-dynamical but by electromagnetic forces. The Reynolds number $\mathrm{Re} = v\,d/v_{th}\,\lambda$ is small: if $v =$ 50 km/sec, $d = 5000$ km (diameter of a prominence knot), $v_{thermal} = 150$ km/sec, and the mean free path $\lambda = 6000$ km, one obtains $\mathrm{Re} = 0.3$ so that the streaming is laminar, and its motion should be described by STOKES's law. Equalizing the accelerating force (which is equal to the acceleration of gravity) to the resisting force gives:

$$\frac{4\pi}{3}\,\varrho_{prom}\,r^3\,g = 6\pi\,r\,\nu\,\varrho_{cor}\,v$$

where ν is the kinematic viscosity, $\nu = v_{therm}\,\lambda/3$. With this equation UNSÖLD calculated an upper limit of the descending velocity of 30 km/sec. This might perhaps explain the observed phenomena of matter condensing in the high regions and falling toward the corona with velocities of this order of magnitude. Observations show that even quiescent prominences continuously exhibit internal motions.

The above result was obtained neglecting the influence of a supporting magnetic field. If the magnetic field is stronger than $H \approx 0.2$ Gauss, the matter cannot move but very slowly in a direction tangential to the field. The force influenced

[1] H. ZIRIN: Astrophys. Journ. **124**, 451 (1956).
[2] Y. ÖHMAN: Ark. Astr. **2**, 1 (1957).
[3] U. ERICSSON: Ark. Astr. **2**, 97 (1957).
[4] Cf. footnote 3, p. 214.
[5] A. UNSÖLD: Physik der Sternatmosphären, 2. Aufl., p. 702. Berlin: Springer 1955.

by a magnetic field on matter in which an electrical current flows is

$$\boldsymbol{K} = \frac{1}{c}(\boldsymbol{j} \cdot \boldsymbol{H}) \tag{69.1}$$

and, assuming the displacement current to be zero, hence $\sigma = \infty$:

$$\operatorname{curl} \boldsymbol{H} = \frac{4\pi}{c}\boldsymbol{j}. \tag{69.2}$$

Thus, Eq. (69.1) may also be written as $K = -\frac{1}{4\pi}(\boldsymbol{H} \cdot \operatorname{curl} \boldsymbol{H})$ (cf. KIPPENHAHN and SCHLÜTER[1]). The force exerted by the field is perpendicular to the field and can only compensate gravity when the lines of force are parallel to the Sun's surface. This may perhaps explain the position of some filaments in the centre of a *BM* region, dividing this region into two parts, or at the high latitude side of *UM* or *BM* regions, where the combination of the field and the Sun's poloidal field might locally cause the lines of force to be parallel to the surface. (At the low latitude side this cannot occur.) For a detailed discussion of the stability problem of a prominence in such a field and of the prominence's influence on the field reference is made to the original paper[1].

The assumption $\sigma = \infty$ is certainly not wholly justified. If the prominence also contains neutral particles, matter will move slowly downward through the field, but it can be shown that the velocity of fall is negligibly small, also because, if matter would tend to fall, an electrical field will be induced opposing the fall motion (LENZ's law). This compensation explains the long-lived stationarity of some prominences.

The greater velocities, also occurring in prominences and surges might perhaps be explained as a manifestation of shock waves[2] or hydromagnetic (ALFVÉN) waves. The former move with a velocity slightly greater than the thermal velocity, 150 km/sec. The motion of the front can be made observable since cooling occurs at it, heating behind it. The origin of the front may e.g. be due to a sudden partial collapse when, by a local decrease of temperature, recombination sets in producing a decrease of pressure.

The ALFVÉN waves have a velocity[3] $v_A = H/\sqrt{4\pi\varrho}$. For a particle density of 10^{10} cm^{-3}, $v_A = 50\,H$ km/sec, so that with $H = 1$ Gauss a considerable velocity of these waves is obtained. The waves might arise by the same kind of collapse, which in a field compresses the lines of force and produces a moving disturbance.

The passage of the Alfven waves may produce a fluctuation in the local kinetic temperatures, equal to the variation in the local field-strengths, which thus may explain the occurrence of moving knots.

C. The corona.

I. Optical observations.

70. The corona, survey and problems. The corona, like the chromosphere, has been discovered during total solar eclipses as a pale aureole around the Moon's disk. The surface brightness of the inner corona is about 10^{-6} of that of the Sun's disk, and it decreases rapidly outward with a factor 10^{-3} over a distance

[1] R. KIPPENHAHN and A. SCHLÜTER: Z. Astrophys. **43**, 36 (1957).
[2] J. HOUTGAST, in: Relations entre les phénomènes solaires et géophysiques, p. 21. Lyon 1948.
[3] H. ALFVÉN: Ark. Mat. Astronom. Fys. B **29**, No. 11 (1942).

of one solar diameter. The corona is visible to distances many times the solar diameter, merging gradually into the zodiacal light. The kinetic temperature of the coronal gas is high, between 5×10^5 and 2×10^6 °K. Disturbed regions may attain higher temperatures, up to several millions of degrees.

Three components can be distinguished:

(a) The L corona shows an emission line spectrum, produced by a highly ionized gas (CaXII–CaXV; FeX–FeXV; NiXII–NiXVI; and AX). Such degrees of ionization can be explained only by assuming ionization temperatures between 5×10^5 and 10^6 °K. The explanation of these high temperatures presents a difficult problem.

(b) The K corona: the light is partially polarized; the continuous spectrum consists of photospheric light scattered by free electrons, close to the Sun, with high kinetic temperatures: the photospheric Fraunhofer lines are washed out in this scattering.

(c) The F corona: the light is less polarized, the continuous spectrum presents the absorption lines of the photospheric Fraunhofer spectrum, mainly produced by diffraction of sunlight to the zodiacal dust particles, at great distances to the Sun, which have relatively low velocities.

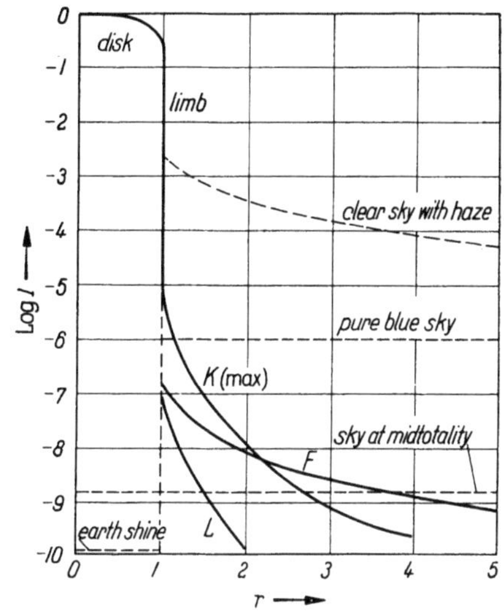

Fig. 73. Schematic representation of the relative surface intensities in the corona and around the Sun. K, F and L denote the K, F and L coronas. H. C. VAN DE HULST: in: The Sun (ed. G. P. KUIPER). Chicago 1953.

The K and F coronas form the "*white corona*". Their light is of nearly the same colour as that of the Sun.

A schematic representation of the intensities and the extension of the various parts of the corona is given in Fig. 73, after VAN DE HULST[1]. The total intensity of the light emitted in the L corona is only 1% or less of that of the total continuous radiation of the corona, but the great importance of the L corona is that it yields a wealth of information on temperature and pressure in the lowest parts of the corona.

In the first part of this chapter we deal mainly with the *quiet minimum* corona; only rarely observed in pure form. The minimum corona shows as typical structures the two systems of polar rays and a nearly uniform atmosphere in the equatorial region.

The *activity* of the corona is exhibited in various forms: the active coronal regions, corresponding with Centres of Activity, the coronal condensations, and the coronal rays of which it is suggested that they correspond to the C regions.

Up to 1930, the total amount of time that the corona had been under observation since its discovery in 1851 was about one hour. So it was very important that after 1930 and 1945 two new techniques made it possible to observe the corona also outside eclipses.

LYOT's *coronagraph* enables one to observe the L corona between limb distances of 0.07 and 0.4 R. The instrument is mainly useful at high altitude stations. The Poulkovo coronagraph[2] permits to photograph the L corona at sea level.

[1] H.C. VAN DE HULST, in: The Sun (ed. G.P. KUIPER), p. 257. Chicago 1953.
[2] I.A. PROKOFYEVA: Isw. Gl. Astr. Obs. Poulkovo **160**, 170 (1957).

Also with LYOT's[1] photoelectric L coronagraph it is observable at sea level.

The K (electron) corona was seen outside eclipses for the first time in 1930 by LYOT at the Pic du Midi, by visual observations. With a new photoelectric K coronagraph[2], in use since 1956 at the High Altitude Observatory (Boulder) and since 1957 at Meudon the streamers of the K corona can be detected up to limb distances of 1.0 to 1.5 $R_{\odot}$.

The coronagraph shows the corona beyond the limb, and information on the structure of the corona before the disk can only be obtained by interpolating between observations of the same region at E and W limb passages, a fortnight apart. Such a procedure may give rise to errors, especially because certain coronal structures have rather short lifetimes.

Since 1945 *radio-observations* have regularly been made at many stations. Because the solar radio-radiation emerges mainly from the corona, this new observing technique is very important for increasing our information on the corona. The observations at various wavelengths refer to emission processes at different height levels, but it is only with great technical difficulties that a sufficient resolution can be obtained to record the disk distribution of the intensity. Daily radiomaps of the Sun at 21 cm were made for the first time in June 1957 by CHRISTIANSEN *et al.*[3] (see Fig. 100).

In the first part of this chapter the optical observations of the corona are discussed; the radio observations, especially important for the understanding of the disturbed parts of the corona will be treated in the second part of this chapter.

General references (apart from the books listed at p. 362):
SIEDENTOPF, H.: Die Sonnenkorona. Ergebn. exakt. Naturw. **23**, 1 (1950).
SHKLOVSKI, I. S.: Solnetshnaia Korona. Moskva 1951.
BUGOSLAVSKAIA, E. J.: Struktura solnetshnoi Korony. Trudy Gos. Astr. Inst. Sternberga **19** (1950).
ALLEN, C.W.: Rep. Progr. Phys. **17**, 135 (1954).
WALDMEIER, M.: Die Sonnenkorona. Zürich: Birkhauser, Bd. I, 1951; Bd. II, 1957.

a) The quiet minimum corona; photometry, ionization and excitation.

71. The continuous spectrum; F and K corona. The designation "white corona" initially given to the F and K corona is only partly right. New observations, up to greater limb distances and in a bigger spectral region than the early ones, have confirmed K. SCHWARZSCHILD's suggestion that the outer corona is reddish with respect to the inner corona and to the photosphere. ALLEN[4] (1940) found that the intensity ratio $I(6500\,\text{Å})/I(4000\,\text{Å})$ increased by a factor 1.5 between $\varrho(r/R) = 1.2$ and 2.4. It is valuable to extend the measurements to the infrared. At the eclipse of 25 February 1952 BLACKWELL[5] found that the ratio $I(19000\,\text{Å})/I(4300\,\text{Å})$ increased by a factor 2.17 between $\varrho = 1.5$ and 2.5. At the eclipse of 30 June 1954 ALLEN[6], from direct photometry, could not find a colour difference (4500 to 6200 Å) between the Sun and the corona up to $\varrho = 10$ to 12, but from spectrographic observations of the same eclipse[7] he found a redden-

[1] B. LYOT: C. R. Acad. Sci., Paris **226**, 25 (1948); **226**, 137 (1948).
[2] G. WLÉRICK and J. AXTELL: Astrophys. Journ. **126**, 253 (1957); cf. also L'Astronomie **71**, 409 (1957).
[3] W.N. CHRISTIANSEN, D.S. MATHEWSON and J.L. PAWSEY: Nature, Lond. **180**, 944 (1957).
[4] C.W. ALLEN: Monthly Notices Roy. Astronom. Soc. London **106**, 137 (1946).
[5] D.E. BLACKWELL: Monthly Notices Roy. Astronom. Soc. London **112**, 652 (1952).
[6] C.W. ALLEN: Monthly Notices Roy. Astronom. Soc. London **116**, 69 (1956).
[7] C.W. ALLEN: Monthly Notices Roy. Astronom. Soc. London **116**, 413 (1956).

ing increasing with distance from the Sun. However, the corona was observed through thin clouds, so that the sky correction was considerable and uncertain. MICHARD[1] confirmed BLACKWELL'S and ALLEN'S (1940) observations at the eclipse of 1955 (Fig. 74).

The "white" colour for small ϱ values agrees with the fact that the inner coronal emission is due to photospheric light, mainly scattered by free electrons (Thomson scattering). The reddening for greater ϱ values shows the increasing contribution of light diffracted by small solid particles, for which diameters of the order 10^{-3} cm are estimated.

The (rare) photo-electric *absolute brightness measurements*[2] show that the total radiation emitted in the blue and violet, in a zone with $1.03<\varrho<6.00$,

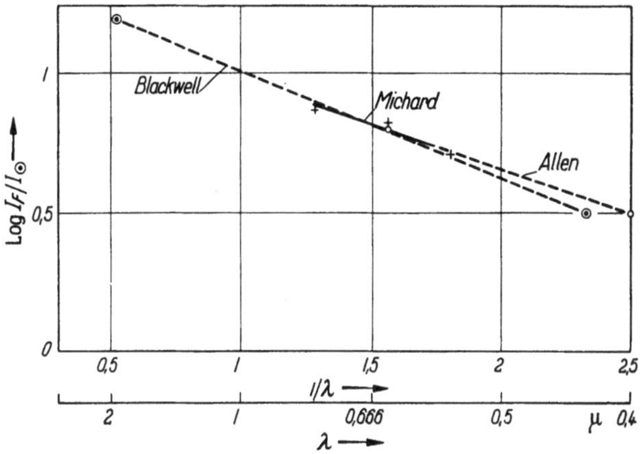

Fig. 74. Spectral distribution of light in the *F* corona compared to the *K* corona or to the Sun, showing the relative reddening of the *F* corona. Only the slopes of the lines are known, the three observational curves are made to coincide at $\lambda = 6400$ Å. The influence of the *K* corona was eliminated. R. MICHARD: Ann. d'Astrophys. **19**, 229 (1956).

ranges from 0.32×10^{-6} to 1.06×10^{-6} times the intensity of the total solar light. There are clear indications for a dependence on the solar phase; the intensity seems to be greatest (1.0×10^{-6}) at sunspot maximum; smallest (0.5×10^{-6}) at minimum phase (VAN DE HULST[3]).

The *relative* distribution of coronal intensities is determined at almost any total eclipse. The photometric calibration of the exposures, in which the intensity ranges over factors up to 10^3 and sometimes more in a short interval, offers great difficulties. Fig. 75a shows a typical minimum corona (1954). The maximum corona (Fig. 75d) is distinguished from the minimum corona in that it has nearly circular isophotes.

Characteristic for the corona as it appears between minimum and maximum (example: 25 February, 1952; Fig. 75c) are the long rays, which may suggest how to explain the maximum corona: when the rays increase in number, also at higher latitudes, and cover the limb almost homogeneously, the corona becomes more or less circular. The coronal rays are transient phenomena which do not occur when the Sun is wholly quiet. It should be stressed that an understanding of the minimum corona is of primordial importance for the explanation of all observed coronal forms and intensities. The maximum corona shows a greater density than the minimum corona and is moreover subjected to the disturbances, like coronal rays, helmets and arches, which mark periods of greater

[1] R. MICHARD: Ann. d'Astrophys. **19**, 229 (1956).
[2] V. B. NIKONOV and E. K. NIKONOVA: Isw. Krymsk. Astrophys. Obs. **1**, 83 (1947).
[3] H. C. VAN DE HULST: Bull. astronom. Inst. Netherl. **11**, 135 (1950).

activity. So, as throughout in this review, we start with a discussion of the *undisturbed minimum corona.*

The best studied and most typical specimen is the one of 30 June 1954. This corona was the first one photographed that was nearly completely free of any

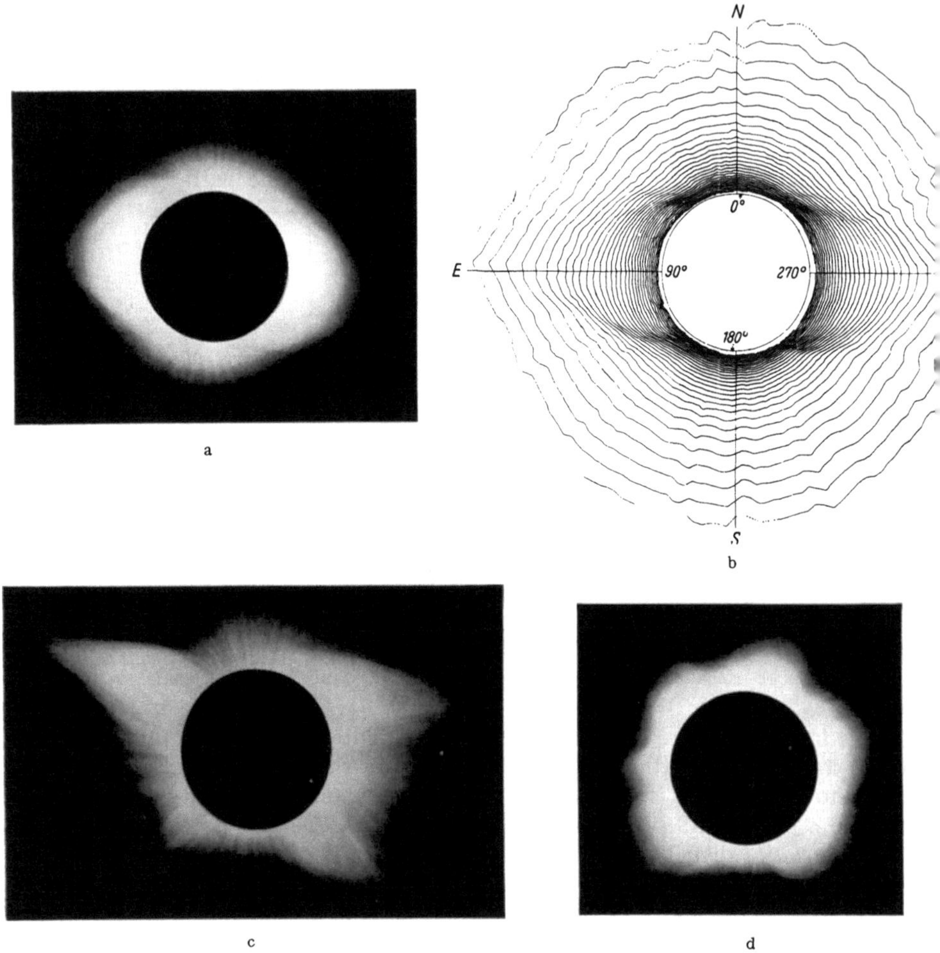

Fig. 75a—d. The corona. a) and (b) The minimum corona (Gotland, 30 June, 1954). (b) gives isophotes. Courtesy Å. WALLENQUIST, Uppsala. (c) An intermediate type (Khartoum, 25 February, 1952). Courtesy G. VAN BIESBROECK, Yerkes Observatory. (d) An example of a maximum corona (Sora, USSR, 19. VI. 1936). Courtesy G. RIGHINI, Arcetri.

disturbed regions. No sunspot was seen for 8 days preceding or 12 days following the eclipse. At that eclipse the brightness of the white corona was measured by ALLEN[1], BLACKWELL[2], WALDMEIER[3] and WALLENQUIST[4]. Fig. 75b gives some isophotes (WALLENQUIST[4]). Remarkable are the beautiful symmetry, the absence of streamers, the variation of the isophotal oblateness with ϱ and the intensity minimum of the inner corona at $b=65°$ in each of the four quadrants. Fig. 76 shows equatorial and polar surface intensity, plotted against ϱ and ob-

[1] C.W. ALLEN: Monthly Notices Roy. Astronom. Soc. London **116**, 69 (1957).
[2] D.E. BLACKWELL: Monthly Notices Roy. Astronom. Soc. London **115**, 629 (1955).
[3] M. WALDMEIER: Z. Astrophys. **36**, 275 (1955); **43**, 289 (1957).
[4] A. WALLENQUIST: Uppsala Ann. **4**, No. 4 (1957).

tained from a compilation of all observations and calibrated in absolute intensity units. The agreement with the curve derived earlier by VAN DE HULST[1] for a characteristic "minimum corona" is good for small ϱ values but at great ϱ values the differences between polar and equatorial intensities are greater at the 1954 eclipse.

The "*ellipticity*" ε is according to the usual definition (LUDENDORFF): $\varepsilon = (A - P)/P$ where A and P are the average equatorial and polar diameters

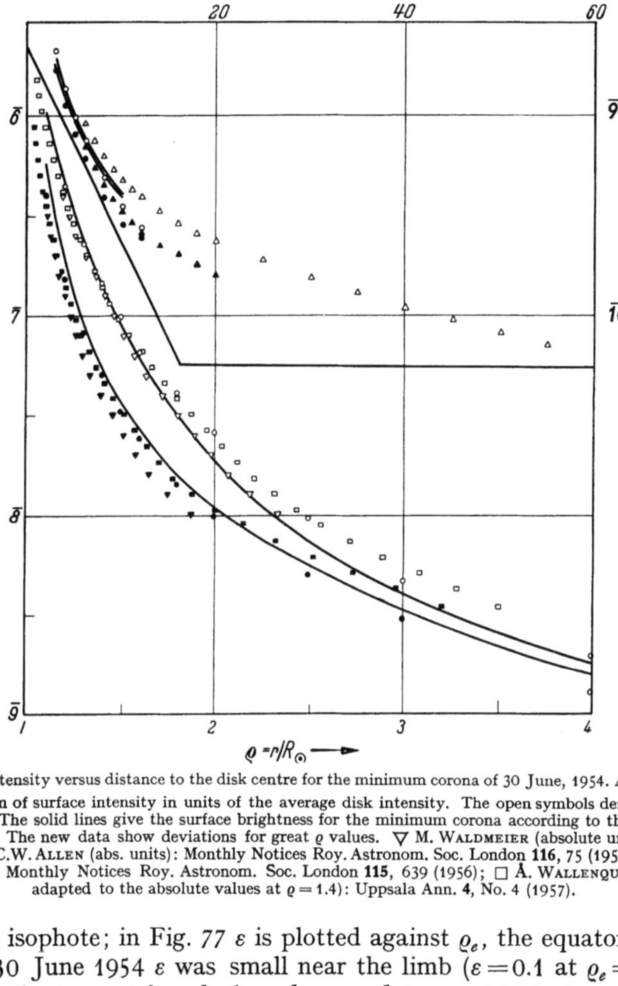

Fig. 76. Surface intensity versus distance to the disk centre for the minimum corona of 30 June, 1954. Abscissa: $\varrho = r/R_\odot$; ordinate: logarithm of surface intensity in units of the average disk intensity. The open symbols denote equatorial, the filled polar data. The solid lines give the surface brightness for the minimum corona according to the often used model of VAN DE HULST. The new data show deviations for great ϱ values. ▽ M. WALDMEIER (absolute units): Z. Astrophys. **43**, 289 (1957); ○ C.W. ALLEN (abs. units): Monthly Notices Roy. Astronom. Soc. London **116**, 75 (1956); △ D.W. BLACKWELL (abs. units): Monthly Notices Roy. Astronom. Soc. London **115**, 639 (1956); □ Å. WALLENQUIST (rel. intensities, adapted to the absolute values at $\varrho = 1.4$): Uppsala Ann. **4**, No. 4 (1957).

for a certain isophote; in Fig. 77 ε is plotted against ϱ_e, the equatorial isophotal radius. On 30 June 1954 ε was small near the limb ($\varepsilon = 0.1$ at $\varrho_e = 1.2$). It increased to 0.18 at $\varrho_e = 1.8$ and then decreased to $\varepsilon = 0.075$ at $\varrho_e \approx 3.5$. It was shown for the first time in 1952 and was confirmed later that for still greater ϱ_e values the ellipticity again increases, showing the influence of the highly flattened zodiacal cloud. The point of minimum ellipticity was at $\varrho_e = 4.2$ in 1952; at $\varrho_e = 3.2$ to 4.4 in 1954 and at 3.8 in 1955. This picture shows how both constituents of the corona intermix, the F corona being the most important at the greatest distances. BLACKWELL[2] and SAITO[3] showed the continuous transition between the coronal ellipticity and that of the zodiacal light (Fig. 77).

[1] H.C. VAN DE HULST: Bull. astronom. Inst. Netherl. **11**, 135 (1950).
[2] Cf. footnote 5, p. 255.
[3] K. SAITO: Publ. Astr. Soc. Japan **8**, 126 (1956).

The next step is to separate the contribution of the K and the F part to the total light. There are two methods. The most direct method is to make use of the *Fraunhofer spectrum:* if the F corona was the only constituent, the Fraunhofer lines would have the same profiles as in the photospheric spectrum. If d is the line depression $I_F/[I_F + I_K(1-\eta)] = d_{\text{obs}}/d_{\text{phot}}$, where η is the fraction of photospheric light absorbed by the Fraunhofer lines; in the blue spectral region $\eta \approx 0.1$ (cf. [1]).

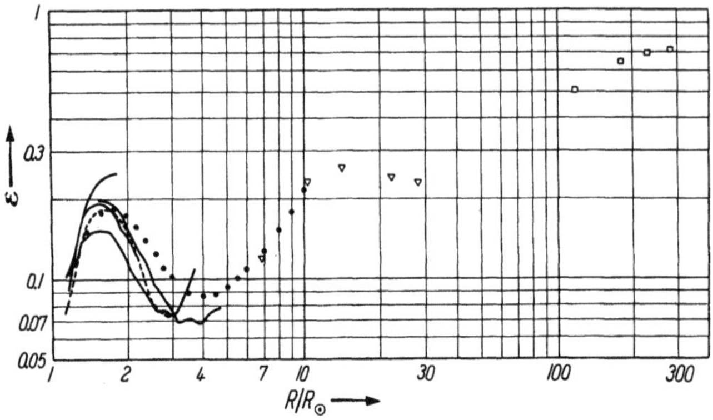

Fig. 77. The ellipticity of the minimum corona (1954) and of the zodiacal light. Solid lines: M. WALDMEIER: Z. Astrophys. **36**, 275 (1955) and M. WALDMEIER and H. BACHMANN: Z. Astrophys. **38**, 125 (1955). Dashed: Å. WALLENQUIST: Uppsala Ann. **4**, No. 4 (1957). Dots: K. SAITO: Publ. Astr. Soc. Japan **8**, 126 (1957). Triangles: D. E. BLACKWELL: Monthly Notices Roy. Astronom. Soc. London **115**, 648 (1956). Squares: A. BEHR and H. SIEDENTOPF: Z. Astrophys. **32**, 19 (1953).

Spectral observations have only been made at the eclipse of 1954 (ALLEN [2]), of 1923 (GROTRIAN [3], very close near a deep minimum with a monthly mean: $R=5.7$) and 1940 (ALLEN [4], at the decreasing branch, $R=62$). GROTRIAN'S

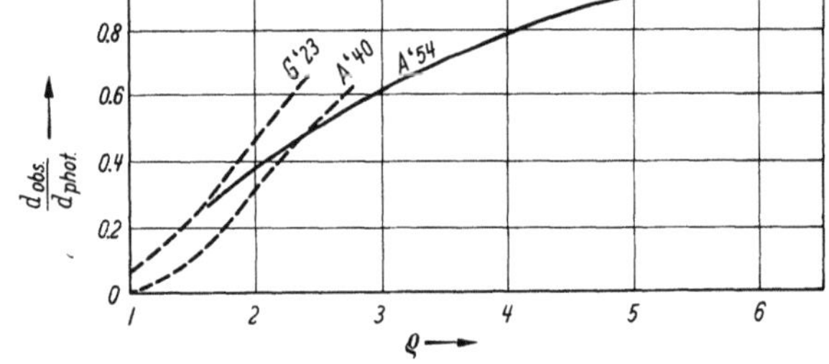

Fig. 78. Ratio of line depths in the coronal and photospheric spectra versus ϱ $(=r/R_{\odot})$. G = GROTRIAN; A = ALLEN'S observations of 1940 and 1954. After C.W. ALLEN: Monthly Notices Roy. Astronom. Soc. London **116**, 413 (1956).

depth ratios are larger than ALLEN'S 1940-values by about 0.10 to 0.15 for all ϱ values, which shows the greater relative intensity of the K corona in 1940 (Fig. 78). ALLEN'S 1954-values extend to much greater ϱ-values than the earlier determinations, but are less accurate because of the low dispersion of the spectrograph and the bad sky.

For *greater ϱ values* a way to disentangle the F and K coronas is to make use of the observed *polarization* of the light, but the method offers greater difficulties because of the uncertain polarization of the dust particles.

[1] M. WALDMEIER: Z. Astrophys. **46**, 17 (1958).
[2] C.W. ALLEN: Monthly Notices Roy. Astronom. Soc. London **116**, 413 (1956).
[3] W. GROTRIAN: Z. Astrophys. **8**, 124 (1934).
[4] C.W. ALLEN: Monthly Notices Roy. Astronom. Soc. London **106**, 137 (1946).

If one assumes, as is often done, that the light of the dust component is unpolarized, the observed polarization must come from the electron corona. Since the amount of polarization of a pure electron corona with a given density distribution can exactly be computed, the observed degree of polarization $p(\varrho)$ may be used to determine the relative importance of the two constituents.

Since a scattering element receives light from all parts of the Sun's disk the degree of polarization of a pure electron corona should be small when close to the limb (because radiation is received from many directions) and approaches to a constant value at greater distances.

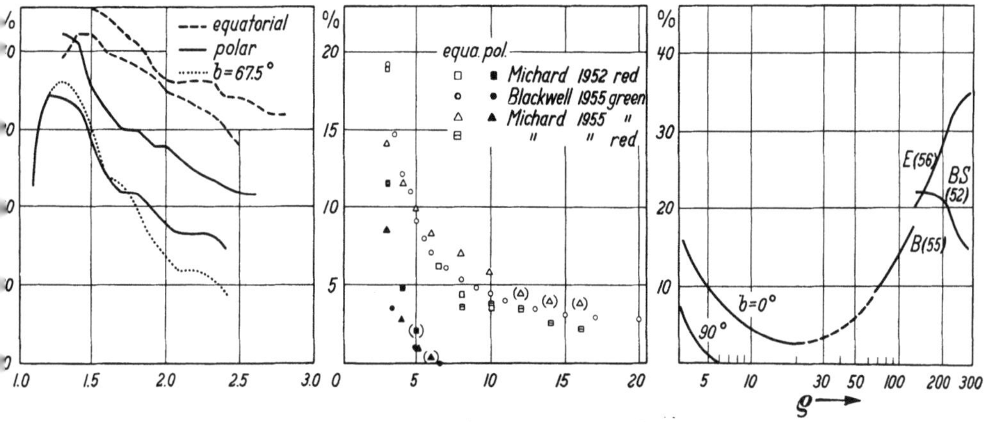

Fig. 79a–c. Polarization of the corona and the zodiacal light. Abscissa: $\varrho = r/R_{\odot}$; ordinate: polarization in percents. (a) Close to the Sun; eclipse of 30 June, 1954. Solid lines: polar data; dashed: equatorial ($b=0°$); dotted: $b=67°.5$. M. WALDMEIER: Z. Astrophys. **40**, 120 (1956). (b) $\varrho = 3$ to 20. Eclipses of: 1952: R. MICHARD, Ann. d'Astrophys. **17**, 320 (1954); 1954: D. E. BLACKWELL, Monthly Notices Roy. Astronom. Soc. London **115**, 629 (1955); 1955: R. MICHARD, Ann. d'Astrophys. **19**, 229 (1956). (c) Greater elongations. Solid line: Average from equatorial eclipse observations shown in Figs. 79a and b. Dashed: Outside eclipse zodiacal light observations taken in the ecliptical plane, 1955. D. E. BLACKWELL: Monthly Notices Roy. Astronom. Soc. London **116**, 365 (1956). Solid: Zodiacal light observations. A. BEHR and H. SIEDENTOPF: Z. Astrophys. **32**, 19 (1953); H. ELSÄSSER: Mitt. astr. Ges. **1957**, II, 61; Sterne **34**, 166 (1958).

As a rule the various observed polarization-curves differ greatly, not only from one eclipse to the other, but there are also great differences between various observations of the same eclipse, obviously because of the difficulty to eliminate scattered sky light; in this procedure both under- or overcorrection may appear: also the far F corona is sometimes identified with the sky background. Fig. 79 shows the observed polarization of the coronas of 25 February, 1952 and 30 June, 1954 (MICHARD *et al.*[1,2], WALDMEIER[3], BLACKWELL[4]) where also the polarization of the zodiacal light is shown (BLACKWELL[5], BEHR and SIEDENTOPF[6], ELSÄSSER[7]). If E_t and E_r are the intensities of the observed radiations for which the electric vector vibrates along the tangent and the radius of the solar image, we define $p = (E_t - E_r)/(E_t + E_r)$. The polarization falls to a minimum at $\varrho \approx 20$, because of the increasing importance of the F corona, which produces only weak polarization. There is a steep increase in polarization at greater distances. The question is still open, whether this peculiar character of the p-curve is caused by an increased polarization of dust particles at great position angles or by an increase

[1] R. MICHARD, A. DOLLFUS, J.C. PECKER, M. LAFFINEUR and M. D'AZAMBUJA: Ann. d'Astrophys. **17**, 320 (1954).
[2] R. MICHARD: Ann. d'Astrophys. **19**, 229 (1956).
[3] M. WALDMEIER: Z. Astrophys. **40**, 120 (1956).
[4] D.E. BLACKWELL: Monthly Notices Roy. Astronom. Soc. London **115**, 629 (1955).
[5] D.E. BLACKWELL: Monthly Notices Roy. Astronom. Soc. London **116**, 365 (1956).
[6] A. BEHR and H. SIEDENTOPF: Z. Astrophys. **32**, 19 (1953).
[7] H. ELSÄSSER: Mitt. astr. Ges. **1957**, II, 61; Sterne **34**, 166 (1958).

of the relative importance of electrons over dust particles at greater distances to the Sun (assuming that dust causes no polarization). Some authors[1] favour the second alternative and find an N_e curve falling to $N_e \approx 5 \cdot 10^2$ cm^{-3} at 1 astronomical unit (ELSÄSSER[2]), others[3,4] remark that also dust particles can produce a measurable polarization, which, however, is difficult to compute theoretically, but which might under certain circumstances[4] lead to negligibly small electron densities in interplanetary space. THIESSEN[5] showed that in the Sun's vicinity dust particles will be directed by the Sun's general magnetic field thus causing an additional polarization.

The absence of Fraunhofer lines in the spectrum of the K corona shows that the Doppler motions of the electrons must be high, so that in the scattered light the Fraunhofer lines are wholly washed out. Once, a trace of the influence of the lines has been found: on a low dispersion spectrogram of the photosphere the accumulation of Fraunhofer lines produces a wide and shallow depression near 3800 Å. GROTRIAN[6] could recognize this depression in the spectrum of the K corona but widened to about 60 Å, which would correspond with an electron temperature $> 10^5$ °K. That the depression was not found again by other observers, shows how difficult it is to find a shallow and very wide depression in a low dispersion spectrum of which the spectral intensity varies rather irregularly.

The theoretical calculation of the line profile obtained after scattering by a high temperature electron gas is straight-forward but elaborate from the mathematical point of view. The theory was given by VAN HOUTEN[7] for a strong line with 100% depression and $W = 10$ Å, for $\varrho = 1.03$, $\lambda = 4000$ Å, $T_{\text{el}} = 10^6$ °K and for the electron densities given by VAN DE HULST's model. The scattered profile appears to have a central depression of 6% and a half-width of 120 Å, hence it is only observable in very accurate photometry.

72. The *L*-corona; excitation and ionization. For an understanding of the coronal emission line spectrum and the X-ray spectrum it is necessary first to discuss the excitation and ionization conditions. Therefore the question is: what are the physical conditions in a gas with a temperature near 10^6 °K, and with electron densities between 10^9 and 10^6 electrons per cm^3? What we principally want to know is the degree of ionization. Computations have been made by various authors among which may be mentioned MIYAMOTO (1943), BIERMANN (1947), WOOLLEY and ALLEN (1948, 1950), SHKLOVSKI[8] and ELWERT[9]. We mainly follow the latter's discussion.

Let there be N_r and N_{r+1} ions in the r-th and $(r+1)$st degree of ionization (atoms having lost r and $r+1$ electrons). Then the numbers of possible transitions per cm^3 are:

$$\text{photo-ionizations:} \quad N_r Q_r^{r+1}, \tag{72.1}$$

$$\text{photo-recombinations:} \quad N_{r+1} Q_{r+1}^r, \tag{72.2}$$

$$\text{collisional ionizations:} \quad N_r N_e S_r^{r+1}, \tag{72.3}$$

$$\text{three-body recombinations:} \quad N_{r+1} N_e^2 S_{r+1}^r. \tag{72.4}$$

[1] F.L. WHIPPLE and J.L. GOSSNER: Astrophys. Journ. **109**, 308 (1949).

[2] Cf. footnote 7, p. 255.

[3] H.C. VAN DE HULST, in: Les particules solides dans les astres. Liège Colloqiuum 1954, p. 89, 1955.

[4] D.E. BLACKWELL: Observatory **77**, 187 (1957).

[5] G. THIESSEN: Z. Astrophys. **32**, 173 (1953).

[6] W. GROTRIAN: Z. Astrophys. **3**, 199 (1931).

[7] C.J. VAN HOUTEN: Bull. astronom. Inst. Netherl. **11**, 160 (1950).

[8] I.S. SHKLOVSKI: Isw. Krymsk. Astrophys. Obs. **6**, 105 (1950).

[9] G. ELWERT: Z. Naturforsch. **7a**, 202, 432 (1952); **9a**, 648 (1954); with references to earlier papers.

The stationary state requires (72.1) + (72.3) = (72.2) + (72.4). However, it can be shown, that in normal coronal conditions (72.4) $\ll$ (72.2): the probability of a three-body collision is evidently very small, and that the number of collisional ionizations (72.3) exceeds the number of photo-ionizations (72.1) (but only by about two orders of magnitude). So in the quiet corona:

$$N_{r+1}/N_r = S_r^{r+1}/Q_{r+1}^r. \tag{72.5}$$

It is of importance that the ionization depends only on T and not on the density, since the functions S and Q do not depend on the density.

Only when the total solar radiation increases more than a hundred times the "normal" value—as seems the case for $\lambda < 15$ Å, when a coronal condensation with $T \approx 6 \times 10^6$ °K occurs in the corona—, and when $N_e > 10^8$, is photo-ionization no longer negligible. Application of a generalized ionization formula shows that in such a case ions with ionization energies as high as 800 ev (e.g.: Ca XV, $\chi_i = 814$ eV) can still be formed[1], without appreciably changing the Fe ionization.

The next step is to compute the values of the probability functions S and Q.

(a) The probability Q_{r+1}^r for *photo-recombination* is derived from the photo-ionization probability Q_r^{r+1} using the assumption of thermodynamic equilibrium:

$$N_r Q_r^{r+1} = N_{r+1} Q_{r+1}^r.$$

Since the ratio N_r/N_{r+1} is known for thermodynamic equilibrium, the function Q_{r+1}^r is easily derived (KRAMER's formula). In computing Q_r^{r+1} it is assumed that the atoms are hydrogen-like, and the Gaunt factor is unity. Further, the radiation field is approximated by WIEN's law.

The result is subject to two restricting conditions, viz. the validity of MAXWELL's law of velocity distributions, and, much more significant, the validity of the atomic approximations (hydrogen-like ions) used. The resulting expression is (cf. also UNSÖLD, Physik der Sternatmosphären, 2nd ed., p. 653f.):

$$Q_{r+1}^r = \frac{64\sqrt{\pi}}{3\sqrt{3}}\, c\, r_0^2 \left(\frac{\chi_H}{kT}\right)^{\frac{1}{2}} \frac{Z^2}{n} \frac{\gamma}{1 + u_n^{-1}} \tag{72.6}$$

where the symbols have their usual meaning; r_0 is the classical radius of the electron (2.82×10^{-13} cm); $\chi_H = 13.595$ eV; $n = n_0$ for the ground level of the ion; $u = h\nu/kT$; Z is the charge of the ion and γ is the ratio of all recombinations to the number of recombinations to the ground level, $\gamma \approx 3$ to 5.

(b) The probability for *collisional ionization* is computed on the basis of an expression for the cross section for ionization by electron collisions, derived by THOMSON (1912):

$$\pi d^2 = \frac{\pi e^4}{E}\left(\frac{1}{\chi_r} - \frac{1}{E}\right)$$

with d = collision cross section,

χ_r = ionization energy of the ion,

E = kinetic energy of the colliding electron.

Defining $u = E/\chi_r$ and writing $e^2 = 2a_0\chi_H$; a_0 is the first Bohr radius:

$$\pi d^2 = 4\pi a_0^2 (\chi_H/\chi_r)^2 \frac{u-1}{u^2} K \tag{72.7}$$

where

$$K = \tfrac{1}{2} n^* [1 + 0.3(u-1)]$$

[1] G. ELWERT: Z. Astrophys. **44**, 112 (1958).

is a semiempirical correction term, valid for $1<u<2$, derived by comparing the computed πd^2 with experiments and theoretical calculations[1].

n^*, the effective quantum number, is the number of electrons in the outer shell.

The total number of collisional ionizations is obtained after integrating the electron velocities over the Maxwellian velocity distribution. Finally, after some further calculations:

$$S_r^{r+1} = e^4 \sqrt{\frac{2\pi kT}{m}}\, \frac{n_r^*}{\chi_r^2}\, \mathrm{e}^{-u_r} \left(1 + \frac{0.3}{u_r}\right) \tag{72.8}$$

with

$$u = \chi/kT; \qquad \chi/\chi_H = Z^2/n^{*2}.$$

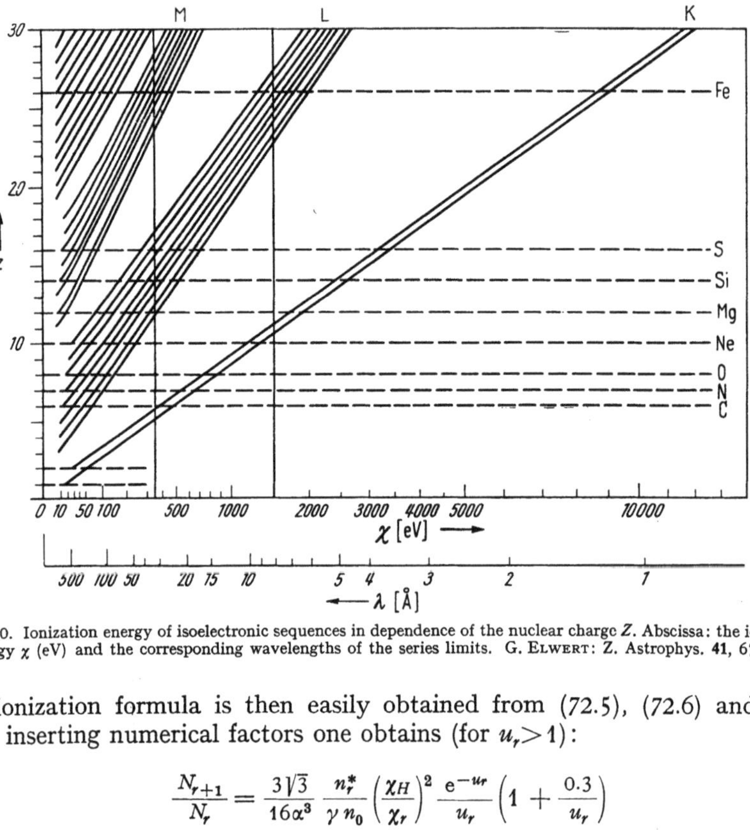

Fig. 80. Ionization energy of isoelectronic sequences in dependence of the nuclear charge Z. Abscissa: the ionization energy χ (eV) and the corresponding wavelengths of the series limits. G. ELWERT: Z. Astrophys. **41**, 67 (1956).

The ionization formula is then easily obtained from (72.5), (72.6) and (72.8). After inserting numerical factors one obtains (for $u_r>1$):

$$\frac{N_{r+1}}{N_r} = \frac{3\sqrt{3}}{16\alpha^3}\, \frac{n_r^*}{\gamma\, n_0} \left(\frac{\chi_H}{\chi_r}\right)^2 \frac{\mathrm{e}^{-u_r}}{u_r} \left(1 + \frac{0.3}{u_r}\right) \tag{72.9}$$

with $\alpha = \frac{2\pi e^2}{hc} = \frac{1}{137}$; the other symbols have been introduced before. If a high degree of precision is not needed, the expression (72.9) may be approximated by[2]:

$$\frac{N_{r+1}}{N_r} = 5\times 10^5 \left(\frac{\chi_H}{\chi_r}\right)^2 \frac{\mathrm{e}^{-u_r}}{u_r}.$$

A survey over the ionization conditions is obtained with the aid of Fig. 80 showing the ionization energies of iso-electronic series. Fig. 81 gives the state of ionization of Fe for some temperatures after ELWERT (solid) and after the earlier

[1] Cf. E. R. HILL: Austral. J. Sci. Res. **4**, 437 (1951). — A. WERNER: Austral. J. Phys. **7**, 25 (1954).

[2] G. ELWERT: Z. Astrophys. **41**, 67 (1956).

computations of SHKLOVSKI. The agreement is good for the high temperatures, less so for the lower temperatures. The theory obviously explains the observation that one kind of atom can occur simultaneously in various stages of ionization.

73. The emission line spectrum. Some thirty emission lines have been found in the spectrum of the innermost part of the corona in the wavelength interval between 3328 and 10798 Å. With the coronagraph many of these lines, even up to the far infrared region, can be observed outside eclipses. A summary of the lines and some of their most important properties is given in the Table 16. All identified lines are transitions to the ground level except the line 7059.62 Å.

The lines are divided into four classes[1] according to their response on solar activity (lines of type IV occur in the most active regions[2]); these classes differ mainly in their ionization potential. The equivalent width W is expressed in the intensity of the continuous spectrum adjacent to the line (unit 1 Å). This definition has the drawback that the continuous spectrum decreases greatly with limb distance, so that a constant $W(\varrho)$ signifies a decreased emission intensity.

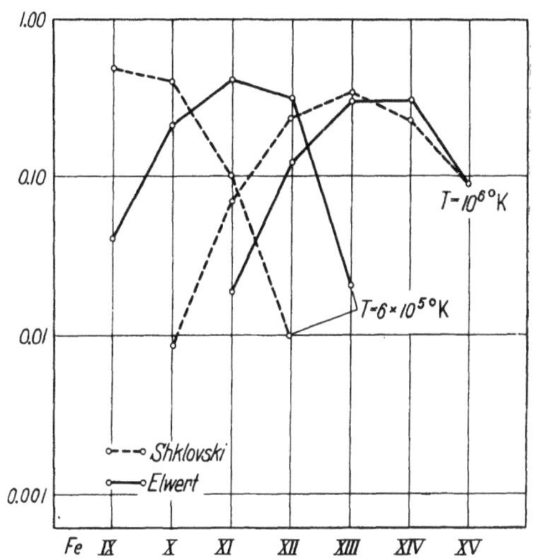

Fig. 81. Ionization in the corona according to SHKLOVSKI and ELWERT. The ordinate gives the fraction of iron ions in the various states of ionization.

The identification of the lines as transitions in highly ionized atoms has been started by GROTRIAN[3] and principally performed by EDLÉN[4,5]. The lines are due to magnetic dipole transitions: forbidden transitions between two even or two odd terms (selection rule: $J=0, \pm 1$ with $0 \rightarrow 0$ forbidden). The occurrence of these lines in the coronal spectrum shows that the lifetimes of the upper states are long and that the probabilities are small for a collisional downward transition or for an (upward) transition under absorption of radiation. The next problem is whether the upper levels of the transitions are excited by absorption of radiation or by collisions. The answer varies for the various transitions, depending among others on the collisional cross sections, on the absorption coefficients for radiation, and on the radiation flux, but generally the following rules may be given:

(a) In the case of excitation of the upper level by *absorption of radiation*, the number of emissions is equal to the number of absorptions, since there is no other way to leave the upper state, as remarked above. In turn, the number of absorptions is proportional to $N_{0,s}$, the number of ions in the ground-level of the ion, if the lower level is the ground-level as it is the case for many coronal lines.

The equivalent width, as defined in the usual way, is the ratio of this emission to the emission in the continuous spectrum of the K corona (the emission lines all occur in the inner corona, where the K part is dominant). The continuum

[1] B. LYOT and A. DOLLFUS: C. R. Acad. Sci., Paris **237**, 855 (1953); cf. also A. DOLLFUS: C. R. Acad. Sci., Paris **245**, 2011 (1957).
[2] B. EDLÉN: Monthly Notices Roy. Astronom. Soc. London **114**, 700 (1954).
[3] W. GROTRIAN: Naturwiss. **27**, 214 (1939).
[4] B. EDLÉN: Ark. Mat. Astronom. Fys. **28** B, 1 (1941). — Z. Astrophys. **22**, 30 (1942).
[5] Some new identifications: CH. PECKER: C. R. Acad. Sci. Paris, Dec. 1958.

Table 16.

λ (Å)	Class	Ion	Multiplet	Ionization- and excitation-potential (eV) χ_r	upper term $\chi_{r,s}$	Transition probability A (sec⁻¹)	Average equivalent width (Å)
3170		Cr XI	$^3P_1-{}^1D_2$				
3329.6	III	Ca XII	$^2P_{\frac{3}{2}}-{}^2P_{\frac{1}{2}}$	589	3.72	488	0.7
3388.0	II	Fe XIII	$^3P_2-{}^1D_2$	325	5.96	87	10
3454.3	I						1.4
3534.0	I	V X	$^3P_1-{}^1D_2$				
3600.9	III	Ni XVI	$^2P_{\frac{1}{2}}-{}^2P_{\frac{3}{2}}$	455	3.44	193	1.1
3642.8	II	Ni XIII	$^3P_1-{}^1D_2$	350	5.82	18	0.7
3800.7	I						0.4
3885		Mn XII					
3987.3	I	Fe XI	$^3P_1-{}^1D_2$	261	4.68	9.5	0.4
3998		Cr XI	$^3P_2-{}^1D_2$				
4086.5	III	Ca XIII	$^3P_2-{}^3P_1$	655	3.03	319	0.3
4232.2	II	Ni XII	$^2P_{\frac{3}{2}}-{}^2P_{\frac{1}{2}}$	318	2.93	237	0.7
4256.4		K XI					
4311							
4351.0	III	Co XV					
4358.9	III						
4412.4	IV	A XIV	$^2P_{\frac{1}{2}}-{}^2P_{\frac{3}{2}}$	682	2.84	108	
4566.6							0.3
4586							
5116.0	II	Ni XIII	$^3P_2-{}^3P_1$	350	2.42	157	0.8
5302.9	II	Fe XIV	$^2P_{\frac{1}{2}}-{}^2P_{\frac{3}{2}}$	355	2.34	60	20
5444.5	IV	Ca XV	$^3P_2-{}^3P_1$	814	4.46		0.15
5536		A X	$^2P_{\frac{3}{2}}-{}^2P_{\frac{1}{2}}$	421	2.24	126	0.4
5694.5	IV	Ca XV	$^3P_0-{}^3P_1$	814	2.18	95	0.3
6374.5	I	Fe X	$^2P_{\frac{3}{2}}-{}^2P_{\frac{1}{2}}$	233	1.94	69	5.0
6535.4		Mn XIII					
6701.9	III	Ni XV	$^3P_0-{}^3P_1$	422	1.85	57	1.2
7059.6	III	Fe XV	$^3P_1-{}^3P_2$	390	31.7		0.8
7891.9	I	Fe XI	$^3P_2-{}^3P_1$	261	1.57	44	6.0
8024.2	III	Ni XV	$^3P_1-{}^3P_2$	422	3.39	22	0.3
10746.8	II	Fe XIII	$^3P_0-{}^3P_1$	325	1.15	14	48
10797.9	II	Fe XIII	$^3P_1-{}^3P_2$	325	2.30	9.7	30

intensity is proportional to N_e; hence $W \propto N_{0,s}/N_e$. According to the ionization formula (72.5) this ratio depends only on the temperature, not on the pressure. Since T is virtually constant in the corona, W should be *independent of height* if the upper level is *excited by radiation*.

(b) Suppose next that the upper level is excited by *electron collisions*. The number of downward transitions is equal to the number of exciting collisions, hence equal to $N_e N_{0,s}$. Since the ionization is constant (it depends only on T, which does not vary) $N_{0,s} \propto N_e$, and hence the number of emissions is proportional to N_e^2 and $W \propto N_e$. In this case, W should decrease with increasing height.

Observations of the λ 5303 line by GROTRIAN[1] shows that W decreases to $\varrho \approx 2$ and remains more or less constant for greater distances. This indicates

[1] W. GROTRIAN: Z. Astrophys. 7, 26 (1933).

that the first mechanism prevails in the outer parts of the corona, where the density is too low to produce an appreciable number of collisional excitations. The second mechanism, collisional excitation, is more important in the inner corona.

The same result is obtained after more quantitative computations which are compared with outside eclipse observations of the strength of the λ 5303 line (WALDMEIER[1]). So the coronagraph observations which refer to the innermost part of the corona should be treated with the second mechanism[2]. For this case WALDMEIER[3], following EDLÉN, derived quantitative relations for the energy emitted in the red and the green lines (cm^{-3} sec^{-1} solid $angle^{-1}$):

$$\left.\begin{aligned} E_g &= 15.2\times 10^{-6} N_e^2\, T^{-\frac{1}{2}}\, 10^{-1.19\times 10^4\, T^{-1}}\ \mathrm{Fe\,XIV/Fe}\,, \\ E_r &= 6.3\times 10^{-6} N_e^2\, T^{-\frac{1}{2}}\, 10^{-0.988\times 10^4\, T^{-1}}\ \mathrm{Fe\,X/Fe}\,. \end{aligned}\right\} \tag{73.1}$$

Integration over the line of sight gives the observed emission; self-absorption is unimportant. It is the importance of these two relations that their simultaneous solution yields values for N_e and T for any part of the corona, as soon as the two $E(h)$ functions can be found from the observed $E(\varrho)$ values. Useful tables and graphs for the practical solution of these equations were published by WALDMEIER[4]. The tables are based on ionization computations of PETRI[5]; the use of ELWERT's tables would change the results.

74. The X ray spectrum of the Sun. *Observations* of the far ultraviolet and X ray spectrum of the Sun have been made by means of the rocket technique. The most detailed *calculations* are made by ELWERT[6], they are in reasonable agreement with the observations.

α) Calculations. For $\lambda < 912$ Å a part of the *continuous radiation* is produced by bound-free transitions of hydrogen and for $\lambda < 227$ Å and 504 Å also by recombinations in the helium continuum. For wavelengths smaller than 100 Å the continuous radiation is mainly emitted in free-free and free-bound transitions of the electrons in the fields of the highly ionized metals of the corona. In both cases the radiation is proportional to $N_e \cdot N_i$, i.e. $\propto N_e^2$ (because the ionization depends only on T, which is constant). The computations by ELWERT were made according to standard methods, assuming the ions to be hydrogen-like and introducing proper correction factors ("Gaunt factors") if necessary. The corona was assumed to be isothermal. The electron density $N_e(h)$ as deduced by BAUMBACH was adopted. In fact it suffices to compute $\int N_e^2\, dh$ and to multiply the computed emission rates per electron by this value. The results are expressed as I/Q' (erg cm^{-2} sec^{-1} $Å^{-1}$) at the Earth's distance, where Q' is a factor, tentatively introduced to take account of the uncertainty of the cross sections and of the density fluctuation in the corona; $1 < Q' < 3$ (cf. also Sect. 85).

The contribution of the *line emission* is more important, but the computation is a formidable task. First the wavelengths of the possible coronal lines have to be determined from a critical investigation of the energy levels of the strongly ionized ions of the corona. Next it is necessary to determine the mechanism of line emission. Like the lines in the visible wavelength region the upper levels

[1] M. WALDMEIER: Z. Astrophys. **38**, 143 (1955); especially his Fig. 1. Cf. also A. UNSÖLD: Physik der Sternatmosphären, 2. Aufl., S. 660, Table 94.

[2] A later investigation by WALDMEIER (footnote 3, p. 267) showed that for the *red* line radiative excitation is not wholly negligible.

[3] M. WALDMEIER: Z. Astrophys. **30**, 137 (1952).

[4] M. WALDMEIER: Z. Astrophys. **35**, 95 (1954).

[5] W. PETRI: Forschungsber. Obs. Wendelstein **2** (1952).

[6] G. ELWERT: Z. Naturforsch. **9**a, 637 (1954). — Z. Astrophys. **41**, 67 (1956).

appear to be excited by collisional excitation (by electrons), rather than by recombination of ions with electrons. The energy differences between the various levels are, as a rule, so great that only the first level above the ground state can be excited.

With the aid of estimated values for the oscillator strengths of the lines, with an adopted coronal model, and taking self-absorption into account, the emitted radiation is given for five temperatures between 5×10^5 °K and 6×10^6 °K (temperature of a coronal condensation[1]). Although the strength of each line varies with T the total selective emission of the corona remains constant between certain limits. It is of the order of $2.5\times10^3\,Q'$ erg cm^{-2} sec^{-1}, corresponding at the

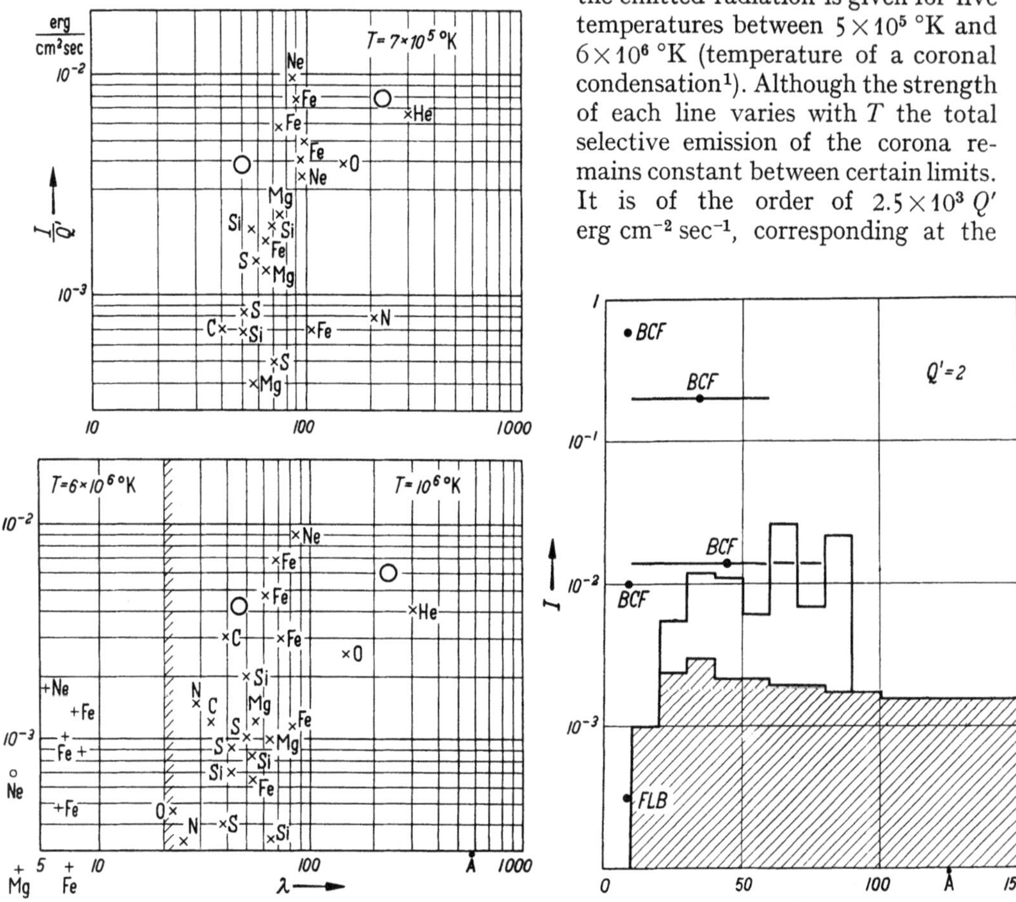

Fig. 82a and b. (a) Spectrum of the corona for $T = 7\times10^5$ and 10^6 °K and for a coronal condensation of 6×10^6 °K (left part of lower figure). The lines are shown by crosses, the series limit continua by circles. The ordinate gives the intensity at the Earth's distance divided by an uncertainty factor Q'. G. Elwert: Z. Naturforsch. **9**a, 637 (1954) and in Solar Eclipses and the Ionosphere, p. 167, Pergamon Press 1956. (b) Comparison of computed intensities with various observations showing the great intensity fluctuations in the short wavelength region. Shaded: continuous emission; open: line emissions. C. de Jager: Ann. Géophys. **11**, 330 (1955).

Earth's distance to an incident radiation of $6\times10^{-2}\,Q'$ erg cm^{-2} sec^{-1}. The wavelength of maximum emission shifts from 80 to 90 Å for 6×10^5 °K to about 60 Å for 10^6 °K (Fig. 82a).

A comparison of the computed continuous and selective emission of the corona is given in Fig. 82b in which the selective emission in intervals of 10 or 20 Å has been summed up. The same has been done with the continuous emission for $\lambda < 100$ Å; for greater wavelengths the steps in the wavelength intervals

[1] Cf. footnote 6, p. 261.

have been made larger. The figure shows the greater importance of the line emissions over the continuous emission, which is caused by the fact that in the high-energy tail of the Maxwell curve there are more electrons with excitation energies than electrons which are able to produce recombinations, while the continuous and selective f-values are more or less of the same order.

The X ray radiation of the *quiet corona* is mainly emitted at wavelengths above 20 Å. Radiation of shorter wavelengths is emitted by *highly excited regions* like coronal condensations. ELWERT[1] showed that very *hot regions* ($\approx 10^7$ °K) may emit radiation down to $\lambda \approx 1.5$ Å.

β) Comparison with the observations. The observations[2–5] show a considerable non-instrumental scatter at the short wavelength end (Fig. 82b); the main cause is the variability of the solar radiation, especially in the region of hard X-rays. It seems reasonable to assume that the lowest observational points correspond to the quiet corona, whereas the higher points reflect the coronal emission at more active moments. The measured lowest intensities above 20 Å agree with ELWERT's computations for $T_{el} = 750000°$ K, a value about equal to the colour temperature of the radiation measured[4] between 8 and 100 Å: $T_c \approx 7 \times 10^5$ °K.

The deviating points in the short wavelength region show either that some mechanism emitting X-rays between 8 and 15 Å at normal coronal temperatures has been neglected, or that the observations have not been made when the Sun was completely quiet, or that the coronal layers producing the radiation between 8 and 15 Å do continuously contain many small high-temperature condensations. Arguments in favour of the second suggestion were given by FRIEDMAN[3]: there is some correlation between the intensities measured between 8 and 20 Å and the central disk intensities of coronal lines extrapolated to the same days. The increase of the X-ray radiation appears to be accompanied by an increase in the population of the ions with high ionization potentials.

Observations by FRIEDMAN *et al.*[6] in the course of summer and autumn of 1957 showed that the highly excited Sun can emit X-ray radiation down to wavelengths, as short as 1.5 Å. This radiation must be emitted by very hot coronal regions: The radiation between 6 and 7 Å is emitted at temperatures of about 5×10^6 °K by the relatively rare elements Si and S. A further increase of the temperature would yield radiation by recombination and by discrete transitions of K electrons of elements following to Si and S in the periodical system. These, however, have a very low abundance and only by increasing T to above 10^7 °K, ions of FeXXV and FeXXVI are formed, which emit radiation between 1 and 2 Å (see Fig. 80). ELWERT suggested that this radiation is emitted during or after flares (footnote 6, p. 261).

The—observational and computational—results of this section and of Sect. 19α are summarized in Fig. 83 giving the ultraviolet and X-ray spectrum of the Sun. The curve is based partly on the observations (solid lines); the interpolation between them is based on theoretical calculations. The short wavelength end of the spectrum (<50 Å) is variable (shaded).

---

[1] Cf. footnote 6, p. 261.
[2] H. FRIEDMAN, S.W. LICHTMAN and E.T. BYRAM: Phys. Rev. **83**, 1025 (1951).
[3] E.T. BYRAM, T. CHUBB and H. FRIEDMAN: Phys. Rev. **92**, 1066 (1954).
[4] H. FRIEDMAN: Ann. Géophys. **11**, 174 (1955).
[5] E.T. BYRAM, T. CHUBB and H. FRIEDMAN: J. Geophys. Res. **61**, 251 (1956).
[6] Cf. footnote 1, p. 209.

The centre-limb variation of the X-ray radiation was computed by SHKLOVSKI[1] and later more detailed, by ELWERT[2]. The radiation shows two crescent-like brightenings in the equatorial region (Fig. 84). The greatest part of the radiation

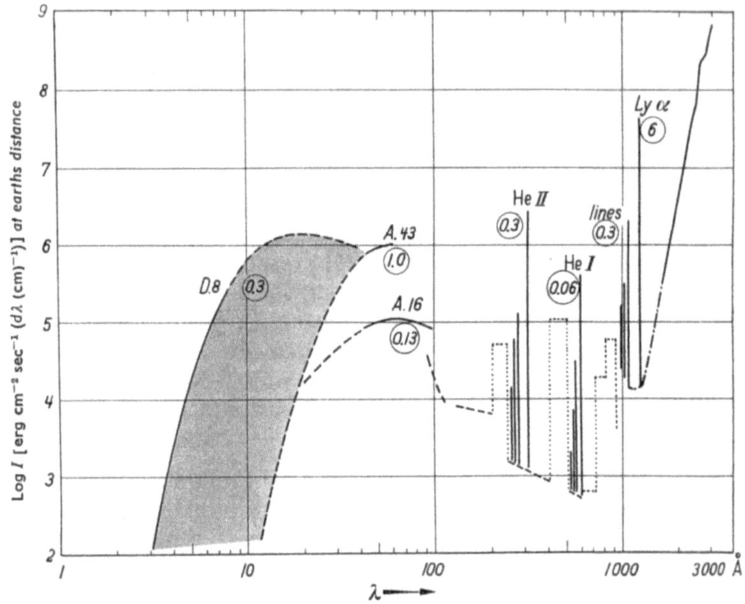

Fig. 83. Ultraviolet and X-ray spectrum of the Sun. Solid lines: observations. Dashed and dotted: computed or interpolated: — · — · — photosphere; .... chromosphere, summed up in intervals of 100 Å (OSTER: Z Astrophys. **40**, 28 (1956); ---- corona [ELWERT: Z. Naturforsch. **7a**, 432 (1952), and FRIEDMAN, 9ième Rapport Comm. Relations Solaires Terrestres, p. 41, 1957]. The total flux emitted in the various groups of lines or under the solid curves, in the X-ray region is indicated (erg · cm⁻² · sec⁻¹). The fluxes emitted in the helium lines are computed, the others are observed. The emission of the active corona is shaded.

appears to be emitted by the limb regions; the computed residual intensity during a total eclipse is about equal to the value deduced from *E*-layer observations (see Chap. E).

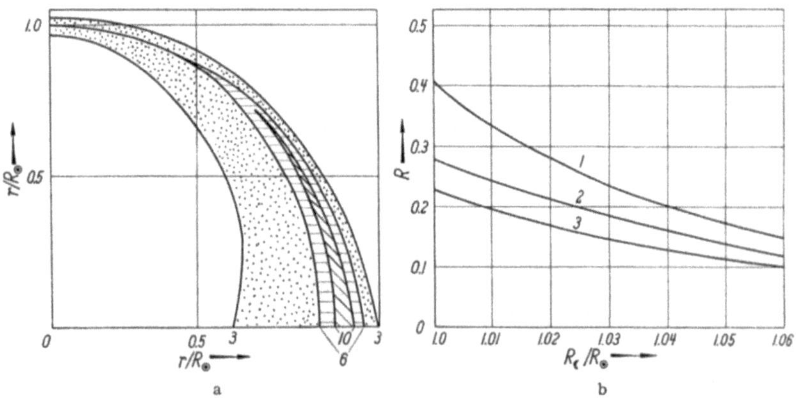

Fig. 84a and b. (a) Intensity distribution of the X-ray radiation over the Sun's disk. It has been assumed that the optical depth in the line centres in the disk centre ($\tau_{0,0}$) is equal to unity. (b) Residual intensity R computed for total solar eclipses, as a function of the projected lunar radius ($R_☾/R_\odot$). 1. for $\tau_{0,0} = 0.2$; 2. for $\tau_{0,0} = 0.5$; 3. for $\tau_{0,0} = 1.0$. G. ELWERT: J. Atmos. Terr. Phys. **12**, 187 (1958).

[1] I. S. SHKLOVSKI: Isw. Krymsk. Astrophys. Obs. **4**, 80 (1949).
[2] G. ELWERT: J. Atmos. Terr. Phys. **12**, 187 (1958).

b) The structure of the quiet minimum corona; temperatures and densities.

75. Variation of density with height in the *K* corona. The problem to determine electron densities in the corona can only be solved if the relative contribution of the K corona to the total coronal light is known. In Sect. 71 it was shown that this problem does not offer fundamental difficulties for the inner corona where use is made of the observed depressions of lines in the scattered Fraunhofer spectrum. In the outer corona, where use is made of the (faint) polarization, not much reliance can be placed on the existing solutions because of our insufficient knowledge of the polarization of the dust particles.

α) After having obtained the surface brightness and its variation with height in the K corona, it is necessary to transform these values into electron densities. Following SCHWARZSCHILD and BAUMBACH[1] it is assumed that the emission $\mathcal{E}$ per cm³ at a distance d (in solar radii) to the Sun's centre may be written as a polynomial: $\mathcal{E} = \sum_n a_n d^{-n}$.

The contribution of one of the polynomial terms to the surface brightness $E(\varrho)$ at a projected distance $\varrho\,(=r/R)$ to the centre of the disk is obtained after integrating along the line of sight (cf. Fig. 85):

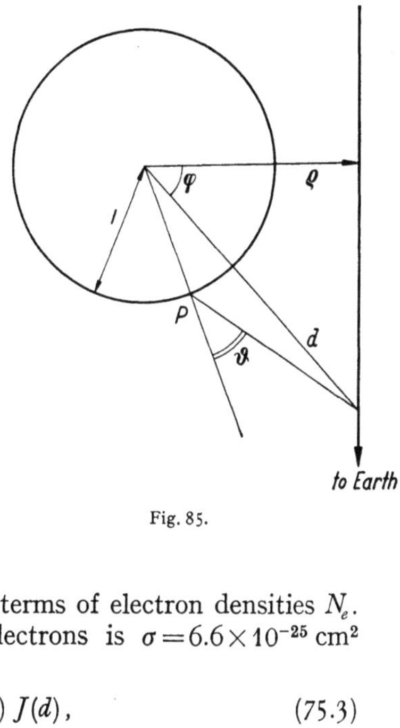

Fig. 85.

$$\left.\begin{aligned}\frac{E_n}{a_n} &= 2\int_0^\infty d^{-n}\,dx = 2\int_0^\infty (\varrho^2 + x^2)^{-\frac{n}{2}}\,dx\\ &= \frac{2}{\varrho^{n-1}}\int_0^{\pi/2} \cos^{n-2}\varphi\,d\varphi.\end{aligned}\right\} \tag{75.1}$$

Hence, if the observed surface brightness is written as $E = \sum_n b_n \varrho^{-n}$, the source function is

$$\mathcal{E} = \sum_n a_n d^{-n-1}. \tag{75.2}$$

β) The next step is the interpretation of $\mathcal{E}$ in terms of electron densities N_e. The (THOMSON) scattering coefficient of free electrons is $\sigma = 6.6\times 10^{-25}$ cm² independent of λ. Hence

$$\mathcal{E} = \sigma N_e(d)\int \frac{I\,d\omega}{4\pi} \equiv \sigma N_e(d)\,J(d), \tag{75.3}$$

where I is the intensity of radiation in a point P at the Sun's surface. The integration of I, leading to J is made over the whole disk, assuming that the disk intensity is given by the simple formula:

$$I(\vartheta) = I(0)\,\frac{1+\beta\cos\vartheta}{1+\beta}$$

with $\beta = 4.0$, a value valid at 4300 Å, a wavelength where most observations are made. Then

$$J(d) = \frac{I(0)}{2+2\beta}\left\{1 - \left(1 - \frac{1}{d^2}\right)^{\frac{1}{2}} + \frac{\beta}{2}\left[1 - \frac{1}{2}\left(d - \frac{1}{d}\right)\mathrm{Log}\,\frac{d+1}{d-1}\right]\right\} \tag{75.4}$$

and for $\varrho > 1.2$; after developing (75.4) into powers of d^{-2}:

$$J(d) = \frac{I(0)}{2+2\beta}\left[\frac{1}{2d^2} + \frac{1}{8d^4} + \cdots + \beta\left(\frac{1}{3d^2} + \frac{1}{15d^2}\cdots\right)\right],$$

[1] S. BAUMBACH: Astronom. Nachr. **263**, 121 (1937).

yielding with (75.3):

$$N_e(d) = \frac{2.18 \times 10^8 \,\mathcal{E}(d)}{\frac{1.833}{d^2} + \frac{0.392}{d^4} + \cdots}. \tag{75.5}$$

In this way, the electron densities can be found directly from the observations, by means of Eqs. (75.2) and (75.5).

γ) VAN DE HULST[1] introduced a refinement in solving for N_e by using MINNAERT'S expressions[2] for the intensity of radiation E_t and E_r emitted in the two directions of polarization, in which the anisotropy of radiation is taken into account. In each of the two integral equations thus obtained $N_e(d)$ occurs, whereas the observations yield $E = E_r + E_t$. The two integral equations must therefore be solved simultaneously in such a way that $E_r + E_t = E$, and that both equations yield the same $N_e(d)$ function. The solution was obtained by trial and error. In the final $N_e(d)$ curves distinction is made between the polar and equatorial density curves, but as is known these two curves do not give the extremes: in the inner corona the intensity and hence the density reaches a minimum value for b between 60° and 70° (cf. Fig. 75).

A somewhat other method to determine N_e was applied by BLACKWELL[3] to his aeroplane observations of the 1954 eclipse, which give the observed polarization at great distances. At a particular distance to the Sun let the degrees of polarization observed of the electron corona and of the F corona be p_0, p_K and p_F respectively. Assuming $p_F = 0$, then

$$p_0/p_K = K/(K+F), \tag{75.6}$$

where K and F denote the intensities of the K and F coronas. The computation was made in approximations. (i) Assume plausible values for $p_K(\varrho)$, then (ii) $K(\varrho)$ can be computed from p_0 with (75.6). (iii) With this derived brightness distribution of the K corona the polarization $p_K(\varrho)$ can be computed. (iv) If this one and the adopted $p_K(\varrho)$ functions differ, a second approximation is attempted etc.

A difficulty is that at $\varrho \approx 20$ about one quarter of the total brightness comes from parts of the corona more than 40 $R_\odot$ distant from the Sun, and at greater elongations this distant contribution becomes even more serious. Hence the analysis of the observed distribution of brightness and polarization necessitates assumptions on the distribution of electron densities up to very great distances ($d = 100$ to 500) to the Sun; for these values use should be made of zodiacal light observations.

The resulting electron densities are given in the Table 17 on p. 271. Slightly greater N_e-values were obtained by ELSÄSSER[4] from a discussion of zodiacal light observations. The gradient of the (Log N_e, Log R)-curve appears to be constant for $d < 20$, and is about 0.3 of that value for $d > 30$. The zodiacal electron densities were obtained assuming that the light of the dust component is unpolarized. That this assumption need not be true was already indicated in Sect. 71.

Although the F corona is observed even close to the limb, the solid particles cannot be expected to occur at so small distances because they would evaporate. For one particular case (SiO_2 crystals) it has been shown[5] that they can only occur at distances $d > 4\,R_\odot$ from the Sun's centre.

[1] H. C. VAN DE HULST: Bull. astronom. Inst. Netherl. **11**, 135 (1950).
[2] M. MINNAERT: Z. Astrophys. **1**, 209 (1930).
[3] D. E. BLACKWELL: Monthly Notices Roy. Astronom. Soc. London **116**, 56 (1956).
[4] H. ELSÄSSER: Z. Astrophys. **37**, 114 (1955); cf. also footnote 7, p. 255.
[5] J. OVER: Circ. Amsterd. **12** (1958).

76. The structure of the minimum corona in monochromatic light: temperatures. Characteristics for the minimum corona of 30 June, 1954 are the extended structureless equatorial atmosphere, the polar rays and the intermediate region of minimum intensity near $b = 65°$. The distribution of intensity of the emission lines during 1954 was described by WALDMEIER[1], on the basis of observations made in the first four and the last six months of the year (no observations were made during the two months of nearly complete inactivity: May and June). Only the red line at 6374 Å appears to be well visible (Fe X). The green line at 5303 Å, which as a rule has an intensity superior to that of the red line, was mostly invisible in 1954 and does only appear in disturbed regions, in the two sunspot zones, for the old and new cycles[2].

So the red line is presumably characteristic of the minimum corona. The assumption that the temperature of the minimum corona is such that the ionization curve for Fe has its maximum at Fe X (Fig. 81), corresponds to an electron temperature of about 5×10^5 °K. In that case the intensity of the line is principally determined by $\int N_e^2\, dy$, where the integration is made over the line of sight. In the foregoing section it was found that the brightness of the *white corona* is determined by $\int N_e\, dy$. The conclusion drawn there, that $(N_e)_{\text{pole}}/(N_e)_{\text{equator}} \approx 2$, does not wholly agree with the observations, showing for the *red* (λ 6374) *corona* an intensity ratio of about 2 to 3. This difference is due[3] to the fact that the red line is partly also excited by radiation ($\propto N_e$).

An attempt to determine T and N_e simultaneously in the corona was made by WALDMEIER[4] and later by TRELLIS[5] on the basis of the method described in Sect. 73. The computations were based on a simultaneous discussion of the intensity of the red and the green line; WALDMEIER's data refer to the maximum year 1947 and to 1953. TRELLIS discussed the observations of 1943 to 1955 including the minimum year 1954. For 1947 T appears to be nearly constant $= 735000° \pm 10000°$ along the whole limb; for the minimum year 1954 TRELLIS found[2] at the equator: $690000° \pm 10000°$, the poles have 730000°. WALDMEIER[3] finds 680000° and 670000°. The electron densities, found for 1953 at $d = 1.05$ are $N_e = 1.0 \times 10^8$ at the pole and 2 to 2.5×10^8 at the equator, while VAN DE HULST, from white light studies, found at the same height 1.3 to 1.8×10^8 for pole and equator and at $d = 1.15$: 7×10^7 ($b = 0$), 1×10^7 ($b = 70°$) and 4×10^7 ($b = 90°$).

The corona before the disk. With the aid of the known density distribution in the corona WALDMEIER[6] has attempted to deduce the form of the corona in front of the disk. The monochromatic emission of the coronal lines is proportional to $\int_{-\infty}^{\infty} N_e^2\, dy$, where y is the coordinate along the sightline. Since the corona is optically thin, the upper limit H of the integral is ∞ or is to be found at the photosphere if the sightline passes or does not pass along the disk. The disk values of $\int N_e^2$ were obtained by interpolating between coronal brightness measurements at the east and west limbs of the Sun at successive limb passages. The resulting intensity distribution greatly resembles that of the radio-brightness distribution at 20 cm (cf. Fig. 97).

---

[1] M. WALDMEIER: Z. Astrophys. **37**, 233 (1955).
[2] At Pic du Midi the green line, though faint, was visible also during the minimum phase.
[3] M. WALDMEIER: Z. Astrophys. **45**, 155 (1958).
[4] M. WALDMEIER: Z. Astrophys. **35**, 95 (1954).
[5] M. TRELLIS: Ann. d'Astrophys. Suppl. **5** (1957).
[6] M. WALDMEIER: Z. Astrophys. **38**, 143 (1955).

As to the *transition to the chromosphere* coronagraph observations do not show any decrease of coronal line intensity towards the disk for $\varrho > 1.02$ or 1.03, that is 14000 to 20000 km above the photosphere. This signifies that the lower border of the corona must be below 14000 km, whereas eclipse observations[1] push this limit still lower; the maximum emission certainly occurs below 10000 km and the lines can perhaps still be distinguished as low as 5500 km.

77. The polar rays. One of the most striking appearances of the minimum corona are the two systems of polar rays. They are very clearly visible on white light exposures (cf. Fig. 75a), where they can be followed from the poles downward till they seem to merge into the more amorphous brightness distribution of the equatorial zone. WALDMEIER[2], in subtle and patient observing, has succeeded in recording them also in monochromatic light of the red line, 6374 Å. He observed a mean number of ten (monochromatic) rays over 50° of the limb in the polar regions. The number of rays depends on the visibility and also on the extension of the polar zone, and this latter in turn depends on the occurrence of activity centres at high latitudes. Rays are invisible near maximum (BUGOSLAVSKAYA[3]). On visual inspection of white light photographs, the rays have widths of the order of 7000 km at the basis of the corona; the width increases with increasing distance to the limb (VAN DE HULST[4]; BUGOSLAVSKAYA[3]). When studying isophotes from photographs, the images seem much coarser; the fine details seem to have disappeared: the eye is known to be sensitive to small irregularities in the intensity distribution. The same somewhat coarser structure appears in WALDMEIER's diagrams in which the intensity is plotted against the position angle along the limb (measures spaced by 1°). Isophotal studies by VAN DE HULST showed that the average intensity ratio between rays and inter-ray region is 1.10, it may amount to 1.20. Since the measured intensities in white light are proportional to $\int N_e\,dy$ and since the width of one single ray is about $\frac{1}{40}$ times the "thickness" of the polar region, the local electron density may be $1 + 40 \times 0.1 = 5$ times greater in the rays than in between. A more detailed determination by DZUBENKO[5] (eclipse of 30 June, 1954) yielded a ratio of electron concentrations of 9, independent of height.

The structural variations and the lifetimes of the rays were studied outside eclipse (coronagraph) by WALDMEIER[2]. No great differences were perceptible at two and even more consecutive measurements about one hour apart, but as a rule it was not possible to find the rays again after one night. In the stationary case the number of rays $N = n\,l$, where n is the number of rays originating per hour, and l is the mean lifetime. From counts of n and N at four days l was determined; it ranged between $12^{\rm h}.9$ and $18^{\rm h}.7$ with $15^{\rm h}.6$ as an average value. Even the distribution function of lifetimes could be determined:

lifetime (hours) . .	0—6	7—17	18—41	42—65
Relative number .	0.05	0.68	0.25	0.02

The structure and the situation of the rays suggest a magnetic explanation; it seems as if the rays have something to do with the Sun's general magnetic field, which is known to have in the polar regions a sightline component of the

[1] R. G. ATHAY and W. O. ROBERTS: Astrophys. Journ. **121**, 231 (1955).
[2] M. WALDMEIER: Z. Astrophys. **37**, 233 (1955).
[3] E. J. BUGOSLAVSKAYA: Astron. Zhurn. **34**, 233 (1957).
[4] H. C. VAN DE HULST: Bull. astronom. Inst. Netherl. **11**, 150 (1950).
[5] N. I. DZUBENKO: Astron. Zhurn. **34**, 379 (1957).

order of 1 Gauss; the field might be 1.5 to 3 Gauss. Such a field corresponds with a magnetic energy density $H^2/8\pi = 0.1$ erg cm^{-3}. On the other hand at $\varrho = 1.2$, the region where the rays can best be seen, $N_e = 2 \times 10^7$ giving a mean kinetic energy per cm³ of $\frac{3}{2} N k T = 0.005$, twenty times less than the magnetic energy (at $\varrho \approx 1.03$ the magnetic and thermal energies are about equal). So in this part of the corona magneto-hydrodynamic laws should be applied. When somewhere a density maximum occurs, the magnetic field will prevent its sideward diffusion and the electrons will mainly diffuse into the direction of the lines of force, hardly at right angles to them. This explains the long extensions of the polar rays. The angles between the rays and the normal to the Sun's limb may give information about the direction of the field. VAN DE HULST found that the rays, when extrapolated backwards, intersect at a point on the solar axis at a distance $q = 0.55\, R_\odot$ to $0.70\, R_\odot$ from the centre. For a dipole field one would expect $1/q = 3 \sin b$, so for $b = 80°$: $q = 0.34$. Hence the field displayed by the ray system is not a dipole field. Virtually the same result was found for the 1954 minimum corona[1,2] ($q = 0.7\, R_\odot$), but at $\varrho > 1.3$ the directions of the rays approach those of the dipole field lines[2]. BACHMANN[3] showed that the observed forms of the polar rays could be described as due to a superposition of such dipole fields, either resembling the field of a bar magnet with a length of $1.2\, R_\odot$, or of two inner solar ring currents, symmetrical around the axis at a mutual distance of $1.0\, R_\odot$.

The density gradient in the rays as measured by VAN DE HULST, BUGOSLAVSKAYA and DZUBENKO (l. c.) does not differ from the mean density gradient in the adjacent parts of the corona. DZUBENKO determined the temperature of the rays from the density gradient, the values ranged between 0.8×10^6 and 1.3×10^6 °K.

78. Summary of temperature determinations; line widths; turbulence. In this section a review is given of the various temperature determinations of the minimum corona.

α) The non-appearance of Fraunhofer lines in the K corona indicates an electron temperature, in any case greater than 10^5 °K.

β) From the ionization of Fe in the quiet minimum corona an *electron-* (*not:* ionization-) temperature was found of about 600000°; Fe X is near maximum ionization (Sect. 76). However, higher ionized Fe atoms occur also. SHKLOVSKI[4] found that the relative abundances of the various Fe ions can be explained by introducing an inhomogeneous corona, with $T = 600000°$ and 1200000°. The first temperature may apply to the undisturbed parts of the corona, and it is tempting to assume that the other applies to the disturbed parts (cf. Sects. 79 and 80).

γ_1) From the *intensities* of the red and green Fe lines of the minimum corona an *electron temperature* of 700000° was found in Sect. 76.

γ_2) The measured total intensities in the X-ray spectrum of the Sun correspond to an *electron temperature* of 750000° K (Sect. 74).

δ) Anticipating part II of this chapter we notice that the radioradiation of the quiet Sun at metre wavelengths indicates an *electron temperature* of 700000° K in the lower part of the corona.

---

[1] S. K. VSESSVIATSKY and G. M. NIKOLSKY: Astron. Zhurn. **32**, 354 (1955).
[2] M. WALDMEIER: Z. Astrophys. **41**, 115 (1956).
[3] H. BACHMANN: Z. Astrophys. **44**, 56 (1957).
[4] I. S. SHKLOVSKI: Isw. Krymsk. Astrophys. Obs. **4**, 109 (1950).

ε_1) The density determinations of the K corona (Sects. 71 and 75) yield $N_e(h)$, which is proportional to the density. This function may be compared to the law of hydrostatic equilibrium. For an isothermal gas the generalized barometric formula is $N_e \propto \exp\,(G M_\odot\, \mu\, m_{\mathrm{H}}/\varrho R_\odot\, kT)$ (G = gravitational constant; m_{H} = proton mass; μ = molecular weight; $\varrho = r/R_\odot$). For $\varrho < 3$ Log N_e appears to depend linearly on ϱ^{-1}. So up to that distance the corona is practically isothermal. VAN DE HULST obtained for the minimum equator and pole respectively: $T_{\mathrm{kin}} = 1.6 \times 10^6$ and $1.1_5 \times 10^6$ °K.

ε_2) Similarly, intensity gradients in the monochromatic corona[1] yield *kinetic temperatures* of 1.8×10^6 °K (λ 5303) to 2×10^6 °K (λ 6374) with a considerable scatter, especially in greatly disturbed regions (displaying λ 5694 emission).

ζ) The widths of the emission lines indicate similar kinetic temperatures. Neglecting for the moment the widths determined in condensations, coronal knots, etc. and the various width determinations in the quiet corona with low resolving power-spectrographs, there remain LYOT's and DOLLFUS's determinations of the width of the red coronal line in slightly excited coronal regions[2] (resolving power: 20000). Between $1.04 < \varrho < 1.24$ the line has a Gaussian profile described by a *kinetic temperature* of $(2.0 \pm 0.3) \times 10^6$ °K or by $\eta = 25$ km/sec. This value seems independent of the phase of the solar cycle.

η) JARRETT and VON KLÜBER[3], from Fabry-Pérot interference fringes of the 5303 Å line (resolving power 2600) at the eclipse of 30 June, 1954, found T_{kin} between 2.2×10^6 and 5×10^6 °K in the range $1.05 < \varrho < 1.3$ (neglecting turbulence). Most values scatter around 2.5×10^6 °K. BILLINGS[4], HIRSCH and VARSAVSKY[5] from coronagraph observations at the Climax Observatory found *kinetic temperatures* of about 2.0×10^6 °K.

Hence, there seems to be some disagreement between the electron temperature of the corona, yielding values between 6 and 8×10^5 °K and the kinetic temperatures determined from line widths and density gradients. A part or perhaps the whole of this difference might be ascribed to the fact that the measurements referred to in points β, γ, δ and ε_1 refer to the quiet minimum corona, whereas the measurements of ε_2, ζ and η refer to more excited parts of the corona. It should further be borne in mind that the temperatures, deduced from relative and absolute line strengths are still somewhat uncertain because they are influenced by the inaccurate present knowledge of the collisional cross sections involved. However, WERNER's result[6], based on new collisional cross sections, yielding $T = 2 \times 10^6$ °K from the equal intensities of the red and the green lines is not confirmed by ELWERT, who finds $T = 7 \times 10^5$ °K. So we are inclined to accept for the quiet parts of the minimum corona an electron temperature of about $(7 \pm 1) \times 10^5$ °K.

The question might be posed whether a part of the difference between the two groups of temperatures could be ascribed to a certain amount of microturbulence in undisturbed parts of the corona. Large-scale mass motions have only been observed in excited coronal regions (values up to 15 and 20 km/sec, cf. Sects. 79 to 81 below) whereas the undisturbed corona does not show velocities greater

[1] D. E. BILLINGS and R. H. COOPER: Z. Astrophys. **43**, 218 (1957).
[2] A. DOLLFUS: Convegno Volta **1952**, 184 (1953). — C. R. Acad. Sci., Paris **236**, 996 (1953).
[3] A. H. JARRETT and H. VON KLÜBER: Monthly Notices Roy. Astronom. Soc. London **115**, 343 (1955).
[4] D. E. BILLINGS: Astrophys. Journ. **125**, 817 (1957).
[5] D. E. BILLINGS, S. HIRSCH and C. VARSAVSKY: Astrophys. Journ. **123**, 532 (1956).
[6] A. WERNER: Austral. J. Phys. **7**, 25 (1954).

than 5 km/sec[1]. Very high microturbulent velocities would be needed to explain the observed kinetic temperatures of $\approx 2\times 10^6$ °K in terms of thermal velocities corresponding to 0.7×10^6 °K *plus* turbulence.

Reference is further made to the correlation found by BILLINGS[2] between the local kinetic temperature of excited regions and the ionization temperature (determined from line strength ratios); this may be explained by a true increase of the local kinetic temperature in "hot regions" rather than by an increase of the local microturbulence.

Table 17. *The structure of the quiet corona.*

$\varrho = \frac{r}{R_\odot}$	h km	Log N_e (cm^{-3}) and plasma frequency f_0 (MHz) min (equator)	MHz	min (pole)	MHz	log N_e (cm^{-3}) maximum equator	pole
1.011	7500	9.05	280	8.95	235	—	—
1.014	10000	8.6	180	8.45	150	8.9	8.6
1.022	15000	8.3	125	8.1	100	—	—
1.03	21000	8.2	110	8.0	90	8.8	8.5
1.06	42000	8.1	100	7.85	75	8.7	8.4
1.1	70000	7.9	80	7.6	56	8.55	8.25
1.2	140000	7.6	56	7.1	30	8.2	7.9
1.4	280000	7.1	30	6.4	14	7.7	7.4
1.6	420000	6.75	21	5.9		7.3	7.0
2.0	700000	6.25	12	5.3		6.8	6.45
3.0	1.4×10^6	5.6		4.2		5.6	
5.0	2.8×10^6	4.8		3.4		4.6	
10	6.2×10^6	4.0					
20	13×10^6	3.4					
65	44×10^6	3.0					
215	1 A.U.	2.8					

Sources. $\varrho < 1.03$: C. DE JAGER: Paris symposium 1958 on Radio Astronomy, (1959). $1.03 < \varrho < 3$ (*min.*): H. C. VAN DE HULST: Bull. astronom. Inst. Netherl. **11**, 135 (1950), slightly changed in accordance with Fig. 76.
$1.01 < \varrho < 2$ (*max.*): G. NEWKIRK: I.A.U. 1958 Radiosymposium at Paris (1959).
$\varrho > 3$: D. E. BLACKWELL: Monthly Notices Roy. Astronom. Soc. London **116**, 56 (1956), and H. ELSÄSSER: Mitt. Astr. Ges. **1957**, II, 61.

c) The active parts of the corona and the "maximum corona".

As in the other parts of the Sun coronal activity is displayed mainly in the centres of activity (CA). It was noted (Sect. 47) that the first (spot) phases of a CA correspond with the active coronal regions which show short coronal rays, arches and condensations, whereas the end phase of a CA is rather correlated with the long and broad coronal rays. At the top of coronal activity (the "maximum corona"), the number of CA's is so great that the corona is nearly circular and the position of its axis can not or hardly be determined from photographs. The polar rays are invisible, perhaps occulted by the numerous other rays. The shape and extension of the maximum corona varies from case to case, depending on the intensity of the maximum. We are not justified to speak of *the* maximum corona, whereas the notion "minimum corona" has a well defined significance. VAN DE HULST (footnote 1, p. 253) found that the equatorial brightness distribution of the average *quiet maximum corona* exceeds that of the minimum corona

[1] A. DOLLFUS: C. R. Acad. Sci., Paris **244**, 1880 (1957).
[2] See footnotes 4 and 5, p. 270.

by the factor 1.8 for $\varrho<3$. For greater ϱ values the intensity seems to be independent of the solar activity, showing the preponderant contribution of the F corona

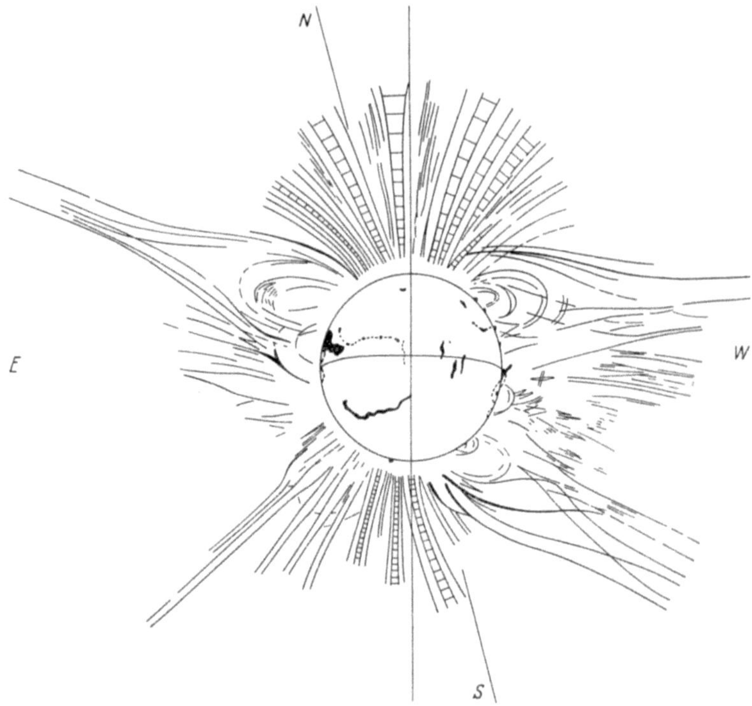

Fig. 86. Schematic drawing of the corona of 25 February, 1952 showing rays, polar plumes and arches. G. M. NIKOLSKIJ: Astron. Zhurn. **30**, 286 (1953).

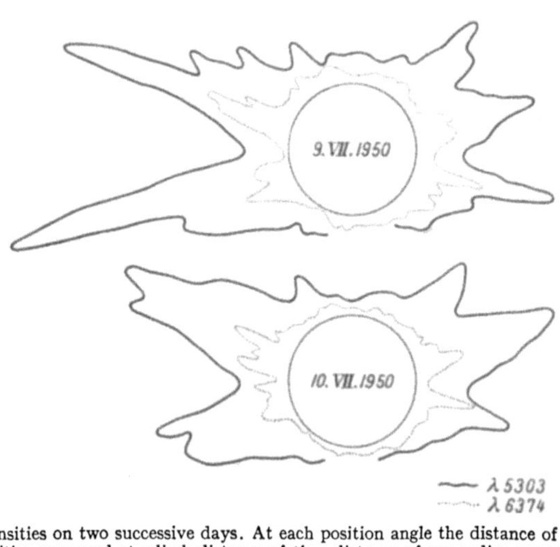

Fig. 87. Coronal line intensities on two successive days. At each position angle the distance of the curves to the limb is proportional to the intensities measured at a limb distance of 1′; a distance of one radius corresponds with 50×10^{-6} the intensity of 1 Å of the continuous spectrum of the disk's centre. The brightness maxima are often called "jets" ("Koronastrahl") by coronagraph observers. Observer M. TRELLIS. Courtesy J. RÖSCH, Pic du Midi.

at great distances. For a description of the coronal structures observed during eclipses between 1887 and 1950 reference is made to BUGOSLAVSKAIA's paper (see general list of references), where the relation between the coronal structures and the surface phenomena is also shown. The relative intensities of coronal formations at various limb distances are given by BUGOSLAVSKAIA for two eclipses[1] (1914 and 1941).

79. The coronal activity regions and the coronal condensations. Fig. 88 shows the relations between, on the one hand, the photospheric and chromospheric phenomena and, on the other the *monochromatic (L) corona*[2]. Figs. 88a and b give the photospheric and chromospheric phenomena (faculae, filaments, spot-groups), Figs. 88c to e the intensity of the lines $\lambda\lambda$ 6374, 5303 and 5694. As LYOT found, the line λ 5694 shows the best correlation with active spot areas; WALDMEIER[3] and DOLDER *et al.*[4] noticed the close correlation between the emission of this line and flares. Also the line 5303 is intensified above CA's. At latitudes below 40° each CA is correlated with a brightening of the line, at higher latitudes the line can also be strong above quiet chromospheric regions and above some filaments, though sometimes its intensity—like that of the other coronal lines—is slightly decreased above quiescent filaments[5]. The line is faint or invisible during sunspot minimum. The red line λ 6374 is visible over a greater part of the solar surface, mainly in the spot belts. The intensity changes only little during the sunspot cycle. The line is sometimes strong on the sides of the regions where the green line is abnormally enhanced, especially on the equatorial side[6]. Its intensity is decreased near quiescent filaments. Above filaments the ionization temperature seems to be enhanced, but the electron density tends to decrease. The greatest densities occur in regions where the yellow line is enhanced, that is above CA's at the top of their development.

The monochromatic corona above CA's is characterized by relatively stable ray and arch structures, well known from eclipse observations. The rays are relatively short, about 150000 km long, often curved and radiate from the CA's. According to J.W. EVANS' description[7] of motion picture films taken by DUNN in 5303 Å, "arches often occur in the form of several concentric parabolic or semicircular filaments, rather sharply outlined, usually decreasing in brightness from the inner to the outer filaments. They tend to expand gradually over periods of hours". Short-periodic fluctuations were first described by LYOT[8]. In one case[7] an arched streamer was observed which was initially strongly curved, then stretched within 5 minutes, to become nearly vertical simultaneous with the appearance of a flare. The velocity at the end of this "whip" was ≈ 600 km per sec.

However, "most of the dynamic activity consist of rapid brightness variations, rather than of motions of material ... A common feature is the sudden development of a sharply outlined dark area, which may be a roughly circular spot perhaps 3×10^4 km in diameter"[7]. Also DOLLFUS[9] remarks that the observed motions reflect variations in the excitation conditions, rather than mass displacements. Often a disturbed structure re-originates the next day. The forms suggest an electromagnetic explanation.

[1] E. J. BUGOSLAVSKAIA: Astr. Circ. SSSR. **157**, 8 (1955).
[2] M. WALDMEIER: Z. Astrophys. **38**, 219 (1956).
[3] M. WALDMEIER: Astr. Mitt. Zürich **146** (1946).
[4] W.O. ROBERTS, D.E. BILLINGS and F.D. DOLDER: Astrophys. Journ. **119**, 120 (1954).
[5] M. WALDMEIER: Experienta **2**, 220 (1946).
[6] W. PETRI: Naturwiss. **42**, 119 (1955).
[7] J.W. EVANS: Publ. Astronom. Soc. Pacific **69**, 421 (1957).
[8] B. LYOT: Ann. d'Astrophys. **7**, 31 (1944).
[9] A. DOLLFUS: C. R. Acad. Sci. Paris **244**, 1880 (1957).

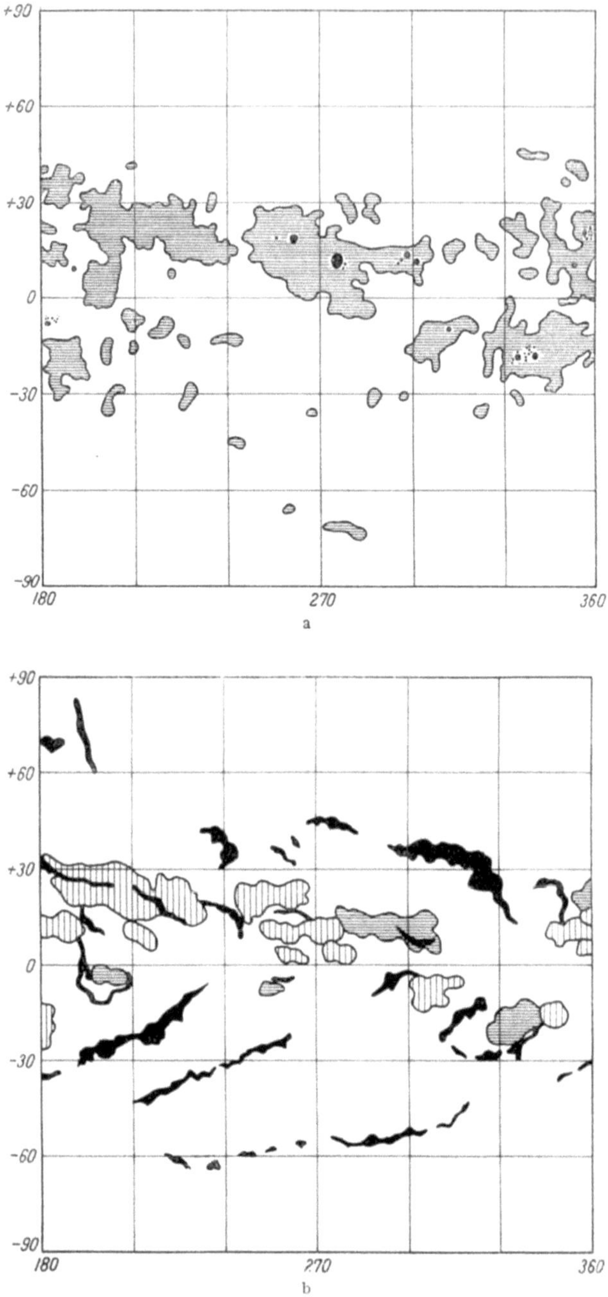

Fig. 88a—e. Relation between photospheric, chromospheric and coronal phenomena derived from observations from 5 to 19 March, 1950. (a) Photospheric phenomena: black: spots; shaded: faculae. (b) Chromospheric phenomena: black: filaments.

GENTILI and TRELLIS[1] have reported rapid intensity variations of the green line (5303 Å): hourly measurements show that the intensity in a bright coronal region may change by a factor 2 in two hours, or even less. *White light observations*

[1] M. GENTILI DI GIUSEPPE and M. TRELLIS: C. R. Acad. Sci., Paris **243**, 1724 (1956).

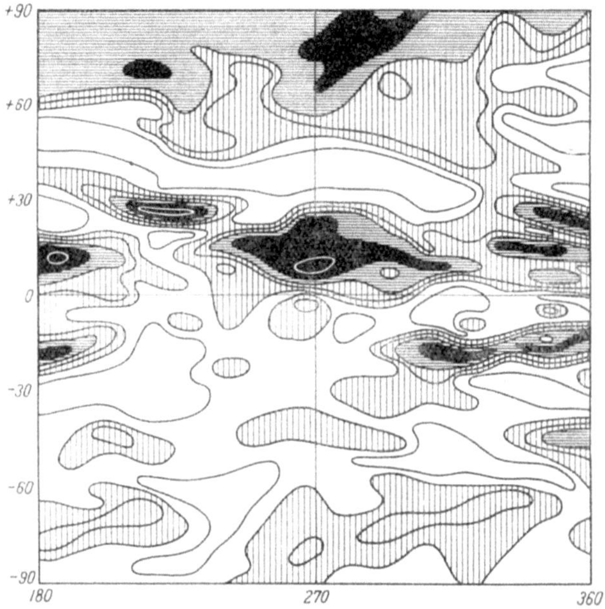

c

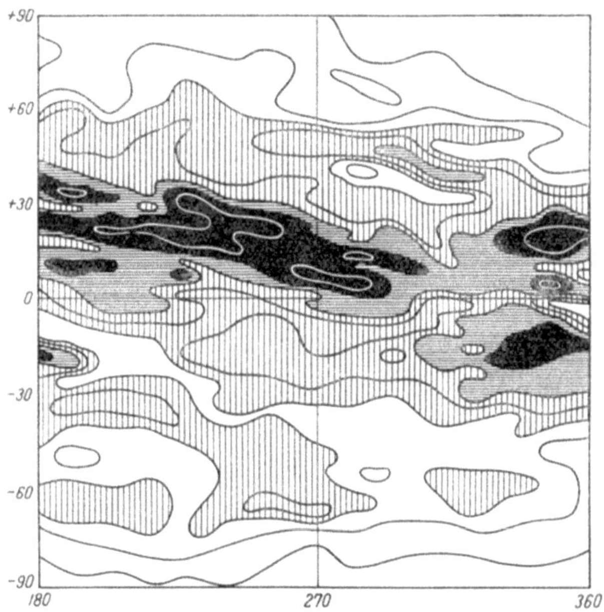

d

Fig. 88 c and d. (c) Intensity of the 6374 line. (d) Intensity of the 5303 line.

of the regions above faculae and CA's show complex systems of short thin rays, widths $\approx 0.03\,R$ or more. These rays sometimes take arch-like forms in the neighbourhood of spots. Motions of 10 km/sec may occur in these "bushes", helmet-like structures, especially studied by BUGOSLAVSKAYA[1]. NEWKIRK, from

[1] E. J. BUGOSLAVSKAYA: Comm. Sternberg State Astr. Inst. **96** (1954).

K coronameter observations finds that the average active region condensation has (for $\varrho < 2$) about twice the electron density of the quiet corona. The diameter of the condensation is 25° at the basis; a pinching in at $\varrho \approx 2$ is evident from the records.

It is clear that we may schematize: the coronal active regions are characterized by motions and magnetic fields and by enhanced 5303 Å intensity; moreover the most active centres show the 5694 Å line. There are, however, grades in the activity of the coronal active regions, grades that may be best characterized by the intensity and the frequency of occurrence of the yellow line 5694 Å.

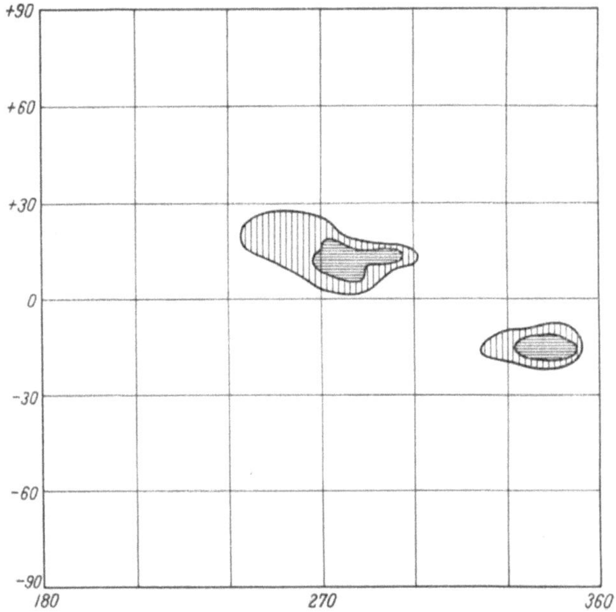

Fig. 88e. Intensity of the 5694 line. M. WALDMEIER: Z. Astrophys. **38**, 219 (1956).

For the solar cycle 1944 to 1954, the table below gives the probability of observing the line λ 5694 in connection with spot-groups (WALDMEIER[1]):

Type of spot. .	no spots	A	B	C	D	E	F	G	H	I
Probability . .	0.04	0.02	0.03	0.12	0.21	0.37	1.00	0.32	0.17	0.07

This again shows the correlation with flare activity. The temperatures derived for yellow active regions are usually determined from line widths[2]; they range between 10^6 and 6×10^6 °K. The high values can only be reduced to 10^6 or 2×10^6 °K by introducing a very intense microturbulence, of the order of 100 to 200 km/sec.

The highest temperatures occur in the *permanent coronal condensations*. They are (WALDMEIER[3]) the coronal CA's at the top of their development; mainly characterized by great densities (a bright continuous spectrum[4,5] indicates electron densities of the order of 0,5 to 2×10^{10} cm^{-3}), by the occurrence of the 5694 Å line

[1] M. WALDMEIER: Z. Astrophys. **39**, 219 (1956).
[2] CH. PECKER, D.F. BILLINGS and W.O. ROBERTS: Astrophys. Journ. **120**, 509 (1954).
[3] M. WALDMEIER: Z. Astrophys. **40**, 221 (1956).
[4] M. WALDMEIER: Z. Astrophys. **20**, 172 (1940).
[5] D. E. BILLINGS: Astrophs. Journ. **125**, 817 (1857).

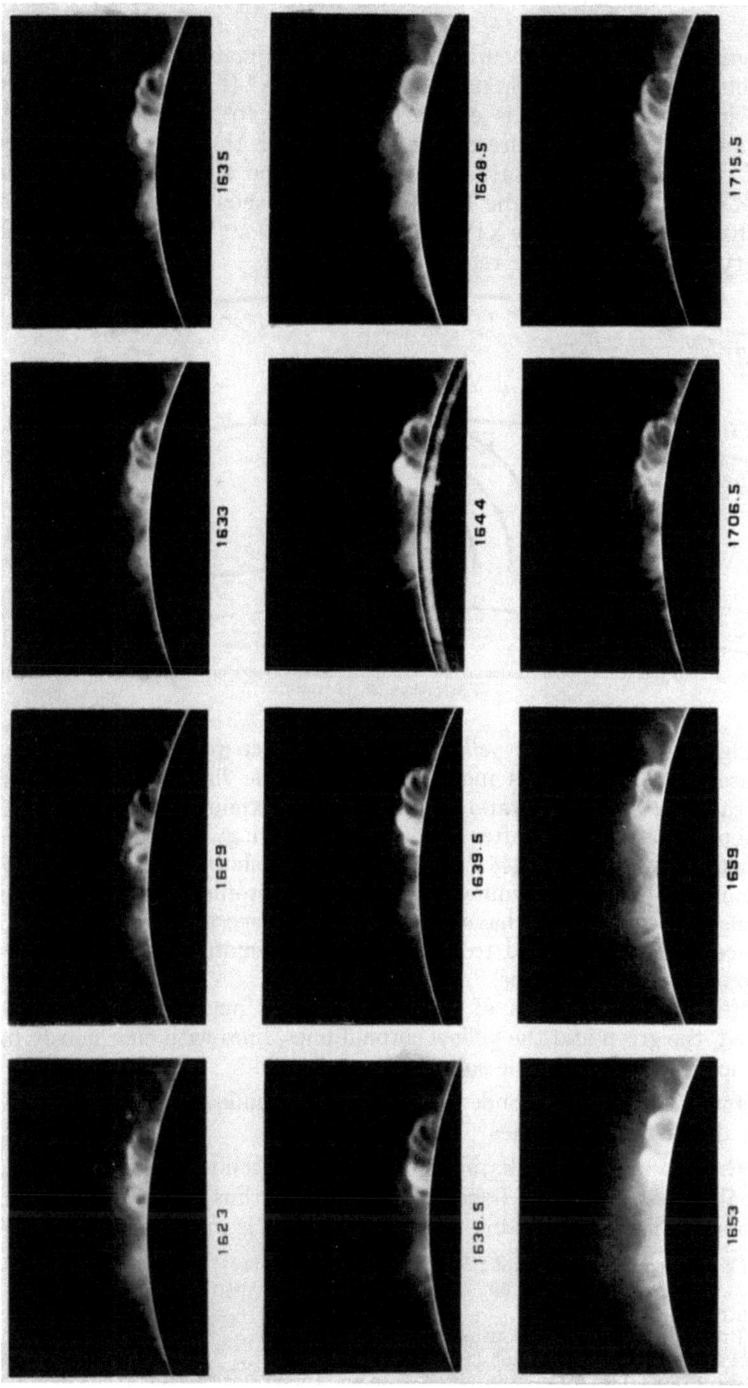

Fig. 89. Short-periodic fluctuations of the coronal structure. Observations of the 5303 Å line, made 22 November, 1956, at Sacramento Peak Observatory with a 15 cm coronagraph. Note the sudden changes between 16.44, 16.48 and 16.53 UT. The lowest part of the 16.44 observation contains the spectrum of the line obtained through a circular slit. Courtesy R. B. DUNN, Sacramento Peak Observatory.

and the (rare) occurrence of the 5445 Å line with half the intensity[1] of λ 5694. They have semispherical or elliptical forms without much internal structures and with lifetimes of the order of several days; they often accompany the *F* spot-groups.

A schematized model of a coronal condensation, mainly derived from radio-observations, was given by WALDMEIER and MÜLLER[2] (Fig. 90). It was supposed that the kinetic temperature is of the order of 5×10^6 °K; later WALDMEIER[3] found an ionization temperature of 2×10^6 °K. With the aid of high dispersion spectra of permanent condensations BILLINGS[4] found a kinetic temperature of 4.2×10^6 °K. The profiles of the line 5303 Å often show very broad wings, indicating that a part of the FeXIV emission from along the line of sight comes from a very high temperature region.

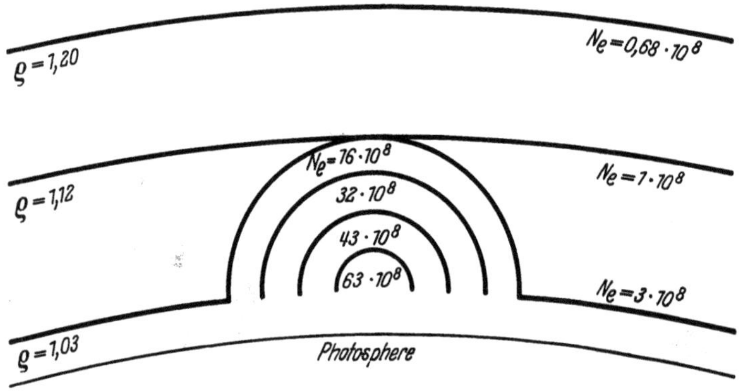

Fig. 90. Sketch, of a "permanent coronnl condensation". The diameter is 125 000 km. M. WALDMEIER and H. MÜLLER. Z. Astrophys. **27**, 58 (1950).

The brightening up of the *yellow line* occurs often together with *flares* but the increased emission decays more slowly than the flare emission. A sharp decline of the 5684 Å emission rarely occurs[5]. The maximum intensity is generally reached about half an hour after the flare's beginning when the Hα emission has vanished[6]. Also the line 5445 Å may appear for a short moment, when 5694 Å is at maximum intensity. Asymmetric coronal line profiles resulting from sight-line motions of coronal clouds have rarely been observed and only in the vicinity of limb flares[7]. So, flares tend to increase both the motions and the intensities of neighbouring coronal regions.

The difficulty that spectra of the condensations may present at the same time the red, the green and the yellow coronal lines, lines with enormously differing ionization potentials, can be solved in two ways[8]:

1. It is possible that the condensations are surrounded by cool regions which emit the red and the green lines.

2. By the strong X-ray radiation emitted by the condensation in wavelength regions smaller than 15 Å the *M* shell of Ca is ionized thus enabling the formation of CaXV ions. Since at the same time the intensity is insufficient to ionize the

[1] M.C. BRETZ and D.E. BILLINGS: Astronom. J. **62**, 138 (1957).
[2] M. WALDMEIER and H. MÜLLER: Z. Astrophys. **27**, 58 (1950).
[3] Cf. footnote 3, p. 276.
[4] D.E. BILLINGS: Astrophys. Journ. **125**, 817 (1957).
[5] W. PETRI: Z. Astrophys. **34**, 68 (1954).
[6] M. WALDMEIER: Transact. Internat. Astronom. Union **9**, 672 (1957).
[7] W.O. ROBERTS: Transact. Internat. Astronom Union **9**, 659 (1957).
[8] G. ELWERT: Z. Astrophys. **44**, 112 (1958).

L shell of *iron* atoms ($\chi_i > 1300$ eV) the ionization of Fe is not fundamentally changed.

The X-ray emission of the coronal condensations, assumed to be built up according to the Waldmeier-Müller model with adopted temperatures of 3×10^6 and 6×10^6 °K was calculated by ELWERT[1] (cf. Sects. 54 and 74).

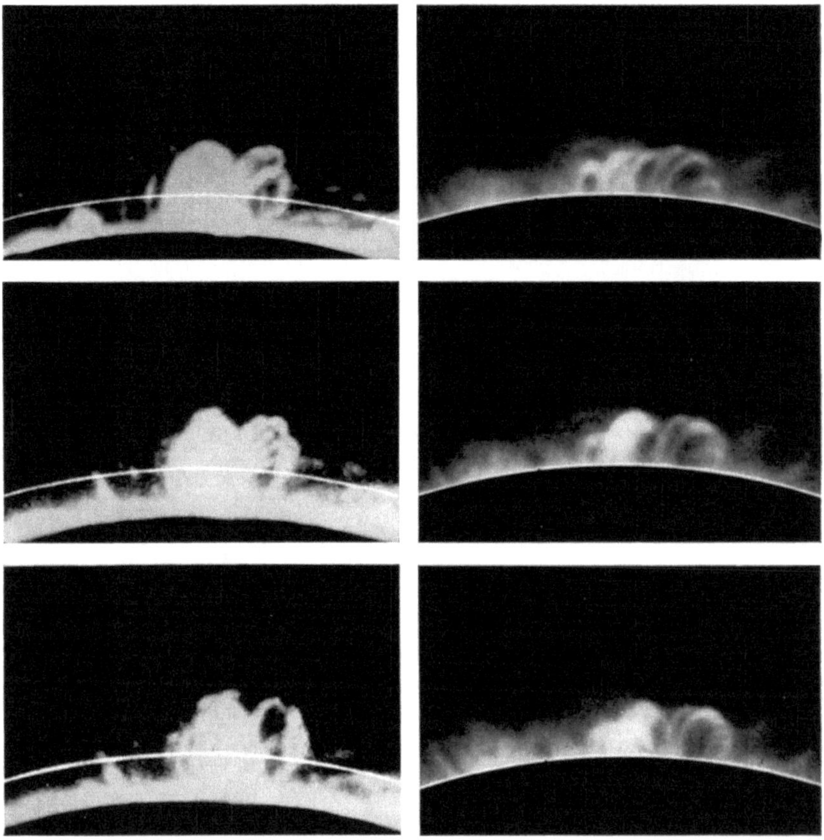

Fig. 91. Sporadic coronal condensations and their correlation with loop prominences (22 September, 1955). Simultaneous pictures obtained in H α (left) and the coronal line 5303 Å. From above to below: 16.33; 16.44 and 16.54 UT. The H α pictures are overexposed, and show only a general loop structure; on a lighter exposure they show multiple loops. Courtesy R. B. DUNN, Sacramento Peak Observatory.

The *sporadic condensations* occur in the permanent ones, are short lived, and coincide with the Hα prominences observed as knots, loops and funnels. They are observed mainly in the lines 5303 Å (see Fig. 91) and 6374 Å. They are obviously caused by a local decrease of the electron temperature in the permanent condensations: in the prominences this temperature decreases to below 10^5 °K (cf. also Sect. 68).

Coronal Doppler motions have been observed in those sporadic condensations that contain active loop prominences[2]. The velocities are generally smaller than those observed at the same place in the prominences, but they have the same signs. It is suggested that the expansion or contraction of the magnetic field defining the loop gives rise to the Doppler motions.

[1] G. ELWERT: Z. Astrophys. **41**, 67 (1956).
[2] G. NEWKIRK: Ann. d'Astrophys. **20**, 127 (1957).

80. The coronal streamers. A beautiful specimen of a streamer (fan; ray) was shown by the 1952 corona (Fig. 75c). The long streamer at position angle 70° could be traced out to a distance of at least 12 solar radii.

MICHARD *et al.*[1], and HEPHURN[2], investigating the correspondence between the "white" coronal structures and the brightness variation of the green line (5303 Å) in the *L* corona, found a good correlation as far as the inner white corona is concerned, but VON KLÜBER[3] found that the maxima of the green line did not coincide with the positions of the long streamers of the outer corona. There were indications rather of shifts between them, showing that the streamers have not

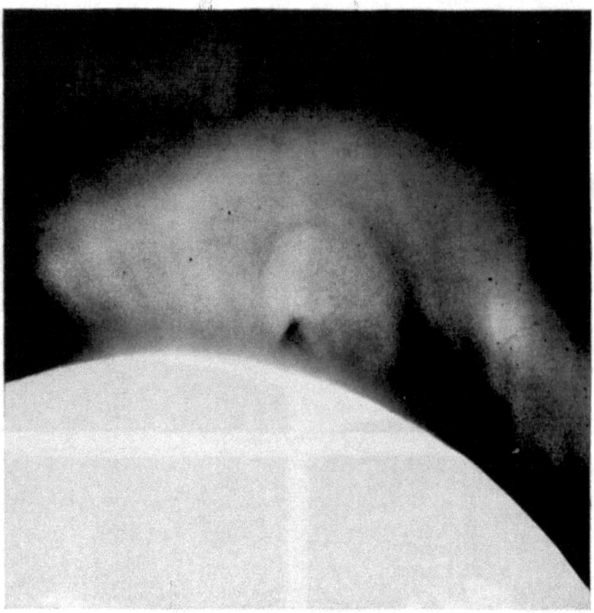

Fig. 92. A small prominence (invisible on the picture) occured in the dark hole in the middle of the photograph; it was accompanied by a coronal arch and by short rays; eclipse of 30 June, 1954. In the reproduction the Sun's disk has been masked. Courtesy Å. WALLENQUIST, Uppsala.

to do with regions of enhanced line emission. Mention has been made of ALLEN'S suggestion that the coronal rays are correlated with the quiescent prominences (p. 190). The other suggestion, that they are also correlated with the *M* and *C* regions is not confirmed[4]. There is no indication that the streamers represent outstreaming motions. Appreciable velocities have not been observed in them.

The streamers of the maximum corona are only slightly inclined to the normal, the inclinations go to $+20°$ and $-20°$. Those of the corona near minimum have an average inclination of 30° towards the equator with a scatter of 10 to 15° (BUGOSLAVSKAYA[5]).

The bases of the fans may have diameters up to one solar radius and are often composed of a system of concentric more or less elliptical arches with tops reaching to $\varrho=1.5$ to 1.8. There is a peculiar relation with the prominences: the brighter arches often border a dark region around the prominences. The filaments

[1] Cf. footnote 1, p. 255.
[2] N. HEPHURN: Astrophys. Journ. **122**, 445 (1955).
[3] H. VON KLÜBER: Observatory **72**, 207 (1952).
[4] C.W. ALLEN, in: 9ième Rapport Commission Relations Solaires Terrestres, p. 79, 1957.
[5] E.Y. BUGOSLAVSKAYA: Astron. Zhurn. **34**, 233 (1957).

occur in these arches, but not necessarily exactly in the centre (Fig. 92). The arch-structures might reflect the decreased coronal density near filaments, as if the corona is locally rarified by the condensation of matter into the prominence. The ratio of densities in this dark "dome" and in the "normal" corona is not known. Slow upward rising motions of 2 km/sec may occur in the arches.

Often, instead of one big arch system, the basis of the fans may contain a greater number of smaller arches. They consist of one or more concentric arches; their base width may amount to 20° and their top height to $\varrho = 1.25$. The ratio between the height a and the half-width b at the basis is:

a/R	0.2	0.4	0.6—0.8
a/b	1.5	2	2.5

Hence, the smallest arches are nearly circular; greater ones are more elliptic. Small prominences may occur in the central parts of these arches. Sometimes such arches are located above sunspots; these obviously form a transition structure between the quiet-region fans and the arches above active regions.

The intensity decreases less quickly in the streamers than in the surrounding coronal regions, but the decrease of N_e with distance is equal[1,2]. The polarization is greater than in the surroundings, which is due to the greater electron density which exceeds that in the streamers by a factor 7. Here we are up against an analogous case as was met in the polar rays: a high density region extends over a great distance with the same density *gradient* as the surrounding less dense parts, and it seems tempting to ascribe the streamers also to the influence of a magnetic field. Some cases were reported in which the polarization angles in rays were not perpendicular to the limb[1,3]. Deviations up to 35° were measured, but VON KLÜBER[4] (1952 eclipse) did not find deviations exceeding 10°.

81. Coronal theories. In this section a short review of coronal theories is given. For further details the reader is referred to UNSÖLD, Physik der Sternatmosphären, pp. 666seq. where the thermal equilibrium of the corona is dealt with, and to VAN DE HULST's contribution in KUIPER, The Sun, pp. 302seq., where especially the mass balance is treated in detail.

The heating of the low corona. In Sect. 29 it was shown that the flow of mechanical energy coming from the photosphere, though dissipating partly in the chromosphere, is still important enough when arriving in the regions where $N_e \approx 4 \times 10^9$ cm^{-3}. At that level the dissipated mechanical energy is $\varrho \xi^3/H$ erg cm^{-3} sec^{-1} (since $\xi \approx v_S$) and the radiation per cm³ and sec (the total of bound-free and bound-bound emissions) is $1.8 \times 10^{-21}\, T^{-\frac{1}{2}} N_e^2$ erg cm^{-3} sec^{-1}. With $\xi = 15$ km/sec, $H = 1000$ km, $\varrho = 1.6\, m_{\mathrm{H}} N_e$, $T = 6000°$ equality between dissipation and radiation occurs for $N_e = 4 \times 10^9$ cm^{-3}. Higher up, the temperature increases because the emission of radiation becomes relatively less important. The radiation of the corona is of the order 5×10^3 erg cm^{-2} sec, whereas the inflow of mechanical energy is about a factor 10^2 greater.

A part of the mechanical energy absorbed by the low corona is transported back downward to be re-radiated in the chromosphere (the upper chromosphere and corona are good heat-conductors). This flow of energy determines the temperature gradient of the transition zone chromosphere-corona[5]. The conduction

[1] M. SCHMIDT: Bull. astronom. Inst. Netherl. **12**, 59 (1953).
[2] See footnote 2, p. 280.
[3] G. NEWKIRK, G. WLERICK and J. AXTELL: Astronom. J. **62**, 95 (1957).
[4] H. VON KLÜBER: Monthly Notices Roy. Astronom. Soc. London **118**, 201 (1958).
[5] A. UNSÖLD: Physik der Sternatmosphären, 2. Aufl., p. 672. 1955.

equation is $W=-K\,dt/dh$, with $K=C_{v,\mathrm{at}}\,v/4Q$; $C_{v,\mathrm{at}}$ is the specific heat per constant volume per particle, and Q is the average collision cross section for ions and electrons with velocity v. UNSÖLD estimated $K=1.65\times10^{-6}\,T^{\frac{5}{2}}$. (For more detailed calculations cf. OSTER[1].) Hence

$$\pi F_{\mathrm{cond}}=1.65\times10^{-6}\,T^{-\frac{5}{2}}\,dT/dh.$$

By integrating this expression, and taking $\pi F_{\mathrm{cond}}=10^6$ erg cm^{-2} sec^{-1}, and assuming an "observed scale height" for T of 4000 km, we find a coronal temperature of 0.96×10^6 °K, quite an acceptable value.

An other part of the mechanical energy absorbed by the corona is lost by *evaporation*[2]. That evaporation must certainly be important is clear from the observations (1) that the thermal velocity of the coronal gas is close to the velocity of escape (the maximum possible coronal temperature, 1.5×10^6 °K, is determined by the condition that the thermal energy of the protons should be about 2 to 3 times less than the gravitational energy if the electrostatic field is taken into account—this field reduces the parabolic energy to half its value), and (2) that the size of the corona is compatible with the Sun's radius. This evaporation mechanism explains the coronal cooling (mainly the fastest particles disappear) and it shows also that hot coronal regions must necessarily be a source of particle emission.

BIERMANN and TEN BRUGGENCATE[3] have suggested that the energy necessary to maintain the high temperature of the *outer parts of the corona* comes from the gravitational energy of accreted interstellar matter. This suggestion has been elaborated by BONDI, HOYLE and LYTTLETON[4], but it appears that the theory requires too great a density (45 atoms cm^{-3}) of the interstellar medium to explain the observed coronal temperature and density. BLACKWELL and DEWHIRST[5] showed moreover that the interplanetary density *distribution* predicted on the basis of the accretion mechanism differs from the observed distribution. So the theory of accretion of interstellar matter fails to explain the principal features of the corona. However, an effect that has been neglected hitherto is that of the accretion of the inter*planetary* dust, which drifts towards the Sun under the influence of the Poynting-Robertson effect. (cf. [6]).

The problem of the *mass balance*, already touched here, was discussed by VAN DE HULST[7], partly on the basis of PICKELNER'S[2] discussion of the evaporation of the corona. The quantities of evaporated and accreted matter are of the same order and are too small to have any influence both on the evolution of the Sun and on its energy balance, as far as the undisturbed Sun is concerned. Nevertheless the Sun practically continuously emits corpuscular radiation, and the observations of comets show that this radiation produces a mean acceleration of the tail ions of about 100 $g_\odot$, the values ranging between 20 $g_\odot$ and 1000 $g_\odot$. BIERMANN[8] estimated that the active Sun emits more than 10^{36} particles per second (mostly protons and electrons). The quiet Sun seems to emit 10^{33} protons

[1] L. OSTER: Z. Astrophys. **42**, 228 (1957).
[2] S.B. PICKELNER: Isw. Krymsk. Astrophys. Obs. **5**, 34 (1950).
[3] L. BIERMANN and P. TEN BRUGGENCATE: Veröff. Göttingen **83** (1946).
[4] H. BONDI, F. HOYLE and R.A. LYTTLETON: Monthly Notices Roy. Astronom. Soc. London **107**, 184 (1947).
[5] D.E. BLACKWELL and D.W. DEWHIRST: Monthly Notices Roy. Astronom. Soc. London **116**, 637 (1956).
[6] M. MINNAERT, in: Les Particules solides dans les Astres, Liège colloquium 1954, p. 15, 1955.
[7] H.C. VAN DE HULST, in: The Sun (ed. G.P. KUIPER), p. 302f. 1953.
[8] L. BIERMANN: Z. Astrophys. **29**, 274 (1951). — Z. Naturforsch. **7**a, 127 (1952). — Observatory **77**, 109 (1957).

per second[1,2]. ALFVÉN[3] concluded from this fact that owing to the existence of large-scale solar magnetic fields not only the inner corona must rotate with the Sun, as is observed, but also an appreciable part of the emitted particle stream, perhaps to near Mercury's orbit (cf. also [4]). SHIMOODA[5] has attempted to describe the interplanetary field as it would have been built up by the material ejected from the Sun. Mention should also be made of CHAPMAN's finding[6] that a steady conductive heat flow from the coronal gas would make the corona extend far beyond the Earth, the coronal temperature near the Earth being 2×10^5 °K.

II. Radio emission from the Sun.

The solar radio emission is composed of a semi-permanent part referring to thermal radiation by the quiet Sun (it varies only slowly and slightly in the course of the solar cycle), and a superimposed variable part, mainly emitted by centres of activity. The latter consists partly of radiation of thermal origin, partly of non-thermal radiation.

First the quiet Sun will be discussed, next the disturbed one, starting with the thermal radiation from centres of activity. After that the various kinds of non-thermal radiation are considered.

References (see also the *general references* on p. 362):

Daily flux measurements are published in the Quarterly Bulletin on Solar Activity, ed. Sternwarte, Zürich.

HULST, H.C. VAN DE: A Course in Radio Astronomy. Mimiographed, Leiden 1951.

LOVELL, A.C.B., and J.A. CLEGG: Radio Astronomy. London: Chapman & Hall 1952.

PAWSEY, J.L., and R.N. BRACEWELL: Radio Astronomy. London: Oxford University Press 1955.

COUTREZ, R.: Radio Astronomie. Monographie No. 5, Obs. Roy. Belgique, 1956.

Many authors: Radio Astronomy. I.A.U. symposium No. 4. Ed. H.C. VAN DE HULST, Cambridge University Press 1957; I.A.U. symposium at Paris. Ed. R.N. BRACEWELL, Stanford Univ. Press 1959.

a) The quiet Sun.

82. Theory of the emission of thermal radio-radiation by a plasma. In this section we summarize the formulae, relevant to the absorption and emission of radiation in a highly ionized gas such as the corona. For more detailed discussions see the references listed above and at the end of this contribution.

α) The emission of thermal radio-radiation in the corona is caused by free-free transitions of electrons in the field of ions, mainly protons. It is easy to ascertain that other radiation mechanisms play no role. The computation of the emission is based on the evaluation of the radiation emitted by an electron moving in the x direction and passing an ion (charge Ze) at a minimum distance p. The electron is accelerated and this leads to isotropic emission of radiation, the intensity integrated over the spectrum being $Q = 2e^2(\ddot{x})^2/3c^3$. The further computation involves the finding of the spectral distribution of the emitted radiation first for one single collision and next by performing the integration over all possible collision parameters p. If this latter problem is treated classically divergences would occur if p would be permitted to take all values between 0 and ∞,

[1] S. PICKELNER: Dokl. Akad. Nauk. SSSR. **72**, 255 (1950). — Isw. Krymsk. Astrophys. Obs. **7**, 87 (1951).

[2] Cf. footnote 7, p. 282.

[3] H. ALFVÉN: Ark. Mat. Astronom. Fys. A **28**, No. 6 (1942).

[4] Cf. footnote 8, p. 282.

[5] H. SHIMOODA: Publ. Astr. Soc. Japan **8**, 95 (1957).

[6] S. CHAPMAN: Smithson. Contrib. Astrophys. **2** (1), 1 (1957).

but these can be avoided by considering that deflections greater than—say—$\pi/2$ do hardly occur. This is in fact the point where approximations are introduced; the lower limit of p has been defined differently by different authors. From a detailed *quantum mechanical discussion*, in which this cut-off problem is treated more precisely, ELWERT[1] derived the following expression

$$4\pi\,\mathcal{E} = N_i N_e \frac{32\sqrt{2\pi}\,e^6 Z^2}{3c^3 m^{\frac{3}{2}} (kT)^{\frac{1}{2}}} \ln \frac{kT}{1.44\, Z e^2 N_i^{\frac{1}{3}}}.$$

$\mathcal{E}$ is the emitted radiation per unit solid angle, Z is the nuclear charge, the other symbols have their usual meaning. The logarithmic term varies only slightly with temperature and density.

The absorption coefficient $\varkappa_\lambda$ is then found with KIRCHHOFF'S law, $\mathcal{E}_\lambda = \varkappa_\lambda B(T)$ where B can be replaced by the asymptotic Rayleigh-Jeans law, which is a good approximation in the range of radio wavelengths:

$$B(T) = \frac{2k}{c^2} f^2 T. \tag{82.1}$$

Here f is the frequency in Hz[2]; $f = c/\lambda$. In local thermodynamic equilibrium

$$\varkappa_\lambda = \mathcal{E}_\lambda \lambda^2 (2kT)^{-1}. \tag{82.2}$$

Since, roughly, the corona consists of 80% H, 20% He and is wholly ionized:

$$N_i N_e Z^2 = 1.6\, N_e^2$$

and hence

$$\varkappa_\lambda = \frac{\zeta N_e^2}{f^2 T_e^{\frac{3}{2}}} \quad (\text{cm}^{-1}) \tag{82.3}$$

where

$$\zeta = 1.32 \times 10^{-2} \ln (367\, T N_e^{-\frac{1}{3}})$$

is a slowly varying function of N_e and T and, hence of λ. Average ζ-values, sometimes useful, are found by inserting in this expression the T and N_e-values at the depth where the emitted radiation mainly comes from:

λ	1	3	10	30	100	300	600 cm
ζ	0.085	0.097	0.124	0.15	0.17	0.18	0.18

The above theory holds no longer when the gas has a refractive index $n \ll 1$, as occurs in the corona (region of metre waves). In that case one has to consider the anomalous dispersion of the radiation near the "resonance frequency", f_0, being the natural oscillation frequency of the plasma:

$$f_0 = (e^2 N_e/\pi m)^{\frac{1}{2}} = 8.98 \times 10^{-3} \sqrt{N_e}\ [\text{MHz}]$$

(cf. also Sect. 91 and the Table 17 on p. 271). The theory (MARTIN; SHKLOVSKI; SMERD and WESTFOLD) shows that in that case the absorption coefficient $\varkappa_\lambda$ becomes nearly identical to the expression (82.3) but for a factor $1/n$. So in the *general case* one may assume

$$\varkappa_\lambda = \frac{\zeta N_e^2}{n f^2 T_e^{\frac{3}{2}}} (\text{cm}^{-1}) \tag{82.4}$$

[1] G. ELWERT: Z. Naturforsch. 3a, 477 (1948).

[2] Instead of the somewhat cumbersome notations c/s and Mc/s, generally used in the English literature, we will use throughout Hz and MHz.

β) The intensity of the emergent radiation is

$$I_\lambda(0) = \int_0^\infty S_\lambda(\tau_\lambda)\, e^{-\tau_\lambda}\, d\tau_\lambda \tag{82.5}$$

where S is the source function, and

$$\tau_\lambda = \int_0^s \varkappa_\lambda\, ds.$$

Often S is identified with the black-body function B [expression (82.1)].

γ) If $n \neq 1$ the ray paths are curved and the optical depth has to be computed along these curved rays, with the usual Snellius equation:

$$n\, r \sin i = a,$$

where i and the constant a are defined in Fig. 93a; the expression transforms for a spherically symmetric corona to:

$$i = a \int_0^\infty \frac{d\varrho}{\varrho\sqrt{n^2\varrho^2 - a^2}}. \tag{82.6}$$

The index of refraction n in the absence of a magnetic field is

$$n = (1 - f_0^2/f^2)^{\frac{1}{2}} \tag{82.7}$$

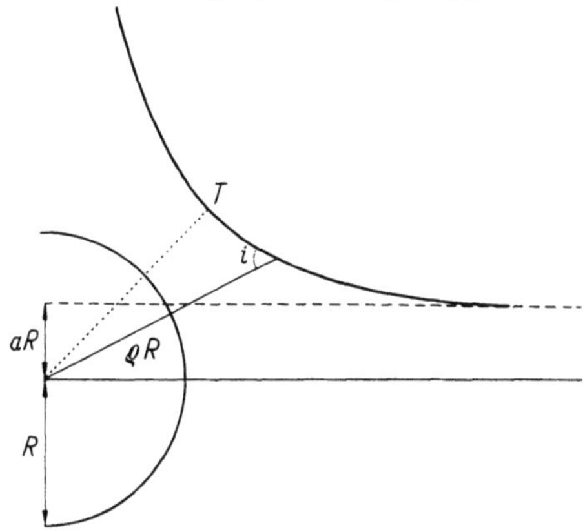

Fig. 93 a.

where f and f_0 are the frequency and the plasma frequency. The ray trajectories thus computed assuming a model of the corona in which N_e decreases outward

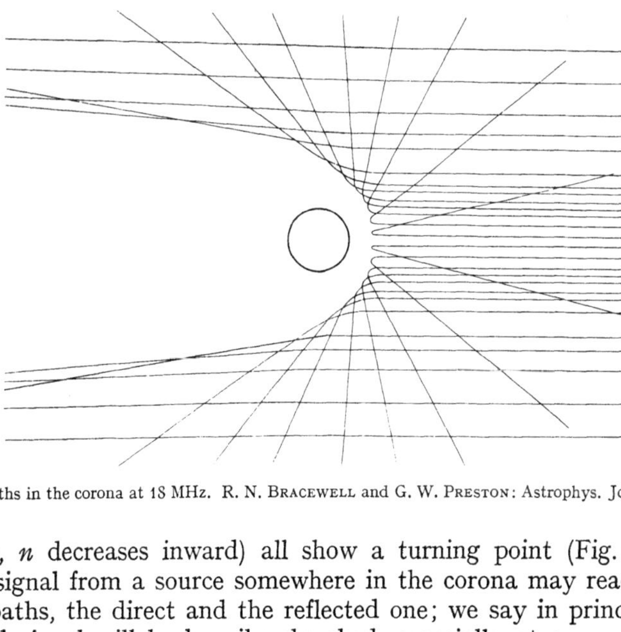

Fig. 93b. Ray paths in the corona at 18 MHz. R. N. BRACEWELL and G. W. PRESTON: Astrophys. Journ. **123**, 14 (1956).

(and, hence, n decreases inward) all show a turning point (Fig. 93b). So in principle a signal from a source somewhere in the corona may reach the Earth along two paths, the direct and the reflected one; we say in principle, because the reflected signal will be heavily absorbed, especially at cm and dm waves. For $f < f_0$, n is imaginary; radio-waves originating in a layer below the layer with $n = 0$ are strongly damped and cannot escape outward.

δ) In the presence of a magnetic field H the electrons, vibrating under the influence of an electromagnetic radiation field, are influenced by a force perpendicular to the direction of motion. In that case the emitted radiation becomes elliptically polarized; it is circularly polarized for the gyro-frequency:

$$f_H = \frac{e\,H}{2\pi\,m\,c} = 2.80\,H\ (\mathrm{MHz}).$$

Moreover, in an oscillating plasma the phase velocity of the waves will be modified, due to the combination of the incident radiation field and that re-emitted by the electrons. In the presence of a magnetic field two modes of propagation will occur: along the paths of the ordinary and of the extraordinary ray, with converse polarization, and with different absorption coefficients[1]. Each of these has its own refractive index n_0 and n_e: for the ordinary ray the refractive index n_0 is determined by (82.7); for the extraordinary ray

$$n_e = \sqrt{1 - \frac{f_0^2}{f(f - f_H)}}.$$

ε) The state of polarization of an electromagnetic wave received at the Earth need not be equal to the state of polarization in the region where the radiation is mainly emitted, it can be considerably changed on the ray-path, even in the ionosphere (for metre waves).

ζ) Finally some definitions. It is customary to express the solar *flux density*, usually measured in watt per m² per Hz, into the *apparent disk temperature* T_d. This is the temperature of a black body with the same angular diameter as the optical Sun, which emits in the same wavelength region the observed amount of radiation. This definition is the more easy to handle since for these long wavelengths PLANCK'S law reduces to that of RAYLEIGH-JEANS (82.1), making T_d proportional to the emitted radiation.

Analogously the *brightness temperature* T_b represents the surface brightness in a point of the disk. It is connected to the local kinetic temperature T_k by:

$$T_b = \int_0^\infty T_k(\tau)\,e^{-\tau}\,d\tau.$$

The integration is made along the ray trajectory.

83. The radio-spectrum of the quiet Sun. Records of the solar radioemission show various components (Fig. 94): the *background radiation*, constant for years and showing only a small variation with the solar cycle, and a transient component consisting of

(i) a *slowly varying component*, period about 27^d (=synodical rotation period), discovered on decimetre waves, but also faintly visible on short metre waves;

(ii) *isolated bursts*, with an average duration of one second or less, mainly on metre waves but also on decimetre waves;

(iii) *outbursts*, on all frequencies, but mainly on metre waves, duration of minutes to some tens of minutes;

(iv) *noise storms* (restricted to metre waves) consisting of numerous short-lived bursts ($\approx 0^s.5$), and lasting for hours or days;

(v) an *enhancement of the base level* often occurs shortly after an outburst or together with a noise storm.

[1] See e.g. T. TAKAKURA: Publ. Astr. Soc. Japan **8**, 204 (1957).

The intensity of the *quiet Sun* is obtained by eliminating the transient components, presumably due to the centres of activity. In the region of the *decimetre wavelengths* the elimination of the slowly varying component due to thermal emission of facular centres may present difficulties when there are many sources. The difficulties are overcome by plotting (DODSON[1]) the intensity of the Sun's

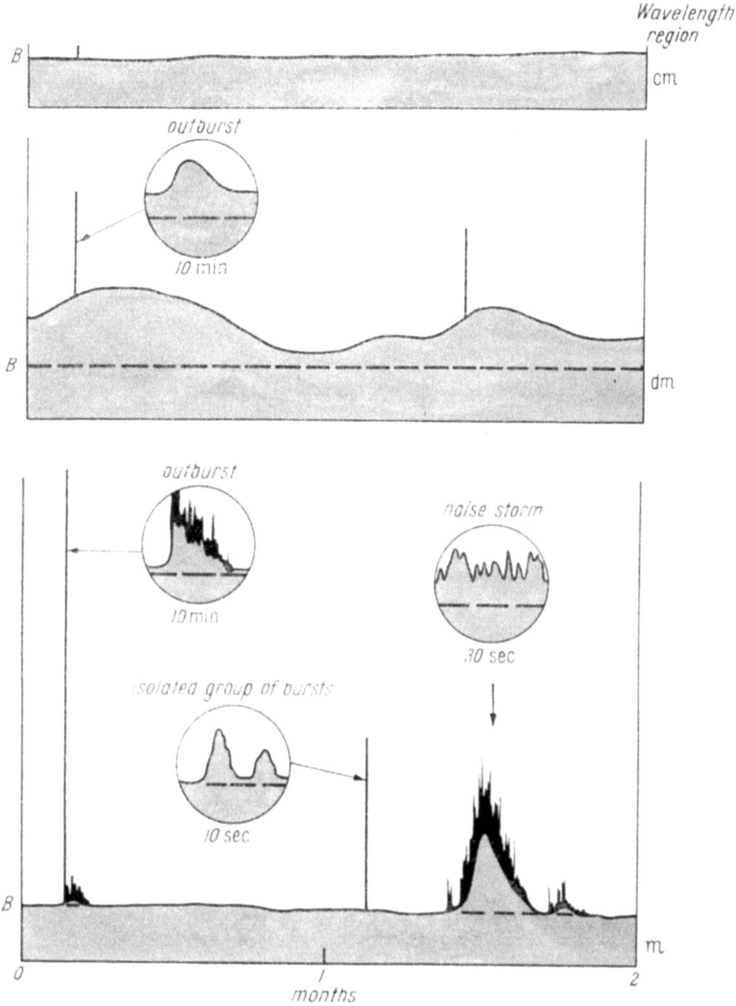

Fig. 94. Idealized types of solar radio transients at centimetre, decimetre and metre waves. J. L. PAWSEY and S. F. SMERD: in: The Sun (ed. G. P. KUIPER), p. 479. Chicago 1953.

radiation at decimetre wavelengths against the area of the $K(Ca^+)$ faculae insofar as these are brighter than a certain, arbitrarily fixed limit (1.5 times the radiation of the undisturbed Sun in the centre of the K line). The curve thus obtained is extrapolated to zero facular area. The good correlation, so far found for radiation at λ 10.7 cm (Fig. 95a) indicates that the enhanced radiation at decimetre wavelengths is emitted by regions having the same disk distribution as the bright faculae. The same good correlation was found earlier by CHRISTIANSEN and

[1] H. W. DODSON: McMath-Hulburt Obs. Reprint **41** (1955).

WARBURTON[1] (21 cm) and confirmed later by CHRISTIANSEN *et al.*[2] in a detailed investigation of the distribution of radiosources on the disk at 21 cm. DENISSE

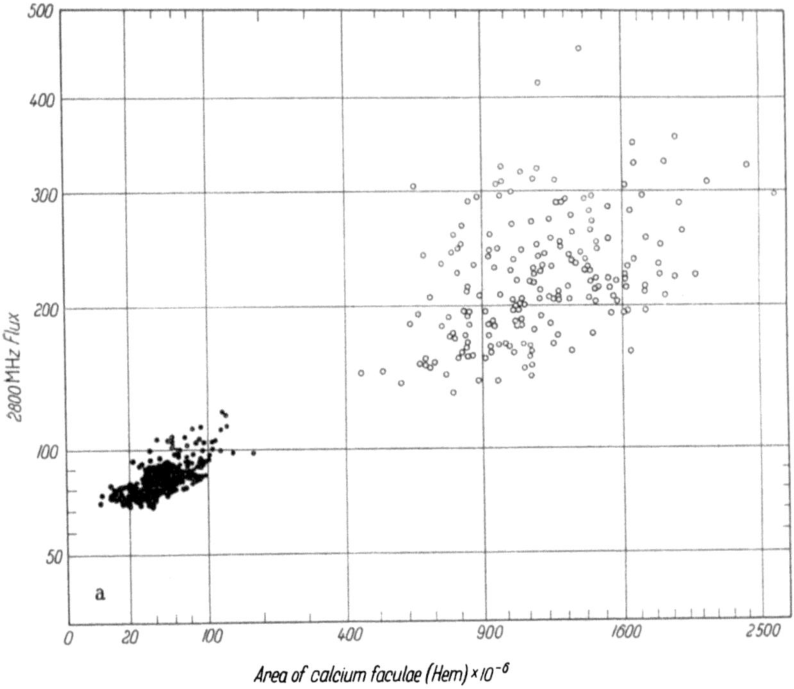

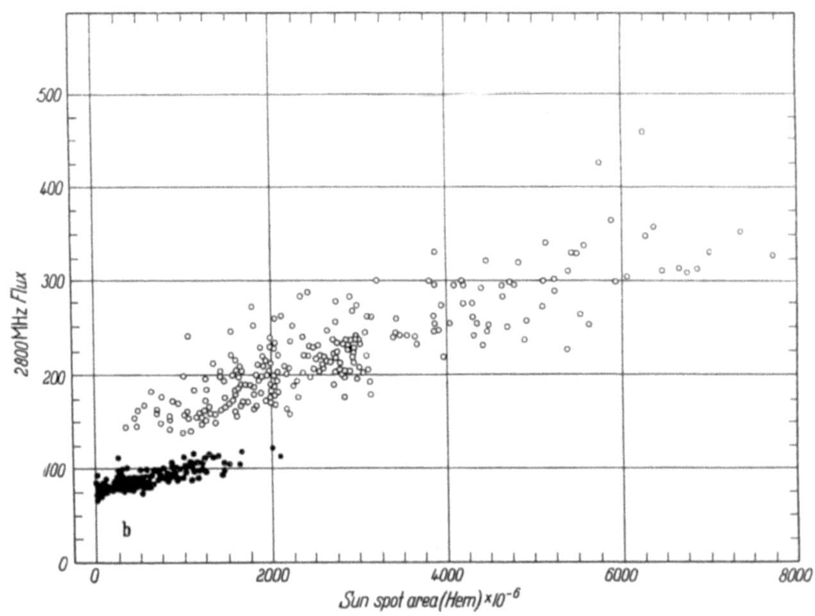

Fig. 95a and b. 2800 MHz flux versus area of calcium plages (above) and versus sunspot area (below) for 1947 (○) and 1952 (●) H. W. DODSON: McMath-Hulbert Observatory. Reprint **41** (1956).

[1] W. N. CHRISTIANSEN and J. A. WARBURTON: Austral. J. Phys. **6**, 190 (1953).

[2] W. N. CHRISTIANSEN, J. A. WARBURTON and R. D. DAVIES: Austral. J. Phys. **10**, 491 (1957).

and KUNDU[1] found a good correlation between the 10.7 cm radiation and the E layer ionizing flux; this relation permits to define a quiet Sun.

Earlier it had been tried to find the background component by correlating the radio radiation and the visible sunspot area, but this correlation was less good. It could, however, be improved[2,3] by correlating the radio radiation with the spot area observed two rotations earlier. This result again suggests the primary correlation with faculae (longer lifetime of faculae).

In the region of *metre waves* the correlation with spots is still worse. There are active and inactive spot-groups. An often used way to find the intensity of the undisturbed Sun is by making a histogram of measured intensities and to look for the lowest intensities, or—perhaps better—for the lowest maximum in the histogram.

Thus the radio-spectrum of the quiet Sun may be obtained as shown in Fig. 96 (after ALLEN[4], with additions[5-7]). It then appears that the quiet Sun intensity still differs for the sunspot maximum and minimum, at least at wavelengths greater than about 1 cm, notwithstanding the reduction to activity level zero (cf. also [8-10]). This result is compatible with that found from optical coronal observations: the electron density of the quiet corona differs in minimum and maximum.

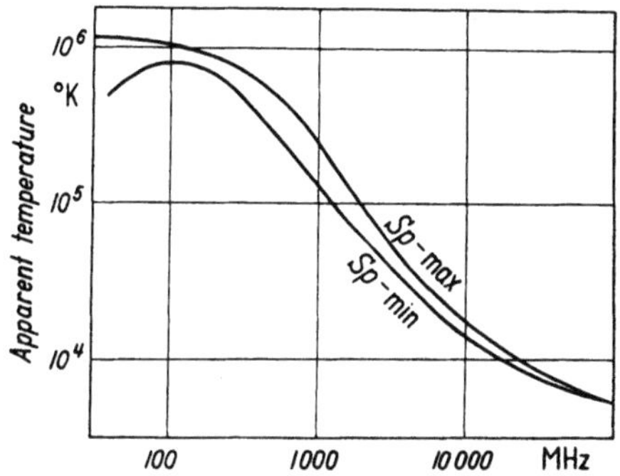

Fig. 96. Radiospectrum of the Sun for spot maximum and minimum. C. W. ALLEN: Monthly Notices Roy. Astronom. Soc. London **117**, 174 (1957).

Also the brightness temperature T_b of the Sun's centre, as deduced from high-resolution observations (Figs. 96 and 97) increases with increasing wavelength; this is due to the increase of $\varkappa_\lambda$ with λ: the corona is transparent for small wavelengths and absorbing for longer ones.

The spectrum described here is the *continuous* spectrum of the Sun. The problem whether the spectrum might also contain absorption or emission *lines* has been discussed by various authors. The 21 cm line of ground-state neutral hydrogen does not seem to occur in the solar spectrum because the number N_1 of ground-state neutral H atoms is too small in the transparent parts of the upper chromosphere and corona.

In Sect. 27 mention has been made of the probability to find the line at 3.03 cm ($2P \rightarrow 2S$ transition of H).

84. The disk distribution of the quiet Sun's radio radiation. The expression "disk distribution" is not wholly correct since at metre wavelengths an appreciable

[1] J. F. DENISSE and M. R. KUNDU: C. R. Acad. Sci., Paris **244**, 45 (1957).
[2] J. H. PIDDINGTON and R. D. DAVIES: Monthly Notices Roy. Astronom. Soc. London **113**, 582 (1953).
[3] C. W. ALLEN, in: Radio Astronomy I.A.U. symposium No. 4, p. 253, 1957.
[4] C. W. ALLEN: Monthly Notices Roy. Astronom. Soc. London **117**, 174 (1957).
[5] R. J. COATES: Astrophys. Journ. **128**, 83 (1958) (at 8.6 and 4.3 mm: $T_b = 8500°$ and $7000°$).
[6] R. N. WHITEHURST and F. H. MITCHELL: Astronom. J. **61**, 192 (1956) (at 7.5 mm: $T_b = 6000° \pm 500°$).
[7] R. N. WHITEHURST, F. H. MITCHELL and J. COPELAND: Astronom. J. **62**, 38 (1957) (at 6 mm: $T_b = 4500° \pm 400°$).
[8] Cf. footnote 2, p. 288.
[9] H. TANAKA: Proc. Res. Inst. Atmosph. **3**, 117 (1955).
[10] A. D. FOKKER and L. D. DE FEITER: Ionosphere and Radio Astron. Section, Netherl. P.T.T. rep. No. **4** (1954).

intensity is still observable beyond the position of the optical limb. Another characteristic is that the quiet radio Sun is not spherically symmetric; at least for wavelengths greater than 10 cm the isophotes show a polar flattening.

The variation of the radiation intensity across the disk can be determined by interferometric and by eclipse observations. Such observations permit (a) to discover localized regions of enhanced intensity, (b) to determine the distribution of undisturbed radiation. Especially on metre waves, where the Sun is virtually never calm, the eclipse observations may often be useful for the first purpose, not so for the latter.

α) Eclipse observations. When the various parts of the Sun are covered by the Moon this enables one to discriminate between the contribution from different parts of the disk. The limit of resolution is imposed by diffraction; it is of the order of $\sqrt{\lambda/d}$ where d is the distance of the Moon to the Earth: at $\lambda = 10$ cm it is 3″. For the scanning of the Sun's disk partial eclipses are just as useful as total ones (except for $\lambda \lesssim 3$ cm), but observations from one site alone are as a rule insufficient to determine the whole two-dimensional intensity pattern of the Sun. More complete results may be obtained from simultaneous observations from two or more stations, but in no case can eclipse observations give unambiguous data on the total solar brightness distribution. Moreover, never has one surveying discussion been given of all radio observations of the same eclipse.

Nevertheless, some of the fundamental discoveries were made during eclipses: on 23 November, 1946 COVINGTON found that centres of activity may emit an enhanced radiation at decimetre waves (cf. Sect. 87). That the radio corona at 178 cm is elliptical appears from the work of BLUM, DENISSE and STEINBERG[1], who could show that their eclipse observations at 178 cm of 1 September 1951 and 25 February 1952 were incompatible with a spherical symmetric corona. Their derived axis ratio was 0.65 to 1.

A complete list of all eclipse observations up to 14 February 1953 is given by HEY[2]. Later observations, especially referring to the eclipse of 1954 are found in the report of the 1955 Jodrell Bank symposium.

β) Interferometric observations. High resolution aerials are gradually coming into regular use at various observing stations. Four types are used:

(a) At cm wavelengths a sufficiently large receiver yields a fairly high resolution, e.g. at 1 cm the diameter of a paraboloid should be 10 metres to yield a resolving power of 3′ (HAGEN's parabolic receiver at 8.5 mm has a resolving power of 4′).

(b) The cliff interferometer, mainly of historical interest, consists of an aerial mounted on a cliff and makes use of the interference between the direct signals and those reflected against the sea surface.

(c) The spaced-aerial interferometer makes use of two or 2^n separate aerials connected on the same receiver [e.g. at Nancay (France) a resolving power of 3′.8 at 177 cm is obtained by means of 16 5-metre paraboloids over a base length of 1550 metres). At Mitaka (Tokyo) a half-power width of 4′.5 at 7.5 cm is obtained with an interferometer consisting of 8 parabolic reflectors of 1.5 m diameter.

By means of phase-shifting devices the pattern of an interferometer aerial may be swept over the Sun, thus permitting rapid scanning, and this procedure may be repeated for various mutual distances of the aerials (Cambridge, England), thus enabling the observer to find the centre-limb distribution of the brightness temperature $T_b(r)$.

[1] E. J. BLUM, J. F. DENISSE and J. L. STEINBERG: C. R. Acad. Sci., Paris **234**, 1597 (1952). — Ann. d'Astrophys. **15**, 184 (1952).

[2] J. S. HEY: Vistas in Astronomy **1**, 526 (1956).

Let n be the distance between the two antennas, expressed in units of λ, then in the process of phase shifting the amplitude A is:

$$A(n) = 2\pi \int_0^\infty T_b(r)\, r\, J_0(2\pi n r)\, dr,$$

i.e. the Fourier-Bessel transformed of T_b. Hence,

$$T_b(r) = 2\pi \int_0^\infty A(n)\, n\, J_0(2\pi n r)\, dn.$$

(d) Finally there are systems making use of two such aerials, perpendicular to each other, permitting two-dimensional interferometry. A first version was CHRISTIANSEN and WARBURTON's system consisting of a 32 paraboloid aerial and one of 16 perpendicular to it. The most elaborate specimen is CHRISTIANSEN's new system[1] consisting of two 32 aerial interferometers arranged in cross formation (64 aerials each 6 metres in diameter; length of each arm 380 metres; frequency 1430 MHz). These interferometers can be used separately or else coupled together in "Mills Cross" fashion to produce pencil beams 3′ in diametre and spaced 1° apart. The fringes may be moved by means of phase shifters and the Sun can therefore be scanned line by line analogous to the formation of a television picture.

The interpretation of interferometric observations in which a good resolving power is obtained in one direction only, offers an interesting obliteration problem which has been discussed by various authors, cf. [2-5]. As in all such kinds of problems one has to reach a very high degree of observational accuracy in order to infer with reasonable accuracy the energy distribution of the Sun. The accuracy attained with interferometric observations is often overestimated; for a detailed discussion of the uncertainty caused by the finite distance of the elements, cf. [5].

The accompanying Fig. 97 shows for wavelengths between 0.85 and 790 cm the Sun's intensity profile for the quiet minimum Sun (years around 1954) as derived by various methods. When a two-dimensional intensity distribution is known, both the polar (dashed) and equatorial (solid) profiles are given. The interesting form of the system of isophotes is shown in detail for 21 cm and 60 cm in Fig. 98. Notable is the bright (equatorial) ring (segments), perceptible at $\lambda \approx 1$ cm but impressively strong near 10 cm, broadening and shifting towards the disk centre at wavelengths around 50 cm and, finally, disappearing at metre waves. This shows how the optical thickness of the corona, being very small at 1 cm, becomes greater for longer wavelengths, finally turning into complete opacity for metre waves. The shift of limb maximum towards the disk centre for increasing wavelengths may partly be of instrumental origin but shows also the increasing importance of the ray curvature. Also the increasing flattening of the corona for increasing wavelength is an interesting phenomenon.

A slight polarization of the quiet Sun's thermal radiation, due to the poloidal magnetic field is predicted (SMERD[6]), with a differential polarization of 0.2% for an assumed magnetic field of 2 Gauss. The existing observations[7] are not accurate enough to detect this polarization[8].

[1] Footnote 3, p. 250.
[2] S.F. SMERD and J.P. WILD: Radio astronomy. I.A.U. colloquium No. 4, p. 290, 1957.
[3] J. ARSAC: Austral. J. Phys. **10**, 16 (1957).
[4] S.F. SMERD and J.P. WILD: Phil. Mag. **2**, 119 (1957).
[5] G. ELWERT and H. SIEDENTOPF: Z. Naturforsch. **11**a, 769, 961 (1956).
[6] S.F. SMERD: Austral. J. Sci. Res. **3**, 265 (1950).
[7] P.A. O'BRIEN and E. TANDBERG HANSSEN: Observatory **75**, 11 (1955).
[8] R.G. CONWAY: Observatory **76**, 106 (1956).

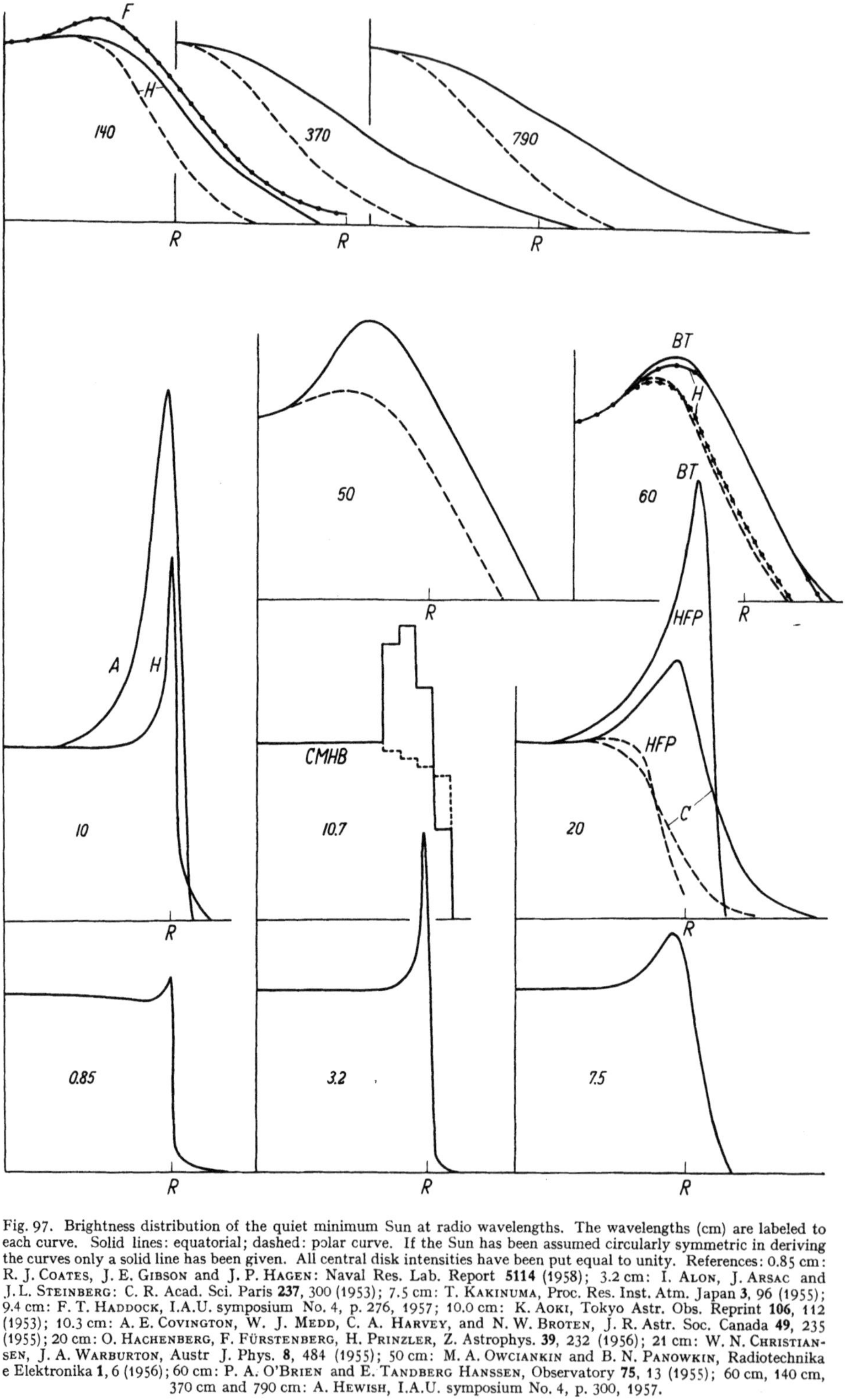

Fig. 97. Brightness distribution of the quiet minimum Sun at radio wavelengths. The wavelengths (cm) are labeled to each curve. Solid lines: equatorial; dashed: polar curve. If the Sun has been assumed circularly symmetric in deriving the curves only a solid line has been given. All central disk intensities have been put equal to unity. References: 0.85 cm: R. J. COATES, J. E. GIBSON and J. P. HAGEN: Naval Res. Lab. Report **5114** (1958); 3.2 cm: I. ALON, J. ARSAC and J. L. STEINBERG: C. R. Acad. Sci. Paris **237**, 300 (1953); 7.5 cm: T. KAKINUMA, Proc. Res. Inst. Atm. Japan **3**, 96 (1955); 9.4 cm: F. T. HADDOCK, I.A.U. symposium No. 4, p. 276, 1957; 10.0 cm: K. AOKI, Tokyo Astr. Obs. Reprint **106**, 112 (1953); 10.3 cm: A. E. COVINGTON, W. J. MEDD, C. A. HARVEY, and N. W. BROTEN, J. R. Astr. Soc. Canada **49**, 235 (1955); 20 cm: O. HACHENBERG, F. FÜRSTENBERG, H. PRINZLER, Z. Astrophys. **39**, 232 (1956); 21 cm: W. N. CHRISTIANSEN, J. A. WARBURTON, Austr J. Phys. **8**, 484 (1955); 50 cm: M. A. OWCIANKIN and B. N. PANOWKIN, Radiotechnika e Elektronika **1**, 6 (1956); 60 cm: P. A. O'BRIEN and E. TANDBERG HANSSEN, Observatory **75**, 13 (1955); 60 cm, 140 cm, 370 cm and 790 cm: A. HEWISH, I.A.U. symposium No. 4, p. 300, 1957.

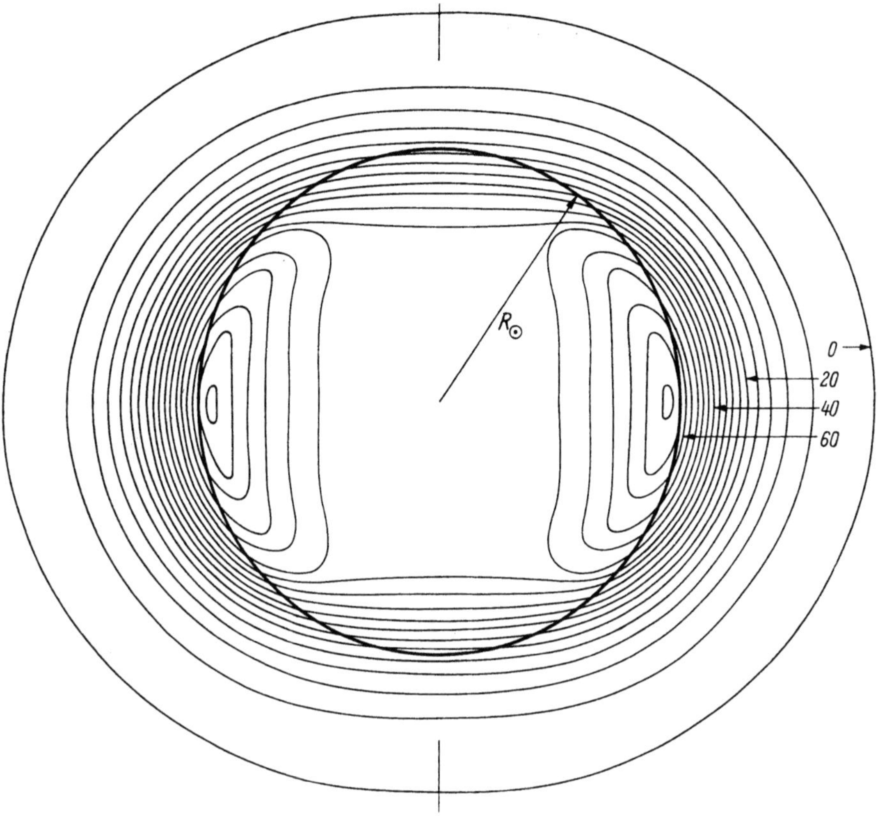

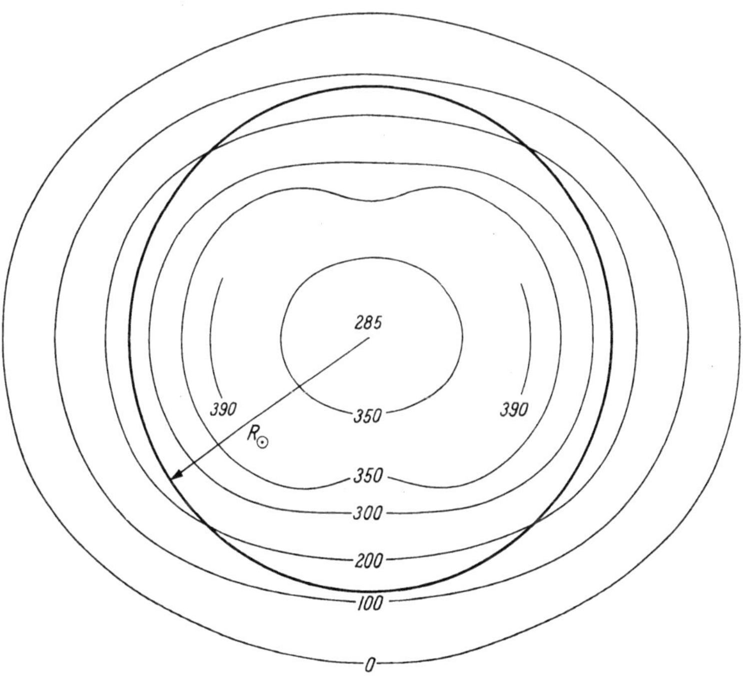

Fig. 98. Average brightness distribution of the quiet minimum Sun at 21 cm (above) and at 60 cm (below). The lines are labeled with the brightness temperature (unit: 10^3 °K). 21 cm: W. N. Christiansen and J. A. Warburton: Austral. J. Phys. **8**, 474 (1955). 60 cm: R. G. Conway and P. A. O'Brien: Monthly Notices Roy. Astronom. Soc. London **116**, 386 (1956).

85. Absorption and scattering of point source radiation by the corona. Eclipse and interferometric observations show that the corona extends much further than is generally assumed on the basis of current coronal models. Also, less limb brightening at metre wavelengths is observed than is computed. These two effects indicate that the corona may scatter the radio-radiation. This conclusion is supported by the observation of the scattering of the radiation from a point source when being occulted by the corona[1,2]. The radio-source Taurus A (the Crab nebula) is occulted each year by the corona near the 14th or 15th of June. The outer parts of the corona at $\varrho > 4.7$ pass over the point source, the influence of the eclipse is visible from about 9 to 22 June. Observations are made at metre wavelengths where the intensity of the Crab nebula is comparable to that of the Sun. The diameter of Taurus A is smaller than that of the Sun, but its radiation can be isolated from the solar background by arranging an interferometer of considerable spacing in such a way that the maxima of the interference pattern subtend an angle small as compared to the solar diameter.

On the basis of the *refraction mechanism*, applied to a spherical homogeneous coronal model in which N_e decreases outward monotonously, one would expect (MACHIN[1], LINK[3]) that the rays of the radiosource would be refracted sidewards so that during the eclipse an initial increase of the intensity of the radiostar would be followed by a reduction at mid-eclipse. Computations based on N_e values deduced from optical estimates of the coronal brightness[4] predicted a rather weak effect at $\lambda \approx 8$ m and no effect at $\lambda \approx 4$ m. Observations from 1950 onward had at first no success because the Sun was too disturbed but later observations, since 1952 showed

(i) a considerable decrease of source brightness at 3.5 and 7.9 metre. In 1952 the intensity went down to 0.25 at mid-eclipse; in 1953 it was even zero[5]. The observations[1,4] showed the influence of the corona up to 15 $R_\odot$ and even to 20 $R_\odot$ [5]. An increase of intensity did not occur at long metre-waves.

This result is incompatible with the homogeneous refraction theory, but suggests *scattering* due to refraction by coronal density irregularities. However at 178 cm an increase of intensity has been observed[6] in 1957 and 1958; this again would plea for the homogeneous refraction theory.

(ii) The position of the source does not change systematically within a range of 0′.5 (argument against refraction).

(iii) Some interferometric observations showed that the apparent angular diameter of the source, being about 4′, increases during the occultation to 12′ at 1.78 m[6], to 18′ at 3.5 m and to 27′ at 5.8 m[5]. Other observers find no or smaller effects[7]. The increase of the diameter indicates the influence of scattering elements in the corona. HEWISH[8,5] assuming scattering elements of various diameters, found $\delta N_e \approx \frac{1}{3} N_e$ for an assumed element diameter of 10^5 km; $\delta N_e \approx 0.1 N_e$ for elements of 10^4 km. Similar results were obtained by VITKEVITCH[4] and SCHEFFLER[9]; there is no unique solution. From the centre-limb variation at

[1] K.E. MACHIN and F.G. SMITH: Nature, Lond. **168**, 599 (1951).
[2] V.V. VITKEVITCH: Dokl. Akad. Nauk SSSR. **11**, 585 (1951).
[3] F. LINK: Bull. Astr. Inst. Czech. **3**, 6 (1952).
[4] V.V. VITKEVITCH: Radio astronomy. I.A.U. symposium No. 4, p. 313, 1957. — Astron. Zhurn. **32**, 150 (1955) and **35**, 52 (1958).
[5] A. HEWISH: Proc. Roy. Soc. Lond., Ser. A **228**, 238 (1955); comm. at 1958 (Paris) Symposium on Radio astronomy (1959).
[6] E.J. BLUM and A. BOISCHOT: Observatory **77**, 205 (1957); comm. at 1958 (Paris) Symposium on Radioastronomy (1959).
[7] O.B. SLEE: Observatory **76**, 228 (1956).
[8] A. HEWISH: Vistas in Astronomy **1**, 603 (1956).
[9] H. SCHEFFLER: Mitt. Astr. Ges. **1956**, 22 (1957). — Z. Astrophys. **45**, 113 (1958).

metre waves, and assuming $L_0 = 100$ km for the diameter of the scattering elements, SCHEFFLER found $[\overline{(\delta N_e)^2}/\overline{(N_e)^2}]^{\frac{1}{2}} = 0.15$ at $\varrho = 5$. Assuming that the inhomogeneities are related to the coronal rays, thus taking $L_0 \approx 10^5$ km, this value is 5 at $\varrho = 5$. In this latter case the far corona would consist of dense regions sourrounded by almost empty ones. HEWISH (footnote 5, p. 294) found that the scattering is greater in directions parallel to the Sun's axis than perpendicular to it. This result could be interpreted assuming that the inhomogeneities have a ray-structure similar to the solar magnetic lines of force. The scattering is smallest near sunspot minimum (1953 to 1954).

We finally note that VITKEVITCH[1] has suggested annually to investigate the outer corona by means of the occultations of Jupiter.

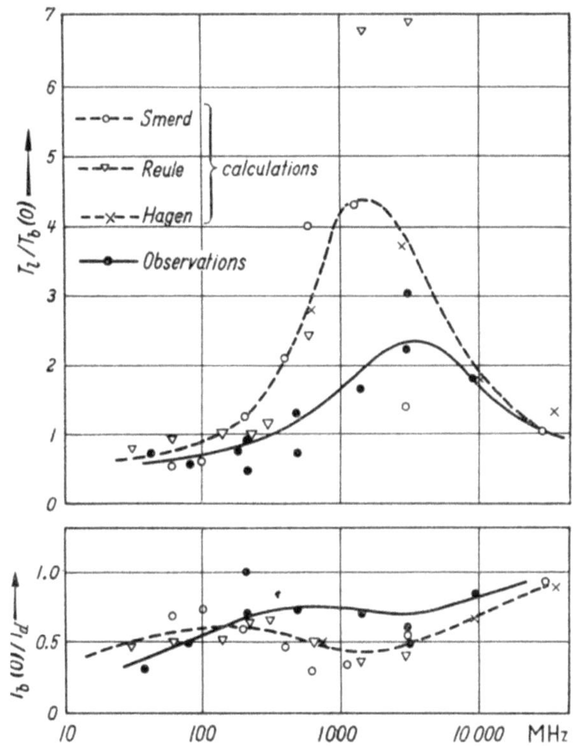

Fig. 99. Comparison of observed and computed parameters of the solar intensity distribution. Upper graph: $T_l/T_b(0)$; lower graph: $T_b(0)/T_d$. C.W. ALLEN: in: Radio astronomy (ed. H. C. VAN DE HULST). I.A.U. symposium No. 4, Cambridge 1957.

86. The structure of the quiet corona from radio-observations. The preceding sections show that the computation of the emergent radiation for a given chromospheric and coronal model is straightforward and offers no difficulties, except for the influence of scattering of radiowaves by inhomogeneities. So, especially in the years before 1952, the computation of the emergent radio-radiation as a function of wavelength and heliocentric angle has been a favourite subject for many authors. (i) Generally, they calculated the radiation for an isothermal chromosphere and an isothermal corona with a discontinuous intensity jump in between. (ii) The earlier computations even neglected the curvature of the rays[2], which is only permissible in the range of cm and short dm waves. (iii) Since moreover the scattering of radio-waves was neglected throughout and (iv) spherical symmetry of the Sun was assumed, it is clear that these early computations could only lead to a rough survey; the refined sets of two-dimensional centre limb observations, now available, should be interpreted in terms of an axial symmetric corona and chromosphere, the latter being non-isothermal, in which the density- and temperature-fluctuations are fully incorporated. This has not yet been done.

---

[1] V.V. VITKEVITCH: Astron. Zhurn. **34**, 217 (1957).

[2] The ray curvature was perhaps considered for the first time by G. BURKHARDT and A. SCHLÜTER: Z. Astrophys. **26**, 295 (1949).

In this section we confine ourselves, following ALLEN[1] to a comparison of some of the main particularities of the observed radiation with those computed for various coronal models.

ALLEN[1] has compared the observed and computed values of the function $T_b(0)/T_d$ (Fig. 99), provisionally assuming that this function does not depend on the solar cycle. Where the observations show a polar flattening, an average of polar and equatorial values has been taken. The observations and the computations apparently have the same trend, but the calculated curve generally is somewhat lower than the observed one, especially near $\lambda = 30$ cm, indicating that the calculated limb brightening is greater than the observed one.

In the same way $T_l/T_b(0)$ was plotted, T_l being the brightness temperature of the intensity maximum at the limb, or, if there is no evidence of limb brightening, T_l being equal to T_b at the limb. Especially near $\lambda = 30$ cm:

$$\left(\frac{T_l}{T_b(0)}\right)_{\text{comp}} > \left(\frac{T_l}{T_b(0)}\right)_{\text{obs}} .$$

Generally, the calculated limb brightening is more pronounced than the observed one.

b) Thermal radiation from centres of activity.

At centimetre and short decimetre waves the solar enhanced radiation consists of short lived *bursts* and a *slowly fluctuating component* which appears roughly to have a period equal to the synodic rotation period of the Sun. Both are correlated with the occurrence of activity centres on the disk. These variable components are caused by thermal radiation of centres of activity. First we discuss the slowly varying component. The bursts shall be dealt with in Sect. 89.

87. Investigation of individual activity centres. *Eclipse observations* were the first to give the clue that centres of activity, usually recognized by their sunspots, emit an enhanced background radiation, especially at decimetre wavelengths. A classical example is COVINGTON's observation[2] made at the eclipse of 23 November 1946 at 2800 MHz = 10.7 cm. At the moment that a large spotgroup was eclipsed or uncovered by the Moon, the solar radiation decreased or increased abruptly. Similar observations have been made later at nearly each eclipse. A year later, COVINGTON[3] reported to have found a close correlation between his daily measurements of the solar radiation at 10.7 cm and the total visible sunspot area.

The forms and sizes of the emission regions can only be studied well when many observations are combined, made at the same wavelengths by various observers at different sites; such observations *sometimes* permit to trace the isophotes of a centre of activity with a refinement that cannot always be reached by interferometric observations. Some cases are mentioned below.

α) The partial solar eclipse of 20 June 1955 was observed in Japan at 10 cm wavelength from three sites differing by four degrees in latitude[4]. There was one strong emission centre, the profile of which could be defined fairly accurately due to the circumstance that at the three observing stations the Moon either wholly covered the emission centre, or let it just free, or again eclipsed it only

[1] Cf. footnote 3, p. 289.
[2] A.E. COVINGTON: Nature, Lond. **159**, 405 (1947).
[3] A.E. COVINGTON: Proc. Inst. Radio Engrs. **36**, 454 (1948).
[4] T. HATANAKA, K. AKABANE, F. MORIYAMA, H. TANAKA and T. KAKINUMA: Publ. Astr. Soc. Japan **7**, 161 (1955).

partly. The contours of the emission centre closely agreed with that of a Ca^+ facular field. The top brightness of the field had a radiation temperature of 3×10^6 °K, the outer contours of the Ca^+ faculae corresponded with $T_b = 10^6$ °K.

β) Sometimes an emission centre emitting at decimetre waves may be located high in the corona. CHRISTIANSEN, YABSLEY and MILLS[1] (1 November, 1948; 600 MHz 50 cm), found eight sources of which seven corresponded with facular areas; three of them contained spots. The eighth source was in the corona above a prominence. This shows also that radiation from disturbed regions may originate from fairly high coronal regions although for this radiation the level for $\tau_\lambda = 1$ or the plasma frequency level are chromospheric. The mean brightness temperature of these areas was 5×10^6 °K, the average area was 4×10^{-3} of the solar disk.

γ) As a rule the position and the contours of the emission fields are found by trial and error, but a more straightforward mathematical method was twice applied by COUTREZ *et al.*[2,3] who represented the solar intensity distribution by an arbitrary function $B(x, y)$, of m parameters with $\int B(x, y)\, dx\, dy = 1$. When the Moon covers the area $L(t)$, the residual flux is $\varphi(t)$. The system of $n (\geq m)$ equations:

$$\int_{L(t)} B\, dx\, dy = 1 - \varphi(t)$$

can be solved.

All eclipse observations suffer from the drawback that eclipses are rare and that they are insufficient for a general insight into the intensities and areas of the disturbed regions at decimeter wavelengths. *Interferometer observations*[4] obtained at $\lambda = 21$ cm with a $32 + 16$ element interferometer (resolving power of 3′ in east-west direction) and later with a 64-elements cross (cf. Sect. 84 β d) permitted not only to detect the individual sources but also to trace their shape (for the largest) and their variations across the disk and during their life. The results are:

1. The positions and the shapes of the radio-sources agree with the associated *faculae*, not with the associated *sunspot-groups*.

2. The flux varies from small values up to 10^{-21} Watt m^{-2} Hz^{-1}. Assuming that in the latter case the emitting region has an area of 10 square minutes of arc, and that the emitting region is homogeneous and does not consist of smaller brighter dots, the average brightness temperature is 0.6×10^6 °K. The maximum value is 1.5×10^6 °K.

3. It seems that the radio-sources move slightly more rapidly over the disk than the corresponding faculae; the interpretation is that the sources are in the average case 40000 km higher than the faculae (maximum height: 10^5 km). Furthermore the centre-limb variation follows a cosine law, indicating that the emitting source must be rather flat for this wavelength.

4. After two rotations the radio emission has in the average reduced to $^1/_6$ of its maximum value, whereas the spot area decreases much faster. So the sources, as the faculae, are more persistent than the spots. The average life time is four months.

---

[1] W. N. CHRISTIANSEN, D. E. YABSLEY and B. Y. MILLS: Austral. J. Sci. Res. A **2**, 506 (1949).

[2] R. COUTREZ, A. KOEKELENBERGH and E. POURBAIX: Observatory **73**, 239 (1953).

[3] R. COUTREZ: Ann. d'Astrophys. **20**, 23 (1957).

[4] W. N. CHRISTIANSEN, J. A. WARBURTON and R. D. DAVIES: Austral. J. Phys. **10**, 491 (1957); and 1958 (Paris) Symposium on Radio astronomy (1959).

88. The spectrum and centre-limb variation of the enhanced radiation of activity centres. Polarization. For those cases in which the radiation of the individual activity centres cannot be determined (e.g. by lack of resolving power) the centre-limb variation of the enhanced radiation has been studied statistically[1-3] by measuring the intensity when the source—assumed characterized by spot-groups—drifts over the disk.

The results are not very reliable but they suggest that the "limb darkening" of the enhanced thermal radiation decreases with increasing wavelength. At

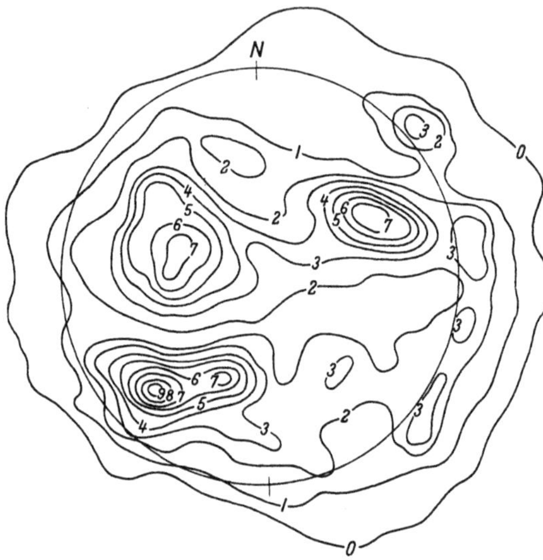

Fig. 100a. Radiobrightness distribution over the Sun at 21 cm (1423 MHz) obtained 14 October, 1957; 02h15 UT, by CHRISTIANSEN et al. The brightness temperature is specified by the number on each contour times 10^5 °K. Courtesy W. N. CHRISTIANSEN, Sydney.

metre waves (e.g. $\lambda = 150$ cm) the intensity of the sources seems to be reduced to 0.3 or 0.2 of its central disk value two days after central meridian passage.

For shorter wavelengths the limb darkening is weaker: the intensity decreases to 0.5 at $r/R = 0.57$ for $\lambda = 50$ cm and at $r/R = 0.75$ at $\lambda = 25$ cm. This agrees with CHRISTIANSEN's finding (Sect. 86) that at 21 cm the centre-limb variation follows a cosine law. Similar results were found at 7.5 cm[4] and at 3.2 cm[5]. Also at these wavelengths the radio-spots appear one to two days before the sunspots and disappear later, indicating that the radio-spots are located at about 35000 km above the photosphere for $\lambda = 7.5$ cm.

λ (cm)	$R=100$	$A=2\times10^{-3}$	$F=2\times10^{-3}$	$T_b\ (\times10^{-3})$		Height ($R_\odot$)	
				E	F	E	F
3				100			
7.5					700	0.03	0.04
10.7	55 (33)	72 (44)	58 (36)	800	3000	0.05	0.07
21		20 to 48			1500	0.10	0.15
25	20 (66)	26 (86)	19 (36)				
50	13 (180)	16 (220)	13 (180)	1600	3200	0.14	0.22

[1] M. WALDMEIER: Z. Astrophys. **32**, 116 (1953).
[2] H. MÜLLER: Z. Astrophys. **39**, 160 (1956).
[3] K.E. MACHIN and P.A. O'BRIEN: Phil. Mag. **45**, 973 (1954).
[4] T. KAKINUMA: Proc. Res. Inst. Atmos. **4**, 78 (1956).
[5] M. GUTMAN: In Paris Radio-Astronomy Symposium (1959).

In order to determine the *spectrum* of an activity centre the observations should first be reduced to the same size of the emitting region. The subsequent table (ALLEN[1], KATAYAMA[2] with some values added by us from other sources) gives for five wavelengths the flux of the 27-day component, respectively for a sunspot

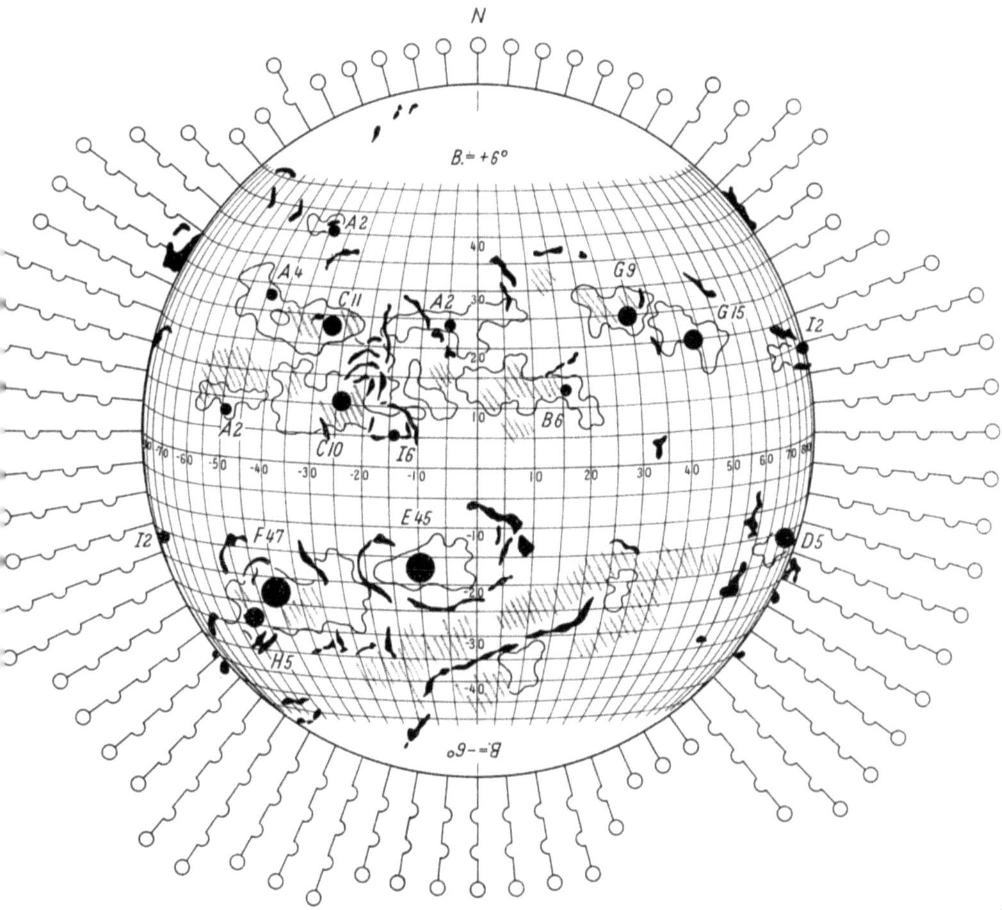

Fig. 100b. Corresponding daily map of the Sun based on optical observations, issued by the Fraunhofer institute (Freiburg, Germany). Black dots: sunspots shaded areas: faculae; black ribbons: filaments and prominences. The coronal intensities (5303 Å) are indicated by the radial lines outside the Sun's limb.

number $R = 100$, a sunspot area $A = 2 \times 10^{-3}$ visible hemisphere and an equal (photospheric) facular area F. The flux data are expressed in 10^{-22} W m^{-2} Hz^{-1}. Between brackets the table gives in the 2nd, 3rd and 4th columns the corresponding increases of the apparent disk temperature ($\times 10^{-3}$) T_d. It further gives the effective heights and brightness temperatures of activity centres belonging to E and F type spots.

Fig. 101 shows the spectrum of the long-lived spot component, thus determined for a projected spot area of 0.002 of the Sun's disk. This kind of radiation has its

[1] C. W. ALLEN: Monthly Notices Roy. Astronom. Soc. London **117**, 174 (1957).

[2] A. KATAYAMA: Paper at meeting Astr. Soc. Japan, 1954, quoted by T. TAKAKURA: Publ. Astr. Soc. Japan **8**, 191 (1956).

greatest importance for wavelengths between 10 and 75 cm, the effect is small outside this region. At the shortest wavelength (0.43 cm) COATES[1] who scanned the Sun with a 3 m paraboloid found that spot areas contributed only a few percent to the total radio emission. At short metre waves a weak and only shortlived thermal radiation is found in activity centres, bright centres at 178 cm only sometimes emit a continuous radiation, which however seems not to be related to CA's.

COVINGTON[2] and PIDDINGTON and MINNETT[3] have found a small and variable circular polarization up to 15% at $\lambda = 10$ cm. At 7.5 cm the value usually does not exceed 10%[4]; at 20 cm no polarization was found (footnote 4, p. 297). The observations for June to August 1948 are to be interpreted as follows: Sunspots in the NE-SW quadrants show an excess of left-handed circular polarization (the reverse in the NW-SE quadrants), but for the cycle starting in 1954, the polarization turned to right-handed in these quadrants ($\lambda = 7.5$ cm; TANAKA and KAKINUMA[5]). So the sense of this (small) circular polarization generally reverses when the source passes the central meridian. (The sense of rotation of a circularly or elliptically polarized wave is defined as right-handed if the electric vector in a fixed plane in the wave front rotates clockwise when viewed into the direction of propagation; the source being behind the observer[6]). These observations suggest a relationship with the magnetic fields of the spots or faculae.

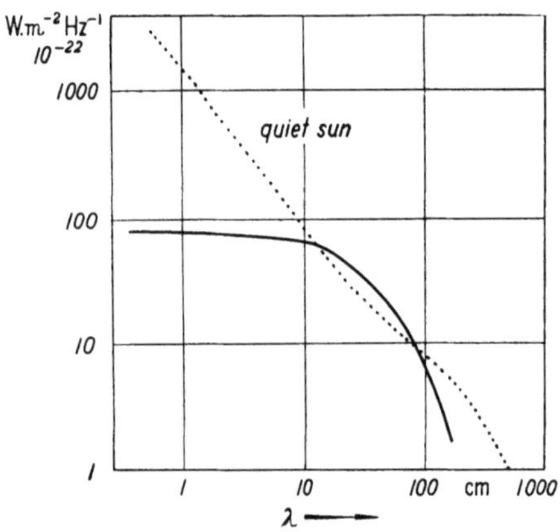

Fig. 101. Solid line: Tentative spectrum of the 27 days component, for a sunspot area $A = 2 \times 10^{-3}$ visible hemisphere. For comparison the quiet Sun spectrum is given by a dotted line.

89. The sources of the enhanced thermal radio-radiation; the cm-bursts. The interpretation of the slowly varying component, as being due to the thermal radiation of centres of activity, is quite natural because the radiation temperatures of these emission sources never exceed values of 10^7 °K, and because of the rather long lifetimes of these sources. The relation with the regions of enhanced activity of the λ5303 line was noted by LAFFINEUR *et al.*[7]. Earlier, WALDMEIER and MÜLLER[8] already suggested that "permanent coronal condensations" are the source of the slowly varying component. Computations based on a detailed model of such a condensation[8,9] (cf. also Fig. 91 in Sect. 79) showed the best agreement with the observed spectrum, (Fig. 101), for $T_{el} = 6 \times 10^6$ °K, higher than is found from the photometry of the coronal lines. A similar result was found by PIDDINGTON and MINNETT[3], who used a homogeneous model of a condensation ($T = 10^7$ °K; $N_e = 10^9$ cm^{-3}).

However, also in this connection a clear distinction should be made between a coronal condensation and a simple active coronal region. *K*-coronameter

[1] R. J. COATES: Proc. Inst. Radio Engrs. January **1958**.
[2] A. E. COVINGTON: Proc. Inst. Radio Engrs. **37**, 407 (1949).
[3] J. H. PIDDINGTON and H. C. MINNETT: Austral. J. Sci. Res. **4**, 131 (1951).
[4] See footnote 4, p. 298.
[5] H. TANAKA and T. KAKINUMA: Proc. Res. Inst. Atmos. **3**, 102 (1955).
[6] Cf. Trans. Internat. Astronom. Union **9**, 576 (1957).
[7] M. LAFFINEUR, R. MICHARD, J. C. PECKER and B. VAUQUOIS: Ann. d'Astrophys. **17**, 358 (1954).
[8] M. WALDMEIER and H. MÜLLER: Z. Astrophys. **27**, 58 (1950).
[9] M. WALDMEIER: Z. Astrophys. **32**, 116 (1953).

observations[1] showed that the average density in the latter regions is only twice that of the quiet corona, and that the temperature is near 10^6 °K. With such a model NEWKIRK[1] could quantitatively explain the radiobrightness of active regions at the limb and their radiospectrum. So, the *coronal activity regions* have to be considered as the *sources of the slowly varying component*. Often the statement is read that the sources should be flat ($\lesssim$100000 km) since they are not observed at metre wavelengths. This conclusion is not correct: a local *density* increase over the whole range of heights of an isothermal corona will not result in a local increase of the brightness temperature, but a *temperature* increase does. So the only conclusion can be that the coronal activity regions do not differ much in temperature from the quiet corona.

The *centre-limb observations* (Sect. 88) reveal that the directivity of the radiation increases with increasing wavelength. At decimeter waves the observations show a cosine curve and can be explained by the effect of perspective. At longer waves the lower parts of a condensation at the limb are obscured by the low corona: the reflection level is not spherical, and its distance to the photosphere is greater for greater λ-values (see also Fig. 93). If a source, situated relatively close to the layer $n=0$, moves over the disk by the solar rotation, it might become invisible from a certain value of f upward, because of its disappearance behind the reflection layer. A complicating factor is that the ray paths are curved at metre wavelengths, thus apparently shifting the limb condensation towards the disk centre.

In these model considerations the influence of the magnetic field is not yet incorporated; its rôle might be important in a final solution. The observations at 10 cm exhibit some circular polarization, indicating the influence of the magnetic fields of the BM regions. So in this case the influence of the splitting of the rays into an ordinary and an extraordinary component has to be taken into account (cf. Sect. 82δ). Because of the magnetic fields the layer $n_e=0$ is higher than the layer $n=0$ in the field-free case. The extraordinary ray follows a shorter path and is at the same time more heavily absorbed at the same height than the ordinary ray, so that the resulting radiation is polarized (PIDDINGTON and MINNETT[2]).

Finally a short note on the *bursts* at centimetre-wavelengths. The observations[3,4] show the greater part (90%) of the centimetre burst, for $\lambda < 20$ cm, to be rather smooth, in great contrast to the bursts observed at metre waves. Often the tail part seems to decay exponentially with a decay time of 10 to 20 min[4]. Their top intensities are of the same order of magnitude as the quiet Sun's level. About half of them show circular polarization. The hypothesis that most of these bursts are caused by thermal radation emitted by flare regions is supported by (1) the decay time, being of the same order as that of the flares; (2) the relatively low top intensities; (3) their diameters, being small ($\lesssim 1'$); (4) the observation that many centimetre bursts correspond with a flare (cf. footnote 4, p. 308), and (5) the fact that the solar centimetre radiation is emitted by the same height levels where the flares tend to occur.

c) Non-thermal radiation: type III and type *U* bursts.

The type III and U type bursts are discussed in this subdivision before we shall treat the more complicated burst phenomena in subdivision ***d***, since the

[1] G. NEWKIRK: In I.A.U. Radio-Symposium at Paris (1959).
[2] J.H. PIDDINGTON and H.C. MINNETT: Austral. J. Sci. Res. **4**, 131 (1951).
[3] O. HACHENBERG: Z. Astrophys. **46**, 67 (1958).
[4] A.E. COVINGTON, M.R. KUNDU and H. TANAKA and T. KAKINUMA: papers presented at 1958 I.A.U. Radio-Symposium at Paris (1959).

Table 18. *Classification of non-thermal radio-*

Class	Duration	Polarization	Associated optical feature
Type III burst	Some seconds or less	Mainly polarized	Often first phase of a flare Flare-puff
U type burst	Between 3 and 10 seconds		Flare (80%)
Type II or outburst (generally very complex in character)	Minutes or tens of minutes (individual lifetimes of components: from some tenths of a second to some seconds)	Variable	Flare Ejections at limb
Type I or storm-burst (occurring in noisestorm)	Individual lifetime: one second or less. Lifetime of noisestorm: hour or days	Mainly polarized	Emitted in "*R* centres", not clearly correlated with CA's
Type IV phenomenon	Order of one hour	100% polarized	Preceded by flare ($0^h.0$ to $1^h.5$)
Type V phenomenon	Some minutes		Often flare

latter can hardly be understood without an understanding of the burst phenomenon in its uncomplicated form. For a classification of the non-thermal radio-phenomena reference is made to the accompanying table; cf. also the review by DENISSE [1].

90. Spectra and polarization, time profiles. The type III burst may occur at random intervals as well as in groups at decimetre and metre wavelengths. Originally they have also been called isolated or unpolarised bursts, denoting that they do not occur in groups and are not circularly polarized in contrast to the other (storm-) bursts. Later investigations showed that the type III bursts are as a rule strongly polarized[2]. Moreover, although the type III bursts may occur sporadically, they often cluster together, at a rate of 10 to 100 per hour (type III storm), and they sometimes occur in groups in the first phase of an outburst.

The spectra of type III bursts were described for the first time by WILD and MCCREADY[3] (Fig. 102) and later by WILD *et al.*[4-7]; they are characterized by a very fast frequency drift.

The spectrum, if extended over a sufficient frequency range, sometimes shows the second harmonic (Fig. 102a—c)[5,6]. The rate of frequency drift decreases con-

[1] J.F. DENISSE: In 1958 (Paris) symposium on Radio astronomy (1959).
[2] M. KOMESAROFF: Austral. J. Phys. **11**, 338 (1958).
[3] J.P. WILD and L.L. MCCREADY: Austral. J. Sci. Res. **3**, 387 (1950).
[4] J.P. WILD: Austral. J. Sci. Res. **3**, 541 (1950).
[5] J.P. WILD, J.A. ROBERTS and J.D. MURRAY: Nature, Lond. **173**, 532 (1954).
[6] J.P. WILD, J.D. MURRAY and W.C. ROWE: Austral. J. Sci. Res. **7**, 439 (1954).
[7] J.P. WILD: Radio Astronomy. I.A.U. symposium No. 4, p. 321, 1957.

phenomena (mainly observed in the metre range).

Dynamic spectrum		
Bandwidth	Spectral classification and description	Corresponding speed through corona
Overall spectrum broad (some tens of MHz (outburst) and some tens to some hundreds of MHz (type III bursts) but sometimes composed of narrow individual peaks; Fig. 102d). Outbursts are composed of many narrow individual peaks	Rapid frequency-drift, as a rule towards the lower frequencies; rate: $(f/4.5)$ MHz/sec	About 10^5 km/sec
	Initial frequency drift towards low frequencies followed by drift towards high frequencies. Turning point mostly between 100 and 150 MHz. Detailed microstructure	As type III bursts
	Type II: frequency-drift towards the lower frequencies; rate 0.2 MHz per sec. Fairly sharp cut-off at lower frequencies	About 10^3 km/sec
Some MHz. Spectrum of noise storm: some tens of MHz	Type I: nearly stationary spectrum but centre-of-gravity of the storm may drift slowly; rate 0.1 MHz/sec	
Hundreds of MHz	Very strong enhancement of base level cut-off at $f < 500-1000$ MHz	Standstill after type II motion
Some tens of MHz	$f < 100$ MHz	$v \approx 3000$ km/sec, after type III motion

tinuously with time[1], approximately according to $df/dt = -f/4.5\ \text{sec}^{-1}$. At low frequencies the drift decreases so rapidly that one or both bands can, in some cases, become almost stopped at a fixed frequency. The curves for the first and second harmonics are similar in shape but the relative band-width of the former tends to be narrower. The ratio of peak frequencies is not exactly equal to 2 but somewhat less: in the majority of cases it is between 1.85 and 2.00.

Roughly one in ten type III bursts shows the first and the second harmonics; the third harmonic has once been observed (HADDOCK). It is not known whether, in the cases in which only one component appears, one is dealing with the first or the second harmonic.

Interferometric[2] and optical[3] observations show that the source of the radiation moves upward through the corona with speeds sometimes exceeding 10^5 km/sec. The positions at which the individual frequencies are emitted do not change during the burst.

These observations suggest that the radiation is emitted by periodic oscillations of the coronal plasma. To produce an overtone the oscillations must be non-linear. The observations that the frequency shifts towards lower frequencies with time is in agreement with the observed displacement of the sources, but the velocity deduced from the frequency shift on the basis of the model of the quiet corona is smaller than the observed velocities. This suggests that the high parts of the coronal activity regions have a greater density than the quiet corona.

[1] See footnote 6, p. 302.

[2] J.P. WILD, K.V. SHERIDAN and G.H. TRENT: In 1958 (Paris) Symposium on Radio astronomy (1959).

[3] R.G. GIOVANELLI: Austral. J. Phys. **11**, 350 (1958).

Sometimes the spectrum is complex. The individual bursts have band widths of 5 to 10 MHz (Fig. 102d).

The precise time profiles of these rapid phenomena cannot be read from the common slow intensity records, but should be studied by means of fast recording instruments. At long metre waves (50 to 100 MHz; 6 to 3 m) each individual type III burst has a half-width of about 2 to 5 seconds at any one frequency and no fine structure is evident within each burst (WILD). At 200 MHz (150 cm) monochromatic observations[1] showed that the singly occurring bursts are often asymmetric, having a steep ascending and a less steep descending part. The longest lived ones (half-width ≈ 1.5 sec) are perhaps to be identified with type III bursts.

At 75 cm the type III bursts have a half-width of 0.2 sec[2].

A type of bursts, possibly related to the type III bursts, are the so-called *U-type bursts*[3]. Their spectrum is *U*-shaped, with the opening towards high frequencies. So it consists of a fast burst which first decreases in frequency, then increases again, after a turning point. The turning point is mainly between 100 and 150 MHz, in some cases at greater frequencies. The greater part (75%) of the bursts coincided with flares. They last for some seconds.

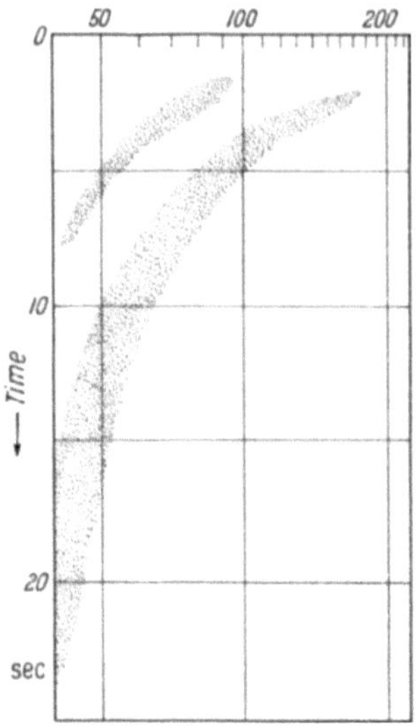

a

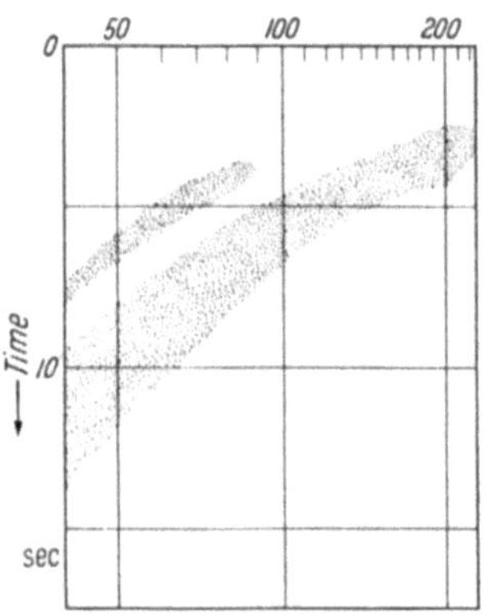

b

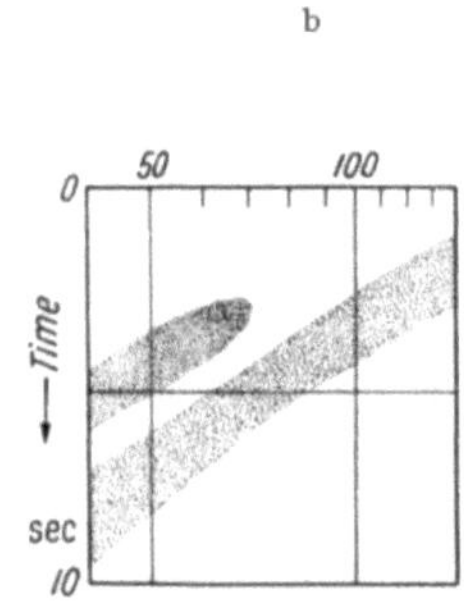

c

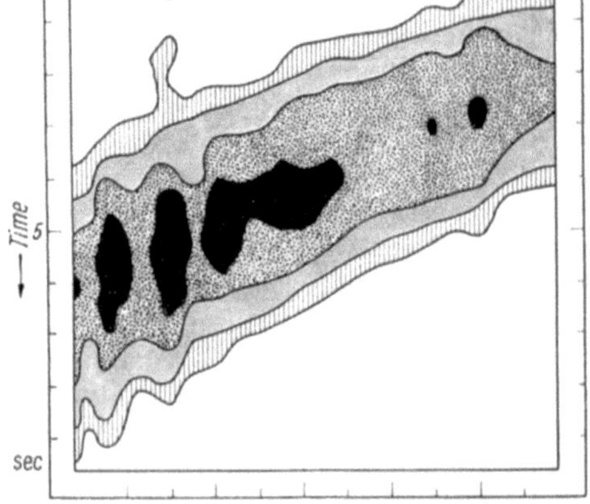

d

Fig. 102a–d. Spectra of type III bursts. Abscissa: frequency in MHz; ordinate: time in seconds. Fig. 102d shows, on a narrower frequency scale, the fine structure of a type III burst. J. P. WILD, J. D. MURRAY and W. C. ROWE: Austral. J. Phys. 7, 439 (1954). J. P. WILD: Austral. J. Sci. Res. 3, 541 (1950).

[1] C. DE JAGER, F. VAN'T VEER and CH. L. SEEGER: Rech. Obs. Astr. Utrecht **14** (1) (1958).
[2] T. DE GROOT: Nature, Lond. **180**, 382 (1957); **181**, 1676 (1958), and priv. comm.
[3] A. MAXWELL and G. SWARUP: Nature, Lond. **181**, 36 (1958).

There seems to be no essential difference between U type and type III bursts, but for the inversion; also the value of $|df/dt|$ is of the same order. The inversion might be due to an exciting agent moving along a curved path in the corona, e.g. along magnetic lines of force, which are known to be curved in active regions.

Also the *"herring-bone structures"*, fine structures, sometimes observed in type II bursts (ROBERTS) are perhaps type III phenomena: the type III bursts originate from a type II band (cf. Sect. 92) and go towards low and high frequencies.

Another type of burst, related to the type III burst, but only observed at the longer metre waves was described by J. A. ROBERTS[1]: the frequency *increases* with time (2 to 8 MHz/sec), the bursts are of short duration and contain two elements; the second being a repetition of the first after a delay of 1.5 to 2 seconds. Occasionally they occur within and form part of the structure of a type III burst. It is suggested that the second elements of the bursts are echoes of the first, reflected from lower levels of the corona. The bursts are assumed to occur at the second harmonic of the coronal plasma frequency (in that case they are emitted *above* the reflex frequency).

91. Theories of type III bursts. The observation that type III spectra may show both the fundamental frequency and the second harmonic, is an indication that the emission may be caused by macroscopic oscillations of the coronal gas, and suggests (SHKLOVSKI 1946) that certain radio-bursts are due to plasma oscillations. Since the ascending parts of the bursts near $\lambda = 100$ cm last only between 0.1 and 0.3 sec the time of *excitation* must be short. It seems reasonable to assume that the emission *decays* by collisions in the corona.

The observations also show that in the whole wavelength range the wavelength of emission of a particular burst increases with time, thus suggesting an exciting agent moving through the corona, in the direction of a decreasing density gradient and exciting its consecutive parts.

A plasma is a quasi-neutral (partly or totally) ionized gas. In a plasma high frequency oscillations can occur with the natural frequency of the gas which is[2]

$$\left.\begin{aligned} \omega_0 &= (4\pi e^2 \overline{N}_e/m)^{\frac{1}{2}} \\ \text{or}\qquad f_0 &= (e^2 \overline{N}_e/\pi m)^{\frac{1}{2}} = 8.98\times 10^{-3}\sqrt{\overline{N}_e}\,\text{MHz}. \end{aligned}\right\} \tag{91.1}$$

The problems arising next are (1) how these oscillations are excited and (2) how they can lead to the emission of radiation. The answer to the first problem is involved in the observation that at least a number of bursts and all type III and type II bursts seem to be excited by systematic gas motions. It is felt almost intuitively that longitudinal plasma oscillations can easily be excited by gas motions. Longitudinal oscillations, in contrast to transverse ones, are always connected with space-charge separation (it can be shown that space-charge separation is essential for the emission of radiation): LARENZ[3] showed that high-frequency plasma oscillations connected with a nearly complete space-charge separation occur when the systematic gas motions are equal to or greater than the "electron-sound-velocity" $a_e = (5kT/3m_e)^{\frac{1}{2}}$ (of the order of 10^3 km/sec), where m_e is the electron mass. The corresponding electrostatic potential differences are of the order of 10^4 or 10^5 eV, thus producing intense oscillations. It is true that space-charge separation *occurs* already at the "ion-sound-velocity" $a_i = (5kT/3m_i)^{\frac{1}{2}}$

[1] J. A. ROBERTS: Austral. J. Phys. **11**, 215 (1958).

[2] Cf. e.g. G. JOOS: Lehrbuch der theoretischen Physik, 9. Aufl., S. 716. Leipzig: Akademische Verlagsgesellschaft 1956.

[3] R. W. LARENZ: Z. Naturforsch. **10a**, 766 (1955).

(m_i is the ion mass) but the corresponding ion oscillations do not occur at the high radio-frequencies.

Velocities of the order of the electron-sound-velocity described above can occur in the corona in the form of shockwaves[1] (expansion of dense matter into a less dense medium). A detailed discussion of magnetohydrodynamic shock waves in the corona in connection with type II and type III bursts, was published by WESTFOLD[2].

The second problem is how the excited plasma oscillations can lead to the emission of radiation, and this problem again consists of two parts:

(a) The longitudinal plasma oscillations should be coupled with simultaneous *transverse* oscillations since for longitudinal oscillations of density and space charge the sum of the current density and of the displacement current, and hence, the Poynting vector is zero. Otherwise stated: the curl-free longitudinal plasma waves and the divergence-free transverse radiation waves are not coupled.

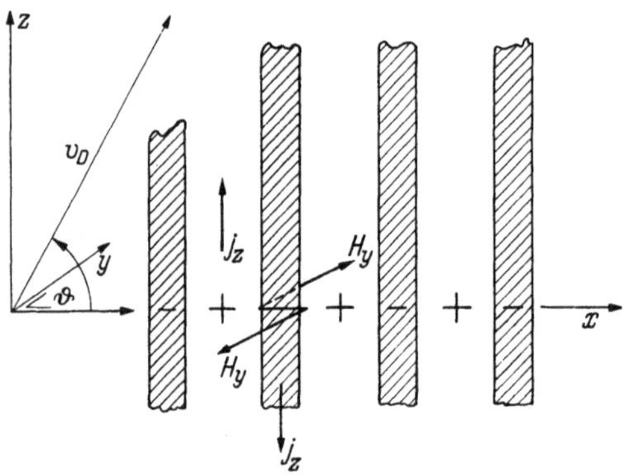

Fig. 103. Emission of electromagnetic radiation by coronal plasma oscillations after LARENZ (for details see text).

(b) Even if the Poynting vector is not zero, radiation can only propagate if the local refraction index is positive (for pure longitudinal or transverse plasma waves, $n=0$). However, in the absence of a magnetic field, $n=0$ for $f=f_0$. Thus, electromagnetic radiation cannot propagate. These difficulties disappear in a magnetic field when the waves progress perpendicular to the field, like in ionospheric reflections (cf. also LÜST[3]). Then two wave components originate, parallel and perpendicular to the direction of motion, hence giving rise to a transverse component of the electrical field. Thus the current density is no longer compensated by the displacement current, leading to a non-zero Poynting vector. (The Appleton-Hartree dispersion formula shows that $n \neq 0$ in a narrow frequency region around the plasma-frequency.)

But, also in absence of a field, the difficulty can be solved in various ways, e.g. (i) in the case that fast systematic motions occur (LARENZ[4]). Suppose e.g. a space-charge wave expanding into the x direction and superimposed by a general plasma motion into another direction (angle ϑ with the x direction). Then the following will happen (Fig. 103): At one point—not moving with the gas—the successive positive and negative space charges produce an alternating current containing, because of the systematic plasma drift, only a transverse component j_z, proportional to $|v_D| \sin\vartheta$. This transverse component is connected with a transverse magnetic field H_y, perpendicular to the x-z plane, defined by the drift vector and the wave vector. This variable magnetic field in turn influences the transverse current, yielding a coupled electromagnetic longitudinal and transverse wave; a further coupling between the two directions of motion is effectuated by the Lorentz force. Since the magnetic field and the divergence-free com-

[1] R.W. LARENZ, in: Tagung über Plasmaphysik, 1956, Leipzig: Akademie-Verlag p. 16, 1958.

[2] K.C. WESTFOLD: Phil. Mag. **2**, 1287 (1957).

[3] R. LÜST: Z. Astrophys. **37**, 67 (1955).

[4] R.W. LARENZ: Z. Naturforsch. **10**a, 901 (1955).

ponent of the electric field are perpendicular to the direction of motion of the space-charge waves, the resulting Poynting vector of the electromagnetic radiation is parallel to the direction of the density waves.

The calculations were extended by LARENZ[1] to the case that a field of compressible turbulent motions is superimposed on the general motion field. In both cases the dispersion formula $n = n(f)$ appears to consist of two branches; for one of them $n = 0$ for $f = f_0$ as usual, but for the other, for which the longitudinal oscillations are more important than the transverse ones, $n > 0$ for $f = f_0$; hence plasma radiation originates *and* it can be emitted.

(ii) Another successful attempt to explain how plasma oscillations in the absence of a magnetic field may lead to the emission of radiation is due to FIELD[2]. In an *inhomogeneous* medium plasma waves and radiation waves may be coupled

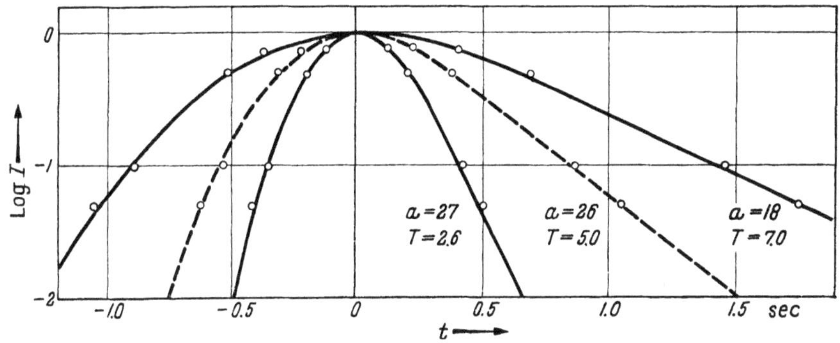

Fig. 104. Observed profiles of radiopips at 150 cm (dots) compared with profiles computed on the basis of a type III mechanism (lines). The profile parameters a (MHz/sec) and T (10^6 °K) are specified by the numbers on each profile.

because the longitudinal plasma waves are refracted by the density gradient, producing a bending of the lines of force and thus a curl. The amount of radiation thus emitted is sufficient to account for the observed intensities. TAKAKURA[3], discussing the case of inhomogeneities *in* a magnetic field showed that an extraordinary wave, emitted *below* the region $n_e = 0$ is split at a discontinuity into an ordinary and an extraordinary ray (analogous to optical double refraction). The ordinary ray component may subsequently be emitted.

The next problem is that of the size l_0 of the oscillating gas masses:

Upper limit. The bursts are excited in 0.1 sec; the excitation moves at a speed of 10^4 to 10^5 km/sec; then $l_0 \approx 10^3$ or 10^2 km. In that case the oscillation energy of the electrons must be small ($\approx 10^{-6}$) as compared with the kinetic energy of the electrons, since otherwise the emitted energy of the bursts would be much greater than the observed values[4].

Lower limit. Assume that the electron cloud vibrates in a strictly ordered way[3], then it must contain about 10^{11} electrons in a sphere with a diameter of the order of 10 cm.

If one assumes that $l_0 \gg 10$ cm, then the kinetic energy of the electrons is greater than the oscillation energy. In that case (i) the emission pattern shows a Doppler widening along the frequency axis[4]. Assume further (ii) that the coronal density decreases monotonically outward; let (iii) the plasma be excited by an exciting front moving upward such that $df_0/dt = a$ (constant); f_0 is the plasma

[1] See footnote 1, p. 306.
[2] G. B. FIELD: Astrophys. Journ. **124**, 555 (1956).
[3] T. TAKAKURA: Publ. Astronom. Soc. Japan **8**, 182 (1957).
[4] Cf. footnote 1, p. 304.

frequency; assume (iv) that the excited elements decay according to $\exp(-\lambda t)$ where λ is defined by the collision frequency of the gas: $\lambda = 4.2\times 10^3\, N_e T^{-\frac{3}{2}}$ cf. [1]). Then the time profile is (cf. Fig. 104)

$$I_{f_0}(t) = \text{const}\, e^{-\lambda t}\left[1 \pm \Phi(|Y|)\right] \tag{91.2}$$

where Φ is the error function, $Y = \frac{at}{\Delta} + \frac{\lambda\Delta}{2a}$, and Δ is the Doppler width: $2.26\times 10^5 f\sqrt{T}$. The comparison of the observed profiles at 200 MHz with those calculated with (91.2) shows the best agreement for a values between 18 and 27 MHz per sec and for temperatures ranging from 2.6×10^6 °K to 7.0×10^6 °K. The broadest and rare pips are emitted before flares and occur in regions with the highest temperatures; the a values are of the order 20 MHz/sec, of the same order as the value found observationally for type III bursts.

d) Other non-thermal radiophenomena, often connected with flares.

92. Description of radio-phenomena, connected with flares. There is a bewildering diversity in the phenomena that may occur at the moment of flare-outbreak. At first sight the only possible inference seems to be that strong flares are often associated with the emission of radio-radiation; the observed intensities are as a rule greater when the flare is more important[2]. There seems to be only moderate correlation between radiobursts and flares if the time over which the statistics are taken is restricted to a few minutes around the beginning of the flare. However, DODSON, HEDEMAN and OWREN[3], discussing 200 MHz events occurring within half an hour of a flare's commencement, found more correlation: at $\lambda = 150$ cm, 78% of the flares correspond with radio-phenomena. At $\lambda = 10.7$ cm distinctive events are associated with 57% of the observed flares and 19% of the subflares; each outstanding radio-event at 10.7 cm corresponds with a flare[4]. The radio-phenomena may differ from one case to another. At first DODSON *et al.* distinguished eleven different cases, but it appeared possible, as a working scheme, to draw a schematized picture of what an "ideal" flare would, or should do:

On a *slow record* at metre waves, e.g. at 200 MHz (pen deflection time of about 1 second), one would observe first a short but intense burst of radiation, lasting one to five minutes (Fig. 105). (Actually this happens in about one case in twenty.) After a short decrease the radiation increases again, and often reaches a higher and longer lasting intensity. These two events are the first two phases of the *outburst*. The start of the first phase may coincide with the onset of the optical flare, the second phase may occur during the decreasing part of the light curve. When the optical flare has nearly ended, it may be followed by a *noise storm* consisting of a great number of storm bursts, and/or by an *enhancement of the base level*. (This latter event can sometimes be classified as a "type IV burst".) Both may last for hours and sometimes for days, but in the latter case it seems uncertain whether the whole storm is still the consequence of the same flare.

This is the ideal case, but it rarely occurs in this standard form. Actually one, two, or three of the described radio-phenomena may remain absent and sometimes even nothing happens, especially during the occurrence of weak

[1] J.C. JAEGER and K.C. WESTFOLD: Austral. J. Sci. Res. A **2**, 322 (1949).
[2] E.V. APPLETON and J.S. HEY: Phil. Mag. **37**, 73 (1946).
[3] H.W. DODSON, E.R. HEDEMAN and L. OWREN: Astrophys. Journ. **118**, 169 (1953).
[4] H.W. DODSON, E.R. HEDEMAN and A.E. COVINGTON: Astrophys. Journ. **119**, 541 (1954).

flares. In a period when the Sun is rather flare-active, it often happens that the noise storms from various flares overlap, so that a flare may originate while a noise storm is in progress.

Conversely, some of the radio-phenomena described above may occur, without flares being observed optically, but one is of course never wholly certain that flares did not actually occur.

This complicated picture becomes simpler if one regards the *radiospectrum*[1] at metre wavelengths at the moment of a flare (Fig. 106). This spectrum is obtained with a receiver which sweeps through the frequency range 100 to 600 MHz

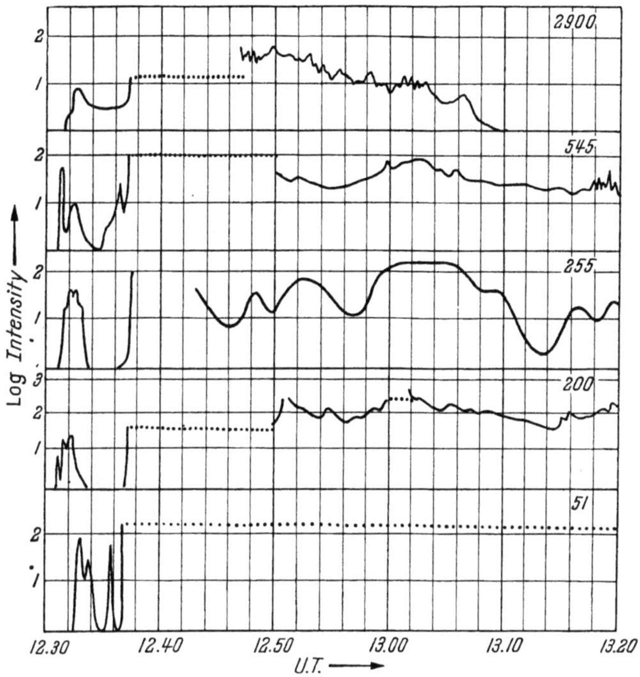

Fig. 105. Record of the outburst of August 31, 1956. The curves are labeled with the frequency (MHz). (Compilation made by A. D. Fokker, based mainly on observations at Nera, Holland and Humain, Belgium.)

every half a second. Each spectrum consists of an intensity-modulated line. The successive spectra are photographed on a continuously moving film, showing the "dynamic spectra".

Such spectrographs are now (1958) in use at Sydney, Freiburg observatory (Germany), the Naval Research Laboratory (Washington), the Harvard Radio Astronomical Station (Fort Davies, Texas), the University of Michigan, the Crimean Observatory, the Tokyo Observatory. A "microspectrograph" is in use at the Dwingeloo Observatory (Holland) which obtains spectra in a narrow band (12 MHz around 330 MHz) but with a high time resolution (0.01 sec).

Spectrographic observations of this kind show that the first part of an outburst like the one shown in Fig. 105 often consists of a cluster of type III bursts; the second part consists of type II bursts (cf. Fig. 106).

In Wild's description[2]: "In its most complete form the outbursts in this frequency range (38 to 240 MHz) appear to consist of two discrete phases followed by prolonged storminess. The first phase starts near the very beginning of the

[1] Cf. J.P. Wild: Trans. Internat. Astronom. Union 9, 661 (1957).

[2] Cf. footnote 7, p. 302.

optical flare and lasts for about one minute ... The onset of the second phase is heralded by the sudden appearance of one or more intense emission bands at a frequency of the order of 100 Mc/s. These bands, which are sometimes superposed on a diffuse background continuum, may be accompanied by their harmonics at double the frequency. With the passage of time they gradually drift in emission towards the lower frequencies—the 'type II' dynamic spectrum."

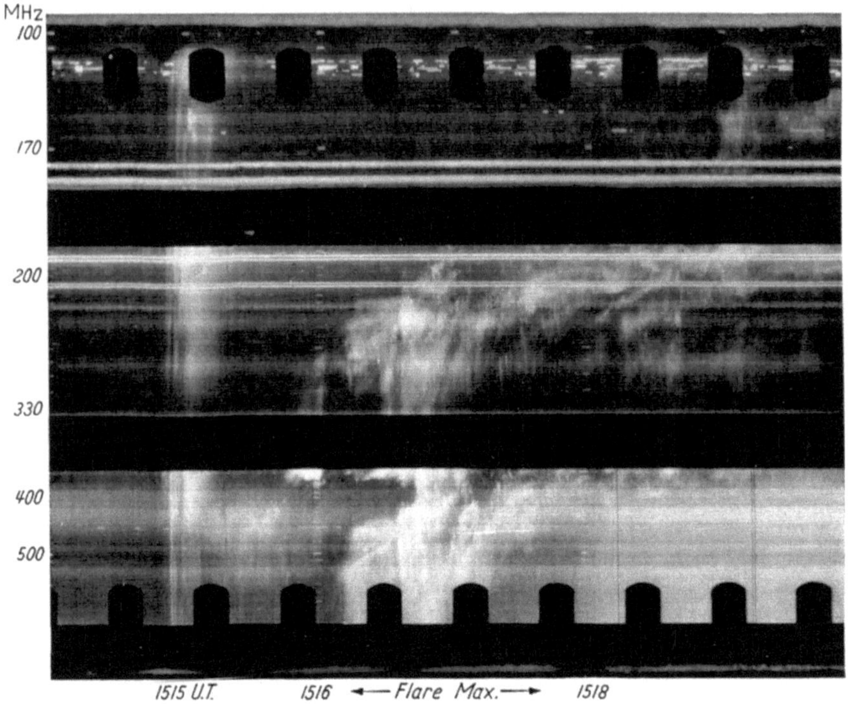

Fig. 106. The dynamic spectrum of the completely developed outburst of 9 December, 1957, observed at frequencies between 100 and 600 MHz, showing (a) at 15.15 UT a cluster of type III bursts; (b) between 15.16 and 15.20 UT two complex type II bands at the fundamental and second harmonic frequencies (e.g. at 15.17½ UT the maximum intensities are at 210 and 420 MHz); the bands show a detailed microstructure; (c) a type IV burst starting at 15.18 UT. Its intensity increased in the following part of the spectrum (not shown here) and it lasted for 2½ hours. The cut-off of the type IV burst was between 330 and 370 MHz (not shown). The horizontal lines between 170 and 200 MHz and the short bars near 120 MHz are caused by interference. Courtesy F. T. Haddock, University of Michigan.

Of 308 flares, mostly microflares, observed[1] between November 1955 and July 1956, 71 (23%) were accompanied by type III events (within ± 2 min of flare time) and two were accompanied by type II events. Reversely, during periods of optical observations, 204 type III events were recorded of which 133 (65%) occurred during flares (± 2 min) and 3 type II events were observed, of which 2 accompanied flares. Furthermore, there are indications that the behaviour of flares occurring within a particular centre of activity may depend on the activity centre itself. Some centres of activity appear to produce flares associated in all cases with type III events, and conversely. Flares occurring in other centres of activity show little or no correlation with radio-bursts.

So, "nearly all activity on radio waves associated with flares large and small, consists, apart from subsequent storminess, either of a group of type III bursts

[1] R. E. Loughhead, J. A. Roberts and M. K. McCabe: Austral. J. Sci. Res. **10**, 483 (1957).

(common), or a type II burst (much rarer), or a compound burst in which the type II follows a type III group. That spaced single-frequency records often show effective simultaneity in the starts at different frequencies is readily understood because (1) type III groups can start simultaneously within a second or less over widely spaced frequencies, and (2) type II bursts frequently have complex spectra in which the drift can only be recognized when continuous spectra data are available" (WILD, private communication).

The flares with associated outbursts on $\lambda = 410$ cm are asymmetrically distributed over the disk[1], more occur at the east-half. This asymmetry is assumed to be due to particles expelled by the flare and absorbing non-isotropically because of the curved particle trajectories, due to the solar rotation. The difference in optical depth, west compared with east, is of the order of 0.5 at $\lambda = 4.1$ m.

A flare is a short-lived event in a Centre of Activity. Another category of short-lived transients in a CA are the active prominences, but except for the very high-velocity ejections, observed in the early stages of some limb flares, there is only in some cases (spot-groups with looped prominences and downward streaming) a correlation between prominence activity and radio-events (cf. [2–7]).

93. Spectra and polarization of outbursts and type II bursts. Outbursts were discovered on monochromatic records as occasional and sudden large increases of the solar radio flux. The radiation may increase a thousandfold (depending on the wavelength), the period of enhancement may last for some to tens of minutes, or more. The radiation of an outburst has been claimed to be unpolarized, but this is certainly not always so. As a rule, the first part of an outburst (type II burst) may show only slight or random polarization, but the storm part is often wholly (circularly) polarized in one sense (Fig. 107).

Outbursts have been reported at frequencies as low as 9.2 MHz (SHAIN) and as high as 35000 MHz (HAGEN and HEPBURN[8]), hence at wavelengths from 0.85 cm to 33 m, but they do not always occur at all frequencies. They occur more often, are intenser and last longer at metre wavelengths than at shorter wavelengths. In the centimetre wave region they are virtually only observed when a flare or subflare is in progress[9]. Fig. 108 shows the peak intensities of the largest outbursts on record. This curve should not be confused with the "spectrum" of an outburst. UNSÖLD[10] has drawn attention to the similarity between this curve and that of the radiospectrum of the bright radiopoint sources and has suggested that both kinds of radiation might have a similar origin. Near 10 m, equivalent disk temperatures of 10^{12} °K may be reached, and since the radiating area is surely smaller than 0.01 of the disk, the radiation temperature must in that case be higher than 10^{14} °K. Hence the radiation cannot be of thermal origin.

Slow records show big intensity fluctuations (Fig. 105) and a comparison of records, made on neighbouring frequencies, shows how the detailed aspect rapidly changes with frequency.

---

[1] J. S. HEY and V. A. HUGHES: Monthly Notices Roy. Astronom. Soc. London **115**, 605 (1956).
[2] A. K. DAS and K. SETHUMADHAVAN: Nature, Lond. **172**, 446 (1953).
[3] H. W. DODSON and R. DONSELMAN: Astrophys. Journ. **113**, 519 (1951).
[4] O. E. H. RYDBECK: Convegno Volta **11**, 290 (1953).
[5] J. P. WILD and H. ZIRIN: Austral. J. Phys. **9**, 315 (1956).
[6] S. ROSSELAND and E. TANDBERG-HANSEN: Astrophys. Norv. **5**, 279 (1957).
[7] Cf. footnote 5, p. 212.
[8] Cf. J. P. WILD: Vistas in Astronomy **1**, 573 (1957). With list of references.
[9] Cf. footnote 4, p. 308.
[10] A. UNSÖLD: Z. Astrophys. **26**, 176 (1949).

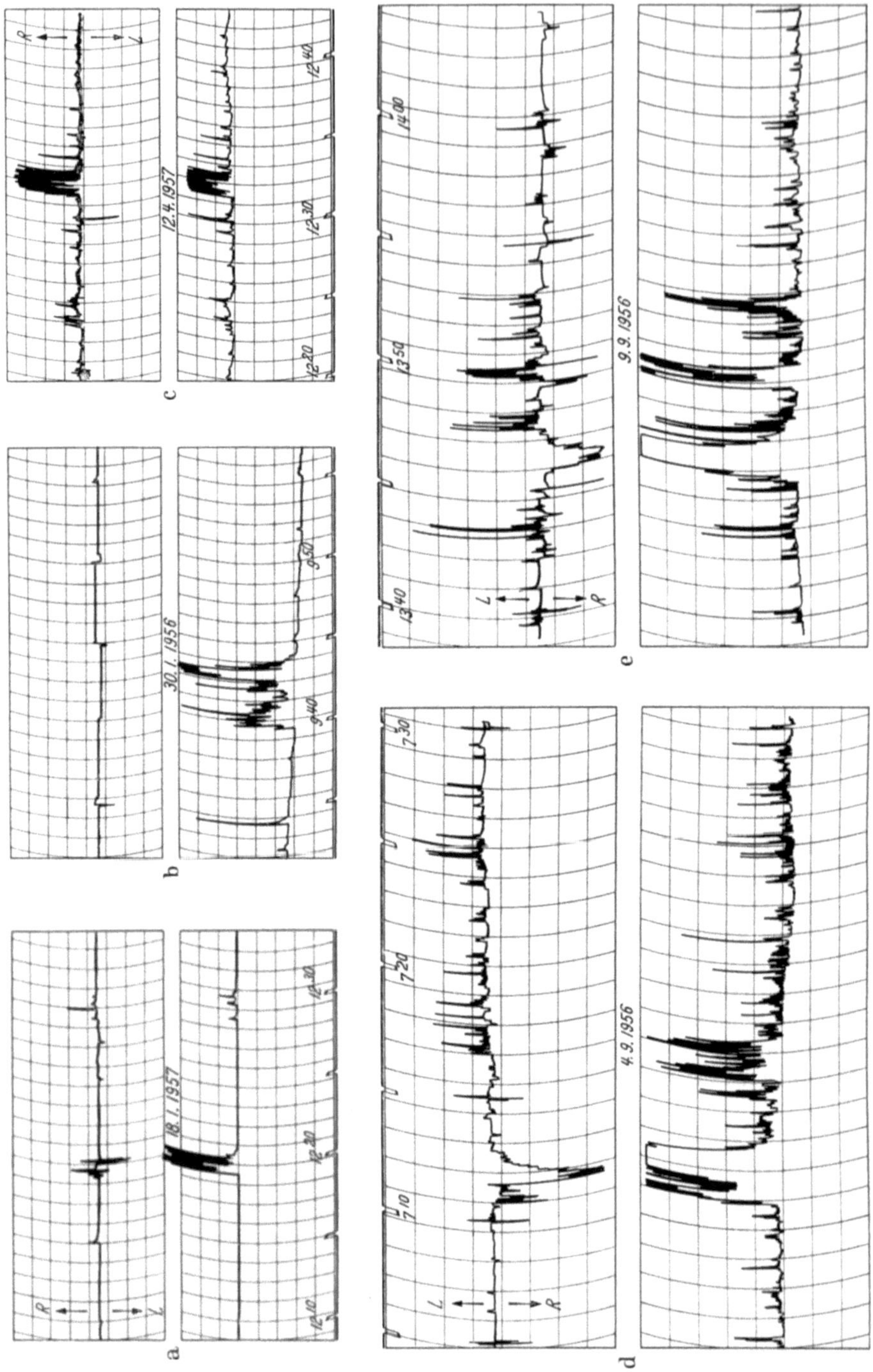

Fig. 107a–e. Outbursts at 200 MHz and their polarization. Each diagram consists of two parts: the lower one shows the recorded intensities; in the upper diagram the upward deflections give the intensity of the righthanded circular polarization component, the downward ones give the intensity of the lefthanded component. (a) and (b) Two unpolarized outbursts. (c) A righthanded polarized outburst. (d) and (e) These outbursts show a slight righthanded polarization, whereas the noise storm is rather lefthanded polarized (especially on 4 September, 1956). Courtesy A. D. FOKKER, Nera Observatory, Holland.

In Sect. 92 it was shown that fully developed outbursts start with type III bursts which precede the (rare) type II bursts. The type III bursts were described in subdivision **c**. The spectra of type II bursts show three characteristic features: (1) the wavelength drift, (2) the occurrence of the fundamental mode (frequency f) together with the first overtone (frequency $2f$), (3) the cut-off at lower frequencies.

(1): The wavelength drift has been studied for a small number of outbursts leading to $df/dt = -0.22$ MHz/sec, with relatively small scatter. The corresponding velocities are in the range 300 to 700 km/sec (cf. Fig. 102, Sect. 89). In one well-studied case the velocity remained virtually constant, equal to 475 km/sec between heights of 0.5×10^5 and 5×10^5 km[1]. These velocities were computed assuming the density distribution of the *quiet* corona. The active corona, however, has about twice or more that density (Sect. 79) so that the true velocities are also greater than the initially given values. This could be confirmed both by direct optical observations (cf. footnote 5, p. 212) and by interferometry (footnote 2, p. 303), thus confirming WILD's[2] suggestion to associate this velocity with that of the particles producing intense magnetic storms[3].

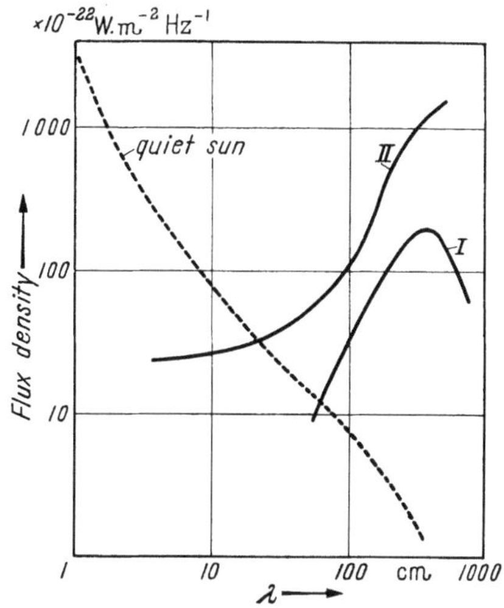

Fig. 108. Tentative spectra of the peak intensities of outburst (type II burst) and of stormbursts (type I bursts) as compared with the spectrum of the quiet Sun. Slightly changed, after a compilation by R. COUTREZ (Radio astronomie).

When the first and second phases of the outburst are present (Fig. 109), the phenomenon can be explained fairly simply (WILD[2]) in terms of the ejection of matter at two different velocities from a common explosion near the transition region chromosphere-corona. In an alternative explanation (TAKAKURA[4]) the frequency drift is attributed to the decrease of density at the front of an expanding gas cloud.

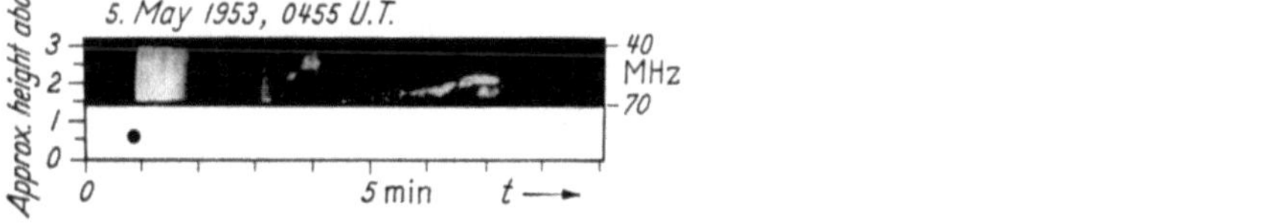

Fig. 109. Explanation of the two phases of an outburst in terms of an outbreak from one common height. In each case the black dot indicates the time and height of a hypothetical initiating "explosion". Courtesy R. P. WILD, Sydney.

(2) and (3): Approximately 60% of the type II bursts show the two harmonics[3]. The first overtone has about the same intensity as the fundamental mode, and has about the same relative spectral intensity distribution (the

[1] Cf. footnote 6, p. 302.
[2] Cf. footnote 7, p. 302.
[3] J. A. ROBERTS: In 1958 (Paris) Symposium on Radio astronomy (1959).
[4] T. TAKAKURA: Proc. Astr. Soc. Japan 6, 185 (1954).

wavelength scale being of course twice that of the fundamental mode). The frequency ratio is not exactly equal to 2 but somewhat smaller (1.90 to 1.99). This can be explained, in a similar manner as for the type III bursts, by the assumption that the amount of transmitted radiation is much greater at the frequency $2f$ than at f.

The remarkable splitting of each of the two outburst components, which is sometimes observed (see Fig. 109) has tentatively been ascribed by WILD[1] to the influence of the spiral motion of the electrons in a magnetic field. The distance between the two components might be put equal to $2f_H$, where f_H

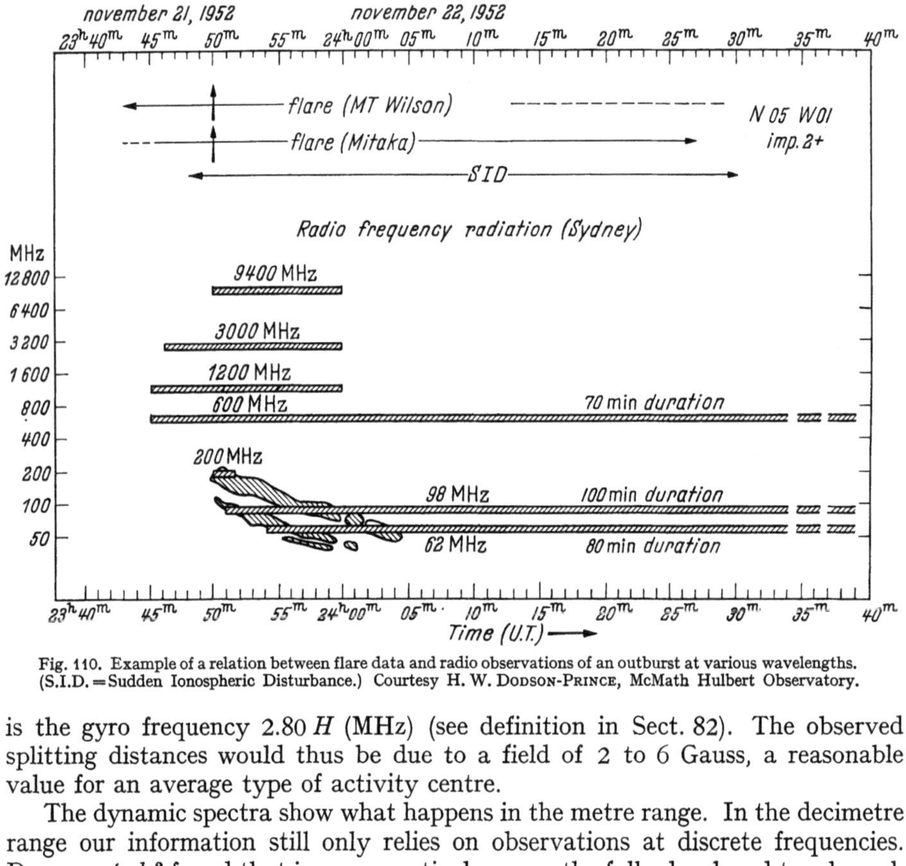

Fig. 110. Example of a relation between flare data and radio observations of an outburst at various wavelengths. (S.I.D. = Sudden Ionospheric Disturbance.) Courtesy H. W. DODSON-PRINCE, McMath Hulbert Observatory.

is the gyro frequency 2.80 H (MHz) (see definition in Sect. 82). The observed splitting distances would thus be due to a field of 2 to 6 Gauss, a reasonable value for an average type of activity centre.

The dynamic spectra show what happens in the metre range. In the decimetre range our information still only relies on observations at discrete frequencies. DODSON *et al.*[2] found that in some particular cases the fully developed two-branch type II burst, when combined with observations at other frequencies, indicates that the first radio-emission takes place at frequencies near 500 to 1000 MHz, followed by emissions at shorter and longer wavelengths (Fig. 110). This observation qualitatively agrees with a finding of DAVIES[3] who, from a statistical investigation of observations made at seven frequencies between 60 and 10000 MHz, concluded that in an average case—there are many exceptions, and the conclusion is only statistically correct—bursts come first at 3000 MHz, $\lambda = 10$ cm, leading to the postulate of a simultaneous up and downward motion of the excitation, starting from a region somewhere in the top part of the chromosphere. This

[1] J. P. WILD: Austral. J. Sci. Res. A **3**, 399 (1950).

[2] H. W. DODSON: In Radio astronomy. I.A.U. symposium No. **4**, 327, 1957.

[3] R. D. DAVIES: Monthly Notices Roy. Astronom. Soc. London **114**, 74 (1954).

result should be compared with what is known about flares; these too are upper chromospheric phenomena, often occurring in the low corona, leaving unaffected the lower lying photospheric layers (the weak and medium strong photospheric lines are not influenced by flares).

94. The noise storms; type I (or polarized) bursts. A noise storm occurs mainly in the range of metre waves. It may last for hours or even for days and it consists of numerous stormbursts of spectral type I, of greatly varying intensity and all of short duration ($\lesssim 1$ sec), often occurring in so great a number that they overlap, thereby producing an extremely complicated intensity pattern, in which an enhancement of the base level is often suggested by the accumulation of bursts. On the other hand there is a loose correlation between noise storms and enhancements of the base level. Both may occur during the second phase of a flare or after its disappearance, but correlation with flares or outbursts is not clear in all cases. This is partly due to the fact that the noise storms may last for so long a time that the connection with one particular flare is difficult to trace when more than one flare originates during the lifetime of a noise storm.

A peculiar type of "noise storm" was observed[1] on 4 November 1957; at 200 MHz it consisted of an enhancement of the base level up to 900 times the quiet Sun level showing short periodic intensity fluctuations, period 0.2 to 0.3 sec, somewhat resembling ionospheric scintillation of point sources.

In this section we discuss: (α) the duration and intensities of the bursts, (β) the relation between noise storms and enhanced radiation, (γ) the structure of the storms, (δ) the spectra, (ε) the polarization, (ζ) the correlation with sunspots and other disk phenomena and (η) the centre-limb variation.

α) Duration and intensity of bursts. The duration of the stormbursts, as determined with the aid of high speed recorder pens (deflection time smaller than 0.05 sec) is short. At 164 MHz their lifetime is of the order of 0.2 sec[2]. At 200 MHz (150 cm) they have a half-width of 0.4[3] to 0.3 sec[4]. There does not seem to be an appreciable difference in form between the stormbursts and the other bursts observed at that wavelength (cf. Sect. 89), yet the stormbursts may have the shortest lifetimes. At 400 MHz (75 cm) the half-width of the (rare) stormbursts is of the order of 0.2 sec[5].

The distribution of top intensities of the bursts is roughly exponential[6] with at 200 MHz

$$N = \text{Const} \times \exp(\alpha A^{\beta}) \tag{94.1}$$

where N is the number of pips with fluxes between A and $A + 0.25 \times 10^{-22}$ W m^{-2} Hz^{-1}. The constant C is a measure for the "activity" of the noise storm. The values of α range between 0.8 and 0.9; those of β between 0.75 and 0.85.

β) Relation between noise storms and enhanced radiation. Whether or not the enhanced radiation is due to accumulation of many stormbursts has been a matter of controversy since the first years of solar radio astronomy, but high speed records show that, as a rule, the enhanced radiation (the so-called "type I continuum") is not due to an accumulation of individual pips (Fig. 111). That the reverse case can also occur, was claimed by BLUM[2] *et al.*; FOKKER and ROOSEN[7], investigat-

[1] Staff of I.R.A. section of the Neth. Telecommunications Services. Nature, Lond. **181**, 542 (1958).
[2] E. J. BLUM, J.F. DENISSE and J.L. STEINBERG: C. R. Acad. Sci., Paris **232**, 387 (1951).
[3] Cf. footnote 1, p. 304.
[4] Ö. ELGARÖY: Nature, Lond. **180**, 862 (1957).
[5] T. DE GROOT: Nature, Lond. **180**, 382 (1957).
[6] O. HAUGE: Astrophys. Norv. **5**, 227 (1956).
[7] A.D. FOKKER and J. ROOSEN: Ionosphere and Radio Astr. Section Netherl. P.T.T. report **12** (1957).

ing high-speed records ($\lambda = 150$ cm), found that the pip density during noise storms is so high that the observed intensity fluctuations are superpositions of a great number of elementary pips. (The term "pip" introduced by REBER is used to denote the short-lived peaks that form the fundamental emission event in solar radioastronomy; the bursts are often complexes of pips.) Similar results were obtained by VITKEVITCH[1].

Spectral observations often show the (true) continuum to extend over a fairly broad band (several tens of MHz), often providing the background for the stormbursts.

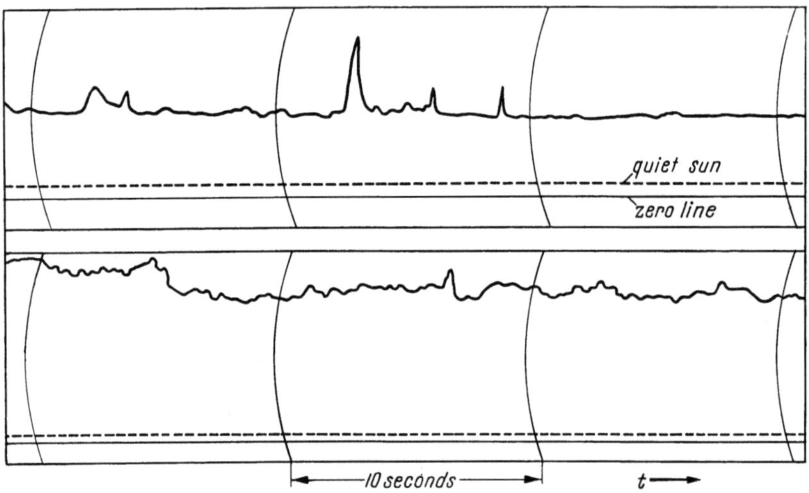

Fig. 111 a and b. Two cases of an enhancement of the base level, most probably not being due to accumulation of many small bursts. The quiet Sun level (dashed) is far below the actual intensity level. Wavelength 150 cm; the pen deflection time was 0ˢ02. (a) 10 November, 1956; 10.55 UT; also showing a group of "isolated" bursts. (b) 3 August, 1956; 11.51 UT; showing variability of the base level. Courtesy A. D. FOKKER; Nera Observatory Holland.

These results all refer to well developed noise storms. For the weak storms, being a transition phenomenon between the quiet Sun and the developed noise storms, there seems to be hardly any correlation between the activity constant C of formula (94.1) and the mean height B of the base level[2] if the base level intensity B is smaller than 10^{-21} W m^{-2} Hz^{-1}; above that value C seems to increase but slightly with increasing B. Also FOKKER[3] has found only a slight correlation between the background level and the stormburst intensity.

γ) The structure of a noise storm is by no means homogeneous. Weak or moderate storms consist of short-lived groups (lifetimes of the order of five to ten seconds at 200 MHz[2]; 6 sec at 169 MHz[4]). In turn these groups are not wholly independent of each other but cluster together into groups with a lifetime of the order of 10 or 20 minutes[2]. The detailed structure of intense storms is much more difficult to investigate, also because the individual pips cannot be distinguished.

δ) The spectra. The overall spectrum of noise storms and enhanced radiation was determined by DENISSE[5] from a preliminary compilation of storm observa-

[1] V.V. VITKEVITCH: Radio astronomy. I.A.U. symposium No. **4**, p. 363, 1957. — Dokl. Akad. Nauk. SSSR. **101**, 229 (1955).

[2] Cf. footnote 1, p. 304.

[3] A.D. FOKKER: Ionosphere and Radio Astr. Section Netherl. P.T.T. report **5** (1954).

[4] Cf. footnote 2, p. 315.

[5] J.F. DENISSE: Ann. d'Astrophys. **13**, 181 (1950).

tions at various wavelengths. The observations can be represented by

$$T_e \approx 10^{11} \times f^{-1.66} \quad (f \text{ in MHz}). \qquad (94.2)$$

Another, more modern spectrum, composed by COUTREZ[1], is represented in Fig. 108.

The detailed analysis of the metre wave spectrum, by the Australian dynamic spectrograph, has revealed that the individual bursts are nearly monochromatic; their halfwidth Δf, measured at about 100 MHz is 4 MHz. At 160 MHz, $\Delta f \approx 5$ MHz[2]. At 200 MHz VITKEVITCH[2] found a spectral bandwidth of 6 and 11 MHz for peaks of small and great amplitude respectively. Other observations by the same author yielded 4 to 7 MHz. At the same frequency DE JAGER and VAN 'T VEER[3] found $\Delta f = 6$ MHz. FOKKER (private comm.) found at 200 MHz: $\Delta f =$ 4 to 6 MHz. At 400 MHz, Δf as measured by DE GROOT[4] is 7 MHz. If this widening is interpreted as due to Doppler widening[3], these observations yield electron temperatures of 1.7×10^6 °K (100 MHz), 1.05×10^6 °K (160 MHz), 0.43×10^6 to 3.2×10^6 °K (200 MHz), and 3.4×10^5 °K (400 MHz), all reasonable values for the corona.

The Australian spectra show that big stormbursts are as a rule stationary (no frequency shift). This was confirmed for weak bursts by DE JAGER and VAN 'T VEER[3], who could not find a systematic time difference between peaks recorded simultaneously at 200 and 205 MHz and by DE GROOT[4] (394 to 406 MHz).

It often appears that the stormbursts cluster in a restricted but variable spectral frequency range[5]. This effect may sometimes be so pronounced that a noise storm is observed at one frequency, whereas at a neighbouring frequency,

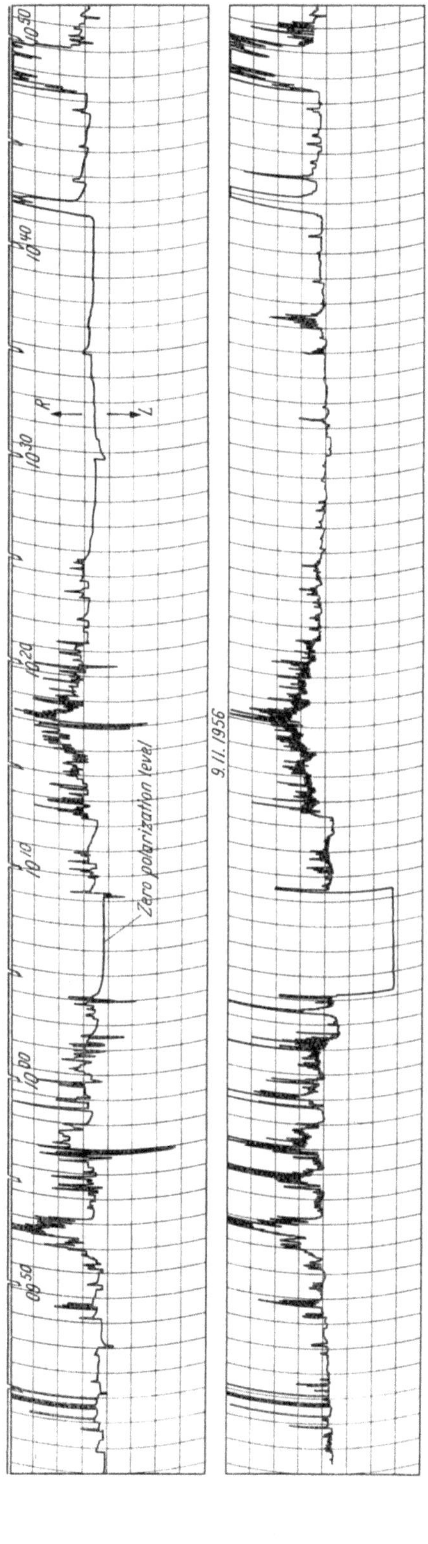

Fig. 112. Noise storm at 200 MHz showing nearly completely righthanded circular polarization. Courtesy A. D. FOKKER Nera, Observatory, Holland.

[1] R. COUTREZ: Radio Astronomie 1957.
[2] Cf. footnote 1, p. 316.
[3] Cf. footnote 1, p. 304.
[4] T. DE GROOT: In 1958 (Paris) Symposium on Radio astronomy (1959).
[5] J.P. WILD: Trans. Internat. Astronom. Union 9, 661 (1957).

differing by 20 MHz or so, one does not observe any appreciable storm[1]. In the dynamic spectra one observes how the stormbursts fill up small bands in the spectrum which, however, may drift with a rate of 0.1 to 0.2 MHz sec^{-1}, corresponding with an upward velocity of ≈ 100 km/sec. Occasionally also downward movements are observed.

ε) The polarization. The stormbursts and the enhanced continuum are often elliptically or circularly polarized. As a rule the sense of polarization of bursts and continuum is equal, but the often quoted statement that both are 100%

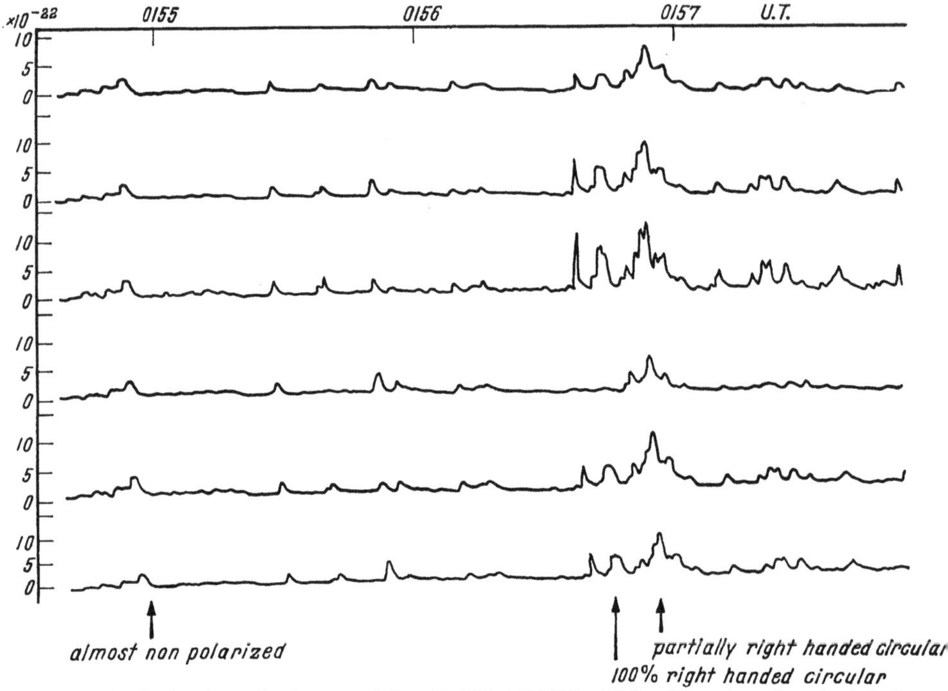

Fig. 113. Polarization of stormbursts measured June 29, 1957 at 200 MHz at Tokyo Observatory. From top to bottom records (1), (2), (5) and (6) show the intensities of four linear components, (1) in the direction NS, (2) in the direction EW, (6) between them at 45°, and (5) perpendicular to (6). The records (3) and (4) show the intensities of the *R* and *L* circular components. Cf. T. HATANAKA, S. SUZUKI and A. TSUCHIYA: Publ. Astr. Soc. Japan **7**, 114 (1955).

polarized, is not always confirmed. In broad lines stormbursts, produced by a spot-group with a preceding *N* polarity spot, are left-handed polarized, reversely for a *S* polarity preceding spot-group. So the polarization seems connected with the magnetic field of the spots, or of the *BM* regions, of which it is known that they follow the same polarity rules as the spots. As a rule, therefore, the sense of polarization of a noise storm does not change during the passage of the storm over the disk. (According to the usual radio-astronomical definition, a ray is right-handed circularly polarized when the electric vector turns clockwise for an observer looking into the direction of propagation. The above conclusion shows that we are generally dealing with the ordinary ray.)

The above rule, however, is valid only in broad lines; many exceptions occur. It may happen that the sense of polarization of a source suddenly changes[2]. A possible interpretation is that the major region of emission of the stormbursts migrates from above the leading spot to above the following spot.

[1] E. J. BLUM and J. F. DENISSE: C. R. Acad. Sci., Paris **231**, 1214 (1950).

[2] A. D. FOKKER: Paper at Symposion on Radionavigation, Ionospheric Research, Radio astronomy, Essen 1957.

A very detailed analysis of the polarization of stormbursts was made by HATANAKA, SUZUKI and TSUCHIYA[1], who measured six components of the polarization—two circular and four linear (Fig. 113). Four components are sufficient for a complete determination of the sense and direction of the polarization. [Similar equipments were developed at Cornell (USA)[2] and at Nera observatory (Holland).] It was found that type I bursts are partly elliptically, partly randomly polarized. On most days the degree of polarization exceeds 90%, but the ellipticity may vary from about 100% (i.e. circular polarization) when the emitting source is near the disk centre, to 10% (nearly linear polarization) when the source is close to the limb.

The ellipticity, the degree of polarization and the tilt angle usually do not change considerably on the same day. The tilt angle changes from day to day, but this effect may be due to the Faraday effect of the Earth's ionosphere. Also the corona produces a certain Faraday rotation, dependent on the frequency. Furthermore, an instrumental depolarization is caused by the dispersion of ellipse orientation angles due to differential Faraday rotation within the pass band of the receiver. Hence the ellipticity and the tilt of the polarization ellipse cannot give direct information on the solar emission mechanism.

ζ) Correlation with sunspots and other disk phenomena. The noise storm emission comes from the neighbourhood of large sunspot groups[3], hence from centres of activity in their first (active) phase, the height of the source being 0.3 to 1.0 $R_\odot$ above the photosphere.

Statistically, noise storms at metre wavelengths are not so much associated with the total area of the spot group, but rather with the area of the largest spot of the group: if the latter is of the order of 300×10^{-6} of the Sun's disk ($H_0\approx1500$ Gauss) short-lived storms occur at 200 MHz. At 97 MHz the treshold value is about 2800 Gauss[3] and 400×10^{-6} visible disk. During the minimum year 1954—when the mutual relations were easy to trace—enhanced 200 MHz radiation was emitted from all regions in which flares occurred and from all regions with spot areas $\geqq170\times10^{-6}$ of the visible hemisphere and magnetic fields $\geqq1300$ Gauss. On the other hand there were relatively large, bright faculae and regions with relatively strong 5303 Å coronal emission which produced no noise storms when passing over the disk[4]. Great noise storms occur only if the largest spot has an area exceeding 400 to 500×10^{-6} of the disk. FOKKER[5] found that the noise storms are mainly emitted by *D*, *E*, *F* and *G* spots. According to SIMON[6] the *F* spots have the greatest probability to emit noise storms, although the spots with the most important flares or those with many flares are not necessarily those producing the strongest noise. DENISSE[7] and BARABASHEV and GORDON[8] showed that the greatest number of bursts at metre waves occurs at moments of maximum variation of the spot areas, just as for the flares. It is important in this connection, that all storms seem to be preceded by an outburst (HATANAKA and MORIYAMA[9]; OWREN[10]). Hence both storms and flares may be connected with induction phenomena in the field of the spots. This is

[1] T. HATANAKA, S. SUZUKI and A. TSUCHIYA: Publ. Astr. Soc. Japan **7**, 114 (1955).
[2] M.H. COHEN: Cornell research report E E 278, 1957.
[3] R. PAYNE-SCOTT and A.G. LITTLE: Austral. J. Phys. A **4**, 508 (1951).
[4] H.W. DODSON and F.R. HEDEMAN: Astrophys. Journ. **125**, 827 (1957).
[5] A.D. FOKKER: Document for comm. 5. URSI meeting Boulder 1957.
[6] P. SIMON, in: Radio Astronomy. I.A.U. colloquium No. 4, p. 334, 1957.
[7] J.F. DENISSE: C. R. Acad. Sci., Paris **228**, 1571 (1949).
[8] N. P. BARABASHEV and I. M. GORDON: Dokl. Akad. Nauk Ukraina **1**, 6 (1954).
[9] T. HATANAKA and F. MORIYAMA: Tokyo Astr. Obs. Reprint **99** (1952).
[10] L. OWREN: Radioastr. Report Cornell **15** (1954).

confirmed by SEVERNY[1], who finds that radiobursts at 50 cm not only correspond with flares but also with rapid motions of prominences in the fields of spots and with other rapid processes in active regions. WILD and ZIRIN[2] found that radio-noise is mainly emitted by activity centres containing looped and downward streaming prominences, whereas DODSON[3] once noticed the relation between a 200 MHz burst and the appearance of an active dark flocculus. The source of a noise storm can be small, but it is enlarged by coronal scattering[4].

η) *The center-limb variation* of the enhanced radiation or of the noise storms has been investigated by various authors. Statistical methods yielded conflicting results. The initial opinion that the emission cone of the radiation is sharp and increases for increasing frequencies has not been confirmed by later investigations. Indeed, interferometric observations have shown that noise-active centres may also occur close to the Sun's limb; an interferometric investigation of the distribution of radio-noise active spots[5] (255 MHz) shows that there is hardly any pronounced centre-limb effect, although in some cases the noise storm emission ceases before the corresponding activity centre has reached the Sun's limb (disappearance of the source behind the plane of total reflex).

Other interferometric observations[6,5,7] showed that low-frequency stormbursts originate from higher levels than high-frequency radiation.

95. The type IV and type V phenomena; tangential motions of sources. In a number of cases outbursts are accompanied or followed by an enhancement of the base level at metre waves; sometimes this enhancement may attain considerable values, with apparent temperatures up to 10^{10} and 10^{12} °K. BOISCHOT[8] has shown that some of these enhancements have particular properties. They were called type IV bursts:

(a) Type IV bursts always occur after a flare; the intensity increases during 20 to 40 min, it decreases slower (one or some hours); there is no clear correlation with the flare intensity.

(b) The radiation is circularly polarized.

(c) Spectrographic observations show that the enhanced emission is remarkably stable and does not present type I bursts. The spectra are continuous and occur in the frequency range 100 to 600 MHz, with a cut-off at the side of lower frequencies (cf. Fig. 106).

(d) Source diameters up to 150000 and 350000 km have been measured.

(e) The emission centres often show tangential motions. Observations of type IV bursts emitted after limb flares show that these extended sources attain maximum heights ranging between 0.3 and 4 or 5 R above the photosphere.

Tangential motions of sources have been measured soon after the introduction of the interferometer technique. Since most interferometers consist of a one-dimensional aerial, only one of the motion components is traced. Moreover outbursts may last shorter than the time necessary to pass through one fringe of the interferometer pattern. This difficulty was overcome by LITTLE and

[1] A.B. SEVERNY: Isw. Krymsk. Astrophys. Obs. **15**, 111 (1955).
[2] J.P. WILD and H. ZIRIN: Austral. J. Phys. **9**, 315 (1956).
[3] H.W. DODSON: McMath-Hulburt Obs. reprint **41** (1955).
[4] J.A. HÖGBOM: In 1958 (Paris) Symposium on Radio-Astronomy (1959).
[5] A.D. FOKKER: Ionosphere and Radio Astron. Section, Netherl. P.T.T. report **12** (1957).
[6] Cf. footnote 10, p. 319.
[7] Y. AVIGNON, E.J. BLUM, A. BOISCHOT, R. CHARVIN, M. GINAT and P. SIMON: C. R. Acad. Sci., Paris **244**, 1460 (1957).
[8] A. BOISCHOT: C. R. Acad. Sci., Paris **244**, 1326 (1957).

PAYNE-SCOTT[1] with an interferometer at 97 MHz of which the fringe pattern was artificially swept over the Sun's disk at a rate of 15 times per second (the "swept lobe" interferometer).

The tangential displacements, if assumed to be due to radially ejected matter, usually[2] lead to outward velocities between 400 and 7000 km/sec (normal range 500 to 3000 km/sec). The sources start at or above the zero refractive index level and, when moving upward, may eventually reach a height of $2\,R_\odot$.

Greater velocities are sometimes reported. Interferometric observations of a source at 26 July 1956 by FOKKER[3], showed how a source initially situated near the disk's centre, displaced over the disk to close to the limb in three minutes time. If this is the projection of a radial motion, the upward velocity would be 8500 km/sec, but also in this case the source stopped abruptly at the end of its activity. Some other cases of source motions have been listed by BOISCHOT[4].

In some cases drifts are reported that apparently cannot be reconciled with radial motion, e.g. a source moving from the limb towards the disk centre[3]. WILD[5] suggested that in such a case one is observing the image of the source, reflected against the nearly globular Sun, which acts like a convex mirror. By some reason the source itself should not be observed (anisotropic emission towards the Sun?). In another case[3] a source appeared to move at nearly constant height above the photosphere, travelling over an angle of 40° in 2.5 minutes.

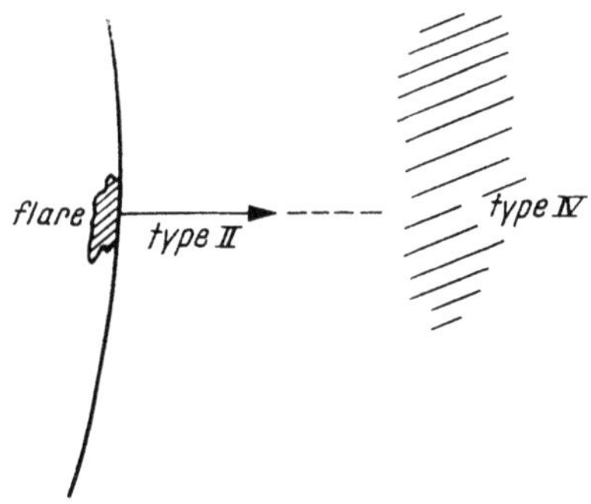

Fig. 114. Suggested relation between a flare and the corresponding type II and type IV radiobursts, after BOISCHOT and DENISSE.

BOISCHOP and DENISSE[6] have shown that the emission of the type IV bursts cannot be explained by plasma oscillations (at the coronal heights of several solar radii plasma oscillations would correspond with frequencies much lower than 169 MHz), but the energy

$$P = 6\times 10^{-27}\,H^2\,(\text{Gauss})\,E^2\,(\text{eV})$$

emitted by synchotron radiation of relativistic electrons with energy E at the frequency $f = 1.6\times 10^{-5}\,H E^2$, can explain the observed emission ($P \approx 10^{-19}$ W m^{-2} Hz^{-1}) by taking $N_e = 4\times 10^{32}$, $E = 3$ MeV, $H = 1$ Gauss. Also the observed duration of the type IV bursts ($\approx 1^{\text{h}}$) is of the same order as the mean free lifetime (3×10^4 sec) of electrons of 3 MeV in a region with an electron density of $2\times 10^8\,\text{cm}^{-3}$. TAKAKURA[7] advocates a field of the order of 200 Gauss; his argument being that the sharp cut-off at low frequencies should occur at the gyrofrequency, where resonance absorption is very strong.

Therefore, the type IV bursts show, consistently with optical coronal observations, that some time after flares a higher lying coronal region can be excited to high temperatures. It is most probably these regions that are responsible for the

[1] A. G. LITTLE and R. PAYNE-SCOTT: Austral. J. Sci. Res. A **4**, 489 (1951).
[2] R. PAYNE-SCOTT and A. G. LITTLE: Austral. J. Sci. Res. A **5**, 32 (1952).
[3] A. D. FOKKER: Ionosphere and Radio Astron. Section, Netherl. P.T.T. report **12** (1957).
[4] H. BOISCHOT: C. R. Acad. Sci., Paris **244**, 1326 (1957).
[5] Cf. footnote 8, p. 311.
[6] A. BOISCHOT and J. F. DENISSE: C. R. Acad. Sci., Paris **245**, 2194 (1957).
[7] T. TAKAKURA: Comm. at Paris 1958 Radio astronomy symposium (1959).

emission of the type IV bursts, and perhaps also of the cosmic ray particles which (as is known) also reach the Earth some 15 min to 1.5 hour after a flare.

Finally a remark on the *type V phenomena*: these show some similarity to type IV bursts. Type V bursts occur together with or shortly after type III bursts; they last for 0.5 to 3 minutes and reach great intensities, up to 10^{-18} watt. m^{-2} Hz^{-1}. The bandwidth is some tens of MHz, with $f < 100$ MHz. Interferometric observations show type V bursts to have a proper motion of about 3000 km/sec (WILD *et al.*, footnote 2, p. 303).

D. Solar rotation and the solar cycle.

a) The solar cycle.

96. The Wolf number and the 11-year period. The numbers and areas of the spots and the spot groups vary from day to day. A useful quantity to measure the sunspot activity is the Wolf number (sunspot number):

$$R = K(10g + f),$$

where f is the number of spots on the visible hemisphere, and g is the number of spot groups. K is an adaptation coefficient to reduce the R values to those derived for the Fraunhofer refractor with 8 cm aperture and magnification 64× with which WOLF's original countings were made in 1849. For this instrument, still in use at the Eidgenössische Sternwarte in Zürich, $K = 1$. If there is only one spot on the Sun, then $R = 11K$. Since all efforts should be made to keep the long series of observations as homogeneous as possible, K is determined for each observer by comparing a long series of observations with the standard observations made in Zürich. Great variations may occur when comparing simultaneous observations made at various observatories; the seeing is an important factor. The probable error of one R determination is about 15%; that of a monthly mean is 3 to 5%.

The sunspot numbers R are published among others in the "Astronomische Mitteilungen der Eidgenössischen Sternwarte Zürich" and in the "Quarterly Bulletin on Solar Activity". Independent R determinations by the Solar Division of the A.A.V.S.O. (U.S.A.), published in "Sky and Telescope" and in "Solar-Geophysical Data" initially disagreed with the Zürich determinations for great R values. Presently the agreement seems to be better. Since 1950 the U.S. Naval Observatory publishes in its *Circulars* daily position areas, counts of spots and sunspot numbers R, as determined at this observatory. For 1956 e.g. the American mean monthly R-values are by about 6 ± 4% greater than the Zürich values, the largest differences occurring for the smallest R-values. For that year the annual mean values were 152 (U.S.) and 142 (Zürich).

At some observatories *spot areas* are measured (Greenwich since 1874, collaborating with Kodaikanal and Cape Town). It is true that the area might be in some respects a better measure for the solar activity than the Wolf number in which small and large spots are counted in the same way, but the area determination requires a great amount of labour. Statistically there appears to be a correlation between the spot area A (in 10^{-6} of the visible hemisphere) and the spot number R; for 1878 to 1953: $A = 16.1\ R$; there is a great scatter in the values of A/R[1].

[1] W. GLEISSBERG: Naturwiss. **43**, 196 (1955).

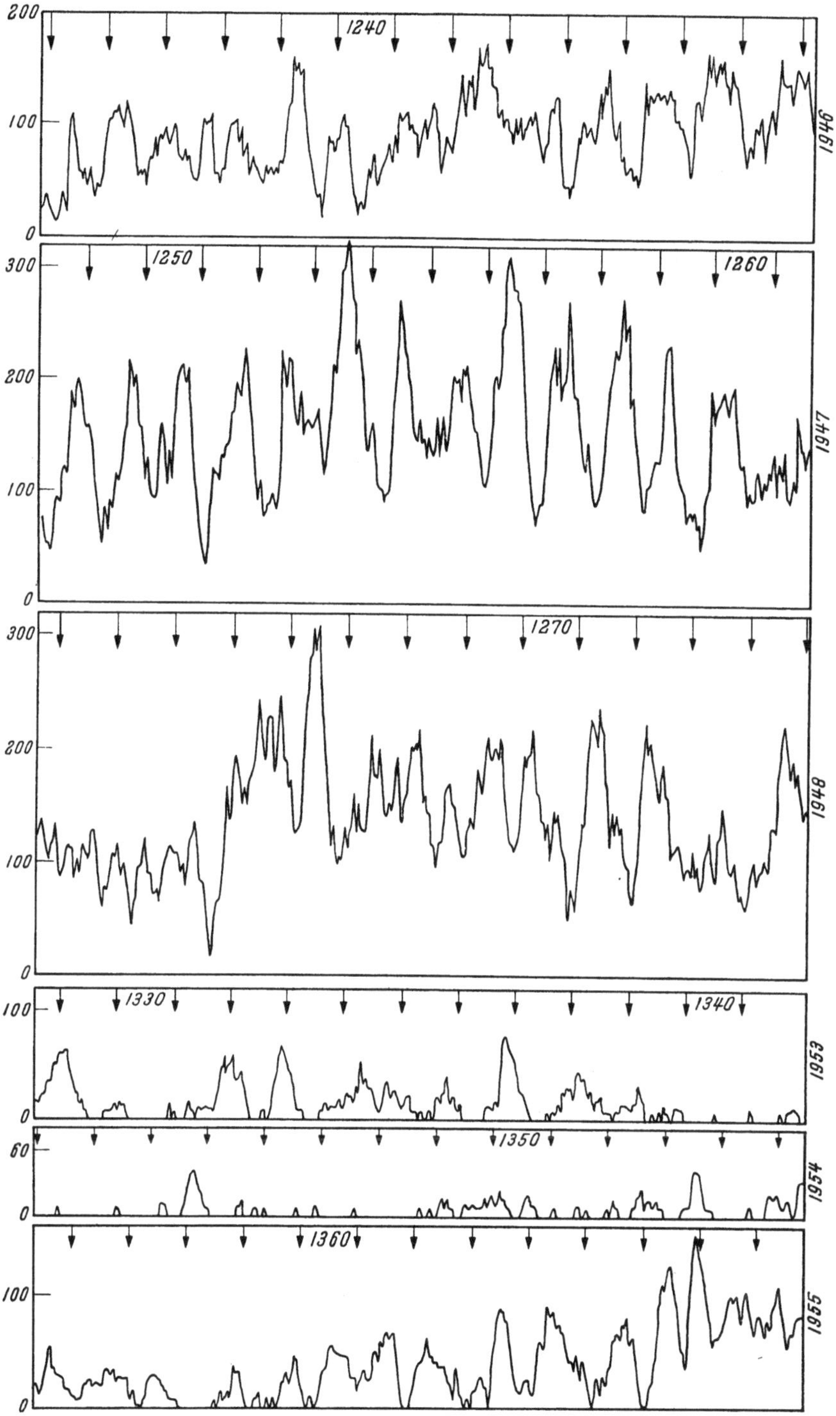

Fig. 115. Daily sunspot numbers for the maximum years 1946 to 1948 and the minimum years 1953 to 1955, taken from WALDMEIER'S annual reports on the solar activity (Astron. Mitt. Eidgen. Sternwarte Zürich). Each tenth solar rotation number is indicated along the upper side of the diagrams. The ordinates give the sunspot numbers.

There are some defects to the Wolf number as it is defined and used:

(1) It does not measure the activity of the *whole* Sun, since it does not account for the spots at the invisible hemisphere, and in this way the 27-day period (= the average synodic rotation period), important to geophysicists but less to astrophysicists is introduced (Fig. 115). This effect can be eliminated by determining R for the whole Sun by extrapolating the development of spot groups to the invisible side[1].

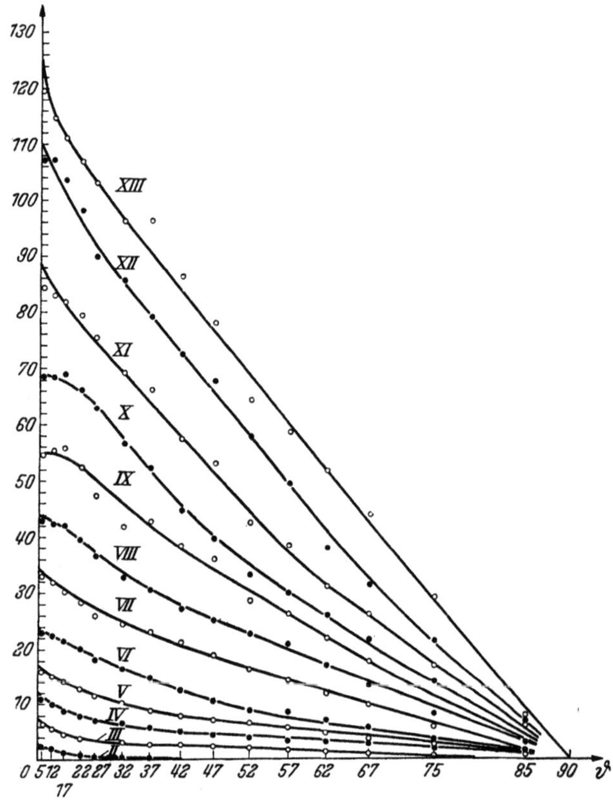

Fig. 116. The visibility of sunspots. Abscissa: the heliocentric angle ϑ. Ordinate: the numbers of spots in groups which at central disk passage contained the numbers of spots found at the intersection of the solid lines with the ordinate axis. M. ROGGENHAUSEN: Z. Astrophys. **30**, 249 (1952).

(2) In determining the Wolf number the counting results are not corrected for the difference in visibility of spots near the centre and limb of the Sun. The visibility of a spot should be proportional to $\cos\vartheta$ if the spots were flat structures on the disk (ϑ is the heliocentric angle), but, as it appears from a statistical analysis, it is proportional to $\cos\vartheta\,(1-\mu\tan\vartheta)$ where μ depends on the size of the spot; the average value is 0.075 (see also Fig. 116). The specific form of these visibility curves was suggested to be due to a ring of (elevated) faculae surrounding the spot[2].

(3) The areas of chromospheric faculae correlate better with other signs of solar activity (coronal line emission; radio-radiation at decimetre waves; ionospheric observations) than sunspots do.

[1] U. BECKER and K.O. KIEPENHEUER: Z. Astrophys. **33**, 132 (1953).

[2] M. HOTINLI: Publ. Istanbul Univ. Obs. **42** (1951); **56** (1955). — Ann. d'Astrophys. **20**, 45 (1957).

The periodicity in the number of spots on the disk was discovered in 1843 by SCHWABE after 20 years of observations. The period, 11.0 years, was later determined by WOLF, who started systematic sunspot observations. With the aid of scattered observations since the discovery of the telescope, the sunspot activity can be traced back till 1610 (see Fig. 117). The sunspot curve is by no means a regular, strictly periodic curve. High and low maxima occur, often alternately, and also the time intervals from minimum to minimum or from maximum to maximum spread around 11 years.

Superimposed on the slow changes, the spot number may vary enormously in the course of only two weeks (Fig. 115). In this short-periodic variation the 27-day average synodic period of solar rotation can often be found back over a longer interval of time than the average lifetime of the spots, indicating that

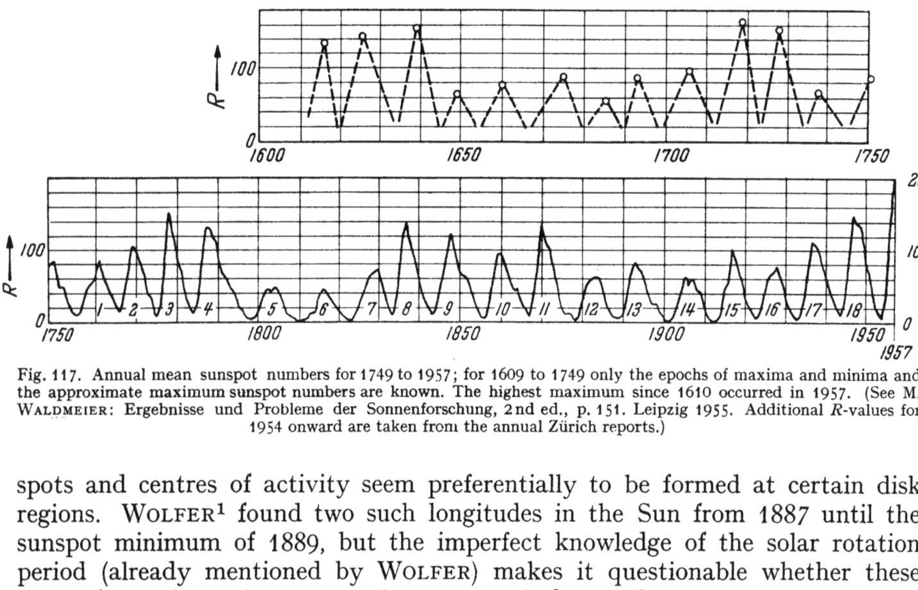

Fig. 117. Annual mean sunspot numbers for 1749 to 1957; for 1609 to 1749 only the epochs of maxima and minima and the approximate maximum sunspot numbers are known. The highest maximum since 1610 occurred in 1957. (See M. WALDMEIER: Ergebnisse und Probleme der Sonnenforschung, 2nd ed., p. 151. Leipzig 1955. Additional R-values for 1954 onward are taken from the annual Zürich reports.)

spots and centres of activity seem preferentially to be formed at certain disk regions. WOLFER[1] found two such longitudes in the Sun from 1887 until the sunspot minimum of 1889, but the imperfect knowledge of the solar rotation period (already mentioned by WOLFER) makes it questionable whether these seats of spot formation were real: an error of 0°.1 per day makes 75° in the two years lifetime of WOLFER'S seat. Indeed, U. BECKER[2], studying the spot-groups occurring between 1880 and 1940 found 46 seats of spot formation with lifetimes up to 21 rotations with an average value of perhaps ten rotations. The solar rotational velocity deduced from these seats turns out to be 0°.1 per day faster (for $0° < |b| < 25°$) than the velocity deduced from long lived sunspots.

The prediction of the future development of the R-function has often been attempted, but extrapolations by means of a Fourier analysis of the R-curve failed, probably because the curve cannot be considered as a superposition of strictly periodic functions. Progress was made when WALDMEIER introduced his "eruption hypothesis" (1935)[3], according to which the length and character of each cycle from minimum to minimum is independent of the others. A new cycle starts already when the old cycle is still decaying, a fact that is clear from the laws on the latitude variation of the spots (Fig. 119 in Sect. 97). So, each cycle, defined by its number in the Zürich standard numbering (cf. Fig. 117) must be treated separately; it has its own character, defined by parameters such as:

[1] A. WOLFER: Mem. Soc. Spettr. Ital. **29**, 152 (1901).
[2] U. BECKER: Z. Astrophys. **37**, 47 (1955).
[3] M. WALDMEIER: Astr. Mitt. Zürich **133** (1935).

R_M: the maximum sunspot number (smoothed monthly mean);
T: the time to rise from minimum to maximum (years);
Θ: the fall time, from maximum to $R=7.5$, etc.

The parameters T, Θ and the other cycle parameters all depend on R_M (WALDMEIER[3]), proving that the spot curves $R(t)$ form a *one-dimensional set of curves* with parameters R_M.

Essential features of the sunspot curves are (see Fig. 118):

(1) The time, necessary to reach maximum intensity decreases with increasing R_M, so that approximately $\int\limits_0^{t_M} R\,dt$ is the same for each cycle.

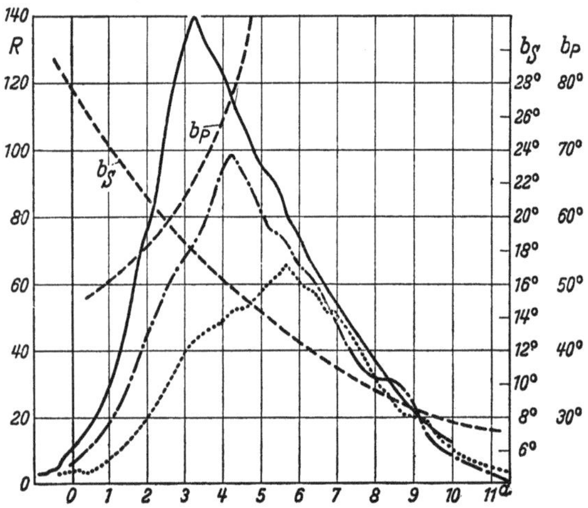

Fig. 118. Characteristic sunspot curves with respectively low, mean and high maxima. The curves b_s and b_p show the variation of the mean latitude of the sunspot zone and of the polar prominence zone. M. WALDMEIER: Z. Astrophys. **42**, 34 (1957).

(2) After maximum the $R(t)$ curves all have a similar, more or less exponential form.

Property (1) advocates for the eruptive character of the sunspot cycles, whereas (2) suggests the exponential decrease of the solar activity after maximum.

KSANTAKIS[1] drew attention to the secondary fluctuations in the sunspot numbers. For cycles with equal T-values these fluctuations appear at the same moment, if reckoned from the time of maximum activity.

97. The butterfly diagram of sunspots. Spots occur in two zones parallel to the equator at about equal but opposite latitudes; during each solar cycle the mean latitude of the spots decreases continuously. This law was discovered by CARRINGTON (1858) and put in a precise form by SPÖRER. The mean latitude is 35° in the very first part of the cycle, although occasionally greater values are reported. Short-lived faculae occur sometimes in high latitudes without being able to develop spots. Spots occur rarely above 45° and have short lifetimes; pores, living only for some hours are sometimes reported up to latitudes of 75°. In 1915 a well developed spot was observed at 60°; in August 1953 a short-lived spot-group consisting of two small spots occurred at 58°. In August 1956 a single spot at $b=48°$ was observed during 17 consecutive days. The more active a cycle is, the higher the latitude at which spots and faculae can occur. A list of high latitude spots and spot-groups ($b=41$ to 43°) with lifetimes greater than 2 days was given by WALDMEIER[2].

At the end of each cycle the mean latitude of the spots has decreased unto about 8°. Then the new cycle has already started with small short-lived spots at high latitudes. This is illustrated by the "butterfly" diagram, made for the first time by MAUNDER (cf. Fig. 119). The law of zones can be traced back to 1621 (SPÖRER).

[1] Z. KSANTAKIS: Astron. Zhurn. **34**, 391 (1957).
[2] M. WALDMEIER: Z. Astrophys. **43**, 29 (1957).

So, in fact, a spot cycle lasts longer than 11.0 years, the average time from minimum to minimum (or maximum to maximum). The true duration of a cycle, determined as the whole time interval during which spots of a certain cycle are visible on the Sun, ranges between 11.8 and 14.4 years; mean value 13.2 years. Even this is not yet the "true" period of solar activity, since high latitude coronal activity starts still earlier: the coronal activity of the 1957 cycle is thought to have started in 1951 at $b \approx 70°$ for $\lambda = 5303$ Å[1]. A similar result is shown by the prominence activity (Fig. 121): both the *prominences* and the *corona* present a high latitude activity zone beside the main equatorial zone. The former starts before sunspot minimum. The picture is completed by a finding of U. Becker[2] that also *sunspot* have a high latitude zone, starting near spot minimum a $\bar{b} \approx 28°$ and lasting, with increasing latitude till spot maximum (the beginning part of this branch is faintly visible in some of the "butterflies" of Fig. 119).

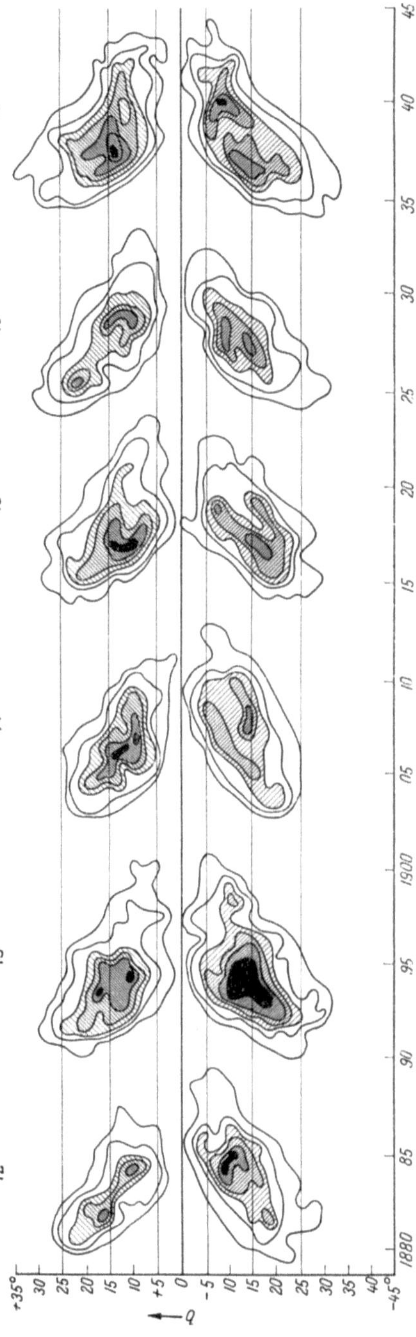

Fig. 119. Butterfly diagram of sunspots for 1880 to 1945. The lines connect latitudes of equal annual spot numbers. U. Becker: Z. Astrophys. **37**, 47 (1955).

Waldmeier[3] found that the mean latitude of the sunspot-zone during the cycle depends on R_M. In an active cycle the mean latitudes are greater than in an inactive one. Fig. 120 shows for five different R_M values the mean migration of the spot zones. The abscissa gives the time, reckoned from the sunspot maximum, measured in years; the ordinate gives the mean latitude. It appears that this set of curves can be combined into one if the time is counted from the beginning of the sunspot cycle[3]. At the time of a high maximum, which comes early in the sunspot cycle, the latitude of the spot belt is higher than at the time of a low, and hence late maximum.

There is a certain north-south asymmetry of solar activity; the

[1] M. Trellis: Suppl. Ann. d'Astrophys. **5** (1957).
[2] U. Becker: Priv. comm. 1958.
[3] M. Waldmeier: Astr. Mitt. Zürich **138**, 470 (1939).

activity in the northern cycle tends to develop somewhat earlier than in the southern cycle. Sometimes also the southern spot activity extends somewhat beyond the equator. The spottedness is not always equal in the northern and southern hemispheres[1]. In very active cycles (high R_M) the southern hemisphere is the more spotted at the beginning of the cycle, and the northern at the end, and conversely for very inactive cycles (low R_M).

In Sect. 36 it was remarked that the polarity of the spot-groups changes with the solar cycle. The sunspot curve of Fig. 117 shows further that in many cases the maxima of the R-curves are alternately high or low. These are reasons why it seems physically more appropriate not to speak of an 11-year but of a 22-year cycle. Thus, two successive cycles belong to each other, and there still remains the problem which two of the successive spot cycles form a pair. To investigate this, GNEVISHEV and OLJ[2] plotted ΣR for the even cycles against (a) ΣR for the following odd cycle and (b) against ΣR for the preceding odd one. The first diagram gave a good correlation (correlation coefficient 0.91 ± 0.11), whereas the latter's correlation was bad (correlation coefficient 0.50 ± 0.24). Hence every even cycle (cf. numbering in Fig. 117) forms a pair with the following odd one.

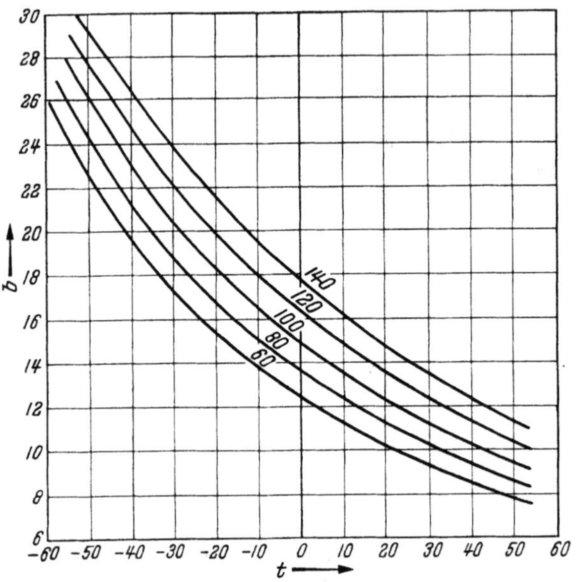

Fig. 120. Dependence of the latitude motion of the sunspot zone on the maximum R-value for each cycle. Abscissa: time difference from the epoch of spot maximum, measured in rotation periods. Ordinate: mean heliographic latitude. The curves are labeled with the R_{max}-value of the cycle. M. WALDMEIER: Astr. Mitt. Zürich **138**, 470 (1939).

98. Long scale fluctuations of the solar activity. As early as 1862 WOLF suggested a long cycle superposed on the 11-year solar activity. This long period of about 80 years has proved to be real. GLEISSBERG[3,4] could trace it back in the sunspot observations to 1610 and in SCHOVE's[5] list of aurorae to A.D. 300. Maxima of the long wave occurred in 1639, 1719, 1778, 1860 and 1957. Evidently the N-S asymmetry[6] found in Sect. 97 is thus a property of the long cycle, as was stressed by WALDMEIER[7]: very active cycles occur when the southern hemisphere is the more spotted at the beginning, very inactive cycles occur in the reverse case. However, the average N-S asymmetry, integrated over one cycle is zero for very active and very inactive cycles, whereas Σ (N-S) >0 or <0

[1] H. W NEWTON and A. S. MILSON: Monthly Notices Roy. Astronom. Soc. London **115**, 398 (1955).
[2] M.N. GNEVISHEV and A.I. OLJ: Astron. Zhurn. **25**, 18 (1948).
[3] W. GLEISSBERG: Publ. Istanbul Univ. Obs. **27** (1944).
[4] W. GLEISSBERG: Naturwiss. **42**, 410 (1955). — Reviews in Sterne **34**, 117 (1958), and J. Brit. Astr. Ass. **68**, 148 (1958).
[5] D. J. SCHOVE: J. Geophys. Res. **60**, 127 (1955).
[6] H.W. NEWTON and A. S. MILSON: Monthly Notices Roy. Astronom. Soc. London **115**, 398 (1955).
[7] M. WALDMEIER: Z. Astrophys. **43**, 149 (1957).

during the ascending or descending parts of the long cycle. GLEISSBERG[1] could find the 80-year period in the relative geomagnetic effects produced by the northern and southern solar hemispheres.

Attempts have been made to find still longer periodicities, but these can no longer be based on sunspot records but only on recordings in old chronicles of visual observations of big sunspots, the observations of aurorae (these are more frequent during sunspot maxima), the detection of comets[2] (the sky is thought to be more transparent at sunspot minimum). With such very unreliable data LINK[2] believes it possible to detect a 410-year period which he traces back till the year 2000 B.C. A longer period, of 600 to 700 years, is suggested by the asymmetry in the sunspot curves[3] and by tree-rings in Californean Sequoia gigantea[4].

99. The prominences during the solar cycle. Also the distribution of filaments and prominences on the disk reflects the 11-year cycle, but the Maunder diagram for the prominences and filaments is more complicated than that for the sunspots and does not show the typical butterfly structure. Fig. 121 demonstrates the latitude distribution of prominences between 1905 and 1952[5]; confirming and extending smilar investigations by D. and M. D'AZAMBUJA[6]. In each hemisphere there are at least two zones where prominences are frequent, but the zone of high latitude is less important. The major zone in turn sometimes shows two peaks, one corresponding to the mean latitude of the spots (the active spot filaments). The other, at a slightly greater latitude, corresponds with the stable quiescent filaments, which often occur at the high latitude side of activity centres. The mean latitude of the major zone which is about 30° at the beginning of the sunspot cycle, decreases to about 17° at the end, thus being higher than that of the spot zone. Shortly before sunspot minimum the high "polar" prominence zone becomes visible (latitude between 40 and 50°); it remains visible till the next sunspot maximum. The poleward drift of this prominence belt ends near sunspot maximum when $\bar{b}_p \approx 90°$. The $\bar{b}_p(t)$ curves have the same forms in different cycles, and they can be made coinciding if the time t is counted from the beginning of each cycle[7]. The latitude is low at the time of a high sunspot maximum; it is high at the time of a low maximum; this is a reflection of the fact that a high maximum occurs earlier than a low one. At the time of a high maximum ($R_M =$ 140) $(\bar{b}_p)_{\max} \approx 65°$; for $R_M = 80$ $(\bar{b}_p)_{\max} \approx 90°$ and for lower maxima the polar prominence zone has disappeared at the time of maximum activity (see also Fig. 118, Sect. 96).

Another aspect of the distribution of filaments with latitude is their orientation[8]. The high filaments are orientated nearly parallel to the equator; those at lower latitudes have greater inclinations.

Fig. 121 already made clear that the number of prominences depends on the solar cycle, but upon closer examination this appears to apply mainly to the high prominences[9], those exceeding the "natural platform height", 30000 km. Just

[1] W. GLEISSBERG: Z. Astrophys. **43**, 144 (1957).
[2] F. LINK: Bull. Astr. Inst. Czech. **6**, 122 (1955).
[3] W. GLEISSBERG: Publ. Istanbul Univ. Obs. **51** (1954).
[4] I.W. MAXIMOW: Dokl. Akad. Nauk SSSR. **92**, 1149 (1953).
[5] R. ANANTHAKRISNAN: Nature, Lond. **170**, 156 (1952). — Proc. Indian Acad. Sci. **40**, 72 (1954).
[6] L. and M. D'AZAMBUJA: Ann. Obs. Meudon **6**, 52 (1948).
[7] M. WALDMEIER: Z. Astrophys. **42**, 34 (1957).
[8] See e.g. D.H. MENZEL and B. BELL: Convegno Volta **11**, 143 (1953).
[9] D.E. BILLINGS and C. KOBER: Astronom. J. **62**, 242 (1957).

like the spicules the number of low prominences changes only slightly during the solar cycle.

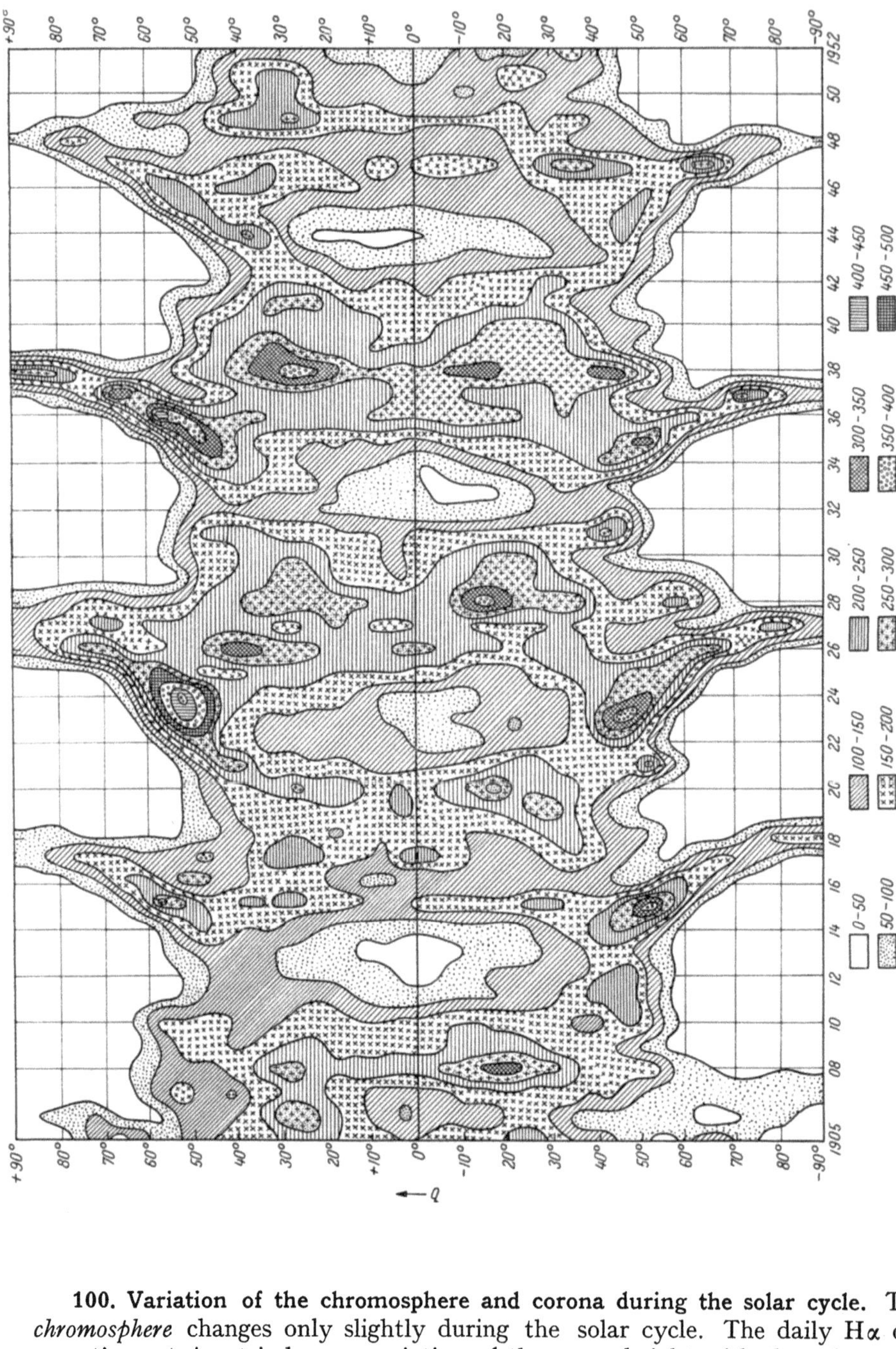

Fig. 121. Isopleths of mean daily prominence areas from 1905 to 1952. The contour lines are drawn at 50×10^{-3} sq. minutes of arc of prominence area. Remarkable are the equatorial motion of the major prominence zone and the poleward motion of the polar prominence zone. R. ANANTHAKRISHNAN: Proc. Indian Acad. Sci. 40, 72 (1954).

100. Variation of the chromosphere and corona during the solar cycle. The *chromosphere* changes only slightly during the solar cycle. The daily Hα observations at Arcetri show a variation of the mean height with the solar cycle (Fig. 122). From a discussion of chromospheric emission gradients (flash spectra)

ATHAY *et al.*[1] found that the average emission gradient does not show any significant dependence upon the solar cycle.

The corona. We have already referred to the brightness variation of the white corona (Sect. 71); at sunspot minimum the integrated brightness between $\varrho = 1.03$ and 6 is half the sunspot maximum value. The variation of the surface brightness for different limb distances was investigated by W.N. ABBOTT[2], the variation is greatest at $\varrho \approx 1$ to 1.5. The 11-year variation of the height of the emission at 5303 Å is demonstrated in Fig. 123a (WALDMEIER[3]). The variation of the coronal temperature during the sunspot cycle was deduced by WALDMEIER[4] from the average intensity ratio of the coronal lines 5303 and 6374 Å, with the aid of formulae (73.1); cf. Fig. 123b. Similar results were found by TRELLIS[5].

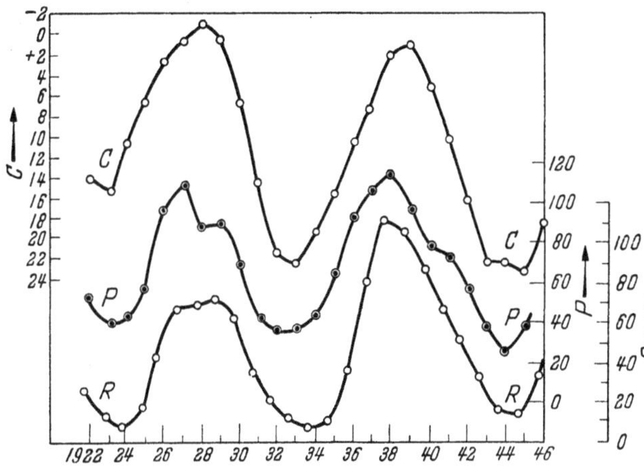

Fig. 122. The height variation of the Hα chromosphere (*C*), defined as the sum, in absolute units of the four differences: north pole minus equator, equator minus south pole, south pole minus equator, equator minus north pole, in seconds of arc the frequency of prominences (*P*) and the sunspot number (*R*). M. G. FRACASTORO: Oss. e Mem. Arcetri **64**, 33 (1947).

The latitude of the monochromatic coronal "rays" (intensity maxima of the monochromatic corona) changes during the cycle. Coronal activity, measured with the 5303 Å line, starts at $b = \pm 50$ to 70° one to four years before sunspot minimum[5]. The activity zone then splits into a major zone having the same latitude of the spot-zone and a polar zone moving rapidly polewards to reach $b = 90°$ at the maximum of solar activity and to disappear afterwards[6].

Radio observations made during solar eclipses show that from 1947 to 1954 the residual intensity measured at mid-eclipse decreased at all wavelengths, indicating a greater contraction of the coronal isophotes[7].

The most striking aspect of the white corona is its ellipticity. At sunspot minimum the corona is flattened (example: the minimum corona of 1954) and at sunspot maximum if has a more circular shape. In order to define the ellipticity by one single parameter, LUDENDORFF introduced the parameter $(a + b)$,

[1] R.G. ATHAY, D.H. MENZEL and F.Q. ORRALL: Smithson. Contrib. Astrophys. **2** (4), 35 (1957).
[2] W.N. ABBOTT: Ann. d'Astrophys. **18**, 81 (1955).
[3] M. WALDMEIER: Sonnenkorona **2**, 142 (1957).
[4] M. WALDMEIER: Z. Astrophys. **30**, 137 (1952).
[5] M. TRELLIS: Ann. d'Astrophys. Suppl. **5** (1957).
[6] R. MÜLLER: Z. Astrophys. **38**, 212 (1955).
[7] B.M. TCHIKHATEV, in: Radio astronomy. I.A.U. symposium No. 4, p. 311, 1957.

being the ellipticity at $\varrho = 2.0$ (the ellipticity ε is written as $\varepsilon = a + b(\varrho - 1)$). VAN DE HULST[1] plotted $(a + b)$ against the phase φ of solar activity (cf. also SAITO[2], Fig. 6). Here φ is -1, and $+1$ at two consecutive maxima; it is 0 at

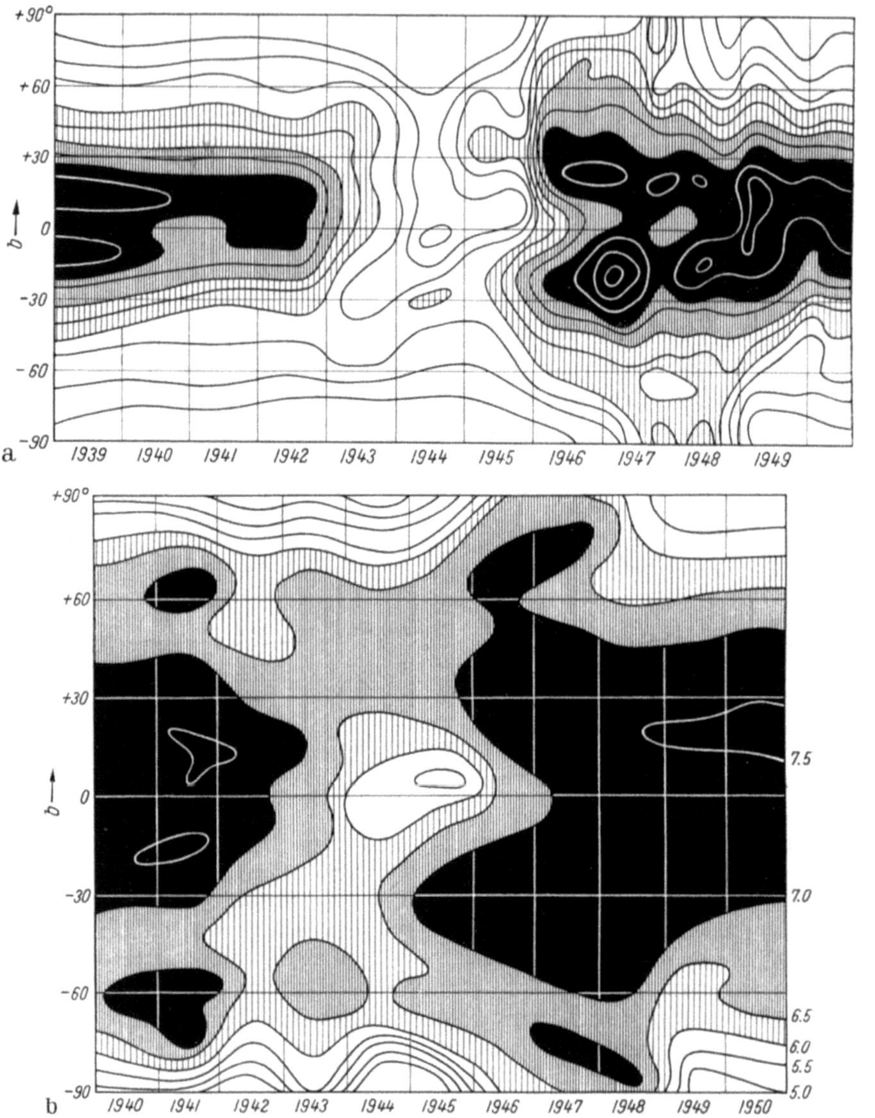

Fig. 123a and b. (a) Isohypses of the corona in 5303 Å light from 1939 to 1950. The lines correspond with distances of 0'.5. Shaded are heights above 2'.0, black those above 4'.0. M. WALDMEIER: Die Sonnenkorona, II, p. 142. 1957. (b) Isotherms of the corona from 1940 to 1950; temperatures are given in units of 10^5 °K. M. WALDMEIER: Z. Astrophys. **30**, 137 (1952).

the minimum between, and proportional to the time, according to LUDENDORFF'S definition. However, the form of the solar corona is greatly defined by the long rays[3], which, in turn, are related to the quiescent filaments. So the flatten-

[1] H.C. VAN DE HULST: Bull. astronom. Inst. Netherl. **11**, 135 (1950).
[2] K. SAITO: Publ. Astr. Soc. Japan **8**, 126 (1956).
[3] M. WALDMEIER, H. ARBER and H. BACHMANN: Z. Astrophys. **42**, 204 (1957).

ing of the corona will depend on the number of polar prominences, and therefore it seems more appropriate to define a phase φ_1 in such a way that equal phases correspond to equal positions of the prominence zones (see Sect. 99);

$$\varphi_1 = \frac{T - m_1}{m_2 - m_1}$$

where m_1 and m_2 are the epochs of the preceding and following minima; T is the time of observation. The variation of $(a+b)$ with φ_1 is shown in Fig. 124. It appears that the corona has its greatest flattening at minimum. It is nearly spherical at $\varphi_1 = 0.32$, when the polar prominence zone at $b = 70°$ has its maximum development.

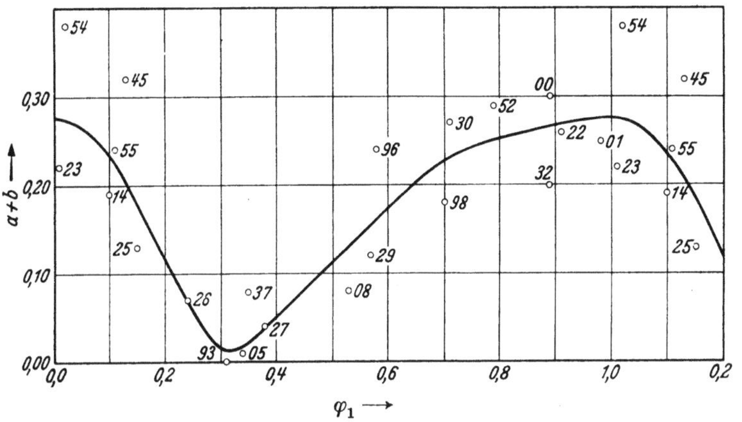

Fig. 124. Coronal ellipticity during the solar cycle. The points are labeled with the number of the year, omitting the century number. M. WALDMEIER, H. ARBER and H. BACHMANN: Z. Astrophys. **42**, 156 (1957).

101. Other parameters measuring the solar activity. The main parameters measuring the solar activity are the sunspots (numbers and areas), the areas and the numbers of faculae, the numbers of prominences and filaments, the numbers of flares, the height of the Hα-chromosphere, the form of the corona, the intensity of the coronal lines, the radio radiation of the Sun. Most of these quantities are measured daily at various observing stations.

1. Daily maps of the Sun are made at the observatory of Meudon, based on spectroheliographic disk observations made at Meudon and coronal observations at the Pic du Midi. The maps contain the spots, faculae, coronal line intensities and data on flares if observed on one of these observatories. The maps are weekly dispatched.

2. The Fraunhofer Institut prepares daily maps of the Sun from observations of a dozen institutes covering the whole Earth. The maps contain essentially the same data as the Meudon Map, but as a matter of course, they are more complete. The maps are bimonthly dispatched.

3. The Poulkovo Observatory and the Commission for solar phenomena of the Astronomical Council of the Sovjet Union publish daily maps of the Sun[1], containing faculae, spots, filaments, prominences and coronal intensities, as well as the H, D and Z components of the geomagnetic field.

[1] Solnetchnye Dannye, Bull. **1** (1956).

Fig. 125 shows for the same day the solar maps published by these three institutes. The comparison shows the relatively good agreement, the minor differences, the relative accuracy and completeness of each of the maps.

Fig. 125a—c. Daily maps of the Sun for 28 April, 1957. On all maps the facular areas are shaded. Filaments and prominences are black, spots are indicated in various ways. The way of giving coronal intensities differs. (a) Pulkovo, USSR. Solid line around Sun: intensity of 5303 Å

4. The Boulder Solar Activity Maps, published quarterly, contain for each solar rotation data on the various centres of activity as measured by means of spots, faculae, coronal line intensities and flares, with additional data on the prominence activity at the limb.

5. The Quarterly Bulletin on Solar Activity, edited by the Zürich observatory with financial help of the International Astronomical Union, contains daily data on most kinds of solar activity (spots, faculae, prominences, flares, radio-noise, corona). The work is published quarterly but with a considerable delay.

6. The Solar-Geophysical data issued quarterly by the U.S. National Bureau of Standards contain R numbers, and data on faculae, coronal intensities, flares, radio-intensities, and geomagnetic activity indices.

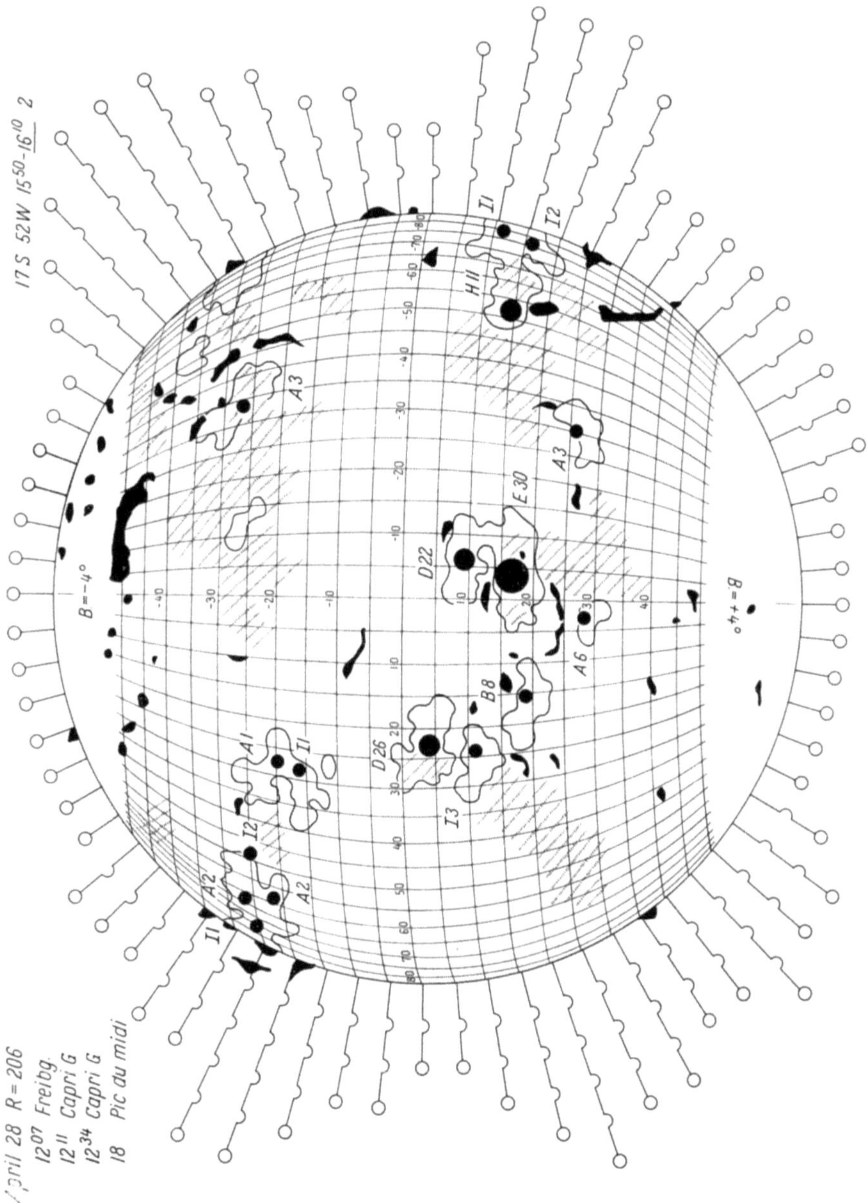

Fig. 125 b. Fraunhofer Institut, Germany; the lengths of the radial lines are proportional to the intensity of 5303 Å line.

7. Finally mention should be made of the Cartes Héliographiques de la photosphère (Zürich), the Synoptic Maps of the chromosphere (Meudon observatory), and summarizing maps on chromospheric phenomena published monthly like those of Mrs. M. D'AZAMBUJA (l'Astronomie; Bulletin de la Société Astronomique de France), those of the Kanzelhöhe Obervatory (published in the Documenta-

tion des Observateurs; Observatoire de Haute Provence), those of the Tokyo Observatory, published in the Japanese Bulletin of Solar Phenomena, etc.

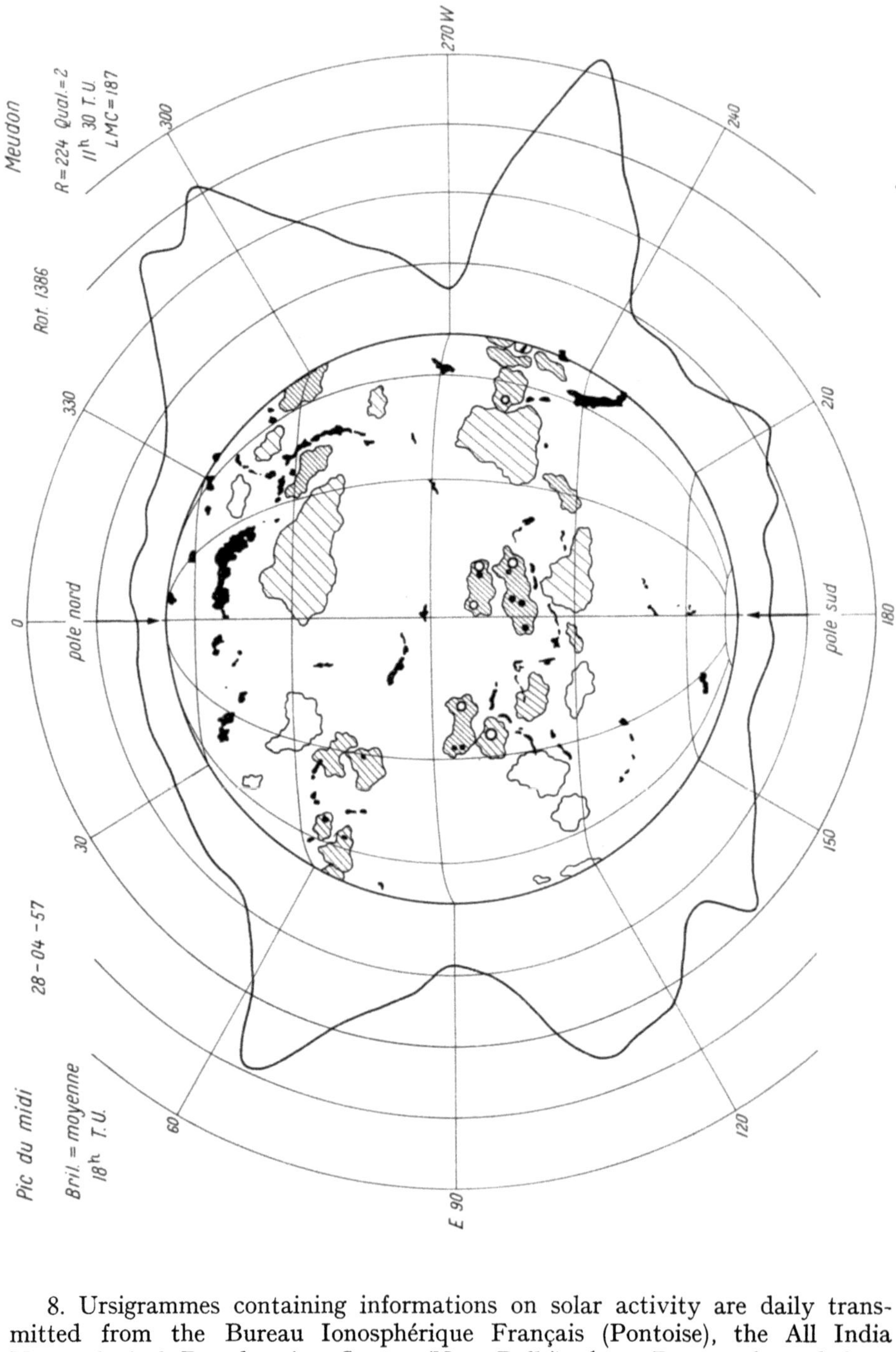

Fig. 125c. Meudon, France. The distance of the solid line to the limb is proportional to the intensity of the 5303 Å line.

8. Ursigrammes containing informations on solar activity are daily transmitted from the Bureau Ionosphérique Français (Pontoise), the All India Meteorological Broadcasting Centre (New Delhi), from Darmstadt and from Tokyo.

Fig. 126 (partly after ALLEN[1]) shows the variation of the monthly mean values of the sunspot number R, the sunspot area A, the facular area F, the prominence area P, the coronal 5303 Å intensity, the E layer ionizing flux I, and the radio-radiation at 50, 25 and 10.7 cm, from 1947 to 1954. Some of these phenomena show a time lag behind the sunspot number variations, presumably

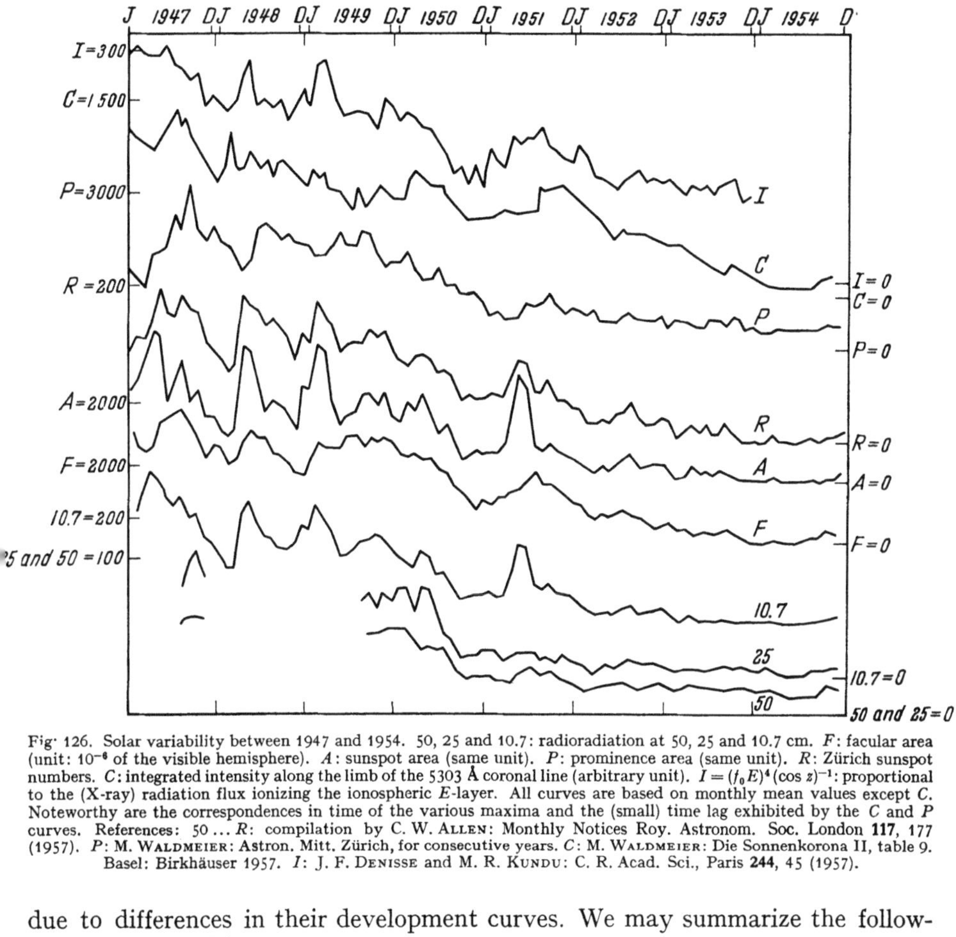

Fig. 126. Solar variability between 1947 and 1954. 50, 25 and 10.7: radioradiation at 50, 25 and 10.7 cm. F: facular area (unit: 10^{-6} of the visible hemisphere). A: sunspot area (same unit). P: prominence area (same unit). R: Zürich sunspot numbers. C: integrated intensity along the limb of the 5303 Å coronal line (arbitrary unit). $I = (f_0 E)^4 (\cos z)^{-1}$: proportional to the (X-ray) radiation flux ionizing the ionospheric E-layer. All curves are based on monthly mean values except C. Noteworthy are the correspondences in time of the various maxima and the (small) time lag exhibited by the C and P curves. References: 50 ... R: compilation by C. W. ALLEN: Monthly Notices Roy. Astronom. Soc. London **117**, 177 (1957). P: M. WALDMEIER: Astron. Mitt. Zürich, for consecutive years. C: M. WALDMEIER: Die Sonnenkorona II, table 9. Basel: Birkhäuser 1957. I: J. F. DENISSE and M. R. KUNDU: C. R. Acad. Sci., Paris **244**, 45 (1957).

due to differences in their development curves. We may summarize the following delay times[1,2]:

radioemission at 10.7 cm:	1 to 2 days[2]
bright H flocculi:	5 days
Ca^+ flocculi:	10 days
faculae:	20 days
5303 Å corona line:	26 days.

b) Solar rotation and the Sun's general magnetic field.

102. The law of solar rotation. The westward drift of spotgroups over the disk shows the solar rotation, the average synodic period being about 27 days. The rotation period further depends on the latitude: the equatorial region rotating

[1] C. W. ALLEN: Terr. Magn. Atm. Electr. **53**, 433 (1948).
[2] M. WALDMEIER: Astron. Mitt. Zürich **215** (1958).

fastest (CARRINGTON 1858). Observations of one single object may, however, be influenced by longitude drift of that object, and have to be verified by statistical studies on many kinds of solar phenomena.

The most extended series of observations refer to sunspots; other observations refer to the photospheric faculae, the chromospheric Ca^+ faculae, the filaments and prominences and even to coronal activity centres. All observations refer to medium latitudes (0 to 60°), only some filaments and the polar faculae reach greater latitudes. A number of objects can be followed during one disk passage or longer, others can only be observed for a short time, like the polar faculae. The latter are known to have short lifetimes, and only eight of the most persistent ones, with lifetimes of 40 to 50 hours, could be used for the determination of the rotation velocity of the Sun. These latter determinations made it possible to extend the rotational velocity curve to 80° latitude; before that time it was only known up to 65°.

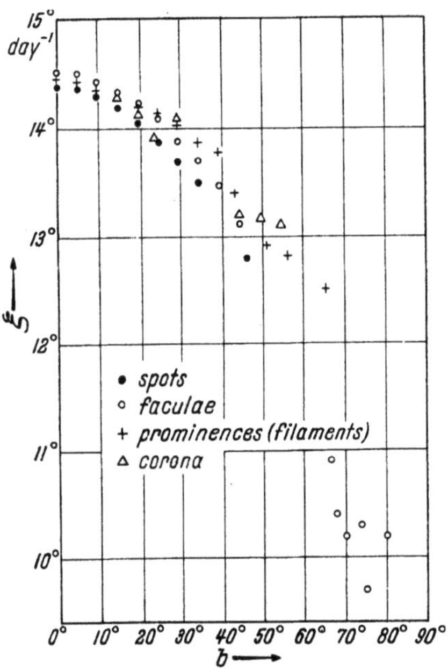

Fig. 127. Rotational velocity of the Sun in dependence of heliographic latitude b. The sources are: *Sunspots:* mean values are taken from a compilation of spot data given in M. WALDMEIER: Ergebnisse ... der Sonnenphysik, 2nd ed., p. 52, 1955. The $b=45°$-value is a mean value from the observation of two high-latitude spots by M. WALDMEIER: Z. Astrophys. **43**, 29 (1957) and M. KOPECKY, J. KVICALA and J. PLACEK: Bull. Astr. Inst. Czech. **8**, 106 (1957). *Faculae:* M. WALDMEIER: Mitt. Zürich **147** (1946); Z. Astrophys. **27**, 24 (1950). *Polar faculae:* M. WALDMEIER: Z. Astrophys. **38**, 37 (1955). *Prominences:* L. and M. D'AZAMBUJA: Ann. Obs. Meudon **6**, 7 (1948); U. BECKER: Z. Astrophys. **42**, 1 (1957). *Corona:* W. PETRI: Z. Astrophys. **34**, 68 (1954); M. WALDMEIER: Z. Astrophys. **27**, 24 (1950). The data of M. TRELLIS [Ann. d'Astrophs. Suppl. **5** (1957)] have not been included in this survey.

Fig. 127 assembles rotation velocity measurements, determined by means of the five kinds of objects mentioned above. The decrease of velocity with increasing latitude is clear. It is often tried to represent the observed daily displacement ξ (degrees per day) by a polynomial in $\sin b$ or/and $\cos b$ (b is the latitude), like[1]:

$$\xi = 14^\circ\!.38 - 2^\circ\!.77 \sin^2 b. \qquad (102.1)$$

It is clear that such interpolation formulae have only a formal meaning.

The various layers and objects all seem to have the same angular velocity at the same latitude. The small systematic residuals between the various objects, previously ascribed to an increase of rotational velocity with height, might rather be ascribed to the imperfectly eliminated influence of proper motions of the various structures. The rotational velocity determined by means of spots, is somewhat below the mean curve (but the most pronounced difference, near $b=45°$ is the effect of two well-studied spots only), and thus is perhaps due to such an effect. U. BECKER[2] has shown that long-lived seats of spot formation have a rotational velocity greater by $0^\circ\!.1$ per day than that of the spots. A similar result was found from the study of long-lived facular fields (M. D'AZAMBUJA, unpublished).

The rotational velocity can also be determined by means of radial velocity observations of Fraunhofer lines or emission lines. These measurements show a considerable scatter. At $b=0$ the ξ-determinations by means of metal absorption

[1] H. W. NEWTON and M. L. NUNN: Monthly Notices Roy. Astronom. Soc. London **111**, 413 (1952).

[2] U. BECKER: Z. Astrophys. **37**, 47 (1955).

lines of the photosphere range between 13°.2 and 14°.8. The mean scatter is 0°.35, hence 2.5%, whereas the mean scatter derived from the observed motions of various objects (Fig. 127) is about 0.4%, nearly one order smaller. At $b=40°$ the scatter is still greater, nearly twice that at $b=0$. This can easily be explained: the mean equatorial velocity is 2 km/sec; an error of 2.5% corresponds to an error in the measured velocity of 0.05 km/sec, hence with $\Delta\lambda = \pm 0.0008$ Å. This is about the limit of accuracy of wavelength measurements. Another reason for this great scatter is the inhomogeneity of the photospheric velocity field (cf. also HART[1]). The internal motions in prominences and the spicular motions in the chromosphere reach values even ten to forty times greater than in the photosphere; there is no good reason to expect to obtain the rotational velocity with some precision from radial velocity measurements of these objects.

It should, finally, be noted that no measurable variation of the Sun's rotation period could be detected, at least not between 1878 and 1944 (NEWTON and NUNN, l. c.).

103. The latitude drift of sunspots and its variation in the solar cycle. The fore-going section dealt with the motion component parallel to the equator and its differential effects. Besides, the latitude drift of spots indicates the occurrence of a motion component perpendicular to the equator. The proper motion in latitude is always much smaller than the motion in longitude so that the effect is only observable for long-lived spots.

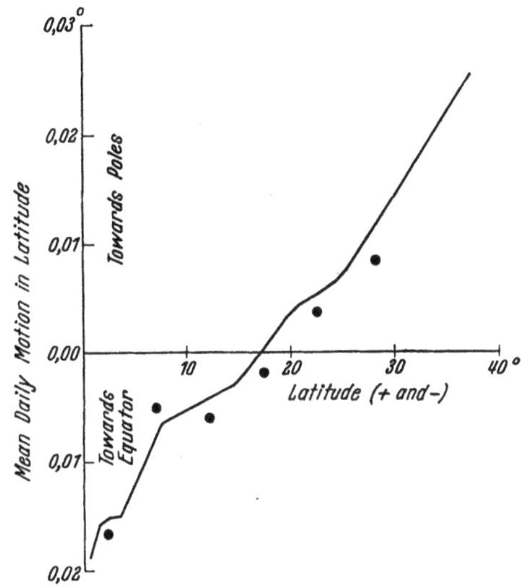

Fig. 128. Average daily latitude drift of sunspots, with periodicities in time and latitude eliminated. Solid line: J. TUOMINEN. Z. Astrophys. **37**, 145 (1956); Dots: R. S. RICHARDSON and M. SCHWARZSCHILD: Convegno Volta **11**, 228 (1953).

Such investigations have first been made in 1941 by TUOMINEN[2] and later by RICHARDSON and SCHWARZSCHILD[3]. TUOMINEN'S discussion is based on the Greenwich sunspot observations 1874 to 1941 and 1945. All groups with lifetimes longer than one rotation have been investigated. The total number of spot-groups was divided into subgroups, according to their latitude (zones of 1° width), and mean latitude drifts were determined.

In RICHARDSON and SCHWARZSCHILD'S investigation about 1800 spots have been used, occurring in the Greenwich material from 1874 to 1938 and the Mount Wilson material from 1939 to 1950. The total material was divided into six groups, according to the absolute value of the latitude (there are no systematic differences between the two hemispheres). The main conclusion emerging from Fig. 128 is that spots with $|b| < 16°$ have an equatorward motion, whereas spots at greater latitudes have a poleward motion, increasing with increasing latitude. The small oscillatory character of the curve, initially found by TUOMINEN, was smoothed out in a later analysis. U. BECKER'S[4] investigation throws doubt to the reality of these small deviations from the mean curve.

[1] A. B. HART: Monthly Notices Roy. Astronom. Soc. London **114**, 17 (1954).
[2] Cf. J. TUOMINEN: Z. Astrophys. **37**, 145 (1955).
[3] R. S. RICHARDSON and M. SCHWARZSCHILD: Convegno Volta **11**, 228 (1953).
[4] U. BECKER: Z. Astrophys. **34**, 129 (1954).

TUOMINEN[1], RICHARDSON and SCHWARZSCHILD[2] further found that the latitude drift depends on the phase in the 22-year solar cycle. If the mean latitude drift (Fig. 128) is subtracted from the observed motions, there remain small residuals, depending on the phase of the 22-year cycle: In odd cycles, the spot-groups show a preference for moving equatorward (apart, of course, from the general latitude motion of Fig. 128 which remains valid throughout the whole solar cycle). The average amplitude of this new 22-year superimposed latitude drift curve is 0.005 degrees/day according to RICHARDSON and SCHWARZSCHILD. It depends on the latitude b according to TUOMINEN, with an average daily displacement of $0\overset{d}{.}0005 \times b$. The mean value is of the same order as the former one. For suggested explanations of the effect, cf. RICHARDSON and SCHWARZSCHILD[2], and PLUMPTON and FERRARO[3].

U. BECKER[4] found indications that the 22-year effect is not physically real but is produced by variations in the sunspot frequency curve from cycle to cycle.

104. The Sun's general magnetic field. After many previous attempts to measure the Sun's general magnetic field, yielding values up to 50 Gauss, a decisive determination of its value has only been made after 1948, when the visual and photographic methods used up to that time were generally replaced by photoelectric methods (cf. Sect. 32), which make use of the fact that a small line shift produces great intensity variations at the steep parts of the line profiles. The effect can be doubled by using two slits, at the wings of the profile. (HALE, in 1930, was the first to apply the photoelectric method.)

In this way THIESSEN[5], observing at 45° latitude, found in 1949 a field which, assuming a dipole field, would yield a pole field strength of $+(1.5 \pm 0.75)$ Gauss. In 1951 the pole field strength H, found in the same way, was $+(2.4 \pm 0.5)$ Gauss[6]. The positive sign signifies that the direction of the field relative to that of the solar rotation is opposite to that of the Earth.

The definition of positive magnetic polarity as used generally for the Sun and stars is that a magnetic vector pointing towards the observer is positive in sign. Hence the north geographic pole would have a negative magnetic polarity; the Sun's north pole has a positive magnetic polarity. As seen from the north pole of rotation, a spin in the positive direction of a sphere, having a positive electric charge, will induce a field of positive magnetic polarity.

On the other hand, KIEPENHEUER, measuring in 1950 and 1951 with an apparatus having a sensibility limit of 1 Gauss, could not detect a field component above 1 Gauss.

Technical innovations introduced by the BABCOCKS[7] permitted the recording of fields as weak as 0.3 Gauss. They detected the weak poloidal "general" field, restricted to latitudes within about 30° from the poles (Fig. 46b). The poloidal or "general" field during 1953 to 1956 has been consistently positive in the north, negative in the south. Its intensity varies from place to place, and its boundaries are irregular. If allowance is made for limb-darkening, the mean field component in the line of sight is about 1 Gauss in very high latitudes. These results are inconsistent with THIESSEN's observations, in which it was assumed that the field extends at least down to 45°. THIESSEN's results are attributed by the BABCOCKs to the influence of the so-called BM regions.

[1] J. TUOMINEN: Z. Astrophys. **30**, 261 (1952).
[2] Cf. footnote 3, p. 339.
[3] C. PLUMPTON and V. C. A. FERRARO: Astrophys. Journ. **121**, 168 (1955).
[4] Cf. footnote 4, p. 339.
[5] G. THIESSEN: Observatory **69**, 228 (1949).
[6] G. THIESSEN: Z. Astrophys. **30**, 185 (1952).
[7] H. W. and H. D. BABCOCK: Astrophys. Journ. **121**, 349 (1955).

Assuming that the magnetic field at the surface is $H = H_p(b - 46°)/4\pi$, where H_p is the polar field strength, and that the direction of the lines of force is as if diverging from a point at the Sun's axis, 0.6 of the radius out of the centre, one finds $H_p = +2.6$ Gauss, in accordance with THIESSEN'S result.

The angle between the rotational and the magnetic axes of the Sun (as determined from the directions of the polar rays)[1] is smaller than 0°.7. The field does not seem to change in the eleven-year cycle.

No features on spectroheliograms appear to be related to this weak field. Regions of polar faculae do not show any observable enhancement of the field. In any case the observations support the suggested relation (SCHUSTER, Report B.A.A.S. 1892) between the polar rays and the general magnetic field. VAN DE HULST[2] showed that the polar rays can easily be explained by a field smaller than 0.2 Gauss.

The total magnetic flux F_t of the poloidal field (above 46°) is

$$F_t = \int_{b=46°}^{90°} H \cos(b - \psi)\, ds = 8 \times 10^{21} \text{ Maxwell},$$

where ψ is the projection angle in the meridional plane. As to the order of magnitude this field is five times as great as the magnetic flux of an average bipolar magnetic region or of a strong unipolar magnetic region.

We should further remark that the total mean magnetic flux of the Sun obtained by summing up with their proper signs all magnetic fluxes of the sunspots, ranges between values of 0.5×10^{21} to 8×10^{21} Maxwell[3]. (If, however, various BM regions are caused by the same poloidal field looping up through the surface and back, this value should be reduced by an unknown factor of the order 2 to 8.)

Time of decay of the general field. COWLING[4] has shown that the decay time of the general solar magnetic field is very long. In matter at rest, the variation of the magnetic field with time is slowed down by induction leading to the equation

$$4\pi\sigma \frac{\partial \boldsymbol{H}}{\partial t} = \nabla \boldsymbol{H}^2,$$

which is approximated to

$$4\pi\sigma \frac{H}{t_0} = \frac{H}{l^2}$$

(cf. also Sect. 39), where t_0 is the time of decay and l is a characteristic "scale length" of the field; σ is the conductivity. For high temperature T

$$\sigma = 2 \times 10^{-14} \frac{T^{\frac{3}{2}}}{Z}$$

where Z is the mean charge of the particles. Assuming $l = 2 \times 10^{10}$ cm; $\sigma = 10^{-4}$ e.m.u. we obtain $t_0 = 10^{10}$ years. Thus, variations in the field seem to occur extremely slowly, and the presently existing field might be of the same order as that occurring when the Sun was just formed.

105. The internal rotation and circulation of the Sun. The information collected in Sects. 102 and 103 on the rotation and latitude motion of the solar surface layers, suggest that the latitude drift of the sunspots may be interpreted

[1] M. WALDMEIER: Z. Astrophys. **41**, 115 (1956).
[2] H.C. VAN DE HULST: Bull. Astronom. Inst. Netherl. **11**, 410 (1950).
[3] W. GROTRIAN and H. KÜNZEL: Z. Astrophys. **28**, 28 (1950).
[4] T.G. COWLING: Monthly Notices Roy. Astronom. Soc. London **105**, 166 (1945).

as real large-scale mass motions. The sunspot drift suggest the occurrence of meridional currents, a schematic representation of which can easily be made, being given that matter streams towards the equator for $|b| < 16°$, and that it streams poleward for $|b| > 16°$.

A theorem derived by VON ZEIPEL and elaborated by VOGT and EDDINGTON predicts the occurrence of meridional currents: a star in radiative equilibrium does not rotate uniformly. VOGT[1] and EDDINGTON[2] suggested that with an acceptable law for the energy generation the poles would become hotter than the equatorial regions, leading to meridian circulation. At the surface ΔT would be $< 1°$.

In these considerations the influence of the magnetic field is neglected. The influence of the magnetic field, if sufficiently great, would be to change the rotation pattern in that of isorotation[3,4]. In that case the angular velocity is constant along cylinders formed by the rotation of lines of force around the rotation axis (because otherwise the lines of force would be continuously wound round the axis of rotation).

From the observational side, internal circulation is suggested by the observed relative abundance deficiencies of light atomic nuclei like H^2 and Li, as compared to their terrestrial and meteoritic relative abundances. Assuming that the latter abundances are still of the same order of magnitude as the abundances in the solar system at the time of its origin, a great part of the photospheric light nuclei seem to have been destroyed since, e.g. by a slow internal circulation. The ratio between the solar and the terrestrial relative abundances of lithium (both abundances compared with those of elements like Si or Fe) is about 0.01. Further it is known that the lifetime of lithium is short for $T \approx 3 \times 10^6$ °K. Hence 99% of the initially occurring photospheric lithium has at least once been brought to the level with $T \approx 3 \times 10^6$ °K. Assuming a total solar lifetime of 5×10^9 years, we obtain for this part of the Sun a circulation time roughly of the order of 10^9 years.

This circulation refers to rather deep layers: the region with $T = 3 \times 10^6$ °K is about halfway the Sun's centre. For still deeper layers the rotation time is much longer, because the viscosity of the solar matter increases quickly with depth. SWEET[5] found that the time to be taken for material carried by meridional currents, to go from the centre to the surface of the Sun, is 8×10^{12} years. Qualitatively this result is confirmed by MESTEL[6]. It is natural that the circulation time becomes much shorter toward less deep layers and takes values of the order of one or several months in the solar convection zone, a layer with a depth[7] of about 100000 km. At a depth of 10^5 km the temperature is nearly 10^6 °K; there nuclear reactions are still unimportant. The upward velocity of the convection elements is 1 km/sec in the upper part of the convection zone, whereas it is extremely small at its very basis. The time, necessary to traverse the main part of the convection zone may be of the same order as the lifetime of the spots.

This discussion makes it clear that *the solar interior up to the lower limit of the convection zone should be regarded as a perfectly rigid body*, as far as transient phenomena are considered with a life time shorter than a year. Important mass displacements occur only in the regions above that limit.

[1] H. VOGT: Astronom. Nachr. **223**, 229 (1925).
[2] A. S. EDDINGTON: Monthly Notices Roy. Astronom. Soc. London **90**, 54 (1930).
[3] V. C. A. FERRARO: Monthly Notices Roy. Astronom. Soc. London **97**, 458 (1937).
[4] H. ALFVÉN: Ark. Mat. Astronom. Fys. **29**a, 11 (1943).
[5] P. A. SWEET: Monthly Notices Roy. Astronom. Soc. London **110**, 548 (1950).
[6] L. MESTEL, in: Les Processus Nucléaires dans les Astres. Liège colloquium 1953, p. 213, 1954.
[7] E. BÖHM-VITENSE: Z. Astrophys. **46**, 108 (1958).

The problem whether meridional currents might occur in the convection zone, i.e. the analogue of the von Zeipel-theorem for a convective region in a star, has been discussed by BIERMANN[1]. Taking into account the non-isotropic form of turbulent exchange of energy and by considering the momentum balance in a meridional plane, it could be shown that the theorem applies also to the convection zone: a state of pure rotation without meridional currents is impossible for the convection zones of stars.

It seems quite natural that this outer convection zone is the very *basis of the long-scale transient phenomena*: the solar cycle, the sunspots. The core of the Sun has a rigidity, many orders greater than that of the more viscous outer convection zone. (The *central* convection zone, originally assumed, and connected with the carbon cycle, seems hardly to be developed in the Sun.) So it seems that the sunspots, their latitude drift, the migration of the activity zones, the differential rotation of the Sun—all transient phenomena with a relatively great time scale—should be explained by the dynamics or perhaps the magneto-hydrodynamics of this zone, treated in connection with the solar rotation and the Sun's general magnetic field. Only quite recently has the close connection been stressed between the various aspects of solar activity and the convection zone (BIERMANN[1]).

Since the Sun's general magnetic field is small, its influence will be neglected in a first survey. The state of motions will be discussed by means of considerations given in basical form already by JEANS[2], and adapted to modern ideas.

Let the inner, nonconvective part of the Sun rotate with a constant angular velocity ω_0 and let it have the shape of a rotational ellipsoid, characterized by the meridional equations

$$r_0^2 = \frac{a^2(1-e^2)}{1-e^2\cos^2 b},$$

a, e, r_0 and b respectively being the semi major axis, excentricity, radius vector and heliographic latitude. Convection tends to transport momentum into a radial direction. Since the angular momentum of rotation $\omega^2 r$ of any volume of unit mass should remain constant, ω has to decrease outward as r^{-2}. Hence for $r=R$ the angular velocity is

$$\omega = \omega_0 \frac{r_0^2}{R^2} = \omega_0 \frac{a^2}{R^2}\left\{1 - \frac{e^2}{1-e^2}\sin^2 b + \cdots\right\}.$$

A comparison of this expression with the observational formula (102.1) yields: $e=0.40$, hence $b/a=0.92$ and assuming a thickness of 10^5 km for the convection zone, so with $a/R=\frac{6}{7}$: $\omega=0.74\,\omega_0$ so that the uniform angular rotational velocity of the viscous inner part should be about 19°/day. This value will decrease by viscosity, but this happens extremely slowly. COWLING[3] estimates an equalizing time of $1.5\times10^{13}\,\varrho$ years; hence the effect is only appreciable in the very outer part of the convection zone, where ϱ takes sufficiently small values. So the actual rotational velocity of the Sun's central part should still be about equal to its original value existing when the Sun came into being.

A consequence is that the non-uniform solar rotation cannot supply the magnetic energy of the sunspots[4]. The latter is $W=\frac{H^2}{8\pi}\cdot V$, where V is the volume of a bipolar spot-group, tentatively put equal to the product of spot-area (adopt

[1] L. BIERMANN: I.A.U. symposium **6**, 248 (1958). With further references.
[2] J. JEANS: Monthly Notices Roy. Astronom. Soc. London **86**, 328 (1926).
[3] T.G. COWLING, in: The Sun (ed. G.P. KUIPER), p. 532. 1953.
[4] H. ALFVÉN: Tellus **8**, 274 (1956).

3×10^{10} cm²) and the distance between the two spots (say 10^{10} cm). With $H=3000$ Gauss, one obtains $W=10^{35}$ erg, an energy that is perhaps acquired in 10^3 years (Sect. 39). Assume that there are 100 new spots per year, then a comparison with the kinetic energy of differential rotation (10^{39} to 10^{40} erg) shows the insufficiency of explaining the spots as originating from the non-uniform rotation (lifetime of the Sun $\approx 10^{10}$ years). BIERMANN[1], however, was able to show that the magnetic flux created by impressed electro-motive forces has roughly the order of magnitude observed in big spots and BM regions.

E. Solar and terrestrial relationships.

The Sun enables and regulates life on Earth; its influence on human behaviour, on biological, meteorological and other geophysical processes is enormous. Nevertheless, the astrophysical notion of solar and terrestrial relationships does not refer to this influence but rather to the much fainter and for human life much less important effects produced by the solar high energy wave and particle radiation. Because these waves and particles are absorbed in the upper terrestrial atmosphere, this chapter deals mainly with the high atmosphere. An extensive treatment of the upper atmosphere appears in the geophysical part of this Encyclopedia (Vol. XLIX); for that reason the present chapter is restricted to the solar aspect of the problem. The Earth's atmosphere is used here as a spectrum analyser, both for the wave-radiation frequency spectrum and for the particle velocity spectrum. The relation between solar activity and chemical and biological processes[2] is not discussed here.

We first discuss the "normal" solar ultraviolet and X-ray electromagnetic radiation; next the wave-radiation emitted together with flares. There are four types of particle emission: (1) the cosmic ray particles emitted by great flares, (2) the "slower" particles emitted together with or after flares (and from regions near certain types of sunspots), (3) the recurrent particle streams emitted by M regions, (4) the continuous emission of the Sun. The cosmic ray particles were treated in Sect. 72, the particles of headings (2) to (4) will be discussed in Sect. 110 and 111.

106. Solar ultraviolet and X-ray radiation and the ionosphere. Fig. 84 in Sect. 74 shows the energy distribution in the solar ultraviolet and X-ray spectrum for $\lambda < 3000$ Å. Fig. 129 gives a schematic review of the absorption by the Earth's atmosphere. These two figures provide two of the basic data for the computation of the ionization in the Earth's atmosphere.

For a thorough description of the theory of the formation of the ionospheric layers (CHAPMAN 1931) the reader is referred to Vol. XLIX of this Encyclopedia or to modern monographs on the ionosphere[3,4]. In this section only the aspects will be discussed which are of direct heliophysical importance.

Assume ionization by the process:

$$X + h\nu \rightarrow X^+ + e^- . \tag{106.1}$$

Assume further that the ionization is due to a monochromatic radiation with density Q (quanta cm^{-2} sec^{-1}) and Q_∞ outside the atmosphere. The probability for the process (106.1) is given by the cross section σ of the molecule X. The

[1] L. BIERMANN: Z. Naturforsch. 5a, 65 (1950).
[2] G. PICCARDI: Geophys. e Meteorol. 6 (1958); and U. BECKER: Ark. Meteorol., Geophys. u. Bioklim., Ser. B 6, 511 (1955).
[3] K. RAWER: Die Ionosphäre. Groningen: Noordhoff 1953.
[4] S.K. MITRA: The Upper Atmosphere. Asiatic Society, Calcutta, 2nd ed. 1953.

density is N (cm^{-3}). Hence the electron production is

$$P = N\sigma Q. \tag{106.2}$$

Since N usually increases with depth and Q decreases, there will be an optimum electron production at a certain height. The number of quanta absorbed along the path ds is

$$dQ/ds = -N\sigma Q \quad \text{and} \quad dQ/dz = -N\sigma \sec\chi\, Q$$

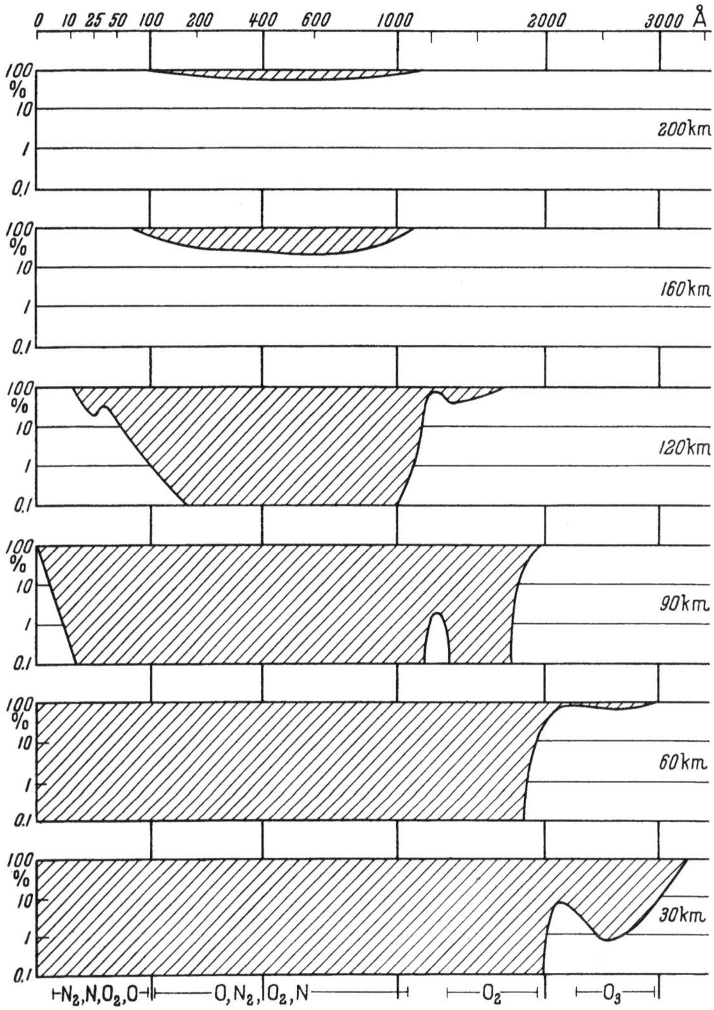

Fig. 129. Transmission of the Earth's atmosphere at various heights. The six graphs are logarithmic. The approximate wavelength regions of the molecular absorptions are indicated along the lower abscissa.

where z is the height measured in vertical direction, and χ is the zenith-angle of the incident radiation. Hence

$$Q = Q_\infty \exp\left[-\sigma \sec\chi \int_z^\infty N(z)\, dz\right]. \tag{106.3}$$

To evaluate the integral the validity of the barometric formula is assumed; this assumption is roughly correct. Hence

$$N = N_0 \exp(-z/H), \quad H \text{ is the scale height}. \tag{106.4}$$

Combination of (106.2) to (106.4) yields

$$P = \sigma\, Q_\infty\, N_0 \exp\left[-\frac{z}{H} - \sigma N_0 H \sec\chi\, e^{-z/H}\right]. \tag{106.5}$$

This expression shows that one well-defined layer is formed when one ionizing reaction occurs throughout the upper terrestrial atmosphere. The height of this layer increases with increasing solar zenith angle. Since different reactions would produce different layers, the ionosphere has often been pictured as consisting of various more or less parabolic layers of electron densities. The precise values of the final electron densities still depend on the values of the recombination coefficient; recombination counteracts ion formation.

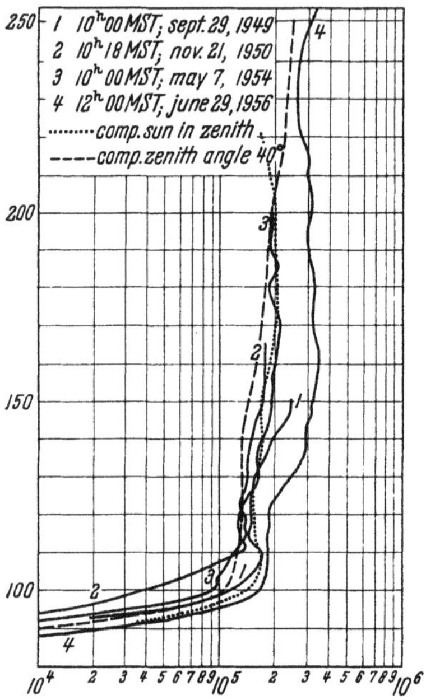

Fig. 130. Observed and computed electron densities in the upper atmosphere. Observations: cf. J. E. JACKSON and J. C. SEDDON: References in Naval Res. Rep. **5087** (1958); Computations: H. K. KALLMANN: footnote 1, p. 347.

Ionospheric research with radio-sounding at vertical incidence did at first suggest the existence of various "layers" each with its own maximum density $(N_e)_{\max}$, determined by the (plasma) frequency of total reflection. The heights of the various layers, F_2, F_1 and E were determined by means of the group time of flight of the transmitted waves.

Rocket observations (N.R.L. firings made after 1949)[1] permitted to determine N_e by measuring the phase shift $\Delta f = f v n e^{-1}$ of a constant frequency signal ($f = 4$ MHz), where v is the known velocity of the rocket, usually fired almost vertically, and n is the refractive index of the medium, directly related to the electron density. The observations demonstrated that in fact there is only one ionized region, above 90 to 100 km (Fig. 130); the various "layers" correspond with slight irregularities in the $N_e(h)$ profile. Moreover it appeared that the F_2 region is much more constant in altitude and substantially lower than is commonly believed on the basis of ground observations. The best average heights and maximum N_e-values seem to be

for the E-region:	$h = 100$ to 110 km;	$(N_e)_{\max} = 1.5 \times 10^5$ cm^{-3}
F_1	160 km	2×10^5 cm^{-3}
F_2	260 km	2 to 5×10^5 cm^{-3}.

The height of the F_2-layer determined by ionospheric sounding varies between 250 and 400 km. The "virtual height" is still about 100 km higher. These differences are due to the considerable group-velocity retardation experienced by the radiosignals in the continuously dense ionosphere where over a great range of altitudes the density differs only slightly from the density of the reflection layer. The conclusion from this paragraph is that if in the following part of this section the classical layer concept will be used, this is only to roughly indicate an ionospheric height level.

[1] Cf. J. W. TOWNSEND, H. FRIEDMAN and R. TOUSEY: Naval Research Lab. report **5087**, 11 (1958) contains complete list of references. Further: J. C. SEDDON and J. E. JACKSON: I.G.Y. Rocket Report Series **1**, 149 (1958).

The "D" region is not observable by means of rockets. It was discovered by its absorption effects at very long (20 km) radiowaves. There is as yet no wholly satisfactory evidence for a separate existence of this "region", although reflections at 1 to 5 MHz from a height of about 80 km are regularly observed at stations like De Bilt and Slough. The most important property of the D-region is that in it the product $N_e \nu$ (ν is the electron collision frequency) reaches a maximum, and this causes the (often observed) high absorption of radiowaves in this region.

It seems that the observed $N_e(h)$ profiles can be explained by theoretical calculations. However, there are great uncertainties in the cross sections (as a function of the wavelength) of the individual air particles, as well as in the solar $Q_\infty(\lambda)$ curve. So the same observed $N_e(h)$ profile might be explained by different combinations of $Q_\infty(\lambda)$ and of the cross sections. The dotted lines in Fig. 130 shows the $N_e(h)$ curve calculated[1] on the basis of an extension of the Chapman theory, with (i) a continuous spectrum of solar radiation entering into the Earth's atmosphere, (ii) molecular and atomic densities varying with altitude, (iii) absorption cross sections, continuously varying with wavelength, (iv) recombination constants continuously varying with height and wavelength. The calculations, obviously much more complicated than for the simple Chapman theory, confirm the observations.

In the calculations described here the absorption cross sections assumed for the X-ray region were such that the solar X-rays hardly contribute to ionosphere formation. Other authors[2-4], however, adopting other cross sections, found that the E and F_1 layer ionization can be due to X-rays. It is one of the chief problems of ionospheric research to solve for this ambiguity; a better knowledge of absorption cross sections is of primordial importance. A decision to which λ region the ionizing radiation is due, should now be taken with the help of other observations (e.g. centre-limb variation of ionizing radiation, cf. Sect. 107).

Thus, the conclusion to be drawn from this section may be summarized in the following Table 19. The attachment of one ionization reaction to one "layer" is too schematic but serves only to roughly indicate the various ionization reactions and the approximate regions where these occur.

Table 19.

h(km)	Region	Principal ionization due to	Suggested main wavelength region of ionizing radiation	Alternative possibility
80	D	NO	Lyα (1215 Å); <20 Å during flares	
110	E	O_2, O, (N_2)	$\lambda \leq 80$ Å	≈ 1000 Å
160	F_1	N_2, O	$\lambda \leq 200$ Å	≈ 800 Å
260	F_2	N_2	200 to 900 Å	

The recombination coefficient varies with height[3]. The calculated *radiative* recombination coefficient is small, of the order of 10^{-12} cm^3 sec^{-1} [if α is the recombination coefficient, the number of recombinations (cm^{-3} sec^{-1}) is $\alpha N_e N_{\text{ion}}$], but this value seems to occur only in the high ionosphere (above 300 km), whereas

[1] H. K. Kallmann: In "Les molécules dans les astres", Liège Colloque 1956, p. 31 (1957) and Ann. de Géophys. **14**, 140 (1958).

[2] R. J. Havens, H. Friedman and E. O. Hulburt: Physics of the Ionosphere, Cambridge 1954, p. 237, 1955.

[3] Review by D. R. Bates, in: Solar Eclipses and the Ionosphere, ed. by W. J. G. Beynon and G. M. Brown, p. 191. London: Pergamon 1956.

[4] G. Elwert: J. Atmos. Terr. Phys. **4**, 68 (1953).

in the lower and denser regions the effective recombination coefficient α' is greater, to reach values of the order of 2×10^{-8} in the E region. Various processes have been suggested (cf. [1-5]) to explain these high values.

In the classical layer theory a very simple approximation for $(N_e)_{\max}$ may be obtained from the equation expressing that the electrons are formed by radiative ionization and disappear by recombination (rate $\alpha' N_e^2$, where α' is the effective recombination constant):

$$dN/dt = P_0 \cos\chi - \alpha' N_e^2 \tag{106.6}$$

P_0 is the rate of maximum electron production for the Sun in the zenith. Hence:

$$N_{\max} = (P_0 \cos\chi/\alpha')^{\frac{1}{2}} \quad \text{and} \quad (P_0/\alpha') = N_{\max}^2/\cos\chi. \tag{106.7}$$

This relation can be used to interpret the observed eleven-year variation of the ionospheric reflection frequencies. Ionospheric sounding yields the following values[6] for the quantity P_0/α' (see confirmation by[7]):

Layer	Sunspot max	Sunspot min	Ratio	Dependence on yearly mean solar R value[8]
F_1	17×10^{10}	6.9×10^{10}	2.4	$1+0.0124\,\bar{R}$
E	4.9×10^{10}	2.15×10^{10}	2.25	$1+0.0097\,\bar{R}$

(For the extremely low and high sunspot minima and maxima of June 1954 and October 1957 this ratio reached the record values 3.4 and 4.7 for the E and the F_1 layers[9] respectively.) So, if α' remains constant, the radiation flux in the wavelengths responsible for the ionospheric ionization varies by a factor greater than two between sunspot maximum and minimum. The intensity of the E-layer ionizing radiation is very closely correlated with that of the 10 cm radio-radiation[10], suggesting the same region of origin for the two radiations (cf also Sect. 107). This is perhaps one of the closest daily solar-terrestrial relations.

107. The ionosphere during solar eclipses[11]. Additional information about the origin and the wavelength of the ionizing radiation is obtained from eclipses. The effect of an eclipse on a Chapman layer has been summarized by RATCLIFFE[12]. During an eclipse the quantity Q in formula (106.5) becomes time-dependent, hence (106.5) becomes

$$P(t,h) = N_0\, Q(t) \exp\left[-\frac{z}{H} - \sigma N_0 H\, e^{-z/H} \sec\chi(t)\right]. \tag{107.1}$$

[1] Cf. footnote 3, p. 347.
[2] D.R. BATES and H.S.W. MASSEY: Proc. Roy. Soc. Lond., Ser. A **192**, 1 (1947).
[3] R.J. HAVENS, H. FRIEDMAN and E.O. HULBURT, in: Physics of the Ionosphere, Cambridge Conference, p. 237, 1954.
[4] J.A. RATCLIFFE: Phil. Trans. Roy. Soc. Lond. **248**, 621 (1956).
[5] R. MICHARD, in: 9ième Rapport commission Relations Phénomènes Solaires Terrestres, p. 64, 1957.
[6] E. APPLETON and W.R. PIGGOT: J. Atm. and Terr. Phys. **5**, 141 (1954).
[7] D.H. MENZEL, J.G. WOLBACH and H. GLAZER, in: Solar Eclipses and the Ionosphere, p. 279. London: Pergamon Press 1956.
[8] C.W. ALLEN: Terr. Magn. Atm. Electr. **53**, 433 (1948).
[9] C.M. MINNIS: Nature, Lond. **181**, 543 (1958).
[10] J.F. DENISSE and M.R. KUNDU: C. R. Acad. Sci., Paris **244**, 45 (1957).
[11] Cf. Solar eclipses and the Ionosphere (ed. W.G.J. BEYON and G.M. BROWN). London: Pergamon Press 1956. With complete list of references.
[12] J.A. RATCLIFFE, in: Solar Eclipses and the Ionosphere, p. 1. London: Pergamon Press 1956.

The electrons may disappear by a recombination-like process (rate $\alpha' N_e^2$) or by an attachment-like process (rate γN_e). So the number $N_e(t, h)$ at any given time or height will be determined by

$$d N_e(t, h)/dt = P(t, h) - \alpha' [N_e(t, h)]^2 \tag{107.2}$$

or

$$d N_e(t, h)/dt = P(t, h) - \gamma N_e(t, h). \tag{107.3}$$

There are two ways of determining α' or γ from the observations. (i) With Eqs. (107.2) and (107.3) in combination with (107.1) and assuming a certain disk distribution of the ionizing radiation, the variation of N_e can be computed for any adopted value of α' or γ; the results may be compared with the observations. Reversely, with an assumed value of α'(or γ) the disk distribution of the ionizing radiation can be derived. Another, more direct but less precise way is by making use of the fact that the time of minimum electron density depends on the value of α' or γ: for very small α' or γ values $(N_e)_{\min}$ occurs a considerable time after totality (see e.g. Fig. 131). Clearly, the great difficulty in interpreting eclipse results is that both α' and the disk distribution of ionizing radiation are unknown and can vary during the eclipse. A short review of eclipse results is given below.

E region. Observations of the minimum eclipse of 1932 suggested[1] that in the E region roughly $(N_e)_{\max} \propto$ (luminous area)$^{\frac{1}{2}}$, corresponding with the *recombination* mechanism. Here, a uniform disk distribution of the ionizing radiation was assumed. Observations at later eclipses permitted to derive quantitative information on the value of α' or γ and on the brightness distribution on the disk. During the 1952 eclipse it was found[2] that activity centres can emit enhanced ionizing radiation. A narrow strip at the western limb had a surface brightness at least 30 times that of the disk centre. That a part of the ionizing radiation comes from activity centres was also proved from outside eclipse observations (footnote 10, p. 348). Further[2] $\alpha'_E = 1.2 \times 10^{-8}$. Also at the 1955 eclipse[2] a centre of activity emitted 20% of the ionizing flux.

The distribution of the sources of ionizing radiation over the *quiet Sun's* disk could be nicely investigated during the eclipse of 30 June, 1954 when the Sun was unusually inactive. There was an intensification near the east and west limbs ($I \approx 1.5\, I_c$), whereas the poles showed a weakening ($I \approx 0.8$ to $0.9\, I_c$), a result that resembles the radio-intensity distribution at decimetre waves and the distribution of X-rays over the disk[3] (cf. Figs. 85 and 98). About 10 to 15% of the ionizing radiation seems to remain uncovered during totality[4], whereas at the 1952 eclipse the residual intensity at totality was virtually zero[2].

Whether the E-region ionizing radiation is due to X-rays (≈ 100 Å) or to radiation of the order of 800 to 1000 Å cannot yet be decided: the former radiation has a residual intensity at totality of the order of 15%, the latter of the order of some percents.

F layers. For the F_1 layer α' is about half of that of the E layer[2], but things are less simple for the F_2 layer because of the occurrence of polarization fields and since violent horizontal and vertical movements prevent the development of the ideal Chapman layer, as described by Eqs. (106.1) to (106.5).

---

[1] S. S. Kirby, L. V. Berkner, T. R. Gilliland and K. A. Norton: J. Res. Nat. Bur. Stand. **11**, 829 (1933).
[2] C. M. Minnis: J. Atmos. Terr. Phys. **6**, 91 (1955); **9**, 36, 201 (1956); **10**, 229 (1957).
[3] C. M. Minnis: Observatory **77**, 94 (1957).
[4] See footnote 12, p. 348.

$F_{1\frac{1}{2}}$ *layer*. In sites near the geomagnetic equator with dip angles less than 20° a new critical frequency, the $F_{1\frac{1}{2}}$ "layer" is sometimes formed, especially during eclipses. In general, the tendency to form stratifications is enhanced during eclipses (during the eclipse of 25 February, 1952 the F_1 critical frequency was observed very clearly in Djibouti—as a rule it is there hardly visible in that time of the year[1]). Also the E layer sometimes shows a splitting resembling the formation of the $F_{1\frac{1}{2}}$ critical reflex. The explanation of the $F_{1\frac{1}{2}}$ critical frequency was suggested by RATCLIFFE. Let, before the eclipse, $N_e(h)$ be linear and let during the eclipse $P(t)$ be constant for all heights and have the form as described in Fig. 131 a. Then, at different heights, the variation $N_e(t)$ of electron density will be as shown in Fig. 131 b (slow recombination at great altitudes produces a retardation of the ionospheric eclipse in the F_2 region). The cross sections at constant t indicate the occurrence of a steeper $N_e(h)$ gradient at moments 1 and 2, thus suggesting the formation of a new "layer".

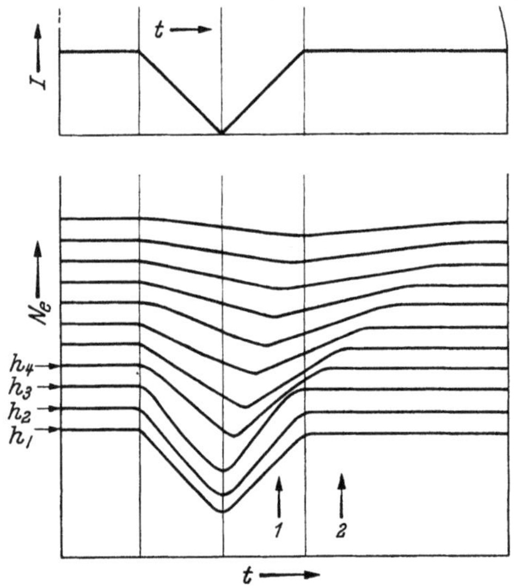

Fig. 131. Explanation of the extra critical frequency $f_0F_{1\frac{1}{2}}$ during eclipses. J. A. RATCLIFFE: in: Solar eclipses and the ionosphere, p. 1. 1956.

D layer. The existing eclipse observations are difficult to interpret since they only rely on intensity measurements of long and medium waves obliquely reflected or on measurements of the ionospheric absorption of the field intensities of remote transmitters. (Vertical D layer sounding has not yet been made during eclipses.) There is some evidence that the sources of D and E ionization are located in the same regions of the Sun[2,3].

108. Radiation effects of flares; the problem of the *D* layer. Flares are associated with an increase of the Sun's ultraviolet, X-ray, and corpuscular radiation. The ultraviolet wave radiation acts at the very moment of visibility of the optical flare; the influence of the corpuscular radiation is perceptible only after the travel time of the particles. The precise moment of X-ray emission is not yet certainly known.

In this section the influence of the *wave radiation* is treated; if manifests itself on the sunlit side of the Earth by various kinds of "Sudden Ionospheric Disturbances" (S.I.D.'s):

(a) the Mögel-Dellinger effect (Short Wave radio Fade-out; S.W.F.);

(b) the disappearance (Fade Out) of ionospheric reflections from all layers (E, F_1, F_2);

(c) the increase of absorption of cosmic radio-noise at $\lambda \approx 10$ to 15 m (Sudden Cosmic Noise Absorption; S.C.N.A.);

(d) the magnetic crochets;

[1] K. RAWER: Personal communication.
[2] Cf. footnote 12, p. 348.
[3] W. R. PIGGOT, in: Solar Eclipses and the Ionosphere, p. 106. London: Pergamon 1956.

(e) the increase of intensity of atmospheric noise, or of signals from remote stations, at very long wavelengths >10 km (Sudden Enhancement of Atmospherics; S.E.A.);

(f) the phase change of signals at very long wavelength (Sudden Phase Anomalies; S.P.A.) and

(g) the Sudden Fieldstrength Anomalies (S.F.A.) of distant transmitters (≈ 15 kHz) recorded on radio-circuits.

These effects, which usually start together with the onset of the optical flare and of the radio-noise at decimetre waves, or some minutes later, can be explained by an increase of ionization in the regions below 80 km.

a) The short wave radio fade-outs (S.W.F.) are *absorption effects*; they consist, as a rule, of a sudden cease of the radiosignals observed at frequencies of 5 to 20 MHz (15 to 60 metre) and over long distances (>1000 km) along the sunlit hemisphere, but some of the disturbances may start more gradually than the common "sudden fade-outs", and are less easily recognized as flare associated phenomena. The onset of the fade-out is virtually simultaneous with the onset of strong flares—weak flares do not cause appreciable effects. In strong "Dellingers" the fade-out may be complete for five or ten minutes, then a gradual recovery sets in.

In 1936 fade-outs occurred for 2.6, 14 and 43.4% of the flares of importance 1, 2 and 3 respectively. Normally the short radio-waves travel by reflecting against the E and F layer, but at the time of a flare the reflecting ceiling decreases (by 15 km at the time of strong flares) to a region where the gas density and, therefore, the collision frequency are much higher. So radio-waves (but for those with $\lambda>10$ km) loose their energy by damping; they are not reflected but absorbed in the low ionosphere.

b) and c) The ionospheric fade-outs and the S.C.N.A. (Figs. 132a and d) are explained in the same way as the Dellinger effect, by increase of the damping in the regions below the E layer. The higher ionospheric layers seem hardly influenced by the effects, so the origin of the ionization has to be looked for in a kind of penetrating radiation, mainly influencing the region below 90 or 80 km.

d) Crochets are displacements of small amplitude in the continuous records of the H, D and V components of the Earth's magnetic field. A sudden onset and a rapid rise is followed by a gradual return to normal (Fig. 132b). The crochets are generally small and only well observable when the magnetogram is for the rest fairly undisturbed. Their magnitude is very small when the Sun's zenith angle exceeds 75°. The increase of ionization causes an *increase of the conductivity*, principally in the lowest ionospheric layers. The corresponding increase of the ionospheric dynamo current system influences the Earth's magnetic field. The sudden impuls of the crochets is correlated in time with the flash phase of the flare.

The dynamo current referred to was initially proposed by STEWART (1882), the theory is developed by CHAPMAN *et al.* It occurs in normal days mainly in the E layer, and is caused by the daily periodical motions of the ionized regions of the Earth under the influence of the Sun's heat radiation and of the atmospheric tides. By the influence of these motions Foucault currents are induced in the ionosphere by the Earth's permanent magnetic field. These currents are naturally accompanied by magnetic fields, which become superimposed on the permanent geomagnetic field. They could also be detected[1] (from a statistical study of E layer sounding observations) by the distortion of the E layer produced by their effect.

e) Sudden enhancements of atmospherics (S.E.A., Fig. 132c). While at short wavelengths radio-waves are absorbed during a solar flare, and at medium

---

[1] E.V. APPLETON, A. J. LYON and A. G. TURNBULL: Nature, Lond. **176**, 897 (1955).

wavelengths nothing happens, at very long wavelengths the signals obtained from remote stations are strengthened during a flare. The increase of the noise

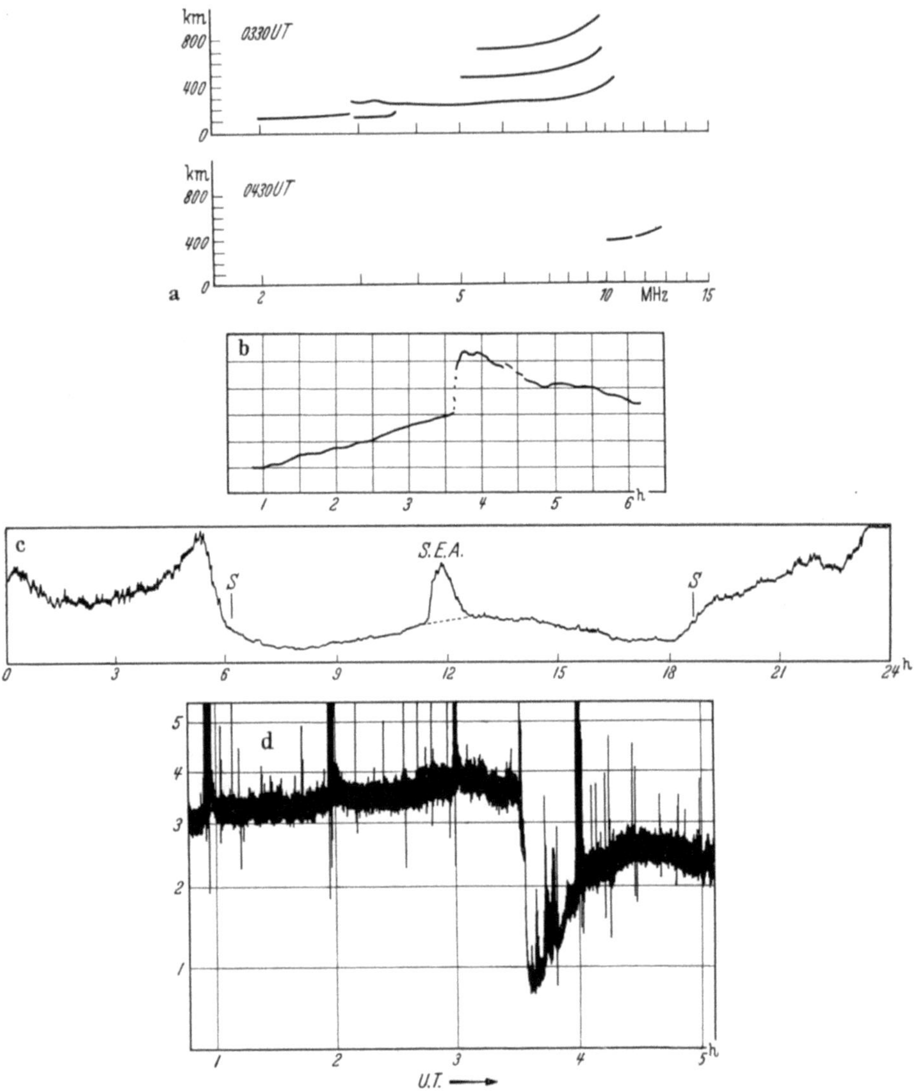

Fig. 132a—d. Some types of S.I.D.'s (sudden ionospheric disturbances) mainly observed during the great flare of 23 Feb., 1956 (03.32 to 04.15). (a) Disappearance of ionospheric reflexes (ionospheric Fade Out) on 23 Feb., 1956. Upper figure: $03^h\!.00$ UT. Lower figure: $04^h\!.00$ UT. Courtesy K. R. RAMANATHAN, Ahmedabad, India. (b) A magnetic crochet in a H magnetogram at $03^h\!.35$ UT on 23 Feb., 1956. Scale: 29.5 γ/division. Courtesy A. K. DAS, Kodaikanal, India. (c) A Sudden Enhancement of Atmospherics, on 24 March, 1951. Observations at 27 kHz; S-S mark the times of local sunrise and sunset at ground level. M. A. ELLISON: J. Atmos. Terr. Phys. **6**, 226 (1953). (d) A Sudden Cosmic Noise Absorption at 03.35 UT on 23 Feb., 1956, frequency 25 MHz. Courtesy K. R. RAMANATHAN, Ahmedabad, India.

at long wavelengths (>10 km), principally emitted by distant tropical thunderstorms, can be used as a flare indicator. At a long-wave receiver this noise, summed up over the whole tropical zone, is rather constant and shows only a regular daily variation. Its enhancement during flares is due to the *improved reflectivity* for long waves, at oblique incidence against the basis of the D layer during a flare. As a rule, all flares of importance classes 2 and 3 and 50% of the importance 1 type flares produce S.E.A.'s[1].

f) The sudden phase anomalies (S.P.A.) have been measured in a wide frequency range (16 kHz to 2.5 MHz)[2]. Normally, when the sky wave and the ground wave are received from a nearby long-wave transmitter, there will be a certain phase difference between the two waves. When, during a flare, the reflecting ceiling falls, there occurs a relative phase shift. The S.P.A. offers a fairly direct method to measure the lowering of the reflection height during flares.

g) The sudden field strength anomalies (S.F.A.) are due to interference between the ground wave and the sky wave from the *D* layer.

There is some controversy whether the onsets of the S.I.D.'s occur at the same moment as the optical flare. In any case ELLISON[3] has made the following list:

					No. of cases
Flare starts	-10^{m}	SPA	SEA	start	
	-6^{m}	Crochet		start	
	-2^{m}	SWF		start	
Maximum of flare	0				
	$+2\overset{m}{.}5$	Crochet		maximum	13
	$+5^{m}$	SPA		maximum	8
	$+5^{m}$	SFA		maximum	21
	$+7^{m}$	SEA		maximum	8

According to DODSON[4] the S.I.D. starts as a rule during the ascending branch of the flare light curve and before maximum. Very many exceptions occur. After very intense flares (3 or 3^{+}) the *D* layer ionization may not return to its pre-flare value for a period of the order of 1^{h} after all visible radiation has ceased.

The above features refer rather to "normal" flares, but a unique feature displayed by the great flare of 23 February, 1956 was that it also produced pronounced ionospheric effects on the night side. The onset of these effects was at 03.45 ± 0.01 UT, later than that of the optical flare but simultaneously with the onset of the cosmic ray burst; it is obviously due to the influence of the cosmic ray burst associated with the flare (cf. Sect. 109).

The problem how to explain the increase of ionization in the lowest ionospheric layers was already touched in Sect. 54. The normal "*D* layer" is produced by the ionizing influence of Lyα, which penetrates to about 80 km through the accidental occurrence of a sharp (1 Å wide) absorption window at 1215 Å, but no flare can produce a Lyα radiation sufficiently intensive to reduce the *D* layer ceiling by as much as 15 km, as was observed during strong flares. Since furthermore there is an imperfect correlation between the time duration of the Hα emission of flares (supposed to go more or less parallel to the Lyα emission) and the duration of the anomalous *D* layer ionization, SIEDENTOPF[5] and later FRIEDMAN and CHUBB (see Sect. 54) assumed that the lowering of the reflecting layer was due to hard X-rays of wavelengths near 2 Å. The X-rays are so much more efficient in producing the *D* layer ionization, because they ionize all constituents of the ionosphere, whereas Lyα ionizes only NO molecules, present as a trace constituent in the upper atmosphere. The prediction was beautifully confirmed by the rocket observations during flares when X-ray radiation in a sufficient quantity was observed above 2 Å (cf. Sect. 54). However, not all flares

[1] M. A. ELLISON: J. Atmos. Terr. Phys. **4**, 226 (1953). — Comm. Obs. Edinburgh **6** (1953).
[2] K. WEEKES: Observatory **75**, 65 (1955).
[3] M. A. ELLISON: Vistas in Astronomy **2**, 799 (1957).
[4] H. W. DODSON: Trans. Internat. Astronom. Union **9**, 650 (1957).
[5] H. SIEDENTOPF: Arch. elektr. Übertragung **4**, 97 (1950).

that emit X-rays produce S.I.D.'s (WARWICK and ZIRIN[1]). This is due (CHUBB *et al.*[4]) to the steep increase of the atmospheric absorption with depth: the amount of ionizing radiation should *greatly* increase to produce an appreciable lowering of the "*D* layer" ionization. The observations of limb flares[2] show that no S.I.D.'s are produced by flares with heights lower than 15000 km, thus suggesting that the ionizing radiation comes mainly from regions above that height.

109. Corpuscular effects associated with flares or "active sunspots". Although flares themselves have little to do with high velocity particle emission, particles may be emitted by other mechanisms, associated with flares. The solar evidences for the emission of particles of various energies have been summarized in Sect. 59. The arrival of the solar particles near the Earth is shown by a magnetic storm, especially manifest in the *H* component of the magnetic field. As a rule, the storm starts with an *onset*, followed first by a positive deflection, remaining positive during about 3 to 6 hours (average deflection 10 γ), and then by a negative deflection (-20 to $-30\,\gamma$) with a very slow recovery, sometimes taking more than 2 or 3 days. Rapid fluctuations up to $+500\,\gamma$ and $-1000\,\gamma$ are superimposed over this mean curve (Fig. 133)[3]. This figure shows a "sudden commencement storm" (s.s.c.); besides, there occur also non-s.s.c. (cf. Sect. 110).

Observations show that 80, 30 and 20% of the *flares* in the disk centre of importance 3^+, 3 and 2 are followed by a geomagnetic storm within about two days[4]. The average flare storm interval is 34^h for relatively "small" flares (imp. 3); it is 22^h for the most important ones (3^+), the shortest interval on record[5] is 17^h. These travel times correspond with mean velocities of 1200, 1900 and 2400 km/sec, respectively. With the exception of great flares, it is often difficult to find out to what flare a certain storm is due. This causes an uncertainty in the determination of travel times. The early determinations might give incorrect travel times due to the incompleteness of the early flare statistics. Furthermore, the restriction to the disk centre was not justified: geomagnetic storms can be emitted from flares all over the disk. Finally a distinction should be made between geomagnetically active and inactive flares (see below).

It was noticed by BECKER and DENISSE[6] that the activity centres emitting magnetic storm particles often also emit *radio-noise storms*. In connection with the foregoing this might be said in another way: flares followed by a radio-noise storm tend to emit particles. This suggests that the radio-noise storms bear some connection with disturbances moving through the Sun. A similar result was found by SIMON[7]: the geomagnetic activity is more closely correlated with radio features than with optical features, and in this respect a distinction should further be made between "active" and "inactive" sunspots.

The geomagnetic activity tends to increase[8, 9] one to four days after the passage of a radio-noise active spot-group (the so-called "*R* spots"); in the same time interval the critical F_2 frequency decreased at night. The central meridian passage

---

[1] J.W. WARWICK, H. ZIRIN, T.A. CHUBB, H. FRIEDMAN, R.A. KREPLIN and J.E. KUPPERIAN: Nature, Lond. **180**, 500 (1957).

[2] C.S. WARWICK: Astrophys. Journ. **121**, 385 (1955).

[3] CHAPMAN: Mixed Commission of the Ionosphere; 3rd meeting, p. 84, 1953.

[4] H.W. NEWTON: Vistas in Astronomy **1**, 666 (1956).

[5] H.W. NEWTON and W. JACKSON: 7ième Rapport Commission Relations Phénomènes Solaires-Terrestres, p. 107, Paris 1951.

[6] U. BECKER and J.F. DENISSE: J. Amer. Terr. Phys. **5**, 70 (1954).

[7] P. SIMON: Radio Astronomy. I.A.U. symposium No. 4, p. 334, 1957.

[8] J.F. DENISSE: Ann. Géophys. **8**, 55 (1952). — C. R. Acad. Sci., Paris **236**, 1856 (1953).

[9] P. SIMON: Ann. d'Astrophys. **19**, 122 (1956). — Ann. Géophys. **12**, 167 (1956); 9ième Rapport Commission Relations Phénomènes. Solaires Terrestres, p. 69, 1957.

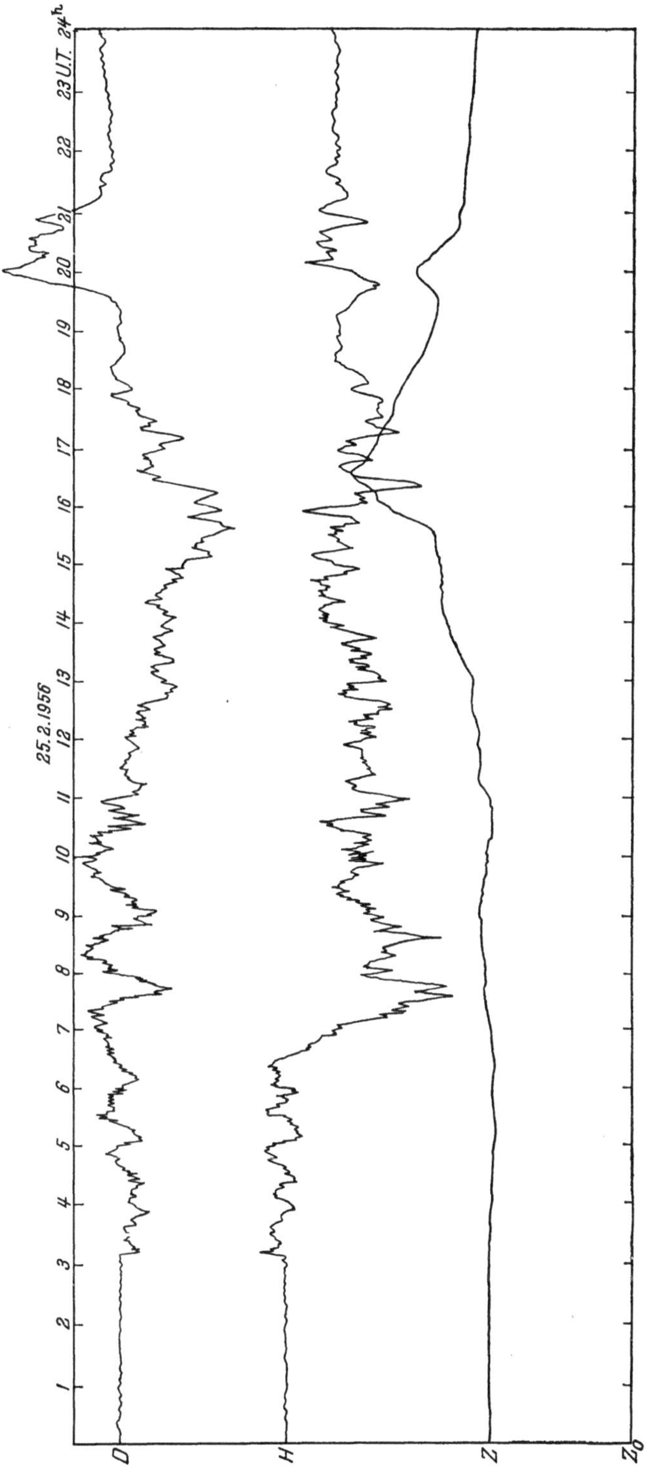

Fig. 133. Magnetic storm of 25 February, 1956, following the flare of 23 February, 1956. Courtesy J. VELDKAMP, De Bilt, Netherlands.

of activity centres that do not emit radio-radiation ("*Q* spots") correspond with a *de*crease of geomagnetic activity at about the same time interval. Even if the centre contains many flares, which, however, do not emit radio-radiation, the geomagnetic activity does not increase. The first observation suggests a relation between the radio-noise emitted by some spots and the emission of fast particles.

DODSON and HEDEMAN[1] looking for a correlation, not with noise storms but with *outbursts* found a close association between flares with major early bursts on 200 MHz (example: Fig. 105; flare of 31 August, 1956) and magnetic storms with sudden commencements. The average time interval is 56 to 60^h, longer than assumed in NEWTON's statistics (but the new results also apply to weaker flares); the extreme values are 18^h and 4 to 5 days (upper limit uncertain). The particle emission can be produced from all points on the disk but seems strongest close to the central meridian. On the basis of these statistics one calculates an average velocity of 700 km/sec for the "flare storms". The geomagnetic storm may last for many hours, it is strongest in the polar region, weaker near the geomagnetic equator. ROBERTS (footnote 3, p. 313) found strong correlation between flares producing *type II bursts* and geomagnetic storms.

SINNO[2] (cf. also[3]) found that the probability of a geomagnetic storm depends on the importance of a flare and the intensity of the radionoise (200 MHz), and also on the position of the source on the disk (more storms are produced by important flares, emitting strong radionoise, in the disk centre).

In *conclusion* sudden-commencement geomagnetic storms (s.s.c.) may be related to the c.m.p. of radio-noise active sunspots or be emitted together with flares with major premaximum bursts or with type II bursts; about 25% of the s.s.c.'s remain unexplained[4]. The non-s.c. storms which seem mainly to be emitted by *M* regions are treated in Sect. 110.

If the observed average particle velocity (found from the travel time) is smaller than 10^3 km/sec it can no longer be put equal to the velocity of ejection. OLJ[5], taking the Sun's gravitational field, and the Sun's and the Earth's rotation into account computed that average travel velocities of 750, 500 and 250 km/sec correspond to velocities of ejection of 870, 770 and 670 km/sec. Hence the "flare storm particles" seem to be ejected with $v \approx 850$ km/sec.

The theory of the phenomena occurring around the Earth during and after the approach of the interplanetary particle-cloud (cf. Fig. 134) is mainly due to CHAPMAN and FERRARO[6,7]. (For a summary of other theories cf. [8].) The cloud of charged particles emitted by the Sun, being ionized but neutral, behaves as if it were perfectly conducting; it carries with it a part of the solar magnetic field. This field will be unobservable when it arrives near the Earth because of the geometrical broadening during the travel-time Sun-Earth, but its effect is that the cloud acts as a diamagnetic body, impermeable to the Earth's magnetic field. So, upon approach to the Earth the terrestrial magnetic field is pushed back, causing surface currents to be induced in the cloud's front part at

[1] H.W. DODSON and E.R. HEDEMAN: In the press.
[2] K. SINNO: Rep. Ion. Res. Japan **11**, 195 (1957).
[3] Y. HAKURA: J. Radio Res. Japan **5**, 57 (1958).
[4] J. ROOSEN: Ionosphere and Radio Astr. Section, Netherl. P.T.T., report **14** (1957).
[5] A.I. OLJ: Solnetshnye Dannye **1**, 79 (1955).
[6] S. CHAPMAN and V.C.A. FERRARO: Papers in Terr. Magn. **36** (1931); **37** (1932); **38** (1933); **45** (1940).
[7] V.C.A. FERRARO: Ann. Géophys. **11**, 284 (1955).
[8] V.C.A. FERRARO: 9ième Rapport Commission Relations Phénomènes Solaires Terrestres, p. 91, 1957; also: E.N. PARKER: J. Geophys. Res. **61**, 625 (1956).

the expense of the kinetic energy of the cloud. This front part is extremely thin, having a thickness of the order of $\sqrt{\tau/\sigma}$, where τ is the time to increase the field by a factor e and σ is the electric conductivity of the matter. For a particle density 1 cm^{-3} the front has a thickness of 5 km; if $N = 10^3$ cm^{-3} it is 160 metres.

Because of this shielding effect, protons and electrons at the other side of the cloud are not deflected; they overtake the braked front part, thus enabling the cloud particles to penetrate somewhat further into the Earth's magnetic field, each time again forming a new front until the cloud is finally stopped.

The initial compression of the magnetic lines of force of the Earth's field causes a sudden increase of this field, thus producing the observed sudden onset of the storm. This is the first phase of the storm, starting already when the cloud is at an average distance of 5×10^5 km.

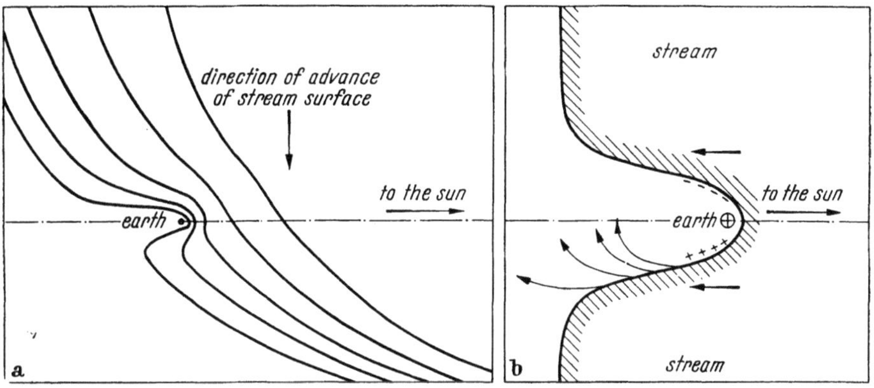

Fig. 134a and b. (a) Illustration of the counteraction between a particle stream coming from the Sun and the Earth's magnetic field. (b) Unstable surface charges are supposed to be formed on the walls of the curved hollow space. The arrows indicato the paths described by the escaping charges. These might form a ring current around the Earth. V. C. A. FERRARO: Ann. de Géoph. **11**, 284 (1955).

The second, main phase of the storm appears to be associated with the formation of a ring current around the Earth. The Earth's magnetic field produces a hollow cavity in the approaching cloud, open to the side of the Earth, away from the Sun (Fig. 134). The opposite deflecting forces of the magnetic field will produce a charge separation on the inner side of the hollow: it becomes positively charged on the side facing the morning side of the Earth. The origin of the ring current, for which the evidences mainly rely on observational data, not so much on theory, can be explained qualitatively with Fig. 134.

The inner radius of the ring is differently estimated by different authors; an average value is 40000 km; the outer radius might be about 65000 km. Once this ring having been set up, it will be very stable due to its electromagnetic inertia. Its slow decay, estimated to take some days, agrees well with the slow recovery of the geomagnetic field during and after the storm. Some of the particles may obtain a subsequent acceleration in this ring current; they may reach the Earth and produce the aurorae.

A detailed discussion of the *auroral problem* is far outside the scope of this section[1]. Important for the understanding of the aurorae is the occurrence in their spectra of the H_α and H_β lines, which are moreover Doppler displaced. This shows that protons coming from outside the Earth play a role in the formation of the aurorae.

[1] Summary of auroral observations and theories: J.W. CHAMBERLAIN, in Vol. 4 of Advances in Geophysics (ed. H.E. LANDSBERG and J. VAN MIEGHEM). New York: Acad. Press 1958.

The radial velocities of the incoming particles have been determined with auroral spectra of Hα and Hβ[1-3]. The profiles of the lines due to the incoming particles (spectrograph pointed at magnetic zenith) are displaced and asymmetric, the broadening and displacement (up to 10 Å) shows a mean velocity of approach of 350 to 450 km/sec; the maximum velocities range between 1400 and 3000 km per sec. The velocities *outside* the Earth's atmosphere are certainly much greater than the above quoted values since the incoming particles are slowed down in the atmosphere.

The essential auroral problem is to explain the observed intensity $I_{\mathrm{H}\alpha}$ and the observed distribution of velocities v by one and the same distribution of initial velocities v_∞. Moreover these initial velocities should not be in disagreement with the observed travel times of the particles. The problem has not completely been solved (see BATES' paper[4]): A beam of incident protons is mainly stopped near the 100 km level: a proton of 5 keV penetrates down to $h=135$ km; one of 25 keV comes to 120 km, one of 200 keV to 105 km, protons of 1000 and 4000 keV come to 90 respectively 75 km. The main auroral region is between 90 and 120 km. The computation of the rate of recombination in a beam of incident protons shows that Hα emission virtually only takes place when the proton energy has been reduced to about 100 keV, and that more than half of the emission occurs at energies less than 50 keV. The Hα emission is naturally followed by a Lyα emission with estimated intensities 16 to 1600 erg cm^{-2} sec^{-1}. (In an average case the Lyα emission following the Hα emission might double the "D layer" ionization[5] by ionization of NO.)

The range in which emission would take place if a mono-energetic beam of particles would enter the atmosphere is small, of the order of 3 km, in contradiction with the observations. The contradiction could be solved by the hypothesis that there is a great spread in the velocities of the incoming protons, or that the mono-energetic protons have a certain dispersion in direction[6]. CHAMBERLAIN[3] found the best agreement between observed and calculated luminosity curves and line profiles by assuming a velocity spread in the incident flux of particles varying as v^{-2}; the angular dispersion is probably proportional to $\cos^2\vartheta$ (ϑ is the angle with the normal).

However, a velocity spread cannot be reconciled with the long travel times: there would be an observable time difference between the arrival of slow and fast particles. This suggests the existence of an acceleration mechanism close to the Earth.

Another, stronger argument for the existence of such an accelerating mechanism is based on the inference that to be able to penetrate down to 100 km, the initial velocity v_∞ of the protons must have been as great as 10^4 km/sec. This great velocity cannot be reconciled with the travel times (1 day corresponds with 1750 km/sec).

MARTYN[7] showed that a potential difference of 10^5 to 10^6 eV, set up in the ring current, might initiate a discharge, giving such velocities that the accelerated particles may reach the polar zone down to 100 km.

[1] A.B. MEINEL: Astrophys. Journ. **113**, 50 (1951).

[2] J.I. GALPERIN: Astr. Zhurn. **34**, 131 (1957); **35**, 382 (1958).

[3] L. VEGARD: Géophys. Publ. Oslo **18**, No. 5 (1952); **19**, No. 9 (1956). — J.W. CHAMBERLAIN: Astrophys. Journ. **120**, 360 (1954).

[4] D.R. BATES: Ann. Géophys. **11**, 253 (1955).

[5] A. OMHOLT: J. Atmos. Terr. Phys. **9**, 28 (1956); **10**, 320 (1957).

[6] A. OMHOLT: J. Atmos. Terr. Phys. **9**, 18 (1956).

[7] D.F. MARTYN: Nature, Lond. **167**, 92 (1951).

An effect, accompanying magnetic storms, has been mentioned earlier, that of the decrease of the general cosmic ray intensities. The explanation of this modulation effect has not yet been given.

The ionospheric effects of solar *cosmic rays* have been remarked only in the case of the flare of 23 February, 1956 by the influence exerted on the *night side* of the Earth. The various kinds of then observed Sudden Ionospheric Disturbances[1,2] are explained by a sudden decrease by about 8 km of the low ionosphere. The onset of these phenomena was at 03.45 $\pm$ 0.02 UT, still somewhat depending on the place on Earth, and are obviously caused by the increased ionization produced by the cosmic ray particles. The effects were greater at higher latitudes, and reached a maximum at geomagnetic latitudes near 60°; no effects were observed below 30°, corresponding with a spectral distribution $\sim \exp(-7E)$ above $E = 10^9$ eV. It was suggested that the increased ionization is produced by singly ionized He atoms with energies of some GeV: protons cannot produce the effect, their minimum energy should be 2 GeV to produce the observed latitude distribution, but in that case they would penetrate too deeply.

110. Corpuscular effects of *M*-regions; continuous corpuscular radiation. The preceding section described the strong magnetic storms. These occur only in connection with flares (hence "sporadically") and usually start suddenly. Many of the moderate and weak magnetic storms are not related to flares, and they show a strong tendency to recur with a period of 27 days. Such a periodicity has been determined by the daily magnetic character figures in which a distinction is made between days of different grades of disturbedness. In Fig. 135 (left) the white squares represent the quietest days[3]. The days are arranged in 27-day periods. The tendency for the same region to remain active during many solar rotations is clear. The years preceding sunspot minimum are most favourable for the appearance of long recurrencies of *M* storms because of the occurrence of old activity centres, which are not destroyed by the formation of new centres. Another way to trace the 27-day recurrence period is to select a great number of "zero dates", being days of high character figures, and to determine the mean character figures for 30 or more following or preceding days. It then, of course, appears that the column for day zero has a high mean, but another top occurs after 27 or 28 days. Naturally autocorrelation or similar techniques could also be applied. That the recurrency is better developed near minimum than near sunspot maximum is also shown by Fig. 136 (FOKKER[4]).

The recurrence tendency is a property of moderate and weak magnetic storms. The relation between the recurrence tendency of the magnetic storms and their intensities is given by the Table 20 below:

Table 20.

Intensities of storms	Number of storms	Number of recurrences	Ratio
$< 130\gamma$	73	18	0.25
130 to 149 γ	91	24	0.26
150 to 179 γ	91	19	0.21
$> 180\gamma$	88	11	0.12
Very intense storms	60	3	0.05

[1] M.A. ELLISON and J.H. REID: J. Atmos. Terr. Phys. **8**, 291 (1956). — Comm. Obs. Edinburgh **11** (1956).

[2] A. EHMERT and K. REVELLIO: Z. Geophys. **23**, 113 (1957).

[3] J. BARTELS: Terr. Magn. **39**, 201 (1934).

[4] A.D. FOKKER, in: Commission Mixte Phénomènes Solaires Terrestres, Réunion Dublin, p. 17, 1956.

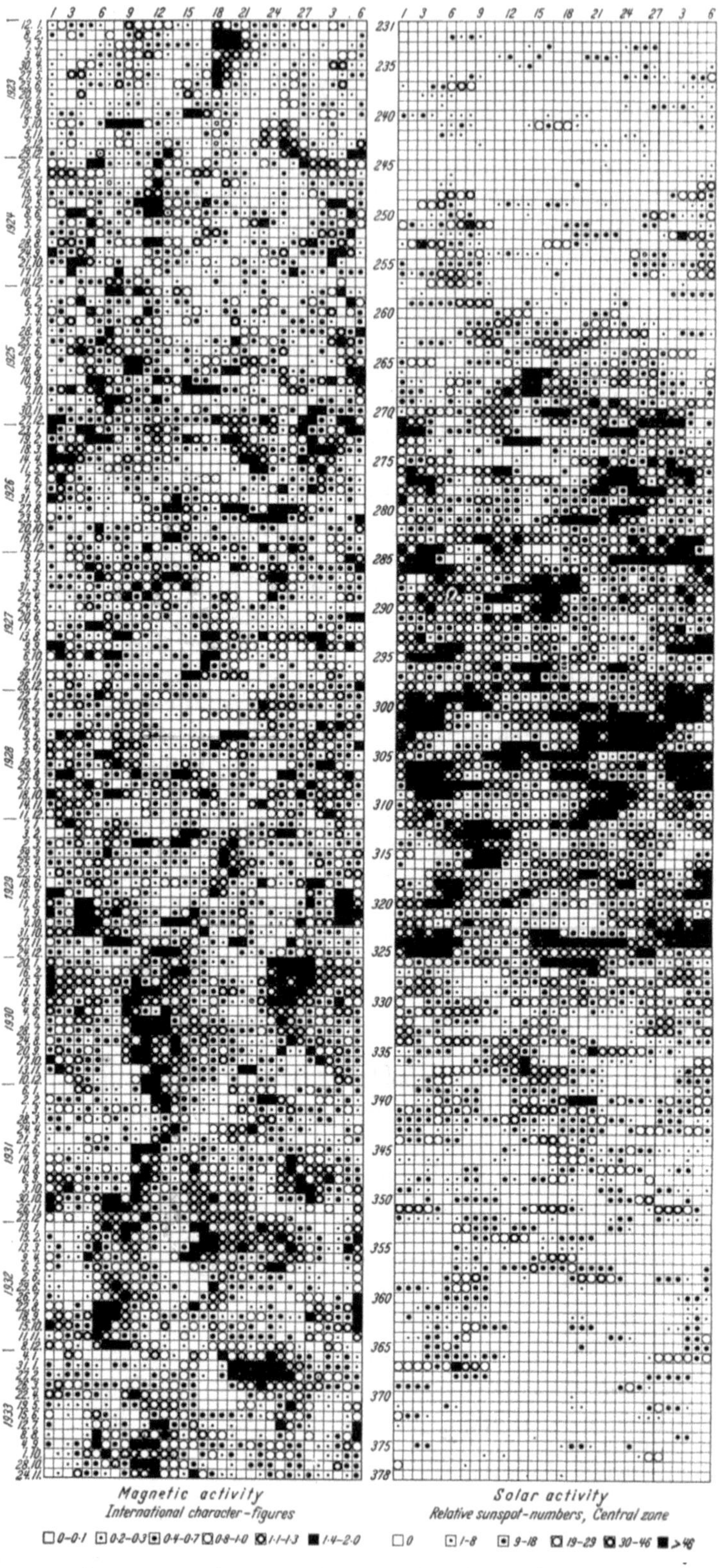

Fig. 135. Twenty-seven-day recurrence time-patterns for geomagnetic activity (left) and sunspot numbers R (right). The diagram covers one solar cycle (1923 to 1933) and shows the occurrence of long lived M regions, especially after sunsopt maximum. S. Chapman and J. Bartels: Geomagnetism, I, p. 411. Oxford: Clarendon Press 1940.

That very intense storms recur only rarely confirms the knowledge that they are produced simultaneously with flares.

The fact that no definite solar source could be detected, responsible for these particle beams, has led BARTELS to the introduction of the M-regions, hypothetical regions assumed responsible for the emission of particles during periods, sometimes lasting for several solar rotations. In Sect. 47 reference was made to the relation between the M-regions, BABCOCK'S UM-regions and WALDMEIER'S C-regions: the particles are mainly emitted by regions where the coronal 5303 Å intensity and also the number of prominences may be small but still greater than one would expect on the basis of the local number of sunspots or faculae. There is a negative correlation with bright areas of 5303 Å emission ($\approx$ centres of activity and facular fields). No M region does extend over the equator: there is a relatively sharp equatorial cut-off of the emitted particles.

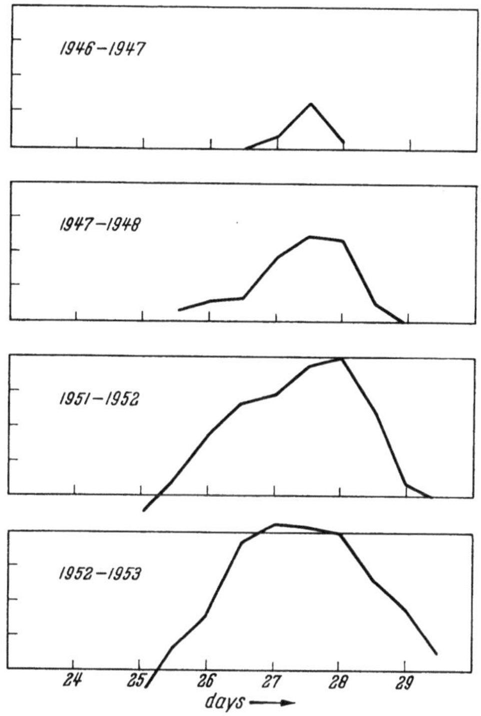

Fig. 136. Recurrency of magnetic activity for two years near sunspot maximum and two years shortly before minimum. The graphs are derived by a modified method of superimposed 27-day rows. The recurrence tendency is greatest near the minimum of solar activity. A. D. FOKKER: Comm. Mixte Phén. Solaires Terrestres, Réunion Dublin p. 17, 1956.

Although the general corpuscular activity follows the same trend as the sunspot activity, reaching high maxima during active cycles and even showing the same N-S asymmetry as the spots[1], the average geomagnetic character figures show better correlation with the *prominence areas*[2]: the years 1945 to 1949 had very high sunspot numbers R but a relatively low total prominence area; accordingly the geomagnetic activity was less in that cycle than in the three preceding ones in which R_{max} was less than in 1947. This would support the assumption that M regions have some relation to coronal rays: these too are related to (quiescent) prominences.

MUSTEL[3] suggested the non-s.c. storms to be connected with the central meridian passage of faculae. His explanation for the difference between s.s.c. and non-s.c. storms is that the first are emitted by activity centres, carrying with them a "frozen" magnetic field. The entrance of this field into that of the Earth would produce the sudden commencement (see also Sect. 47).

As a rule the passage of an M region over the disk does not produce important or even notable disturbances on a radio-record, but in three cases relating to the same disk passage a pronounced synchronism and similarity in form was observed[4] between intense deflections on a magnetogram and radiobursts at 300 cm; dura-

[1] W. GLEISSBERG: Z. Astrophys. **43**, 144 (1957).
[2] R. ANANTHAKRISHMAN: Nature, Lond. **172**, 854 (1953).
[3] E. R. MUSTEL: Astron. Zhurn. **34**, 120 (1957); **35**, 351 (1958). — Isw. Krymsk. Astrophys. Obs. **18**, 162 (1958). — E. R. MUSTEL and A. S. DVORIASHIN: Astron. Zhurn. **35**, 3 (1958).
[4] A. K. DAS and B. N. BHARGAVA: Nature, Lond. **172**, 855 (1953).

tion 5 to 10 min. No optical phenomena were recorded, but the crochetlike magnetic disturbance is likely to be due to a flash of ultraviolet or X-ray emission whereas the radiobursts indicate a simultaneous coronal disturbance.

The travel times of the M type disturbances are longer than those of the great storms; they range between one and four days. Still longer travel times may perhaps also occur, but they are very difficult to trace.

An important property of the radiation producing the weak magnetic storms is their sharp directivity in contrast to the flare-storms: the weak storms seem to occur only when the source is close to the central meridian of the Sun. The emission is confined within an angle of 10 to 15°. The short duration of the storms ($\approx$ one day) is assumed to be due to the fact that the corpuscular fan sweeps over the Earth in this time interval.

Apart from the discrete emissions of solar particles the existence of comets' tails indicates the occurrence of a *continuous stream of particles* streaming radially from the Sun (BIERMANN[1]). The estimates[2] yield a solar loss of mass of 3×10^{10} to 3×10^{12} g sec^{-1}; the resulting streamer density near the Earth may range between 0.1 and 100 atoms/cm^3. BIERMANN[3] estimates a loss of 7×10^{10} g sec^{-1}; the energy loss of the quiet Sun is 10^6 ergs cm^{-2} sec^{-1}, hence about 2×10^{-5} of the solar constant. With an assumed particle velocity of 500 km/sec, and assuming an anisotropic particle radiation the two latter values yield a particle density at the Earth's distance of 150 and 2 cm^{-3}, respectively.

General references.

The three books [*1*], [*2*], [*3*] listed below appeared in 1953 and 1955. They deal extensively with the subject of this article. For that reason it has not been tried to reach completness as to the literature before 1953; only some of the most important papers are mentioned. It was attempted to make the later references more complete.

[*1*] Various authors: The Sun. Vol. I of "The Solar System"; ed. G. P. KUIPER. Chicago, Ill.: Chicago University Press 1953.
[*2*] UNSÖLD, A.: Physik der Sternatmosphären, 2. Aufl. Berlin: Springer 1955.
[*3*] WALDMEIER, M.: Ergebnisse und Probleme der Sonnenforschung, 2. Aufl. Leipzig: Akademische Verlagsgesellschaft 1955.

Some further references:

[*4*] ABETTI, G.: The Sun. London: Faber & Faber 1956.
[*5*] ALFVÉN, H.: Cosmical electrodynamics. Oxford: Clarendon Press 1950.
[*6*] BRUHAT, G., and L. D'AZAMBUJA: Le Soleil, 2nd ed. Paris: Presses Univ 1951.
[*7*] COWLING, T. G.: Magnetohydrodynamics. New York: Interscience Tracs on Physics and Astronomy 1957.
[*8*] ELLISON, M. A.: The Sun and its influence. London: Routledge & Kegan Paul Ltd. 1955.
[*9*] KIEPENHEUER, K. O.: Die Sonne. Berlin: Springer 1957.
[*10*] MUSTEL, E.: Chapter III of "Theoretische Astrophysik", von V. A. AMBARZUMJAN, E. MUSTEL u. a. Berlin: VEB Deutscher Verlag der Wissenschaften 1957.
[*11*] Many authors: Convegno Volta **11**, 1952. Roma: Accad. Naz. d. Lincei 1953.
[*12*] Many authors: Chapters 4, 6, 7 and 8 of "Vistas in Astronomy", ed. A. BEER. London & New York: Pergamon Press 1955/56.

[1] L. BIERMANN: Z. Astrophys. **29**, 274 (1951).
[2] S.K. VSEKHSVJATSKY *et al.*: Astron. Zhurn. **32**, 165 (1955). — E.R. MUSTEL: Astron. Zhurn. **32**, 177 (1955). — G.M. NIKOLSKIJ: Astron. Zhurn. **33**, 588 (1956).
[3] L. BIERMANN: Observatory **77**, 109 (1957).

The Atmospheres of the Planets.

By

HAROLD C. UREY.

With 3 Figures.

I. Introduction.

1. General introduction. The atmospheres of the planets have markedly different compositions from that of the Sun, and the origin of these differences is a fascinating study of the chemistry and physics of the elements which has been pursued by many scholars. The planetary atmospheres consist of hydrogen, helium and the other inert gases and compounds of hydrogen, carbon, nitrogen and oxygen to the exclusion in a high degree of any other elements. Moreover, these atmospheres in some cases have surely changed their chemical composition during the time since the solar system can be said with some precision to have originated. These chemical changes depend upon the chemical properties of the compounds of these elements, the electromagnetic and particle radiation from the Sun, the relative proportions of the elements, the temperatures of the planets and other factors. The terrestrial planets have atmospheres which are very thin, while the atmospheres of the major planets are very deep and hence the two groups present quite different problems. Because of our better understanding of the Earth's atmosphere it will be discussed in detail first and then other atmospheres will be compared with that of the Earth.

It is immediately evident that much of the gaseous compounds of the atmospheric elements were lost from the terrestrial planets early in their history and there are important lines of evidence indicating that such loss also occurred in the case of the major planets. This loss must have occurred while the mass of the planet was distributed within a much larger volume than is the case at the present time (UREY 1951, 1952a). Once an abject as large as Mars accumulated within a gas mass of solar composition, no rapid loss of gas could be expected and it would be difficult to believe that the Earth would be left with a very thin atmosphere and, in particular, such gases as krypton and xenon could not have escaped so completely as they have. The fractionation was not determined by atomic and molecular weights, but by chemical properties, i.e. by the different volatilities of their compounds or by the formation or non-formation of stable non-volatile compounds. Thus in the case of the Earth water and nitrogen were retained, while neon, having a similar atomic weight, was lost in a high degress [ASTON (1924); JEFFREYS (1924); RUSSELL and MENZEL (1933); SUESS (1949); BROWN (1949)].

In order to discuss the early histories of the atmospheres some conclusions in regard to the temperatures at which the planets were formed are desirable. It is well to look for evidence in regard to this matter rather than to trust hypotheses. As mentioned above, a certain fractionation of the elements has occurred in the case of the terrestrial planets and the meteorites and the character of this fractionation puts limitations on the processes to which they have been subjected. Briefly, the terrestrial planets have lost those elements which are volatile at present terrestrial temperatures. The meteorites have not lost the elements which are next to these in volatility, namely, arsenic, cadmium, and zinc, and the Earth has not lost these, and mercury as well, in any important degree (UREY 1954). Other elements may have high volatilities under similar conditions, but no great loss of volatile elements of this class has occurred. We, of course, know nothing about this question for any other planetary body. Also, these elements are not concentrated noticeably in the Earth's surface as might be expected if the Earth had been formed in a molten condition. (The possibility that the Earth's surface, after the origin, was completely mixed with the Earth's interior is not excluded.) The substances concentrated at the Earth's surface are those which are volatile at low temperatures, namely, water, nitrogen, carbon dioxide, the inert gases, and those which are soluble in water at low temperatures, namely, chlorides, bromides, iodides and borates[1] (UREY 1953b). The evidence favors moderate temperatures for the formation of the meteorites and the Earth except for local and temporary high temperatures which produced no fractionation of the elements. In this chapter we shall assume generally low temperatures for the formation of the terrestrial planets.

Atmospheres have developed with time since the formation of the planets through three well recognized possible processes as observed on the Earth, the gravitational escape of gases into interplanetary space, the escape of constituents from the planet's interior into the atmosphere and chemical reactions of planetary gases with the solid materials of the planet. Also the capture of gaseous constituents has been considered from time to time, and limited amounts of certain constituents are produced by radioactive processes. These possibilities will be discussed in the case of each of the planets, but at this point a brief summary can be given. The major planets at present will lose no gases to space, not even atomic hydrogen. The Earth loses hydrogen and helium but no other constituents so far as we know. This is probably true of Venus as well. Mars surely loses hydrogen and helium and may even lose molecules in the molecular weight range of 16 to 20. The Moon and Mercury have lost their atmospheres completely or nearly so, in so far as our observations can determine, and this is expected on theoretical grounds. The satellites of the major planets and Pluto, with the exception of Titan, have no observable atmospheres and presumably they have been lost or condensed on their surfaces.

Since hydrogen will always escape from planets more readily than any other atom or molecule, and since the loss of hydrogen is always equivalent to an oxidation process, the escape of atmospheres produces a more oxidized condition of the planet. Given a mixture of reduced gases such as CH_4, NH_3, H_2O, etc., the escape of hydrogen causes the appearance of more highly oxidized substances such as C, CO, CO_2 and N_2, the many carbon compounds which are all more oxidized than the initial molecules, and even free molecular oxygen, O_2 (POOLE 1941). Thus, the atmospheres of the Earth, Venus and Mars are all highly oxidized and undoubtedly owe this character to the more rapid escape of hydrogen.

It is a well observed fact that certain atmospheric constituents are coming to the surface of the Earth from its interior, and it is a plausible idea that the oceans and atmosphere of the Earth have originated in this way. RUBEY (1955) has advanced many arguments for this point of view. Hence, it is possible that similar processes have occurred on other planets.

[1] The halogens are often referred to as volatiles following GOLDSCHMIDT, but this is in error. Halogen salts are only slightly hydrolized to the hydrogen halide at high temperatures and halogen salts are not markedly volatile.

Two important chemical reactions may remove constituents from atmospheres. Free oxygen reacts with ferrous iron of silicates to produce ferric iron and thus free oxygen would be removed from the atmosphere. The ferric iron of the Earth may have been produced in this way, but oxidation by water may have produced much of it without the necessity of contact with the atmosphere. Chemical reactions between gases and solids are usually slow and often imperceptible because of lack of contact between the gas and the interior of the solid particles. On Earth the oxidation of ferrous iron by oxygen has been promoted by the disintegration of rocks by water as well as by the diverse activities of living organisms.

Carbon dioxide reacts with magnesium, calcium and ferrous silicates to form carbonates and silicon dioxide. The equilibrium pressure of carbon dioxide is very low at terrestrial surface temperatures, e.g. 10^{-8} atmospheres for the reaction,

$$CaSiO_3 + CO_2 = CaCO_3 + SiO_2 .$$

Perhaps the thermodynamic properties of $CaSiO_3$ are not representative of igneous rocks, but there is no doubt that the absorption of CO_2 by silicate rocks is a favored reaction and that it will proceed, if effective contact is made in the presence of water and active erosion. No high pressure of carbon dioxide could be expected to remain for millions of years on the surface of the Earth under conditions of erosion such as those that obtain at the present time (UREY 1952a, p. 148ff). Thus the presence of large amounts of carbon dioxide on Venus require the absence of such erosion as will be discussed later.

2. Table of planetary constants. Certain planetary constants are necessary for numerical calculations and these constants are summarized in Table 1 with

Table 1. *Summary of planetary constants.*

Planet	Distance from Sun $R_♁=1$	Mass $M_♁=1$	Radius $a_♁=1$	Density	Albedo	Color index	g	T_M	T_A
Mercury . .	0.3871	0.0543	0.39	5.05	0.063	1.00	350	626	443
Venus . . .	0.7233	0.8136	0.973	4.88	0.59	1.00	842	370	262
Earth . . .	1.	1.	1.	5.52	0.36	0.37	980	354	250
Mars . . .	1.5237	0.1080	0.520	4.24	0.15	1.30	391	308	218
Jupiter . .	5.2028	318.35	10.97	1.33	0.47	0.88	2592	148	105
Saturn. . .	9.5388	95.3	9.03	0.714	0.46	1.07	1145	110	78
Uranus . .	19.19	14.54	3.72	1.56	0.45	0.50	1070	78	55
Neptune . .	30.07	17.26	3.50	2.22	0.49	0.36	1381	61	43
Pluto . . .	39.52	0.033(?)	0.45(?)	2.0(?)	0.16(?)	0.63	160(?)	60	42
Moon . . .	1.	0.012304	0.2727	3.35	0.073	1.10	162	389	275
Io.	5.2028	0.0122	0.275	3.24	0.50	1.06	158	146	103
Europa . .	5.2028	0.0079	0.241	3.12	0.56	0.77	133	141	100
Ganymede .	5.2028	0.0260	0.435	1.74	0.35	0.64	135	156	110
Callisto . .	5.2028	0.0162	0.392	1.48	0.15	0.68	103	167	118
Titan . . .	9.5388	0.0236	0.384	2.30	0.30	1.17	157	117	83
Triton . . .	30.07	0.022(?)	0.35(?)	2.8(?)	?	0.72	176(?)	?	?

Distance of Earth from the Sun $= 1.4968 \times 10^{13}$ cm
Mass of the Earth $= 5.975 \times 10^{27}$ g
Mean radius of the Earth $= 6370.97$ km
Solar constant $= 1.395 \times 10^{6}$ ergs cm^{-2} sec^{-1}.

References: RUSSELL, DUGAN and STEWART (1945); G. P. KUIPER (1952); H. CAMICHEL (1944); K. STUMPFF (1948); A. DANJON (1948); A. DOLLFUS (1954); G. P. KUIPER (1950); A. DANJON (1954). Relative to the radius of Mars, see H. C. UREY (1952a), p. 61, and this paper. — The solar constant is that given by F. S. JOHNSON (1954).

references to sources. In general, data which has been published in the literature have been selected, rather than summarized tables with no supporting details. One is usually prejudiced in favor of more recent data, but in some cases these are presented without any arguments indicating the reasons for the discrepancy between the new and older data, even when substantial differences exist. No difference in conclusions would result from other choices of data except that KUIPER's value for the albedo of Venus, i.e. 0.76, would make a marked difference in the energy received by Venus.

3. Definition of symbols used.

a Planetary radius.
a_c Planetary radius at critical escape level.
AE Aeon equal to 10^9 years.
c_i Mole fraction of i-th constituent.
C Heat capactiy per gram molecule at constant pressure.
D, D_0 Diffusion coefficient.
g Gravitational acceleration.
G Constant of gravitation $= 6.670 \times 10^{-8}$ erg cm g^{-2}.
h Vertical altitude in the atmosphere.
H Scale height $\left(=\frac{RT}{\mu g}\right)$.
ΔH Change of heat content per gram molecule at constant pressure, heats of vaporization and of chemical reactions.
K Equilibrium constant.
M Planetary mass.
n Number of molecules or atoms per cm^3.
n_0 Number of molecules of atoms per cm^3 at some specified level.
p Polarization of light.
p_i Partial pressure of i-th constituent.
P Total pressure.
P_0 Pressure at some specified level.
r Variable for distance from planet's center.
R Gas constant per gram molecule.
t Variable time, or $\frac{1}{T_0}\frac{dT}{dh}$.
T Absolute temperature.
T_A Average temperature of planet (calculated for black body).
T_c Absolute temperature at the critical escape level.
T_M Temperature of mid-solar point of planet (calculated).
x Variable.
X_i The escape constant $\left(=\frac{GM\mu_i}{a_c R T_c}\right)$.
Γ Adiabatic lapse rate, a positive quantity.
μ Molecular or atomic weight.
$\bar{\mu}$ Mean molecular weight.
ϱ Density.
ϱ_0 Density at some specified level.

II. The Earth's atmosphere [1].

4. Composition. The chemical composition of the Earth's atmosphere is given in Tables 2 and 3 prepared from GLAUCKAUF's discussion (1951). Some constituents are labelled stable and others unstable on the basis of readily available spontaneous reactions which may destroy the constituent. All except the inert gases take part in rapid or slow chemical cycles involving the energetic processes

[1] The Earth's atmosphere is reviewed here only very briefly since it is discussed in great detail in Vols. XLVIII and XLIX of this Encyclopedia. The objective here is to call attention to points which are more pertinent to a discussion of the atmospheres of the other planets.

Table 2. *Non-variable constituents of atmospheric air*[1].

Constituent	Content	Stability	Spontaneous reactions
N_2	78.084%	Stable	
O_2	20.946%	Unstable	Reacts slowly with FeO and carbon compounds
CO_2	0.033%	Unstable	Reacts slowly with silicates
A	0.934%	Stable	
Ne	18.18×10^{-6}	Stable	
He^4	5.24×10^{-6}	Stable	
He^3	6.55×10^{-12}	Stable	
Kr	1.14×10^{-6}	Stable	
Xe	0.087×10^{-6}	Stable	
H_2	0.5×10^{-6}	Unstable	$2H_2+O_2=2H_2O$
CH_4	2×10^{-6}	Unstable	$CH_4+2O_2=CO_2+2H_2O$
N_2O	0.5×10^{-6}	Unstable	$2N_2O=2N_2+O_2$

Table 3. *Variable constituents of dry air.*

Constituent	Content	Origin	Probable destructive reaction
O_3.	0 to 0.07 ppm (summer) 0 to 0.02 ppm (winter)	Radiation	$2O_3=3O_2$
SO_2	0 to 1 ppm	Industrial	$SO_2+H_2O+\frac{1}{2}O_2=H_2SO_4$
NO_2	0 to 0.02 ppm	Industrial	To nitrates or oxygen and nitrogen
CH_2O . . .	Uncertain	Biological or oxidation of CH_4	Oxidation to CO_2 and H_2O
I_2	0 to 10^{-10} g cm^{-3}	Industrial	Many reactions
NaCl . . .	Order of 10^{-10} g cm^{-3}	Sea spray	Solution in water
NH_3	0 to trace	Industrial	$2NH_3+\frac{3}{2}O_2=N_2+3H_2O$ Solution in water
CO	0 to trace (0.8 cm atm)[2]	Industrial	$CO+\frac{1}{2}O_2=CO_2$

initiated by the Sun's light, the activities of living organisms or the heating processes taking place within the Earth, and the inert gases escape from the Earth's interior and may be accumulated from space and lost from the Earth. It would not be possible to determine from any fundamental principles what constituents would be present in the Earth's atmosphere. Given water and oxygen on the surface of the Earth, nitrogen should certainly be a constituent. Oxygen owes its presence to the slowness of its reaction with ferrous iron of the rocks especially and to some process or processes which have produced it in the past and which continue to produce it. Carbon dioxide reacts with the silicate rocks to form limestone and dolomites and it must be continuously supplied either from new sources or by reversal of the reactions to form carbonates by heating processes in the Earth's crust. All the minor constituents except the inert gases are highly unstable under terrestrial conditions. The table indicates some important destructive reactions for these constituents. It is to be expected that all unstable and highly reactive constituents of all atmospheres will be rare.

Chackett, Paneth and Wilson (1950) found only slight variations in the concentrations of helium, neon and argon in samples collected by a V-2 rocket at altitudes of 50 to 70 km. This indicates no diffusive separation of these altitudes. Townsend, Meadows and Pressly (1954) have found no diffusive separation of argon and nitrogen up to 140 km.

---

[1] The classification of H_2, CH_4 and N_2O as non-variable constituents is uncertain.
[2] See Migeotte and Neven (1950).

Particles are suspended in the atmosphere. The most numerous of these are liquid and solid water particles of clouds. Dust particles from the Earth's sea and land surfaces and from the disintegration of meteors are present in detectable amounts. Such particles are confined mostly to the low regions of the atmosphere and because of their efficient removal by rain and other precipitation of water are very rare.

5. Temperature, pressure and density. An exponential variation of pressure and density following the equations.

$$P = P_0 \, e^{-\frac{\bar{\mu} g h}{R T}} \quad \text{and} \quad \varrho = \varrho_0 \, e^{-\frac{\bar{\mu} g h}{R T}}, \tag{5.1}$$

are very often used because of their simplicity and approximate validity. The quantity $RT/\bar{\mu}g$ is called the scale height and will be indicated by the symbol H. However, observations made in recent years show that the Earth's atmosphere deviates markedly from these laws, principally because both the temperature, T, and the mean molecular weight, $\bar{\mu}$, are not constant. Since World War II much observational data has been secured by the use of rockets, particularly by groups working for the U.S. Army. These results have been summarized by "The Rocket Panel" (1952) and will be taken as the basis of discussion in this chapter. Up to 30 km the data have been secured by balloons and the temperatures are measured directly. Above 30 km the basic data give a determination of pressure and density and the temperature must then be derived from these data which requires a knowledge of the mean molecular weight to be derived from other considerations. Up to 80 km height, the molecular weight can be taken as that of surface dry air with confidence. Above this altitude the evidence on this point is indirect and uncertain. Table 4 gives a combination of data due to the Rocket Panel and a revision of such data by JOHNSON (1956) on the basis of X-ray observations and calculations based on them. Oxygen becomes dissociated between 80 and 120 km, but nitrogen remains mostly in the molecular state to the top of the atmosphere.

NICOLET (1954) for purposes of calculation assumes that there is a constant rate of change of scale height, H, with altitude. He extends his table to 500 km and a temperature of 2080° K. He considers the effects of diffusive separation of molecular nitrogen and atomic oxygen. Since this is the most detailed consideration at present, his table is reproduced in Table 5. As altitudes increase, the tabulated values become very uncertain and above 120 km they depend entirely on a few rocket data secured by the Naval Research Laboratory and on only one observation at 220 km.

6. The absorption and loss of energy by the atmosphere. The spectrum of the Sun approximates that of a black body with a mean temperature near 5800° K. The intensity in the far ultraviolet is but poorly known, though there may well be a considerable intensity at very short wavelengths (NICOLET 1952).

In the region between 2200 to 3400 Å the intensity is distinctly less than that corresponding to a black body at 5800° K (HULBURT 1947) and at lower wavelengths the deviation may be greater. Considerable fluctuations in intensity in the ionizing radiation below 1000 Å are shown by the density of ionization in the high atmosphere. Also very considerable variation in intensity occurs in the radio wavelength region. These variations are important in considerations of the high atmosphere of the Earth but the effects will not be discussed in detail in this chapter.

Table 4. *Properties of the terrestrial atmosphere.*

The data of the Rocket panel are used in the first part of the table and the revisions due to Johnson are used for the second part. Some interpolations have been made and the data recast to make entries uniform and comparable.

Height h km	Mol. wt. $\bar{\mu}$	Temperature T °K	Scale height H km	Pressure $\log_{10} P$ dynes cm^{-2}	n cm^{-3}
			Rocket panel		
0	28.97				2.55×10^{19}
5	28.97	276.8	7.83	5.743	1.51×10^{19}
10	28.97	230.8	6.78	5.446	8.77×10^{18}
15	28.97	209.1	6.16	5.107	4.42×10^{18}
17	28.97	208.5	6.14	4.966	3.21×10^{18}
20	28.97	212.8	6.28	4.755	1.94×10^{18}
25	28.97	223.0	6.59	4.418	8.51×10^{17}
30	28.97	231.7	6.85	4.095	3.89×10^{17}
35	28.97	244.5	7.24	3.786	1.81×10^{17}
40	28.97	262.6	7.79	3.497	8.67×10^{16}
45	28.97	271.3	8.06	3.224	4.48×10^{16}
50	28.97	270.8	8.06	2.955	2.41×10^{16}
			Johnson		
50	28.97	271	8.1	2.955	2.4×10^{16}
60	28.97	245	7.3	2.400	7.4×10^{15}
70	28.97	210	6.3	1.760	2.0×10^{15}
80	28.97	185	5.7	1.020	4.1×10^{14}
90	28.97	183	5.6	0.220	6.6×10^{13}
100	28.97	200	6.1	0.465−1	1.06×10^{13}
110	27.0	230	7.4	0.801−2	2.0×10^{12}
120	25.3	280	9.8	0.296−2	5.1×10^{11}
130	25.3	423	14	0.918−3	2.1×10^{11}
140	25.0	594	20	0.674−3	5.8×10^{10}
150	24.5	737	26	0.493−3	3.0×10^{10}
160	23.7	846	32	0.345−3	1.9×10^{10}
180	22.3	979	39	0.101−3	9.3×10^{9}
200	21.1	1043	44	0.893−4	5.4×10^{9}
220	20.0	1074	48	0.618−4	3.4×10^{9}
240	19.2	1088	52	0.451−4	2.3×10^{9}
260	18.2	1095	55	0.371−4	1.6×10^{9}
280	17.5	1098	58	0.217−4	1.1×10^{9}
300	16.8	1100	61	0.070−4	7.7×10^{9}
310	16.5	1100	62	0.999−5	6.5×10^{8}

Table 5. *Atmospheric model with diffusion beginning at 160 km.*

Altitude km	Temperature °K	Scale height H km	Pressure $\log_{10} P$ dynes cm^{-2}	Scale height for O km	n(O) (cm^{-3})	Scale height for N_2 km	$n(N_2)$ cm^{-3}
160	482	18.0	0.38−3	27	1.24×10^{10}	15.4	2.47×10^{10}
180	586	23.4	0.95−4	33	5.21×10^{9}	18.9	6.17×10^{9}
200	688	29.3	0.63−4	39	2.54×10^{9}	22.3	2.03×10^{9}
220	789	35.5	0.36−4	45	1.38×10^{9}	25.7	7.68×10^{8}
240	888	41.9	0.13−4	51	8.05×10^{8}	29.1	3.28×10^{8}
260	987	48.4	0.94−5	57	5.00×10^{8}	32.6	1.54×10^{8}
280	1085	55.0	0.77−5	63	3.26×10^{8}	36.0	7.85×10^{7}
300	1181	61.6	0.62−5	69	2.21×10^{8}	39.4	4.24×10^{7}
350	1416	77.9	0.31−5	84	9.48×10^{7}	48.0	1.12×10^{7}
400	1645	93.9	0.06−5	99	4.77×10^{7}	56.6	3.70×10^{6}
450	1866	109.7	0.83−6	114	2.52×10^{7}	65.1	1.43×10^{6}
500	2080	125.3	0.65−6	129	1.54×10^{7}	73.6	6.14×10^{5}

The energy below 1000 Å is absorbed strongly by O, O_2, CO_2, N_2 and N in the high atmosphere. This produces ionization of these molecules and produces the ionosphere at high altitudes. The height at which this absorption occurs extends from 100 km to the top of the atmosphere. Two regions of intense ionization occur at 200 to 300 km (*F* layer) and at 90 to 110 km (*E* layer). Each of these layers breaks up into two during the day time. Ionization below 80 km also occurs during the day time and is due to Lyman α. These layers are due to the ionization of the molecules present in this region and the heights must be determined by the concentrations of these molecules at the various heights, the intensity distribution in the Sun's spectrum and the absorption coefficients of the molecules. All these effects are interrelated and no general agreement in regard to details has been secured. The subject has been discussed by BHAR (1938), MITRA, BHAR and GHOSH (1938), MOHLER (1940), NICOLET (1945, 1949), MITRA (1952), BATES and MASSEY (1946, 1947) and BATES (1954). The densities of ions in the *E* and *F* regions are of the order of 10^5 and 10^6 and the number of molecules per cm^3 are 4×10^{13} and 10^9 respectively.

The total solar radiation below 1000 Å is very uncertain because it is probable that a very high excess radiation over that expected for a black body is present, at least for brief periods of time. A black Sun at 6000° should give 0.4 erg cm^{-2} sec^{-1} below 1000 Å. Much of the energy in this region may come from the corona (see BYRAM, CHUBB, and FRIEDMAN 1956).

The dissociation energy for N_2 is 9.76 ev and thus radiation lying below 1160 Å is required to produce dissociation. Absorption coefficients for nitrogen in this region are very small and no continuum has been observed. Predissociation from the Birge-Hopfield bands occurs, as HERZBERG and HERZBERG (1948) have pointed out. Since the radiation in this region has small intensity, the rate of production of nitrogen atoms is low (WATANABE, ZELIKOFF and INN 1953). It is most probable that little dissociation of nitrogen occurs. BYRAM, CHUB, FRIEDMAN and KUPPERIAN (1956) have estimated the energy of the Lyman α (λ 1215) line alone as $\sim$6 ergs cm^{-2} sec^{-1}.

If high temperatures are characteristic of the high atmosphere, it is of interest to consider the thermodynamic equilibria for the dissociation of nitrogen and oxygen. The equilibrium constants can be calculated from available thermochemical data[1] (ROSSINI et al. 1952; KELLEY 1949). Some equilibrium constants so calculated are:

$$O_2 = 2O; \quad K_{1000} = 10^{-19.85}, \quad K_{1500} = 10^{-11}, \quad K_{2000} = 10^{-6.5};$$

$$N_2 = 2N; \quad K_{2000} = 10^{-18.25}, \quad K_{2500} = 10^{-13.25}.$$

Use of these constants in the equation $\frac{x^2}{1-x} = \frac{K}{P}$, where x is the faction of the partial pressure of oxygen (or nitrogen) due to dissociated atoms and P is the total pressure in atmospheres due to molecular and atomic oxygen (or nitrogen), shows that even at the pressures characteristic of the escape layer nitrogen is not appreciably dissociated at 2000° K under equilibrium conditions, while oxygen is almost completely dissociated. At temperatures above 3000° K, nitrogen will be dissociated at pressures obtaining at the escape layer. Since the atmospheric conditions at the outer surface of the atmosphere are determined by diffusion from lower levels, we may conclude that nitrogen would be in atomic form only

[1] The dissociation energy of nitrogen is taken as 9.76 ev and not from the reference cited.

if absorption of dissociating radiation occurs, but at high levels oxygen would be highly dissociated, if in any way a temperature of 1500° K or above is maintained. Rapid diffusion of oxygen atoms to lower levels should result in recombination and hence a lower dissociation than calculated.

Particle radiation from the Sun is recognized as important in understanding the dissipation of comet tails. BIERMANN (1951) estimates the intensity as 10^{11} to 10^{12} ions and an equal number of electrons $cm^{-2}\ sec^{-1}$ at the distance of the Earth, and estimates the velocities as 3×10^{8} cm sec^{-1}. This energy amounts to 10^4 to 10^5 ergs $cm^{-2}\ sec^{-1}$, assuming that the particles are hydrogen atoms. If these particles can reach the Earth through its magnetic field at least near the magnetic poles, this energy must be absorbed at the escape layer, namely, within one or two free paths from the surface of the atmosphere. Such amounts of energy would raise the temperature of this layer to millions of degrees in a day if it were not lost. Much of this energy is converted most probably to radiation which is lost rapidly to space or to lower layers of the atmosphere. However, if such large amounts of energy are available, it is not surprising that the surface of the terrestrial atmosphere is at a high temperature. If such particles reach the Earth, they must not have solar composition, for in this case neon would be accumulated by the Earth in much larger quantities than are observed. It is likely that much of this particle radiation does not reach the Earth. CHAPMAN (1956) has suggested that the corona of the Sun envelops the Earth and other terrestrial planets and heats their high atmospheres.

Absorption by molecular oxygen in the Schumann-Runge continuum from 1300 to 1760 Å is very intense (WATANABE, ZELIKOFF and INN, 1953), and leads to oxygen atoms in the ^1D and ^3P levels, and possibly also to some atoms in the ^1S level. (The maximum absorption is at 1420 Å, where k is 380 cm^{-1} N.T.P.) This absorption occurs strongly in the region between 80 and 120 km and causes the region of transition in which oxygen becomes dissociated. BATES and WITHERSPOON (1952) have made estimates of this dissociation, but rate constants are not sufficiently well known to make final calculations possible. The total energy absorbed in this process is 100 ergs $cm^{-2}\ sec^{-1}$, if the effective solar temperature is 5000° K. Since the thermal conductivity of the gas is only some 2×10^{3} in cgs units, this energy cannot be conducted from this region and must be radiated away (SPITZER 1952). A black body at the temperature of this region, about 250° K, would radiate 2×10^{5} ergs $cm^{-2}\ sec^{-1}$, but the gas is far from being a black body radiator. The only infrared radiator is carbon dioxide and this should be present in this altitude, since no important separation of atmospheric gases has occurred up to these levels. According to NICOLET's estimates on diffusive separation, it would be greatly decreased in quantity at the 200 to 300 km level. Radiation by radiative combination of ions is not probable. Carbon dioxide absorbs strongly in the region below 1600 Å (WATANABE, ZELIKOFF and INN 1953), and this must result in dissociation. Carbon monoxide will radiate in the far infrared due to a rotation spectrum and this may be an effective cooling agent in this region. CURTIS and GOODY (1956) have shown that carbon dioxide would be a very efficient radiator in the 80 to 120 km region were it not for the long relaxation time for the vibrational degrees of freedom which prevent the molecule from accepting thermal translational energy into its vibrational degrees of freedom during molecular collisions.

Oxygen absorbs the Schumann-Runge bands from 1760 to 1925 Å much less strongly than it does the continuum and this absorption occurs mostly in the ozone region between 17 and 50 km. Secondary reactions produce ozone which then absorbs all wavelengths between 3000 and 2100 Å. The processes are,

$$O_2 + h\nu = O,$$
$$O_2^* + O_2 = O_3 + O,$$
$$O + O_2 + M = O_3 + M,$$

where M indicates any molecule. Also oxygen absorbs the weak Herzberg continuum at the same level

$$O_2 + h\nu = 2O\,(^3P),$$

and again the secondary reactions produce ozone. The ozone is destroyed by photochemical reactions or by reaction with atomic oxygen. The concentration of ozone is determined by the steady state of these opposing reactions. Its distribution in the atmosphere was first determined by DOBSON (1930) and his co-workers. CHAPMAN (1931) first explained the distribution of ozone in the atmosphere as the result of these photochemical reactions. More recent calculations were made by WULF and DEMING (1938). Observations show that there is more ozone at high latitudes than at low latitudes and more in winter than in summer. SCHRÖER (1949), DUTSCH (1946) and CRAIG (1950) all conclude that theory and observation do not agree. A slow circulation of the atmosphere from equator towards the pole is indicated by the studies on ozone concentrations (see also PAETZOLD's article in Vol. XLVIII of this Encyclopedia).

Such a circulation is also indicated by the great dryness of the stratosphere and the mixed character of the atmosphere relative to the permanent gases, as has been shown by BREWER (1949). The water content above the tropopause falls rapidly to a value corresponding to a frost point of about 190° K within a few kilometers above the tropopause in southern England. BREWER accounts for this by postulating a slowly rising current at the equator from which water snows out at about 190° K or less. This requires heating at the equator due to sunlight and cooling at the latitudes at which sinking occurs. This cooling must occur by radiation. It should be noted that this process provides a fairly sharp boundary with respect to water concentration with very little above the tropopause and much below. It was recognized by BREWER that this current cannot be a simple one because of the Coriolis forces due to the Earth's rotation. ELSASSER (1942) finds that a cooling of 1.1° C per day is possible and this would permit a sinking of 100 meters per day.

The energy absorbed in the ozone region amounts to 10^4 ergs cm^{-2} sec^{-1} and this energy, as well as energy received from deeper layers of the atmosphere, is lost by the radiation of ozone and carbon dioxide in the infrared, as has been shown by GOWAN (1947) and by GOODY (1949). Their calculations made arbitrary assumptions in regard to water content above the tropopause but, nevertheless, are essentially correct within the probable limits of error in such complicated problems.

Visible and near ultraviolet light to 3000 Å is not absorbed by any strong bands by the terrestrial atmosphere. There are weak oxygen bands in the red which remove little energy from the solar radiation. This is also true of near infrared bands of carbon dioxide and water. Clouds reflect much light and have albedos ranging from 50 to 70%, and they diffuse sunlight, but absorb little, so that most of the unreflected light reaches the Earth's surface. The albedo of the Earth's surface lies in the range of 15% and, naturally, varies greatly. About 64% of the solar radiation in the visible is absorbed by the Earth's surface. About 86% of the solar energy lies in this spectrum and hence 7.7×10^5 ergs cm^{-2} sec^{-1}

reach the surface of the Earth and are absorbed as heat energy and ultimately must be radiated from the Earth as infrared heat radiation. When clouds are not present, radiation between 8 to 14 μ is transmitted by the atmosphere and on the average some 10% of the radiation may be lost directly from the surface. Radiation in limited bands from the ozone layer at a temperature from 210 to 270° K contributes to the loss and the radiation by water in the far infrared at the tropopause at 190 to 220° K must make another important contribution. The temperature of a black body radiating the total energy from the entire surface and receiving the total energy given in Table 6 only over $\frac{1}{4}$ of its surface would be 250° K, which is not an unlikely average over the sources mentioned.

The absorption in the infrared extends throughout the atmosphere, but must be most important at the tropopause because of the strong absorption of water. Only through a window between 8 and 14 μ does infrared reach the Earth and there is an ozone band at 9.65 μ which absorbs considerable energy within this window. The total energy absorbed in the infrared is small and of the order of 1.74×10^5 ergs cm^{-2} sec^{-1}.

Table 6 summarizes the amounts of energy absorbed, the altitude at which it occurs and the important absorber at each region.

Table 6.

Energy source	Energy absorption erg cm^{-2} sec^{-1}	Altitude km	Absorbing mechanism
Particle radiation	10^4 to 10^5 (?)	300 to ∞	Collisions
<1000 Å	0.4	200 to 300	Ionization of molecules
1000 to 1760 Å	10^2	80 to 120	$O_2 + h\nu = 2O$
1760 to 3000 Å	10^4	17 to 50	$O_2 + h\nu = 2O$; $O_3 + h\nu = O_2 + O$
3000 to 8000 Å	7.7×10^5	0	Black body absorption
Infrared	1.74×10^5	0 to 350	Absorption by CO_2, O_3, H_2O, OH, CO, etc. but mainly by H_2O between 0 to 15 km

7. The transfer of energy within the atmosphere. The most striking method of energy transfer is by convection in the troposphere. To a close approximation the troposphere is adiabatic in temperature distribution and this is maintained by convection. Energy reaches the Earth's surface and is transferred to the air and throughout the troposphere energy is radiated to space and partly to the overlying layers. This constitutes a thermodynamic engine operating between different temperatures, and it supplies the mechanical energy for the movement of the atmosphere in the troposphere. The tropopause may be a very effective radiator if condensation of water occurs, as is apparently true for the Earth, since a sharp radiating boundary can be produced in this way. The adiabatic temperature, pressure and density gradients can be calculated for the dry and moist atmosphere by usual thermodynamic methods. For cloudless air the adiabatic temperature lapse rate, Γ, is $\frac{\mu g}{C}$, where μ is the molecular weight, g the acceleration of gravity which is constant for all problems considered here, and C is the heat capacity per gram molecule at constant pressure. (All constants will refer to molar quantities and not to grams and the gas constant is the usual physical constant and not that per gram of dry air as commonly used in meteorological publications.) The pressure and density are then

$$P = P_0 \left(1 - \frac{\Gamma h}{T_0}\right)^{\frac{C}{R}}, \qquad \varrho = \varrho_0 \left(1 - \frac{\Gamma h}{T_0}\right)^{\frac{C}{R} - 1}, \tag{7.1}$$

where T_0 is the surface temperature and h the height above the surface. Following BRUNT's discussion (1944) for the moist atmosphere we find that

$$\frac{dT}{dh} = -\Gamma \frac{\left(1+\frac{p_2}{p_1}\frac{\Delta H}{RT}\right)\left(1+\xi\frac{\mu_2}{\mu_1}\right)}{1+\frac{(p_1+p_2)\,p_2}{p_1^2}\frac{R}{C}\frac{\Delta H}{RT}}, \tag{7.2}$$

where p_1 and p_2 are the partial pressures, and μ_1 and μ_2 the molecular weights of the permanent and condensible gases respectively, and ΔH is the heat of vaporization of the condensible gas per gram molecule. ξ is the molar ratio of the total amount of condensible gas in the condensed and gas phases to the permanent gas. If the condensed phase is removed as condensation occurs, ξ equals p_2/p_1. In this equation all heat capacity terms are neglected in comparison with the $\Delta H/RT$ terms since they are always negligible. ΔH for sublimation of ice is nearly constant with temperature and is 12200 cal/mole. Its value for vaporization of liquid water varies slightly, but for present purposes can be taken as 10640 cal per mole. The heat capacity per gram molecule of dry air is 6.957 calories per mole-degree. The dry lapse rate is 9.9° C per km. The calculated lapse rates for saturated air at a total pressure of 1 atmosphere and at 245, 273.2 and 300° K are 9.2, 6.6 and 3.9 deg./km respectively[1]. Near the surface the observed lapse rates are larger, due to the air not being saturated, but at higher altitudes they are smaller than calculated values. In fact the atmosphere is intermediate between an adiabatic one and an isothermal one. This is due partly to the air currents not rising vertically because of the rotation of the Earth, so that the local ground temperature is not the proper zero temperature for the adiabatic process, but this does not explain the lapse rate at the equator. The effect is caused by an involved combination of causes, i.e. radiative cooling, convection and advection of air.

Molecular conduction is not an important mechanism for the transport of heat because of the low conductivity of gases. Energy may be transferred in the high atmosphere by the diffusion of atoms which are produced in one level to other levels where recombination is more rapid.

The tropopause height varies with latitude. GOODY (1949) for purposes of calculation has summarized data and calculates the temperature of the tropopause assuming an average lapse rate of 6.5 deg./km (Table 7).

Table 7. *Estimated temperatures at the tropopause and surface of the Earth* (GOODY).

Latitude	Height km	Ground temperature °K	Tropopause temperature °K	Tropopause pressure atm
0	17.8	302	186	0.079
10	16.5	300	193	0.098
20	14.8	296	200	0.127
30	12.9	291	207	0.167
40	11.2	284	211	0.210
50	9.7	278	215	0.260
60	8.5	272	218	0.307
70	7.6	267	219	0.339
80	7.0	264	220	0.375

[1] Extensive tables calculated by more exact formulae are given in the Smithsonian Meteorological Tables, Smithsonial Miscellaneous Collections, Pub. 4014 (1951). See Tables 79 and 80. They are not easily converted to calculations for other atmospheres.

Such variations may be expected in the case of other planets. It is interesting to note that the tropopause at the equator has a lower temperature than it has at high latitudes. The solar energy at the equator is partly converted to heat and partly to the potential energy of the rising atmosphere. This energy is partly radiated at the equatorial tropopause and partly carried to the high latitudes, where it is radiated at the warmer tropopause as the air sinks. A similar effect is observed on Venus in that the day side is cooler than the night side of the planet (see below).

8. The escape of hydrogen and helium. JEANS (1916) gave a derivation for the rate of escape of planetary atmospheres and the analysis was extended by JONES (1923). The escape layer can be defined as that level above which there is only one free path for a very fast molecule moving vertically away from the planet, i.e.

$$\frac{1}{n_t \pi \sigma^2} = 1, \tag{8.1}$$

where σ equals the molecular diameter. Using σ equal to 3.15×10^{-8} cm, this gives n_t equal to 3.2×10^{14} molecules cm^{-2}, as the total number of molecules above the escape layer. Since the total number of molecules above such a layer is the number per cm^3 multiplied by the scale height, H, equal to $\frac{RT}{\mu g}$ the number of molecules per cm^3 at the escape layer is[1],

$$n_c = 3.2 \times 10^{14}/H. \tag{8.2}$$

The rate of escape in molecules per sec and cm^2, L, is equal to the number of molecules per cm^3 at the escape layer multiplied by one-fourth the velocity and the Maxwell distribution factor integrated over all velocities greater than the escape velocity, namely,

$$v_c = \sqrt{\frac{2GM}{a_c}}, \tag{8.3}$$

that is,

$$L = n_c c_i 4\pi \left(\frac{\mu_i}{2\pi RT}\right)^{\frac{3}{2}} \int_{v_c}^{\infty} \frac{v}{4} e^{-\frac{\mu_i v^2}{2RT_c}} v^2 \, dv, \tag{8.4}$$

where T_c is the temperature of the escape layer, n_c is the total number of molecules per cm^3, c_i is the fraction of molecules of the i-th kind at the escape layer and μ_i is the corresponding molecular weight. Integration gives,

$$L = n_c c_i \sqrt{\frac{RT}{2\pi \mu_i}} (1 + X_i) e^{-X_i}, \qquad X_i = \frac{GM \mu_i}{RT_c a_c}. \tag{8.5}$$

If the value of T_c is known, or is estimated, the scale height can be estimated and the total number of molecules per cm^3 at the escape layer can be calculated. An approximate value for a_c can be estimated. However, the value of c_i depends on the level above which diffusive equilibrium exists and usually this is assumed to be above about 150 or 160 km (SPITZER 1952, NICOLET 1954). SPITZER has pointed out that a large correction must be made to these formulae, if the escape layer is at a markedly higher temperature than that of the lower atmosphere. His calculations show that the escape of He^4 from the Earth cannot be expected to occur in the proper amounts unless the temperature of the escape layer is 1500 or 2000° K for 2% of the time.

[1] This is true exactly for an isothermal atmosphere and one for which the scale height varies linearly with temperature (NICOLET 1954, p. 650).

Hydrogen will not be lost at impossibly high rates because of the Brewer effect (see p. 372). Using GOODY's estimate of 186° K for the temperature of the equatorial tropopause, the vapor pressure of water at this point is 0.16 dynes per cm^{-2} and the total pressure is 0.079 atmosphere, which gives 2×10^{-6} for the ratio of water to dry air. BATES and NICOLET (1950) have discussed the dissociation of water vapor using about ten times the concentration used above, and they show that only atomic hydrogen should be present at high altitudes. NICOLET's estimate (Table 5) of 2.47×10^{10} molecules of N_2 per cm^3 then requires that there be 1.23×10^5 atoms per cm^3 of hydrogen at this level[1].

HARTECK and JENSEN (1948) have given a discussion of diffusion in the Earth's atmosphere and have argued that diffusion in the stratosphere is the slow mechanism for the loss of water to the high atmosphere. Though the evidence for mixing to higher levels has invalidaded their conclusions, it is desirable to apply their calculations to the region above the 160 km level. The diffusion equation is,

$$J = n\,D_i\left[\left(\frac{1}{H} - \frac{1}{H_i}\right)c_i - \frac{d c_i}{d h}\right], \tag{8.6}$$

where J is the current in atoms per cm^2 and sec, D_i is the coefficient of diffusion of the i-th kind of molecule in the mean gas, H and H_i are the mean scale height and the scale height for the i-th gas, i.e., $\frac{RT}{\mu g}$ and $\frac{RT}{\mu_i g}$, and h is the distance above the 160 km level measured vertically. (Thermal diffusion is neglected, since it is a small effect for observed temperature gradients.) $n D_i$ is approximately a constant with respect to pressure variations. H and μ depend on the temperature and on the mean molecular weight which varies because of the diffusive separation of N_2 and O in the high atmosphere. H_i is linear in the temperature and D varies as its square root. Using Table 5 we can estimate these variations and they are important in the high atmosphere. The mean molecular weight varies from 16 to 24 and the temperature from 482 at 160 km to 1416 at 350 km, which is the altitude of the escape layer. NICOLET (1957) has shown that even higher temperatures must be assumed in order to account for the observed rates of escape of He^4 and He^3. We therefore assume a linear variation of the temperature with altitude and that this variation is more important than the variation in atomic weight. Accordingly the diffusion equation is

$$J = \frac{n_0 D_0}{(1+th)^{\frac{1}{2}}}\left[\left(\frac{1}{H_0} - \frac{1}{H_{i0}}\right)\frac{c_i}{(1+th)}\,\frac{1}{\left(1+\frac{h}{r_0}\right)^2} - \frac{d c_i}{d h}\right], \qquad t = \frac{1}{T_0}\,\frac{dT}{dh}. \tag{8.7}$$

The quantity, t, has a value between 1×10^{-7} from temperature and 1.75×10^{-7} deg. per cm^{-1} from the scale heights from the data of NICOLET's table and the values will be larger, if the top of the atmosphere is 3000 or 4000° K, and the escape layer will be higher also. For these higher temperatures we use t as $2\times 10^{-7}\,cm^{-1}$. The integral of this equation, setting $\left(1+\frac{h}{r_0}\right)^2$ equal to a constant mean value, is,

$$J = n_0 D_0 t\left[-c_i Y + c_{i0} Y(1+th)^B\right], \tag{8.8}$$

where

$$Y = \frac{B-\frac{3}{2}}{(1+th)^B - (1+th)^{\frac{3}{2}}}, \tag{8.9}$$

---

[1] J. H. HOUGHTON (private communication) on the basis of new observations estimates the amount of water at high altitudes as about double that used here.

and

$$B = \frac{1}{t}\left(\frac{1}{H_0} - \frac{1}{H_{i0}}\right)\frac{1}{\left(1+\frac{h}{r_0}\right)^2}, \tag{8.10}$$

and h is the height of the escape layer above the 160 km level where diffusion begins. In a steady state this diffusion current, J, must equal the escape current, L. Combining these equations, c_i can be calculated in terms of its value, c_{i0}, at the 160 km level and then the escape current becomes[1]

$$J = L = \frac{n_0 D_0 t Y c_{i0}(1+th)^B n_c \sqrt{\frac{RT}{2\pi\mu_i}}(1+X_i)\,e^{-X_i}}{n_0 D_0 t Y + n_c \sqrt{\frac{RT}{2\pi\mu_i}} X_i\,e^{-X_i}}. \tag{8.11}$$

For the Earth the mean value of $\left(1+\frac{h}{r_0}\right)^2$ can be taken as unity, but allowance must be made for this factor when escape from Mars is considered.

If the first of the two terms in the denominator is much larger than the second, the escape rate is determined by the molecular escape and if the reverse is true, the escape is controlled by diffusion. At lower temperatures the former condition obtains and at very high temperatures the latter is the case. The coefficients of diffusion of H, He^3 and He^4 in air are 1, 0.68 and 0.59 $cm^2\,sec^{-1}$ at N.T.P. and the escape radius of the Earth can be taken as about 6.8×10^8 cm in all cases. The rates of escape calculated for these molecules are given in Table 8.

Table 8. *Escape of hydrogen and helium from the Earth's atmosphere.*

	T °K				
	1500	1500+∞	2500	3000	4000
t cm^{-1} . . .	1.5×10^{-7}	1.5×10^{-7}	1.75×10^{-7}	2×10^{-7}	2×10^{-7}
$160+h_c$ . .	300 km	300 km	400 km	420 km	525 km
			Escape rates		
H	1.1×10^7	1.37×10^7	1.1×10^7	9×10^6	9×10^6
He^3	0.028	27.2	9	13	19
He^4	166	1.8×10^7	6.63×10^5	2.3×10^6	9.75×10^6
			Time of escape (years)		
He^3	1.6×10^8	1.6×10^5	5.0×10^5	3.4×10^5	2.3×10^5
He^4	2.16×10^{11}	2.0×10^5	5.4×10^6	1.5×10^6	3.7×10^5
Ratio $\left(\frac{He^3}{He^4}\right)$	8×10^{-4}	0.8	0.09	0.23	0.62

The first three lines give the assumed values for the temperatures at the escape layer, t, and the heights of the escape layer. The ratios of He^3 and He^4 to air

[1] As t approaches zero this reduces to a similar formula derived by using HARTECK and JENSEN's formula, namely

$$J = L = \frac{n_0 D_0 c_{i0}\left(e^{h\left(\frac{1}{H}-\frac{1}{H_i}\right)}-1\right)^{-1} e^{h\left(\frac{1}{H}-\frac{1}{H_i}\right)} n_c \sqrt{\frac{RT}{2\pi\mu_i}}(1+X_i)\,e^{-X_i}}{n_0 D_0\left(e^{h\left(\frac{1}{H}-\frac{1}{H_i}\right)}-1\right)^{-1} + n_c \sqrt{\frac{RT}{2\pi\mu_i}}(1+X_i)e^{-X_i}}.$$

Substituting approximate mean values for the variable quantities in this equation, approximately the same numerical values for the rate of escape as those given in Table 8 are secured.

are the observed values 6.55×10^{-12} and 5.24×10^{-6} respectively. The hydrogen ratio is taken as 2×10^{-6} assuming that air at the equatorial tropopause is dried by cooling to 186° K. The times of escape for He^3 and He^4 are the total numbers of these atoms in the atmosphere per cm^2 divided by the calculated rates of escape. The second column gives the rates of escape for essentially the atmosphere given in Table 4. The third column contains rates of escape calculated on the assumption that diffusion is rate-controlling and that a very high temperature exists in the escape layer only but that otherwise the atmosphere is again that of Table 4. This temperature distribution is labelled as $1500 + \infty$. The last three columns assume a linear increase in temperature and scale height from the 160 km level at 482° K to escape layers at the temperature indicated. The use of another temperature for the top of the mixing level such as that given in Table 5 would make little difference in the results. The rates of escape may not be uniform over the Earth's surface and hence exact agreement between estimated rates of production of He^3 and He^4 and calculated rates of loss cannot be expected. It is interesting that only some 10^6 to 10^7 atoms of He^4 and some 10 or 20 atoms of He^3 can be lost per cm^2 sec regardless of how high the temperature of the high atmosphere may be.

The calculated escape rates may be correct to better than a factor of 10. As NICOLET concludes, only at high temperatures for the escape layer do the rates of escape of He^3 and He^4 and their ratios become comparable to the estimated rates of production, which he estimates as 2 and 2×10^6 atoms cm^{-2} sec^{-1}. These estimates may be in error by small factors. When the rate is diffusion-controlled, as it is at the higher temperatures, it is not necessary that the atmosphere be continuously at the high temperature, because an increase in concentration could occur which could then be "flashed off" quickly by a temporary high temperature. The high temperature periods need last only some 0.01 to 0.001 of the total time. Since at equilibrium about 2×10^{11} atoms of He^4 will be above the escape layer, and since the mean rate of escape must be some 2 to 5×10^6 atoms per second, these high temperature flashes must be spaced less than one day apart. As SPITZER suggests, these high temperatures require a very considerable excess in the Sun's light in the region below 1000 Å or the high temperatures are due to an intense particle radiation as discussed by BIERMANN.

The loss of hydrogen is equivalent to $\sim$20 g of water per cm^2 during 4.5×10^9 years and this is equivalent to only one-tenth of the atmospheric oxygen. However, the atmospheric methane of 2 ppm. supplies twice as much hydrogen to the high atmosphere as has been assumed, and the uncertainty in the temperature of the tropical tropopause and its efficiency as a drying mechanism is somewhat uncertain. Again, there is probably a considerable production of reduced gases, i.e. H_2, CH_4, and H_2S, and ammonia salts from the Earth's interior and these can only become oxidized without diminishing the atmospheric oxygen by an escape of hydrogen from the Earth or an increase of reduced substances, i.e. carbon compounds, etc. in the Earth's surface regions. Also the arrival of some 10^{11} to 10^{12} atoms of hydrogen and $\sim$0.1 as many helium atoms cm^{-2} sec^{-1} means that these atoms must escape immediately from an intensely high temperature atmosphere, for otherwise hydrogen and helium should accumulate on the Earth and the oxygen in the atmosphere would disappear[1]. These make

[1] It should be noted that these particles cannot have the mean solar composition and hence the accelerating mechanism must favor low mass atoms and ions. For example, solar proportions would lead to an estimated arrival of 10^8 to 10^9 neon atoms cm^{-2} sec^{-1} and in geologic time this would supply 10^{25} to 10^{26} cm^{-2}, while the number in the atmosphere is only 4×10^{20} cm^{-2}.

any quantitative comparison of calculated rates of escape meaningless. We can conclude that the rate of escape need not be excessive in any way, because of the effectiveness of the Brewer effect at the tropical tropopause and the slow diffusion rate above the 160 km level.

9. Probable history of the atmosphere[1]. The most remarkable feature of the terrestrial atmosphere as compared to that of other planets is the presence of elementary oxygen which is not present in detectable amounts, in other atmosphere. It was early suggested by KOENE that this was due to photosynthesis of plants. Also it has been recognized ever since JEANS derived formulae for the escape of hydrogen from atmospheres that the escape of hydrogen from atmospheres leads to an increase in the oxidized character and probably to free oxygen, though this latter point is not certain (ARRHENIUS 1910, POOLE 1941). Many details of the history of the atmosphere have been presented, most of which are erroneous, chiefly on the basis of elementary chemical and physical principles, though in other cases because the great complexity of atmospheric problems cannot be unravelled without extensive observational results on the present atmosphere which have been obtained only during this century. Balloons, aeroplanes and rockets have proved to be the indispensable tools for this investigation and they cannot be applied to the historical problems.

In order to discuss the development of the atmosphere, it is necessary to secure a correct beginning, and this is difficult. We do not know whether the Earth was formed at high or low temperatures, but certain limitations can be placed on this problem[2]. Unless other sources of energy are discovered, it can be stated firmly that the only source of heat for producing a melted primitive Earth was its own gravitational energy of accumulation. If the Earth accumulated sufficiently rapidly so that this energy could not be radiated into space, it could have been melted. On the other hand, accumulation during some tens or hundreds of millions of years would surely have produced an Earth at low temperatures. With the iron of its present core distributed more or less uniformly throughout its mass, a very considerable amount of potential energy over that of an Earth with a core existed, and this energy, together with the greatly increased radioactive heating some 4.5×10^9 years ago may have caused an active reorganization

[1] The literature on the origin of the atmosphere contains several erroneous concepts which are referred to and discussed repeatedly, and which we wish to eliminate from the discussion immediately. (1) It is not necessary for free oxygen to be present in order that life shall exist. Vast amounts of living substance exist now under anaerobic conditions. (2) Ozone absorbs actinic rays fatal to present terrestrial organisms, but such rays do not penetrate deeply into natural water so that living organisms would be protected even if the actinic rays reached the Earth's surface. In the absence of oxygen volatile carbon compounds would be common and these would surely absorb the rays. On many grounds, the ozone blanket is not necessary to preserve life. (3) Carbon dioxide will not remain in any atmosphere for a long period of time providing erosion of land surfaces is possible, since it will react with the silicates of calcium, magnesium and ferrous iron to form carbonates of these elements and silicon dioxide. The equilibrium pressure of carbon dioxide is very low $\sim 10^{-8}$ atm, even lower than the partial pressure now in the atmosphere, and hence this gas would be rapidly removed from the atmosphere, rapidly, that is, in terms of geologic time. (4) The oceans and surface of the Earth could not have been noticeably at higher temperature because of the primitive heat of the Earth for more than thousands of years. We cannot define our meaning of the termination of the Earth's origin within such a small time interval. It is probable that the accumulation of the Earth required tens or hundreds of millions of years during which time the composition of the Earth's atmosphere was determined by the process of accumulation. Such accumulation must have been accompanied by such violent processes that it is meaningless to refer to a mean temperature of the oceans or the atmosphere in describing them.

[2] H.C. UREY (1956b) has reviewed the evidence on this problem which will not be repeated here.

and considerable heat generation. Plutonic activity may have been much greater than now. It is interesting that the oldest known uranium minerals are only 2.6×10^9 years old, while the oldest meteorites are 4.4 to 4.55×10^9 years old by the uranium, Rb^{87}—Sr^{87} and K^{40}—A^{40} dating methods. Complete melting (JACOBS and ALLAN 1956) and a complete convectional reorganization (UREY 1956a, b) during this period have been suggested. Such processes occurring during 1.5×10^9 years would not have changed the general climatic and atmospheric conditions.

We picture the Earth as having been formed by the accumulation of meteorite-like bodies or bodies such as the Imbrian planetessimal which fell on the Moon (GILBERT 1893)[1]. Such objects contained metallic iron and some water. These substances, which would be heated during their fall on the Earth, react to produce hydrogen at high temperatures. Also carbon and iron oxides will produce carbon monoxide and carbon dioxide. These oxides of carbon would react with hydrogen at lower temperatures to produce methane and other carbon-hydrogen compounds. Nitrogen would react less readily to form ammonia which would dissolve in the oceans.

Hydrogen would escape readily from an atmosphere containing large amounts of hydrogen. Above some high level, corresponding to the present 160 km level, diffusive equilibrium would occur and the high atmosphere would be nearly pure hydrogen. Simple calculations show that the time of escape would be some 10000 to 100000 years if present temperature conditions of the high atmosphere obtained. Such high temperatures depend on the luminosity of the Sun, its effective temperature for ultraviolet light emission, and the intensity of particle radiation. The primitive Sun at the time being considered may have been quite different from what it is now. However, hydrogen probably escaped so rapidly that this occurred during what can be designated as the time of formation. Thus immediately we must consider an atmosphere containing some amounts of methane, other hydrocarbons, carbon-hydrogen-nitrogen-oxygen compounds in small amounts, nitrogen and water vapor. As long as methane, ethane, ethylene, acetylene, or other such gaseous compounds were present, they would carry hydrogen to the high atmosphere, would be decomposed by ultraviolet light and hydrogen would escape at a rate determined by the amounts of such gases in the atmosphere and the temperature of the high atmosphere. However, a time would come when the rate became small. Long before oxidation of carbon to its maximum valence state in carbon dioxide had occurred, carbon would have remained dissolved in the oceans as well-known soluble compounds to a great extent, and its abundance in the high atmosphere would have been low.

The calculated rate of escape of hydrogen at the present time equivalent to 20 g cm^{-2} of water during 4.5×10^9 years is quite obviously insufficient to account for the oxidation of carbon, nitrogen, sulphur, and ferrous iron from their low value states, as they are observed in meteorites, to the oxidized states observed in the surface regions of the Earth. If such oxidation was supplied by the escape of hydrogen—and it is difficult to find other sources of oxidation—large amounts of water must have been decomposed. UREY (1953a) has argued for a maximum value of 3.8×10^4 g cm^{-2}. The low rate of escape of hydrogen at the present time is due not only to the arrival of only small amounts of water in the high atmosphere, where it is photochemically dissociated, but also to the fact that oxygen absorbs the light which dissociates water, namely, in the Schumann region, so that it does not penetrate below the tropopause. This occurs in the 80 to 120 km

[1] See R. BALDWIN (1949) and H.C. UREY (1956a) for more recent discussions.

level where dissociation of oxygen takes place. In the absence of atmospheric oxygen, only carbon dioxide will absorb this radiation and its coefficient of absorption is only about $\frac{1}{30}$ that of molecular oxygen though the absorption curves have very similar trends with wavelength (WATANABE, ZELIKOFF and INN 1953). Above the 80 km level of the Earth there are about 2 cm atmospheres of oxygen calculated as though it were not dissociated. It is partly dissociated, of course, but possibly 1 cm atmosphere is a reasonable estimate for undissociated oxygen. In order for carbon dioxide to produce a similar absorption in this region of the spectrum, there must be some 30 cm atmospheres. However, above the tropical tropopause there are only 20 cm atmospheres at the present time and in the absence of oxygen it would also be partly dissociated and ineffective. The calculation is approximate, but it shows that, in the absence of oxygen in the Earth's atmosphere, decomposition of water below the tropopause would probably occur and thus hydrogen would escape to the high atmosphere in large quantities and would escape from the Earth at a very greatly increased rate. The Earth appears to be much more oxidized than Venus, as will be evident from the discussion of that planet, and this absence of the absorbing layer due to an inadequate amount of carbon dioxide probably accounts for this circumstance.

Evidence in regard to the time of the appearance of an oxidizing atmosphere is obscure. The oldest rocks appear to be as oxidized as the more recent ones. Even now under an oxidizing atmosphere the sediments are highly reduced due to the activity of living organisms. Also ferrous iron can be oxidized by water to the ferric state under appropriate conditions and thus elementary oxygen is not a necessary condition for the production of some oxidation. Two pieces of evidence exist. (1) THODE, MACNAMARA and FLEMING (1953) have observed that S^{34} is concentrated in the sulphate relative to the sulphides and they have given evidence that this has occurred by the repeated oxidation of sulphides by elementary oxygen, and repeated reduction by bacteria. They find that this concentration process began about 10^9 years ago with a very considerable uncertainty in the date. This may mark the period at which atmospheric oxygen appeared. (2) The great deposits of ferric oxide belong to geologic formations of about 2×10^9 years ago. MACGREGOR (1927) suggested that these were deposited under a non-oxidizing atmosphere before free oxygen appeared. This is a most remarkable geologic period. For example, the original iron ore deposit of Venezuela, of which only Cerro Bolivar and a few other smaller deposits remain, is estimated to have been some 100 meters thick and to have extended some 200 km or more in linear dimensions. Iron ore deposits of other geologic periods exist, but not on the scale of these Pre-Cambrian deposits. Under a non-oxidizing atmosphere ferrous bicarbonate would be more soluble than calcium bicarbonate and iron could be weathered from the igneous rocks just as calcium is now. If primitive algae were able to oxidize ferrous iron, ferric oxide could have been precipitated in bogs or lakes and in this way formed these deposits. In order for this process to be feasible, it is necessary that the plants oxidized the iron, but did not produce free oxygen, for free oxygen would interfere with the solution of ferrous bicarbonate and thus this method of erosion would not be possible. It is not improbable that such plants may have evolved at an early period.

It may well be that the photodecomposition of water and methane is only able to remove hydrogen as fast as it or other reducing substances are removed from the Earth's interior due to a decrease in the rate of loss of hydrogen when the atmosphere becomes oxidized to a state in which carbon dioxide is abundant and oxygen is not, and that the existence of elementary oxygen is due to photosynthetic activity of present day plants which maintain a non-equilibrium steady state between the elementary oxygen, the reduced carbon compounds and the carbon dioxide of the Earth's surface. We do not have sufficient information to decide this question, but neither Mars nor Venus has any observable free oxygen. If plants exist on these planets, they must be similar to those which were assumed above to have deposited the Pre-Cambrian iron ores.

10. The origin of life. Life, as we know it, is a complex set of chemical and physical processes characteristic of the element carbon and the complex chemical compounds that it forms with hydrogen, nitrogen, oxygen, sulphur and a great many other elements in small but essential amounts. Two types of metabolism occur, the fermentative or anaerobic and the respirative or aerobic. Biochemists recognize that the former is the more fundamental and that it was probably the primitive one. Thus the observation of OPARIN (1938) that life originated under reducing conditions agrees with the outline of events presented above. BERNAL (1949, 1951) has recognized that such a history is consistent with OPARIN's ideas, and UREY (1952) showed that appreciable amounts of free energy could have been available under conditions of a reducing atmosphere. MILLER (1953, 1955) in a brilliant chemical investigation showed that amino acids and other carbon compounds could have been produced under conditions which can reasonably be postulated to have existed during the non-oxidizing atmosphere of the primitive Earth. Chemical investigations designed to produce complex carbon compounds such as are found in living organisms from carbon dioxide have been uniformly unsuccessful and unconvincing. We can conclude that life evolved under anaerobic conditions and that it would probably evolve whenever conditions closely approximating those on the Earth have occurred anywhere in the universe.

It is of interest to specify what the properties of a planet must be in order that life could thrive on it. If the planet were so large that it retained hydrogen, the intermediate state of oxidation characteristic of living organisms would be maintained only with great difficulty for then CH_4, NH_3, H_2O and H_2S would be the abundant compounds. (The mean states of oxidation of carbon and nitrogen in living organisms are about 0 and -2 respectively.) Thus many reactions of carbon and nitrogen are probably taking place within the atmospheres of the major planets, but it is doubtful if anything remotely resembling living organisms exists on these planets. Again, the planet must not be much smaller than the Earth, for in this case the atmosphere would be lost completely at the required temperatures. Thus Mars has closely the minimum mass required. The temperature must lie within narrow limits. Life always exists in liquid water on Earth. Substances which lower the melting point of water are synthesized and utilized by plants and these are important in their ability to withstand low temperatures. However, reactions of carbon compounds become imperceptible at temperatures below 0° C, and at $-100°$ C and below they would be imperceptible even during geological times. Algae thrive in hot springs at 70° C and if such springs existed for long periods of time, complex forms of both animals and plants capable of living at such a temperature would probably evolve. Yet above 100° C reactions of carbon become very rapid and this slow mild series of reactions, so characteristic of living things, could hardly be maintained. The extreme temperature limits for active living processes probably lie between -10 and 100° C, with the certainty that dormant forms could exist for limited times outside these limits. Living organisms are able to adapt themselves to quite remarkable variations in their chemical environment. Thus some bacteria grow in sulphuric acid, e.g., Thiobacillus thio-oxidans, in the presence of high concentrations of carbon dioxide and alcohol, e.g., yeast, in the absence of oxygen, i.e., the anaerobic bacteria, some worms and molluscs, or in the presence of greatly increased concentrations of oxygen, e.g., man and other mammals. In fact, many chemical substances which are violent poisons for some forms of life are tolerated, or even utilized, by others. It would be difficult to conclude that the presence of any chemical compound would definitely prevent the evolution of a form of life; some living organism would probably evolve which would tolerate it or even use it in its metabolic processes. It is evident that both Mars and Venus conform to these broad requirements and that no other planet does so.

III. The atmosphere of Venus.

11. Chemical composition. The surface of Venus is covered by white clouds which prevent any clear observation of its surface. These clouds are responsible for the high albedo of the plante. The horns of the planet near inferior conjunction are often extended and broadened and, in fact, may form a faint, but distinct complete ring about the planet. (For example, see SLIPHER 1939 and WHIPPLE 1946.)

The first constituent of the atmosphere to be detected was carbon dioxide, which was first observed by ADAMS and DUNHAM (1932). They observed bands at 7820.2, 7882.9 and 8688.7 Å. These are the $\nu_1+5\nu_3$, $2\nu_2+5\nu_3$ and the $5\nu_3$ bands as designated by HERZBERG and HERZBERG (1953), who give their relative intensities as 0.3, 0.1 and 1.0 respectively. The moment of inertia of the absorbing molecule calculated from the bands is 70.4×10^{-40} g cm² in close agreement with the accepted moment of intertia of the CO_2 molecule (see HERZBERG 1945). Using the same spectrograph as that used for the observations on Venus, DUNHAM found that a 42.6 meter path through carbon dioxide at 10 atmospheres pressure produced a faint absorption of the $\nu_1+5\nu_3$ band. The band in the Venus spectrum is stronger and in view of the pressure effect which would increase the absorption in the laboratory observation, and recognizing the double path in the Venus atmosphere, DUNHAM estimated the amount of CO_2 above the cloud layer of Venus as 400 meter atmospheres. Since the $2\nu_2+5\nu_3$ band is well observed in the Venus spectra, and since the intensity of this band is only $\frac{1}{3}$ that of the $\nu_1+5\nu_3$, it seems probable that the correct amount is three times that given by DUNHAM, or 1.2 km atmospheres. ADEL and SLIPHER (1934)[1] estimated the carbon dioxide as 3.2 km atmospheres (2 mile atmospheres) from experiments made with a Hilger E-1 glass prism spectrograph which would be likely to give less reliable results than the method used by DUNHAM. KUIPER (1950) from his infrared observations gives 1 km atmosphere, an estimate very close to the Dunham value as corrected by the subsequent work of the HERZBERGS.

HERZBERG (1949) has devised very long paths and has photographed the bands at 8689, 7883 and 7820 Å with carbon dioxide pressures of one atmosphere and less and concludes "that the amount of carbon dioxide above the cloud layer is of the order of 1000 meter-atmospheres". We shall adopt this value for this quantity.

KOZYREV (1954a) has detected many bands in the night sky of Venus and finds evidence for the second positive and the Vegard-Kaplan bands of nitrogen and for the $B\,^2\Sigma\rightarrow X\,^2\Sigma$ bands of N_2^+. The evidence for the second positive band is particularly good. UREY and BREWER (1957) point out that a number of the bands observed by KOZYREV can be assigned to the comet tail bands of CO^+ and that the absorption bands observed by KOZYREV (1954b) may be due to

[1] Pressure broadening may be larger than estimated by DUNHAM and possibly this is the minimum value as he infers in his reference to ADEL and SLIPHER's previous work.

CO^+ and CO_2^+. They observe that the general color of Venus can reasonably be explained as due to the general absorption of these ions or other gaseous constituents to be expected in an atmosphere which contains little or no oxygen. KUIPER (1952, p. 374) was unable to detect carbon monoxide in the infrared spectrum of Venus and placed its abundance as less than 100 cm atmospheres. These observations are not necessarily in contradiction since UREY and BREWER only maintain the presence of fluorescing molecules in the high atmosphere and, as will be shown subsequently, only some 60 cm atmospheres of CO can be expected.

ST. JOHN and NICHOLSON (1922a) attempted the detection of water and oxygen with a dispersion of 3 Å per mm at Mt. Wilson with negative results. Observations on these two constituents are difficult because of their presence in the terrestrial atmosphere. The atmospheric oxygen bands (${}^1\Sigma_g \rightarrow {}^3\Sigma_g$ transition) are due to a forbidden transition and hence are very weak and unsuitable for a sensitive test in another atmosphere and so small amounts of oxygen would not be detected. In order to avoid the difficulty of the Earth's atmospheric absorption, it is necessary to use high dispersion and observe at such times that the Doppler shift due to relative motion of the Earth and planet displaces the lines. ADAMS and DUNHAM (1934) conclude that not more than $\frac{1}{20}$ as much either oxygen or water vapor is present in the Venus atmosphere as is present on Earth. However, only 8% of the Earth's oxygen is above the tropical tropopause and their estimate applies only to the Venus atmosphere above its clouds; hence, if the clouds are near the tropopause, it is possible that Venus has one-half as much oxygen in this region as does the Earth in its atmosphere above the tropopause.

Since the tropopause is probably at about 233° K (see below) the partial pressure of water in the upper atmosphere must be very low and hence any absorption bands could be detected only with difficulty and none have been observed. (In fact, water has never been detected in any extra-terrestrial spectrum because of these difficulties.)

The large amounts of carbon dioxide indicate that the conditions on Venus are fairly far toward the oxidized condition, though the apparent absence of free

Table 9.

Reaction	ΔF^0_{298} kcal	K_{298}
$2CO = CO_2 + C$	-28.64	$\frac{p_{CO_2}}{p^2_{CO}} = 10^{23.1}$
$CO_2 + CH_4 = 2C + 2H_2O$	-2.87	$\frac{p^2_{H_2O}}{p_{CO_2} p_{CH_4}} = 10^{2.31}$
$2N_2O = 2N_2 + O_2$	-49.52	$\frac{p^2_{N_2} p_{O_2}}{p^2_{N_2O}} = 10^{40}$
$3CO_2 + 4NH_3 = 2N_2 + 3C + 6H_2O$	-29.14	$\frac{p^2_{N_2} p^6_{H_2O}}{p^3_{CO_2} p^4_{NH_3}} = 10^{23.4}$
$CO_2 + 2H_2 = C + 2H_2O$	-15.01	$\frac{p^2_{H_2O}}{p_{CO_2} p^2_{H_2}} = 10^{12.1}$

oxygen shows that it is not as oxidized as the Earth. It is of interest to eliminate certain gaseous compounds as possible major constituents of the Venus atmosphere. The reactions of Table 9 are spontaneous in the direction as written at 25° C, as shown by the standard free energies, ΔF^0_{298}. The conclusions would not be changed if temperatures somewhat above or below 298° K prevail on the planet.

C is taken as graphite and all other substances are in the gaseous state. Thus the molecules CH_4, N_2O, NH_3 and H_2 cannot be expected to be important constituents of the atmosphere of Venus once we know that large amounts of carbon dioxide are present and since nitrous oxide and, in fact, all the oxides of nitrogen are unstable with respect to oxygen and nitrogen, they cannot be expected to be present in large quantities in any atmosphere. The oxides of sulphur are very acidic and will react with basic oxides always present in the solid substances expected in terrestrial planets and thus they would be most surprising constituents of atmospheres. No spectra of any of these molecules have been observed on any terrestrial planet except the Earth, and they are present in this case only as very minor constituents and owe their presence to some special photochemical and kinetic chemical process or to industrial sources in each case. Yet both oxygen and carbon monoxide must be minor constituents in the Venus atmosphere, since carbon dioxide absorbs light in the Schumann region with dissociation into atomic oxygen and carbon monoxide. Recombination will take place, but a steady state must be reached with some amounts of O_2, CO, and indeed both O and O_3 in the atmosphere. At high temperatures CO becomes stable and hence no precipitation of carbon in the high atmosphere can be expected.

The clouds of Venus are and have been a source of much speculation and argument. If our estimates of the amounts of carbon dioxide above the cloud layer and the temperature of the clouds are approximately correct, these clouds cannot be condensed carbon dioxide. The condensation temperature for 0.165 bar of CO_2 is 175° K and for 1 bar pressure it is 194.7° K.

LYOT (1929) has studied the polarization of the clouds. If we let $I_\perp$ and $I_\parallel$ be the intensities of light polarized perpendicular and parallel respectively to the plane through the Sun, Earth and the planet, LYOT defines the polarization as

$$p = \frac{I_\perp - I_\parallel}{I_\perp + I_\parallel} \tag{11.1}$$

Observation shows that this varies with the angle between the Sun and the Earth, but not markedly with the orientation of the scattering surface. LYOT explained this as due to the fact that nearly all polarization occurs at single scattering by small particles and that multiple scattering increases the total light, but not the polarization. Hence the observed polarization is smaller than expected from the theory of single scattering. LYOT's curve for the polarization of light from Venus is reproduced in Fig. 1, together with his experimental curve for water droplets of diameter 2.5 μ decreased by a factor of 3.5. The agreement is fair for large values of φ but not for small values. Considering that the tops of the clouds of Venus are probably solid water rather than liquid water, the agreement is probably as good as could be expected. The polarization characteristics of carbon dioxide clouds are unknown.

MENZEL and WHIPPLE (1955) have presented effective arguments for the view that the clouds are due to water[1]. They point out that LYOT's polarization measurements are in reasonable agreement with the water hypothesis, that these measurements require uniform droplets of large size, i.e. 2 μ, and that these are not likely to be dust and that no likely substance other than water has been suggested. On the Earth only very limited quantities of white solids are known, e.g., chalk or silica, and these owe their presence to the leaching action of water. They estimate the amount of water in the terrestrial atmosphere above Mt. Wilson as 0.85 g cm^{-2} of precipitable water, and that above the cloud layers of Venus

[1] The arguments for and against the water hypothesis were presented many years ago by ST. JOHN and NICHOLSON (1922b).

as $\frac{1}{40}$ as much, so that failure to detect water is understandable. Their estimate of water above the cloud layer may be too low since they assume an adiabatic non-mixing atmosphere. The vapor pressure of water at $-38°$ C, which is a possible, though uncertain, value for the temperature at the cloud layer (see below), is 0.158 millibars. If the atmosphere above the clouds is mixed at least for some distance comparable to the 160 km level of the Earth, there are 100 cm atmospheres of water vapor above the clouds or 0.08 g cm^{-2} of water, or about $\frac{1}{10}$ of that above Mt. Wilson. As will be shown later, the amount above the Venus clouds may be less than this, and hence probably cannot be detected.

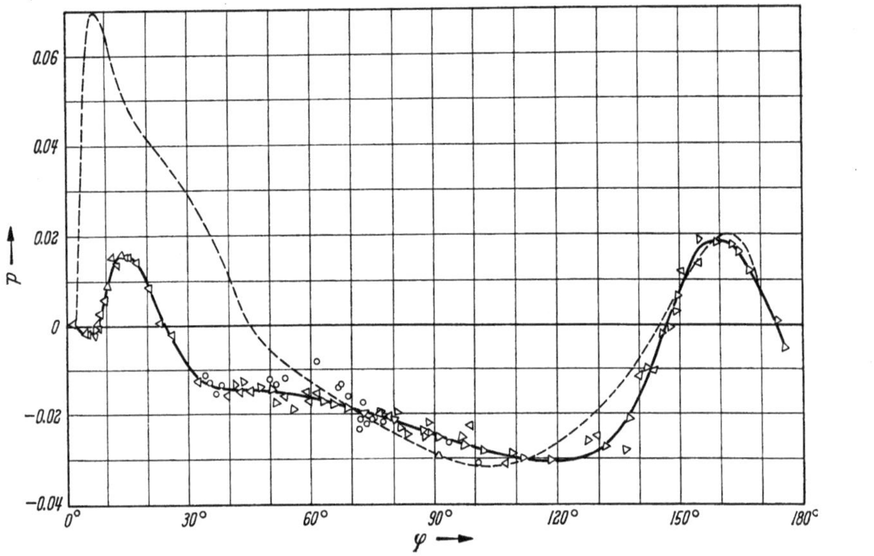

Fig. 1. Polarization of the planet Venus as a function of phase angle (LYOT 1929). The solid curve is the observational curve with the observed points. The broken curve is the observed curve for liquid water droplets of 2.5 μ diameter with the values of the polarization divided by 3.5.

DUNHAM's estimate is that not more than $\frac{1}{20}$ as much water is present above the Venus clouds as that of the Earth's atmosphere.

KUIPER (1952, p. 308) gives the color index of Venus as 0.8 magnitude and DANJON (1948) reported 1.0 magnitude. Since water clouds are whiter than this, some doubt is cast on MENZEL and WHIPPLE's conclusion. KOZYREV (1954b) has measured the reflecting power of Venus at different wavelengths and finds that it increased from 0.15 at 3800 Å to 0.94 at 6500 Å. He also finds some indication for absorption bands at (4120 ± 4) Å and (4372 ± 3) Å. The decrease in reflection is especially marked at wavelengths below 4500 Å. ÖPIK (1956) has discussed this problem and favors the view that the clouds of Venus cannot be due to water because of the distinctly yellow color of the clouds as compared with water clouds of the Earth which are *white*. However, the Earth's atmosphere contains large amounts of elementary oxygen and even ozone and these rapidly destroy any carbon compounds. In the absence of free oxygen, or even in the presence of small amounts of oxygen, carbon compounds would surely be much more abundant in the Earth's atmosphere. Such compounds would contaminate the clouds and would probably produce absorption in the near ultraviolet and blue just as is observed in the clouds of Venus.

UREY and BREWER (1957) have called attention to the absorption of light by CO^+ and CO_2^+ in the high atmosphere of Venus. In the highest regions any absorption by these ions would be radiated and hence would be part of the scattered light. The radiated light would be redder than that absorbed since the emitting transitions would leave the molecule in higher vibrational levels and this energy would be radiated in the red. The absorption bands of CO^+ and CO_2^+ lie in the absorption region observed by KOZYREV and the 4120 Å band is at least close to the wavelength of a group of emission bands of CO_2^+ and both the 4120 and 4372 Å absorptions lie close to the (3, 1) and (4, 1) bands for the $A\,^2\Pi \rightarrow X\,^2\Sigma$ transition of CO^+. The presence of absorption from the $v''=1$ level indicates a high temperature in the high atmosphere of Venus. Only some 10^{13} or 10^{14} molecules cm^{-2} are required for an observable absorption of light if the transition is a permitted one. UREY and BREWER present strong arguments in favor of this interpretation of KOZYREV's data. In the absence of oxygen, a number of molecules such as CN, CH, CH^+, C_2, N_2^+, OH and OH^+ and many compounds of carbon may be present and many of these absorb in the blue and near ultraviolet. These many possibilities make it impossible to exclude the possibility that the clouds of Venus are water on the grounds of the yellowish color of Venus.

UREY (1952a) has shown that in the presence of water, carbon dioxide would react with silicates to form limestone or dolomite and sand, but MENZEL and WHIPPLE point out that this objection would be met if the planet were completely covered by oceans so that effective contact through erosion were prevented. Other possibilities exist. If the initial amount of carbon dioxide were very large, all surface rocks may have reacted and still a surplus could exist. Or mountain building may have ceased and all land areas may be awash so that erosion has ceased. DANJON (1943) made drawings of the dark areas of Venus as observed at slightly different times. He superimposed these drawings and found that a regular pattern appeared. He showed that successive studies of this kind produced the same pattern. He concluded that these are due to permanent surface markings and he found that these markings keep a constant orientation toward the Sun so that the planet rotates only once in its orbital period. These observations have been confirmed and greatly extended by DOLLFUS (1956). We conclude that oceans can hardly cover the entire surface and that either an excess of carbon dioxide exists or that the continents are awash with some coloring material, plants perhaps, in these regions. MENZEL and WHIPPLE's suggestion of water clouds is probably correct.

Alternatives to water as the cloud-forming substance have been proposed, such as quartz particles. But such a substance does not exist in pure form in available quantities on the Earth, and no feasible process for its formation on Venus is evident, either in the absence of water or its presence. It can be stated with confidence that, in the absence of water, dust clouds would be of some very dark brown color such as those of the Sahara dessert or the western United States under "dust bowl" conditions or of the yellow clouds of Mars. The postulate of dust in the atmosphere implies strong convection below the tropopause and this in turn requires that light energy penetrates well below the tropopause and that effective cooling occurs at the tropopause. Light of all wavelengths would be absorbed by finely divided material if it had the composition of igneous rocks or the meteorites, and no great penetration of light into such a layer would occur and hence the whole condition for establishing a dusty atmosphere could not exist. No suitable substance other than water or carbon dioxide can be expected to exist in large quantities on a planet having an oxidation state corresponding to carbon dioxide.

KUIPER (1956) has called attention to the work of HARTECK (1955) on the production of suboxides of carbon ionizing radiation and its possible application to the color of Venus. HARTECK finds that high energy radiations acting on carbon dioxide produce suboxides of carbon and since such radiations reach the high atmosphere of Venus, it can be expected that some amounts of these suboxides may be present in the atmosphere[1]. It is very difficult to estimate the kinetic processes of the high atmosphere of a planet and this has not been done for this problem up to the present time. Because of the well-known instability of these suboxides and because of the high temperature to be expected in the high atmosphere of Venus, it is doubtful if more than small amounts of these substances are present in the atmosphere of Venus. Still only very small amounts of colored substances are required to produce the color of Venus and these compounds may well be the source of the yellowish color. It is very doubtful whether these compounds could constitute the clouds of Venus as KUIPER suggests. Indeed the slight clearing noted by various observers indicates that the clouds precipitate as rain or evaporate, and neither could be expected if they consisted of these resinous suboxides. The clouds of Venus are most probably water with some carbon compounds present and absorption by CO_2^+ and CO^+ must be present to some extent and probably contributes to the color.

12. The temperature and structure of the atmosphere. Early measurements of temperatures show that there is little difference between the day and night sides of the planet (PETTIT and NICHOLSON 1924, COBLENTZ and LAMPLAND 1924). PETTIT and NICHOLSON found 241° K for both the bright and dark hemispheres[2]. Recently, STRONG and SINTON (1956) from the infrared radiation between 8 and 14 μ estimate the temperature of both hemispheres as about −40° C in very close agreement with the observation of PETTIT and NICHOLSON. Recent data of PETTIT and NICHOLSON (1955) indicate 233 and 238° K for the light and dark hemispheres respectively. As ADEL (1941) has shown, this radiation is partly due to carbon dioxide bands in the 10 and 14 μ regions arising from transitions from the $(\nu_1, 2\nu_2)$ resonance levels and the ν_3 level at 9.16 and 10.4 μ, and from the $(\nu_1, 2\nu_2)$ resonance level and the $(\nu_1+\nu_2)$ level at 12.66 and 13.50 μ. Such radiation could hardly originate at the cloud level, for it would be absorbed by the 10^5 cm atmosphere of carbon dioxide above that level since these bands are absorbed by about 500 and 25 cm atmospheres respectively. This suggests that this radiation originates partly high in the atmosphere and gives an indication of temperatures in this region, and possibly only partly at the clouds. From the maximum of intensity P and R branches of the 7820 and 7883 Å bands of CO_2, ADEL (1937a) estimated the temperature as greater than 50° C, and this should be an average temperature for the region above the cloud layer. CHAMBERLAIN and KUIPER (1956) have measured the intensities of these bands more precisely and from the distribution of intensities in the rotational transitions, they find a temperature of $(286 \pm 13)°$ K from the 8689 Å band and $(292 \pm 16)°$ K and $(277 \pm 15)°$ K from the 7820 Å band, using two sets of plates. These bands are comparable in intensity and arise by absorption from the lowest vibrational level and should indicate temperatures at comparable levels of the atmosphere. A mean temperature of $(285 \pm 12)°$ K is thus indicated. The Chamberlain-Kuiper higher temperature is characteristic of the average temperature above the clouds. Their plates were taken at a phase angle of 75 to 80°, and hence the temperatures refer to regions at a considerable angle from the mid solar point.

[1] C_3O_2 is a gas but at moderate temperatures it polymerizes to non-volatile brown higher molecular weight suboxides containing less oxygen.

[2] MENZEL, COBLENTZ and LAMPLAND (1926) reported 330° K but regarded the measurement as doubtful.

MAYER, STOANAKER and McCOLLOUGH (1956) have reported a temperature greater than 350° K by the observations on the 3.15 cm radiation. This radiation presumably comes from the surface of the planet, though this is not certain.

The existence of the cloud layer and evidence for considerable movement of the clouds indicate that the atmosphere is adiabatic below this layer with a tropopause at the top or above the clouds. Since the planet rotates only slowly, a slowly rising current at the mid-solar point similar to the Brewer equatorial rising current in the Earth's atmosphere should exist and we can expect that the temperature of the tropopause at this point is lower than it is at any other point (see p. 372). With an albedo of 0.59, the radiation absorbed from the Sun is 1.3 times that absorbed by the Earth. It must radiate all this energy above the tropopause since the surface of the planet is completely covered by clouds, in contrast to the Earth where radiation from the surface through the window from 8 to 14 μ is possible and effective. We do not understand atmospheric circulations sufficiently to be able to estimate the tropopause temperature but it is probable that a Brewer drying effect would exist and that the tropopause temperatures may be similar to those of the Earth's atmosphere. An albedo of 0.59 requires that the effective radiating temperature is 262° K if the entire surface of the planet had the temperature of a black body[1]. The observed temperatures of 233 and 238° K for the sunlit and dark hemisphere respectively are not inconsistent with this very approximate estimate. If intense drying of the high atmosphere occurs and the clouds are indeed water, as seems very probable, radiation in the far infrared from the cloud layer to space must occur. This radiation will be absorbed only partly by the carbon dioxide above, which is radiating its characteristic bands at 285° K according to CHAMBERLAIN and KUIPER.

Weather bands have been photographed by ROSS (1928) in the ultraviolet. More recently these have been observed by RICHARDSON (1955) and KUIPER (1954). These bands are approximately perpendicular to the terminator. The latter authors disagree on the direction of the pole, KUIPER securing an inclination of the pole by 32° to the orbital plane and RICHARDSON giving 14° for the same datum. SLIPHER (1903) could find no Doppler effect indicating no rotation. This was confirmed by HAMY and MILLOCHAU (1913). The Doppler effect is less sensitive than the method of DANJON and DOLLFUS (see p. 387). At least the weather bands indicating rotation are so faint that agreement in regard to the position of the pole has not been secured.

The absorption of energy in the atmosphere must follow a pattern similar, but not identical, to that in the terrestrial atmosphere. At the top of the atmosphere intense particle radiation must be present and probably produces a very high temperature. The transitions of the second positive bands of nitrogen observed by KOZYREV (1954a) and those of the comet tail bands (UREY and BREWER 1957) indicate that molecules are present in high vibrational levels. The presence of such excited molecules probably does not indicate a true temperature corresponding to the excitation energies, but only that very energetic processes are taking place. Light in the region below 1000 Å will also ionize molecules as occurs on Earth. The Schumann region of the spectrum will be absorbed by carbon dioxide on Venus. This radiation is absorbed by oxygen and penetrates to the 80 km level approximately on Earth, above which level

[1] G.P. KUIPER (1952, p. 304) gives 0.76 for the albedo, but presents no details of measurements nor any explanation of the disagreement with previous observations. This higher albedo would lead to a lower temperature of 234° K.

there are some 4.7×10^{19} oxygen molecules per cm^2. The coefficients of absorption by carbon dioxide follow a similar curve with wavelengths as do those for oxygen but are about $\frac{1}{30}$ as large. Hence this radiation should penetrate through 1.4×10^{21} molecules of carbon dioxide in the case of Venus, and with a scale height of 6.6×10^5 cm the number of molecules at this level would be 2.6×10^{15} cm^{-3}. (The scale height is calculated for a temperature of 285° K and a molecular weight of 42.4.) If carbon dioxide were all dissociated into carbon monoxide and atomic oxygen, there would be 62 cm atmospheres of carbon monoxide in the atmosphere. All the carbon dioxide above the level is not dissociated and, in fact, is probably less dissociated than oxygen is in the Earth's atmosphere since the higher total pressure at the dissociation layer causes more rapid recombination. The bands of CO^+ are the only evidence for its existence. A small amount of molecular oxygen should result from this absorption. These problems have not been studied n detail as have similar terrestrial problems.

There is no known absorber for the region between 2000 Å and the infrared except carbon dioxide in its overtone bands at 7820.2, 7882.9 and 8688.7 Å, where the solar radiation has high intensity. However, the absorption noted by KOZYREV (1954b) in the violet and near ultraviolet shows that absorption does occur and the discussion of UREY and BREWER (1957) shows that a number of molecules may well be present which would absorb through much of the region between 2000 and 4500 Å. The presence of small amounts of carbon suboxides would produce intense absorption as well.

The particle radiation must be about twice as intense at Venus as it is at the Earth, and high temperatures at the escape layer are most probable.

13. Models of the Venus atmosphere. The lower atmosphere of Venus can be constructed on the assumption of the presence of water vapor, but definite conclusions cannot be drawn without more knowledge than we possess. Thus we know the following constants: the pressure of CO_2 at the top of the clouds is 0.170 bars at unknown temperature; the temperature at the surface is possibly ~350° K at an unknown pressure. The lapse rate for a saturated atmosphere is given by Eq. (7.2). The total pressure gradient is

$$-\frac{dP}{dh} = \frac{g\mu_1 p_1}{RT}\left(1 + \frac{p_2}{p_1}\frac{\mu_2}{\mu_1}\right). \tag{13.1}$$

These equations can be solved by successive numerical approximations. The Earth's atmosphere is not saturated with water at any level for most of the time. Venus, with a thicker cloud layer, may be more nearly saturated. HOUGHTON and RADFORD (1938) show that a visibility of 10^5 cm exists when there is only 10^{-8} g cm^{-3} of condensed water. Since the vapor pressure of water at 230° K corresponds to about 6 times this amount of water, it is evident that the top of the cloud layer of Venus may be at some very low temperature and in fact may extend completely to the tropopause. It is evident that we are unable to construct any model of the atmosphere with confidence. For illustration only and as a guide to thinking about the conditions which may exist below the cloud level of Venus, we assume a temperature of −48° C or 225.2° K at the mid-solar tropopause and that the atmosphere is saturated. Also we assume an atmosphere containing 10 mole percent of nitrogen, i.e., an amount of nitrogen approximately the same as that on earth, since this gas should be present to some extent and a mean molecular weight of 42.4. An atmosphere of 1.11×10^5 cm atmospheres (N.T.P.) exerts a pressure of 0.176 bars at the cloud level[1]. Fig. 2 shows a plot of the calculated temperatures vs. altitude. The pressures at h equal to 17 and 20 km below the tropopause, which is assumed to be the cloud level approximately, are 2.66 and 4.10 bars and the temperatures are 326 and 338° K respectively. These temperatures are not unreasonable for a planet receiving 1.3 times as much energy from the Sun as the Earth. The masses of the atmosphere in the two cases are 3160 and 4870 g cm^{-2}. From the solubility of carbon dioxide in water we can conclude that only a few hundred grams cm^{-2} would dissolve in oceans 4 to 5 km deep. The calculations indicate that the total carbon dioxide per cm^2 is comparable to the amount in the sediments of the Earth, estimated as 8300 g cm^{-2}.

The calculated curve (Fig. 2) is for the mid-solar point. For points nearer to the edge of the illuminated hemisphere the tropopause may be at a higher temperature, as is true for the Earth at higher latitudes. At higher levels the curves are drawn so that the temperatures observed in the 8 to 14 μ window are reached at higher altitudes. At still higher altitude the temperature rises to some thousands of degrees as shown by the CO^+ bands indicated by the observations of KOZYREV and their interpretation by UREY and BREWER. It should be noted that a Brewer current rising at the mid-solar point would dry the atmosphere and though higher temperatures as observed by CHAMBERLAIN and KUIPER would be observed, the moisture content of the high atmosphere would be low. An entirely analogous situation exists in the Earth's atmosphere.

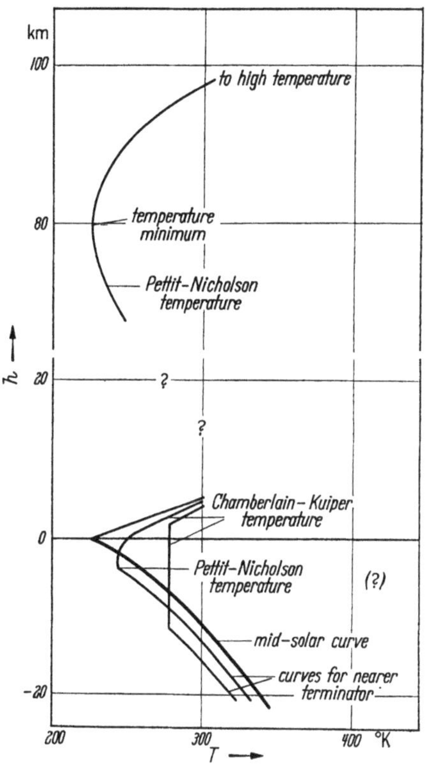

Fig. 2. The temperature-altitude curve for Venus. The heavy curve is that calculated as mentioned in the text. The lighter curves are estimated.

A surface temperature much higher than 350° K makes great difficulty in constructing a model for the lower atmosphere, especially if liquid water exists on the planet, because of the very low lapse rate of a moist atmosphere at higher temperatures. Much higher temperatures would require a very deep atmosphere, if it is moist. If the oceans are completely evaporated, which seems improbable, the lapse rate will be higher, i.e., 4.2° km^{-1} for water vapor and higher than this for mixtures of water and carbon dioxide, and hence the atmosphere need not be quite so deep. Lower surface temperature would decrease the height to the tropopause and the total amount of carbon dioxide on the planet.

Due to the opacity of water and carbon dioxide to infrared light, the surface of the dark hemisphere is highly insulated. The heat capacity of the model atmosphere is 8×10^{10} erg cm^{-2} degree^{-1}. The thermal radiation of a black body

[1] For numerical calculation purposes Eq. (13.1) can be put in a more convenient form

$$\frac{d \ln p_1}{dh} = -\frac{g\mu_1}{RT} - \frac{p_2}{p_1}\left(\frac{\mu_2 g}{RT} - \frac{\Delta H}{RT^2}\frac{dT}{dh}\right). \qquad (13.1')$$

The second term on the right is small as compared to the first.

at 235° K is 1.7×10^5 ergs sec^{-1} cm^{-2} and thus a fall of one degree in temperature would supply the radiation for the dark hemisphere for 5.5 days. The higher temperature of this hemisphere is due to the fall of the atmosphere. A mean fall of 20 meters per day would supply the energy radiated. It is quite reasonable that a moderate circulation of the atmosphere between the two hemispheres would maintain them at nearly the same temperature as PETTIT and NICHOLSON (1955) have observed. The higher temperature of the dark hemisphere is also to be expected, and is similar to the higher temperature of the terrestrial tropopause at high latitudes.

CURTIS and GOODY (1956) have shown that carbon dioxide will not radiate effectively below a total pressure of about 40 dynes cm^{-2} because the relaxation time for the transfer of energy from the translational to the vibration degrees of freedom is too low. Hence, we conclude that far below the escape layer the cooling due to radiation by carbon dioxide will become ineffective and that the high atmosphere of Venus can acquire a very high temperature, quite sufficient to permit the escape of hydrogen. The bands at $\lambda\lambda$ 4372 and 4120 Å observed by KOZYREV, if due to the (3, 1) and (4, 1) transitions of CO^+, indicate high temperatures in the high atmosphere.

14. The escape of gases from Venus. Since Venus is nearer the Sun and has a smaller mass than the Earth, we expect that the escape of gases will be much more rapid from Venus than from the Earth. Since carbon dioxide is dissociated into carbon monoxide and atomic oxygen by radiation in the Schuman region, we must expect that at least carbon monoxide, atomic and molecular oxygen exist in the atmosphere. It is of particular interest to estimate the velocity of escape of hydrogen, for if hydrogen escapes rapidly and the clouds are water, then oxygen should be abundant on the planet. It should be noted that no photodecomposition of water in the lower atmosphere will occur, since the large amount of carbon dioxide will absorb the light of the Schumann region completely and hence no production of hydrogen will occur except in the high atmosphere. Hence, only the transport of water through the tropopause to the high atmosphere and its diffusion to the escape layer need be considered.

As concluded above, the high atmosphere of Venus is probably at 1500° K or higher and again the rate of loss of hydrogen must be determined by the rate of transfer of water through the tropopause and the diffusion of hydrogen to the escape layer. The temperature of the mid-solar tropopause is probably lower than the mean value, just as the terrestrial tropical tropopause has a lower temperature than it has at high latitudes. We shall take this temperature as before to be −48° C or 225.2° K. The vapor pressure of water at this temperature is 0.0498 mb and taking the pressure of carbon dioxide and nitrogen as 178 mb at the same level, we have a mixing ratio of water to the other constituents of 2.9×10^{-4}, which is 145 times that used for the Earth. Obviously, this estimate may be considerably in error. Calculations on rates of diffusion of hydrogen in the atmosphere of Venus are highly uncertain, but the rate should be of similar magnitude to that for the Earth, which was 1×10^7 atoms cm^{-2} sec^{-1} and hence we expect a loss of 1.45×10^9 atoms cm^{-2} sec^{-1} from Venus. This is equivalent to 3000 g of water per cm^2 during 4.5×10^9 years or it is sufficient to have oxidized 1000 g of carbon per cm^2 from the zero oxidation state to carbon dioxide. We have assumed an atmosphere containing $\sim$1100 g of carbon per cm^2 and easily within the limits of the calculations these two quantities could be the same.

The rate of oxidation of carbon on Venus appears to be lower than that for the Earth since some 3000 g of carbon per cm^2 have been oxidised to carbon dioxide on Earth, and this appears to have been substantially complete 2 or 3×10^9 years ago. As mentioned previously, the water of Venus below the tropopause has been protected by its carbon dioxide atmosphere ever since some 100 cm atmospheres of this gas appeared above the tropopause. In the case of the Earth this protection was absent until the elementary oxygen appeared in its atmosphere, since carbon dioxide was removed from its atmosphere by reaction with silicate rocks, a process not possible on Venus due probably to oceans covering the entire surface of the planet or to other causes. Hence, the more rapid oxidation of the Earth's atmosphere is understandable.

When the chemical behavior of carbon dioxide relative to silicate rocks, the effect of *no erosion* on *Venus*, the absorption of light by oxygen, carbon dioxide and water, and the drying of the upper atmosphere due to the Brewer effect are all considered, there is no difficulty in understanding the differences in the atmospheres of the two planets.

15. Evolution of the atmosphere and life on Venus. The evolution of the Venus atmosphere should follow the general course of events as outlined for the terrestrial atmosphere except for those circumstances which have caused the retention of the large quantity of carbon dioxide in the atmosphere. As long as any hydrogen compounds of great volatility, such as methane, remained in the atmosphere, they would provide a method for the transport of hydrogen to the high atmosphere. The concentration of atomic hydrogen at the escape layer would be increased greatly and the rate of escape would be increased in proportion, if the temperature were the same. Following such escape of hydrogen the oceans should have contained large quantities of carbon compounds. Oxidation of these compounds has proceeded during all geologic time and is probably still proceeding.

No positive evidence for the existence of life on Venus is known, though the conditions postulated for the planet would not make it impossible (see discussion for Earth, Sect. 10).

IV. The atmosphere of Mars.

16. Total pressure. The determination of the Martian surface pressure has been attempted by numerous investigations. The numerical values reported are remarkably concordant in spite of the many doubtful assumptions introduced. The problem is to distinguish between the amounts of light scattered by the Martian surface and the atmosphere. The latter is far smaller than the former and hence only a very small effect must be separated from a much larger one. MENZEL (1926) attempted to distinguish between the total amounts of light reflected. LYOT (1929) tried to distinguish between the polarization of light from the atmosphere and that from the solid surface. He assumed RAYLEIGH'S formula for the polarization of the light from the atmosphere and used the experimentally determined polarization of volcanic ash for the surface of Mars. Other studies have been made by BARABASCHEFF and SEMEJKIN (1934), BARABASCHEFF and TIMOSHENKO (1940), SCHARONOW (1941), SYTINSKAYA (1944), FESSENKOFF (1944), HESS (1948) and DE VAUCOULEURS (1945). These are reviewed by DE VAUCOULEURS (1954), who concludes that all are doubtful and that his own determination is approximate only. Only the very recent work of DOLLFUS (1951), which has not yet been published in detail, appears to be physically reliable.

DOLLFUS bases his measurement on the following argument. If B_a and B_s are the surface brightness of the atmosphere and surface at the mid-point at opposition respectively, then at the visual angle, V, and angle, ϑ, from the center of the disk these are

$$B_a \frac{1+\cos^2 V}{2\cos\vartheta}\left(\frac{\lambda_0}{\lambda}\right)^4 \quad \text{and} \quad B_s\,\varphi(V)\,\psi(\vartheta),$$

where $\varphi(V)$ and $\psi(\vartheta)$ are unknown functions. The polarizations of the light are $\frac{\sin^2 V}{1+\cos^2 V}$ and $P_s(V)+\Delta(\vartheta)$ respectively, where $P_s(V)$ and $\Delta(\vartheta)$ are unknown functions again. The total polarization is secured by taking the sum of the products of intensities and polarizations and dividing by the total intensity (the intensity of the atmosphere can be neglected in comparison with that of the surface). Then

$$P(V,\lambda,\vartheta) = P_s(V) + \Delta(\vartheta) + \frac{B_a}{B_s}\left(\frac{\lambda_0}{\lambda}\right)^4 \frac{\sin^2 V}{2\cos\vartheta}\,\frac{1}{\varphi(V)\,\psi(\vartheta)}\,. \tag{16.1}$$

He has used this formula in four ways to deduce the value of B_a/B_s. (1) The values of ϑ and V were varied for one value of λ. In this it is assumed that $P_s(V)$ is constant. (2) With constant V and ϑ and two values of λ ($\sim$5100 and 6200 Å) he secured another value of B_a/B_s. (3) The polarization of surface markings were compared in green and red light for $V=26°$ for which the polarization of atmospheric clouds is small. (4) The fourth determination was made by making measurements in seven wavelengths of light at constant visual angle. All these measurements gave nearly the same value of B_a/B_s with an average of 0.037. From this DOLLFUS concludes that the reduced thickness will be 0.22 times that of the Earth or 227 g cm^{-2}. He gives the surface pressure as 83 millibar[1]. Using the somewhat higher value for g (Table 1) the surface pressure becomes 89 mb. In this calculation it is assumed that the scattering properties of the Martian atmosphere are the same as those of the terrestrial atmosphere per gram of mass.

17. Chemical composition. Only carbon dioxide has been definitely identified as present in the Martian atmosphere. KUIPER (1952) has identified its bands in the Martian spectrum and compared their intensity to the absorption in the terrestrial atmosphere. He constructed a curve of growth by plotting absorption in the terrestrial atmosphere against the path length which was varied by observing with the Sun at varying angles above the horizon. He reports that the observed absorption in light reflected from Mars is the same as that through 8.9 Earth atmospheric masses above McDonald Observatory, and since the path in the atmosphere of Mars averaged over the disk is two times the vertical thickness and the light traverses the atmosphere twice, this gives a factor of 4. He divides the 8.9 by this factor to secure 2.2 times as much carbon dioxide on Mars as on Earth which, corrected to sea level, gives 1.8 times the absorption in a vertical atmosphere of the Earth. KUIPER makes only a very small correction for pressure broadening. GRANDJEAN and GOODY (1955) have made a theoretical correction for the pressure effect and estimate the CO_2 content per cm^2 as 13 times that of the terrestrial atmosphere and this is the more reliable estimate.

[1] DOLLFUS' value of $B_a/B_s = 0.037$ and the albedo can be used to estimate the mass of the atmosphere. Thus,

$$\frac{B_a}{B_s} A \cdot \frac{2}{3} = \frac{(1-a_\lambda)}{2}\,\frac{x}{1033}\,,$$

where A is the albedo at the center of the disk at opposition, and $\frac{1}{2}(1-a_\lambda)$ is the light scattered vertically by the Earth's atmosphere. The factor $\frac{2}{3}$ on the left corrects for the increased Rayleigh scattering of the atmosphere in the backward direction. The value of A is greater than the mean albedo. Values of a_λ can be calculated from RAYLEIGH's law (see Smithsonian Meteorological Tables, Vol. 114, Smithsonian Miscellaneous Collections, 1951, p. 431). Using WOOLLEY's (1953) value for the albedo at 5980 Å, i.e. 0.21, gives 160 g cm^{-2}. The correct value should be larger than this.

The composition of the remaining gas is unknown, but it is probably mostly nitrogen with some radiogenic argon[1]. The atmospheric mass is 227 g cm^{-2}, and assuming nitrogen as the only constituent this is 1.8×10^5 cm atmospheres N.T.P. The carbon dioxide is 1.9% by volume, and assuming a mixed atmosphere the surface partial pressure is 1.5 mb or 1.1 mm of mercury. The higher pressure does not invalidate KUIPER's conclusion that the polar caps of Mars cannot be carbon dioxide since the carbon dioxide frost point for this pressure is $-135°$ C.

KUIPER (1952) reports that he has identified the polar caps as ice by a comparison of the reflection spectrum of frost at low temperatures and that of the polar caps, though no report of the quantitative measurements has been published. DOLLFUS (1951) reports that the polarization from the caps is the same as that of frost at liquid air temperatures. The quantity of water can be estimated from the observation that the temperature of the dark side of the planet is -60 to $-70°$ C, together with the observation of a morning haze, presumably water, which clears during the forenoon (MENZEL 1926, COBLENTZ and LAMPLAND 1926, PETTIT and NICHOLSON 1924b, 1925a, 1930, STRONG and SINTON 1956). This requires that the surface frost point is greater than $-70°$ C, and thus the surface partial pressure of water is 0.00194 mm of mercury or 2.6 dynes cm^{-2} or greater. The fraction of water in the atmosphere is thus 2.9×10^{-5} by volume and 1.8×10^{-5} by mass, and the total water is 4.2×10^{-3} g cm^{-2}. Comparison with terrestrial fogs shows that a fog containing this amount of water in the line of sight would give an easily detectable haze (GEORGE 1950). The amount of water is certainly somewhat higher than that calculated, since the calculation refers only to the uncondensed water vapor. The haze is suspended in the atmosphere above the surface to which the temperature refers and the atmosphere has been cooled at least partly by the surface. It is probable that the actual amount of water may be higher by a factor of 10, namely 0.043 g cm^{-2}. This amount is not markedly different from other estimates that have been made by DUNHAM (1949), who estimated water as less than 5 cm atm. N.T.P., which is equivalent to 0.004 g cm^{-2}.

Early observations to detect oxygen in the Mars spectrum were inconclusive (VERY 1908, CAMPBELL and ALBRECHT 1910, ADAMS and ST. JOHN 1926). ADAMS and DUNHAM (1934) photographed the oxygen spectrum at two times at which the relative velocities of Mars and the Earth were -13.8 and $+12.6$ km sec^{-1} and looked for an assymmetry of the lines due to the Doppler effect. From a study of the displacements of the lines they concluded that the upper limit for the ratio of Martian to telluric oxygen in the atmospheres is not greater than 0.0015, that is, the total thickness is less than 240 cm atm. N.T.P. It will be concluded below that atomic oxygen can escape from Mars and in this case the amounts of molecular oxygen must be very low. Photochemical decomposition of CO_2 must produce unknown amounts of carbon monoxide.

The presence of other components can only be inferred from the total atmospheric pressure and these are generally assumed to be nitrogen and argon, and calculations will be based on the assumption that the important constituent is nitrogen with a molecular weight of 28. The carbon dioxide raises the mean molecular weight to 28.3.

18. Temperature. The temperature of Mars has been studied for many years without much improvement in precision. All measurements agree as to the

---

[1] The total amount of radiogenic argon generated by Mars during 4.5×10^9 years, assuming that the amount of potassium is the average of the chondritic meteorites, is 50 g cm^{-2}. Thus the atmosphere cannot be more than 20% A^{40} even if complete degassing has occurred and this is most improbable.

approximate value of the surface temperatures. Also they agree that the dark areas have temperatures some 8 to 10° C higher than those of the light areas. This shows that the temperatures are those of the surface and not of the atmosphere, or at least that the surface radiation is an important fraction of that observed. As ADEL (1952) emphasizes, the presence of carbon dioxide makes all measurements of temperatures, and especially those near the limb of the planet, doubtful. The effect of some radiation from the atmospheric carbon dioxide will make the observed radiation less since the atmosphere is surely colder than the surface.

DE VAUCOULEURS (1954) has summarized the very extensive measurements of COBLENTZ (1925a, b, c, 1927), COBLENTZ and LAMPLAND (1925), PETTIT and NICHOLSON (1924a, b, 1925b) and COBLENTZ, LAMPLAND and MENZEL (1927). These data were secured during the 1924 and 1926 oppositions. They show considerable variation, and it is difficult to distinguish between observational error, and true variations in temperature. Recently STRONG and SINTON (1956) have reported briefly new measurements which do not disagree with previous estimates.

The temperature of the equatorial region reaches a maximum of about 30° C, and some measurements indicate slightly higher temperatures. The dark regions have higher temperatures than the light regions by some 8° C. The maximum temperature occurs slightly after local noon, but the data on this point are not conclusive. The temperature of the morning limb may be −20° C and of the evening limb 5° C, both at the equator. These values are very uncertain as this is written. The night temperature as reported by MENZEL, COBLENTZ and LAMPLAND is −85° C, and by STRONG and SINTON is −70° C. This latter temperature has been used above to estimate the amount of water in the atmosphere. The variation of temperature with latitude is very marked. The polar caps in winter may have temperatures of −100° C, but in summer their temperatures may be in the neighbourhood of 0° C. The permanent snow cap at the north pole indicates that the general polar temperatures must remain at very low values, i.e. −70° C. All statements in regard to temperatures are subject to considerable probable error, but the temperatures mentioned here give a fair approximation to the temperatures of the planet. It is most likely that these are the surface or near-surface temperatures of the planet and that no observations of temperatures in the higher parts of the atmosphere have been recorded. Thus, there is very little temperature data to use in constructing models of the atmosphere.

19. Clouds and the violet layer. Photographs of Mars in blue, violet and ultraviolet light show less detail than those taken in yellow, red or infrared light, and during a large fraction of the time the surface markings are very obscure when viewed by the shorter wavelengths. At certain periods this violet or blue layer clears and the surface becomes visible. The scattering of light by this layer does not follow RAYLEIGH'S law, i.e. it is not proportional to λ^{-4}. The structure of this layer has been discussed by WRIGHT (1925), SLIPHER (1937), MENZEL (1926), and FESSENKOFF (1926). Certain facts are well established. The violet layer clears occasionally over the entire planet. SLIPHER and WILSON (1954) note that the periods of clearing coincide with times of opposition when the Sun, the Earth and Mars are closely in line and, when clearing at opposition has not occurred, Mars has been markedly out of the plane of the ecliptic. The time required for clearing and reappearance of the violet layer is a few days and a patchy appearance occurs during these times[1].

[1] However, SLIPHER (1954) records a period of two months during which time the blue haze was somewhat transparent.

WRIGHT (1925) found that photographs of Mars in ultraviolet light show about 3% larger radii than those in infrared light. This indicated an atmospheric thickness of 100 km, assuming that the ultraviolet light was scattered by the atmosphere and infrared light was not. MENZEL, FESSENKOFF and others pointed out that the atmosphere was too tenuous to account for the facts in this way. Ross (1926) investigated this and confirmed WRIGHT's results, though he found an even larger difference. SCHARONOW (1941) and BARABASCHEFF and TIMOSHENKO (1940) showed that the difference is at least partly due to the limb darkening in the red and infrared and thus that photographs in these wavelengths are in error on the low side. The theory of the increase in apparent diameter due to scattering by the gaseous atmosphere was investigated by FESSENKOFF, who found that such scattering could not account for the effect. KUIPER (1952) reports that he was unable to detect any difference in the diameter when he viewed the planet through different color filters. He estimates that the violet layer is at an altitude of 10 km. DOLLFUS (1952) reports a height of 60 km for the violet layer[1].

Two types of clouds are observed visually on Mars which can be designated as white and yellow. The former are easily photographed and the latter not. WRIGHT (1927) classified the clouds as types I and II respectively. Type I is visible in shorter wavelengths and becomes invisible as the wavelength increases and thus may be called blue clouds; Type II is readily visible in light of long wavelength but fades as the wavelength of the viewing light is decreased and hence these are yellow. DOLLFUS (1956) classifies these effects as white clouds, blue mists, yellow fogs and the ultraviolet veil[2].

DOLLFUS (1948, 1951) has studied the polarization of the white clouds which he finds to be identical with that of clouds of ice crystals. These clouds are observed at various levels. They are sometimes observed on successive days and even for weeks and in the same location. They move with velocities up to 35 km hr^{-1}. He concludes that local circumstances must be responsible for their formation even in lower latitudes. The white polar winter haze belongs to this class of cloud. MCLAUGHLIN (1954) has pointed out that the surface markings of the planet appear to follow the wind pattern to be expected and he suggests that volcanoes are expelling volcanic ash which is carried by the winds and distributed over the surface. Other possibilities for producing the surface colors exist, as he points out, but he makes a strong case for the winds being effective in establishing the surface pattern in some way. The wind pattern has been estimated by HESS (1950) and reference should be made to his paper. He finds a system of trade winds very analogous to those on Earth. The surface temperatures are not uniform but exhibit warmer and colder regions

The polarization of these clouds decreases with visual angle until V is about 10° and then increases. The shape of the curve is very similar to that of ice crystals and they must be similar to cirrus clouds. The polar winter cloud shows a slightly different polarization curve with a minimum near 16° indicating the presence of larger ice crystals. Whitish mists are observed repeatedly at the same localities indicating that some relief of the surface exists.

---

[1] Such great differences as this made by skilled observers are disturbing. It seems likely that KUIPER and DOLLFUS were not observing the same phenomenon and that subjective effects may be important.

[2] Many colors of clouds are mentioned in the literature and it is difficult to decide whether continuous shading of colors occurs or whether the differences reported in some cases are not due to the inexactness of descriptive words.

The blue mists are observed near the morning and evening limbs. Their polarization increases with visual angle to a maximum at $V = 14°$ and then decreases. They are probably composed of water particles of small dimensions.

The yellow clouds are probably due to dust raised by the surface winds of an arid planet. They are sometimes reported as low-level phenomena, i.e., a few kilometers in height, and at other times as high-level phenomena rising to altitudes estimated up to 30 km. They are more often seen at perihelion at which time storms should be more violent.

GOODY (1957) argues that the violet layer is due to solid particles of small size of the same composition as the dust particles which are responsible for the yellow clouds. In this way he accounts for the absorption of light and hence the obscuration of the surface. The sudden disappearance of the blue haze over the entire planet is not plausibly explained by non-volatile solid particles of any kind. HESS (1950) proposed that the haze was due to solid carbon dioxide, but more recently (HESS 1957) he revised these views and suggested that it is due to crystals of ice. However, ice clouds on Earth are very white with little indication of blue color. KUIPER (1952, pp. 387—396) has emphasized this point of view and by applying the Mie theory of scattering to the polarization measurements of DOLLFUS (1948) he finds that the particles must be 0.15 to 0.20 μ radius. HESS (1957) found that ice clouds could cause the blue haze only if the surface frost point lies near $-90°$ C. This is considerably lower than has been estimated by other methods. Such a low frost point allows very little water for the formation of the polar caps. GOODY'S non-volatile particles do not permit a ready explanation of the clearing. HESS' condensation of water makes it difficult to understand the uniform character of the blue haze over the planet. One would surely expect it to exhibit latitudinal weather bands, and a nearly uniform temperature in the high atmosphere over the entire planet is most improbable, as is a uniform change in temperature by some 10° C within a few days. UREY and BREWER (1957) point out that the CO^+ ion absorbs in the blue and ultraviolet and fluorescences in longer wavelengths and also that some fluorescence of N_2^+ and CO_2^+ may be expected and that the violet haze may be partly due to these effects. They suggest that the ions are produced by particle radiation from the Sun and that the disappearance of the haze at opposition is due to the deflection of these particles by the Earth's magnetic field. They have not considered the problem of the polarization of the blue light.

20. Structure of the atmosphere. The existence of an adiabatic region in an atmosphere such as the troposphere of the Earth requires that heating shall occur at the lower layer and that cooling shall occur at the top. This produces a thermodynamic engine in which heat is converted to the mechanical energy of the winds and storms. Heating at the surface of Mars cannot be as effective as at the Earth's surface since there is little water to absorb the infrared radiations. BREWER and HOUGHTON (1956) have shown that cooling due to radiation from water vapor occurs at the terrestrial tropopause and this shows that similar cooling will occur on Mars, if similar conditions obtain. The preservation of water vapor on the Earth and Venus and also on Mars requires that a cold region occurs, where water is condensed from the atmosphere. We can be confident that a tropopause which is colder at the equator and warmer at high latitudes exists and that the Brewer effect (cf. p. 312) preserves the water. We do not know the temperature of this tropopause, but it is probably at a lower temperature than that of the Earth, since it must radiate less energy. With an albedo of 0.15, Mars receives only 61% as much energy as the Earth and does radiate a substantial amount of energy directly from its surface. If the equatorial tropopause is less than 186° K, namely the temperature of that of the Earth, the loss of hydrogen will be less than that of the Earth at the present time, i.e., 10^7 atoms cm^{-2} sec^{-1}, and this is not so large that it could not be produced continuously from the planet's interior. GOODY (1957) reports the results of calculations using the amounts of carbon dioxide given above and a water content of 10 cm of precipitable water, slightly more than estimated above. He concludes that, including the effects of radiation by carbon dioxide alone, the tropopause would be at 8.5 km but the temperature would fall above this to a minimum of 134° K at 90 km. Including water vapor alone, the tropopause would be at 25 km and the stratosphere would reach a minimum temperature of 153° K. It is to be expected that the inclusion of both would raise the tropopause considerably and might lower the minimum temperature somewhat. However, a minimum of 113° K as required by HESS (1950) for the condensation of CO_2 seems to be impossible. For purposes of calculation we assume a minimum temperature of 130° K at 90 km[1]. The maximum altitude estimated for clouds is near 30 km, and this seems to be a likely altitude for the equatorial tropopause. With a surface temperature of 300° K as a maximum and a lapse rate of 3.75° K, this gives 187° K for the temperature at the equatorial tropopause. A smaller rate above the tropopause probably exists for some distance, but there appears to be no evidence for convection in this region and hence no evidence for another tropopause as suggested by MENZEL (1950). It is likely that a Brewer effect occurs with a slowly rising equatorial current with water falling out, as in the case of the Earth. The calculations of UREY and BREWER indicate that the fluorescing layer shall be at about 80 to 100 km above the surface. In order that ions shall be produced, the total mass above the layer must not be too large. Since Mars can hardly have an appreciable magnetic field, the softer and more intense components of particle radiation should reach the surface and penetrate deeper into its atmosphere and hence produce directly or by secondary radiation a more intense ionization. The most probable ion is CO_2^+, but N_2^+, CO^+ and CO_2^+ all absorb in the violet and near ultraviolet and at sufficiently low pressures should fluoresce.

The great height of the predicted effect would be in accord with the great variation in the radius of Mars as reported in the literature. DE VAUCOULEUR (1954) has summarized the data which justifies his conclusion that the optically measured diameter of the planet is generally observed to be 9″.4 at unit distance or 0.534 Earth radii. Much lower values have been reported by TRUMPLER (1927), KUIPER (1952) and DOLLFUS (1952), i.e., 0.520, 0.523 and 0.530 respectively. CAMICHEL (1954) has made studies of the Martian radius extending over the years from 1941 to 1950 using both the microphotometer method and the movement of surface markings with rotation (TRUMPLER's method) and finds considerable variation in his observations at successive oppositions with minimum values for the equatorial radius near 0.525 and a mean value of 0.53 Earth radii. UREY (1951, 1952) pointed out that this difference might be due to some atmospheric phenomenon and that the smaller values should represent the radius of the planetary surface and that the smaller radius, together with the oblateness calculated from the precession of the satellites, namely 0.0052, requires that Mars contain appreciable amounts of metallic iron and yet has nearly a uniform density. This is a most interesting point and a decision as to its validity depends

[1] In a private communication GOODY expresses the view that the altitude of the minimum may be about 130 km. No important change in the conclusions given here would result from this change in attitude.

on an understanding of the atmospheric phenomena as well as the observational data on the diameter of the planet.

The directly observed oblateness of Mars due to several observers and summarized by DE VAUCOULEURS (1954) is about 0.013 in definite disagreement with that calculated from the precession of the satellites. It seems likely that the discrepancy is due to some atmospheric phenomenon at a lower level over the poles than over the equator by 26 km. This may be a lower tropopause or a lower fluorescent layer, but no definite quantitative theory for the effect has been propsed. KUIPER (1952, p. 389) explains the discrepancy as due to different altitudes of the tropopause, but his very low tropopause (17 km) is unable to explain the difference. The large value favors the view that the atmospheric effect is a high level one. However, CAMICHEL (1954) secured 0.012 from observations on the movements of surface markings with rotation and this indicates that the effect may be due to the solid surface.

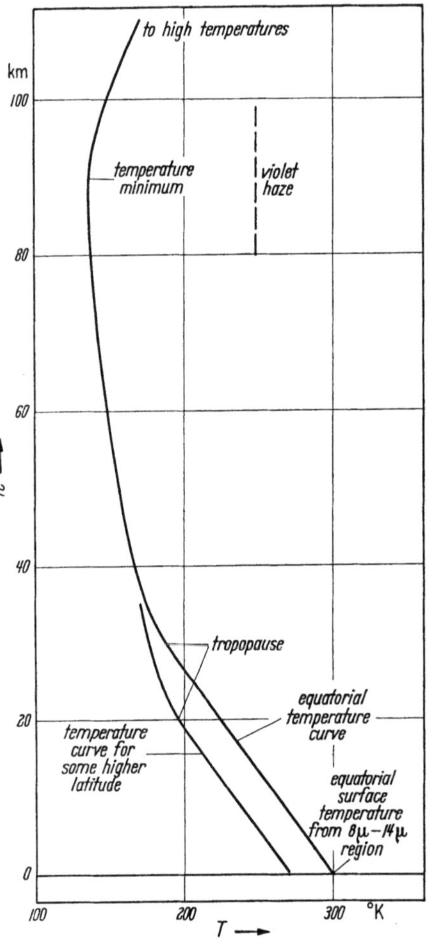

Fig. 3. The temperature-altitude curve for Mars.

Is it possible that great masses of ice mixed with glacial debris or glacial debris alone form this very large elevated region at the equator? In this case the atmospheric pressures at the poles are very much higher than at the equator. This raises questions in regard to the circulation of the atmosphere and the surface temperature distributions and also the difficult question as to how the material became distributed in this way.

Fig. 3 is a diagram of the possible structure of the atmosphere of Mars.

21. The escape of gases from Mars. Because of the small mass of Mars the escape problems are markedly different from those of the Earth and Venus. It is possible that not only hydrogen and helium escape from this planet, but also that oxygen and nitrogen atoms escape as well. It is probable that a high temperature exists at the escape layer of Mars. The intensity of particle radiation should be half as great at Mars as on the Earth. Conduction to lower layers should be less effective than in the terrestrial case because of the greater height of the escape layer. For the purposes of this discussion we assume a temperature of 2000° K at the escape level and conclude that carbon dioxide and nitrogen can remain on the planet and that water, i.e., hydrogen and oxygen, may have been removed in large quantities during past times.

If hydrogen is produced in the Martian atmosphere, it will escape with a velocity that is determined by diffusion only. This will be approximately the same as it is for the terrestrial atmosphere. The presence of carbon dioxide in the high atmosphere prevents the dissociation of water below the tropopause by light in the Schumann region and hence water is preserved on Mars. With a tropopause temperature of 186° K or lower, water will not escape to the high atmosphere any more rapidly than it does on Earth, and with a very low temperature at some higher level the escape will be orders of magnitude less.

Carbon dioxide will be dissociated by light below $\lambda\lambda$ 2000 Å into carbon monoxide and atomic oxygen and no other gas is present to protect it against dissociation by this radiation. The rate of escape of atomic oxygen depends on the temperature of the escape layer, the rate of diffusion to this layer, and also on the structure of the atmosphere in detail, particularly on the temperature of the lower atmosphere and the height to which it is mixed. Low temperatures in the lower atmosphere lower the escape layer and decrease the rate of escape and mixing to higher altitudes decreases the accumulation of atomic oxygen at the escape layer. Table 10 gives the results of calculations on the rate of escape, for different models of the atmosphere. In making the calculations the following temperatures are assumed: (1) a temperature gradient of $-3.75°$ km^{-1} from a surface at 300° K to 30 km and temperature 186.5° K; (2) a gradient of -0.96°K from 30 to 90 km at a temperature of 130° K; (3) different gradients to the escape layers. We have no proof that these temperatures are correct, but it is evident after some experience that no important modification of the results will arise from other temperature distributions or other estimates of the height of the diffusion layer. The mixing ratio, c_{i0} of Eq. (8.11) has been taken as 0.0196, i.e.,

Table 10. *Escape of elements from Mars.*

Maximum mixing level		Escape level		Loss of O atoms	Time (years) to lose observed		Loss of O in 4.5×10^9 years
Altitude km	Temperature °K	Altitude km	Temperature °K	cm^{-2} sec^{-1}	CO_2	H_2O	g cm^{-2}
90	130	1400	1500	6.5×10^9	4.7×10^5	7000	2.4×10^4
300	350	1400	1500	6.1×10^7	5.0×10^7	7.5×10^5	2.3×10^2
90	130	1800	2000	2.0×10^{10}	1.53×10^5	2300	7.5×10^4
300	360	1800	2000	2.0×10^9	1.53×10^6	2300	7.5×10^3

The minimum temperature was taken as 130° K at 90 km. It was assumed that a uniform temperature gradient from this point to the escape level exists in each case.

The calculations were made as follows: we have $T_c = T_0(1+th_c)$ where T_0 is the temperature at the bottom of the diffusion layer where h equals 0. For an atmosphere having a linear temperature gradient we have

$$\frac{d\ln P}{dh} = -\frac{\mu G M}{R T_0}\,\frac{1}{(1+th)(r_0+h)^2}$$

Integrating and substituting n for P from the gas equation gives

$$\ln\frac{n_c}{n_2} = \frac{h_c}{H_0}\,\frac{1}{r_0 t-1}\,\frac{1}{1+\frac{h_c}{r_0}} - \left(\frac{1}{H_0 t}\,\frac{(tr_0)^2}{(tr_0-1)^2}-1\right)\ln(1+th_0) + \frac{1}{H_0 t}\,\frac{(tr_0)^2}{(tr_0-1)^2}\ln\left(1+\frac{h_c}{r_0}\right).$$

We estimate n_0 as 3.2×10^{16} cm^{-3} at the 90 km level. Assuming a value for T_c, we calculate by trial consistent values of h_c and n_c such that $H_c n_c$ equals 3.2×10^{14} cm^{-2}. Escape is now calculated from equation (8.11), using an arithmetic mean for $\left(1+\frac{h}{r_0}\right)^2$ in the value for the constant B.

The assumption of a linear temperature is arbitrary, but a few calculations show that the results are dependent mostly on the mixing height and the temperature assumed for the escape layer.

the fraction of CO_2 in the atmosphere. This means that all CO_2 above the mixing layer is assumed to be dissociated. Two effects are important: namely, increased height of mixing and decreased temperature of the escape layer. Both decrease the loss of oxygen atoms. Only if the escape layer is lower than about 1500° K will the escape of oxygen be prevented.

The loss of atomic oxygen would leave carbon monoxide in the atmosphere. This gas is unstable with respect to graphite and carbon dioxide at low temperatures, but stable at high temperatures, i.e.,

$$2CO = CO_2 + C, \quad F^0_{298} = -28430 \text{ cal}, \quad K_{298} = 10^{20.8},$$
$$F^0_{2000} = 32850 \text{ cal}, \quad K_{2000} = 10^{-3.59}.$$

The reaction at low temperature is slow by usual time scales, but probably not by a geologic time scale. However, the surface of Mars is not noticeably dark as would be expected if graphite were present. Another possible reaction is

$$CO + H_2O = CO_2 + H_2, \quad F^0_{298} = -3990 \text{ cal}.$$

The decrease in standard free energy is not large, but the reaction should proceed. In this case, hydrogen is lost and the net effect is the loss of water. Since water has a very small abundance on Mars, this would be even more catastrophic, as shown in the table. If the high atmosphere is at a temperature of 1500 or 2000 °K, water must be escaping from the rocks of Mars or the present water is the last trace of water from previous oceans and it will be gone in some 10^5 years. This latter conclusion is most improbable. It is interesting to note that CO will be stable at the escape layer and no graphite smoke at a high temperature escape layer should be observed.

The most reasonable interpretation at the present time seems to be as follows: (1) the high atmosphere is at a high temperature somewhat lower than that of the Earth; (2) escape of oxygen occurs at significant rates; and (3) water is escaping from the surface rocks of Mars. A more quantitative statement is not possible.

Nitrogen does dissociate to some extent and hence may also escape from a high temperature atmosphere. At 2000° K the equilibrium constant for the reaction

$$N_2 = 2N$$

is 5×10^{-19}, with concentrations in atmospheres. The total pressure at the escape layer is 2.45×10^{-12} atmosphere. Hence from the equation

$$\frac{x^2}{1-x} = \frac{K}{P} = 2 \times 10^{-7}, \tag{21.1}$$

where x is the fraction of the total pressure due to nitrogen atoms, x equals 4.5×10^{-4}, and this is the mixing ratio, c, as used above. The concentration of atomic nitrogen is probably higher than this. Diffusion of atomic nitrogen from some lower level probably occurs and nitrogen is dissociated by light in the Lyman region and recombination at low pressures is slow. We use this calculated equilibrium value, but recognize that a lower limit for the escape is secured in this way. The escape is not diffusion limited and hence the gravitational escape formula can be used. Using $T_c = 2000°$ K, and an escape layer at 1800 km, the rate of escape of nitrogen is 1.2×10^6 atoms cm^{-2} sec^{-1}, equivalent to 4 g cm^{-2} in 4.5×10^9 years. Thus the rate of escape of nitrogen from Mars can hardly be excessively large even if the escape layer is at 2000° K.

The escape of hydrogen is prevented by the low temperature of the tropopause, and by the absorption of light in the Schumann region by carbon dioxide. Carbon dioxide will be decomposed into carbon monoxide and atomic oxygen, and the atomic oxygen will be lost at a high rate unless mixing to high levels occurs, or unless the temperature of the escape layer is lower than 1500° K. Probably some water is supplied from the surface rocks of the planet. Nitrogen will be preserved due to the great stability of the nitrogen molecule relative to atomic nitrogen even if the escape layer is at 2000° K.

22. The possibility of life and the history of the atmosphere. The positive proof that life in any form depending on the remarkable chemistry of carbon compounds exists on Mars would give us one of the most fascinating facts of all science. Life has appeared on Earth through some process of evolution from inanimate matter and, given similar conditions, we can confidently expect that life bearing considerable similarity to that on Earth would evolve. Proof of life on Mars would confirm this most reasonable postulate.

The most definite evidence for the presence of some plant life is that derived from the greyish or greenish darker areas in the equatorial regions of the planet (SLIPHER 1927). These dark areas change with the seasons, becoming darker in the spring and lighter in the autumn, and ANTONIADI (1930) reports the presence of brown and red colors in different areas in the autumn. V.M. SLIPHER (1924) showed that this green color did *not* have the reflection spectrum of chlorophyl. SCHARONOW (1941) observed the reflection spectrum of the dark areas of Mars and found only a very slight maximum in intensity near 6400 Å between 3800 and 8000 Å. KUIPER studied the infra-red spectrum of Mars between 0.6 and 1.6 μ and found no evidence for any variation in reflectivity through this region. He concludes that chlorophyll is absent and also that water is practically absent. He suggests that lichen-like or moss-like plants may be present since the reflection spectra of these plants are similar to those observed in the dark areas of Mars. Also these plants often grow in comparatively cold and arid regions on Earth. The very great capacity of living organisms to adapt themselves to very diverse conditions suggests that similar organisms on Mars may have adapted themselves to the rigorous conditions of that planet. No suggestion has ever been made of any inorganic substance likely to be present on an arid and lifeless planet that would supply the observed colors and their seasonal variations. Also ÖPIK (1950) has pointed out that drifting dust should have covered these areas, if they had no regenerative powers such as those possessed by living organisms. Many arguments in regard to this question have been presented (TROITSSKAYA 1952, TIKHOFF 1952).

It is possible to give some rather definite conclusions in regard to the past history of the Martian atmosphere if it is to be consistent with the existence of life on the planet. No *dry* chemistry is known which might imitate the very complex chemistry of living organisms and we can definitely conclude that life approximating that we know on Earth is only possible in the presence of water, and it is very difficult to believe that life could have evolved on Mars without considerable liquid water being present. If water was once present, a much more extensive cloud cover could be expected, the albedo would have been higher, and glaciers would have covered both poles continuously. The glaciers must have flowed toward the equator and left extensive glacial drift and it can be expected that substantial amounts of undersurface water is now present in such deposits. Our present knowledge of meteorology is insufficient to conclude that unfrozen oceans in the tropics could not exist. Quite independently of the possibilities of life, it is interesting to consider the possibility of such primitive oceans.

If water once existed in great quantities, it must have been lost in some way, for no great quantities are present now. From the discussion on the escape of oxygen above, it is evident that the crucial questions are the mechanisms of transport of oxygen and hydrogen to the upper atmosphere and whether the escape layer is in fact at a high temperature. If this latter condition did not obtain in the past, only reaction of oxygen with the ferrous iron of the surface rocks could dispose of the oxygen while hydrogen escaped. This reaction probably produced the red coloration of the surface rocks as has often been suggested, but extensive reaction of a gas and solid does not occur readily because of the lack of contact between the gas and the interior of solid particles and not much loss of oxygen by this process can be postulated.

A planet covered by ice would present little surface for the reaction of carbon dioxide with silicate rocks and hence a high concentration of carbon dioxide could be expected on such a planet. In the case of Mars this would result in a greatly increased loss of atomic oxygen, since this would increase the fraction of atomic oxygen in the high atmosphere. Such loss would result in the production of hydrogen by the reaction of carbon monoxide and water and the hydrogen would be lost. This net loss of water would continue until erosion became effective and carbon dioxide reacted to form calcium and magnesium carbonate. The rate of loss of water during this initial period could be very high and as the planet became arid the present situation could develop. Also, the source of the present water may be the glacial debris expected in the equatorial regions on the basis of this postulated history. It can confidently be said that the presence of an oxygen atmosphere at any time is not consistent with the destruction of extensive oceans by the escape of oxygen from the planet since oxygen would move to the high atmosphere, would be photochemically decomposed and hence would escape rapidly, and the total pressure of oxygen would be low. The very low estimate of the maximum amount of oxygen on Mars by ADAMS and DUNHAM (1934) is in accord with this conclusion.

The possibility that Mars once had oceans cannot be excluded. The presence of oceans and glaciers would account partly for the absence of mountains on the planet, for they would have eroded them to low levels. Still, plutonic activity must have been low at least since the disappearance of the water for no high mountains are reported.

V. The atmospheres of Mercury and the Moon.

23. Observational data. The albedo of Mercury and the Moon are very nearly the same, namely, 0.063 and 0.073 respectively, as observed by DANJON (1948). He also finds their color indices to be nearly the same, i.e. +1.00 and +1.10 respectively. No extension of the horns of mercury near inferior conjunction have ever been observed nor is there a luminous ring around the planet at this time. Also, no evidence for the refraction of light at the limb of either the Moon or Mercury has been observed (LYOT 1937, 1938). DOLLFUS (1950, 1952) from polarization measurements concludes that the surface density of the lunar atmosphere is not greater than 10^{-9} that of the Earth's atmosphere, and that that of Mercury is about 3×10^{-3} of that of the Earth.

ANTONIADI (1934) observed a transient haze on Mercury. This haze is reported as being more intense than that on Mars though he notes that it is more difficult to observe at the center of the disk than at the limb, probably due to the greater thickness in the line of sight. It is often sufficiently dense to obscure the dark markings and may appear or disappear within one day, or may remain for many days. These observations indicate that some very tenuous atmosphere exists. The descriptions immediately remind one of the auroral displays on Earth and suggest the presence of fluorescing and absorbing gases such as those present in the terrestrial atmosphere, suggested by UREY and BREWER (1957). PETTIT and NICHOLSON (1936) have determined the temperature of the mid-solar point of Mercury to be 610° K from the infra-red radiations. No observation of the temperature of the dark hemisphere has ever been made.

PETTIT and NICHOLSON (1930) found that the temperature of the mid-solar point of the full Moon to be 134° C. The temperature at the limb was found to be 67° C. However at the greater phase the temperature observed at the subsolar point was found to be 81° C. Thus no true temperature exists. We use 100° C for purposes of the approximate calculations given below.

24. The escape of gases from Mercury and the Moon. Xenon has the highest molecular weight of any likely atmospheric substance and it will remain on Mercury and the Moon more readily than any other. There are 1.8×10^{18} atoms of xenon per cm² of the Earth's surface and this is considerably greater than the number of molecules above the escape layer. For purposes of discussion we assume this number of atoms of xenon to be present on Mercury. The scale height for the mid-solar point temperature and for xenon ($\mu = 131.3$) is 1.1×10^6 cm. It is evident that the escape layer would be 200 to 300 kilometers or more above the surface and that it would be at some high temperature. For discussion purposes we assume 5000° K, that is, somewhat higher than that of the Earth. For this temperature the escape formula indicates a half time of escape of 0.5×10^9 years. Under these conditions xenon would have escaped to a large extent during geologic time as would all other atmospheric constituents. The escape layer would be lowered as the atmosphere was lost until the escape layer came sufficiently near to the surface so that the high temperature could not be maintained and the rate of loss would become very low. A residual atmosphere would remain, consisting of more atoms per cm² than there are above the escape layer, which we have taken as 3.2×10^{14} cm^{-2}. The atmosphere would be very tenuous indeed.

Because of the possibility that mercury is immersed in a very tenuous gas at high temperature, it is necessary to consider the equilibrium gas concentrations at the planetary surface and in the surrounding space. UREY and BREWER (1957) assume that the heat conductivity of the surrounding gas is proportional to $T^{\frac{1}{2}}$ and that the flow of heat across each spherical surface is constant and thus that $4\pi r^2 \sqrt{T \frac{dT}{dr}}$ is constant and proportional to the total flow of heat. They derive the following equation for the density of a gas at the surface of the planet,

$$\ln \frac{\varrho_0}{\varrho_\infty} = \ln \frac{T_\infty}{T_0} + \frac{3 G M \mu_i}{R T_0 a} \frac{(T_\infty/T_0)^{\frac{1}{2}} - 1}{(T_\infty/T_0)^{\frac{3}{2}} - 1} . \tag{24.1}$$

Substituting numerical values, we find that this function has a flat minimum value of 200 for ϱ_0/ϱ_∞ near T_∞/T_0 equal to 64 for atomic oxygen and that it varies only by a factor ~ 10 between T_∞/T_0 equal to 10 and 1000. If there are x atoms of oxygen per cm³ in the neighborhood of Mercury there will be more than $200\,x$ atoms cm^{-3} of atomic oxygen at the surface at the mid-solar point. The scale height is 9×10^7 and hence the atmosphere may be said to contain $1.8 \times 10^{10}\,x$ atoms. At the surface atoms will combine to give simple molecules, e.g. CO, O_2, N_2, etc., and because of the higher molecular weights their concentrations will be somewhat higher than calculated for oxygen atoms. Also higher concentrations will be expected on the dark side. Whether condensation would occur here is difficult to say because of the probability that the surface would be continuously bombarded by high velocity particles, at least to some extent. Also, the chemical situation cannot be estimated. In spite of the very slight concentration of hydrogen expected, some hydrogen compounds of oxygen, carbon and nitrogen may be present. A flow of gas to the dark surface, thence to the illuminated surface and out to space, can be expected.

It is evident that a very tenuous atmosphere can exist and that fluorescence of the neutral and ionized molecules of this atmosphere could produce the haze observed on the planet provided x, the number of atoms of C, N, and O per cm^3, is some reasonable number, say, ~ 100. It is clear that molecules of C, N, O, and H and their ions may be present in the atmosphere of Mercury. We conclude that the haze of Mercury is most probably due to the fluorescence of CO^+ and N_2^+ and other molecules and is thus similar to the terrestrial aurora, the radiation from the dark side of Venus and the blue haze of Mars.

The value of the escape constants, $\frac{GM\mu}{RTa}$, for the Moon is less than that for Mercury by a factor of about 2 and the application of all escape calculations show that it will lose atmosphere more readily than Mercury. The equilibrium concentration of atomic oxygen and other molecules will be much less and the concentration of atoms in the neighborhood of the Moon is probably less than that near Mercury. (A reasonable numerical calculation is instructive. For molecular oxygen, with $T_0 = 373°$ K and $T_\infty/T_0 = 10$, the value of ϱ_0/ϱ_∞ is 5000 and since not more than 1 atom of oxygen per cm^3 in the neighborhood of the Moon can be expected, the concentration of molecules with molecular weight near 32 at the Moon's surface must be negligible.) Since the haze of Mercury is just observable, it is not surprising that none can be seen on the Moon. The lack of fluorescence on the Moon places the limit of the number of molecules capable of fluorescence per cm^2 at less than about 10^{13} or 10^{14}. With a scale height of 10^7 cm, we find that only some 10^6 or 10^7 of fluorescing molecules per cm^3 may be present at the lunar surface and hence the pressure will be less than 10^{-13} or 10^{-12} atmosphere. This argument does not exclude the possibility that a tenuous atmosphere of high molecular weight substances, such as Kr and Xe, exists.

VI. The atmospheres of the major planets and their satellites.

25. General introduction. The mean densities of the major planets and their masses show that their mean chemical compositions are not identical. Jupiter and Saturn may have approximately the same composition and this may be true for Uranus and Neptune, but it can be concluded that Jupiter and Saturn differ markedly in composition from Uranus and Neptune. DE MARCUS (this volume, p. 432 seq.) reviews the evidence for the structure of these planets. The extensive studies on Jupiter indicate that it has a composition near that of the Sun with respect to the ratio of hydrogen to the other elements. In particular it should be emphasized that there is no likely process by which hydrogen, helium and neon could become separated from each other since they are so very volatile. Since DE MARCUS' discussion indicates little deviation from solar composition in the case of Jupiter, it can be assumed reasonably that methane, which is very volatile, also has not been fractionated relative to these gases. We cannot be so certain in regard to water, ammonia and the non-volatile fraction, i.e., silicates and other phases. As indicated above, it is reasonable to assume the same composition for Saturn as the best working hypothesis at present.

Uranus and Neptune must have lost large amounts of their hydrogen, helium and neon. Only in this way can we understand their high densities. Without an exact study of their structure no reliable, or even approximate, estimate of their composition can be made. The presence of absorption bands in their atmosphere, tentatively assigned to HNO, indicates a very high state of oxidation relative to solar matter, as will be discussed later. This is consistent with a considerable loss of hydrogen, helium and neon during the formation of these planets. The intense bands of methane observed indicate little loss of methane or the preservation of carbon in the form of non-volatile compounds during their formation. Thus there is evidence for the loss of gases during the formation process in the case of these two planets, similar to that which occurred when the terrestrial planets were formed.

The spectra of these four planets contain many absorption bands. The first important photograph of these bands is due to V.M. SLIPHER (1905, 1906, 1909, 1929, 1933). WILDT (1931, 1932) pointed out that certain of these bands fell in the region of the ammonia and methane bands, as observed in the laboratory, thus discovering the presence of these substances in the atmospheres of these planets. DUNHAM (1933) photographed the spectra of Jupiter and Saturn and made exact comparisons with laboratory spectra. He estimated the amount of ammonia in Jupiter's atmosphere as 5 to 10 meter atmospheres, and that in the Saturn atmosphere as 2 meter atmospheres. The investigation of the spectra has been continued by ADEL (1937, 1938, 1947), ADEL and LAMPLAND (1941), ADEL and SLIPHER (1934) and KUIPER (1947). The results of the last author have been reported as tentative tables without the discussion of the details of the measurements, but they appear to be more extensive than other published data (KUIPER 1952). No indication of correction for possible pressure and temperature effects is given, and reported data must be regarded as very approximate.

26. The atmosphere of Jupiter. KUIPER gives 15000 and 700 cm atmospheres (N.T.P.) as the amounts of methane and ammonia respectively, above the cloud surface. No other gaseous constituents have been reported. If the solar ratio of hydrogen, helium and carbon (SUESS and UREY, Vol. LI of this Encyclopedia) are assumed, the amounts of hydrogen (H_2) and helium in the atmosphere are 2.7×10^7 and 5.6×10^6 cm atmospheres respectively. Neon and nitrogen must be present also and in solar proportions if our assumptions are to be consistent. At the low temperatures prevailing, ammonia is stable with respect to the elements and will be formed at least at a slow rate in the high atmosphere under the excitation conditions existing there. Consistent with these assumptions the mean molecular weight is 2.35 and the pressure at the cloud layer is 8.1 bars. This is a very tentative pressure because KUIPER's data and the cosmic abundance ratios are approximate and because the assumption of no fractionation during planetary formation is uncertain. (Relative to pressure induced vibrational absorption by H_2, see the discussion under the atmosphere of Uranus below.)

MENZEL, COBLENTZ and LAMPLAND (1926) reported two temperatures for Jupiter, i.e., 120 and 140° K, from the intensity of radiation in the 8 to 14 μ region. Jupiter rotates rapidly and has a massive atmosphere and hence must radiate from its entire surface. The mean radiation temperature expected in this case for a body of albedo 0.47 is 105° K and a maximum of 148° K for the mid-solar point. This measurement was approximate and cannot be used with confidence. Since ammonia at low temperatures can radiate only in the region of 100 to 200 cm^{-1} and near 1000 cm^{-1}, while the maximum of the black body radiator at 105° K is at 370 cm^{-1}, and since methane radiates only at higher frequencies than ammonia, no very effective infra-red radiator is present in the high atmosphere, except that unknown rare molecules may be present. Therefore, it is likely that the radiation from this planet originates at deeper layers and is only partly absorbed by overlying colder layers[1].

HESS (1952, 1953) has found that the amounts of ammonia and methane above the clouds of Jupiter decrease with increasing latitude as determined by the absorption of the 6190 Å methane band and the 6474 Å ammonia band. He finds that the height of the cloud layer increases by 10 km between the equator and 60° latitude. Using 2.35 for the mean molecular weight alters the change in height to 17 km. On the assumption that the clouds are solid ammonia, and using the vapor pressure curve for ammonia, he concludes that the temperature at the cloud level decreases by 6° K between 0° and 60° latitude. He also finds that the morning limb at 60° longitude is 7.6 to 9.7° K lower and that the evening limb at 60° longitude is 3.9 to 4.8° K lower in temperature than the central meridian, all at the equator. Hess assumes a well-mixed atmosphere which means that the tropopause is at the cloud level, for otherwise the partial pressure of ammonia would change much more rapidly with height. The partial pressure of ammonia at the cloud level in this case is $A_{NH_3} P/\Sigma A_i$, where the A's are the cm atmospheres of the constituents. This gives 1.7×10^{-4} atmospheres of NH_3, which is the vapor pressure at 150° K. KUIPER (1952) assumes an adiabatic atmosphere saturated with ammonia at the cloud level and above, and estimates the temperature of the cloud layer as 165° K. It is difficult to understand the large changes in temperature deduced by HESS because of the great heat capacity of the atmosphere. The upper part of the cloud level probably consists of a tenuous fog. As a result, the thickness through which the light penetrates is limited by this circumstance and HESS' observations may indicate a lower penetration nearer the limbs of the planet and thus he does not demonstrate a difference in temperature or cloud level.

DOLLFUS (1957) has been unable to confirm HESS' results and finds that his polarization measurements can be explained by a tenuous fog. He also reports that the polar regions appear to be free of clouds in that the light scattered is polarized as expected for a pure gas. This indicates that the atmosphere is sinking at the poles and again we find evidence for a Brewer current (p. 372). In this atmosphere ammonia probably "snows out" in the low latitudes.

SHAPIRO (1953) has shown that the brightness of Jupiter is related to the mean annual and monthly relative sunspot numbers and has suggested that this is due to some secondary effect of the increased ultraviolet light of the Sun during high sunspot activity. UREY and BREWER (1957) point out that ultraviolet light of the Sun should produce ions and free radicals, e.g. NH_2, N_2^+, CN, CH_2, CH^+, CH, and C_2, and that these undoubtedly fluoresce in the visible light of the Sun and that variations in the number of free radicals may be produced during the Sun spot cycle and thus account for the effects observed. We have no evidence in regard to the temperatures in the high atmosphere.

The atmosphere of Jupiter exhibits many colors from time to time. These range through various shades of yellow and red and even distinctly blue regions are observed. It has been suggested that metallic sodium is dissolved in liquid ammonia to produce the blue color. In view of the well-known instability of

[1] Prof. G. HERZBERG has called the author's attention to the ν_4 to ν_2 transition in methane at 230 cm^{-1}. The excitation of the ν_4 level is at 1306 cm^{-1} and thus very few molecules can be effected in this level. However, the layer of methane is very thick and some radiation way arise in this way.

such solutions and that metallic sodium is produced only by the activity of intelligent living organisms, this is a most improbable suggestion. WILDT (1937) gave a brilliant discussion of the role expected for the free radicals produced from methane and ammonia in the atmosphere of the major planets. These free radicals, e.g. CH_3, CH_2, NH_2, and NH, molecules derived from them such as C_2H_6, C_2H_4, N_2H_4 and free radicals of these in turn must be produced in the high atmosphere and must react with molecular and atomic hydrogen. He concludes that predominantly methane, ammonia and molecular hydrogen must be present. RICE (1955) has suggested that the free radical, NH, is formed in the high atmosphere. This condenses at low temperatures to produce a blue solid. He also suggested that condensed NH_2NH may account for the yellow colors. However, the transport of a free radical through a very thick atmosphere to the cloud surface presents very great difficulties. Though his suggestions may not be completely acceptable, he has called attention in this way to the importance of many possible compounds of carbon, nitrogen and hydrogen which could be produced in the high atmosphere from NH_3 and CH_4 and their free radicals. It is not possible to predict the results of the chemical reactions of the high atmosphere of Jupiter, but compounds of this kind appear to be possible and, in fact, probable[1].

The great Red Spot to the south of and near the equator is the only permanent feature of the surface of Jupiter. It has been observed for some 300 years. It is not rigidly attached since it has been observed to oscillate by 850° with respect to its mean period (PEEK 1939). It is usually suggested that it is due to some solid island floating on liquid, but no plausible suggestion has ever been put forward for the substances composing the solid and liquid. At present no satisfactory explanation can be given. It should be noted that the amout of radioactive heat generated is very small. If Jupiter has the mean solar composition, the amount of silicate materials present is only equivalent to one Earth mass. The Earth is now generating about 50 ergs cm^{-2} and thus only 0.4 erg cm^{-2} would be expected from Jupiter. Of course, the assumption on which this is based is arbitrary and the amount of silicate and hence of radioactive substances may be larger, say, by an order of magnitude. The radioactive substances must be deep in the interior of Jupiter and they may produce a convection at some point in order to dissipate this heat. The red spot is probably associated with some deep-seated activity of this kind. If so, the red color is probable due to compounds of carbon with other elements which cannot be predicted without far more data than are available.

A marked feature of the atmosphere is the presence of latitudinal cloud bands which change their appearance markedly even during periods of a few days. The light bands are clouds and indicate regions of rising atmosphere. The dark bands mark regions of descending currents and "clear skies". The edges of these bands are often irregular and sometimes quite regularly crenulated. Also dark and light spots are a common feature. The dark spots are generally regarded as regions of descending currents, and hence clear skies and the light spots as regions of rising currents and cloudy skies. In fact these are regions of cyclonic and anticyclonic storms (SHAPIRO 1952). These features indicate that the atmosphere is stormy and that it has an adiabatic temperature gradient for some region below the surface of the clouds. This gradient for the composition assumed for

[1] A few compounds selected at random are: diazomethane, CH_2N_2, yellow; tetrazine, $C_2H_2N_4$, red; azomethane, $(CH_3)_2N_2$, yellow; cuprene, $(C_2H_2)_\chi$, yellow. If aromatic derivations are assumed, a very great variety of colored compounds become possible. Cuprene is perhaps the most probable substance.

the atmosphere is 2.44 deg/km. At some level below the clouds sunlight is completely absorbed, an effective warm surface is established and the thermodynamic conditions for mechanical motion of the atmosphere exist. We do not know how deep this level is below the visible surface.

Jupiter does not rotate with uniform velocity, but has an equatorial band about 15° in width which rotates eastward with a velocity of 100 m/sec greater than the velocity of neighboring parts of the planet, while the remainder of the surface rotates as a solid body. SCHOENBERG (1949) has presented a tentative explanation of this phenomenon.

27. The atmosphere of Saturn. The problems of the atmosphere of Saturn are similar to those of Jupiter and only a brief summary of the Saturn atmosphere is given here.

KUIPER (1952) estimates methane as 35000 cm atmospheres above the cloud layer and DUNHAM (1952) estimates ammonia as 200 cm atmospheres. No other gas has been identified. Assuming solar composition as in the case of Jupiter, we secure 6.3×10^7 and 1.3×10^7 cm atmospheres of H_2 and He respectively. Also neon and molecular nitrogen are expected to be important constituents and ammonia should be synthesized to some extent in the high atmosphere. The pressure at the cloud level is then 8.3 bars. (Relative to pressure induced vibrational absorption by H_2, see the discussion under the atmosphere of Uranus below.)

MENZEL, COBLENTZ and LAMPLAND (1926) report a temperature of 128° K from measurements in the 8 to 14 μ region. This temperature is very closely the same as that observed for Jupiter, and is much higher than expected for Saturn, namely, 78° K for a planet radiating over its entire surface and 110° K for a black body at the mid-solar point. Just as in the case of Jupiter, the radiation must originate to some extent in the deep layers of the atmosphere and be partly absorbed by the overlying colder layers. HESS (1952) finds a nearly uniform distribution of methane over the surface, while DOLLFUS (1957) observes an increased thickness of methane near the limbs of the planet.

If the atmosphere above the clouds is a mixed atmosphere, and if the cloud layer is at the tropopause, we can calculate the temperature of the cloud layer. The pressure of ammonia is

$$p_{NH_3} = \frac{A_{NH_3}}{\sum A_i} P$$

as mentioned above. Using the above values, we secure 3.4×10^{-5} atmospheres, which is the vapor pressure of ammonia at 140° K. KUIPER (1950) assumes an adiabatic atmosphere with saturated ammonia vapor above the clouds and secures 152° K. These estimates are very doubtful because of the uncertain value for the ammonia abundance and because of our lack of knowledge in regard to the admixture of the atmosphere.

SATURN'S atmosphere has weather bands of less distinct character than those of Jupiter and thus they indicate a less energetic atmosphere, as would be expected. The colors of its atmospheres are less marked than those of Jupiter (WRIGHT 1927). They are doubtless due to similar but equally unknown compounds of carbon, hydrogen and nitrogen. White spots are occasionally observed (DANJON 1946; WRIGHT 1933; PEEK 1934; E. C. SLIPHER 1938), and are probably due to rising currents of gas. Dark spots are not reported.

28. The atmospheres of Uranus and Neptune. As mentioned above, the higher densities of these planets, as compared to those of Jupiter and Saturn and their lower masses, require that their composition must be different and that they contain a much smaller proportion of hydrogen and helium, and no reasonable assumption in regard to the proportions of these two gases relative to others can be made. DOLLFUS and CAMICHEL (1953) report very faint markings on the disk of Neptune.

KUIPER (1952) gives 220000 and 370000 cm atmospheres of methane on Uranus and Neptune respectively. Ammonia has not been observed. KUIPER (1949) observed a diffuse band at 8270 Å and a series of diffuse lines near 7500 Å. HERZBERG (1952) has identified the diffuse bands as due to the $(v', v'') = (3, 0)$ pressure-induced rotation vibration band of the H_2 molecule. He estimates that there are 4×10^6 cm atmospheres of H_2 and three times as much He in the atmosphere. He found that the double transition (2, 0) (1, 0) was more intense relative to the (3, 0) transition in pure hydrogen as compared to the planetary spectrum and attributed this to an excess of helium in the atmosphere. Since it is difficult to understand how helium could have been concentrated markedly relative to hydrogen, because both are very volatile, nitrogen may be a more likely second gas. According to CHISHOLM and WELSH (1954) nitrogen is more effective than helium in inducing the fundamental band of hydrogen by a factor of 1.85 at 200° K. By a comparison of HERZBERG'S observations with an atmosphere containing equal amounts of hydrogen and nitrogen, we find that his observations are consistent with an atmospheric pressure of 6 bars at the cloud level and 4.2×10^6 cm atmospheres N.T.P. of H_2 and N_2[1]. Since diffusive separation should occur, the total amounts of hydrogen relative to both helium and nitrogen in the whole atmosphere should be larger than estimated. The induced transitions may be produced by methane, but calculations cannot be made without observational data. (It should be noted that the pressure required for producing these transitions

* The value of α for the integrated absorption defined by the equation, $\frac{d \ln I}{dx} = -\alpha \varrho$, where ϱ is the density of hydrogen, is proportional to the density of the perturbing gas. Thus the laboratory observations give

$$\ln \frac{I_0}{I} = K x \frac{p^2}{T^2} (1 + a_{\mathrm{He}} R_{\mathrm{He}}),$$

where p is the pressure of hydrogen, R_{He} is the ratio of He to H_2 and a_{He} is the ratio of the coefficients of helium and hydrogen for inducing the transition. For the fundamental at 298° K, $a_{\mathrm{He}} = 0.6$. The constant K increases with the temperature approximately as the first power. For the planetary atmosphere

$$\ln \frac{I_0}{I} = K \frac{H}{2} \frac{p_0^2}{T^2} (1 + aR) \overline{\left(\frac{1}{\cos\vartheta}\right)} \times 2.$$

For the helium hydrogen atmosphere, assuming the same value of $\ln \frac{I_0}{I}$ for the laboratory and the planetary atmosphere gives $x p^2 = H p_0^2 \overline{\left(\frac{1}{\cos\vartheta}\right)}$. The mean of $\frac{1}{\cos\vartheta}$ over the disk is 2, and $H = 1.73 \times 10^6$ for 78° K, $g = 1070$ and a mean molecular weight of 3.5, i.e. $R = 3$. Using $x = 1.2 \times 10^4$ and $p = 0.25 \times 10^8$ dynes cm^{-2} from HERZBERG'S experiment gives $p_0 = 1.5$ atm and the amount of H_2 is 9×10^6 cm atm N.T.P. The amounts of helium are three times as great (HERZBERG'S values are $\frac{1}{3}$ greater).

If nitrogen is assumed, the value of a_{N_2} is about 1.85 and we secure

$$x p^2 (1 + a_{\mathrm{He}} R_{\mathrm{He}}) = 2 H' p_0^2 \frac{(1 + a_{N_2} R_{N_2})(1 + R_{N_2})}{1 + 14 R_{N_2}},$$

where H' is the scale height for an atmosphere of pure H_2. Without further information we cannot solve for p_0. With $R_{N_2} = 1$, p_0 is 3 bars and this is the pressure of both hydrogen and nitrogen. Throughout we use HERZBERG'S estimate of 78° K for the temperature and at this temperature the vapor pressure of N_2 is 1 atm and nitrogen would condense. We do not know the temperature of the cloud level of Uranus or Neptune and they may be higher than assumed. All results are uncertain because of the approximate character of the observations, the assumption that the ratios of the a's for the (3, 0) and (2, 1) (1, 0) transitions are the same as those for the fundamental, etc.

in the case of Jupiter would be about twice that calculated here because of the higher temperatures expected to prevail at the cloud level.)

At 100° K the vapor pressure of NH_3 is 3×10^{-10} atm and this pressure is evidently too low to supply effective clouds. The clouds of Uranus and Neptune must be either methane or nitrogen. From the amounts of the gases and pressures estimated above, it is possible to calculate the partial pressure of methane in the Uranus atmosphere, namely $(A_{CH_4}/\Sigma A_i)\ P$, where the A's are the abundances. We secure 0.037 and 0.16 atm for the two atmospheres considered. The vapor pressure of methane is equal to these pressures at 84 and 93° K, respectively. At the latter temperature the vapor pressure of nitrogen is 6 atm, and in the case of this model atmosphere the clouds may be either liquid nitrogen or solid methane. It is obvious that the uncertainties of the observed and calculated quantities make a decision in regard to this question impossible.

Similar calculations can be made for Neptune, but the data are still more uncertain and the numerical results are of little interest. The greater amount of methane above the clouds would indicate still higher temperatures at the cloud level. Since the calculated temperature for these planets, assuming radiation from the entire surface, are only 55 and 43° K for Uranus and Saturn respectively, it is evident that the cloud levels are not the effective radiating surfaces. Since H_2, He, N_2, Ne and CH_4 all do not radiate in the far infrared at low pressures, it is evident that high pressures must exist at the radiating layer in order that pressure-induced transitions shall be possible. Hence high pressures at the cloud level and above are required. Also some low temperature levels above the cloud levels must exist which are the effective radiating levels.

KUIPER (1949a) has observed bands in the spectrum of Uranus and Neptune, which cannot be due to methane, and he found no other absorber of these bands. The observed Uranus bands are at 7500, 7524 and 7471 Å. In the Neptune spectrum he found these and another at 7546 Å and a doubtful one at 7570 Å. RAMSEY (1953) observed bands near these in laboratory spectra which were thought to be due to HCO, and Herzberg tentatively proposed that the bands in Neptune and Uranus are due to this molecule. Exact agreement was not secured. More recently, RAMSEY (1957) has found that the laboratory bands are due to HNO. If the planetary bands are due to this molecule, a most surprising chemical composition is indicated since all oxides of nitrogen and all oxygen-hydrogen-nitrogen compounds are thermodynamically unstable with respect to the elements or elements and water, and hence it is very difficult to account for their presence in these atmospheres. However, only some 10^{15} or 10^{16} molecules per cm^2 are required in order to secure absorption and hence only a very small fraction of the atmosphere need consist of the absorbing molecule, if the transition is an electronic one.

29. The atmosphere of the Satellites and Pluto. The four Galilean satellites of Jupiter, Titan of Saturn and Triton of Neptune have masses comparable to the terrestrial moon and might retain atmospheres. Of these, only Titan has a detectable atmosphere of 20000 cm atm of methane (KUIPER 1944). (This datum has not been corrected for temperature and pressure effects.) All these satellites are slightly yellow in color. The presence of methane on Titan supports the view that all primitive atmospheres were highly reduced.

It is of interest to explore the reasons for the presence of methane on Titan and the absence of atmospheres on the other satellites. It is evident that both methane and nitrogen and all other gases than H_2, He and Ne would be condensed on Triton and Pluto, and these gases would not show their presence either by

Table 11.

Planet	Substance	Detected	Amount cm atm (NTP)	Basis of estimate	Remarks
Mercury					Probably fluorescing free radicals and ions produce haze
Venus	CO_2	yes	$\mathbf{10^5}$	Spectroscopic	Much below the cloud layer
	H_2O	yes	oceans	Polarization of clouds	Vast oceans are assumed to exist below the clouds
	N_2	yes	?	Spectroscopic	N_2 and N_2^+ bands in the night sky
	CO	yes	$\mathbf{< 100}$	Spectroscopic	CO^+ bands in night sky and absorption bands in day sky. Limit fixed by near infra-red bands
Earth	Many constituents, see Tables 2 and 3				
Mars	CO_2	yes	**3600**	Spectroscopic	
	H_2O	yes	?	Polarization of clouds	Polar caps consist of ice. Some clouds have the polarization of ice crystals
	N_2	no	$\mathbf{1.8 \times 10^5}$	Total pressure measurement	N_2 is accepted as the most likely non-condensable constituent
Jupiter	CH_4	yes	$\mathbf{1.5 \times 10^4}$	Spectroscopic	
	NH_3	yes	**700**	Spectroscopic	
	H_2	no	2.7×10^7	Density of the planet	Assumed to be present in solar proportions relative to methane on the basis of the Marcus' calculations
	He	no	5.6×10^6		
	N_2	no	4×10^3		
	Ne	no	1.7×10^4		
Saturn	CH_4	yes	**35000**	Spectroscopic	
	NH_3	yes	**200**	Spectroscopic	
	H_2	no	6.3×10^7	Density of planet	Assumed to be present in solar proportions relative to methane
	He	no	1.3×10^7		
	N_2	no	9.5×10^3		
	Ne	no	2.7×10^4		
Uranus[1]	CH_4	yes	$\mathbf{2.2 \times 10^5}$	Spectroscopic	He and H_2 are assumed to be effective molecules in producing transitions of H_2
	H_2	yes	9×10^6	Calculated on HERZBERG'S assumption	
	He	no	2.7×10^7		
	H_2	yes	4.2×10^6	Calculated on assumptions made in this paper	N_2 and H_2 are assumed to be effective molecules in producing transitions of H_2. Solar proportions of He and H_2 are assumed
	He	no	8.6×10^5		
	N_2	no	4.2×10^6		
Neptune[1]	CH_4	yes	$\mathbf{3.7 \times 10^5}$	Spectroscopic	
	H_2	yes	Larger than in uranus	Spectroscopic	Assumed to be effective molecules in producing transitions of H_2
	N_2	no			
	He	no			Solar proportions of H_2 and He assumed
Titan	CH_4	yes	$\mathbf{2 \times 10^4}$	Spectroscopic	
	He	no	—		Assumed high diffusion layer for the preservation of H_2

KUIPER (1952) has been unable to detect N_2O, CH_4, C_2H_4, C_2H_6 and NH_3 on Venus; SO_2, O_3, N_2O, CH_4, C_2H_4, C_2H_6 and NH_3 on Mars; O_3 and SO_2 on Saturn and Uranus, CH_4 and NH_3 on the satellites of Jupiter, and NH_3 on Titan. In most cases, the substances could not be expected to be present because of thermodynamic instability under the known conditions of the planet involved.

[1] Methane has not been considered as the molecule inducing the hydrogen transitions, but it should contribute to this effect.

absorption or fluorescence. We therefore discuss the escape of atmospheres from the most massive satellite of Jupiter, i.e. J III or Ganymede, and from Titan, with 0.0260 and 0.0236 Earth masses and 0.435 and 0.384 Earth radii respectively.

Except for their distance from the Sun, these two objects are very similar. The loss of atmosphere from the one and not from the other may be explained in a number of ways. If the temperatures of the high atmospheres were comparable to those existing in the terrestrial atmosphere, the absence of atmospheres would rest on arguments similar to those applied to the Moon. We shall not assume such temperatures. Then, the loss of methane and ammonia is due to the loss of hydrogen and the condensation of non-volatile compounds of carbon, nitrogen, oxygen and hydrogen. The escape problem is then limited to the escape of hydrogen and helium. A pure hydrogen atmosphere of one gram molecule per cm^3, i.e. 22400 cm atm N.T.P., would be lost from Ganymede or Titan in about 10^7 years, if the escape temperatures are 156 and 117 respectively. Lower temperatures by a factor of $\sqrt{2}$ would preserve hydrogen by a significant degree in the case of Titan. The amounts of radiation in the Schumann region are quite adequate to decompose the methane of Titan during 4.5×10^9 years. Hence recombination of the products of photodecomposition must occur and this requires the presence of hydrogen. As was evident from the study of the escape problem for the terrestrial planets, the presence of the diffusion layer will greatly decrease the rate of loss and this could be supplied by a high level helium atmosphere. Helium will escape quite rapidly from Ganymede, i.e. 2×10^5 cm atm in 4.5×10^5 years at 156°, while the escape from Titan would be only $\frac{1}{30}$ as rapid. If some particle radiation of high energy penetrates occasionally to the region of Jupiter, as indicated by the occasional flaring of the Schassmann-Wachmann Comet, the loss of helium from Ganymede would be greatly accelerated. Hence, the presence of methane on Titan is possibly due to a high-helium atmosphere, which retains hydrogen by the Harteck-Jensen diffusion mechanism.

No atmospheric constituent has been observed on Pluto, which is probably an escaped satellite of Neptune, as suggested by LYTTLETON (1936).

The colors of these satellites are probably due to various compounds of carbon, nitrogen, oxygen and hydrogen. It will be impossible to specify the likely compounds producing the slightly yellowish color and the low albedos of these objects in the foreseeable future.

VII. Conclusions.

Table 11 summarizes the constituents of the planetary atmospheres as discussed in this chapter. In this table are included constituents definitely proven to be present and those inferred as being present on various grounds, with the reasons given briefly in the last column. All amounts given in column 4 are very uncertain. Those for which at least semi-quantitative data are available are recorded in bold face type. It is to be hoped that these will be improved in the future. Especially needed are laboratory data on the absorption bands at pressures and temperatures and in mixtures that may exist in planetary atmospheres. If such observations are to be made effectively, students of the subject must have some estimates of what these conditions may be. It cannot be expected that the conclusions reached here are more than tentative, and it is hoped that the present discussion may suggest valuable lines of effort in the future.

Acknowledgment.

The writer is greatly indebted to Dr. A. W. BREWER for many discussions of meteorological problems and for many constructive criticisms of the manuscript.

References.

ADAMS, W. S., and T. DUNHAM jr.: Publ. Astronom. Soc. Pacific **44**, 243 (1932).
— — Astrophys. Journ. **79**, 308 (1934).
—, and C. E. ST. JOHN: Astrophys. Journ. **63**, 133 (1926).
ADEL, A.: Astrophys. Journ. **86**, 337 (1937a); **85**, 345 (1937b); **87**, 198 (1938); **93**, 397 (1941).
—, and C. O. LAMPLAND: Astrophys. Journ. **93**, 391 (1941).
—, and V. M. SLIPHER: Phys. Rev. **46**, 240 (1934).
ANTONIADI, E. M.: La Planète Mars. Paris: Hermann & Cie. 1930.
— La Planète Mercure. Paris: Gauthier-Villars (1934).
ARRHENIUS, S.: Die Atmosphären der Planeten. Ann. d. Naturphil. Lpz. **9**, 22 (1910).
ASTON, F. W.: Nature, Lond. **94**, 786 (1924).
BAHR, J. N.: Ind. J. Phys. **12**, 363 (1938).
BALDWIN, R.: The Face of the Moon. Chicago, Ill.: Chicago University Press 1949.
BARABASCHEFF, N., and B. SEMEJKIN: Z. Astrophys. **8**, 44 (1934).
—, and I. TIMOSHENKO: Astronom. J. Soviet Union **17**, 52 (1940).
BATES, D. R.: The Earth as a Planet, ed. by G. P. KUIPER, Chap. 12. Chicago, Ill.: Chicago University Press 1954.
—, and H. S. W. MASSEY: Proc. Roy. Soc. Lond., Ser. A **187**, 261 (1946); **192**, 1 (1947).
—, and M. NICOLET: J. Geophys. Res. **55**, 301 (1950).
—, and AGNES E. WITHERSPOON: Monthly Notices Roy. Astronom. Soc. London **112**, 101 (1952).
BERNAL, J. D.: Proc. Phys. Soc. Lond. A **62**, 537 (1949); B **62**, 597 (1949).
— The Physical Basis of Life. London: Rutledge and Kegan Paul Ltd. 1951.
BHAR, J. N.: Ind. J. Phys. **12**, 363—386 (1938).
BIERMANN, L.: Z. Astrophys. **29**, 274 (1951).
BREWER, A. W.: J. Roy. Meteor. Soc. **75**, 351 (1949).
—, and J. T. HOUGHTON: Proc. Roy. Soc. Lond., Ser. A **236**, 17 (1956).
BROWN, H. S.: Atmospheres of the Earth and Planets, Chap. 9, ed. by G. P. KUIPER. 1st ed. Chicago, Ill.: Chicago University Press 1949.
BRUNT, D.: Physical and Dynamical Meteorology. Cambridge 1944.
— Quart. J. Roy. Met. Soc. **59**, 351 (1933).
BYRAM, E. T., T. A. CHUBB and H. FRIEDMAN: J. Geophys. Res. **61**, 251 (1956).
— — — and J. E. KUPPERIAN jr.: Astrophys. Journ. **124**, 480 (1956).
CAMICHEL, H.: C. R. Acad. Sci. Paris **219**, 21 (1944).
— Bull. Astronomique **18**, 83—191 (1954).
CAMPBELL, W. W., and S. ALBRECHT: Lick Obs. Bull. No. 180, 1910.
CHACKETT, K. F., F. A. PANETH and E. J. WILSON: J. Atmos. and Terres. Phys. **1**, 49 (1950).
CHAMBERLAIN, J. W., and G. P. KUIPER: Astrophys. Journ. **124**, 399 (1956).
CHAPMAN, S.: Proc. Roy. Soc. Lond., Ser. A **132**, 353 (1931).
— J. Geophys. Res. **61**, 350 (1956).
CHISHOLM, D. A., and H. L. WELSH: Canad. J. Phys. **32**, 291 (1954).
COBLENTZ, W. W.: Astronom. Nachr. **224**, 361—378 (1925a).
— Bureau of Standards, Sci. Papers **20**, 371—397 (1925b).
— Popular Astronomy **33**, 310—316, 363—382 (1925c).
— **35**, 145—157 (1927).
—, and C. O. LAMPLAND: Publ. Astronom. Soc. Pacific **36**, 274 (1924); **38**, 355 (1926).
— — J. Franklin Inst. **199**, 785; **200**, 103 (1925).
— — and D. H. MENZEL: Publ. Astronom. Soc. Pacific **39**, 97—100 (1927).
CRAIG, R. A.: Meteor. Monographs **1**, No. 2 (1950).
CRAWFORD, M. F., H. L. WELSH, J. C. F. MACDONALD and J. L. LOCKE: Phys. Rev. **80**, 469 (1950).
CURTIS, A. R., and R. M. GOODY: Proc. Roy. Soc. Lond., Ser. A **236**, 193 (1956).
DANJON, A.: L'Astronomie **57**, 161 (1943).
— L'Astronomie **60**, 45 (1946).
— C. R. Acad. Sci. Paris **227**, 652 (1948).
— The Earth as a Planet, ed. by G. P. KUIPER, Chap. 15. Chicago, Ill.: Chicago University Press 1954.
DOBSON, G. M. B.: Proc. Roy. Soc. Lond., Ser. A **129**, 411 (1930) and references given there.
DOLLFUS, A.: C. R. Acad. Sci. Paris, **227**, 331—333 (1948a); **227**, 383 (1948b); **231**, 1430 (1950); **233**, 467, 1066 (1951); **234**, 2046 (1952); **235**, 1479 (1952); **238**, 1475 (1954).
— La Météorologie **1956**, No. 42, 81—91.
— Mem. Soc. roy. Sci. Liège. Ser. IV **18**, 141—146 (1956).
— Private communication (1957).
—, and H. CAMICHEL: Astronomie **1953**.

DÜTSCH, H. U.: Diss. Zürich 1946.
DUNHAM jr., T.: Publ. Astronom. Soc. Pacific **45**, 42, 202 (1933).
— Atm. of the Earth and Planets, ed. by G. P. KUIPER. Chicago, Ill.: Chicago University Press 1949.
— Atmospheres of the Earth and Planets, ed. by G. P. KUIPER, p. 302. Chicago, Ill.: Chicago University Press 1952.
ELSASSER, W. M.: Harvard Meteor. Studies, Milton, Mass. **6**, 97 (1942).
FESSENKOFF, B. G.: Astronom. J. Soviet Union **21**, 257 (1944).
FESSENKOPF, B.: Astronom. Nachr. **228**, 25—32 (1926).
GEORGE, J. J.: Compendium of Meteorology. Amer. Meteor. Soc. 1950, p. 1180.
GILBERT, G. K.: Bull. Phil. Soc. Washington **12**, 241 (1893).
GLUCKAUF, E.: Compendium of Meteorology, p. 3. Amer. Meteor. Soc., Boston, Mass. 1951.
GOODY, R. M.: Proc. Roy. Soc. Lond., Ser. A **197**, 487 (1949).
— Weather **12**, 3 (1957).
GOWAN, E. H.: Proc. Roy. Soc. Lond., Ser. A **190**, 219, 227 (1947).
GRANDJEAN, J., and R. M. GOODY: Astrophys. Journ. **121**, 548 (1955).
HAMY and MILLOCHAU: Annual Report of the Obs. Paris, pp. 23—24. 1913.
HARTECK, P.: J. Chem. Phys. **23**, 902 (1955).
— and J. H. D. JENSEN: Z. Naturforsch. **3**a, 591 (1948).
HERZBERG, G.: Infrared and Raman Spectra. New York: Van Nostrand Co. 1945.
— Atmospheres of the Earth and Planets, 1st ed., ed. by G. P. KUIPER. Chicago, Ill.: Chicago University Press 1949.
— Astrophys. Journ. **115**, 337 (1952).
— Trans. Roy. Soc. Canada **46**, Sec. III, 1—22 (1952).
—, and L. HERZBERG: Nature, Lond. **161**, 283 (1948).
— — J. Opt. Soc. Amer. **43**, 1037 (1953).
—, and D. A. RAMSEY: Unpublished.
HESS, S. L.: Publ. Astronom. Soc. Pacific **60**, 289 (1948).
— J. Meteorology **7**, 1—13 (1950); cf. also Sky and Telescope **9**, 155—157 (1950).
— Lowell Observatory Report, The Study of Planetary Atmospheres 21. 1952.
— Astrophys. Journ. **118**, 151 (1953).
— Technical Report No. 8, Florida State University, Dept of Meteorology 1957; Astrophys. Journ. **127**, 741 (1958).
HOUGHTON, H. G., and W. H. RADFORD: Pap. Phys. Ocean. Meteor., Mass. Inst. Tech., Woods Hole Ocean. Inst. **6**, No. 4 (1938).
HOWARD, J. N., D. L. BURCH and D. WILLIAMS: Geophys. Res. Paper, No. 40. Geophysics Research Directorate 1955.
HULBURT, E. O.: J. Opt. Soc. Amer. **37**, 405 (1947).
JACOBS, J. A., and D. ALLAN: Nature, Lond. **177**, 155 (1956).
JEANS, J.: Dynamical Theory of Gases, Chap. XV. Cambridge: Cambridge University Press 1916.
JEFFREYS, H.: Nature, Lond. **96**, 934 (1924).
JOHNSON, F. S.: J. Meteorology **11**, 431 (1954).
— J. Geophys. Res. **61**, 71 (1956).
JONES, J. E.: Trans. Cambridge Phil. Soc. **22**, 535 (1923).
KELLEY, K. K.: Cont. to Data on Theor. Metallurgy, Bull. 476, Bur. of Mines, U.S. Dept. of Interior 1949.
KOENE (1856).
KOZYREV, H. A.: Publ. Crimean Astrophys. Obs. **12**, 169 (1954a); **12**, 177 (1954b).
KUIPER, G. P.: Astrophys. Journ. **106**, 251 (1947); **109**, 540 (1949a); **110**, 93 (1949b); **120**, 603 (1954) (Pole of Venus).
— Rep. Progr. Phys. **13**, 247 (1950).
— Publ. Astronom. Soc. Pacific **62**, 133 (1950).
— Atmospheres of the Earth and Planets, ed. by G. P. KUIPER. Chicago, Ill.: Chicago University Press 1952.
— Smithson. Contr. Astrophys. **1**, 89 (1956).
LYOT, B.: Ann. Obs. Paris (Meudon), **8** (1929).
— C. R. Acad. Sci. Paris **205**, 895 (1937).
— L'Astronomie **52**, 45 (1938).
LYTTLETON, R. A.: Monthly Notices Roy. Astronom. Soc. London **97**, 108 (1936).
MACGREGOR, A.: South Afr. J. Sci. **24**, 155 (1927).
MARCUS, W. C. DE: This Handbook.
MCLAUGHLIN, D. B.: Publ. Astronom. Soc. Pacific **66**, 161 (1954).
MAYER, C. H., R. M. STOANAKER and T. P. MCCOLLOUGH: As reported in Sky and Telescope **15**, 435 (1956).

MENZEL, D. H.: Astrophys. Journ. **63**, 48—59 (1926).
— Observatory **70**, 223 (1950).
— W. W. COBLENTZ and C. O. LAMPLAND: Astrophys. Journ. **63**, 177 (1926).
—, and F. L. WHIPPLE: Publ. Astronom. Soc. Pacific **67**, 161 (1955).
MIGEOTTE, M., and L. NEVEN: Physica, Haag **16**, 423 (1950).
MILLER, S.: Science, Lancaster, Pa. **117**, 528 (1953).
— J. Amer. Chem. Soc. **77**, 2351 (1955).
MITRA, S. K.: The Upper Atmosphere. Asiatic Society, Calcutta 1952.
— J. N. BHAR and S. P. GHOSH: Ind. J. Phys. **12**, 455—465 (1938).
MOHLER, F. L.: J. Res. Nat. Bur. Stand **25**, 507—518 (1940).
NICOLET, M.: J. Geophys. Res. **54**, 373—381 (1949).
— Physics and Medicine of the Upper Atmosphere. Albuquerque N. M.: University of New Mexico Press 1952.
— The Earth as a Planet, ed. by G. P. KUIPER. Chap. 13. Chicago, Ill.: Chicago University Press 1954.
— Ann. Géophys. **13** (1957).
OPARIN, A. I.: The Origin of Life. New York: Macmillan & Co. 1938.
ÖPIK, E.: Irish Astron. J. **1**, 45 (1950); **4**, 37 (1956).
PEEK, B. M.: J. Brit. Astronom. Assoc. **44**, 220 (1934); **50**, 4 (1939).
PETTIT, E.: and S. B. NICHOLSON: Publ. Astronom. Soc. Pacific **36**, 269—272 (1924a).
— — Pop. Astron. **32**, 601—608 (1924b); **32**, 614 (1924c); **33**, 299 (1925a).
— — Astronom. Nachr. **225**, 331—334 (1925b).
— — Astrophys. Journ. **71**, 131 (1930); **83**, 84 (1936).
— Publ. Astronom. Soc. Pacific **67**, 293 (1955).
POOLE, J. H. J.: Sci. Proc. Roy. Dublin Soc. **22**, 345 (1941).
RAMSEY, D. A.: J. Chem. Phys. **21**, 960 (1953).
— Private communication (1957).
RICE, F. O.: Sci. Amer. **194**, No. 6, 119 (1956). See F. O. RICE and M. FREAMO: J. Amer. Chem. Soc. **75**, 548 (1953).
RICHARDSON, R. S.: Publ. Astronom. Soc. Pacific **67**, 304 (1955).
The Rocket Panel: Phys. Rev. **88**, 1027—1032 (1952).
ROSS, F. E.: Astrophys. Journ. **64**, 243—249 (1926); cf. also Astrophys. Journ. **65**, 266—268 (1927); **68**, 57 (1928).
ROSSINI, E. D., D. D. WAGMAN, W. H. EVANS, S. LEVINE and I. JAFFE: Selected Values of Chem. Therm. Prop., Circular Bur. of Stand. No. 500, 1952.
RUBEY, W. W.: Crust of the Earth, ed. by A. POLDERVAART. Geol. Soc. Amer., Special Paper **62**, 631—650 (1955)
RUSSELL, R. H., R. S. DUGAN and J. Q. STEWART: Astronomy. New York: Gin & Co. 1945.
—, and D. H. MENZEL: Proc. Nat. Acad. Sci. U.S.A. **19**, 997 (1933).
ST. JOHN, C. E., and S. B. NICHOLSON: Phys. Rev. **19**, 444 (1922a).
— — Astrophys. Journ. **56**, 380 (1922b).
SCHARONOW, W. W.: Poulkovo Obs. Circ. No. **32**, 62—73 (1941).
SCHOENBERG, E.: Sitzgsber. bayer. Akad. Wiss., math.-naturw. Kl. **1949**.
SCHRÖER, E.: Ber. Dtsch. Wetterd. U.S.-Zone Nr. 11, Bad Kissingen 1949.
SHAPIRO, R.: The Study of Planetary Atmospheres, Lowell Obs. Report by the Geophys. Res. Division of the Air Force, Cambridge Res. Center 1952.
— J. Meteorology **10**, 350 (1953).
SLIPHER, E. C.: Publ. Astronom. Soc. Pacific. **39**, 209 (1927); **49**, 137—140 (1937).
— Proc. Amer. Phil. Soc. **79**, 441 (1938).
— Sky and Telescope **3**, No. 5, 3 (1939).
— Telescope **7**, 102 (1940); **14**, 361 (1954).
—, and A. G. WILSON: Astronom. Soc. Pacific, Leaflet No. 301, p. 7, 1954.
SLIPHER, V. M.: Lowell Obs. Bull. No. 3 and 4, 1903; No. 16, 1905; No. 27, 1906; No. 42, 1909.
— Publ. Astronom. Soc. Pacific **36**, 261 (1924).
— Pop. Astron. **37**, 138 (1929).
— Monthly Notices Roy. Astronom. Soc. London **93**, 657 (1933).
SPITZER jr., L.: Atmospheres of the Earth and Planets, ed. by G. P. KUIPER. Chap. VII. Chicago, Ill.: Chicago University Press 1952.
STRONG, J., and W. M. SINTON: Science, Lancaster, Pa. **123**, 676 (1956).
STUMPFF, K.: Astronom. Nachr. **276**, 121 (1948).
SUESS, H. E.: J. Geology **57**, 600 (1949).
SYTINSKAYA, N. N.: C. R. Acad. Sci. USSR. **43**, 147 (1944).
THODE, H. G., J. MACNAMARA and W. H. FLEMING: Geochim. et Cosmochim. Acta **3**, 235 (1953).

TIKHOFF, G. A. et al.: Astron. Zhurn. SSSR. **29**, 738–741 (1952).
TOWNSEND jr., J. W., E. B. MEADOWS and P. E. C. PRESSLY: Rocket exploration of the upper atmosphere, ed. by R. L. F. BOYD and M. J. SEATON, p. 169. London: Pergamon Press 1954.
TROITSKAYA, O. V.: Astronom. J. Soviet Union **29**, 57–62 (1952).
TRUMPLER, R. J.: Lick Obs. Bull. **13**, 19 (1927).
UREY, H. C.: Geochim. et Cosmochim. Acta **1**, 207 (1951).
— The Planets. New Haven, Conn.: Yale University Press 1952a.
— Proc. Nat. Acad. Sci. U.S.A. **38**, 351 (1952b).
— XII-th Internat. Congress of Pure and Applied Chemistry, Plenary Lectures. p. 188, 1953a.
— Proc. Roy. Soc. Lond., Ser. A **219**, 281 (1953b).
— Astrophys. Journ. Suppl. **1**, 147 (1954).
— Sky and Telescope **15**, Nos. 3 and 4, (1956a).
— Z. angew. Chem. **68**, 533 (1956).
— Phys. and Chem. of the Earth, Vol. 2. London: Pergamon Press 1956b.
—, and A. W. BREWER: Proc. Roy. Soc. Lond. Ser. A **241**, 37–43 (1957).
VAUCOULEURS, G. DE: C. R. Acad. Sci. Paris **220**, 903 (1945).
— Ciel et Terre **61**, 256 (1945).
— Physics of the Planel Mars. London 1954.
VERY, F. W.: Lowell Obs. Bull. Nos. 36 and 41, 1908.
WATANABE, K., M. ZELIKOFF and E. C. Y. INN: Geophys. Res. Papers No. 21, Geophys. Res. Directorate, Air Force Cambridge Res. Center (1953).
WELSH, H. L., M. F. CRAWFORD, J. C. F. MACDONALD and D. A. CHRISHOLM: Phys. Rev. **83**, 1264 (1951).
WHIPPLE, F. L.: Earth, Moon and Planets. Philadelphia: Blakiston & Co. 1946.
WILDT, R.: Naturwiss. **19**, 109 (1931).
— Veröff. Univ. Sternwarte Göttingen No. 22, 1932.
— Astrophys. Journ. **86**, 321 (1937).
WOOLLEY, R. V. D. R.: Monthly Notices Roy. Astronom. Soc. London **113**, 521 (1953).
WRIGHT, W. W.: Lick Obs. Bull. XII, No. **366**, 48–61 (1925); XII, No. **389**, 50–67 (1927a).
— Publ. Astronom. Soc. Pacific **39**, 231 (1927b); **45**, 292 (1933).
WULF, O. R., and L. S. DEMING: Terr. Mag. and Atm. Elect. **43**, 283 (1938), and articles there cited.

Planetary Interiors.

By

WENDELL C. DEMARCUS.

With 3 Figures.

1. Introduction. During the past several decades enormous progress has been made toward understanding the physics of the stars. As is well known, a large amount of this progress is due to the fact that the properties of matter at high temperatures are relatively simple. This relative simplicity can be attributed to the fact that the macroscopic properties of matter at high temperatures can be deduced from the properties of individual atoms, or ions, etc. by means which are in principle (though not always in practice) quite straightforward. As a result, many properties of matter under physical conditions appropriate to the interiors of stars can be predicted with confidence from atomic data which can be measured in the laboratory. The physics of planetary interiors is, in this respect, quite different from stellar physics for in the case of planetary interiors one has to deal with the properties of matter at low temperatures where condensed phases occur.

In the condensed phases, nature exhibits an often embarrassing diversity of phenomena. Many of these phenomena can be readily classified a posteriori according to the general principles of chemical thermodynamics but the theoretical anticipation of experimental results is notoriously difficult whenever condensed phases are involved. However, at sufficiently high pressures the physical properties of cold matter presumably become amenable to quantitative theoretical treatment. This happens when the number of electrons per unit volume becomes so large that on the average the electrons are in states of sufficiently high kinetic energy (on account of the Pauli Principle) that the electrical contributions to the energy may be neglected. In this case, one describes the matter as a Fermi gas. For a given substance, the lower limit of pressure necessary for the applicability of the Fermi gas equation of state is generally too great to have any direct bearing on planetary interiors. However, there are "cold" bodies for which the relativistic Fermi gas equation of state holds throughout most of their volume. These are the white dwarf stars[1] whose mean densities are enormous (6×10^6 gm/cm^3 for VAN MAANEN's star). It may seem somewhat paradoxical to refer to these stars as "cold" since some have surface temperatures two or three times that of the Sun. Nonetheless, the term "cold" is appropriate[2] since the pressure can be shown to increase with depth sufficiently rapidly to override the effects of the accompanying increase of temperature.

[1] For details on the white dwarfs cf. GREENSTEIN's article in Vol. L, and SCHATZMAN's contribution to Vol. LI of this Encyclopedia.

[2] S. CHANDRASEKHAR: An Introduction to the Study of Stellar Structure, Chap. XI. Chicago 1939.

I. The mass-radius relation of cold bodies.

2. The conjecture of HENRY NORRIS RUSSELL. At the 1935 meeting of the International Astronomical Union, H. N. RUSSELL pointed out that among the planets the radius increases with the mass while the opposite trend is exhibited by the white dwarf stars. The planets and the white dwarfs are cold in the sense that densities throughout most of their bulk are essentially determined by pressure alone. RUSSELL concluded, therefore, that somewhere between the masses of the white dwarfs and of the planets there ought to be a maximum radius for bodies consisting of cold matter. Theoretical calculations substantiating RUSSELL's prediction were soon published by D. S. KOTHARI[1] who found the maximum radius of a cold body to be of the order 10^9 to 10^{10} cm, depending on composition. A number of calculations aimed at improving KOTHARI's first analysis have since been published[2-8]. Fig. 1 shows a $\operatorname{Log} M - \operatorname{Log} R$ plot of the masses and radii of the planets, certain of their satellites (for which these two parameters appear to be reasonably well established) and three white dwarf stars. The continuous curve represents a theoretical $\operatorname{Log} M - \operatorname{Log} R$ relation for configurations of pure cold hydrogen[7].

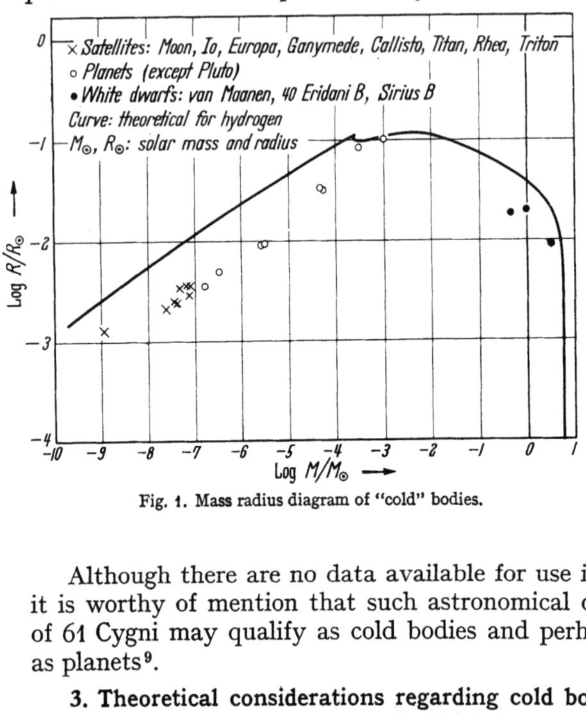

Fig. 1. Mass radius diagram of "cold" bodies.

Although there are no data available for use in the $\operatorname{Log} M - \operatorname{Log} R$ diagram, it is worthy of mention that such astronomical objects as the faint companion of 61 Cygni may qualify as cold bodies and perhaps be appropriately classified as planets[9].

3. Theoretical considerations regarding cold bodies in general. In calculating the internal properties of cosmic bodies it is necessary to assume that the interiors are in hydrostatic equilibrium. This procedure appears to be sufficiently accurate for the larger astronomical objects, but its applicability to cold bodies with masses comparable to Mars is subject to some reservations. It is grossly erroneous for bodies of asteroidal size. For spherical configurations, the assumption of hydrostatic equilibriuml eads immediately to the differential equation

$$\frac{dp}{dr} = -g\varrho. \tag{3.1}$$

[1] D. S. KOTHARI: Monthly Notices Roy. Astronom. Soc. London **96**, 833 (1936).
[2] F. C. AULUCK: Monthly Notices Roy. Astronom. Soc. London **99**, 239 (1939).
[3] D. S. KOTHARI: Proc. Roy. Soc. Lond., Ser. A **165**, 486 (1938).
[4] P. SEN: Z. Astrophys. **16**, 297 (1938).
[5] A. SOMMERFELD: Nuovo Cim. **15**, No. 1 (1938).
[6] W. H. RAMSEY: Monthly Notices Roy. Astronom. Soc. London **110**, 444 (1950).
[7] W. C. DEMARCUS: Ph. D. Thesis, Yale Univ. 1951.
[8] A. A. ABRIKOSOV: Problems of Cosmogony, Vol. III, p. 11—19. Moscow: Academy of Sciences USSR. 1954.
[9] P. VAN DE KAMP: Vistas in Astronomy, Vol. II, p. 1040; A. BEER edit. London and New York 1956.

Here p is the local pressure at distance r from the center of mass of the body, g is the local acceleration of gravity and ϱ the local density. If $M(r)$ denotes the mass within the sphere whose radius is r and G is the constant of gravitation, the acceleration of gravity is given by

$$g(r) = \frac{GM(r)}{r^2}. \tag{3.2}$$

In case the body is rotating it will no longer assume a spherical shape and it is incorrect to employ a single independent variable. However, it will be shown below that the equation of hydrostatic equilibrium, in the case of sufficiently slowly rotating bodies, can still be written to sufficient accuracy in terms of a single independent variable s. For each surface of constant density the value of s is assigned which is numerically the radius of a sphere whose volume would coincide with the volume contained within the surface. The equation

$$\frac{dp}{ds} = -\left[\frac{GM(s)}{s^2} - \frac{2}{3}\omega^2 s\right]\varrho(s), \tag{3.3}$$

where ω is the magnitude of the angular velocity and $M(s)$ is the mass contained within the surface of constant density $\varrho(s)$, correctly defines the condition of hydrostatic equilibrium if terms involving ω^4 may be neglected. The differential equation (3.3) cannot be integrated without further information. For cold bodies this information takes the form of a pressure-density relation[1],

$$\varrho = \varphi(p), \tag{3.4}$$

or its inverse,

$$p = f(\varrho). \tag{3.5}$$

The pressure or the density can be eliminated from (3.3) and a second-order, but generally non-linear, differential equation for the remaining variable results. Explicitly these are

$$\frac{d^2\varrho}{ds^2} + \frac{2}{s}\frac{d\varrho}{ds} + \left[\frac{f''(\varrho)}{f'(\varrho)} - \frac{1}{\varrho}\right]\left(\frac{d\varrho}{ds}\right)^2 + \frac{4\pi G\varrho^2}{f'(\varrho)} = \frac{2\omega^2\varrho}{f'(\varrho)}, \tag{3.6}$$

$$\frac{d^2p}{ds^2} + \frac{2}{s}\frac{dp}{ds} - \frac{\varphi''(p)}{\varphi'(p)}\left(\frac{dp}{ds}\right)^2 + 4\pi G\varphi^2(p) = 2\omega^2\varphi(p), \tag{3.7}$$

where primes denote differentiation with respect to ϱ and p, respectively. There is also an integral form of the equation of hydrostatic equilibrium. Namely, if we set

$$G^* = \int_{\varrho_0}^{\varrho} \frac{dp}{\varrho} \tag{3.8}$$

where ϱ_0 is the density of the external surface of the body (zero pressure) then the integral form of the equation of hydrostatic equilibrium is the Bernoullian integral

$$G^* = \int_{\varrho_0}^{\varrho} \frac{dp}{\varrho} = \int_{s}^{R} \left[\frac{GM(s)}{s^2} - \frac{2}{3}\omega^2 s\right] ds. \tag{3.9}$$

For a homogeneous isothermal planet G^* is the difference between the free enthalpy per gram at the pressure $p(\varrho)$ and the free enthalpy at zero pressure.

[1] If temperature effects are to be considered, one must know the equation of state $\varrho = \varphi(p, T)$ and, in addition, be supplied with the means of obtaining the course of the temperatures throughout the body.

A large amount of past speculations have been based on the heuristic assumption that the march of density with radius conformed to the Laplace Law, viz.,

$$\varrho = \varrho_c \frac{\sin \varkappa r}{\varkappa r} \,. \tag{3.10}$$

The assumption of LAPLACE'S law implicitly is an assumption regarding the compressibility of the matter, namely, that pressure and density are connected by

$$p = A(\varrho^2 - \varrho_0^2) \,, \tag{3.11}$$

with A and ϱ_0 constants. Using LAPLACE'S law for the variation of density within the planet it is also possible to integrate CLAIRAUT'S equation for the ellipticity of the internal strata in closed form. In the absence of any conflicting evidence, this pleasant circumstance made the Laplace law quite popular.

The equation of path (3.11) is a special case of the polytropic equation of path introduced into astronomy by R. EMDEN[1]. EMDEN considered equations of path which, slightly generalized, are of the form

$$p = \frac{k_0}{\gamma} \left[\left(\frac{\varrho}{\varrho_0} \right)^{\gamma} - 1 \right] . \tag{3.12}$$

k_0 and ϱ_0 are the incompressibility and density at zero pressure. The exponent γ is related to the polytropic index n via $\gamma = 1 + \frac{1}{n}$. In case the density at zero pressure vanishes, as would occur for an ideal adiabate or polytrope, k_0 must also be considered to vanish in such fashion that $k_0 \varrho_0{}^{-\gamma}$ remains finite. The "Free Enthalpy" corresponding to the equation of state (3.12) is

$$n \left(\frac{k_0}{\varrho_0} \right) \left[\left(\frac{\varrho}{\varrho_0} \right)^{1/n} - 1 \right] . \tag{3.13}$$

For $\omega = 0$, the Bernoullian integral becomes

$$n \frac{k_0}{\varrho_0} \left[\left(\frac{\varrho}{\varrho_0} \right)^{1/n} - 1 \right] = G \int_s^R \frac{M(s)}{s^2} \, ds \,. \tag{3.14}$$

For a central density ϱ_c put $y = \left(\frac{\varrho}{\varrho_c} \right)^{1/n}$ and differentiate both sides of the preceding equation to obtain

$$n \varrho_c^{1/n} \left(\frac{k_0}{\varrho_0^{\gamma}} \right) s^2 \frac{dy}{ds} = -GM(s) \,.$$

A second differentiation then leads directly to the dimensional form of the Emden equation,

$$\frac{1}{s^2} \frac{d}{ds} \left(s^2 \frac{dy}{ds} \right) + \left(\frac{4\pi G \varrho_c \varrho_0^{\gamma}}{n \varrho_c^{1/n} k_0} \right) y^n = 0 \,. \tag{3.15}$$

Introduction of the unit of length l_E defined by

$$l_E^2 = \frac{n \varrho_c^{1/n} k_0}{4\pi G \varrho_c \varrho_0^{\gamma}} \tag{3.16}$$

and a change to the dimensionless variable x with $s = l_E x$ yields the Emden equation

$$\frac{1}{x^2} \frac{d}{dx} \left(x^2 \frac{dy}{dx} \right) + y^n = 0 \,. \tag{3.17}$$

[1] R. EMDEN: Gaskugeln. Leipzig und Berlin 1907.

The complete mass-radius diagram of all substances obeying the equation of state (3.12) with fixed polytropic index n can then be obtained from a single integration of the Emden equation with the boundary conditions $y(0)=1$, $y'(0)=0$ the integration being carried to the point where y first vanishes. For a configuration with central density ϱ_c the radius of the configuration in units of l_E is determined by the root x_0 of

$$y(x) - \left(\frac{\varrho_0}{\varrho_c}\right)^{1/n} = 0.$$

The mass of the configuration is then

$$M = -4\pi \varrho_c l_E^3 x_0^2 \left(\frac{dy}{dx}\right)_{x=x_0}. \tag{3.18}$$

The form of the mass-radius diagram for very small masses, if $\varrho_0 > 0$, can then be inferred from the fact that for small x, $y(x) = 1 - \frac{1}{3!} x^2 + \cdots$, whence

$$M = \frac{4\pi}{3} \varrho_0 l_E^3 x_0^3 = \frac{4\pi}{3} \varrho_0 R^3, \tag{3.19}$$

as was to be expected. This result, of course, holds for *any* matter forming a condensed phase stable at zero pressure. On the other hand, for central densities large compared to ϱ_0, x_0 tends to the root of $y(x)=0$ and

$$R \to \text{constant}\, M^{\frac{1-n}{3-n}}.$$

If n is less than 1 the radius always increases with mass; for $n=1$ the radius tends to a constant value independent of the mass. For n greater than 1 but less than 3 the radius increases to a maximum value and then decreases or, in case $\varrho_0 = 0$, the radius always decreases with increasing mass. The Fermi gas equation of state has $n=1.5$ in the non-relativistic case but for very high density behaves more like $n=3$. CHANDRASEKHAR[1] has shown that in this case there is a limiting finite mass of zero radius, but nuclear effects preclude the attainment of this interesting limiting case. The existence of a maximum radius for cold bodies is thus seen to be mathematically a consequence of the facts that cold matter has non-vanishing incompressibility at zero pressure and that asymptotically $p \sim \varrho^m$ with $m < 2$.

4. The instability of small cores. The mass-radius relation of chemically homogeneous bodies in general can exhibit some rather intriguing details. For example, it can happen that there is not a single maximum in the values of R as a function of M. The example in mind is the mass-radius relation for chemically homogeneous bodies made up of a substance whose pressure-density relation has discontinuities in density (phase changes). If the density saltus exceeds 50%[2,3] or somewhat less for rotating planets[4], there will be a range of mass values in which the radius is a triple-valued function. The three configurations corresponding to a given mass have respectively a sizeable core of the dense phase, a small core of the dense phase, and no dense phase present. As a general rule, the configuration with the smaller core is unstable. This phenomenon was first noticed by RAMSEY[2]. RAMSEY suggested that the old idea of an exploding planet giving rise to the meteorites can be justified by invocation of this principle.

[1] S. CHANDRASEKHAR: Monthly Notices Roy. Astronom. Soc. London **91**, 446 (1930).
[2] W. H. RAMSEY: Monthly Notices Roy. Astronom. Soc. London **110**, 325 (1950).
[3] M. J. LIGHTHILL: Monthly Notices Roy. Astronom. Soc. London **110**, 342 (1950).
[4] W. C. DEMARCUS: Astronom. J. **59**, 116 (1954).

BULLEN has pointed out also that the existence of this phenomenon might resuscitate DARWIN'S theory of a tidal origin of the Moon, the tides serving merely to trigger a transition from one stable configuration to the other rather than being called upon to effect the actual ejection of the Moon. BULLEN'S suggestion has been followed up by DATTA[1] who has considered the question of the amount of energy which might be released by such transitions.

5. The figure of rotating planets. The interiors of the planets are not open to inspection so that all information regarding their interiors must be inferred from their external features. The most important of the external features are the masses, principal radii and the form of the external gravitational potentials of the planets. To a certain extent the composition of the atmospheres of the planets can be added to the relevant external features. But an extraterrestrial observer would be rather wide of the mark if he concluded that the chemical composition of the Earth was 80% nitrogen and 20% oxygen. The form of the external potential can only be inferred in detail when a planet possesses satellites whose motions can be followed with sufficient accuracy. When such a satellite is lacking, one is forced, by and large, to base all conclusions about the planetary interiors on the radius and mean density.

For the purposes at hand, the term orbit when applied to a planetary satellite will be used to denote a fictitious motion governed solely by the form of the conservative external gravitational potential of the planet; more particularly, that ideal motion which when augmented by the effects of the various relevant perturbing influences, reproduces the observed satellite motions. Except for the possible effects of tidal friction, the orbit in the sense stated can be regarded as well-defined. The conservative external potential of a planet whose internal mass distribution has axial symmetry can always be expanded as a series of Legendre polynomials of argument $\cos\varphi$ where φ is the angle between the axis of rotation of the planet and the radius vector, r, from the center of mass of the planet. The conventional representation of the potential is

$$V = -\frac{GM_1}{r}\left[1 - \frac{2}{3} J a^2 \frac{P_2(\cos\varphi)}{r^2} + \frac{4}{15} K \frac{a^4}{r^4} P_4(\cos\varphi) - \cdots\right], \tag{5.1}$$

where a is the equatorial radius of the planet. M_1 is the total mass of the planet and in general in the sequel the subscript 1 implies evaluation at the planetary surface. This will cause no confusion with the Legendre functions since only even orders of the latter appear. The coefficient J can be expressed explicitly by

$$J = \frac{3}{2}\,\frac{C-A}{M_1 a^2} \tag{5.2}$$

where C is the moment of inertia about the axis of rotation and A is the amount of inertia about any axis perpendicular to the axis of rotation. An expression can also be given for the evaluation of K but the results cannot be expressed in terms of familiar concepts such as the principal moments of inertia. The effects of the terms J and K on the ideal motion of the satellite[2] is to change the period from the Kepler value, and give rise to a secular advance of the line of apsides and a secular regression of the line of nodes. In first approximation the last two effects are of equal magnitudes but in second approximation this is no longer the case, the apse advances at a different rate from that at which the node regresses. Thus, observations of a close satellite in principle yields not only the mass of

[1] A. N. DATTA: Monthly Notices Roy. Astronom. Soc. London, Geophys. Suppl. **6**, 535 (1954).

[2] D. BROUWER: Astronom. J. **51**, 223 (1946).

the primary, but in addition the information embodied in the coefficients J and K. This last information is of little value unless hydrostatic equilibrium is assumed to hold in the interior of the primary[1].

The theory of the figure of a rotating planet with internal hydrostatic equilibrium has been the subject of serveral investigations[2-7]. However, for the most part the emphasis has been on the problem of determining the figure from the density distribution along some radius vector of the planet. One avenue of approach toward the elucidation of the interiors of the planets proceeds by the construction of model planets whose properties are computed from assumed pressure-density relations. Accordingly, a sketch of the theory of the equilibrium of rotating bodies recast into a form suitable for deriving model planets of fixed rotation period solely from a pressure-density relation is given here.

In the following, the theory is carried only through terms of the order ω^4, where ω is the angular speed of rotation. The expansion is basically in powers of centrifugal force, so terms of the order ω^2 will be said to be of the first order, ω^4 of the second order and so forth.

A surface of equal density will be assumed to be approximately an ellipsoid of revolution about the axis of rotation, the equation to the surface being explicitly

$$r = s\left[1 - \frac{4\varepsilon^2}{45} - \frac{2}{3}\left(\varepsilon + \frac{23}{42}\varepsilon^2 + \frac{4}{7}\varkappa\right) P_2(\cos\varphi) + \frac{8}{35}\left(\frac{3}{2}\varepsilon^2 + 4\varkappa\right) P_4(\cos\varphi)\right]. \quad (5.3)$$

Here r is the radius vector from the center of the planet, φ is the colatitude, i.e., the angle between the radius vector and the axis of rotation, the P_n are Legendre polynomials normalized according to the customary convention which takes $P_n(1) = 1$, ε and $\varkappa$ are parameters giving respectively the ellipticity of the surface and the departure of the surface from the true ellipsoidal shape. ε is of the order ω^2 and $\varkappa$ is of the order ω^4. The potential at any point (r, φ) within the planet is then given by

$$\left.\begin{aligned} V = -\frac{4\pi G}{3}\Big[\frac{S_0}{r} - \frac{2}{5}\frac{S_2 P_2}{r^3} + \frac{12}{35}\frac{S_4 P_4}{r^5} + \\ + \frac{3}{2} T_0 - \frac{2}{5} r^2 T_2 P_2 + \frac{32}{105} r^4 T_4 P_4 + \frac{\omega^2 r^2}{4\pi G}(1 - P_2)\Big] \end{aligned}\right\} \quad (5.4)$$

in which the centrifugal potential has been included and

$$S_0 = \int_0^s \varrho(s)\, d(s^3) \qquad T_0 = \int_s^{R_1} \varrho(s)\, d\left[s^2\left(1 - \frac{4}{45}\varepsilon^2\right)\right]$$

$$S_2 = \int_0^s \varrho(s)\, d\left[s^5\left(\varepsilon + \frac{1}{6}\varepsilon^2 + \frac{4}{7}\varkappa\right)\right] \qquad T_2 = \int_s^{R_1} \varrho(s)\, d\left[\varepsilon + \frac{9}{14}\varepsilon^2 + \frac{4}{7}\varkappa\right]$$

$$S_4 = \int_0^s \varrho(s)\, d\left[s^7\left(\varepsilon^2 + \frac{8}{9}\varkappa\right)\right] \qquad T_4 = \int_s^{R_1} \varrho(s)\, d\left(\frac{\varkappa}{s^2}\right).$$

[1] In a sense, hydrostatic equilibrium may be said to be already assumed when the axial symmetry is assumed. However, one can partially defend the axial symmetry assumption on the basis of visual observations.

[2] G. B. AIRY: Phil. Trans. Roy. Soc. Lond., Ser. A **1826**, Part III, 548.

[3] O. CALLANDREAU: Ann. Obs. Paris **19**, E 1 (1889).

[4] G. H. DARWIN: Monthly Notices Roy. Astronom. Soc. London **60**, 82 (1899).

[5] W. DE SITTER: Bull. Astronom. Inst. Netherl. **2**, 97 (1924).

[6] H. JEFFREYS: Monthly Notices Roy. Astronom. Soc. London **113**, 97 (1953).

[7] W. C. DEMARCUS: Astronom. J. **63**, 2 (1958).

The form of the potential (5.4) is due to DARWIN[1] except for the term T_0 which he did not carry because it plays no role when the density distribution is assumed to be given a priori. Inserting the expression (5.3) for r in the expression (5.4) for the potential and using the fact that the Legendre polynomials are linearly independent leads, after some manipulation, to the three equations,

$$-\frac{3V(s)}{4\pi G}=\frac{S_0}{s}\left(1+\frac{8\varepsilon^2}{45}\right)-\frac{4}{25}\varepsilon\frac{S_2}{s^3}+\frac{3}{2}T_0+\frac{8}{75}\varepsilon s^2 T_2+\frac{\omega^2 s^2}{4\pi G}\left(1+\frac{4}{15}\varepsilon\right), \quad (5.5)$$

$$\frac{S_0}{s}\left(\frac{2}{3}\varepsilon+\frac{31}{63}\varepsilon^2+\frac{8}{21}\varkappa\right)-\frac{2}{5}\frac{S_2}{s^3}\left(1+\frac{4}{7}\varepsilon\right)-\frac{2}{5}s^2T_2\left(1-\frac{8}{21}\varepsilon\right)=\frac{\omega^2 s^2}{4\pi G}\left(1+\frac{20}{21}\varepsilon\right), \quad (5.6)$$

$$\frac{S_0}{s}(3\varepsilon^2-8\varkappa)-6\varepsilon\frac{S_2}{s^3}+3\frac{S_4}{s^3}+\frac{8}{3}s^4T_4=0. \quad (5.7)$$

These equations may be solved by successive approximations. In Eq. (5.5) the second order terms are dropped giving

$$V(s)=-\frac{4\pi G}{3}\left[\frac{1}{s}\int_0^s \varrho(s)\,d(s^3)+\frac{3}{2}\int_s^{R_1}\varrho(s)\,d(s^2)+\frac{\omega^2 s^2}{4\pi G}\right]. \quad (5.8)$$

The Bernoullian integral of the equation of hydrostatic equilibrium is

$$V(s)+\int\frac{dp}{\varrho}=\text{const}$$

and when differentiated becomes

$$\frac{dp}{ds}=-\left[\frac{GM(s)}{s^2}-\frac{2}{3}\omega^2 s\right]\varrho(s). \quad (5.9)$$

Before taking up the solution for the parameters ε and $\varkappa$ it is convenient to follow DE SITTER[2] and introduce dimensionless quantities. These are

$$\bar\varrho=\frac{M(R_1)}{\frac{4\pi}{3}R_1^3} \qquad D=\frac{1}{\beta^3}\int_0^\beta \delta(t)\,d(t^3)$$

$$m=\frac{3\omega^2}{4\pi G\bar\varrho} \qquad S=\frac{1}{\beta^3}\int_0^\beta \delta(t)\,d\left[t^5\left(\varepsilon+\frac{1}{6}\varepsilon^2+\frac{4}{7}\varkappa\right)\right]$$

$$\beta=\frac{s}{R_1} \qquad P=\frac{1}{\beta^7}\int \delta(t)\,d\left[t^7\left(\varepsilon^2+\frac{8}{9}\varkappa\right)\right]$$

$$\bar\varrho\,\delta(s)=\varrho(s) \qquad T=\int_\beta^1 \delta(t)\,d\left[\varepsilon+\tfrac{9}{15}\varepsilon^2+\tfrac{4}{7}\varkappa\right]$$

$$Q=\beta^2\int_\beta^1 \delta(t)\,d\left[\frac{\varkappa}{t^2}\right].$$

In terms of the dimensionless variables the Eqs. (5.6) and (5.7) become

$$D\left[\varepsilon+\tfrac{31}{42}\varepsilon^2+\tfrac{4}{7}\varepsilon\right]-\tfrac{3}{5}S\left(1+\tfrac{4}{7}\varepsilon\right)-\tfrac{3}{5}T\left(1-\tfrac{8}{21}\varepsilon\right)=\tfrac{8}{21}m\left(1+\tfrac{21}{20}\varepsilon\right) \quad (5.10)$$

and

$$D\left[3\varepsilon^2-8\varkappa\right]-6\varepsilon S+3P+\tfrac{8}{3}Q=0. \quad (5.11)$$

[1] See footnote 4, p. 425.

[2] W. DE SITTER: Bull. Astronom. Inst. Netherl. 2, 97 (1924).

By comparing (5.1) and (5.4) it is found that

$$S_1 = \tfrac{5}{3} J(1 + \tfrac{2}{3}\varepsilon_1) \tag{5.12}$$

and

$$P_1 = \tfrac{7}{9} K. \tag{5.13}$$

When (5.10) and (5.11) are evaluated at $\beta = 1$, a slight rearrangement yields

$$\varepsilon_1 - J = \tfrac{1}{2}\varepsilon_1^2 - \tfrac{1}{7}\varepsilon_1 m - \tfrac{4}{7}\varkappa_1 + \tfrac{1}{2} m, \tag{5.14}$$

$$K = \tfrac{24}{7}\varkappa_1 + 3\varepsilon_1^2 - \tfrac{15}{7}\varepsilon_1 m. \tag{5.15}$$

Hence, if the mass, equatorial radius and rotational period of a planet are known anf if it has satellites from whose motions the coefficients J and K can be deduced, it is possible to determine the ellipticity of the external surface and the departure of the surface from true ellipticity correct through the fourth power of ω. Values of the ellipticity derived in this way are usually called *dynamical ellipticities* to distinguish them from ellipticities obtained telescopically as, for example, by a filar micrometer. Generally, the dynamical ellipticities are the more reliable.

A certain amount of simplification results if DE SITTER's variable ε' is used instead of ε. ε' is defined by the expression

$$\varepsilon' = \varepsilon - \tfrac{5}{42}\varepsilon^2 + \tfrac{4}{7}\varkappa.$$

By differentiating (5.10) with respect to β and introducing the variable

$$\eta' = \frac{d\ln\varepsilon'}{d\ln\beta}$$

there results, after some rearrangement,

$$\eta'\left[D\left(1 + \frac{4}{7}\varepsilon\right) - \frac{4}{21} m + \frac{4}{7} T\right] = 3D\left(1 + \frac{2}{7}\right)\varepsilon - \frac{3S}{\varepsilon'} \tag{5.16}$$

and differentiating again

$$\beta\frac{d\eta'}{d\beta} + \eta'^2 + 5\eta' - 2\zeta(1 + \eta') - \frac{4}{21}\zeta\xi = 0 \tag{5.17}$$

wherein

$$\zeta = -\frac{d\ln D}{d\ln\beta} = 3\left(1 - \frac{\delta}{D}\right)$$

and

$$\xi = \frac{7m}{D}(1 + \eta') - 3\varepsilon(1 + \eta')^2 - 4\varepsilon.$$

Following a procedure due originally to RADAU[1] the Eq. (5.17) can also be written in the form

$$\frac{d}{d\beta}\left(D\beta^5\sqrt{1 + \eta'}\right) = 5D(\beta)\,\beta^4 F(\eta') \tag{5.18}$$

with

$$F(\eta') = \frac{1 + \frac{1}{2}\eta' - \frac{1}{10}\eta'^2 + \frac{2}{105}\zeta\xi}{\sqrt{1 + \eta'}}$$

RADAU's transformation is quite remarkable in that the function $F(\eta')$ varies very slowly with η'. Neglecting the small term in $\zeta\xi$ the following table exhibits

[1] R. RADAU: C. R. Acad. Sci., Paris **100**, 972 (1885).

the behavior of $F(\eta')$.

η'	0	0.2	0.4	0.6	0.8	1.0	2.0	3.0
$F(\eta')$	1.00000	1.00051	1.00066	0.99928	0.99580	0.98995	0.92376	0.80000

$F(\eta')$ has a maximum at $\eta'=\frac{1}{3}$ at which point its value is 1.00074. Hence, if it appears certain that $\eta'\leqq 0.6$ throughout a planet we may set $F=1$ with an error certainly less than 0.1% and integrate Eq. (5.18) to obtain

$$\sqrt{1+\eta_1'} = 5\int_0^1 D(\beta)\,\beta^4\,d\beta$$

which on partial integration becomes the Radau-Darwin approximation

$$\sqrt{1+\eta_1'} = \frac{5}{2}\left[1-\frac{3}{2}\,\frac{I}{MR_1^2}\right] \tag{5.19}$$

where I is an average moment of inertia defined by

$$I=\frac{8\pi}{3}\int_0^{R_1} s^4\,\varrho(s)\,ds\,.$$

The ellipticity can be computed from the value of η' at the surface, η_1', for eliminating J from (5.14) by (5.12) and (5.16) one obtains

$$\varepsilon_1'(2+\eta_1') = \tfrac{5}{2}\,m + \tfrac{4}{7}\,\varepsilon_1^2 + \tfrac{10}{21}\,m^2 - \tfrac{6}{7}\,\varepsilon_1 m\,. \tag{5.20}$$

In general, the Eq. (5.18) must be solved by successive approximation. Eq. (5.18) is integrated numerically with ξ set equal to zero. This procedure yields an approximate value for ε_1' and $\varepsilon'(\beta)$ is calculated to the same degree of approximation from

$$\frac{\varepsilon'(\beta)}{\varepsilon_1'} = \exp\left\{-\int_\beta^1 \frac{\eta'}{\beta}\,d\beta\right\}.$$

The approximate values for $\varepsilon'(\beta)$ are then used to evaluate the term in ξ and η' recalculated. The acumen displayed by DE SITTER in transforming the dependent variable to ε' is demonstrated in practice by the minuteness of the corrections necessitated by the second approximation.

After the ellipticity is known the solution for $\varkappa$ can be reduced to the solution of a non-homogeneous second order differential equation, but there are many advantages to solutions based on the integral equation directly. Integration by parts can be employed to bring (5.11) to the form

$$\varkappa(\beta) = g(\beta) + \int_0^1 \varkappa(t)\,d\Gamma(\beta,t) \tag{5.21}$$

with

$$g(\beta) = -\frac{3}{8D(\beta)\,\beta^7}\int_0^\beta [\beta^2\,\varepsilon(\beta) - t^2\,\varepsilon(t)]^2\,t^3\,d\delta(t) + \frac{1}{3}\,\frac{\beta^2\,\delta_1\varkappa_1}{D(\beta)} \tag{5.22}$$

and

$$d\Gamma(\beta,t) = \begin{cases} -\dfrac{1}{3}\,\dfrac{t^7\,d\delta(t)}{\beta^7\,D(\beta)} & t<\beta,\\[2ex] -\dfrac{1}{3}\,\dfrac{\beta^2}{t^2}\,\dfrac{d\delta(t)}{D(\beta)} & t\geqq\beta. \end{cases}$$

The contribution of the integral to the surface value of $\varkappa$ is slight for the integrand is largest near $t=1$ so that when $\beta=1$ the integral can be estimated roughly as

$$\int_0^1 \varkappa(t)\, d\Gamma(1,t) \approx -\frac{\varkappa_1}{3}\int_0^1 t^7\, d\delta(t) \leqq -\frac{\varkappa_1}{3}\int_0^1 t^5\, d\delta(t) \leqq \frac{5}{6}\,\frac{I}{M_1 R_1^2}\,\varkappa_1$$

which adequately demonstrates the assertion made. It is therefore to be expected that an iterative solution of the integral equation (5.21) will converge rapidly and this expectation has been confirmed in practice[1].

6. Integral theorems and inequalities. There are a number of instructive bounds which can be placed on various physical features of planets which are based on the assumptions of hydrostatic equilibrium and of monotonic density decrease with radius. Referring to the equation of hydrostatic equilibrium it is clear that the pressure at the center will be smallest for planets of given mass and mean radius if the planet is of uniform density throughout. Thus, P_c, the pressure at the center satisfies the inequality

$$P_c \geqq \frac{2\pi G}{3}\,\bar{\varrho}\, R_1^2\left[1-\frac{2}{3}\,m\right]. \tag{6.1}$$

A less useful upper bound for the density at the center follows from the remark that the pressure at the center will be less than it would be if the density were equal to the central density ϱ_c out to the value s' of the mean radius and vanished otherwise. The value of s' is given by the condition that

$$\varrho_c\, s'^3 = \bar{\varrho}\, R_1^3.$$

Consequently,

$$P_c \leqq \frac{2\pi G}{3}\,\varrho_c^2\, s'^2\left[1-\frac{2}{3}\,\frac{\bar{\varrho}}{\varrho_c}\,m\right]. \tag{6.2}$$

These two inequalities may be combined and given a more symmetrical form

$$\frac{1}{2}\left(\frac{4\pi}{3}\right)^{\frac{1}{3}} G M^{\frac{2}{3}}\,\bar{\varrho}^{\frac{4}{3}}\left(1-\frac{2}{3}\,m\right) \leqq P_c \leqq \frac{1}{2}\left(\frac{4\pi}{3}\right)^{\frac{1}{3}} G M^{\frac{2}{3}}\,\varrho_c^{\frac{4}{3}}\left(1-\frac{2}{3}\,\frac{\bar{\varrho}}{\varrho_c}\,m\right). \tag{6.3}$$

For example, the central pressures of Jupiter and Saturn are bounded below by 11 and 2.1 millions of atmospheres, respectively. Another instructive limit regarding pressures inside the planet refers to the mass weighted average pressure, $\bar{P}$, i.e.,

$$\bar{P} = \frac{1}{M_1}\int^{M_1} P\, dM = -\frac{1}{M_1}\int_0^{M_1} M\,\frac{dP}{dM}\, dM. \tag{6.4}$$

The equation of hydrostatic equilibrium may be written with $M(s)$ as independent variable

$$\frac{dP}{dM} = -\frac{GM}{4\pi s^4} + \frac{\omega^2}{6\pi s}$$

whence,

$$\bar{P} = \frac{1}{4\pi M_1}\int_0^{M_1} \frac{GM^2\, dM}{s^4}\left\{1-\frac{2}{3}\,\frac{\omega^2 s^3}{GM(s)}\right\}. \tag{6.5}$$

If the density never increases with s the following inequality holds

$$\frac{s^3}{M(s)} \leqq \frac{R_1^3}{M_1}.$$

[1] H. Jeffreys: Monthly Notices Roy. Astronom. Soc. London **114**, 433 (1954).

Consequently,

$$\overline{P} \geqq \frac{(1-\frac{2}{3}m)}{4\pi M_1} \int_0^{M_1} \frac{G M_1^{\frac{4}{3}}}{R_1^4} M^{\frac{2}{3}} dM = \frac{3 G M_1^2}{20\pi R_1^4} \left(1 - \frac{2}{3} m\right). \tag{6.6}$$

or

$$\overline{P} \geqq \frac{1}{5} \left(\frac{4\pi}{3}\right)^{\frac{1}{3}} G M_1^{\frac{2}{3}} \bar{\varrho}^{\frac{4}{3}} \left(1 - \frac{2}{3} m\right). \tag{6.7}$$

The average pressure in Jupiter must therefore exceed 4.6 million atmospheres if its density increases monotonically toward the center.

On the assumption that the density does not increase with radius, JEFFREYS[1] has given an upper limit to the surface density of a planet whose average moment of inertia is known,

$$\varrho_s \leqq \tfrac{5}{2} \bar{\varrho}\, (I/M_1 R_1^2). \tag{6.8}$$

Except for the case of the Earth, the moment of inertia of planets has to be obtained from the Radau-Darwin approximation. For the major planets JEFFREYS' inequality implies surprisingly low surface densities. However, DEMARCUS[2] has shown that JEFFREYS' inequality can be sharpened considerably. From (5.12) there results

$$(1 + \tfrac{2}{3}\varepsilon_1)\, J = \tfrac{3}{5} \int_0^1 \delta(t)\, d[t^5 (\varepsilon + \tfrac{1}{6}\varepsilon^2 + \tfrac{4}{7}\varkappa)].$$

Integrating the right hand side by parts this becomes

$$(1 + \tfrac{2}{3}\varepsilon_1)\, J = \tfrac{3}{5} \delta_1 (\varepsilon_1 + \tfrac{1}{6}\varepsilon_1^2 + \tfrac{4}{7}\varkappa_1) - \tfrac{3}{5} \int_0^1 (\varepsilon + \tfrac{1}{6}\varepsilon^2 + \tfrac{4}{7}\varkappa)\, t^5\, d\,\delta(t).$$

Hence, under the same assumption made by JEFFREYS,

$$\delta_1 \leq \frac{\frac{5}{3}(1 + \frac{2}{3}\varepsilon_1)\, J}{(\varepsilon_1 + \frac{1}{6}\varepsilon_1^2 + \frac{4}{7}\varkappa_1)}. \tag{6.9}$$

On dropping the second order term and using (5.14) this last result takes the form

$$\delta_1 \leqq \frac{\frac{5}{3} J}{J + \frac{1}{2} m}. \tag{6.10}$$

In case the second order coefficient K is also known an even sharper inequality can be obtained. For

$$K = \tfrac{9}{7} \int_0^1 \delta(\beta)\, d[\beta^7 (\varepsilon^2 + \tfrac{8}{9}\varkappa)] = \tfrac{9}{7} \delta_1 (\varepsilon_1^2 + \tfrac{8}{9}\varkappa_1) - \tfrac{9}{7} \int_0^1 (\varepsilon^2 + \tfrac{8}{9}\varkappa)\, \beta^7\, d\,\delta(\beta).$$

Table 1. *Superior limits to surface densities* (gm/cm^3).

	JEFFREY'S inequality	Eq. (6.10)	Eq. (6.11)
Mars	4.01	4.00	
Earth	4.60	4.48	4.10
Jupiter	0.86	0.76	0.57
Saturn.	0.37	0.31	0.20
Neptune	1.85	1.78	

[1] H. JEFFREYS: Monthly Notices Roy. Astronom. Soc. London **84**, 534 (1924).
[2] W. C. DEMARCUS: Astronom. J. **63**, 2 (1958).

Hence, as before,

$$\delta_1 \leqq \frac{7}{9}\,\frac{K}{\varepsilon_1^2+\frac{8}{9}\varkappa_1} = \frac{7K}{(2J+m)\,(J+3m)+\frac{7}{3}K}\,. \tag{6.11}$$

In Table 1 the limiting surface densities are calculated from the various inequalities for which sufficient data are available.

It may be argued that these limits are of no importance for the major planets whose outer shells are probably composed of super-critical fluids, but it is possible to obtain inequalities[1] of similar nature for distances beneath the surface corresponding to many tens of scale heights for gaseous hydrogen and *a fortiori* all other gases.

II. Empirical features of the planets.

7. Classification of the planets. The planets fall naturally into three different groupings. These are the Jovian planets comprising Jupiter, Saturn, Uranus and Neptune; the minor planets, the asteroids, meteorites and zodiacal light; and the terrestrial planets, Earth, Venus, Mars and Mercury. The Earth and its satellite, the Moon, are frequently considered as a double planet in which case the Moon should be included among the terrestrial planets. The planet Pluto is not listed above. Its mass is as yet unknown and there are intimations that it is an escaped satellite of Neptune[2-4].

The Jovian planets are characterized by their large masses and radii, *low* values of mean density and the fact that the semi-major axes of their orbits are all greater than five astronomical units from the Sun. The terrestrial planets, on the other hand, are opposite to the Jovian planets in all respects, i.e., they have small masses, small radii, but large mean densities and they are all (unless Pluto should eventually be classified as terrestrial) within or at 1.52 astronomical units from the Sun.

Table 2. *Mechanical data of the planets.*

Planet	Mass (10^{27} gm)	Mean Radius (km)	Mean density (gm/cm³)	m	J	K	$1/\varepsilon$ (dynamic)	ε/m	$I/M_1R_1^2$
Mercury	0.325	2420	5.46						
Venus	4.867	6123	5.06						
Earth	5.977	6371	5.52	0.0034498	0.0016370	0.00001138	297.13	0.9756	0.3335
Mars	0.639	3332	4.12	0.004321	0.003017		192.6	1.199	0.389
Jupiter	1902	69860	1.33	0.0843	0.02206	0.00253	15.3	0.773	0.26
Saturn	569.4	57630	0.71	0.1421	0.02501	0.00386	10.2	0.688	0.21
Uranus	86.9	23700	1.56						
Neptune	102.8	21500	2.47	0.0194	0.0074		58.5	0.88	0.30

Sources. Masses and radii except Jupiter and Saturn; (a) all other data except last column; (b) $I/M_1R_1^2$ for Earth; (c) $I/M_1R_1^2$ for Jupiter and Saturn; (d) $I/M_1R_1^2$ for Mars and Neptune estimated from Radau-Darwin approximation.

(a) G. P. Kuiper: The Atmospheres of the Earth and Planets. Revised Edition. Chicago 1952.

(b) D. Brouwer and G. M. Clemence: The Solar System, G. P. Kuiper, Editor. Chicago (to be published 1958); Vol. 3, Chap. 2.

(c) H. Jeffreys: The Earth. 3rd edit., p. 145. Cambridge 1952.

(d) W. C. DeMarcus: Astronom. J. **63**, 2 (1958).

[1] W. C. DeMarcus: See footnote 2, p. 430.

[2] R. A. Lyttleton: Monthly Notices Roy. Astronom. Soc. London **97**, 108 (1946).

[3] G. P. Kuiper: Astrophys. Journ. **125**, 287 (1957).

[4] E. Rabe: Astrophys. Journ **125**, 290 (1957).

The asteroidal members of the minor planets have their mean distances from the Sun scattered over a rather wide range but seven-eighths (of the first thousand) lie between 2.3 and 3.3 astronomical units from the Sun. It is an extremely interesting fact that the average mean distance of the brighter asteroids is 2.9 astronomical units, a number which is close to the "missing" member, 2.8, of the Titius-Bode sequence. To this interesting fact can be attributed the prevalence of the ideas that the asteroids are the fragments of an exploded planet or else are the building material for a planet the formation of which was never completed.

Some mechanical data concerning the planets are collected in Table 2.

III. Structure and composition of the planets.

a) The constitution of the terrestrial planets.

The similarities of the mean densities displayed by the terrestrial planets leads to the natural assumption that they are similarly constituted. This assumption is seldom, if ever, questioned and the chief point at issue between various authorities seems to be the degree to which the assumed similarities of constitution prevail.

8. Wiechert models. The first model of the Earth which was motivated by physical considerations was given by WIECHERT[1] in 1897. WIECHERT modeled the Earth as having a core of uniform density overlaid by a mantle of constant density. Such a model can nowadays be completely determined from known data. Let R_1 be the radius of the Earth, $x R_1$ the boundary of the core, ϱ_m, ϱ_c, and $\bar{\varrho}$ the mean densities of the mantle, core and the entire Earth. Introduce further the ratios

$$\alpha = \frac{\varrho_m}{\bar{\varrho}} \qquad \beta = \frac{\varrho_c - \varrho_m}{\bar{\varrho}}.$$

Then α, β and x must satisfy

$$\alpha + \beta x^3 = 1 \tag{8.1}$$

in order that the mass of the Earth will be reproduced by the model and

$$\alpha + \beta x^5 = \frac{5}{2} \frac{I}{M_1 R_1^2} \tag{8.2}$$

where I is the average moment of inertia and M_1 is the Earth's mass. The value of x is given by seismology[2] as 0.545 and $I/M_1 R_1^2$ is obtainable from the precession of the equinoxes and is numerically[3] 0.3335. The Eqs. (8.1) and (8.2) may then be solved for α and β. Using 5.53 gm/cm³ for the mean density of the Earth,

$$\varrho_m = 4.22 \text{ gm/cm}^3, \qquad \varrho_c = 12.33 \text{ gm/cm}^3.$$

These densities which were derived in this way by JEFFREYS[4] are somewhat higher than the ones found by WIECHERT who assumed a larger value of x. Nonetheless, WIECHERT'S value for the density of the core, namely 8.2 gm/cm³, when combined with other suggestions such as meteoritic compositions led him to the view that the core was iron. This suggestion has held the field without serious challenge until comparatively recently when it has been questioned by KUHN

[1] E. WIECHERT: Nachr. Ges. Wiss. Göttingen **1897**, 221.
[2] K. E. BULLEN: This Encyclopedia, Vol. XLVII, p. 103.
[3] H. JEFFREYS: The Earth, 3rd edit, p. 145. Cambridge 1952.
[4] H. JEFFREYS: The Earth, 3rd edit., p. 146. Cambridge 1952.

and RITTMANN[1] and by RAMSEY[2]. It cannot be said, however, that WIECHERT's iron core has been abandoned and a lively controversy is presently centered on this point.

9. Earth models based on the Adams-Williamson equation. The most recent work on the density distribution of the Earth's interior is that of BULLEN[3] who has made consistent use of a method due originally to ADAMS and WILLIAMSON[4]. The method makes use of the distribution of the velocities of the two types of bodily earthquake waves, longitudinal and transverse, in order to obtain a first order differential equation for the variation of density with depth. If V_p and V_s are the local velocities of the longitudinal and shear waves at some point within the Earth and if k and μ are the adiabatic incompressibility and adiabatic shear modulus for the local matter whose undisturbed density is ϱ, then

$$V_p^2 = \frac{k + \frac{4}{3}\mu}{\varrho}, \tag{9.1}$$

$$V_s^2 = \frac{\mu}{\varrho}. \tag{9.2}$$

The quantity φ, which is by definition

$$\varphi = V_p^2 - \tfrac{4}{3} V_s^2, \tag{9.3}$$

gives directly the ratio of the adiabatic incompressibility to the density, i.e., $\varphi = k/\varrho$ (adiabatic). Now quite generally we may assume the density to be a function of pressure p, entropy s and the molar numbers n_i of the components of the local material so that

$$\frac{d\varrho}{dh} = \left(\frac{\partial\varrho}{\partial p}\right)_s \frac{dp}{dh} + \left(\frac{\partial\varrho}{\partial s}\right)_p \frac{ds}{dh} + \sum_i \left(\frac{\partial\varrho}{\partial n_i}\right) \frac{dn_i}{dh}, \tag{9.4}$$

h being the depth measured from the surface of the Earth. The equation of hydrostatic equilibrium can also be written (ignoring the centrifugal effects) as

$$\frac{dp}{dh} = g\varrho \tag{9.5}$$

where g is the local gravitational acceleration and, hence,

$$\frac{d\varrho}{dh} = \frac{g\varrho}{\varphi} + \left(\frac{\partial\varrho}{\partial s}\right)_p \frac{ds}{dh} + \sum_i \left(\frac{\partial\varrho}{\partial n_i}\right) \frac{dn_i}{dh}. \tag{9.6}$$

In regions where the composition does not change and the temperature gradient is the adiabatic one, the Adams and Williamson equation is valid

$$\frac{d\ln\varrho}{dh} = \frac{g}{\varphi}. \tag{9.7}$$

The magnitude of the errors arising from use of the Adams and Williamson equation in regions where the temperature gradient is not the adiabatic one has been examined, most thoroughly by BIRCH[5]. He concluded that a departure from the

---

[1] W. KUHN and A. RITTMANN: Geol. Rdsch. **32**, 215 (1941).
[2] H. W. RAMSEY: Monthly Notices Roy. Astronom. Soc. London **108**, 406 (1948).
[3] K. E. BULLEN: Trans. Roy. Soc. New Zealand **67**, 122 (1937). — Bull. Seism. Soc. Amer. **30**, 235 (1940); **32**, 19 (1942). — Monthly Notices Roy. Astronom. Soc. London, Geophys. Suppl. **6**, 50 (1950). — Cf. also BULLEN's contribution to Vol. LXVII of this Encyclopedia.
[4] E. D. WILLIAMSON and L. H. ADAMS: J. Washington Acad. Sci. **13**, 413 (1923).
[5] F. BIRCH: J. Geophys. Res. **57**, 227 (1952).

adiabatic gradient of 1° per kilometer would render the Adams and Williamson equation inaccurate by about 10 to 20%. To put it somewhat differently, Eq. (9.7) will be inaccurate to the extent of 10% if the total change in temperature down to the core boundary differs from the adiabatic one by an amount of order 2000° K. It may be concluded from this that temperature gradients departing from the adiabatic one by amounts large enough to seriously impair the usefulness of the Adams-Williamson equation in chemically homogeneous regions are unlikely.

BIRCH has also put forth an ingenious method for testing the assumption of homogeneity within the Earth. On the basis of BIRCH's homogeneity test and other evidence it appears reasonably certain that the Earth is homogeneous or nearly so for 1000 km $\leqq h \leqq$ 2900 km and for 2900 km $\leqq h \leqq$ 5000 km, the core intervening at 2900 km depth. It appears also reasonably certain that the Earth is definitely inhomogeneous in the range of depths between 400 and 600 km and most certainly heterogeneous from the surface down to depths of 30 km or so. The status of the layer between 30 and 400 km appears uncertain.

To obtain a definite model, BULLEN assumed that the Adams-Williamson equation is valid from 33 to 413 km, from 984 to 2900 km and from 2900 to 5371 km. The crust may be treated in a schematic fashion for it is of little consequence as regards its contribution to the mass and moment of inertia of the planet as a whole. The transition region between 413 and 984 km was joined smoothly onto the solutions for the density of the neighboring regions. The parameters of the inner core of the Earth given in the accompanying table are the result of a compromise between two solutions considered as extreme cases by BULLEN. In any event, its importance for the overall density pattern is not of great moment. The relevant features of BULLEN's model are displayed in Table 3.

Table 3. *Interior of Earth according to Bullen.*

h depth (km)	Density (gm/cm³)	Pressure (10^{12} dynes/cm²)	h depth (km)	Density (gm/cm³)	Pressure (10^{12} dynes/cm²)
33	3.32	0.009	3000	9.57	1.47
100	3.38	0.031	3200	9.85	1.67
200	3.47	0.065	3400	10.11	1.85
300	3.55	0.100	3600	10.35	2.04
400	3.63	0.136	3800	10.56	2.22
413	3.64	0.141	4000	10.76	2.40
500	3.89	0.174	4200	10.94	2.57
600	4.13	0.213	4400	11.11	2.73
700	4.33	0.256	4600	11.27	2.88
800	4.49	0.30	4800	11.41	3.03
900	4.60	0.35	4982	11.54	3.17
1000	4.68	0.39	5121	14.20	3.27
1200	4.80	0.49	5121	16.80	3.27
1400	4.91	0.58	5200	16.85	3.32
1600	5.03	0.68	5400	16.96	3.42
1800	5.13	0.78	5600	17.05	3.50
2000	5.24	0.88	5800	17.12	3.56
2200	5.34	0.99	6000	17.16	3.61
2400	5.44	1.09	6200	17.19	3.63
2600	5.54	1.20	6371	17.20	3.64
2800	5.63	1.32			
2898	5.68	1.37			
	Boundary of core				
	9.43	1.37			

BULLEN has constructed other models of the Earth and a number of important general conclusions can be inferred from his results. Two of these are:

1. The local gravitational acceleration is approximately constant down to the core boundary[1].

2. Pressure as function of depth, down to the core boundary, is insensitive to the model adopted.

A model of the Earth determines a pressure-density relation but it must not be forgotten that the pressure-density relation is not necessarily an isothermal or an adiabatic nor even one of constant composition. By assuming that the model accurately reflects the internal physical features of the Earth one can build model planets of different masses. It must be understood that in so doing one is tacitly, or otherwise, assuming that the temperatures and compositions of the other model planets are the same at corresponding pressures as is the case for the Earth.

There is much to be said for this point of view. For example, the sharp changes in composition presently indicated by essentially all lines of evidence at depths between 400 and 600 km can be attributed to a change of crystal structure of olivine rock as has been pointed out by JEFFREYS and BERNAL[2]. BERNAL's argument was based on the known[3] change in crystal structure of magnesium germanate, Mg_2GeO_4, which is isomorphous with forsterite, Mg_2SiO_4. RAMSEY has gone a great deal further than JEFFREYS and BERNAL and asserted that the large change in density at the core boundary is a further transformation of olivine to a metallic phase. RAMSEY's advocacy of this idea, which is a special case of his assertion that all matter will exhibit *large* discontinuous changes in density before the limiting Fermi-gas status is reached, has such far reaching significance that it seems necessary to give here a brief résumé of the evidence which can be brought to bear on this question.

10. Transitions to metallic phases. In discussing the plausibility of RAMSEY's metallic transition, it is important that the mechanism by which Ramsey supposes these transitions to occur be clearly delineated. If this is not done there is danger of confusion, for experimental results may be held to support or deny RAMSEY's hypothesis when actually other mechanisms are at work. As the writer understands it, RAMSEY's process is solely a supposed "ripping off" of outer electrons, the necessary work being balanced by the product of pressure and volume change. Any change in the lattice structure accompanying the phase change must be viewed as a separate mechanism. Consequently, a catalogue of phase changes known to take place at atmospheric or very modest pressures and the amount of the attendant density changes does not serve to justify RAMSEY's hypothesis. Rather large changes in density are known to occur during which the lattice type changes and there is definitely no "bursting free" of the electrons. The diamond-graphite transition is one example where this is clearly shown. The example also quoted by RAMSEY of the large density difference between the metallic and non-metallic forms of arsenic clearly is not relevant to his mechanism either. It merely shows that for uncompressed solids, a change in lattice type can give rather large changes in density and that at atmospheric pressure metals are usually more dense than non-metals. It would be rather far-fetched to picture this transition as one during which atmospheric pressure rips the valence electrons free.

[1] This fact has been noticed by others, apparently first by A. E. BENFIELD.
[2] J. D. BERNAL: Observatory **59**, 268 (1936).
[3] V. M. GOLDSCHMIDT: Nachr. Ges. Wiss. Göttingen **1931**, 184.

Numerous polymorphic transitions have been found by BRIDGMAN in his comprehensive work in the field of high pressures. Undoubtedly, many of these are attributable solely to electronic rearrangements. Generally, the density changes involved are quite small. One rather famous exception is the transition of cesium which occurs at 45000 atmospheres pressure. There is good reason for discounting the possibility of a change in lattice type here and a theoretical study specifies the nature of this transition[1]. Even this case exhibits a change in density (13%) which must be considered small by RAMSEY's standards.

The final bit of evidence usually quoted in favor of RAMSEY's hypothesis is the presumed phase transition of solid molecular hydrogen to a monatomic metallic phase. (This transition will be more thoroughly discussed below in the section on the Jovian Planets.) Some calculations by KRONIG, DE BOER and KORRINGA[2] have indicated that this transition takes place at 700000 atmospheres and that the density of the metal exceeds that of the molecular phase by a factor of two. More recently, DEMARCUS[3] and ABRIKOSOV[4] have raised grave doubts as to the reliability of the calculations of KRONIG, DE BOER and KORRINGA and find the change in density to be substantially smaller. Moreover, they find that the phase transition occurs at a considerably higher pressure. Here again, the comparison with RAMSEY's mechanism is hindered by the fact that the presumed transition involves a rather thorough going rearrangement of the atomic nuclei. It is also to be noted that ELSASSER[5] has expressed a considered opinion that the change of olivine to a metallic modification, as postulated specifically by RAMSEY to account for the large density change at the boundary of the Earth's core, involves unreasonable changes in the Helmholtz free energy of olivine.

There is now reason to hope that the question may be experimentally decided in the near future. A new technique has recently been reported[6-8] for the measurement of pressure-density relations of solids along the Hugoniot curves which is capable of yielding data at very much higher pressures than the usual techniques. When such measurements are published, current speculations about the interiors of the Earth and the remaining terrestrial planets can be subjected to more meaningful tests than can presently be made.

To sum up: There is no reason to doubt that solids will in general exhibit numerous phase changes in the range of pressures below 10^8 atmospheres. Many of these changes will be attributable solely to the electronic rearrangements without the simultaneous occurrence of a change in the lattice type. The density saltus accompanying the electronic rearrangements may or may not be as large as RAMSEY would have it. The arguments which have been put forth in favor of the large changes in density cannot be regarded as convincing.

11. Models of the terrestrial planets. The assumption that the pressure-density relation of the Earth also holds good in the remaining terrestrial planets is very

[1] R. STERNHEIMER: Phys. Rev. **78**, 238 (1950). But see also F. S. HAM: Advances in Solid State Physics, Vol. I. New York 1955. F. SEITZ and D. TURNBULL Editors, p. 141 where some doubts are expressed regarding the accuracy of STERNHEIMER's calculations.

[2] R. KRONIG, J. DE BOER and J. KORRINGA: Physica, Haag **12**, 245 (1946).

[3] W. C. DEMARCUS: Astronom. J. **63**, 2 (1958).

[4] A. A. ABRIKOSOV: Problems of Cosmogony, Vol. III, pp. 11–19. Moscow: Academy of Sciences, USSR. 1954.

[5] W. M. ELSASSER: Science, Lancaster, Pa. **113**, 105 (1951).

[6] J. M. WALSH and R. H. CHRISTIAN: Phys. Rev. **97**, 1544 (1955).

[7] R. W. GORANSON, D. BANCROFT, B. L. BURTON, T. BLECHAR, E. E. HOUSTON, E. F. GITTINGS and S. A. LANDEEN: J. Appl. Phys. **26**, 1472 (1955).

[8] B. J. ALDER and R. H. CHRISTIAN: Phys. Rev. **104**, 550 (1956).

intriguing[1], for it implies that the seismological properties of the Earth's interior can be used to calculate the internal physical features of all the terrestrial planets. RAMSEY[2] has carried through such calculations. He took the masses of the planets as known data and the pressure-density relation of BULLEN's Model A (Table 3) and computed the radius and mean density of model planets for comparison with observation. His results are given in Table 4. For consistency the

Table 4.

	Relative mass	Observation		Computed	
		mean diameter km	mean density gm/cm³	mean diameter km	mean density gm/cm³
Earth . .	1	12742	5.52	12742	5.52
Venus . .	0.817 ± 0.005	12400 ± 100	4.89 ± 0.12	12570	4.70
Mars . .	0.1078 ± 0.0002	6793 ± 16	3.93 ± 0.03	7020	3.56
Mercury .	0.044 ± 0.008	5000 ± 250	4.0 ± 0.9	5285	3.41
Moon . .	0.01230	3476	3.34 ± 0.01	3480	3.334

observational data used by him are also given although these numbers are no longer up-to-date.

On the basis of the observational material available at the time RAMSEY performed these calculations, the extent of the agreement is quite remarkable. However, the agreement for the case of the Moon was inevitable since the density of the Moon was a datum used by BULLEN to fix the density in the Earth at 33 km depth. It should also be mentioned that the computations were made under the assumption that the crust of the terrestrial planets is in all cases the same fraction of the total mass. This is a departure from the general spirit of the calculation. If this assumption were changed so that the crusts extended in depth to the same pressure level, the agreement for the case of the Moon would be seriously impaired. RAMSEY then shows that better agreement would be secured if BULLEN's densities were subjected to an ad hoc modification intended to reflect the concentration of heavier elements toward the center.

More recently, BULLEN[3,4] has taken up the suggestion of RAMSEY from a somewhat different viewpoint. BULLEN has shown by calculation that the following set of postulates is sufficient to fit the observed masses and diameters of Mars and Venus within the probable errors of the data.

1. The Earth, Venus, and Mars have a common primitive composition.

2. The density distribution of the Earth conforms to BULLEN's Model B[5].

3. The outer core of the Earth is a high pressure modification of the material of the lower mantle.

4. The inner core is chemically distinct from the outer core and mantle.

Fig. 2 shows the mass as function of radius for Earth-like planets as calculated by BULLEN. The empirical points for the terrestrial planets are also

[1] H. JEFFREYS: Monthly Notices Roy. Astronom. Soc. London, Geophys. Suppl. **4**, 62 (1937).

[2] W. H. RAMSEY: Monthly Notices Roy. Astronom. Soc. London **108**, 406 (1948).

[3] K. E. BULLEN: Monthly Notices Roy. Astronom. Soc. London **109**, 457 (1949); **110**, 256 (1950).

[4] K. E. BULLEN and A. H. LOW: Monthly Notices Roy. Astronom. Soc. London **112**, 637 (1952).

[5] K. E. BULLEN: Monthly Notices Roy. Astronom. Soc. London, Geophys. Suppl. **6**, 50 (1950).

included. BULLEN has also calculated the ellipticity of his Martian model and found it to agree satisfactorily with the empirical value.

RABE's[1] revision of the mass of Mercury has definitely precluded its inclusion among the planets conforming to BULLEN's hypothesis. The mean density of Mercury is now believed to be in excess of 5 gm/cm³. Since its mass is about half that of Mars it is obvious that Mars and Mercury cannot be precisely similarly constituted. BULLEN[2] has argued, however, that Mercury's orbit in the past may have been more eccentric and, hence, that temperatures on its surface might have been high enough to volatilize a significant fraction of its mantle. BULLEN points out that on RAMSEY's hypothesis, namely that the internal densities are determined solely by pressure, the evaporation would lower the mean density of Mercury. As a consequence, there would be even more glaring disagreement with the postulate of a common primitive composition. On the other hand, BULLEN remarks that the inclusion of his postulate 4 entails that the evaporation of material would raise the mean density. BULLEN concludes that improved observation of Mercury would appear to be almost an experimentum crucis for the presence, in terrestrial planets, of a quantity of material chemically distinct from that in the Earth's mantle, the proportion being *at least* equal to the proportion of the mass of the Earth's inner core to the mass of the Earth.

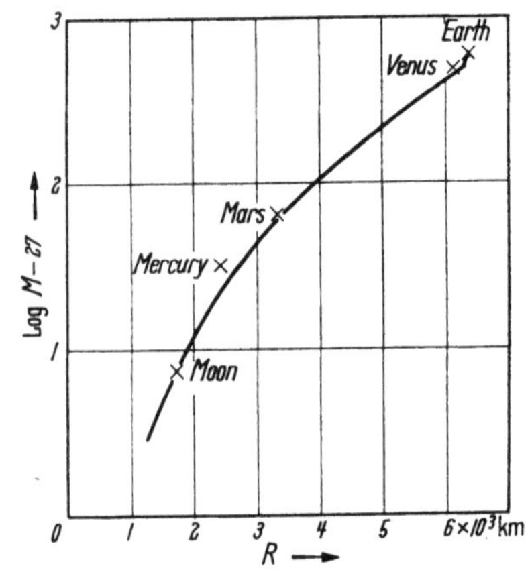

Fig. 2. Mass-radius diagram of Earth-like Planets with distinct inner cores [after K. E. BULLEN and A. H. LOW: Monthly Notices Roy. Astronom. Soc. London **112**, 637 (1952)].

It is not clear, however, to what extent the force of BULLEN's argument might be impaired by consideration of preferential evaporation of light materials.

It is possible to utilize the pressure-density relation of the Earth to assess the composition of the terrestrial planets in a somewhat different fashion. In this case, it is assumed that the outer core of the Earth is chemically distinct from the mantle. By analogy, it is then argued that all of the terrestrial planets are composed of the same materials but not in the same proportions. The terrestrial planets then are viewed as if all have a mantle composed of the same material as the mantle of the Earth and following the same pressure-density relation. Furthermore, they all have a core which is composed of the same material as the Earth's core. The pressures at the core boundaries are now free parameters and may be adjusted to give the correct mean densities. In case this latter pressure is less than the pressure at the boundary of the Earth's core, some extrapolation of thepressure-density relation has to be made. Calculations were carried through along these lines by JEFFREYS[3] in 1937 using a revised version of one of BULLEN' earlier models to determine the pressure-density relation of the Earth. The details of JEFFREY's models are shown in Tables 5 and 6 below. The model of Mercury derived by him is not displayed since the mass of Mercury has since been markedly revised.

[1] E. RABE: Astronom. J. **55**, 112 (1950).
[2] K. E. BULLEN: Physics and Chemistry of the Earth, Vol. 1, L. AHRENS, K. RANKAMA and S. K. RUNCORN Editors, p. 90. London 1956.
[3] H. JEFFREYS: Monthly Notices Roy. Astronom. Soc. London, Geophys. Suppl. **4**, 62 (1937).

Table 5. *Jeffreys' model of Venus.*

r (10^3 km)	p (10^{10} dynes/cm²)	ϱ (gm/cm³)	M (10^{24} gm)
6.15	0	3.29	4910
6.0	4.38	3.41	4677
5.617	16.22	3.69	4101
5.617	16.22	4.23	4101
5	39.38	4.58	3138
4	78.52	5.01	1915
2.911	124.4	5.44	1057
2.911	124.4	9.64	1057
2.0	189.8	10.41	360
1.0	239.2	10.97	47
0.0	255.8	11.15	0

Table 6. *Jeffreys' model of Mars.*

r (10^3 km)	p (10^{10} dynes/cm²)	ϱ (gm/cm³)	M (10^{24} gm)
3.385	0	3.29	643
3.0	4.70	3.42	477.3
2.5	10.68	3.57	310.9
2.034	16.22	3.69	201.3
2.034	16.22	4.23	201.3
2.0	16.69	4.24	193.9
1.5	23.69	4.35	110.7
1.424	24.77	4.37	101.8
1.424	24.77	8.28	101.8
1.4	25.43	8.30	96.8
1.0	34.86	8.45	35.7
0.5	42.15	8.57	4.5
0.0	44.45	8.60	0

If we denote the material of the Earth's mantle by A and that of the core by C then the compositions of Earth, Mars, and Venus given by JEFFREYS are as follows:

	Material A	Material C
Earth . . .	68%	32%
Mars . . .	84%	16%
Venus . . .	78%	22%

JEFFREYS also computed a dynamical ellipticity for Mars on the basis of his model and found $1/e = 205$, whereas H. STRUVE'S value obtained from the perturbation of PHOBOS was $1/e = 192$. He gives arguments that the disagreement is no more than might be expected due to departures from hydrostatic equilibrium.

Another approach leading to a determination of the constitution of the terrestrial planets has been employed by UREY[1]. UREY'S procedure is to estimate the density the planets would have if there were no compression resulting from the forces of gravitation, i.e., he calculates mean densities at zero pressure.

To derive the mean density for the Earth at zero pressure, UREY proceeded in a fashion which will be illustrated by his treatment of a model derived by BULLEN in 1937[2]. The pressure-density relation implied by the core of BULLEN'S model is taken to be applicable to iron. BULLEN'S data for the core were accurately reproduced by the second degree formula

$$\varrho_{\mathrm{Fe}} = 7.227 + 2.068\, P - 0.188\, P^2 \tag{11.1}$$

with ϱ in gm/cm³ and P in units of 10^{12} dynes/cm². UREY assumed that the outer mantle is a silicate of about the composition of olivine and that the inner mantle is a mixture of the silicate phase of the outer mantle and an iron-nickel phase with pressure density relation (11.1). BULLEN'S 1937 model has a discontinuous change in density at 481 km depth from the value 3.69 to 4.23 gm/cm³. By assuming ideal mixing, the weight fraction, w, of iron in the lower mantle can be estimated from the relation

$$\frac{1}{\varrho} = \frac{1-w}{\varrho_{\mathrm{Si}}} + \frac{w}{\varrho_{\mathrm{Fe}}}$$

[1] H. C. UREY: Geochimica et Cosmochimica Acta **1**, 209 (1951).
[2] K. E. BULLEN: Trans. Roy. Soc. New Zealand **67**, 122 (1937).

where ϱ is the density just below the discontinuity, ϱ_{Si} is the silicate density which by the assumptions is equal to density just above the discontinuity, and ϱ_{Fe} is the density of iron at the relevant pressure, $p = 7.561$ gm/cm³ from the formula (11.1) above. The result was $w = 0.25$, i.e., 25% iron phase. Using 3.28 gm/cm³ for the density of the silicate phase (the outer mantle) at zero pressure and 7.227 gm/cm³ for the density of iron at zero pressure the density of the inner mantle at zero pressure then followed immediately as 3.80 gm/cm³. The mean density of the Earth at zero pressure under the assumptions made is then 4.33 gm/cm³. UREY concluded that on the basis of the model Earth used, the Earth was 55% "olivine" by weight and 45% iron nickel. Similar results were obtained for models A and B. UREY remarks also that Venus is so similar to the Earth that its composition will be the same.

Mars, Mercury, and the Moon are only very slightly compressed. Consequently, any reasonable assumption as to the compressibility of the material can be employed to estimate their density at zero pressure, provided no phase changes are involved. UREY's estimates of composition are given in the accompanying table.

	% iron phase
Moon . . .	0
Mercury . .	56%
Venus . . .	45%
Earth . . .	45%
Mars . . .	25%

In some cases UREY gave two values based on different observational values of the planetary radii and the entries in the above table are averages for his cases. The total weight fraction of elemental iron can only be obtained when the composition of the "olivine" is known.

Summary: Opinion as to the material forming the Earth's core is divided. The opposing views are basically that the core is a high-pressure silicate phase of high density and that the Earth's core is iron or nickel-iron. Generally, the silicate camp favors the view that all the terrestrial planets save Mercury are, for practical purposes, chemically identical while the iron camp considers that the terrestrial planets are qualitatively similar in constitution but definitely differ quantitatively.

b) The constitution of the Jovian planets.

The construction of model planets leads to estimates of the pressure-density relations of the materials of which the modeled planets are composed. If it could then be asserted that a given chemical substance or mixture of substances had the same pressure-density relation and that no other chemical substance or mixture could have a pressure-density relation resembling that of the model planet, a solution to the composition problem would have been achieved. This last statement is subject, of course, to reservations about the degree to which the model planets could be safely assumed to reflect the properties of the planet under investigation. In the cases of the terrestrial planets this program could not be carried out, at least by itself. The densities of the terrestrial planets can be matched by a variety of substances when account is taken of the uncertainties that are to be associated with the model planets and in our knowledge of pressure-density relations under the relevant conditions of pressure and temperature In the cases of the two largest planets, the situation is considerably simpler for it turns out that the densities, as function of pressure, calculated from any reasonable model of these two bodies cannot be matched by any matter which is not mainly composed of hydrogen. This is a most fortunate circumstance for hydrogen is the only chemical substance whose pressure-density relation is sufficiently distinctive with respect to all other substances that its presence can be unambiguously detected by the rather broad means at our disposal.

12. Physical properties of hydrogen and helium. Ordinary hydrogen solidifies at a temperature of 14° K. Its density at 4.2° K is 0.089 gm/cm^3. Until recently the only experimental determination of the compressibility of solid H_2 was the determination at 4.2° K for the range of pressures 0 to 100 atmospheres by Miss MEGAW[1]. Recently, however, J. W. STEWART[2] has extended the experimental determination of the 4.2° K isotherm to 20000 atmospheres and has also measured the same isotherm for solid helium. STEWART's measurements are extremely valuable for at the upper limit of his measured pressures the density of hydrogen has been more than doubled. His measurements, therefore, constitute a rather severe test to which theoretical calculations of the equation of state of solid hydrogen and helium can be subjected.

It has been suggested by WIGNER and HUNTINGTON[3] that solid hydrogen may eventually undergo a change to a metallic modification. WIGNER and HUNTINGTON made no attempt to evaluate the pressure at which their suggested phase transition might occur but did give a lower limit to the transition pressure of some 250000 atmospheres. An attempt to derive the complete 0° K isotherm of hydrogen for all pressures was made by KRONIG, DE BOER and KORRINGA[4] in 1946. Implicitly, they assumed that hydrogen would have only two allotropic modifications, the molecular crystal which is the stable phase under conditions accessible to laboratory measurements and the metallic modification. For the molecular phase, KRONIG, DE BOER and KORRINGA assumed that the intermolecular forces in the compressed solid were of the two-body type and that the two-body potential was of the MIE-LENNARD-JONES 6—12 form, i.e., that the potential energy of a pair of hydrogen molecules with mass centers separated by a distance R was expressible in the form

$$\varphi(R) = 4\varepsilon\left[\left(\frac{R_0}{R}\right)^{12} - \left(\frac{R_0}{R}\right)^{6}\right]. \tag{12.1}$$

The adjustable constants ε and R_0 measure the strength and range of the intermolecular force and were evaluated by DE BOER and MICHELS[5] from measured values of the second virial coefficient of gaseous H_2. DE BOER and MICHELS' values are

$$\varepsilon/k = 37° \text{ K}, \qquad R_0 = 2.98 \times 10^{-8} \text{ cm},$$

wherein k is BOLTZMANN's constant. By the use of this interaction potential the total potential energy per mole of hydrogen *atoms* can be evaluated by straightforward summation. On identifying the potential energy thus found with the internal energy U of the solid, the 0° K isotherm could then be calculated from the relation

$$p = -\left(\frac{\partial U}{\partial V}\right)_{T = 0° \text{K}}. \tag{12.2}$$

This relation is only true at absolute zero, the general equation being

$$p = -\left(\frac{\partial F}{\partial V}\right)_T \tag{12.3}$$

where F, the Helmholtz free energy, is the sum

$$F = U - TS \tag{12.4}$$

---

[1] H. D. MEGAW: Phil. Mag. **28**, 129 (1939).
[2] J. W. STEWART: J. Phys. Chem. Solids **1**, 146 (1956).
[3] E. P. WIGNER and H. B. HUNTINGTON: J. Chem. Phys. **3**, 764 (1935).
[4] R. KRONIG, J. DE BOER and J. KORRINGA: Physica, Haag **12**, 245 (1946).
[5] J. DE BOER and A. MICHELS: Physica, Haag **5**, 945 (1938); **6**, 409 (1939).

with S the entropy. Later workers[1,2] have refined these calculations by including the zero-point vibrational energy in the calculation of U. It is an interesting fact that, using the conventional DEBYE[3] theory, the zero point energy is imaginary at the measured volume of hydrogen at zero pressure for the square of the Debye frequency is a negative number[4]. KRONIG et al. followed the same general method of investigating the metallic phase of hydrogen as used by WIGNER and HUNTINGTON, the WIGNER-SEITZ method, but they omitted certain refinements. The phase transition was estimated by them as 700000 atmospheres at which pressure they found the density would change discontinuously from 0.40 gm/cm³ to 0.80 gm/cm³. Inclusion of the zero point energy by later investigators did not lead to a significant change in the estimate of the phase transition pressure or in the magnitude of the estimated density saltus. The magnitude of the density saltus, estimated in the manner sketched by KRONIG et al., has in large measure influenced the present degree of acceptance of RAMSEY'S speculated large changes in density associated with his concept of pressure ionization.

The calculations of KRONIG et al., particularly that portion of their calculations dealing with the molecular phase of hydrogen, are open to objections, however[5,6]. They can be criticized on many grounds, but DEMARCUS has indicated three which he regards as particularly cogent. First and foremost is the fact that the predicted equation of state is "too hard", i.e., the values of the incompressibility are too large as can be seen by direct comparison between the careful calculations of DE BOER and BLAISSE[7] and STEWART'S measures. Such a comparison is only fair at the high pressure end of STEWART'S experimental range because of the breakdown of the zero point energy calculation at the low pressure end. This breakdown would doubtless disappear if anharmonic terms were included in the evaluation of the zero-point energy. Secondly, the equation of state predicted on the basis of the 6—12 potential crossed the equation of state of a perfect Fermi gas of hydrogen for a value of density only slightly beyond the predicted value of the density molecular hydrogen would assume when in equilibrium with the metallic phase. Thirdly, DEMARCUS applies some general results of R. A. BUCKINGHAM[8] concerning the form of the two-body potentials at close distances of approach to show that the correct asymptotic form of the equation of state calculated from a hypothetical exact evaluation of the two-body forces would be too "soft". For the asymptotic form of this hypothetical calculation would have the form $P \sim \varrho^{\frac{5}{3}}$ instead of the correct non-relativistic value of the Fermi gas, $P \sim \varrho^{\frac{5}{3}}$. The Lennard-Jones 6-to-12 law predicts asymptotically $P \sim \varrho^{\frac{15}{3}}$. ABRIKOSOV[6] has calculated an equation of state of molecular H_2 using a two-body potential calculated by him from first principles by the use of wave functions of the WANG type[9] for the hydrogen molecules. However, his equation of state disagrees so markedly with STEWART'S measures at the lower end of the pressure range in which he states his calculations are valid that it does not appear prudent to discuss the presumed hydrogen phase transi-

---

[1] W. H. RAMSEY: Monthly Notices Roy. Astronom. Soc. London **111**, 427 (1951).

[2] J. DE BOER and B. S. BLAISSE: Physica, Haag **14**, 149 (1948).

[3] F. SEITZ: Modern Theory of Solids, Chap. III. New York 1940.

[4] J. DE BOER and B. S. BLAISSE: Physica, Haag **14**, 149 (1948).

[5] W. C. DEMARCUS: Astronom. J. **63**, 2 (1958).

[6] A. A. ABRIKOSOV: Problems of Cosmogony, Vol. III, pp. 11—19. Moscow: Academy of Sciences, USSR. 1954.

[7] J. DE BOER and B. S. BLAISSE: Physica, Haag **14**, 149 (1948).

[8] R. A. BUCKINGHAM: Univ. Wisconsin Naval Research. Lab., Report WIS-AEC-9, "The Repulsive Interaction of Atoms in S States" (Dec. 27, 1956).

[9] Cf. Sect. 5 of KOTANI'S contribution to Vol. XXXVII, Part 2, of this Encyclopedia.

tion on the basis of his calculations. DEMARCUS has concluded that a judicious extrapolation of STEWART'S measures seems to be indicated until our understanding or the quantum theory of molecular crystals is improved and has made two such extrapolations. One of these extrapolations joins smoothly with STEWART'S experimental data and one allows for some error in STEWART'S measures but the magnitude of the assumed error is well within STEWART'S estimated error of 5% in the relative volume changes.

There is very substantial agreement on the pressure-density relation of metallic hydrogen. However, WIGNER and HUNTINGTON'S original paper and KRONIG et al. employ an estimate for the correlation energy which has been revised by WIGNER[1], and DEMARCUS has repeated and extended WIGNER and HUNTINGTON'S calculations using the revised estimate of the correlation energy. Estimates of the phase transformation pressure range from 0.8×10^{12} dynes/cm^2 (RAMSEY) through 1.93×10^{12} dynes/cm^2 and 2.46×10^{12} dynes/cm^2 (DEMARCUS) to 3.6×10^{12} dynes/cm^2 (ABRIKOSOV).

It will appear presently that the equation of state of helium is also of importance for our understanding of the constitution of the Jovian planets. Unfortunately, there are no calculations of accuracy comparable to those on metallic hydrogen for the range of pressures applicable to planetary interiors. RAMSEY has calculated an equation of state from the Lennard-Jones 6—12 potential for helium and finds that the density of He thus calculated is about four times the density values calculated for hydrogen on the same basis. RAMSEY also quotes an unpublished calculation due to BALTENSPERGER who found that the helium would make a transition to a metallic modification at 16 million atmospheres pressure. DEMARCUS has obtained an equation of state for cold helium by joining STEWART'S experimental results smoothly onto a theoretical pressure-density relation which should be valid at high pressures. The theoretical relation was obtained by application of some results of RAIMES[2] who applied the Wigner-Seitz-Fröhlich method to the divalent elements. To adapt RAIMES' results to helium DEMARCUS dropped a term which accounts for the electrons in the ion core since there are no core electrons in helium. The interpolation between the experimental results at low pressures and the theoretical results at high pressures of necessity smooths out any phase transition which might occur. The question of the transition of helium to a metallic state has recently been discussed by TEN SELDAM[3] who has evaluated the density at which the S and P bands cross. He finds that helium should show metallic properties beginning at 190 million atmospheres. He also states, however, that it is not clear whether or not the

Table 7. *Experimental data for solid hydrogen and helium*[4] *at* 4.2° K.

Pressure (kg/cm²)	Molar volume (cm³/mole)		Pressure (kg/cm²)	Molar volume (cm³/mole)	
	hydrogen	helium		hydrogen	helium
0	22.65	(melted)	4000	14.3	9.3
200	21.0	15.8	6000	13.2	8.5
400	20.0	14.5	8000	12.4	8.0
600	19.2	13.5	10000	11.8	7.5
1000	18.0	12.4	12000	11.3	7.2
2043	16.1	10.7	16000	10.6	6.7
3000	15.1	9.9	20000	10.1	6.4

[1] E. P. WIGNER: Trans. Faraday Soc. **34**, 678 (1938).
[2] S. RAIMES: Phil. Mag. (7) **43**, 327 (1952).
[3] C. A. TEN SELDAM: Proc. Phys. Soc. Lond. A **70**, 97 (1957).
[4] J. W. STEWART: J. Phys. Chem. Solids **1**, 146 (1956).

Table 8. *Estimates of equation of state of hydrogen at 0° K.*

Molecular hydrogen			Metallic hydrogen			
P (10^{12} dynes/cm²)	ϱ (gm/cm³) RAMSEY	ϱ DEMARCUS	P (10^{12} dynes/cm²)	ϱ (gm/cm³) RAMSEY	ϱ (gm/cm³) DEMARCUS	Free enthalpy (10^{12} ergs/mole)
0	0.089	0.089	0.8	0.77	0.78	0.664
0.025	0.167	0.205	1.0	0.81	0.82	0.915
0.05	0.195	0.238	2.0	0.98	1.01	2.01
0.1	0.225	0.274	3.0	1.12	1.16	2.93
0.2	0.261	0.324	4.0	1.25	1.29	3.75
0.3	0.284	0.365	5.0	1.36	1.40	4.51
0.4	0.302	0.399	10.0	1.84	1.87	7.56
0.5	0.316	0.428	15.0	2.21	2.24	10.0
0.6	0.329	0.455	20.0	2.53	2.55	12.1
0.7	0.340	0.480	30.0	3.07	3.09	15.7
0.8	0.350	0.503				
1.0		0.548				
1.5		0.643				
2.0		0.732				
2.5		0.819				
3.0		0.903				

Table 9. *Estimated equation of state of 0° K helium*[1].

P (10^{12} dynes/cm²)	ϱ (gm/cm³)	P (10^{12} dynes/cm²)	ϱ (gm/cm³)	P (10^{12} dynes/cm²)	ϱ (gm/cm³)
0.0002	0.234	0.10	0.936	8.00	4.15
0.0010	0.323	0.20	1.15	10.00	4.56
0.0040	0.431	0.40	1.40	20.00	6.21
0.0100	0.534	0.80	1.77	30.00	7.58
0.0200	0.626	1.00	1.89	40.00	8.64
0.04	0.742	2.00	2.45	80.00	11.41
0.08	0.884	4.00	3.16		

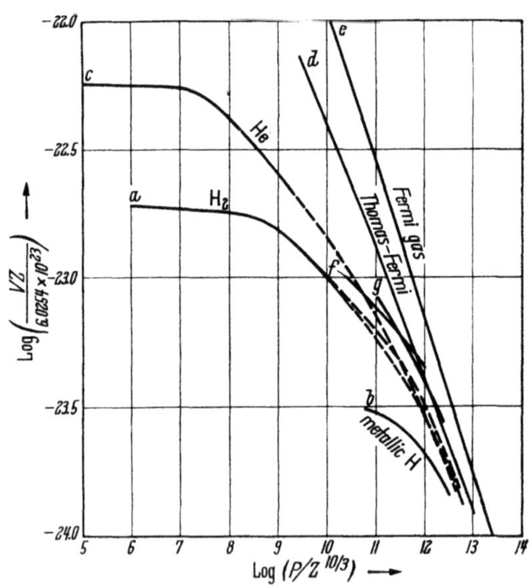

Fig. 3. *a* Equations of state of solid H_2 (after DEMARCUS). Solid portions cover pressure range of STEWART's experiments. *b* Metallic hydrogen (after DEMARCUS). *c* Equation of State of helium (Table 9). Solid portion covers pressure range of STEWART's experiments. *d* Thomas-Fermi atom model (after LATTER). *e* Fermi gas curve. *f* Equation of state of H_2 based on Lennard-Jones 6-12 potential (after RAMSEY). *g* Equation of state of H_2 (after ABRIKOSOV).

[1] W. C. DEMARCUS: Astronom. J. **63**, 2 (1958).

onset of metallic properties is accompanied by *any* discontinuity in density. Numerical results and estimates of the 0° K isotherm of hydrogen and helium are tabulated in Tables 7 to 9. Fig. 3 graphically displays the 0° K isotherms for hydrogen and helium which have been proposed along with the perfect Fermi-gas isotherm and the Thomas-Fermi isotherm[1].

13. Model planets of Jupiter and Saturn. Numerous models of the internal density distributions of Jupiter and Saturn have been proposed[2–7, 8]. The idea that hydrogen played a dominant role in their make-up seems to have crystallized with the publication of WILDT's 1947 review paper[9] on the constitution of the planets. The papers of RAMSEY, FESSENKOV and MASSEVIC, and DEMARCUS have followed up the 1947 suggestion of WILDT. The most careful calculations of models of Jupiter and Saturn on the basis of the equation of state of KRONIG et al. was made by RAMSEY and his collaborator, MILES, in a noteworthy pair of papers[10, 11]. RAMSEY's models have been built on the assumption that Jupiter and Saturn consisted of a uniform mixture of hydrogen and helium surrounding a heavy nucleus. Of this models, the ones he has designated as J_3 and S_2 yield resultant external gravitational potentials which come closest to the ones implied by the constants derived by BROUWER and CLEMENCE[12]. On the basis of the equation of state of hydrogen of KRONIG et al. and the assumption that helium densities are constant multiples of the hydrogen densities at corresponding pressures, these models represent homogeneous mixtures of hydrogen and helium surrounding cores of heavy elements in the amounts

	J_3	S_2
Mantle composition (hydrogen by weight)	0.846	0.818
Mass of mantle: mass of planet	0.948	0.831
Mass of heavy central particle: mass of planet	0.052	0.169
Planetary composition (hydrogen by weight)	0.80	0.68

The internal physical features of these models as function of the mean radius s are given in Tables 10 and 11.

More recently, DEMARCUS[13] has derived new models and abundance estimates based on his recalculation of the properties of metallic hydrogen and on his interpolatory equation of state for molecular hydrogen and helium. The final models used by DEMARCUS are detailed in Tables 12 to 14. Elemental abundance estimates for these two planets can only be performed if several restrictive assumptions are made. For a first estimate DEMARCUS assumed:

1. The density of hydrogen is less than the density of any other chemical substance under the conditions of pressure and temperature existing in the interiors of Jupiter and Saturn.

[1] R. LATTER: J. Chem. Phys. **24**, 280 (1956).
[2] H. JEFFREYS: Monthly Notices Roy. Astronom. Soc. Lond. **84**, 534 (1924).
[3] R. WILDT: Veröff. Univ. Sternwarte Göttingen No. 40, 1934. — Astrophys. Journ. **87**, 508 (1938). — Monthly Notices Roy. Astronom. Soc. London **107**, 84 (1947).
[4] H. BROWN: Astrophys. Journ **111**, 641 (1950).
[5] V. G. FESSENKOV and M. MASSEVIC: Russ. Astronom. J. **5**, 317 (1951).
[6] W. H. RAMSEY: Monthly Notices Roy. Astronom. Soc. London **111**, 427 (1951).
[7] W. H. RAMSEY and B. MILES: Monthly Notices Roy. Astronom. Soc. London **112**, 234 (1952).
[8] See footnote 1, p. 444.
[9] R. WILDT: See footnote 3, p. 445.
[10] W. H. RAMSEY: See footnote 6, p. 445.
[11] W. H. RAMSEY and B. MILES: See footnote 7, p. 445.
[12] See Table 2.
[13] W. C. DEMARCUS: See footnote 1, p. 444.

Table 10. *Ramsey's Jovian model JIII.*

s (10^9 cm)	p (10^{12} dynes/cm²)	ϱ (gm/cm³)	$M(r)$ (10^{27} gm)
6.99	0	0.10	1902
6.9	0.03	0.20	1894
6.7	0.16	0.28	1865
6.5	0.32	0.33	1832
6.3	0.52	0.36	1797
6.1	0.74	0.39	1761
6.05	0.80	0.40	1752
6.05	0.80	0.85	1752
5.5	2.56	1.17	1519
5.0	4.74	1.48	1291
4.5	7.50	1.79	1060
4.0	10.81	2.10	839
3.5	14.7	2.42	640
3.0	19.1	2.73	469
2.5	24.0	3.05	332
2.0	29.9	3.39	229
1.5	37.4	3.78	161
1.0	49.7	4.35	120
0	—	—	99

Table 11. *Ramsey's model of Saturn SII.*

s (10^9 cm)	p (10^{12} dynes/cm²)	ϱ (gm/cm³)	$M(r)$ (10^{27} gm)
5.755	0	0.10	569
5.6	0.02	0.19	560
5.4	0.07	0.24	544
5.2	0.13	0.28	526
5.0	0.20	0.30	507
4.5	0.43	0.35	460
4.0	0.72	0.40	418
3.88	0.80	0.40	409
3.88	0.80	0.86	409
3.5	1.43	0.99	349
3.0	2.47	1.17	277
2.5	3.82	1.38	217
2.0	5.69	1.61	169
1.5	8.63	1.93	134
1.0	14.7	2.46	113
0.5	39.1	3.90	102
0	—	—	96

Table 12. *Calculated properties of the interior of Jupiter* (after DEMARCUS).

β ($=s/R_1$)	ϱ (gm/cm³)	M (10^{27} gm)	P (10^{12} dynes/cm²)	W_{He}
1.000	0.00016	1902.0	0	0
0.994	0.138	1900.31	0.0068	0
0.98942	0.185	1897.15	0.0200	0
0.98942	0.197	1897.15	0.0200	0
0.86	0.593	1707.76	1.06	0.090
0.802	0.777	1591.13	1.93	0.13
0.802	1.08	1591.13	1.93	0.13
0.7	1.56	1270.27	5.07	0.16
0.6	2.12	940.33	9.52	0.23
0.5	2.66	630.96	15.3	0.25
0.4	3.14	379.85	21.9	0.26
0.3	3.58	203.05	29.2	0.26
0.2	4.08	100.37	37.6	0.27
0.15	4.40	72.19	43.2	0.27
0.1	19.09	33.98	63.5	1.0
0.05	27.90	5.19	96.3	1.0
0	30.84	0	110	1.0

W_{He} is the local weight fraction of "helium" assuming that all matter which is not hydrogen is helium.

2. The density[1] of helium is less than that of any element save hydrogen which is likely to be important to the constitution of Jupiter and Saturn.

3. The model planets specified in Tables 12 and 13 are correct in all details.

4. The pressure-density relations of molecular hydrogen and helium as given in Tables 8 (under DEMARCUS) and 9 are correct for whatever temperatures obtain in the two planets.

[1] To avoid excess verbiage, the term density is used rather loosely in this section. An example of a more precise statement is "quotient of molecular weight by partial molar volume".

Table 13. *Calculated properties of the interior of Saturn* (after DEMARCUS).

β $(=s/R_1)$	ϱ (gm/cm³)	M (10^{27} gm)	P (10^{12} dynes/cm²)	W_{He}
1.00	0.00016	569.00	0.000001	0
0.990	0.092	568.20	0.00186	0
0.970	0.185	561.44	0.0200	0
0.970	0.197	561.44	0.0200	0
0.95	0.236	551.86	0.0478	0
0.90	0.293	524.45	0.137	0
0.80	0.397	464.55	0.396	0
0.70	0.498	404.36	0.775	0
0.60	0.611	348.36	1.32	0
0.5227	0.719	309.78	1.93	0
0.5227	0.999	309.78	1.93	0
0.40	1.289	238.64	3.99	0
0.30	4.155	173.00	8.74	0.976
0.20	9.445	75.476	24.0	1.0
0.10	13.92	11.696	45.2	1.0
0	15.62	0	55.5	1.0

Table 14. *Coefficients of external potential.*
(The observed values are those deduced from satellite motions.)

	Jupiter		Saturn	
	model planet	observed	model planet	observed
ε_1	0.0645	0.0652	0.0988	0.0978
$\varkappa_1$	0.00049	0.00052	0.00195	0.00159
J	0.02130	0.02206	0.02594	0.02501
K	0.00250	0.00253	0.00588	0.00386

5. The pressure-density relation of metallic hydrogen and the values of the free enthalpy of Table 8 are correct for whatever temperatures obtain in the two planets.

6. The partial molar volumes of hydrogen and helium when the two occur in mixtures are equal to the molar volumes of pure hydrogen and helium.

7. The empirical data of the planets of Table 2 (p. 431) are correct.

Under these assumptions, it was found that the contribution of hydrogen to the masses of Jupiter and Saturn must exceed 78 and 63% by weight. DEMARCUS then calculates a comprehensive set of "partial derivatives" of the weight fractions of hydrogen of Jupiter and Saturn with respect to relaxation of the preceding assumptions essentially one at a time. His results showed that the lower limits to the abundances are quite insensitive to the relaxation of the assumptions 3 through 7 if we exclude the effects of temperature on the equation of state. The stability of the hydrogen abundances with respect to large variations in the equation of state of molecular hydrogen is perhaps the most surprising feature of the results. The reason can be traced to the effects of the phase transition. If the densities of hydrogen are lowered from the values of Table 8 the phase transition pressure will also be lowered. While the regions of the planets in which the molecular phase is present have their densities lowered, at the same time the volumes of these regions are decreased due to the stability of the metallic form at lower pressures. The cancellation of these two effects is almost complete. Even as regards the free enthalpy of the metallic phase, errors of a few kilocalories per mole lead to errors in the hydrogen abundances of a few percent.

The effects of temperature are not so easily dismissed. On the basis of the experimental data of MICHELS and GOUDEKET[1] for gaseous hydrogen supplemented by theoretical considerations, DEMARCUS concluded that the hydrogen abundances require only slight alterations if the temperatures in the deep interiors are only a few thousand degrees. There are good, but by no means conclusive, reasons for believing that the internal temperatures of Jupiter and Saturn are of this order of magnitude.

14. Uranus and Neptune. The compositions of Uranus and Neptune are much more uncertain than those for Jupiter and Saturn. It is certain that the hydrogen abundances of Uranus and Neptune are considerably lower than those appropriate to Jupiter and Saturn since their masses are smaller but their mean densities are higher. It seems quite plausible that metallic ammonium may be present in substantial amounts in these bodies as has been suggested by RAMSEY. Rough calculations[2] of the 0° K isotherm of this substance have been made and are available for attempts at quantitative models.

---

[1] A. MICHELS and M. GOUDEKET: Physica, Haag **8**, 347, 353, 387 (1941).

[2] M. J. M. BERNAL and H. S. W. MASSEY: Monthly Notices Roy. Astronom. Soc. London **114**, 172 (1954).

Radio Echoes from Sun, Moon and Planets.

By

F. J. KERR.

With 8 Figures.

Introduction. Radio astronomy has two main branches. In the one case, radiation emitted by celestial bodies is studied; in the other, signals are transmitted from the Earth, and echoes received after reflection from celestial bodies. The latter, or radar, method has the advantage that the signals are under the control of the observer, but it is limited in its application to the nearer bodies in the solar system. This restriction is a consequence of the double divergence of the energy which occurs in the two-way propagation path; the received intensity falls off as the fourth power of the distance, whereas in the direct reception method, as in optical astronomy, the second power is involved.

Echoes have so far been received from meteors and from the Moon. The Sun and the nearer planets will probably be reached in the near future, as the problems involved are only those of building sufficiently large equipment. Considerable advances in technique would be needed before echoes could be detected from the outer planets, and it seems unlikely that any of the stars could ever be reached by radar.

The extensive radar investigations of meteors will not be considered in this chapter, as they have been dealt with by LOVELL in Vol. XLVIII and by WHIPPLE and HAWKINS in this volume. A general discussion of the radar method will be followed by an account of Moon echo studies, and then the possibility of detecting echoes from the Sun and the planets will be considered. A brief account is also given of some related phenomena in the immediate neighbourhood of the Earth.

I. Strength and characteristics of echoes.

1. The radar equation. Consider a transmitter whose power output is P_T, and an aerial whose power gain relative to an isotropic radiator is G_T. The flux density (i.e. the power flow per unit area in the transmitted wave) which is incident on a reflecting body at distance d is given by

$$S_I = \frac{P_T G_T}{4\pi d^2} \tag{1.1}$$

provided the target subtends an angle smaller than the aerial beam.

The scattered wave intensity at the receiving aerial is then

$$S_R = \sigma \frac{S_I}{4\pi d^2} \tag{1.2}$$

where σ is the echo cross-section of the reflecting body (the geometric cross-section of an isotropic scatterer giving the same echo intensity).

Then, if the receiving aerial has an effective area A_R and a gain of G_R (commonly equal to G_T), and λ is the wavelength, the power delivered by the aerial to the receiver is

$$P_R = S_R A_R, \tag{1.3}$$

$$= S_R \frac{G_R \lambda^2}{4\pi}, \tag{1.4}$$

$$= \frac{P_T G_T G_R \lambda^2}{(4\pi)^3} \frac{\sigma}{d^4}. \tag{1.5}$$

An additional factor must be introduced if any deviation, absorption, or depolarization is produced by the medium between the Earth's surface and the reflecting body. Such effects are most likely to occur in the ionosphere, and, in the case of the Sun, the solar corona.

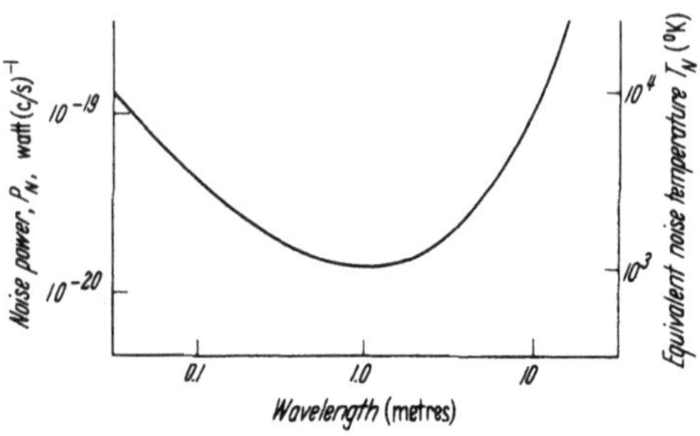

Fig. 1. Approximate magnitude of receiver noise and cosmic noise combined [14].

The ability to detect an echo is limited by the statistical fluctuations of the noise in the receiver output. This noise can be partly external, and partly internal in origin. External types of noise include solar noise, when the Sun is the target, and cosmic noise, when relatively long wavelengths are being used. Internal receiver noise, which is dominant, for example, in short wavelength observations of the Moon, is given by

$$P_N = (N-1)\, k T_0 B \tag{1.6}$$

where N is the receiver noise factor, k is BOLTZMANN'S constant, T_0 the ambient temperature, and B the noise bandwidth. Approximate values for the combined effects of receiver and cosmic noise are given in Fig. 1. This assumes a receiver of good noise factor, and a level of cosmic noise which is representative of the main part of the sky.

The output noise fluctuations can be reduced by integrating after the detector over a period of time containing many noise impulses. Integration over n noise pulses reduces the minimum detectable echo power by a factor $\sqrt{n}$. The effective number of noise pulses per unit time is proportional to the bandwidth B, so that integration for a time t increases the signal-to-noise ratio by a factor $\sqrt{Bt}$. The overall dependence on bandwidth is therefore:

$$\frac{P_R}{P_N} \propto \sqrt{\frac{t}{B}}. \tag{1.7}$$

2. Echo characteristics. Besides the very great distances involved, an important difference between astronomical and terrestrial radar practice lies in the great size of the targets in the line-of-sight direction. To "illuminate" the whole of the visible surface of the Moon or the Sun would require a pulse length greater than 12 milliseconds or 5 seconds respectively. Experimental results indicate that the effective echoing depth of the Moon is only a small fraction of the radius, so that, in this case, shorter pulses (< 1 msec) would be sufficient.

For a smooth, perfectly-reflecting sphere, whose radius is large compared with the wavelength, the incident energy is scattered uniformly in all directions, and the echo cross-section is equal to the projected area of the sphere. When the sphere is partially absorbing, as would normally be the case, the echo cross-section will be nearly equal to the product of the projected area (A) and the reflection coefficient for normal incidence (ϱ):

$$\sigma \approx A\,\varrho. \tag{2.1}$$

If the target is partially rough, the scattering is not isotropic, and the echo cross-section is given approximately by

$$\sigma = A\,\varrho\,D \tag{2.2}$$

where D is the directivity of the scattered energy in the direction of the receiver. To a first approximation, the echo from a large rough body can be taken as made up of specularly reflected components from elementary surfaces normal to the incident radiation. Thus all back scattering occurs at normal incidence, and the normal incidence reflection coefficient can be used.

Echoes from the Moon are surprisingly similar to those to be expected from a smooth reflector, at least at the higher frequencies, with a large proportion of the echo energy coming from the central part of the dish. There is evidence however that the Moon is intermediate between a smooth and a rough body in its echoing behaviour. Other astronomical bodies, in particular the Sun, are likely to be rougher reflectors.

An astronomical target will in general have a component of velocity along the line of sight, due in part to the orbital motions of the Earth and the target object, and in part to the observer's diurnal rotation about the centre of the Earth. In consequence, the returned echo will be displaced in frequency from the transmitted pulse, owing to the Doppler effect, and this must be allowed for in searching for echoes. Over the two-way path, the frequency shift produced by a line-of-sight velocity v is:

$$\Delta f = \frac{2v}{c} f. \tag{2.3}$$

If the target is rotating, the frequency shift will vary over its surface. For a rough reflector, in which the whole disk contributes to the returned echo, the reflected energy will be dispersed over a finite bandwidth; at a frequency of 30 Mc/s, for example, the solar rotation would produce a dispersion of about 1000 c/s in an echo from the Sun.

The power spectrum of the returned energy, or the manner in which the echo fluctuates, will therefore depend on the apparent "brightness" distribution across the disk of the target, which in turn depends on the type of scattering surface. Information on the reflection process can thus be derived from a study of the fading and this method has been employed in the case of the Moon[1]. No aerial has yet been used which is sufficiently directive to derive the brightness distribution directly.

3. Choice of equipment parameters. For maximum sensitivity the pulse length must exceed twice the propagation time across the line-of-sight extension of the echoing region. If the pulse length is greater than this value, integration of the returning echo can be usefully employed, until the pulse length approaches the travel time of the target and back again. In some cases, a practical limitation may operate, in that high transmitter powers become harder to achieve as the pulse length increases.

[1] Rapid fading is also discussed in Sect. 96 of H. Bremmer's article in Vol. XVI of this Encyclopedia.

In normal radar operation, the optimum receiver bandwidth is equal to the reciprocal of the pulse length. If the bandwidth is smaller than this value, part of the band which represents the spectrum of the transmitted pulse will be lost; if it is larger, excessive noise will appear at the receiver output, without any corresponding increases in the echo.

In astronomical radar, a similar argument applies, but this time the optimum receiver bandwidth is equal to the bandwidth of the echo spectrum, as dispersed on reflection[1].

The choice of wavelength will depend on the body concerned, and, to some extent, on the purpose of the investigation. For the Moon and the planets, a short wavelength is to be preferred, so that a high transmitting aerial gain can be more easily obtained, provided the wavelength is not below the limit of high-power transmitters. A longer wavelength, about 10 metres, is however to be preferred for any attempt at Sun echoes, as will be shown in Sect. 10 below.

The discussion so far has been concerned with obtaining maximum sensitivity in echo detection, for which fairly long pulses must be used. Shorter pulses are needed for any precise distance measurement by the radar method, or for a study of the fine structure of an echo. The technical problems are then greater, because the receiver bandwidth must be increased, and also the possibility of integration over the whole reflecting region is lost when excessively short pulses are used; the radio equipment must therefore be correspondingly more powerful. Accurate distance measurement could be carried out with the sharp leading edge of a long pulse, but the bandwidth requirement is just as onerous in this case as it is with short pulses.

II. Moon echoes.

4. Outline of experimental results. Echoes from the Moon were first detected by the U.S. Army Signal Corps in 1946 [*1*], [*2*]; this has historical importance as the first astronomical *experiment* ever carried out. The transmitter operated on a frequency of 111.5 Mc/s (2.7 metres), with a power of 3 kW (later increased to 15 kW), and a pulselength of $\frac{1}{4}$ second; the receiver bandwidth was 57 c/s, which was much higher than the optimum value, but any further reduction would have required too high a frequency stability in the transmitter. The aerial, comprising 64 dipoles and a reflecting screen, was directed horizontally, so that observations could be made only near moonrise and moonset. A view of the station is shown in Fig. 2, and a typical echo trace in Fig. 3.

The calculated echo strength was about 100 times the receiver noise level, but it was found that on several days echoes could not be detected at all, while only rarely did they approach the theoretical strength. In addition, the echoes were subject to rapid fading.

Almost simultaneously, Moon echoes were detected by BAY [*3*] in Hungary at 120 Mc/s, but in this case the equipment sensitivity was lower, and the echoes were only just detectable, even with the aid of a chemical integration system. Consequently, nothing could be found out about variations of echo strength.

The unexpected features in the American results led KERR and SHAIN [*4*], [*5*] in Australia to repeat the experiment at a lower frequency, in order to magnify any ionospheric effects which might be responsible. They used a simple arrangement, involving a short-wave broadcast transmitter, operating near 20 Mc/s. The transmitter was keyed over a landline from the receiving point, 600 km

[1] A suitable definition of bandwidth will allow for any difference between the two band shapes.

away. Both aerials were fixed, and their beams directed at low angles of elevation. The pulse-length was either $\frac{1}{4}$ or $2\frac{1}{4}$ seconds[1], and the bandwidth 70 c/s. At this lower frequency, the receiver sensitivity was limited by cosmic noise.

Fig. 2. The aerial array which was used at the Evans Signal Laboratory in the United States to obtain the first echoes from the Moon.

The echo strength was again found to be well below the expected level on most occasions, and furthermore the echoes always made their first appearance somewhat later than was expected as the Moon rose through the aerial beam. These effects were found to be connected with the F_2 region of the ionosphere. A form

[1] The echo delay-time varies between 2.4 and 2.7 seconds, as the Moon travels round its slightly elliptical orbit.

of short-period fading was shown to be due to the Moon's libration, and it was suggested that a longer-period fading might be related to ionospheric refraction.

Another systematic study of Moon echoes has been carried out by BROWNE, EVANS, HARGREAVES and MURRAY [7] in England, working at 120 Mc/s, with a pulse-length of 30 milliseconds, and a bandwidth of 26 c/s. A fixed aerial was again used, although the beam could be moved in elevation by electrical means; in this case the experiments were all carried out around lunar transit.

The connection between the rapid fading and libration was confirmed, but it was possible to show that the slow fading, at least at this frequency, is due

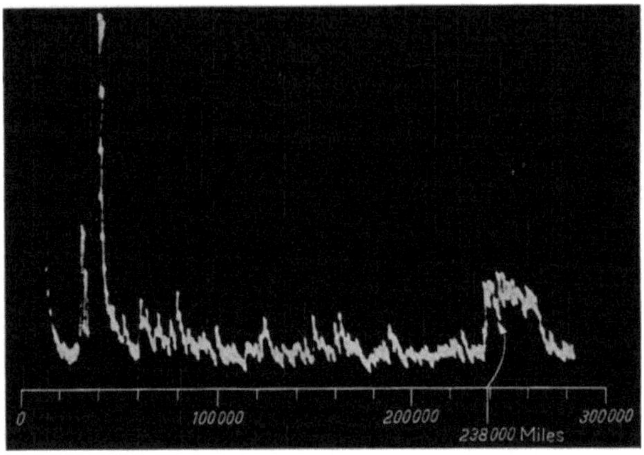

Fig. 3. A typical cathode-ray-tube trace, from the early U.S. Army Signal Corps experiments [1]. The photograph shows the direct signal from the transmitter at the left, followed by receiver noise fluctuations, and a Moon echo pulse at the right.

to the rotation of the plane of polarization of the waves by the ionosphere. This gives rise to a method by which the total electron content of the ionosphere can be measured.

The suggestion has been made at various times that communication circuits could be established between two points on the Earth by reflection from the Moon. The first experiments in this direction were carried out in the United States by the Collins Radio Company and the National Bureau of Standards [10], [11].

A considerable amount of new work on Moon echoes was reported by various groups during 1957 to 1958[1] covering a frequency range from 120 to 3000 Mc/s. Over the whole of this range, it was shown that the Moon reflected in a nearly-specular manner, although the surface was sufficiently rough to extend the echo pulse to some degree and produce the libration-type fading.

5. Echo intensities. In the original experiments of the U.S. Signal Corps at 111.5 Mc/s, the echo strengths were unexpectedly low, and subsequent workers also obtained echo strengths well below the expected values.

In their 20 Mc/s work, KERR and SHAIN compared the median intensity level for each day's test with ionospheric data, and found a good correlation with the critical frequency of the F_2 region (Fig. 4). A similar relationship was found for the altitude at which Moon echoes were first detected in each test.

[1] J. V. EVANS: Proc. Phys. Soc. Lond. **70**, 1105—1112 (1957). — J. H. TREXLER: Proc. Inst. Radio Engrs. **46**, 286—292 (1958). — B. S. YAPLEE, R. H. BRUTON, K. J. CRAIG and N. G. ROMAN: Proc. Inst. Radio Engrs. **46**, 293—297 (1958).

These results show that the anomalies must have arisen in the F_2 region, the departure from theoretical predictions indicating that there are shortcomings in the present theory relating oblique and normal incidence propagation conditions. Rather better agreement with the 20 Mc/s limiting altitude results has been obtained by CHVOJKOVA and LINK[1] in a new treatment of the refraction problem.

Some evidence was obtained in the 20 Mc/s experiments [5] suggesting that vertical deviations of a few degrees occurred during the passage through the ionosphere, but no direct measurements of such deviations have yet been reported.

The earlier comparisons between observed and theoretical echo strengths were all made in terms of a rough Moon, but Evans has now shown that the

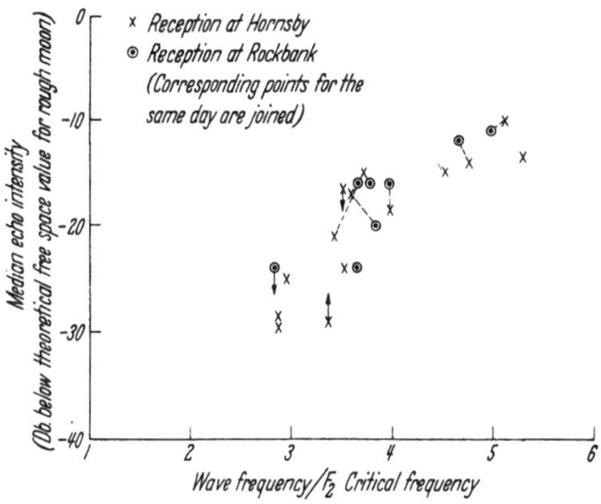

Fig. 4.

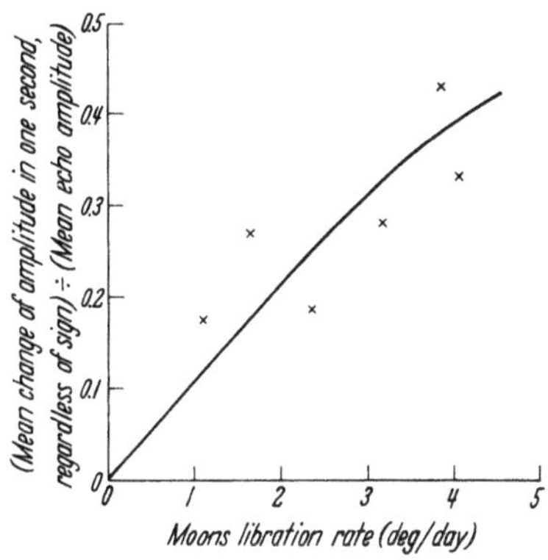

Fig. 5.

Fig. 4. Relation between median echo intensity and the critical frequency of the F_2 layer, in the experiments of KERR and SHAIN near 20 Mc/s [5].

Fig. 5. Relation between speed of rapid fading and libration rate [5]. The ordinate is a measure of the fading rate. The full line is a theoretical curve for a Gaussian approximation to the echo spectrum, while the crosses indicate experimental points for six days.

echo intensities which he observed at 120 Mc/s agree quite well with those calculated for a Moon reflecting in a nearly-specular manner. Agreement between theory and observation is good for higher frequencies also, but Trexler reports that there is still an unexplained difference in the 30 Mc/s region, where ionospheric effects are very important.

6. Doppler shifts and rapid fading. The Doppler shift of Moon echoes [Eq. (2.3)] arises from the rotation of the Earth, which gives an observer a velocity component in the line-of-sight, and also from the ellipticity of the Moon's orbit. This frequency displacement, which amounted to about 300 c/s under the circumstances of the original Signal Corps experiments, gave additional confirmation in the early stages that the echoes were in fact coming from the Moon.

In addition to these systematic motions of the whole Moon, an observer on the Earth sees an apparent oscillation of the Moon at the rate of 1 to 2°/day, about some (variable) axis through its centre. This effect, which is known as *libration*, has four main components, associated with the Moon's orbital motion and the Earth's rotation.

At any instant one half of the Moon is approaching the observer and the other half receding. At the limb, the radial velocity arising from libration amounts to about one metre/second, so that components of an echo coming from the

[1] A. CHVOJKOVA and F. LINK: Bull. Astr. Inst. Czech. 5, 100–103 (1954).

limb and from the centre will differ in frequency by about $0.005 f$ c/s, for a frequency of f Mc/s.

Thus, if the Moon reflects as a partially rough body, with echo components coming from a substantial proportion of the disk, the originally monochromatic transmitted wave will be Doppler-broadened on reflection. Assuming that the receiver bandwidth accepts the whole of the broadened "line", the instantaneous echo intensity will be the resultant of a very large number of components with random amplitudes and phases. Libration will vary the phases of these numerous components, so that the resultant intensity will vary from instant to instant, the rate of intensity variation depending on the rate of libration.

Studies of the rapid fading [5], [7] have in fact shown that the fading rate is proportional to the rate of libration (Fig. 5). This result demonstrates that the Moon is reflecting as a partially *rough* surface.

7. The reflection process. The Moon is approximately uniformly bright at optical wavelengths[1] and in the infrared[2], indicating that the surface is rough in this wavelength range. The early theoretical work on the reflection of radio waves from the Moon also treated the surface as rough. KERR and SHAIN [5] concluded from the connection between the rapid fading and libration that the whole of the Moon's disk was contributing to the echo at 20 Mc/s. They carried out some experiments with short pulses in an attempt to confirm this, but the results were inconclusive.

In 1957, EVANS showed that the rapid fading at 120 Mc/s was several times slower than that calculated for reflection from the whole Moon. It appeared that the power was being returned from a region of the Moon's disk whose diameter is only about a third of the full diameter, implying that reflection was nearly specular. He then showed that short pulses (2 milliseconds) were not noticeably lengthened on reflection, as would have occurred in an echo from the whole disk.

Similar conclusions were reported shortly afterwards by TREXLER, by YAPLEE *et al.*, and by other workers in England and Canada. TREXLER for example found that 50% of the echo energy at 198 Mc/s is reflected in the first 50 microseconds after the leading edge of a short incident pulse contacts the Moon. This means that half the reflected power comes from a circle whose diameter is one-tenth that of the Moon. YAPLEE *et al.* found that the echo at 2860 Mc/s consists of a number of short spikes which fade individually as libration changes the interference between contributions from a number of strong scattering centres near the front edge of the Moon. They obtained a value of 2500 km² for the echo cross-section at 2860 Mc/s with pulses of two microseconds duration.

The occurrence of quasi-specular reflection has been demonstrated at various frequencies in the range 120 to 3000 Mc/s, but the detailed variation of the echoing behaviour with frequency has not yet been worked out. It is clear however that over this frequency range the Moon is intermediate between a smooth and a rough body. No new information is yet available on the lower frequencies.

As explained in Sect. 2, it is convenient to split the echo cross-section of a large, partially-rough body such as the Moon into three factors, the projected area, reflection coefficient and directivity:

$$\sigma = A \varrho D. \tag{7.1}$$

[1] A. V. MARKOV: Astronom. J. USSR. **25**, 172–179 (1948).
[2] E. PETTIT and S. B. NICHOLSON: Astrophys. Journ. **71**, 102–135 (1930).

The directivity (D) can be derived from the distribution of echo energy across the Moon's disk. EVANS obtained a value of 1.8 at 120 Mc/s. The corresponding figures for a smooth and a perfectly rough sphere are 1 and 2.7 respectively [9], while the optical directivity of the Moon to sunlight is 5.7 [5].

The reflection coefficient (ϱ) can only be determined approximately, owing to the other uncertainties in echo strength computations, but the results obtained are consistent with a value of 0.1 to 0.15, which would be expected by analogy with terrestrial materials.

8. Longer-period fading and the magneto-ionic effect. In addition to the fast libration-type fading, KERR and SHAIN [5] observed a slower fading at 20 Mc/s, with a period of the order of several minutes, and suggested that the movement of irregularities in the ionosphere might be responsible. This explanation became untenable, however, when MURRAY and HARGREAVES [6] found that the longer-period fading was slower at 120 Mc/s than at 20 Mc/s, the period now being about 30 minutes. They were able to show that the fading is caused by a magneto-ionic effect[1] in the ionosphere, in which the plane of polarization of the radio waves is rotated in the presence of the Earth's magnetic field.

In the ionosphere, the ray is split into two components, circularly-polarized in opposite directions, which travel at different velocities. From the magneto-ionic theory, it can be shown[2] that the phase difference between the two components, which is produced by a double passage through the ionosphere, is given by:

$$\Omega = \frac{2}{\pi} \frac{e^3 c}{m^2 f^2} \int N\, B \cos\vartheta \sec\chi \, dh \quad \text{radians} \tag{8.1}$$

where f is the frequency, N the electron density, B the geomagnetic induction (gauss), ϑ the angle between the ray and the magnetic lines of force, and χ the angle made by the ray with the zenith at height h.

Thus a change in the total electron content of the ionosphere along the line of sight would produce a slow rotation of the plane of polarization of the returned signals (the resultant of the two components), giving rise to fading of echoes received on a plane-polarized aerial.

The effect was confirmed by receiving alternately on two aerials, linearly polarized in two direction at right angles [6], [7]; differential fading was observed on the two aerials, and the sum of the intensities in the two perpendicular planes was approximately constant, as would be expected in polarization fading. These results confirmed that the long-period fading is produced, at least in part, by changes of polarization in the received echo. The fact that the fading was slower at 120 Mc/s than at 20 Mc/s also supports this interpretation.

During these experiments, it was noticed [7] that the echo strength did not fall to zero at the minima, suggesting that the waves were being depolarized about 10% in the reflection process at the Moon.

Measurements at a single frequency cannot give the total rotation of the plane of polarization, as there is an ambiguity in Ω, equal to an integral number of whole revolutions. The ambiguity can however be resolved by observations at two frequencies. EVANS [8] has carried out such an experiment by switching every three quarters of a minute between 119.28 and 120.72 Mc/s. The signal-to-noise ratio shows a similar variation on the two frequencies, but the fading is more rapid on the lower frequency (Fig. 6).

[1] This effect has often referred to as the Faraday effect in the ionosphere, but this is a misnomer, as the classical Faraday effect involves the polarization of molecular dipoles.

[2] The detailed derivation is given by BROWNE *et al.* [7].

With the fixed aerial system which was used, the Moon could only be observed for about an hour each day near transit. In consequence, there was still an uncertainty in Ω of $\pm\frac{1}{2}$ rotation, but this could be eliminated by observations on successive days, assuming that the ionosphere behaved in the same manner from day to day. The total electron content in the line of sight, as derived from observations of this type at Jodrell Bank in October and November 1955, is shown

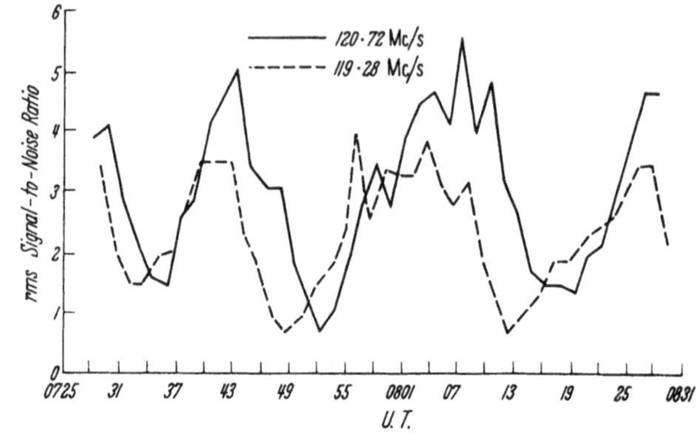

Fig. 6. Comparison between the signal-to-noise ratio of Moon echoes on two adjacent frequencies (EVANS [8], November 9, 1955).

in Fig. 7. The calculated electron content for a parabolic F_2 region is also given, based on critical frequency observations at Slough. The contributions of the E and F_1 regions, and the Earth-Moon space, to the total electron content are believed to be small compared with that of the F_2 region.

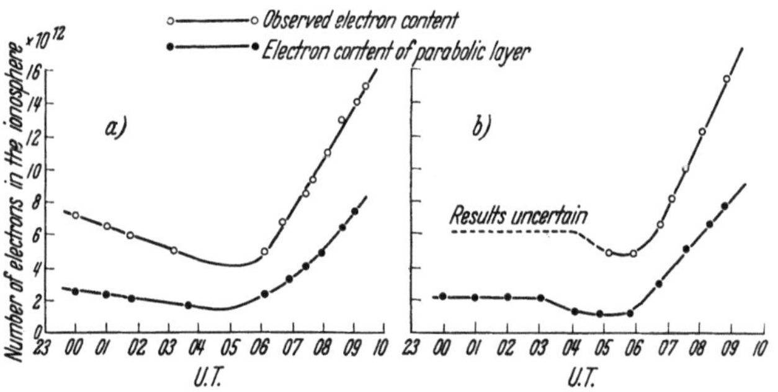

Fig. 7 a and b. Variation of the total electron content of the ionosphere, as determined from Moon echo fading observations, together with estimates from critical frequency data [8]. The ordinates give the number of electrons in the ionosphere per cm² along the line of sight. (a) October 1955. (b) November 1955; the early morning period is uncertain.

In the simple parabolic model it is assumed that there are equal numbers of electrons above and below the level of maximum ionization. The Moon echo results so far indicate that the electron content of the ionosphere is approximately twice that estimated for such a model. Thus about three-quarters of the electrons are above the maximum. A similar result was obtained by SMITH[1] from measurements of the refraction of radiation from discrete radio sources at 81.5 Mc/s.

The long-period fading (of the order of an hour) has been satisfactorily explained through the magneto-ionic effect, but it does not seem possible to account for the day-to-day variations at all frequencies by the same mechanism, as the time scales involved are so very different.

[1] F. G. SMITH: J. Atmos. and Terrest. Phys. 2, 350—355 (1952).

9. Applications of Moon-echo studies. The possibility of measuring the integrated electron content right through the ionosphere is the most important result so far obtained from Moon-echo experiments. The preliminary work described above has been restricted by the use of a fixed aerial, but the method should become of considerable importance when measurements can be made with an aerial capable of following the Moon.

A number of other possible applications have been proposed.

(a) One of the advantages of radar is the ability to measure distances with high accuracy. For this purpose a large receiver bandwidth is necessary, which means a considerable increase in the power requirements above those derived for echo-detection. The most accurate distance measurements made so far have been those by YAPLEE. The accuracy of radar measurements can be expected to surpass that obtainable by classical methods in due course, subject only to the uncertainty in the propagation velocity, which may differ from the velocity of light if the electron density is sufficiently high. Precise measurements of the echo delay-time at two different frequencies should enable the distance to the Moon and the electron density in the Earth-Moon space to be determined separately.

(b) It should be possible to measure the Doppler frequency shift very precisely, provided echoes could be received over a period of several hours. THOMAS[1] made a study of the factors involved, in 1949, following a suggestion that such a measurement might lead to a new determination of the velocity of light, or the radius of the Earth[2]. He concluded however that ionospheric variations and the random changes in the effective centre of reflection at the Moon produce uncertainties which would prevent a precise determination. A reconsideration of the value of such a measurement would be worth while at the present stage. The role of the ionosphere is now much better understood, the velocity of light is more accurately known, and the effect of the reflection variations could be eliminated statistically.

(c) Some of the major increases of solar radio noise might be detectable by reflection from the Moon. This could give the possibility of effectively observing some solar phenomena from two different orientations, at the ends of an Earth-Moon baseline. This type of reflection is not, strictly speaking, radar, but rather similar considerations apply. The flux density received by reflection from the Moon would equal that received directly from the Sun, multiplied by a factor $\frac{\sigma'}{4\pi d^2}$, where σ' is the reflection cross-section of the Moon in the direction concerned, and d is the distance of the Moon from the Earth. This factor is approximately equal to 10^{-6}. A large solar burst, attenuated by this factor, should be detectable with sensitive equipment, but the observation could only be successful at night, when the direct signal from the Sun would not interfere. Reflected solar noise does not yet appear to have been detected. STEINBERG and ZISLER[3] reported the observation of some bursts by reflection, but from the quantitative evidence presented it appears that some other phenomenon, possibly a terrestrial effect, must have led to an erroneous interpretation.

(d) The possibility of communicating between two points on the Earth by reflection from the Moon has often been suggested. Such a circuit would be free of the normal restriction of very high frequencies to short distances, but

[1] A. B. THOMAS: Austral. J. Sci. **11**, 187—192 (1949).

[2] Since the major part of the effect, for low-angle observations, comes from the Earth's rotation.

[3] J. L. STEINBERG and S. ZISLER: C. R. Acad. Sci. Paris **229**, 811—812 (1949).

would be operable only when the Moon was visible from both ends of the circuit. These possibilities were first considered in detail by GRIEG, METZGER and WAER [9]. Because the Moon extends a short pulse to some extent the bandwidth of such a circuit is limited to a few kilocycles; also the echo travel-time of $2\frac{1}{2}$ seconds slows down the exchange of information. However, the operation of experimental circuits by SULZER, MONTGOMERY and GERKS [10], [11] and by TREXLER and others, has shown that, within these limitations, a system of this type is feasible.

III. Future possibilities.

10. The Sun. Although several of the planets approach more closely to the Earth, the Sun presents the best possibility (after the Moon) of returning detectable echoes, because of its much greater size. Also it is a more interesting object for study than the Moon or the planets. An order of magnitude for the power required to produce a detectable echo can be obtained by the following simplified argument:

The angular diameter subtended by the Sun at the Earth is nearly the same as that of the Moon ($\frac{1}{2}°$), so that it will receive about the same total flux from a terrestrial transmitter. Neglecting for the moment the different reflecting properties of the two bodies, the decrease in the Sun's echo relative to that from the Moon will be due solely to the greater distance travelled by the diverging reflected radiation. Since the Sun is about 400 times as distant as the Moon, the Sun's echo will be smaller by a factor of $(400)^2$.

There are however several additional difficulties.

(i) The incident energy would be partially absorbed in the solar atmosphere: at frequencies much above 100 Mc/s, in fact, the reflection coefficient is likely to be vanishingly small.

(ii) The Sun produces a high and variable output of radio-frequency radiation itself and the echo has to exceed this level.

(iii) The solar rotation would disperse the echo energy over a substantial range of frequency, e.g. 500 to 1000 c/s at 30 Mc/s.

As a partial compensation for these factors, reflection would occur high in the solar corona at low frequencies, so that the diameter of the Sun's radio disk would be greater than that of the optical disk.

There are considerable uncertainties in any calculation of echo strength, especially with regard to the reflecting properties of the corona, but it appears [12] that the detection of usable echoes from the Sun would be technically possible, using currently available techniques, though an engineering project of great magnitude would be involved. There is an optimum frequency range, at around 25 to 30 Mc/s, principally determined by the variation of the Sun's reflection coefficient with frequency. The use of a low frequency is indicated on astronomical grounds also, since interest centres largely in obtaining echoes from the outer corona and surrounding clouds or streams of charged particles, because of the difficulty of optical observation in these regions.

The radar method is probably the best way in which turbulence in the corona can be studied. The frequency dispersion of the echo would give information about the internal motions in the scattering region, and different levels in the corona could be picked out by working at different frequencies.

11. The planets. The possibility of detecting echoes from any of the planets can most simply be considered in terms of the requirements for the Moon, as this avoids the necessity of specifying particular values of transmitter power and aerial gain. The relevant quantities are listed in Table 1. The Sun, and the closest bright star, α Centauri, have been added for comparison.

As we are interested in orders of magnitude only, we will neglect the differences in reflection coefficient and directivity between the various bodies. The Sun and some of the planets probably reflect as rough bodies, whereas the Moon appears to be fairly smooth, but the effect of this difference is small compared with the other factors involved.

The primary comparison between two targets depends on their projected area and distance. The echo strength is proportional to $\frac{\pi r^2}{d^4}$ [see Eq. (1.5)], but differences in performance will also arise from the different pulse length and bandwidth requirements. The table gives in each case the length of a pulse which

Table 1. *Comparison of conditions for the detection of echoes from astronomical bodies.*

Target	Distance at near approach d (metres)	Radius[1] r (metres)	Rotation period[2] (days)	"Minimum" pulse length (seconds)	Echo travel time (seconds)	Bandwidth B at 100 Mc/s (c/s)	Echo strength $\frac{\pi r^2}{d^4}\sqrt{\frac{t}{B}}$ relative to Moon
Moon	4×10^{8}	1.7×10^{6}	100 [3]	0.012	2.5	1.7	1
Sun	1.5×10^{11}	7×10^{8}	27	5	1000	2500	4×10^{-6}
Venus	4×10^{10}	6.2×10^{6}	10 [4]	0.04	270	60	2×10^{-7}
Near asteroid	2×10^{9}	2×10^{3}	0.2 ?	0.00001	12.5	1	2×10^{-8}
Mercury	8×10^{10}	2.5×10^{6}	88	0.017	500	3	1.5×10^{-8}
Mars	6×10^{10}	3.4×10^{6}	1	0.023	400	300	8×10^{-9}
Jupiter	6×10^{11}	7.0×10^{7}	0.4	0.5	4000	1700	4×10^{-10}
Saturn	1×10^{12}	5.8×10^{7}	0.4	0.4	7000	1400	5×10^{-11}
α Centauri	4×10^{16}	9×10^{8}	27 [5]	6	$8\frac{1}{2}$ yrs.	3000	7×10^{-25}

would completely illuminate the visible hemisphere. (For a rough body this is the minimum time to avoid loss of sensitivity.) The echo travel time is also quoted, and post-detector integration for the latter period has been assumed. The bandwidth given is the minimum value allowed by the target's velocity of rotation: for definiteness, the values listed are for 100 Mc/s in each case. For a bandwidth B and integration period t, (1.7) indicates that the echo strength will be proportional to

$$\frac{\pi r^2}{d^4}\sqrt{\frac{B}{t}}. \tag{11.1}$$

This quantity is listed in the last column of the table.

These calculations show the magnitude of the problem involved in extending the radar technique beyond the Moon. The possibilities of building sufficiently powerful equipment are considered in Refs. [*12*] and [*14*], for the Sun and planets respectively. These discussions indicate that the detection of echoes from the Sun and Venus, and perhaps Mercury and Mars, would be difficult but feasible.

[1] Optical values. The radio disk will be substantially larger for the Sun and α Centauri, the effective size depending on the frequency used. Planetary ionospheres might similarly increase the values for some of the planets, but the proportional changes would be small.

[2] Sidereal period (except in cases of Sun and Moon). The effective period, as viewed from the Earth, must take into account the orbital motions of the Earth and the planet. The resultant value is variable.

[3] Effective angular velocity arising from libration (order of magnitude only).

[4] Not known. Order of magnitude only.

[5] Solar value taken.

The optimum wavelength for planetary work would depend on technical factors at the time such a project was being undertaken, but is likely to be shorter than the optimum for the Sun. Developments in short-wavelength transmitting tubes may in fact make the nearer planets easier to reach by radar than the Sun.

Little would be gained from an experiment in which planetary echoes were just detectable. Useful information can only be obtained from the characteristics of an echo—its strength, spectrum, variations, etc. A substantial margin of power is therefore needed over that required for mere detection.

The most valuable objective would be an accurate measurement of the distance to a planet. The relative geometry of the solar system is known to a high degree of accuracy from measurements of angular positions of the various bodies, but the absolute scale is known much less accurately. The two best optical determinations of the astronomical unit of distance should be correct to one part in 10000, but they differ by one part in 1000 [*13*]. A radar measurement would be very direct in principle, but to obtain the desired accuracy fairly short pulses must be used.

Other possible future uses of a planetary radar system include:

(i) A measurement of the period of rotation of Venus from the fading statistics of the echo. (The period cannot easily be determined optically.)

(ii) A search for planetary ionospheres, which might show up through slow variations in the echo strength. It has also been suggested[1] that polarization studies might give information on the magnetic field of a planet through the magneto-ionic effect in the planetary atmosphere.

(iii) Echoes might be obtained from some of the smaller bodies in the solar system, such as the asteroids which approach most closely to the Earth. New bodies in this class might possibly be discovered. A study could also be made of the Gegenschein, a concentration of particles on the side of the Earth opposite to the Sun, in the hope of obtaining information about the size of the particles [*13*].

IV. The Earth's immediate neighbourhood.

12. Pulse echo techniques have been used for many years in studies of the ionosphere up to the level of maximum ionization in the F_2 region. Extensions of these techniques have enabled echoes to be detected from meteor trails[2] and from auroral arcs[3]. In all three cases the echo studies yield both geophysical information about conditions in the upper atmosphere, and astronomical information about the material (meteors and charged particles) incident on the Earth from outside.

It has recently been found that a natural form of radar can be used to study the space around the Earth. At very low frequencies (around 10 kc/s) a whistling form of atmospheric disturbance is often heard. These "whistlers", as they are called, are sometimes, but not always, associated with "click" type atmospherics, and they frequently occur in groups, with a constant spacing of several seconds between successive whistlers, and a rate of change of pitch which decreases from one whistler to the next.

Following up a suggestion originally made by ECKERSLEY[4], STOREY [*15*] developed a theory to account for the observed characteristics of whistlers. In

[1] B. BURRELL: Observatory **77**, 118—119 (1952).

[2] A. C. B. LOVELL: Volume XLVIII of this Encyclopedia, p. 427.

[3] K. BULLOUGH and T. R. KAISER: J. Atmos. and Terrest. Phys. **5**, 189—200 (1954); **6**, 198—214 (1955). — H. G. BOOKER, C. W. GARTLEIN and B. NICHOLS: J. Geophys. Res. **60**, 1—22 (1955).

[4] T. L. ECKERSLEY: Nature, Lond. **135**, 104—105 (1935).

this theory, the long time-delays in whistlers are due to propagation along a path extending out from the Earth to a distance of several Earth-radii; the radio waves are guided by magnetic lines of force, striking the Earth again at a point in the opposite hemisphere, where they are reflected back along their original path. The characteristic falling tone results from the variation of propagation time with frequency.

One type of whistler is generated by a lightning flash in the vicinity of the observing station (see Fig. 8). Successive members of a train of whistlers then correspond to repeated trips to the conjugate point in the opposite hemisphere and back again. In the other type, the lightning flash is in the opposite hemisphere, so that no direct atmospheric click is heard. The whistlers in this case are received after 1, 3, 5, 7, ... trips over the long path. STOREY's theory has been confirmed by simultaneous observations in Alaska and New Zealand, at geomagnetically conjugate points [*16*]. Whistler trains with delays 1, 3, 5, 7, ... at one station were shown to correspond to trains with delays (0), 2, 4, 6, ... at the other station.

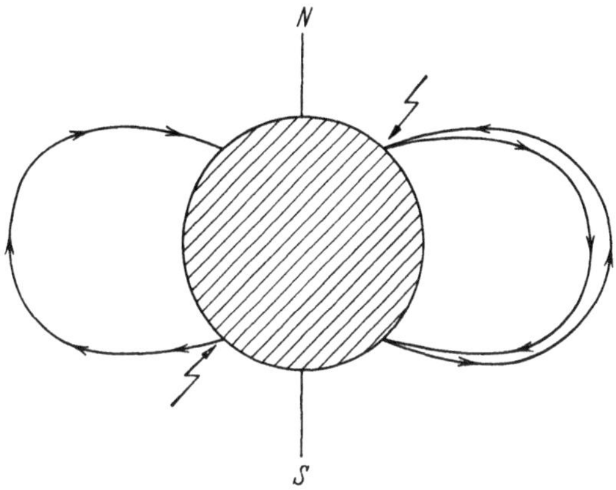

Fig. 8. Probable paths of the two types of whistler for an observer in the Northern Hemisphere. The forked arrow indicates the position of the lightning flash [*15*].

The frequency dispersion of whistlers depends on the electron density and magnetic field along their path. Observations of whistlers can thus provide information about conditions in a region which can be regarded as well out in interplanetary space.

The two parameters, electron density and magnetic field, can be separated by using an extension of the theory, formulated independently by ELLIS [*17*] and by HELLIWELL et al. [*18*]. These authors have shown that the variation of magnetic-field intensity along a line of force produces an upper frequency limit for whistler propagation. This limit can be used to derive the minimum value of magnetic field along the path, i.e. the field in a region far above the equator. Whistlers with such a frequency limit have been detected [*18*], but estimates of the field strength have not yet been published.

Whistler observations have indicated that the electron density in the outer part of the propagation path is of the order of 400 electrons/cm³ [*15*]. This value was unexpectedly high, but subsequent observations of the zodiacal light by BEHR and SIEDENTOPF[1] have yielded comparable values.

A related subject, which remains rather a mystery in the literature, is that of the "long-delay echoes" at about 10 Mc/s, which were reported by STÖRMER, VAN DER POL and others around 1928. These echoes had time delays of the order of 10 seconds. One suggestion was that they came from a ring current around the Earth, produced by charged particles from the Sun, but it is now believed that such a ring would be too close to the Earth. Another was that the long time delays arose from a long low-velocity path through the upper ionosphere. These echoes have never been satisfactorily explained and recent attempts to detect them have been unsuccessful[2].

[1] A. BEHR and H. SIEDENTOPF: Z. Astrophys. **32**, 19–50 (1953).
[2] K. G. BUDDEN and G. G. YATES: J. Atmos. and Terrest. Phys. **2**, 272–281 (1952).

General references.

Moon-echoes—early experiments.

[1] MOFENSEN, J.: Radar echoes from the moon. Electronics **19**, 92—98 (1946).

[2] DEWITT jr., J. H., and E. K. STODOLA: Detection of radio signals reflected from the moon. Proc. Inst. Radio Engrs. **37**, 229—242 (1949).

[3] BAY, Z.: Reflection of microwaves from the moon. Acta phys. Hungar. **1**, 1—22 (1946).

Moon-echoes—systematic studies (to 1956).

[4] KERR, F. J., C. A. SHAIN and C. S. HIGGINS: Moon echoes and penetration of the ionophere. Nature, Lond. **163**, 310—313 (1949).

[5] KERR, F. J., and C. A. SHAIN: Moon echoes and transmission through the ionosphere. Proc. Inst. Radio Engrs. **39**, 230—242 (1951).

[6] MURRAY, W. A. S., and J. K. HARGREAVES: Lunar radio echoes and the Faraday effect in the ionosphere. Nature, Lond. **173**, 944—945 (1954).

[7] BROWNE, I. C., J. V. EVANS, J. K. HARGREAVES and W. A. S. MURRAY: Radio echoes from the moon. Proc. Phys. Soc. Lond. **69**, 901—920 (1956).

[8] EVANS, J. V.: The measurement of the electron content of the ionosphere by the lunar radio echo method. Proc. Phys. Soc. Lond. **69**, 953—955 (1956).

Communication via the Moon.

[9] GRIEG, D. D., S. METZGER and R. WAER: Considerations of moon-relay communication. Proc. Inst. Radio Engrs. **36**, 652—663 (1948).

[10] SULZER, P. G., G. F. MONTGOMERY and I. H. GERKS: An U-H-F moon relay. Proc. Inst. Radio Engrs. **40**, 361 (1952).

[11] Via the moon. Sky and Telescope, **11**, 194—195 (1952).

Future possibilities.

[12] KERR, F. J.: On the possibility of obtaining radar echoes from the sun and planets. Proc. Inst. Radio Engrs. **40**, 660—666 (1952).

[13] LOVELL, A. C. B. and J. A. CLEGG: Radio Astronomy, Chaps. 20—22. London: Chapman & Hall 1952.

[14] PAWSEY, J. L. and R. N. BRACEWELL: Radio Astronomy, Chap. IX. Oxford: Clarendon Press 1955.

The neighbourhood of the Earth.

[15] STOREY, L. R. O.: An investigation of whistling atmospherics. Phil. Trans. Roy. Soc. Lond., Ser. A **246**, 113—141 (1953).

[16] MORGAN, M. G., and H. E. DINGER; M. G. MORGAN and G. McK. ALLCOCK: Observations of whistling atmospherics at geomagnetically conjugate points. Nature, Lond. **177**, 29—31 (1956).

[17] ELLIS, G. R.: On the propagation of whistling atmospherics. J. Atmos. and Terrest. Phys. **8**, 338—344 (1956).

[18] HELLIWELL, R. A., J. H. CRARY, J. H. POPE and R. L. SMITH: J. Geophys. Res. **61**, 139—142 (1956).

a

b

c

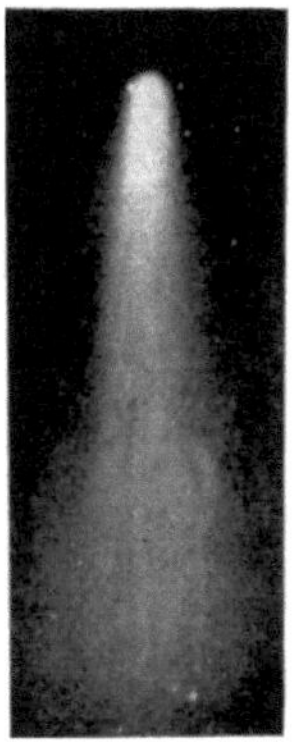

d

e

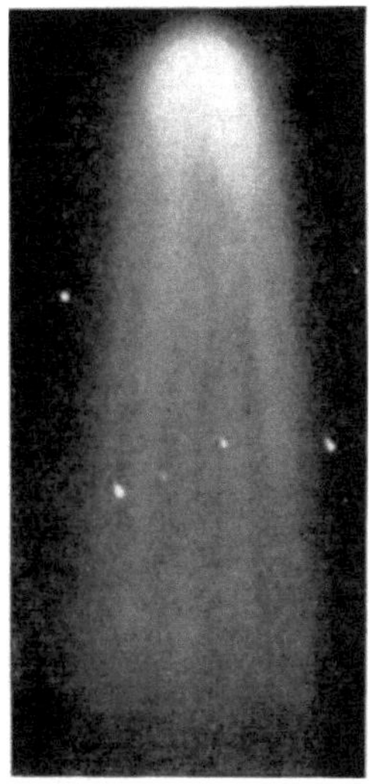

f

Fig. 1a—f. Komet Halley 1910 bei verschiedenen Sonnenabständen. a. $r = 5{,}0$ A.E.; b. $r = 2{,}3$; c. $r = 1{,}5$; d. $r = 0{,}60$; e. $r = 0{,}60$; f. $r = 0{,}60$. [Entnommen N. T. Bobrovnikoff, Lick Obs. Publ. **17**, Part II (1931).]

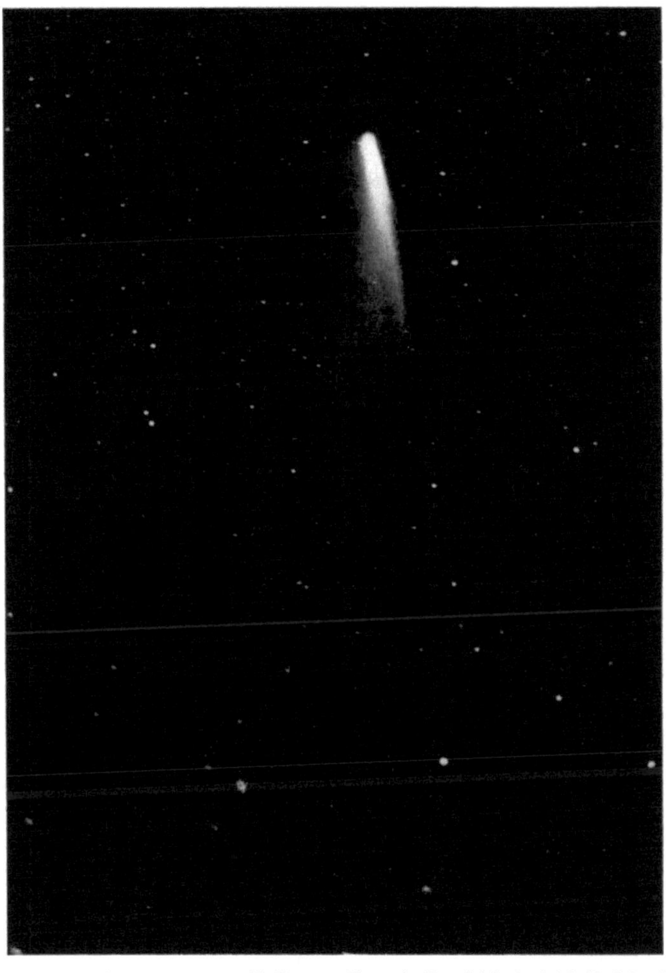

Fig. 2. Komet Mrkos 1957d, 17. August 1957 (G. Romano, Treviso). Das Original zeigt beide Schweife in nahezu doppelter Länge wie in dieser Reproduktion.

Die Kometen.

Von

K. Wurm.

Mit 31 Figuren.

I. Einleitung.

1. Allgemeine Formen und allgemeines Verhalten der Kometen. Im Durchschnitt werden jetzt jährlich sechs bis sieben Kometen entdeckt. Es sind dies größtenteils Neuentdeckungen, da nur der eine oder andere von ihnen zu den schon bekannten periodischen Objekten gehört. Trotz dieser nicht geringen Zahl von Kometenerscheinungen hat die Kometenforschung doch ihre Schwierigkeiten, sich das für ihre Weiterentwicklung geeignete Beobachtungsmaterial zu beschaffen. Die Mehrzahl der Erscheinungen ist außerordentlich schwach und bringt kaum mehr als einige Positionen für eine angenäherte Bahnbestimmung. Große Kometen stehen zur Zeit ihrer größten Helligkeit meist nahe der Sonne, so daß ihre Beobachtung weitgehend auf die Zeit kurz nach bzw. kurz vor der Dämmerung beschränkt bleibt.

Tabelle 1.
Kometenentdeckungen (nach M. F. Baldet, Extrait de l'Annuaire du Bureau des Longitudes pour l'an 1950)

Vor Chr. Geb.	Anzahl
2500—2000	6
—1500	5
—1000	6
— 500	10
0	118
Nach Chr. Geb.	**Anzahl**
0— 500	172
—1000	207
—1500	277
1500—1600	77
—1700	38
—1800	77
1800—1820	31
—1840	35
—1860	82
—1880	79
—1900	124
1900—1910	45
—1920	52
—1930	52
—1940	61
1940—1950	65

Der wesentlichste Teil eines Kometen ist sein *Kern*, der nur einige Kilometer Durchmesser hat und im Fernrohr mit seiner nächsten Umgebung als ein heller, sternförmiger Lichtpunkt erscheint (Fig. 1 b)[1]. In ihm ist die ganze Masse konzentriert, und bei größeren Sonnendistanzen (etwa $r > 5$ Astronomische Einheiten) besteht ein Komet aus nichts anderem. Nähert er sich auf seiner Bahn der Sonne, so beginnt der Kern Gas und Staub auszuwerfen, und diese „Aktivität" steigert sich bis zur größten Sonnennähe, dem Perihel. Um den Kern bildet sich mit dessen Gas- und Staubabgabe erst einmal der *Kometenkopf (Coma)*, eine rundliche Atmosphäre mit einem Durchmesser zwischen einigen zehntausend bis zu einigen hunderttausend Kilometern. Die Flächenhelligkeit in den Köpfen fällt vom Kern aus nach außen ab, die Kopfränder sind meist schwer festzulegen. Der Staub wird erkenntlich durch ein gestreutes oder diffus reflektiertes Fraunhofer-Spektrum, die Gase werden nachweisbar durch Banden und Atomlinien. Die relativen Anteile von Gas und Staub wechseln in weiten

[1] Was im Fernrohr als Kometenkern erscheint ist gewöhnlich nicht der Kern allein, sondern dieser einschließlich des ihm dicht anliegenden Teiles der Gas-Staubatmosphäre des Kometenkopfes. Die Atmosphäre des Kometenkopfes hat ihre größte Dichte in der unmittelbaren Umgebung des Kometenkernes.

Grenzen, vielfach ist der Staub kaum festzustellen, in anderen Fällen fehlt das Gasspektrum.

Die *Schweife* entwickeln sich gewöhnlich erst für heliozentrische Distanzen $r < 1{,}5$. Die normalen Schweife sind stets von der Sonne abgewendet. Sie verdanken ihr Entstehen zum Teil dem solaren Lichtdruck, zum Teil wahrscheinlich der Einwirkung solarer Korpuskelströme. Die Wirkung der letzteren ist im

Fig. 3. Komet Morehouse 1908. (Aufnahme MAX WOLF, Reflektor der Sternwarte Heidelberg.)

eigentlichen Mechanismus noch ungeklärt. Gelegentlich, aber doch relativ selten, werden sog. *anomale Schweife* sichtbar, die in der Projektion auf den Himmelshintergrund mit dem verlängerten Radiusvektor Winkel bis zu 180° bilden, also zur Sonne gerichtet sind (s. Fig. 25 u. 26 auf S. 503).

Größte Helligkeiten und größte Schweiflängen werden gewöhnlich kurz nach dem Perihel erreicht. Gas-Staubkometen zeigen in der Regel Doppelschweife in Richtung des verlängerten Radiusvektors, einen wenig oder gar nicht gekrümmten *Gasschweif* und einen in der Bewegung dem Kometen nachfolgenden stärker gekrümmten *Staubschweif* (s. Fig. 2 und 31). Im Kopf fließen beide ineinander. Beide Schweife schmiegen sich eng an die Bahnebene an und sind deshalb aus perspektivischen Gründen nicht getrennt erkenntlich, wenn die Erde in oder nahe der Bahnebene steht.

Die Gasschweife bestehen fast immer aus einem Bündel von Strahlen (Fig. 3 und 4), die aus der nächsten Umgebung des Kometenkernes hervorbrechen. Ihre Strukturen ändern sich von Nacht zu Nacht meist radikal, häufig schon innerhalb weniger Stunden.

Die Gas- und Staubmassen in Kopf und Schweif gehen dem Kometen ständig verloren und werden vom Kern her erneuert, vielfach in diskontinuierlichen Aus-

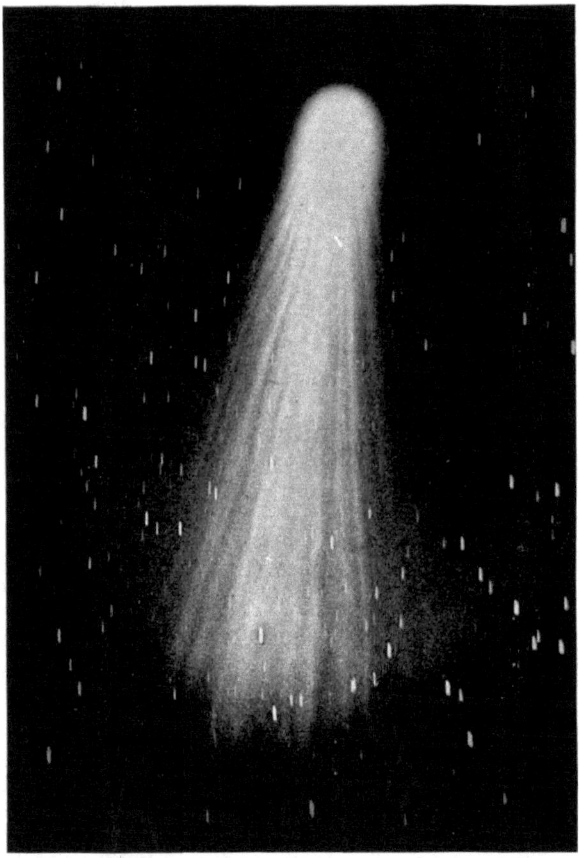

Fig. 4. Komet Daniel 1907. (Aufnahme MAX WOLF, Reflektor der Sternwarte Heidelberg.)

brüchen. Bei den Materie-Ausströmungen, ob diese nun kontinuierlich oder stoßartig erfolgen, wird die der Sonne zugewandte Richtung bevorzugt. Dabei bilden sich nicht selten parabolische Enveloppen um den Kern (Fig. 5).

Helligkeitsmessungen an Kometen *einschließlich* des Schweiflichtes lassen sich kaum durchführen. Zahlreich vorhanden sind dagegen Helligkeitsmessungen für das Kopf- und Kernlicht. Die in der Literatur publizierten Helligkeiten beziehen sich gewöhnlich, wenn nicht anders vermerkt, auf den ganzen Kometenkopf. Bei der Zusammenfassung und Reduktion der Helligkeiten H für ein Objekt wird von den Beobachtern meist durch Ausgleichung eine Darstellung in der Form

$$H = \frac{H_0}{\varDelta^2 \cdot r^n} \tag{1.1}$$

versucht, worin H_0 die Helligkeit für die heliozentrische Distanz $r = 1$ und $\varDelta$ die Distanz von der Erde bedeutet. Der „Aktivitäts-Exponent“ n schwankt — von extremen Fällen abgesehen — zwischen 2 und 6 mit einer Häufung um den Wert 4.

Die bisher vorliegenden Helligkeitsmessungen sind größtenteils visueller Art[1]. Dem allmählichen Anstieg der Helligkeit zum Perihel überlagern sich fast immer unregelmäßige Fluktuationen. Monochromatische Helligkeitsmessungen an einzelnen Banden oder innerhalb des Kontinuums, die physikalisch bedeutsamer wären als die Daten über visuelle oder photographische Totalhelligkeiten, fehlen leider vollständig.

Da jeder Komet in Perihelnähe ständig Materie verliert, so taucht die Frage auf, nach welcher Zeit im Durchschnitt die Fähigkeit zu einer Atmosphären-Bildung erlischt. Ausreichend präzise Helligkeitsmessungen reichen historisch

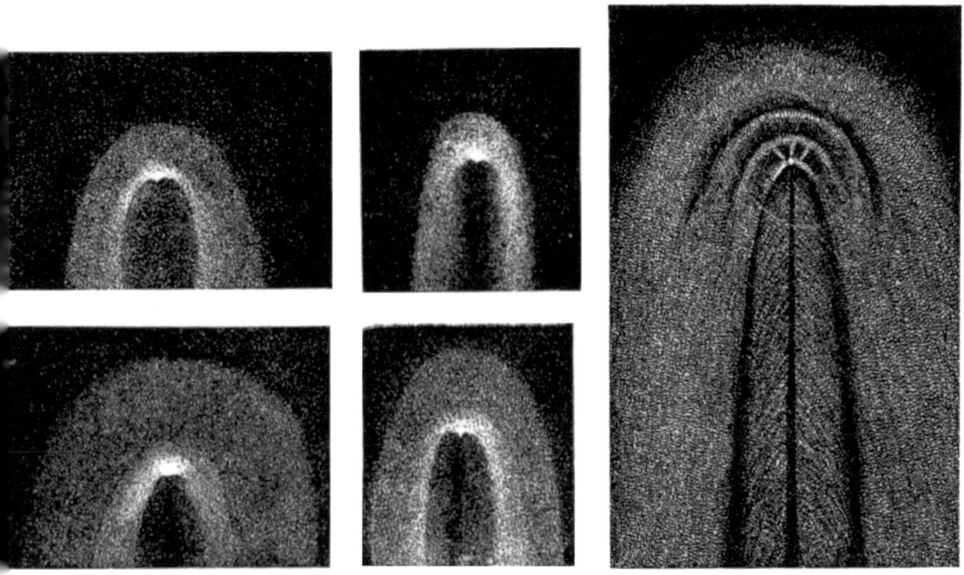

Fig. 5. Ausströmungsbilder beim Kometen Donati, September und Oktober 1858. (Zeichnungen nach dem Anblick im Fernrohr von C. Pape, Hamburg-Altona.)

noch nicht weit genug zurück, um diese Frage beantworten zu können. Für den Halleyschen Kometen mit einer mittleren Periode von 77 Jahren liegen Beobachtungsberichte rückwärts über 28 Perihelannäherungen vor. Aus diesen ergibt sich der Eindruck, daß innerhalb dieses Zeitabschnittes jedenfalls keine beträchtliche Abnahme seiner Perihelhelligkeit eingetreten ist.

Unter den Astronomen ist die Meinung sehr verbreitet, daß die Helligkeiten und insbesondere die Helligkeitsfluktuationen eng mit Vorgängen auf der Sonnenoberfläche korreliert sind (Fleckentätigkeit, Eruptionen), und daß dafür eindeutige Beweise geliefert wurden. Eine eingehende Prüfung der diesbezüglichen Publikationen ergibt jedoch, daß es sich dabei vorläufig noch um Vermutungen handelt[2].

II. Die Kometenbahnen.

2. Allgemeine Charakteristika der Kometenbahnen. Die Bahn eines Kometen um die Sonne wird festgelegt durch sechs Konstante, die sechs *Bahnelemente* (s. Fig. 6):

i die Neigung der Bahnebene gegen die Ebene der Ekliptik.

[1] Systematische Helligkeitsmessungen an zahlreichen Kometen einschließlich vieler schwacher Objekte sind innerhalb der letzten Jahrzehnte insbesondere von M. Beyer durchgeführt worden. Die Resultate und deren Diskussion findet man in einer Reihe fortlaufender Publikationen in den Astron. Nachrichten. Dort auch weitere Literatur.

[2] Vgl. dazu L. Houziaux u. L. Battiau: Bull. Acad. Belg. **43**, 171 (1957).

☊ den aufsteigenden Knoten der Bahn; das ist der Punkt, in dem der Komet von der südlichen auf die nördliche Seite der Ekliptik überwechselt. Das Symbol ☊ wird gewöhnlich auch dazu benutzt, um die Länge dieses Punktes, d.h. den Winkel ♈ S ☊ zu kennzeichnen.

ω die Länge des Perihels P, gerechnet in der Bahnebene vom Knotenpunkt ☊ (Winkel ☊ SP).

q die Periheldistanz SP, meist in astronomischen Einheiten (A.E.) ausgedrückt.

e die Exzentrizität der Bahnkurve.

T den Zeitpunkt des Durchgangs durch das Perihel[1].

Die überwiegende Mehrzahl der Kometen bewegt sich in stark exzentrischen Bahnen mit Exzentrizitäten sehr nahe an 1, ganz im Gegensatz zu den Verhältnissen bei den Planeten und Planetoiden. Neuere Ergebnisse exakter Bahnbestimmungen führen mit großer Wahrscheinlichkeit zu dem Schluß, daß alle Kometen ursprünglich periodisch sind, also elliptische Bahnen beschreiben. Gleichzeitig muß dabei aber bemerkt werden, daß auch alle, soweit sie in das System der Planeten eindringen, früher oder später durch Planetenstörungen in hyperbolische Bahnen geworfen werden und dem Sonnensystem verloren gehen.

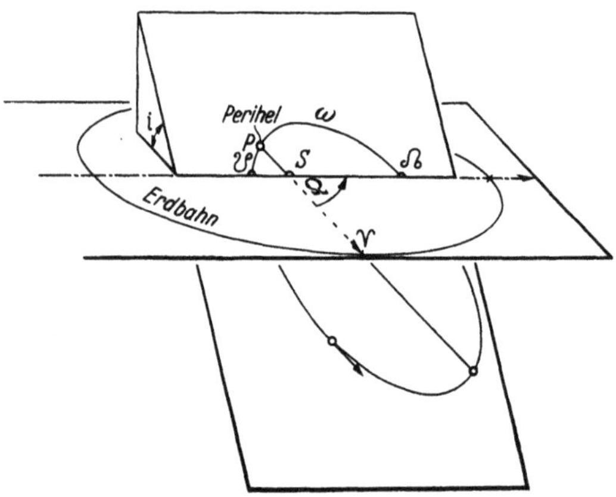

Fig. 6. Elemente einer Kometenbahn.

Ein kleiner Teil der bekannten Kometen bewegt sich in Ellipsen-Bahnen, die vollständig innerhalb des Planetensystems liegen. Man bezeichnet sie als die *kurzperiodischen* Glieder der Gesamtheit der Kometen. Die Abgrenzung dieser Gruppe gegen die langperiodischen Objekte ist wegen eines kontinuierlichen Überganges ein wenig willkürlich. Wir werden hier, wie das neuerdings geschieht, auch diejenigen Objekte mit in die Gruppe einbeziehen, deren Bahnen nur zum größten Teil innerhalb der Neptunbahn liegen. HALLEYs Komet mit seiner Periode von 77 Jahren, dessen Aphel mit 35,31 A.E. zwischen Neptun und Pluto liegt, gehört also auch in diese Gruppe gleichzeitig mit dem Kometen Encke, dessen Periode mit 3,3 Jahren die kürzeste von allen ist.

Die Gruppe der kurzperiodischen Kometen enthält etwa 85 Mitglieder. Innerhalb dieser existiert eine Untergruppe mit rund 50 Individuen, deren Perioden vergleichbar mit der Periode des Planeten Jupiter sind. Die Aphele der Kometen dieser Jupitergruppe liegen größtenteils dicht an der Bahnkurve des Planeten und ihre Bahnneigungen i sind für Kometenbahnen durchweg klein und bleiben unterhalb von 45°. Bemerkenswert ist weiterhin, daß innerhalb der Jupiterfamilie kleine Exzentrizitäten relativ häufig sind. Für den Kometen 1942 VII (Oterma) ergibt die Bahn eine Exzentrizität $e = 0{,}143$ und für den Kometen 1925 II (Schwaßmann-Wachmann) $e = 0{,}136$ (Merkur 0,206). Die Bahnneigungen

[1] Es lassen sich auch andere Sätze von Elementen wählen, die jedoch stets auf obige zurückgeführt werden können.

der gerade genannten Objekte sind 4,0° und 9,4° (Merkur 7°). Unter den Planetoiden sind andererseits Exzentrizitäten $e > 0{,}2$ relativ häufig. Die Bahnneigungen bleiben zwar stets unter 90°, gehen in manchen Fällen aber über 30° hinaus (Hidalgo $e = 0{,}65$; $i = 42{,}5°$).

Die kurzperiodischen Kometen unterliegen ständig in ihrer Bahnbewegung störenden Einflüssen durch die Planeten, insbesondere durch die großen Planeten. Neben langsam verlaufenden Änderungen der Bahnelemente, insbesondere des besonders empfindlichen Elementes e, sind in den letzten 200 Jahren in der Jupitergruppe auch eine Anzahl radikaler Bahnumwandlungen festgestellt worden.

LAPLACE war der erste, der an eine Entstehung der kurzperiodischen Kometen der Jupitergruppe als Folge naher Passagen von langperiodischen an den

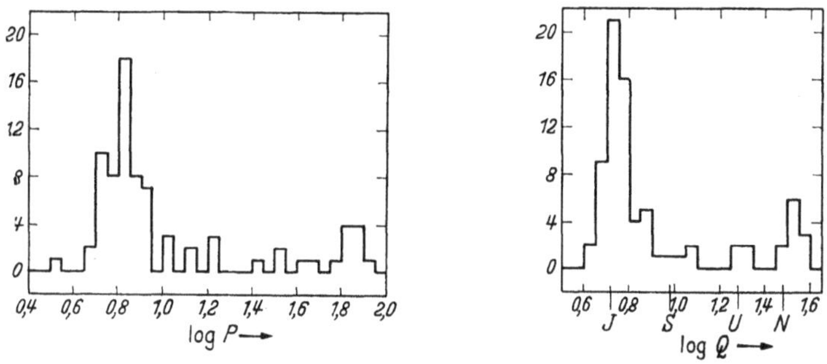

a b

Fig. 7a u. b. Die Verteilung der Perioden P und Aphelienabstände Q bei den kurzperiodischen Kometen. [Nach A. J. J. VAN WOERKOM: Bull. Astronom. Inst. Netherl. **10**, Nr. 399 (1948).]

Planeten dachte. Es wird heute auf Grund späterer Untersuchungen ziemlich allgemein angenommen, daß dieser Gedanke zutreffend ist (s. Ziff. 19).

Bei den *langperiodischen Kometen* liegt die Bahnkurve zur Hauptsache außerhalb des Systems der Planeten und die Beobachtungen, die nur selten über 5 A.E. hinausreichen, bedecken allgemein nur ein winziges Stück der ganzen Bahn. Im Gegensatz zu den kurzperiodischen Objekten findet man hier alle Bahnneigungen zwischen 0 und 180°. Die bisher vorliegenden definitiven Kometenbahnen längerer Perioden, etwa 500 an der Zahl, sind mit prozentual geringen Ausnahmen sog. oskulierende Bahnen, sie gelten nur für die Zeit der Beobachtung, geben aber keine ausreichende Auskunft über die Bahnelemente vor dem Eintritt des Kometen in das System der Planeten oder für die Zeit nach dem Austritt aus demselben. Für den Kometen 1914 V (Delavan), der über ein besonders langes Zeitintervall von 629 Tagen beobachtet worden war, berechnete G. VAN BIESBROECK[1] zunächst eine Exzentrizität $e = 1{,}0001618 \pm 0{,}0000035$, also eine schwach aber entschieden hyperbolische Bahnkurve. Eine rückwärts bis zum Jahre 1908 durchgeführte Berechnung der planetarischen Störungen ergab dann aber für die Zeit vor dem Eintritt in das System der Planeten eine Exzentrizität $e = 0{,}9999781 \pm 0{,}0000035$ und somit eine Ellipse. Die Bahn hat eine Apheldistanz von 170000 A.E. und eine Periode von $24 \cdot 10^6$ Jahren. Um ähnlich exakte ursprüngliche Bahnen berechnen zu können reicht in den meisten Fällen die Qualität der gemessenen Positionen nicht aus. Außerdem sind solche Aufgaben mit einem großen Arbeitsaufwand verbunden.

---

[1] G. VAN BIESBROECK: Publ. Yerkes Obs. V, Teil II, 1927.

Tabelle 2.

Nr.	Komet	oskul. $1/a$	ursprünglich $1/a$	$\Delta(1/a)$
1	1853 III	$-0{,}0008193$	$+0{,}0000829$	$+0{,}0009022$
2	1863 VI	$-0{,}0004949$	$+0{,}0000166$	$+0{,}0005115$
3	1882 II	$+0{,}0118963$	$+0{,}0121488$	$+0{,}0002525$
4	1886 I	$-0{,}0006944$	$-0{,}0000071$	$+0{,}0006873$
5	1886 II	$-0{,}0004770$	$+0{,}0003166$	$+0{,}0007936$
6	1886 IX	$-0{,}0005765$	$+0{,}0000630$	$+0{,}0006395$
7	1889 I	$-0{,}0006915$	$+0{,}0000420$	$+0{,}0007335$
8	1890 II	$-0{,}0002151$	$+0{,}0000718$	$+0{,}0002869$
9	1897 I	$-0{,}0008722$	$+0{,}0000396$	$+0{,}0009118$
10	1898 VII	$-0{,}0006074$	$-0{,}0000157$	$+0{,}0005917$
11	1902 III	$+0{,}0000810$	$+0{,}0000054$	$-0{,}0000756$
12	1904 I	$-0{,}0005040$	$+0{,}0002165$	$+0{,}0007205$
13	1905 VI	$-0{,}0001424$	$+0{,}0006210$	$+0{,}0007634$
14	1907 I	$-0{,}0004991$	$+0{,}0000252$	$+0{,}0005243$
15	1910 I	$+0{,}0002143$	$+0{,}0006921$	$+0{,}0004778$
16	1914 V	$-0{,}0001465$	$+0{,}0000119$	$+0{,}0001584$
17	1922 II	$-0{,}0003806$	$+0{,}0000038$	$+0{,}0003844$
18	1925 I	$-0{,}0005665$	$+0{,}0000540$	$+0{,}0006205$
19	1925 VII	$-0{,}0002730$	$+0{,}0001150$	$+0{,}0003880$
20	1932 VI	$-0{,}0005948$	$+0{,}0000441$	$+0{,}0006389$
21	1936 I	$-0{,}0004870$	$+0{,}0002050$	$+0{,}0006920$

Ursprüngliche Kometenbahnen: nach E. SINDING, Publ. og mindre Meddelelser fra Københ. Obs. Nr. 146, 1948.

Bereits vor VAN BIESBROECK haben jedoch einige Autoren wie insbesondere G. FAYET[1] und E. STRÖMGREN[2] gründliche Versuche unternommen, um über den wahren Charakter der ursprünglichen Kometenbahnen Aufschluß zu bekommen, wobei sie verständlicherweise ihre Aufmerksamkeit vorwiegend den Objekten zuwandten, deren oskulierende Bahnen hyperbolische oder parabolische Elemente ergeben hatten. FAYETs Untersuchung, die ein Material von 150 Objekten umfaßt, hat mehr statistischen Charakter, während E. STRÖMGREN individuelle, exakte Störungsrechnungen durchgeführt hat. Das interessanteste Ergebnis der Strömgrenschen Rechnungen kann der Tabelle 2 entnommen werden, in der die behandelten Kometen (einschließlich des van Biesbroeckschen Objektes) aufgeführt sind. In den theoretischen Untersuchungen zu den Bahnumwandlungen ist die Größe $1/a$ (a große Halbachse) ein praktischeres Element als die Exzentrizität e, weshalb dieses in der Tabelle aufgeführt ist[2]. Von den behandelten 21 Bahnen besaßen 18 oskulierende Bahnen hyperbolischen Charakter, drei waren elliptisch. Die als ursprünglich bezeichneten Werte von $1/a$ charakterisieren die Bahn außerhalb der Region der Planeten und bezogen auf das Gravitationszentrum Sonne plus große Planeten. Mit einer Ausnahme erfolgt beim Übergang zu den ursprünglichen Bahnen eine Verschiebung in Richtung zu

[1] G. FAYET: Recherches concernant les excentricités des comètes. Paris 1906. — E. STRÖMGREN: Über den Ursprung der Kometen. Publ. og mindre Meddelelser fra Københ. Obs. Nr. 19, 1914.

[2] Die reziproke große Halbachse $1/a$ ist ein Maß für die Energie der Masseneinheit in der Bahn. Es folgt dies aus der Energiegleichung der Bahnbewegung im Zweikörperproblem:

$$\frac{1}{2}\,\frac{m\,M}{m+M}\,V^2 - \frac{G\,M\,m}{r} = -\,\frac{G\,M\,m}{2a}\,.$$

Der erste Term links ist die totale kinetische Energie der Körper mit den Massen M und m und der relativen Geschwindigkeit V, der zweite die potentielle Energie der Gravitation. G bezeichnet die Gravitationskonstante.

einer elliptischen bzw. schwächer exzentrischen Bewegung, die Änderung $\Delta(1/a)$ ist positiv. $\Delta(1/a)$ ist von der Ordnung $+0{,}0005$. Man beachte, daß $\Delta(1/a)$ von höherer Ordnung ist als das Mittel der ursprünglichen $1/a$-Werte. In zwei Fällen (1886 I und 1898 VII) bleiben die ursprünglichen Elemente schwach hyperbolisch. Es ist jedoch durchaus wahrscheinlich, daß in diesen Beispielen die Genauigkeit von Beobachtung und Rechnung nicht vollständig ausreicht, um die aufgeführten negativen Werte von $1/a$ zu garantieren, und daß auch hier parabelnahe elliptische Bahnen vorliegen.

Kataloge der Kometen und deren Bahnelemente:

1. J. G. Galle: Verzeichnis der Elemente der bisher berechneten Kometenbahnen. Leipzig 1894.
2. A. C. D. Crommelin: Comet Catalogue: A Sequal to Galle's Kometenbahnen. Mem. Brit. Astronom. Assoc. **26** (1925); **30** (1932).
3. A. S. Yamamoto: Preliminary General Catalogue of Comets. Publ. Kwasan Obs. **1**, Nr. 4 (1936).
4. F. Baldet u. G. de Obaldia: Catalogue Général des Orbites des Comètes de l'An 466 à 1952. Obs. de Paris. Section d'Astrophysique de Meudon 1952.
5. S. K. Vsesviatsky: Physikalische Daten der Kometen. Moskau 1958 (Russisch).

3. Einige besonders bemerkenswerte Kometen. α) *Halleys Komet*[1]. Dieser Komet ist bekanntlich der erste, für den eine Wiederkehr vorausgesagt wurde. In der Tabelle 3 stehen die Elemente der drei Kometen der Jahre 1682, 1607 und 1531, wie sie von Halley berechnet wurden, und aus deren Ähnlichkeit er auf eine Identität schloß. Die Berechnungen für 1607 basierten auf Beobachtungsdaten von Kepler, die für 1531 auf solchen von Apian. Die Differenzen in den Perioden erklärte sich Halley durch Jupiter- und Saturnstörungen, die er jedoch nicht berechnen konnte.

Tabelle 3. *Elemente des Halleyschen Kometen.*

Jahr	ω	Ω	i	q	e	T	Periode
1531	104°18′	45°30′	163°0′	0,5800	0,967391	Aug. 25,799	
							76 Jahre plus 62 Tage
1607	106°45′	47°48′	162°40′	0,5850	0,967391	Okt. 26,912	
							75 Jahre minus 42 Tage
1682	109°12′	50°48′	162°18′	0,5825	0,967391	Sept. 14,896	

Für die nächste Wiederkehr 1758—1759 berechnete dann später Clairaut unter Berücksichtigung der Planetenstörungen den Periheldurchgang auf den 13. April 1759, ±30 Tage. Der Durchgang erfolgte am 13. März 1759. Clairauts Rechnungen litten unter der ungenauen Kenntnis der Massen der Planeten.

Cowell und Crommelin[2] bestimmten ausgehend von den Erscheinungen von 1835 und 1910 die Periheldurchgänge rückwärts bis 240 v. Chr. Für jede der 27 Erscheinungen zwischen 87 v. Chr. und 1910 lassen sich in der astronomischen Literatur bzw. in alten Chroniken Berichte über das Auftauchen eines Kometen in der richtigen Jahreszeit und im richtigen Sternfeld auffinden.

Die Perioden von einem Periheldurchgang zum anderen variieren bis zu nahezu fünf Jahren infolge von Planetenstörungen. Die mittlere Periode beläuft sich auf 77 Jahre.

[1] Ältere Literatur bei J. G. Galle, loc. cit.

[2] P. H. Cowell u. A. C. D. Crommelin: Investigation of the motion of Halley's Comet (Appendix to Greenwich Observations 1909. Edinburgh: H. M. Stationery Office 1910).

β) Der Septemberkomet 1882 II. Kometengruppen. Das vorige Jahrhundert war reich gesegnet mit hellen Kometenerscheinungen, zu denen auch der am Tage sichtbare Septemberkomet 1882 II gehörte, dessen Perihel mit 0,0077 A.E. Distanz vom Zentrum der Sonne in der Sonnenkorona lag (s. Fig. 8). Am 1. September am Morgenhimmel entdeckt ging er am 17. September $6^h 24^m$ nachmittags, Berliner Zeit, durchs Perihel. In der Korona zerteilte sich der Kern des Kometen in vier Teile, die sich allmählich in der Richtung der Bahn voneinander trennten. H. KREUTZ[1] berechnete die Bahnen für jeden der vier Kerne und fand Perioden

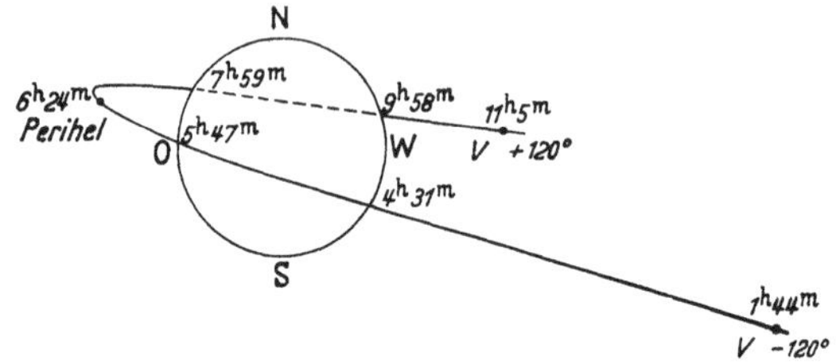

Fig. 8. Bahn des großen September-Kometen 1882 nahe der Sonne.

von 664, 769, 875 und 959 Jahren. Diese Bruchstücke werden also wahrscheinlich als vier verschiedene Kometen in einem zeitlichen Abstand von angenähert einem Jahrhundert wiederkehren.

Tabelle 4.

Komet	$\omega + \Omega = \pi$	Ω	i	q	e
1668	248,8	358,6	144,3	0,0666	(1)
1843 I	278,7	1,3	144,3	0,0055	0,99991
1880 I	279,9	6,1	144,7	0,0055	(1)
1882 II	276,4	346,0	142,0	0,0077	0,99991
1887 I	266,2	324,6	128,5	0,0097	(1)
1945 g	270,7	321,6	137,0	0,0063	(1)

Es ist bemerkenswert, daß bereits fünf Kometen entdeckt wurden, deren Bahnelemente mit denen des Septemberkometen angenähert identisch sind (s. Tabelle 4). Diese Übereinstimmung in den Elementen kann kaum als zufällig angesehen werden, und es erscheint äußerst wahrscheinlich, daß diese ganze Kometengruppe durch einen gleichen Prozeß aus einem Mutterkometen entstanden ist, wie die vier von KREUTZ behandelten Splitter des Septemberkometen. I. G. PORTER[2] führt in seiner Monographie „Comets and Meteor Streams" nicht weniger als 19 verschiedene Kometengruppen auf, deren Daten allerdings nicht alle so überzeugend erscheinen wie die der hier aufgeführten Gruppe, und die außerdem meist nur zwei und drei Mitglieder enthalten.

γ) Der Komet 1889 V (Brooks). Durch einen nahen Vorbeigang an dem Planeten Jupiter erlitt der Komet 1889 V im Jahre 1886 eine radikale Bahnumwandlung. Der Fall ist von dem amerikanischen Astronomen C.L. POOR[3] im Jahre 1894 analytisch-numerisch behandelt worden. Die Fig. 9, welche diese Bahnumwandlung veranschaulicht, wurde nach den Poorschen Daten angefertigt.

[1] H. KREUTZ: Untersuchungen über das Kometensystem 1843 I, 1880 I und 1882 II. Publ. Kieler Sternwarte Nr. 3 u. 4, 1888.

[2] I. G. PORTER: Comets and Meteor Streams. The International Astrophysics Series. Vol. 2. London 1952.

[3] C.L. POOR: Astronom. J. **13**, 123, 127, 177 (1894).

Der Komet lief vor 1886 in einer Bahn mit einer Periode von 27,0 Jahren ($a = 9{,}0$ A.E.). Deren Neigung gegen die Ekliptik war klein und betrug nur etwa 7°. Am 24. März 1886 trat der Komet in die ,,Aktionszone" des Planeten Jupiter ein, d.h., er kam dem Planeten so nahe, daß die Schwereattraktion des Planeten die der Sonne übertraf. Dieser Zustand dauerte mehrere Monate bis

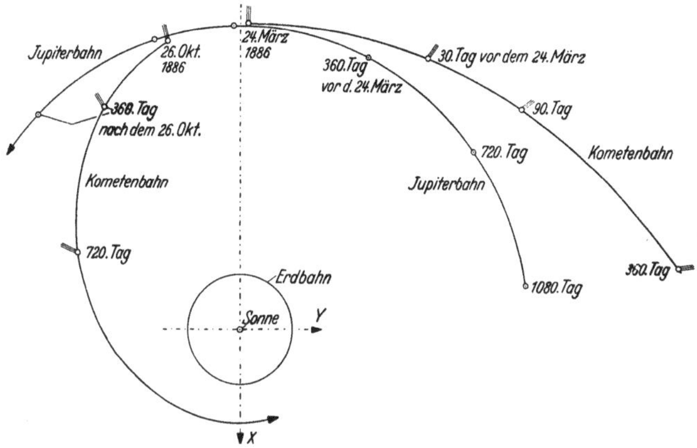

Fig. 9. Bahnumwandlung des Kometen 1889 V.

zum 26. Oktober. Nach dem Austritt aus der Aktionszone ($\varrho \approx 0{,}332$ A.E.) des Planeten lief der Komet in einer Ellipse mit einer Umlaufzeit von etwa 6,8 Jahren ($a = 3{,}6$ A.E.). POOR hat auch die durch die Sonne gestörte Hyperbelbahn

Tabelle 5.

Komet	Zeit	a (A.E.)	q (A.E.)	P in Jahren
Lexell	vor 1767	5,06	2,96	11,4
	1770	3,15	0,67	5,6
	nach 1779	6,37	3,33	16,2
Brooks 2 = 1889 V	vor 1886	9,00	5,44	27,0
	1889—1921	3,59	1,95	6,8
	nach 1921	3,64	1,86	6,95
Wolf 1	vor 1875	4,18	2,54	8,54
	1874—1922	3,59	1,59	6,80
	nach 1922	4,07	2,36	2,20
Comas Sola	vor 1912	4,46	2,15	9,43
	nach 1912	4,17	1,77	8,52
Schwaßmann-Wachmann 2	vor 1921	4,43	3,55	9,30
	nach 1921	3,46	2,09	6,42
Whipple	vor 1922	4,74	3,90	10,30
	nach 1922	3,83	2,50	7,50

um Jupiter genauer berechnet. Danach ging der Kometenkern durch das System der Jupitermonde und zwar innerhalb der Bahnkurve des Satelliten I, dessen mittlere Distanz d vom Zentrum des Planeten nur das 5,9fache des Radius desselben beträgt ($d = 4.2 \cdot 10^5$ km). Durch den Vorbeigang an Jupiter wurde der Kern infolge von Gezeitenkräften in mehrere Stücke aufgebrochen, die später für einige Zeit als getrennte Kometen gesehen werden konnten. Neben dem Hauptkometen gab es vier schwächere Begleiter. Der Hauptkomet ist bis heute regelmäßig beobachtet worden. Er erfuhr übrigens eine zweite, geringe Bahnänderung im Jahre 1921.

In Tabelle 5 sind fünf weitere Kometen aufgeführt, die durch nahe Vorbeigänge an dem Planeten Jupiter eine starke Änderung der Bahn erfuhren. Der Komet Lexell kam noch näher an den Planeten heran als der Komet 1889 V.

δ) Der Komet Encke. Dieser periodische Komet hat mit 3,28 Jahren von allen bekannten Kometen die kleinste Periode. Er wurde zuerst gesehen im Jahre 1786 von den Franzosen MECHAIN und MESSIER. Eine Bahn konnte jedoch mangels ausreichender Beobachtungen erst im Jahre 1818 durch ENCKE errechnet werden. Seit dieser Zeit ist er bei jeder Perihelannäherung beobachtet worden. ENCKE (1791—1865) fand, daß nach Berücksichtigung aller planetarischer Störungen noch ein Einfluß auf die Bewegung des Kometen übrig blieb, der die Periode stetig, wenn auch nur gering, verkürzte. Durch BACKLUND u. a.[1] ist dieser Effekt später bestätigt worden. Derselbe ist jedoch von variabler Größe und hat sich im Laufe der Zeit etwas vermindert. Für die Jahrzehnte von 1819—1865 ergab sich eine Änderung $\Delta\mu$ der mittleren täglichen Bewegung pro Umlauf von $+0\overset{''}{.}1044$, für die Zeit von 1871—1894 betrug die Zunahme nur noch $+0\overset{''}{.}0677$. MATKIEVICZ gibt für 1918—1924 $\Delta\mu = +0\overset{''}{.}046$, für 1921—1928 $\Delta\mu = +0\overset{''}{.}020$, und wiederum einen höheren Wert $\Delta\mu = +0\overset{''}{.}097$ für 1924—1931.

Bei der Deutung des Effektes dachte ENCKE an die Wirkung eines widerstehenden Mittels (Meteorschwärme). Da jedoch bei den anderen Körpern des planetarischen Systems solche Anomalien nicht festzustellen waren, so äußerte BESSEL die Ansicht, daß die Ursache im Kometen selbst läge und mit der Abstoßung von Materie zusammenhinge[2]. Bei der Beobachtung des Halleyschen Kometen hatte BESSEL festgestellt, daß der Kometenkern bevorzugt in Richtung zur Sonne Materie ausstieß. Er argumentierte, daß dieser Vorgang die Bahnbewegung beeinflussen müsse, und daß man nur über die Größe des Effektes noch streiten könne.

Tabelle 6.

Komet	Zeit	$\Delta\mu$ proTag
Encke. .	1819—1865	$+43232''10^{-9}$
Biela . .	1806—1832	+ 7345
Brooks II	1889—1910	+ 5391
Winnecke	1859—1886	— 212
Wolf . .	1884—1919	— 210

Der Enckesche Komet scheint nicht der einzige zu sein, bei welchem die Bahnbewegung durch die Newtonschen Attraktionskräfte im Sonnensystem allein nicht eine vollständige Erklärung findet. A.D. DUBIAGO[3] gibt in einer Publikation noch vier weitere Fälle an (Tabelle 6). Außer DUBIAGO hat sich auch F. WHIPPLE[4] mit der Deutung der Erscheinung beschäftigt. Beide Autoren akzeptieren den Besselschen Gedanken der Massenabstoßung. Sie versuchen eine Abschätzung der notwendigen Massenabgabe ΔM bzw. des Verhältnisses $v\Delta M/M$ (M gleich der gesamten Kometenmasse, v gleich der Abstoßungsgeschwindigkeit). Es scheint, daß für den Kometen Encke die Abgabe von wesentlich weniger als einem Prozent der Masse pro Umlauf ausreicht, um das aus der Bewegung errechnete $\Delta\mu$ zu erklären. Die Verhältnisse benötigen jedoch noch einer besseren Klärung.

---

[1] O. BACKLUND: Astronom. Nachr. **184**, 89 (1910). — A. MATKIEVICZ: Bull. Obs. Poulkova **14**, Nr. 6 (1935). — A.D. DUBIAGO: Russ. Astronom. J. **25**, Nr. 6 (1948).

[2] W. BESSEL: Astronom. Nachr. **13**, 345 (1836).

[3] A.D. DUBIAGO: Russ. Astronom. J. **25**, Nr. 6 (1948).

[4] F. WHIPPLE: Astrophys. Journ. **111**, 375 (1950). Siehe auch SALOH EL-DIN HAMID u. F. WHIPPLE: Astronom. J. **58**, 100 (1953).

III. Spektren und Physik der Kometen.

4. Vorbemerkungen. Kometen erscheinen am Himmel als ausgedehnte, diffuse Objekte, und sie besitzen — abgesehen von der nächsten Umgebung des Kometenkernes — nur eine sehr geringe Flächenhelligkeit. Ihre spektroskopische Erforschung zwingt zur Benutzung von sehr lichtstarken Spektralanordnungen, was die Erreichung großer Auflösung ausschließt[1]. Bevorzugt wurden in der Vergangenheit Objektivprismen benutzt, sowohl wegen ihrer höheren Lichtstärke wie auch aus dem Grunde, daß diese in einer Aufnahme eine spektrale Zerlegung des ganzen Kometenbildes liefern. Ihre schwerwiegenden Nachteile bilden natürlich die geringe Dispersion und die starke Überlagerung der monochromatischen Bilder. Aufnahmen mit Spaltspektrographen bei enger Spalteinstellung blieben bisher fast ausschließlich auf das hellere Kern-Kopfgebiet der Kometen beschränkt.

Die allgemeinen Hauptmerkmale ihrer Spektren sind folgende: 1. das Vorhandensein einer größeren Anzahl von Emissionsbanden (Molekülspektren), 2. ein kontinuierliches Spektrum, welches die Fraunhoferschen Linien der Sonne erkennen läßt, und 3. das gelegentliche Erscheinen von einigen Atomlinien.

Die relative Stärke dieser drei Komponenten schwankt von Komet zu Komet in weiten Grenzen und wechselt ebenfalls häufig für ein und dasselbe Objekt mit der Zeit. Das Erscheinen von Atomlinien ist an kleine heliozentrische Distanzen gebunden (etwa $r < 0{,}7$ A.E.). Das Fraunhofer-Kontinuum ist stets am intensivsten im Kerngebiet und hier fast immer zu erkennen. In den übrigen Partien des Kometenbildes kann das Kontinuum gänzlich verschwinden, insbesondere in den geraden Schweifen. Das Auftreten von Molekülbanden in Emission muß als das Haupt-Charakteristikum der Kometenspektren angesehen werden. Bei großen Sonnendistanzen können diese allerdings fehlen oder noch unsichtbar bleiben und bei sehr kleinen Sonnendistanzen durch das verstärkte Fraunhofer-Kontinuum bis zur Unsichtbarkeit überstrahlt werden. Das Kontinuum entsteht zweifellos durch Streuung des Sonnenlichtes an Staub und freien Elektronen. Ob die Elektronen einen wesentlichen Beitrag liefern ist noch ungeklärt.

Die Banden im Spektrum sind für Kopf und Schweif nicht identisch. Die Köpfe zeigen Banden-Emissionen von neutralen und ionisierten Verbindungen, im eigentlichen Schweif liegen nur ionisierte Moleküle vor.

Eine Bibliographie der Kometenspektren, beobachtet bis etwa zum Jahre 1938, findet man in einer Publikation von N. T. BOBROVNIKOFF[2]. P. SWINGS und L. HASER publizierten kürzlich einen ausgezeichneten Atlas von Kometenspektren[3], der das beste bisher gewonnene Material aller Sternwarten in sehr guten Reproduktionen nun allgemein zugänglich macht. In diesem Werk findet man auf 23 Bildtafeln über 200 Einzelspektren von 66 verschiedenen Kometen. Von besonderer Qualität sind die zahlreichen von P. SWINGS selbst und einigen Mitarbeitern am McDonald Observatorium, USA, hergestellten Spaltaufnahmen einiger Kometen der letzten zwei Jahrzehnte. Neben den Reproduktionen enthält die Publikation ebenfalls Tabellen zur Identifizierung der

---

[1] *Anmerkung bei der Korrektur.* Unter Benutzung des 200 inch-Reflektors auf dem Mt. Palomar hat J. L. GREENSTEIN im August 1957 eine Reihe von ausgezeichneten Spektren von dem Kometen Mrkos gewonnen, die bei einer Dispersion von 18 Å/mm eine bisher nie erreichte Auflösung haben [Astrophys. Journ. **128**, 106 (1958)]. Diese wie auch manche andere Publikation über die beiden hellen Kometen des Jahres 1957 konnten in diesem Artikel leider keine Berücksichtigung mehr finden.

[2] N. T. BOBROVNIKOFF: Phys. Rev. **14**, 172 (1942).

[3] P. SWINGS u. L. HASER: Atlas of Representive Cometary Spectra. Astrophysical Institute of the University of Liège, Liège 1955.

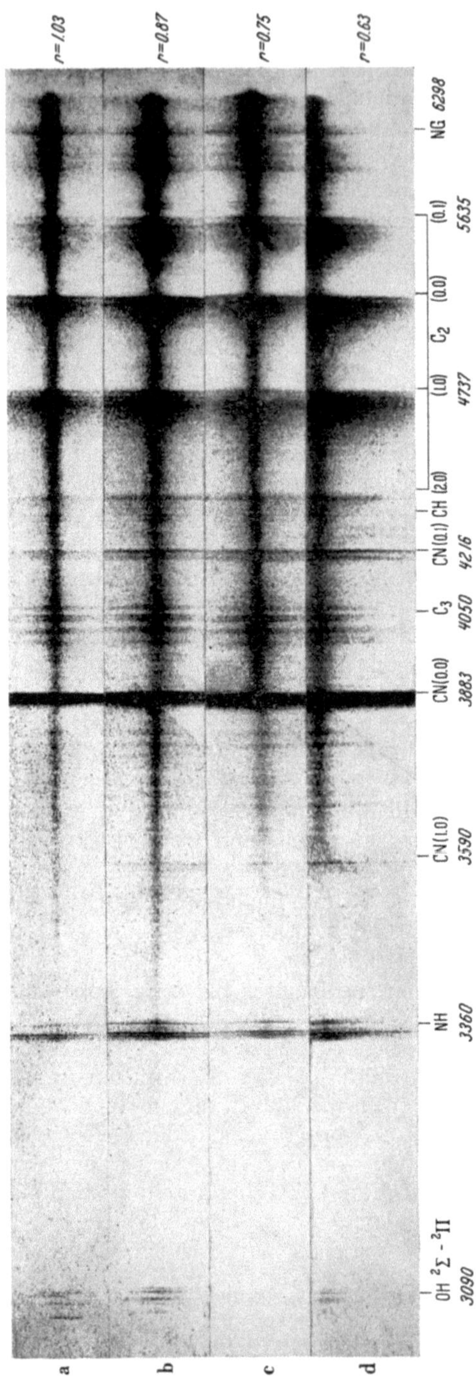

Fig. 10a—d. Spektrum des Kometen Cunningham (1940c). (Aufnahme P. SWINGS. Siehe Atlas von SWINGS und HASER.)

Kometen-Emissionen sowie eine Tabelle der Bahnelemente der Objekte. In einem einleitenden Text sind die Hauptmerkmale der Kometenspektren und die wesentlichsten Grundlagen der Kometenphysik beschrieben. Die Bandenspektren der Kometen können zum größten Teil auch in Laboratoriums-Lichtquellen erzeugt werden. Zum Zwecke eines Vergleichs sind in dem Atlas eine Anzahl von entsprechenden Laboratoriums-Aufnahmen aufgenommen worden.

5. Die Strahlung der Kometenatmosphären (allgemeine Zusammensetzung). Die in Fig. 10 wiedergegebenen Aufnahmen, die dem Atlas von SWINGS und HASER entnommen sind, vermitteln leicht einen Überblick über die Hauptmerkmale eines vollentwickelten Kometenkopf-Spektrums und zwar für den Fall, daß das kontinuierliche Licht gegenüber dem Gasspektrum sehr stark zurücktritt. Der Spalt des Spektrographen (Quarzoptik) war bei den Aufnahmen a bis c so orientiert, daß der Kometenkern auf der Spaltmitte lag, bei der Aufnahme d dagegen an einem Ende des Spaltes. Die Kernposition ist durch die höchste Lichtintensität im Spektrum und durch ein von Rot nach Ultraviolett durchgehendes Kontinuum ausgezeichnet. Fig. 11 zeigt Spektren desselben Kometen, aufgenommen mit einer Glasoptik. Die Identifizierung der kräftigsten Emissionen, wie sie hier vorliegen, hat mit einer Ausnahme (C_3 λ 4050) von vornherein keine besonderen Schwierigkeiten bereitet, da sie mit schon lange wohlbekannten Molekül-Emissionen übereinstimmen. Die (0, 0)-Bande des CN bei λ 3883 bildet stets die hervorstechendste Einzelbande im Spektrum. Abweichend von den Verhältnissen in Laboratoriums-Lichtquellen (Kohlebogen in Luft, Ent-

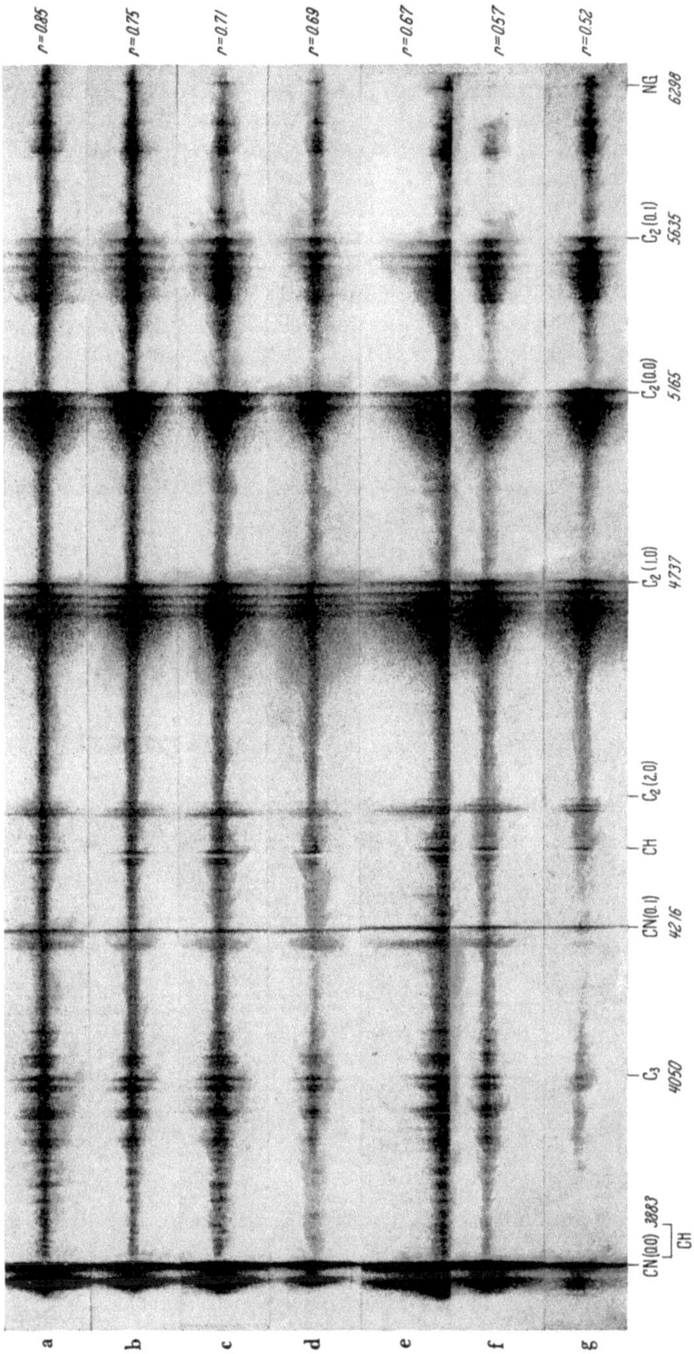

Fig. 11a—g. Spektrum des Kometen Cunningham (1940c). Aufnahme P. Swings, s. Atlas von Swings und Haser.)

ladungen durch Gemische von Kohlenwasserstoffen und Stickstoff) haben die anderen CN-Banden wie CN (1, 1) ..., CN (0, 1) ... und CN (1, 0) ... eine abnorm geringe Intensität. Diese Erscheinung findet in der besonderen Anregungsart der

Kometenspektren (Fluoreszenzanregung durch das Sonnenkontinuum, s. Ziff. 6) eine befriedigende Deutung. Die C_2-Banden, die insgesamt betrachtet im Kometenlicht noch kräftiger erscheinen als die von CN, weisen andererseits eine mehr normale Intensitätsverteilung auf. Die drei Bandengruppen $C_2(0, 0)$, $C_2(1, 1)$ …, $C_2(1, 0)$, $C_2(2, 1)$ … und $C_2(0, 1)$, $C_2(1, 2)$ … sind ähnlich ausgebildet wie in Bogen- und Gasentladungen. Nach der Fluoreszenz-Hypothese ist auch dies zu erwarten und erklärt sich zwanglos auf Grund des Fehlens eines permanenten Dipolmomentes bei dem symmetrischen C_2 (s. Ziff. 6).

Die Auflösung der Rotationsstruktur der Banden ist infolge der geringen Dispersion der Spektralapparate bei den engen und zahlreichen Zweigen der C_2-Banden ($A\,^3\Pi - X\,^3\Pi$-System) überhaupt nicht zu erreichen, für die CN(0, 0)-Bande ($B\,^2\Sigma - X\,^2\Sigma$-System) nur ganz mangelhaft[1]. Bemerkenswert ist, daß die unsymmetrischen Moleküle wie CN, OH, NH und andere nur mit den niedrigsten

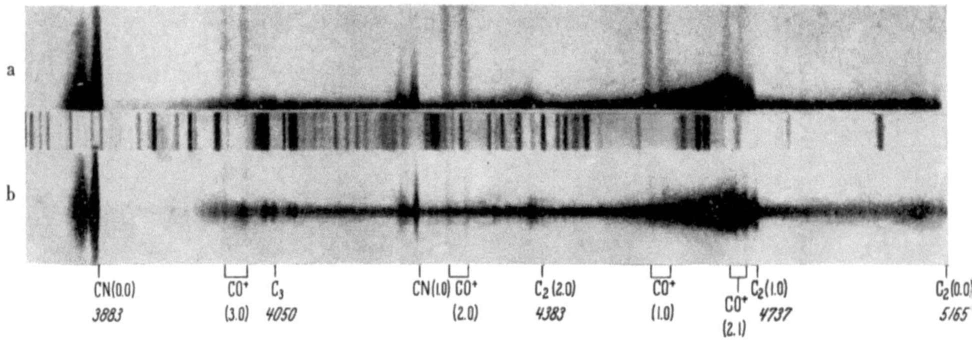

Fig. 12. Komet 1911 V. (Aufnahme Lick Obs. Siehe Atlas von SWINGS und HASER.)

Rotationslinien auftauchen; die Bandenzweige sind sehr kurz. Auch dies wird durch die Fluoreszenz-Hypothese verständlich. Diese Vereinfachung der Bandenstruktur hat zur Folge, daß bei den leichten Hydrid-Verbindungen wie OH und NH (s. Fig. 10), bei denen die natürliche Rotationsaufspaltung infolge der kleineren Trägheitsmomente relativ groß ist[2], trotz der geringen Dispersion doch schon die einzelnen Rotationslinien erkenntlich werden.

Die in den Fig. 10 und 11 angezeigte, mit C_3 identifizierte Banden-Emission hat lange einer Zuordnung widerstanden und konnte erst neuerdings im Laboratorium reproduziert werden[3]. Die Emission besteht aus zahlreichen Linien des Bereiches von λ 3950 bis λ 4140 mit einem Intensitäts-Maximum bei λ 4050.

Das CH-Radikal besitzt in dem Spektralbereich von λ 3000 bis λ 5000 zwei lange bekannte Bandensysteme, die in Entladungen durch Gase bei der Anwesenheit von Kohle und Wasserstoff leicht auftauchen. Beide Systeme treten

---

[1] Vgl. dazu jedoch Fußnote 1 auf S. 477.

[2] Die Folge der Rotationsterme eines zweiatomigen Moleküls ist bekanntlich in erster Näherung durch

$$F(J) = B\,J(J+1)$$

gegeben mit der Quantenzahl $J = 0, 1, 2, \ldots$, der Rotationskonstanten

$$B = \frac{h}{8\pi^2 c^2 I}$$

und dem Trägheitsmoment $I = \mu\, r_0^2$ (h Plancksche Konstante, c Lichtgeschwindigkeit, μ reduzierte Molekülmasse, r_0 Abstand der Atomkerne). B hat für CN den Wert $1{,}89\,\mathrm{cm}^{-1}$, ist jedoch für NH mit $7{,}82\,\mathrm{cm}^{-1}$ mehr als viermal so groß.

[3] Im Zusammenhang mit dieser Emission ist eine umfangreiche Literatur entstanden, die hier nicht aufgeführt werden soll. Siehe Atlas von SWINGS und HASER.

auch in den Kometenspektren auf, jedoch nie mit hoher Intensität. Das eine System ($A\,^2\Delta - X\,^2\Pi$, das bekannte G-Band der Sternspektren) hat seine stärkste Bande (0, 0) bei λ 4315, die (0, 0)-Bande des zweiten Systems $B\,^2\Sigma - X\,^2\Pi$ liegt auf der langwelligen Seite der CN (0, 0)-Bande bei λ 3883. Beide Systeme sind arm an Einzelbanden und gewöhnlich werden nur ihre (0, 0)-Übergänge erkennbar.

Im langwelligeren Bereich der Kometenspektren ($\lambda > 5500$) existieren eine Reihe von Koinzidenzen mit den sog. α-Banden des Ammoniaks, die aber sehr wahrscheinlich nicht NH_3 sondern NH_2 zum Träger haben. Dieses NH_2-Spektrum ist jedoch weder in den Kometen noch im Laboratorium bisher ausreichend entwirrt worden, um die Zuordnung über jeden Zweifel sicher zu stellen.

Wie schon bemerkt ist das Kometenleuchten arm an Atomlinien. Für kleinere heliozentrische Distanzen ist das D-Dublett des Natriums eine normale Erscheinung (s. Fig. 13). Weitere Atomlinien tauchen erst dann auf, wenn ein Komet sich der Sonne bis auf Entfernungen von der Größenordnung des Sonnenradius nähert, was sehr selten zu beobachten ist. Von R. Copeland und J. G. Lohse[1] wurden Linien von Fe und Ni bei dem Kometen 1882 II visuell in einem Spektroskop gemessen, als dieser sehr nahe an der Sonnenoberfläche stand.

Außer den bisher aufgezählten neutralen Partikeln findet man in den Kometenköpfen zeitweise gleichzeitig eine Reihe von ionisierten Molekülen. An erster Stelle sind die Moleküionen CO^+, N_2^+ und CO_2^+ zu nennen. Obwohl diese drei genannten Partikeln charakteristisch für die Kometenschweife sind, findet man dieselben doch auch in den Kometenköpfen. Andererseits treten aber die typischen Partikel der Kometenköpfe, die neutralen Moleküle CN, C_2, OH, NH und CH nie in den Kometenschweifen auf, eine Tatsache, die schon lange bekannt ist. Falls kein Kontinuum vorliegt, so entfällt mehr als 90% des Schweiflichtes auf das Resonanzbanden-System $A\,^2\Pi - X\,^2\Sigma$ des CO^+ (sog. Kometenschweif-Banden). Die Emissionen von N_2^+ und CO_2^+ sind im Vergleich zu der von CO^+ schwach. Die Dublett-Banden des Kometenschweif-CO^+-Spektrums erstrecken sich über den gesamten photographischen und visuellen Bereich des Spektrums,

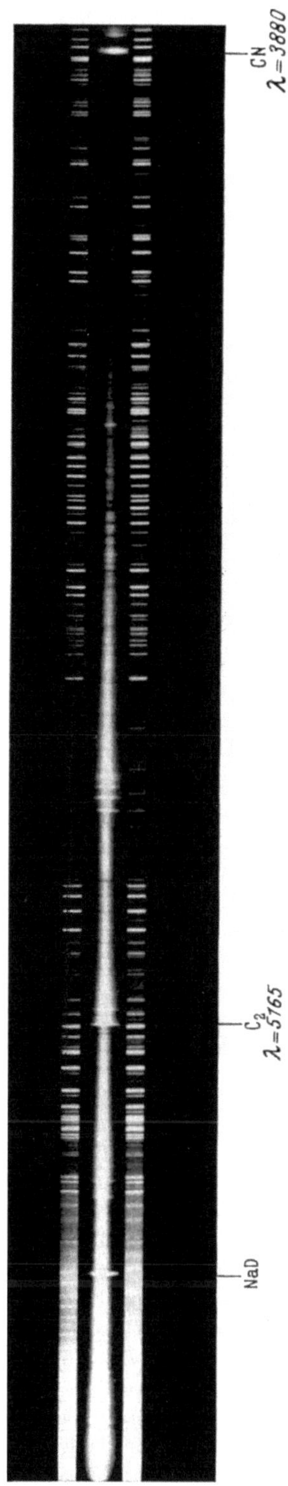

Fig. 13. Spektrum des Kometen Mrkos 1957 d. (Aufnahme G. Mannino, Sternwarte Asiago).

[1] R. Copeland u. J. G. Lohse: Copernicus (Dublin) **2**, 225 (1882).

sind jedoch besonders kräftig im blau-violetten Gebiet. Bei gemischten Schweiftypen (Gas und Staub) kann man häufig beobachten, daß die visuellen und photographischen Schweifstrukturen sehr voneinander verschieden sind, da im Visuellen mehr das Kontinuum, also die Verteilung des Staubes, dagegen im Photographischen mehr die Verteilung des Gases die erfaßte Struktur bestimmt (s. Fig. 14a und b). Die CO^+-Kometenschweifbanden bilden einen $^2\Pi - {}^2\Sigma$-Übergang mit einer weiten Aufspaltung des $^2\Pi$-Terms, so daß jede Einzelbande eine auch bei mäßiger Dispersion erkennbare Doppelkante bildet. Fig. 15 zeigt eine Objektivprismen-Aufnahme des Kometen Morehouse 1908 III (nach F. BALDET, entnommen dem Atlas von P. SWINGS und L. HASER), welche die CO^+-Emission fast vollständig frei von anderen Emissionen erkennen läßt. Der Komet Morehouse besaß nur ganz schwache Andeutungen der Emissionen von CN und C_2 und bestand (soweit es sich um die im Sichtbaren und Photographischen emittierenden Partikeln handelt) fast nur aus CO^+ mit einer Beimengung von N_2^+. In den Spaltaufnahmen des Kometenkopfes des Kometen 1911 V (Lick-Aufnahmen) der Fig. 12 lassen sich die vier intensivsten „Dubletts" der Kometenschweifbanden als weit getrennte „Linien" erkennen. Die Rotations-Struktur dieser Banden, die stark ausgeprägte Kanten bilden, bleibt vollständig unaufgelöst.

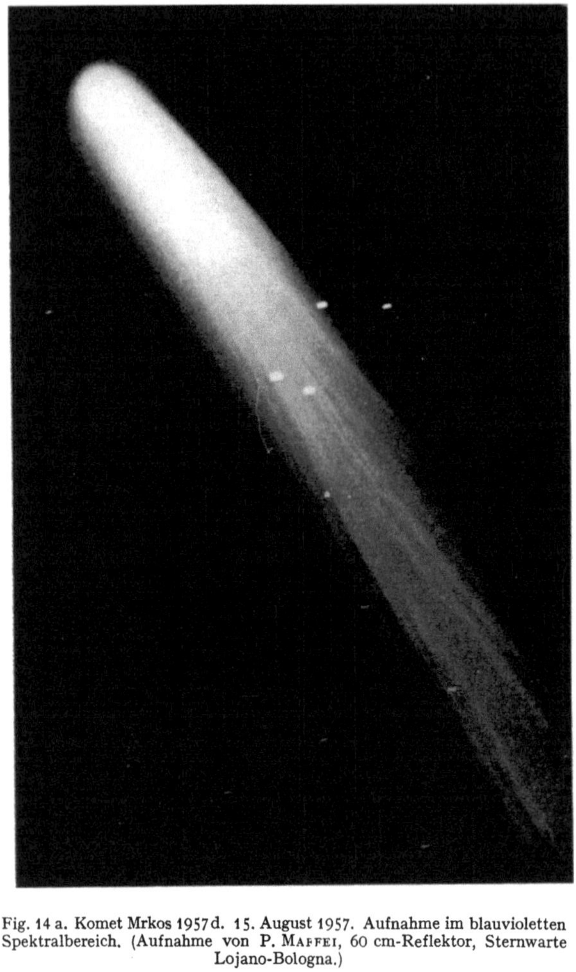

Fig. 14 a. Komet Mrkos 1957d. 15. August 1957. Aufnahme im blauvioletten Spektralbereich. (Aufnahme von P. MAFFEI, 60 cm-Reflektor, Sternwarte Lojano-Bologna.)

Das N_2^+-Molekül ist in den Kometenschweifen im wesentlichen nur mit der (0, 0)-Bande λ 3914 des $B\,^2\Sigma - X\,^2\Sigma$-Resonanzsystems vertreten. SWINGS und PAGE[1] konnten nachweisen, daß auch das CO_2^+-Ion mit einigen schwachen Banden in den Schweifen vorliegt. Für das CO^+-Ion sind außer den starken Resonanzbanden durch SWINGS ebenfalls Banden des Baldet-Johnson-Systems $B\,^2\Sigma - A\,^2\Pi$ nachgewiesen worden. Der untere Elektronenzustand dieses Systems $^2\Pi$ ist identisch mit dem oberen Elektronenzustand der Resonanzbanden.

An ionisierten Teilchen sind dann weiterhin die Ionen CH^+ und OH^+ nachgewiesen. Deren Emissionen zeigen nur eine geringe Ausdehnung in das eigent-

[1] P. SWINGS u. T.L. PAGE: Astrophys. Journ. **111**, 530 (1950).

liche Schweifgebiet und werden meist nur in den Köpfen im Gebiet um den Kern deutlich erkennbar.

Während bei mittleren Sonnendistanzen ($r = 1{,}5$ bis $r = 0{,}5$ A.E.) meist das Gasspektrum vorzuherrschen scheint, findet man für kleine und ebenso für größere heliozentrische Distanzen vielfach einen hohen Anteil an kontinuierlichem Licht. Bei dem Kometen Schwassmann-Wachmann II, der eine wenig exzentrische Bahn besitzt, und dessen Sonnen-Entfernung sich bei einer Umlaufzeit von 16,41 Jahren nur zwischen den Grenzen $r = 5{,}51$ und $r = 7{,}43$ A.E. ändert, ist bisher spektroskopisch ausschließlich nur ein Kontinuum beobachtet worden[1]. Der ausgedehnte, rundliche Kopf dieses Objektes besteht anscheinend aus Staub. Es ist nicht auszuschließen, daß auch das Sonnenlicht streuende Gase vorliegen, die jedoch von der Art sein müssen, daß sie keine sichtbare Fluoreszenz ergeben. Die beiden hellen Kometen des Jahres 1957 (Arend-Rolland, Mrkos) die bei heliozentrischen Distanzen zwischen $r = 0{,}4$ und $r = 0{,}8$ A.E. beobachtet wurden, zeigten beide (insbesondere der erste) ein bedeutendes kontinuierliches Spektrum, das sich über den ganzen Kometen erstreckte und zeitweise das Gasspektrum nur schwer erkennen ließ. Soweit durch Objektivprismen-Aufnahmen feststellbar, besaßen die Kometen 1910a (beobachtet bei $r = 0{,}13$ A.E.) und 1941 I (beobachtet bei $r = 0{,}53$ A.E.) ein rein kontinuierliches Schweifspektrum[2]. Beim Kometen Mrkos 1957 ließ sich erkennen, daß dieser zwei Schweife verschiedener Krümmung ausgebildet hatte (s. Fig. 2 und 31), die in Kopfnähe zusammenflossen. Dieses Phänomen der Doppelschweife bei Staub-Gaskometen ist seit langem bekannt. Dasselbe entsteht dadurch, daß die festen Staubteilchen, die den stärker gekrümmten Schweif bilden, einer geringeren Sonnenrepulsion unterliegen als die CO^+-Ionen des Gasschweifes (s. Ziff. 16).

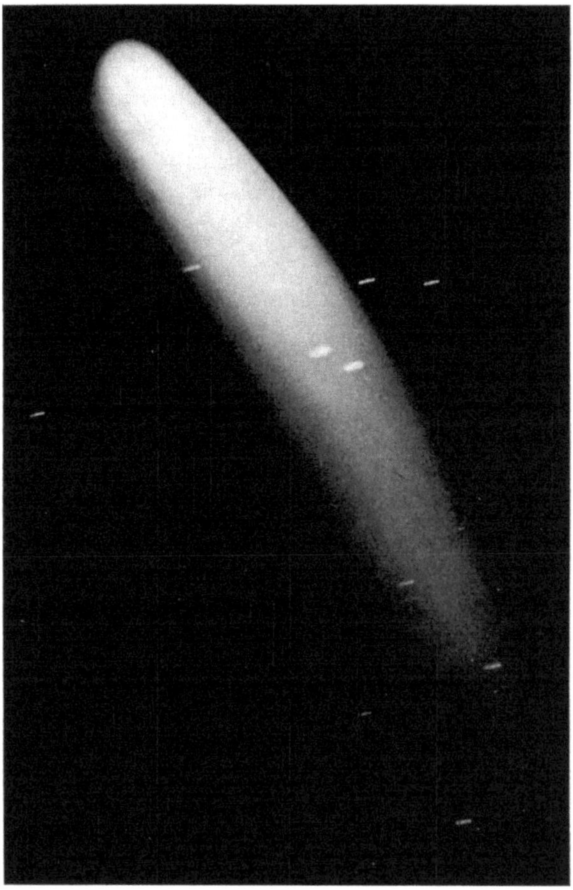

Fig. 14b. Wie Fig. 14a, jedoch Aufnahme im roten Spektralbereich. (Aufnahme von P. Maffei, 60 cm-Reflektor, Sternwarte Lojano-Bologna.

[1] N.U. Mayall: Publ. Astronom. Soc. Pacific **53**, 340 (1941). — G.H. Herbig: Publ. Astronom. Soc. Pacific **58**, 61 (1946).

[2] Siehe Atlas von Swings und Haser.

6. Entwicklung der Bandenintensitäten mit der heliozentrischen Distanz[1]. Die Bandenemission, die bei der Annäherung eines Kometen an die Sonne an erster Stelle erscheint, ist die (0, 0)-Bande des CN bei λ 3883. Sie hat sofort eine Ausdehnung über den ganzen Kometenkopf und behält diese auch weiterhin bei. Die kritische Distanz für das Erscheinen dieser Bande liegt bei etwa 3 A.E. An zweiter Stelle folgen die Emissionen von C_3 und NH_2 ($r \approx 2$ A.E.). Bei Distanzen von $r \approx 2$ A.E. ist die C_3-Bandenemission kräftiger als die dann auch meist sichtbare (0, 1)-Bande des CN, das Intensitäts-Verhältnis C_3:CN(0, 1) verkleinert sich mit abnehmendem r. C_3 und NH_2 dehnen sich weniger weit vom Kern in den Kopf hinein aus als CN.

Die C_2-Swanbanden tauchen bei $r \approx 1{,}8$ A.E. auf. Ihre Ausbreitung in den Kopf hinein ist ebenfalls weniger weit als die von CN, übertrifft aber diejenige von C_3. Die relativ hohen Vibrations- und Rotations-Temperaturen des C_2-Moleküls

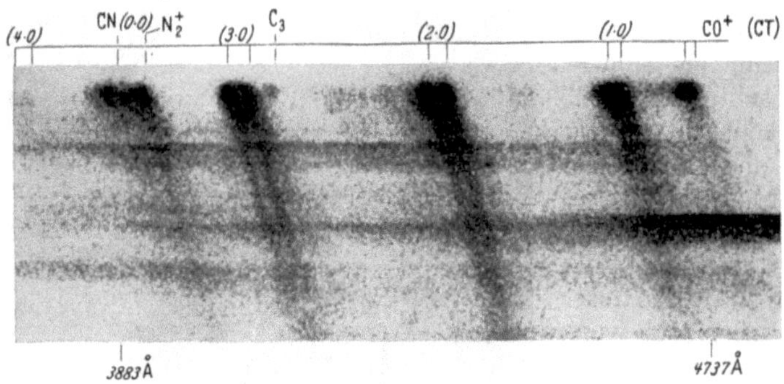

Fig. 15. Komet Morehouse 1908. (Aufnahme F. BALDET, s. Atlas von SWINGS und HASER.)

sind vom Beginn ihres Erscheinens an vorhanden. Die anderen neutralen Verbindungen (OH, CH, NH) lassen sich dann für $r \approx 1{,}5$ A.E. nachweisen, womit das eigentliche Kopfspektrum voll entwickelt ist.

Nach dem Periheldurchgang verlaufen die Veränderungen im Spektrum im umgekehrten Sinne.

Die CO^+-Schweife sind selten früher als für $r = 1{,}5$ A.E. deutlich entwickelt.

Wenn hier von einer allgemeinen, gleichförmigen Entwicklung der Kometenspektren gesprochen worden ist, so darf daraus nicht geschlossen werden, daß wir es bei allen Objekten mit einer genau gleichen Entwicklung und einer einheitlichen chemischen Zusammensetzung zu tun haben. Die Tatsache, daß ihre Mehrzahl immer nur für kleine Bereiche der heliozentrischen Distanz unter Beobachtung bleibt und quantitative Intensitätsmessungen sehr schwierig sind, gestattet kaum quantitative Rückschlüsse auf Unterschiede in der chemischen Zusammensetzung. Allerdings kennen wir zumindest ein Beispiel, für das eine ganz auffallende Abweichung von dem normalen Gasspektrum vorlag. Komet Morehouse 1908 hatte bereits bei $r = 1{,}5$ A.E. einen außerordentlich stark entwickelte $CO^+ - N_2^+$-Schweif. Die $CO^+ - N_2^+$-Emission war zudem besonders kräftig in der Umgebung des Kometenkernes. Im Vergleich dazu traten die Moleküle CN und C_2 nur ganz schwach in Erscheinung. Aus letztem Grunde war ein eigentlicher Kopf bei diesem Kometen überhaupt nicht ausgebildet.

SWINGS u. Mitarb. haben weiter festgestellt, daß auch für die neutralen Kopfmoleküle die relativen Banden-Intensitäten schwanken können. Dies trifft

[1] Vgl. dazu die Zusammenfassung der bisherigen Beobachtungsergebnisse und die Literaturangaben in dem Atlas von SWINGS und HASER.

insbesondere für die relative Stärke der OH- und NH-Emissionen zu. Von einer einheitlichen chemischen Zusammensetzung kann somit keine Rede sein, da es keinesfalls angängig ist, die beobachteten Differenzen auf Änderungen in der Anregung zurückzuführen.

7. Spektrum und Anregung. Schon von SCHWARZSCHILD und KRON[1] wurde vermutet, daß die Banden- und Linienemission in den Kometen ein Fluoreszenzvorgang ist, der durch das Sonnenlicht angeregt wird. H. ZANSTRA[2] und K. WURM[3] haben später die Fluoreszenzhypothese näher verfolgt und durch überzeugende Argumente belegt. Es ist zunächst zu bemerken, daß die kräftigen Emissionen, die vorliegen, ausnahmslos den Resonanzsystemen der Trägerpartikel entspringen. Man benötigt, wie sich zeigen läßt, nur sehr geringe Gasmassen, um die gemessenen absoluten Helligkeiten auf der Basis der Annahme einer Fluoreszenzanregung zu erklären. Von K. WURM wurde dann gezeigt, daß die Fluoreszenzvorstellung und nur diese in der Lage ist, die eigentümlichen Intensitäts-Verhältnisse innerhalb der Bandensysteme und der Bandenzweige zwanglos verständlich zu machen.

Wie in Ziff. 6 schon erwähnt wurde, tritt das symmetrische C_2-Molekül mit einer normalen Ausbildung der Banden und Bandenzweige auf (zahlreiche Einzelbanden, volle Bandenzweige), während das unsymmetrische CN eine anormal kräftige (0, 0)-Bande und nur bruchstückhaft ausgebildete Zweige aufweist. Die Länge der Bandenzweige ist beim CN mit der Sonnendistanz r veränderlich und zwar in der Weise, daß das Intensitätsmaximum in einem Zweig sich mit abnehmendem r zu höheren Rotations-Quantenzahlen verschiebt.

Verfasser hat diese Besonderheiten wie folgt erklärt. Die Fluoreszenzanregung durch das Sonnenkontinuum geht in der Kometen-Atmosphäre ganz ungestört vor sich. Anders als bei Laboratoriumsexperimenten erleiden infolge der extrem niedrigen Dichten die absorbierenden und emittierenden Moleküle keine Stöße durch andere Moleküle, und die optische Dicke der gesamten Atmosphären-Schicht ist gering. Abgesehen vom Anfangszustand beim Eintritt in die Atmosphäre wird wegen der Abwesenheit von Stößen die Verteilung der Moleküle über die Quatenzustände allein durch die Intensität und Intensitätsverteilung der anregenden Sonnenstrahlung und durch die Molekülkonstanten (Übergangswahrscheinlichkeiten, Verweilzeiten) bestimmt. Wesentlich ist dann weiterhin, daß die Absorptionsprozesse an einem individuellen Molekül im Mittel in relativ großen Zeitabständen aufeinander folgen, dies um so mehr, je weiter der Komet von der Sonne steht. Für $r=1$ ist beispielsweise die Zeit zwischen zwei sukzessiven Absorptionsvorgängen von der Ordnung 10^2 sec (s. unten, S. 486), ein Zeitintervall, welches bereits groß ist im Vergleich zu den Verweilzeiten in den Schwingungsniveaus der Grundzustände von Dipolmolekülen und vergleichbar mit den Verweilzeiten in den tiefsten Rotationszuständen. Bei einer Beurteilung der Emissionsstrukturen der Banden ist es deshalb wichtig, zu beachten, daß die unsymmetrischen Moleküle wie CN, CH usw. mit permanentem Dipolmoment nach jeder Anregung Zeit finden, um in das tiefste Schwingungsniveau $v''=0$ und in die tieferen Rotationsniveaus zurückzufallen. Eine neue Anregung kann also immer wieder nur aus diesen tiefliegenden Termen heraus erfolgen.

Für ein symmetrisches Molekül wie C_2 ohne Dipolmoment liegen nun aber die Verhältnisse insofern anders, als jetzt die Verweilzeiten in den Schwingungs-

[1] H. SCHWARZSCHILD u. E. KRON: Astrophys. Journ. **34**, 342 (1911).
[2] H. ZANSTRA: Monthly Notices Roy. Astronom. Soc. London **89**, 178 (1928).
[3] K. WURM: Z. Astrophys. **5**, 10 (1932); **8**, 281 (1934); **9**, 62 (1934). — Mitt. Hamburger Sternwarte **8**, Nr. 51 (1943).

und Rotationszuständen außerordentlich lang sind, und mit einer Ausstrahlung von reiner Schwingungs- und reiner Rotationsenergie nicht mehr gerechnet werden kann.

Zur Abschätzung der Häufigkeit der Fluoreszenzprozesse in den Resonanzsystemen genügt es, wenn wir die Moleküle durch klassische, räumliche Oszillatoren idealisieren. Die Anzahl N_A der Absorptions- und Reemissionsprozesse zwischen den Elektronenzuständen eines Moleküls in der Zeiteinheit ist dann gegeben durch

$$N_A = \frac{1}{h\nu} \frac{\pi e^2 f}{m_e} \varrho'_\nu \tag{7.1}$$

mit e gleich der Ladung des Elektrons (in elektrostatischen Einheiten) und m_e gleich der Masse des Elektrons. f bezeichnet die Oszillatorenstärke des Systems, ϱ'_ν die Strahlungsdichte (für die Absorptionsfrequenz ν) am Ort des Kometen und h die Plancksche Konstante. Angenäherte Werte für f für einige Moleküle (CN, C_2, OH, CN, NH) sind nach einem Verfahren von MULLIKEN[1] durch LYDDANE, ROGERS und ROACH[2] berechnet worden. Für CN und C_2 sind die f von der Ordnung 10^{-2}. Bei Anwendung von Gl. (7.1) auf das C_2-Molekül, dessen Resonanzfrequenzen auf dem Maximum der Sonnenstrahlung liegen, erhalten wir

$$N_A \approx \frac{5 \cdot 10^{-2}}{r^2} \tag{7.2}$$

und für CN einen um einen Faktor 3 bis 5 kleineren Wert. Gl. (7.2) wurde errechnet mit einer Sonnentemperatur $T = 6000°$ und einem Verdünnungsfaktor für die Strahlung gleich $R_\odot^2/4\pi r^2$ ($R_\odot$ gleich dem Sonnenradius). Für die Entfernung $r = 1$ A.E. erfolgen somit die Prozesse an einem einzelnen C_2-Molekül im Durchschnitt in einem zeitlichen Abstand von 10 bis 100 sec, am CN etwa alle 100 sec.

Für das CN-Molekül können wir bei den Schwingungsniveaus mit Übergangswahrscheinlichkeiten der Ordnung $A_{\text{osc}} = 10^1$ bis 10^3 rechnen. Bei den Elektronensprüngen eventuell gewonnene Schwingungsquanten gehen also selbst bei kleineren Sonnendistanzen bis zu $r = 0{,}1$ und darunter bis zum Einsetzen eines neuen Elektronensprunges durch Ultrarot-Ausstrahlung wieder verloren. Jede Anregung in dem $B\,^2\Sigma - X\,^2\Sigma$-System trifft danach die Moleküle immer in dem Schwingungsniveau $v'' = 0$ des Grundzustandes an. Wegen der symmetrischen Lage der Potentialkurven in den Elektronenzuständen $X\,^2\Sigma$ und $B\,^2\Sigma$ wird deshalb ein Sprung $X\,^2\Sigma(v''=0) \rightarrow B\,^2\Sigma(v'=0)$ mit der Reemission $B\,\Sigma^2(v'=0) \rightarrow X\,^2\Sigma(v''=0)$ gegenüber allen anderen Fluoreszenz-Sprüngen stark bevorzugt sein.

Das C_2-Molekül weist für die beiden Elektronenzustände des Swansystems eine ähnliche symmetrische Lage der Potentialkurven auf wie das CN. Infolge der Unfähigkeit des Moleküls zur Ausstrahlung von reiner Schwingungsenergie kann hier aber nicht der gleiche Effekt erwartet werden wie beim CN. Die Unkenntnis des Anfangszustandes der Verteilung über die Quantenzustände und die nur mangelhafte Kenntnis der relativen Übergangswahrscheinlichkeiten für die verschiedenen $A\,^3\Pi(v') \leftrightarrow X\,^2\Pi(v'')$-Sprünge machen es unmöglich, das Zustandekommen der beobachteten relativen Intensitäten der Banden des Systems genauer aufzudecken. Die relativen Intensitäten der Banden sind in etwa dieselben wie man sie in einer Lichtquelle im Temperatur-Gleichgewicht bei 2000 bis 3000° erwarten kann[3].

---

[1] R. MULLIKEN: J. Chem. Phys. **7**, 14 (1939); **8**, 382 (1940).

[2] R.H. LYDDANE, F.T. ROGERS u. F.E. ROACH: Phys. Rev. **60**, 281 (1941). Vgl. dazu auch R.H. DWYER u. O. OLDENBURG: J. Chem. Phys. **12**, 361 (1941).

[3] K. WURM: Handbuch der Astrophysik, Erg.-Bd. VII, S. 305. Berlin 1936.

Die (0, 0)-Bande des CN besteht bei kleiner Dispersion gewöhnlich aus zwei diffusen „Linien" mit etwas Struktur (s. Fig. 11 u. 16). Die eine dieser diffusen Linien wird gebildet von den unaufgelösten Rotationslinien des P-Zweiges (langwellig), die andere von den R-Linien. V. M. SLIPHER[1] und später J. DUFAY[2] haben als erste darauf aufmerksam gemacht, daß die Aufspaltung dieser „Linien" mit der heliozentrischen Distanz der Kometen veränderlich ist. DUFAY bestimmte für neun verschiedene Kometen, beobachtet bei verschiedenen Sonnendistanzen,

Fig. 16. Struktur der CN-Bande (0,0) λ 3883. (Entnommen dem Atlas von SWINGS und HASER.)

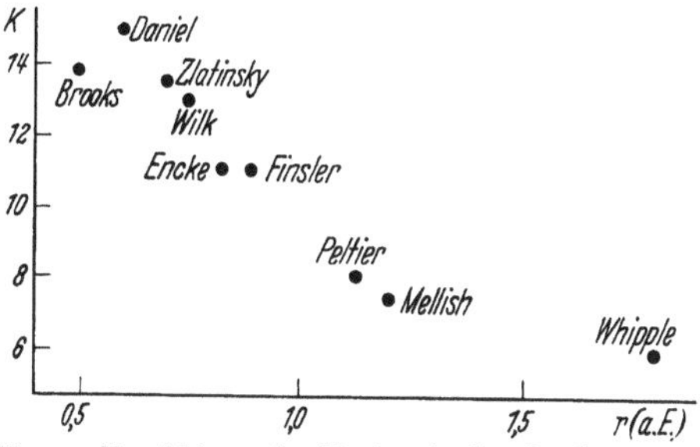

Fig. 17. Verschiebung der Maxima in den Bandenzweigen der (0,0) Bande λ 3883 des CN. (Nach J. DUFAY.)

die Laufzahlen K in den Serienlinien, die am nächsten mit den Positionen der Intensitätsmaxima übereinstimmen. Das Resultat zeigt Fig. 17, in welcher die Abszisse die heliozentrische Distanz zur Zeit der Beobachtung bezeichnet und die Ordinate die genannte Laufzahl K (beide ergeben sich als gleich für P- und R-Zweig). Mit abnehmender Sonnendistanz r verschiebt sich also das Intensitätsmaximum in den Zweigen zu höheren Rotations-Quantenzahlen K.

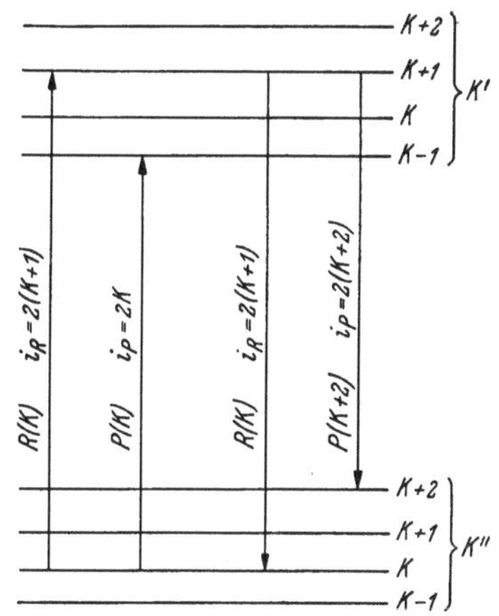

Fig. 18. Intensitäten der Übergänge in einem $^2\Sigma - {}^2\Sigma$-System.

Wie Verfasser gezeigt hat, ist dieser Effekt gerade zu erwarten, wenn eine Fluoreszenzanregung vorliegt, und wenn die Verweilzeiten in den Rotations-Niveaus von derselben Länge werden wie das Zeitintervall zwischen zwei aufeinander folgenden Absorptionsprozessen am Molekül. Man erkennt dies sofort unter Beachtung der relativen Übergangswahrscheinlichkeiten in P- und R-Zweig eines $^2\Sigma - {}^2\Sigma$-Überganges (s. Fig. 18). Nach den quantentheoretischen Intensitätsformeln hat in Absorption die Linie $P(K)$ (Numerierung nach dem Grundniveau) den Intensitätsfaktor $i_{P(K)} = 2K$, die Linie $R(K)$ jedoch den Faktor $i_{R(K)} = 2(K+1)$. In Emission haben die von einem bestimmten Niveau ausgehenden Linien $P(K+2)$ und $R(K)$ die Intensitätsfaktoren $i_{P(K+2)} = 2(K+2)$ und $i_{R(K)} = 2(K+1)$. Bei kleinen K sind also in Absorption die R-Sprünge, in Emission die P-Sprünge bevorzugt. Bei einer störungsfreien Fluoreszenzanregung, erzeugt durch ein gleichförmiges Kontinuum, werden also die Moleküle Rotationsenergie gewinnen, wenn das Zeitintervall zwischen zwei aufeinanderfolgenden

[1] V. M. SLIPHER: Bull. Lowell Obs. Nr. 74.
[2] J. DUFAY: C. R. Acad. Sci., Paris **206**, 1948 (1938).

Absorptionen kleiner wird als die Verweilzeiten in den Rotationsniveaus. Im Prinzip ist damit die Rotationsstruktur beim CN und deren Veränderlichkeit mit r erklärt.

P. Swings[1], A. McKellar[2] und insbesondere J. Hunaerts[3] haben später die Verhältnisse mehr im Detail untersucht und auch andere unsymmetrische Moleküle wie CH, NH und OH bearbeitet. Besonders erwähnenswert aus diesen Arbeiten ist der einwandfreie Nachweis, daß sich in der Feinstruktur der Banden auch eindeutig ein Einfluß der Unregelmäßigkeiten im Sonnenspektrum (stellenweise Schwächung des Kontinuums durch breite Absorptionslinien) widerspiegelt. Hunaerts, der die Rotationsstruktur der Hydride CH, NH und OH sehr ausführlich behandelt hat, unternahm den Versuch, die unbekannten Verweilzeiten in den unteren Rotationszuständen aus der aufgefundenen Zweigausbildung zu berechnen. In Tabelle 7 findet man sein Resultat für NH. Für CH und OH kommt der Autor zu denselben Größenordnungen.

Tabelle 7. *Verweilzeiten in den ersten Rotationsniveaus für den Grundzustand* $^3\Sigma(v''=0)$ *des* NH

Rotations-quantenzahl J''	Verweilzeit τ_{rot}
2	1290 sec
3	355 sec
4	145 sec
5	72 sec

8. Photodissoziations- und Photoionisations-Prozesse[4]. Die Gasatmosphäre eines Kometen muß sich beständig erneuern. Ein Teil der Gase wird durch starke Repulsionskräfte in der Richtung des Schweifes abgetrieben, ein anderer Teil expandiert allseitig vom Kern aus in den Raum. Es kann nicht bezweifelt werden, daß außer den Verbindungen, die wir durch eine Emission erkennen, auch noch zahlreiche andere vorliegen, die durch das Sonnenkontinuum nicht oder nur äußerst schwach angeregt werden können, die also unsichtbar bleiben.

Die Chemie der Atmosphären der Kometen ist sicherlich im Prinzip sehr einfach[5]. Nach einer Abschätzung der Dichten[6] und nach den schon oben auseinandergesetzten stark unterschiedlichen „Schwingungs-“ und „Rotationstemperaturen“ der Verbindungen kann man mit Sicherheit schließen, daß normalen chemischen Reaktionen, die ja an Stoßprozesse gebunden sind, in ihnen keine Bedeutung zukommt. An chemischen Umwandlungen können nur Zersetzungen der Verbindungen durch den photochemischen Primärprozeß und Photoionisation eintreten[7]. Trotz dieses einfachen Gesichtspunktes bleibt es trotzdem äußerst schwierig, für irgendeines der nachgewiesenen Moleküle seine „Zerfallsreihe“ aufzudecken, was seinen wesentlichen Grund in der schwachen Anregung der Atmosphäre hat. Spektroskopisch erfaßbar sind ja nur solche Partikel, deren Resonanzfrequenzen im astrophysikalisch zugänglichen Spektrum liegen. Die Schwierigkeiten werden ausreichend beleuchtet, wenn wir uns einmal

---

[1] P. Swings: Lick Obs. Bull. **19** (1951).

[2] A. McKellar: Rev. Mod. Phys. **14**, 179 (1943). — Astrophys. Journ. **98**, 1 (1943); **99**, 162 (1944).

[3] J. Hunaerts: Ann. Obs. Roy. Belg. **5**, Nr. 1. — Symposium Liège 1952; 1956, S. 82 u. 63. — Ann. Obs. Roy. Belg., Ser. III **6**, 97.

[4] Vgl. dazu K. Wurm: Z. Astrophys. **9**, 62 (1934) und Mitt. Hamburger Sternwarte **8**, Nr. 51 (1943).

[5] Die Chemie der Kerne mag dagegen äußerst kompliziert sein. Vgl. dazu weiter unten, Ziff. 10.

[6] Die Dichten sind im Zentrum des Kometenkopfes bei helleren Kometen von der Ordnung 10^4 bis 10^5 Teilchen pro cm^3. Dieses sind Partialdichten der Molekule wie CN und C_2. Vgl. dazu K. Wurm, loc. cit. Hamburger Mitt.

[7] Neuerdings ist von einigen Autoren ein Einfluß von solaren Korpuskelströmen auf die Kometenatmosphären ernstlich zur Diskussion gestellt worden (s. Ziff. 16). Wir sehen hier zunächst einmal von einem solchen Einfluß ab.

eine hypothetische, verzweigte Umwandlungsreihe ansehen, welcher das beobachtete CO^+-Ion angehören könnte, und die durch folgendes Schema zu charakterisieren wäre:

$$\begin{array}{l} CO_2 \rightarrow CO \rightarrow CO^+ \rightarrow CO^{++} \\ \quad\searrow \\ \qquad O \rightarrow O^+ \ldots \end{array}$$

Das CO^+-Ion geht durch Photoionisation aus CO hervor, und dieses ist durch eine Photodissoziation aus CO_2 entstanden. Das chemische sehr stabile CO_2 fassen wir als ein Entgasungs- oder Verdampfungsprodukt des festen Kometenkernes auf. Das CO^+-Ion soll dann durch eine weitere Ionisation in CO^{++} übergehen. Die drei Partikel, die mit dem CO^+ in einer Reihe stehen, könnten im Kometenspektrum aber nur bei einer intensiven UV-Anregung sichtbar werden, und dasselbe gilt für das von CO_2 abgespaltene Atom O und dessen Ionen. In Wahrheit können wir nicht einen der angegebenen Schritte garantieren. Obwohl die Umwandlung $CO + h\nu = CO^+ +$ Elektr. sehr wahrscheinlich ist, so kann nicht ausgeschlossen werden, daß das CO^+ seine Existenz wesentlich oder doch zum Teil dem Prozeß $CO_2 + h\nu \rightarrow CO^+ + O$ verdankt. Der Prozeß $CO^+ + h\nu \rightarrow CO^{++}$ ist ebensowenig garantiert und könnte durch den Dissoziationsvorgang $CO^+ + h\nu \rightarrow C + O^+$ (oder $C^+ + O$) übertroffen werden.

Von dem CO^+-Molekül läßt sich auf jeden Fall sagen, daß es im Strahlungsfeld der Sonne einer Dissoziation (und Ionisation) sehr widersteht und deshalb eine relativ große Existenzdauer besitzt. Offensichtlich ist dadurch erst die Möglichkeit der Ausbildung langer und intensiver Gasschweife gegeben. Bei mittleren Sonnendistanzen ($r = 0{,}5$ bis 1,0 A.E.) zählt die mittlere Existenzdauer der CO^+-Ionen sicherlich nach Tagen. Es folgt dies einfach aus der Tatsache, daß man bei diesen heliozentrischen Distanzen schon vielfach isolierte CO^+-Schweifwolken bis zu sechs Tagen im Auge behalten konnte. Ob die anderen Schweifmoleküle wie CO_2^+ und N_2^+ eine ähnliche Stabilität dem Sonnenlicht gegenüber aufweisen, hat sich wegen der Schwäche der Banden dieser Ionen noch nicht entscheiden lassen; es scheint dies aber wahrscheinlich.

Verfasser hat schon früh darauf hingewiesen[1], daß die Verschiedenheit von Kopf- und Schweifemission in der verschieden langen Existenzdauer der Molekülsorten (CN, C_2, OH, NH einerseits und CO^+, N_2^+, CO_2^+ andererseits) eine einfache Erklärung findet. Die Moleküle der Kopfgruppe erleiden eine Dissoziation (oder Ionisation) bevor sie in den eigentlichen Schweif abgetrieben werden können. Die mittleren Lebensdauern der Moleküle CN und C_2 liegen bei mittleren Sonnendistanzen in der Größenordnung von einigen Stunden und nicht von Tagen. Dafür sprechen schon die Beobachtungen, daß merkliche und sogar starke Helligkeitsverminderungen des Kometenkopfes sich von einer Nacht zur anderen vollziehen können.

Die Theorie der sukzessiven Umwandlung der Moleküle in den Kometen, deren Grundlage Verfasser schon vor mehr als einem Jahrzehnt entwickelt hat[2], ist in ihrer praktischen Anwendung noch immer ein Problem der Zukunft geblieben.

9. Polarisation des Kometenlichtes. Das Vorliegen einer teilweisen Polarisation des Kometenlichtes wurde zuerst von ARAGO an den Objekten 1819 III und 1835 III (Halley) festgestellt[3]. Nach ARAGO stellte dann WRIGHT[4] Polarisations-Beobachtungen an den großen Kometen 1881 III (Tebbutt) und

[1] K. WURM: Z. Astrophys. **8**, 281 (1934).

[2] K. WURM: Mitt. Hamburger Sternwarte **8**, 51 (1943). Formal hat diese Theorie eine große Ähnlichkeit mit der Theorie des radioaktiven Gleichgewichts.

[3] Siehe BREDIKHIN-JÄGERMANN: Kometenformen. St. Petersburg 1903.

[4] A. W. WRIGHT: Amer. J. Sci. **22**, 142 (1881); **22**, 372 (1881).

1881 IV (Schaeberle) an. Bei dem ersten fand er eine durchschnittliche Polarisation von 23% mit Fluktuation von Nacht zu Nacht, an dem zweiten maß er einen Durchschnitt von 13%. Der Phasenwinkel war in beiden Fällen angenähert gleich. DANJON[1] berichtet über photographische und visuelle Messungen an dem Kometen 1927 II (Pons-Winnecke), die Werte von 10 und 12% ergaben.

Besondere Erwähnung verdienen dann die mehr systematischen Untersuchungen von ÖHMANN[2]. Bei dem Kometen 1939 III fand er bei visueller Messung 6% Polarisation. Wichtiger sind seine darauf folgenden Beobachtungen an den Objekten 1940 c (Cunningham) und 1941 c (Paraskevopoulos), bei denen er eine spektrale Zerlegung vornahm. Diese zeigte, daß im ersten Falle im wesentlichen Bandenlicht und im zweiten fast nur kontinuierliches Licht vorlag. Die instrumentelle Ausrüstung bestand aus einem 40 cm - Astrographen mit einem Objektivprisma kleiner Dispersion und einem am Ende des Lichtweges eingebauten Wollaston-Prisma, mit Kollimator- und Kameralinse. Die Eigenpolarisation der Optik erwies sich im verwendeten zentralen Teil des Bildfeldes als vernachlässigbar klein.

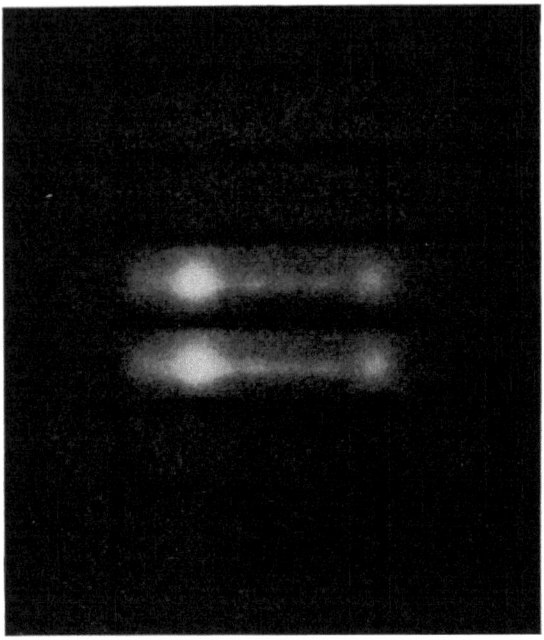

Fig. 19. Polarisations-Aufnahmen im zerlegten Kometenlicht. (Nach I. ÖHMANN.)

Schon der einfache visuelle Anblick der Doppelspektren des Kometen 1941 I (Fig. 19) macht es evident, daß sämtliche Bandenemissionen des photographierten Bereiches eine Polarisation aufweisen. Der Polarigraph war bei den Aufnahmen nach dem Positionswinkel Komet—Sonne orientiert und senkrecht zur Dispersionsrichtung. In Tabelle 8 sind die von ÖHMANN mitgeteilten Ergebnisse für die Bandengruppen C_2 λ 5165; C_2 λ 4737 und CN λ 3881 aufgeführt, die nach einer sorgfältigen Photometrie der Platten gewonnen wurden.

Die Dispersion in den Beträgen von p[3] ist zu einem Teil auf die Änderung des Phasenwinkels zurückzuführen. Die Messungen sind allerdings nicht exakt genug, um einen Gang mit dem Phasenwinkel, der über die Zeit der Beobachtungen von 42 auf 156° anstieg, klar erkennen zu können. Als wahrscheinlichste Werte der Polarisation für einen Phasenwinkel $\alpha = 90°$ gibt ÖHMANN die folgenden an:

$$p_{5165} = 8{,}3\% \pm 1{,}8\%; \quad p_{3881} = 10{,}1\% \pm 1{,}3\%$$
$$p_{4737} = 10{,}6\% \pm 0{,}6\%; \quad p_{\text{total}} = 9{,}5\% \pm 1{,}8\%.$$

[1] A. DANJON: C. R. Acad. Sci., Paris **186**, 1427 (1928).

[2] I. ÖHMANN: Monthly Notices Roy. Astronom. Soc. London **99**, 624 (1939). — Stockholms Obs. Ann. **13**, Nr. 11 (1941).

[3] Es ist $p = \frac{I_1 - I_2}{I_1 + I_2}$, I_1 Intensität des ordentlichen, I_2 des außerordentlichen Strahles.

Die quantentheoretische Interpretation der Polarisations-Verhältnisse eines Molekülspektrums sind recht verwickelt, da der Grad der Polarisation sowohl von Zweig zu Zweig wie auch noch mit der Rotations-Quantenzahl variiert. Eine quantitative Deutung setzt auf jeden Fall eine volle Auflösung der Zweige und Intensitätsmessungen an Einzellinien voraus. Wie ÖHMANN erläutert, ist der beobachtete Polarisationsgrad von rund 10% für die Systeme von C_2 und CN roh qualitativ mit der Theorie in Übereinstimmung.

Bei dem zweiten von ihm behandelten Spektrum (Komet 1941 IV) findet ÖHMANN höhere Polarisationsgrade. Die Messungen wurden in zwei Bereichen im Spektrum mit den Schwerpunkten λ 4700 und λ 4300 durchgeführt. Die Resultate sind in diesem Falle quantitativ weniger genau als für den Kometen 1941 I, und zwar sowohl wegen einer Verschleierung der Platten durch Mondlicht wie auch wegen der geringeren Anzahl der gewonnenen Aufnahmen. Der Autor hält jedoch die höhere Polarisation, die sich jetzt auf 20 bis 27% beläuft, durchaus für gesichert. Unterschiede für die beiden gemessenen Wellenlängenbereiche lassen sich soweit nicht feststellen. Da im Spektrum die CN- und C_2-Banden nicht klar hervortreten, so ist anzunehmen, daß die gemessene Strahlung zur Hauptsache dem Kometenkontinuum angehört. ÖHMANN deutet die vorliegende Polarisation als durch eine Mischung von Streuung und diffuser Reflexion hervorgerufen. Im Falle einer reinen Streuung sollte der zu beobachtende Polarisationsgrad noch wesentlich höher ausfallen. Der Phasenwinkel lag in den Tagen der Beobachtungen zwischen 58° und 98°.

Tabelle 8.

Beobachtungszeit	Polarisation p (Komet 1941 I)			
	C_2 λ 5165	C_2 λ 4737	CN λ 3881	Totales photographisches Licht
Nov. 1940				
26,80				0,046
Dez. 1940				
4,70		0,086		
4,72		0,070		
13,66		0,072		
13,68		0,107		
14,65		0,105		
14,67		0,056		
19,66		0,095	0,113	
19,68		0,050		
20,73		0,130	0,091	
22,66	0,067	0,067		
22,68	0,089	0,092		
22,69		0,106	0,125	
23,72				0,094
31,67		0,112	0,052	
Jan. 1941				
2,69		0,081	0,074	
3,67		0,083		
3,68		0,080		

10. Vorstellungen über die Konstitution der Kometenkerne. Obwohl die Kometenkerne als recht helle Lichtpunkte leicht erkenntlich sind, so erweisen sie sich doch als einer näheren Erforschung sehr schwer zugänglich. Aussagen über ihre Konstitution sind nach wie vor noch sehr spekulativ.

Gewöhnlich wird gesagt, daß sie aus einer Ansammlung meteorischen Materials bestehen. Über die Dimensionen der Einzelkörper gehen die Meinungen sehr auseinander. Sicher ist, daß die Gesamtmasse und die gesamte Ausdehnung eines Kometenkernes — astronomisch gesehen — immer sehr klein ist, sehr klein selbst im Vergleich zu den Planetentrabanten. Erste sichere Anhaltspunkte dafür erhielt man auf Grund von Durchgängen von zwei Kometen durch das System der Jupitermonde (Lexell 1770 I[1]; Brooks 1889 V[2]), in dem die Durchgänge

[1] U. J. LEVERRIER: Théorie de la comète périodique de 1770. Ann. Obs. Paris **131** (1857).
[2] C. L. POOR: Researches upon Comet 1889 V, Part III. 1889. Astronom. J. **13**, 177 (1893).

nicht die geringsten Störungen hinterließen. LAPLACE[1] berechnete dann auf Grund der Tatsache, daß der Komet Lexell bei einem nahen Vorbeigang an der Erde deren Bahnbewegung nicht beeinflußte, als oberen Grenzwert 0,0002 Erdmassen. Eine weitere Gelegenheit zur Abschätzung einer Kometenmasse auf dynamischem Wege ergab sich im Anschluß an die Aufteilung des Kometen Biela (1846) in zwei Komponenten, die sich getrennt weiter bewegten und unmittelbar nach der Teilung in ihrer Bewegung feststellbare Störungen aufeinander ausübten. J. v. HEPPERGER[2] errechnete auf dieser Basis Werte von der Ordnung 10^{-7} Erdmassen.

Zahlreicher sind die Abschätzungen der Massen und Dimensionen nach photographischen Daten. Nach RICHTER[3], der 22 verschiedene Kernhelligkeiten nach Beobachtungen von WIRTZ[4] und BEYER[5] zu diesem Zwecke behandelt hat, schwanken die Massen zwischen 10^{17} und 10^{23} g ($5{,}0\times10^{-11}$ und 1×10^{-5} Erdmassen). Die effektiven photometrischen Durchmesser werden zwischen 40 und 3000 km gefunden. Für den Halleyschen Kometen, der zu den mittelgroßen Objekten zu rechnen ist, errechnete VORONTSOV-VELYAMINOV[6] einen Durchmesser von 30 km und eine Masse von 10^{19} g. Der Autor benutzte dabei die Kernhelligkeit ($15{,}5^m$) für den Zeitpunkt, als der Komet bei einer Sonnendistanz von $r=3{,}4$ vorwiegend in kontinuierlichem Licht strahlte. Unter Annahme des gleichen Albedo-Wertes wie für den Planetoiden Ceres (0,028) und eines Phasengesetzes wie für Merkur ($m=m_0+0{,}015\cdot\alpha$, α Phasenwinkel) ergab sich, daß der Kern dem Sonnenlicht eine reflektierende Fläche von $3\cdot10^{13}$ cm² darbot. Über die Berechtigung der Voraussetzungen in bezug auf Albedo und Phasengesetz läßt sich streiten.

Nach RICHTER[7] sind die Albedo-Werte der Kometenkerne im Durchschnitt auffallend klein, schwanken aber in weiten Grenzen. Einige typische Beispiele aus der Richterschen Untersuchung von 22 Objekten findet man in Tabelle 9. Es sind gleichzeitig die vom Autor unter gewissen Annahmen errechneten Durchmesser der Kerne aufgeschrieben[8]. Allerdings scheinen uns die angenommenen d äußerst unsicher, und da diese d immer obere Grenzen darstellen, so haben die errechneten A als untere Grenzen zu gelten, die nach unserer Meinung sogar um Größenordnungen zu klein sein können. Dem Werte $d=900$ km für den Halleyschen Kometen (1909 c) steht der Wert $d=30$ km von VORONTSOV-VELYAMINOV gegenüber. Letzterer ist natürlich ebenfalls unsicher und mag wesentlich zu klein sein. Diese Diskrepanz beleuchtet jedoch die Unsicherheit der Ergebnisse.

N. RICHTER hat ebenfalls versucht, aus den Helligkeits-Beobachtungen von 14 Kometenkernen genaueres über das Phasengesetz der Kernhelligkeiten in Erfahrung zu bringen. Aus einer Kombination der Helligkeitsmessungen von

[1] P. S. DE LAPLACE: Méchanique céleste **4**, 230 (1805).
[2] J. v. HEPPERGER: Wien. Ber. IIa, 115, 785 (1906).
[3] N. RICHTER: Statistik und Physik der Kometen. Leipzig 1954.
[4] V. WIRTZ: Astronom. Nachr. **177**, 81 (1908); dort auch frühere Literatur.
[5] M. BEYER: Astronom. Nachr. **278**, 217 (1950); dort auch frühere Literatur.
[6] B. VORONTSOV-VELYAMINOV: Astrophys. Journ. **104**, 226 (1946).
[7] N. RICHTER: Astron. Nachr. **276**, 47 (1948).
[8] Die Berechnung der Albedowerte wurde nach der bekannten Lambertschen Formel durchgeführt:

$$A=\frac{3}{2}M_0\frac{\sin^2 S}{\sin^2 S_0\cdot\sin^2\sigma}.$$

M_0 ist das beobachtete Helligkeitsverhältnis in der Opposition des Kometenkerns zur Sonne, S der scheinbare Halbmesser der Sonne von der Erde aus gesehen, S_0 der scheinbare Halbmesser der Sonne vom Kometen aus gesehen, σ_0 der scheinbare Halbmesser des Kometenkernes von der Erde aus gesehen. Da die geringste Anzahl der benützten Kometenhelligkeiten aus der Opposition stammen, mußten diese Helligkeiten größtenteils mittels einer Phasenkurve auf die Opposition reduziert werden.

Tabelle 9.

Komet	Durchmesser in km	Albedo A	Komet	Durchmesser in km	Albedo A
1882 I	500	0,0450	1937 c	130	0,0773
1903 c	150	0,0320	1939 c	40	0,0449
1904 a	480	0,0005	1940 c	180	0,0296
1906 b	700	0,0002	1907 d	3070	0,0034
1932 g	490	0,2087	1909 c	900	0,0113
1932 k	130	0,3020	1927 i	1035	0,0230

14 Objekten, die einzeln für sich genommen nur immer ein kleines Phasenwinkel-Intervall überdecken, leitet er eine Phasenkurve ab, die anfangs bis zu Winkeln von 50 bis 60° sehr steil verläuft und bei diesen Winkeln schon Korrekturen von $+3{,}0^m$ bis $+4{,}0^m$ verlangt[1], was zwei bis drei Größenklassen mehr sind als für die Planeten und den Erdmond. Der Autor sieht in dem besonderen Phasengesetz der Kerne, das dem ähnelt, welches man bei staubförmigen Massen erhält[2], einen Beweis für eine vorwiegend staubförmige Struktur der Kometenkerne. Es läßt sich nicht bestreiten, daß viele Kometenkerne reich an Staub sind, an festen Teilchen von zum Teil so geringer Dimension, daß sie leicht durch den Lichtdruck der Sonne in Richtung des Schweifes abgetrieben werden können. Es ist jedoch schwierig, den Standpunkt zu verteidigen, daß solche Staubwolken die permanenten Bestandteile der Kometenkerne sind, die ihnen ihre Existenz für astronomische Zeiten sichern. Für eine kompaktere Struktur bestehend aus nur einigen wenigen großen Blöcken oder nur einem großen Block sprechen eine ganze Reihe von Erfahrungen und Überlegungen. Die klare Aufspaltung einzelner Kometen in zwei oder mehrere Komponenten, von denen jede Komponente nach zunehmender räumlicher Trennung einen vollständigen Kometen darstellte, läßt sich zwar auf Grund einer Zerstückelung großer Blöcke aber nicht einer Zerteilung und Aufteilung einer Staubwolke verstehen. Ebenso beachtenswert ist die Tatsache, daß nachweislich eine ganze Reihe von Kometen eine Passage der Sonnenkorona überstanden haben. Die am besten bekannten Beispiele sind die großen Kometen 1843 und 1881 II. Bei starker Aktivität erfolgt die Ausströmung von Materie aus den Kernen bevorzugt in Richtung zur Sonne. Diese Bevorzugung einer Richtung weist auf das Vorliegen eines festen, mechanischen Widerstandes. Der Fall des kurzperiodischen Kometen Encke zeigt, daß die Kerne eine hohe Anzahl von Sonnen-Umläufen mit starker Annäherung an die Sonne ($q = 0{,}338$ A.E.) durchführen können, ohne daß sich ihre Fähigkeit zur Gasabgabe erschöpft, was wiederum schwerlich mit kleinen Dimensionen der Kernkörper in Einklang zu bringen ist.

Die Beantwortung der Frage nach der physikalisch-chemischen Konstitution des Kernmaterials, das die Atmosphäre mit Gas und Staub versorgt, hat ihre ganz besonderen Schwierigkeiten. Wie oben schon bemerkt wurde, lassen die im Spektrum erkenntlichen Radikale keine sicheren Schlüsse auf die Natur ihrer Muttersubstanzen zu, aus denen sie entstanden sind außer, daß sie diese Radikale enthalten müssen. Es liegt natürlich nahe, in erster Linie an die stabilen Verbindungen H_2O, NH_3, CO_2, CH_4, C_2N_2 und N_2 zu denken. Nicht ganz leicht ist es, für das beobachtete C_3 ein plausibles Muttermolekül zu nennen. Bis vor kurzem neigte man allgemein zu der Auffassung, daß die gasförmigen Substanzen in okkludiertem Zustand an festen, gesteinsartigen Massen haften und durch

[1] N. Richter: Astronom. Nachr. **276**, 44 (1948).

[2] Vgl. dazu die Experimente von N. Richter, dargelegt in seiner Monographie „Statistik und Physik der Kometen". Leipzig 1954.

Erwärmung nach und nach aus diesen ausgetrieben werden. Gegenwärtig wird jedoch mehr ein Kometen-Kernmodell diskutiert, das von einer anderen Art ist und vor etwa acht Jahren von F. Whipple[1] entworfen wurde.

Nach Whipple besteht ein Kometenkern aus einem Konglomerat von zu „Eis" erstarrten Gasen und zwar aus solchen, die bei Zimmertemperatur in Verbindung mit ihrem Kondensat schon über einen hohen Dampfdruck verfügen, und die im Kometenkern durchsetzt und vermischt sind mit bei normalen Temperaturen festem meteorischem Material. Diese letzteren Bestandteile bilden, obwohl locker aneinander haftend, das mehr beständige Gerüst des Kerns, das seine Quasi-Stabilität aber einzig dem Umstand verdankt, daß nur eine verschwindend kleine Gravitationskraft wirksam ist. Bei der Annäherung eines Kernes an die Sonne beginnen die Eise an den Oberflächenteilen, die der Bestrahlung ausgesetzt sind, zu verdampfen. Schwächer haftende, kleine meteorische Teilchen werden von den expandierenden Gasen mitgerissen. Aus zwei Gründen ist das Eisreservoir gegen eine rasche Erschöpfung geschützt. Die auftreffende und absorbierte Strahlungsenergie wird zum größten Teil als Verdampfungswärme aufgebraucht, so daß — abgesehen von der Erreichung kleinster Sonnendistanzen — die Kerne immer nur mäßig erwärmt werden. Nachdem die Eise der Oberflächenschichten abgedampft sind, bildet sich bald durch die erhalten gebliebenen Strukturen des festen meteorischen Materials eine Isolationsschicht besonders geringer Wärmeleitfähigkeit. Letztere hat ihren Grund in der Kleinheit der Kontaktflächen innerhalb des lockeren Materials.

Das Modell ist von Whipple nach dem Gesichtspunkt konstruiert, daß es in seinen Konsequenzen einigermaßen zwanglos die wesentlichsten, von der Beobachtung gestellten Forderungen befriedigt. Ein Kern stellt eine ergiebige aber auch gleichzeitig beständige Quelle für Gase und Staub dar, die mechanische Festigkeit muß schließlich einen allmählichen aber vollständigen Zerfall und eine Auflösung in kleinste Meteoriten-Körper gestatten.

Die bei den Kometen häufig auftretenden plötzlichen Helligkeitssteigerungen (Gas-Staubausbrüche) können nach den Whippleschen Vorstellungen durch das Zusammenbrechen ausgehöhlter Gerüstteile entstehen, wodurch der Strahlung neue „Eisflächen" zum Abdampfen dargeboten werden. Eine andere Möglichkeit zu plötzlichen Gas-Staubausbrüchen besteht nach Whitney[2], der diese Seite des Kometenproblems auf der Basis des neuen Modells diskutiert hat, in der Verdampfung und explosionsartigen Befreiung von enger eingeschlossenen, starken Konzentrationen der Eise, wobei insbesondere an die niedrig siedenden Verbindungen CH_4 und CO_2 gedacht ist.

B. Donn und H. C. Urey[3] haben die Aufmerksamkeit darauf gelenkt, daß infolge der tiefen Temperaturen, die das Kernmaterial besitzt, den abgesättigten Molekülen der Eise in beachtlichem Prozentsatz freie Radikale beigemischt sein können. Sie verweisen auf Experimente von Rice[4], Broida[5] u. a., in denen es gelang, freie Radikale wie NH, OH, und Gemische aus atomarem und molekularem Sauerstoff und desgleichen Stickstoff und Wasserstoff bei tiefen Temperaturen nahe dem absoluten Nullpunkt zu stabilen Konglomeraten zu kondensieren. Solche Gemische werden hochexplosiv bei etwas gesteigerten Temperaturen. Beispielsweise bildet sich bei der Temperatur der flüssigen Luft

[1] F. Whipple: Astrophys. Journ. **111**, 375 (1949).
[2] C. Whitney: Astrophys. Journ. **122**, 190 (1950).
[3] N. Donn u. H. C. Urey: Astrophys. Journ. **123**, 339 (1953) Symposium, Liège 1956.
[4] F. O. Rice u. M. Freamo: J. Amer. Chem. Soc. **73**, 5539 (1951); **75**, 548 (1953).
[5] H. Broida u. I. R. Pellam: Phys. Rev. **95**, 845 (1954). — J. Chem. Phys. **23**, 407 (1955).

(77° K) aus OH-Radikalen Wasserstoffperoxyd (H_2O_2). Letzteres muß in Verbindung mit Kohlenstoffverbindungen wiederum als hochexplosiv gelten.

Die Autoren halten es für möglich, daß Gas-Staubausbrüche der Kometenkerne durch Reaktionen zwischen freien Radikalen und anderen unstabilen Verbindungen ausgelöst werden. Der Eintritt solcher explosiven Reaktionen kann eingeleitet werden durch Erwärmung oder bei großen Sonnendistanzen auch durch das Auftreffen solarer Korpuskeln. Wie weit der Gedanke der Autoren für die Kometenphysik wirklich von Bedeutung ist, läßt sich gegenwärtig schwer beurteilen.

Wir müssen hier noch erwähnen, daß B. U. LEVIN[1] schon vor längerem eine Näherungsformel aufgestellt hat, welche die Ergiebigkeit des Kometenkernes an Gasen unter der Voraussetzung beschreibt, daß der unterliegende Prozeß entweder eine Desorption oder eine Verdampfung darstellt. Diese Ergiebigkeit beträgt nach dem Autor

$$n = n_0 \sqrt{\frac{k T}{2 \pi m}}\, e^{-L/RT}, \tag{10.1}$$

worin L entweder die Ablösearbeit (Desorption) oder die Verdampfungswärme bezeichnet und zwar pro Mol des Gases, R die Gaskonstante, T die Temperatur des Kernkörpers, m das Molekulargewicht und k die Boltzmann-Konstante. n_0 bezeichnet die Gesamtzahl der vorliegenden adsorbierten bzw. kondensierten Moleküle und schließlich n die der pro Zeiteinheit in den Gasraum eintretenden Moleküle.

Gewöhnlich wird angenommen, daß die Oberflächentemperatur der Kernkörper durch die einfache Formel

$$T = T_0/\sqrt{r} \tag{10.2}$$

darstellbar ist, worin T_0 die Temperatur für $r=1$ bezeichnet. Die Intensität des Kometenleuchtens, soweit es von einer Fluoreszenzanregung herrührt, sollte proportional sein zu n (da die Anzahl der entstehenden fluoreszenzfähigen Moleküle zu n proportional sein sollte), proportional zur Intensität des Sonnenlichtes ($\sim r^{-2}$) und proportional zu Existenzdauer τ der fluoreszierenden Moleküle, die wiederum proportional zu r^2 ist, falls die Moleküle durch Lichtabsorption vernichtet werden. Die beiden letzten Faktoren heben sich gegenseitig auf. Man erhält dann für die Intensität I des Kometenleuchtens als Funktion von r den Ausdruck

$$I = I_0\, r^{-\frac{1}{4}}\, e^{-B\sqrt{r}} \tag{10.3}$$

mit $B = L/R\,T_0$ und I_0 gleich einer Konstanten, die durch die individuellen Eigenschaften des Kernes bestimmt ist.

Die Levinsche Formel ist mehrfach auf ihre Verwendbarkeit hin von verschiedenen Autoren geprüft worden[2], ohne daß jedoch bisher Klarheit darüber gewonnen werden konnte, ob sie den wesentlichsten Vorgang der Atmosphärenbildung erfaßt und zutreffend darstellt. Es ist durchaus wahrscheinlich, daß in ihr ein wahrer Kern steckt. Die bisher vorliegenden Helligkeitmessungen an Kometen beziehen sich fast ausnahmslos auf das gesamte Licht des Kometenkopfes (visuell oder photographisch), und es läßt sich nicht entscheiden, wie weit dabei das reine Fluoreszenzlicht erfaßt worden ist. Man wird erst dann ein besseres Urteil über ihre Bedeutung gewinnen können, wenn ausreichende monochromatische Helligkeitsmessungen vorliegen.

---

[1] B. U. LEVIN: Russ. Astronom. J. **20**, 48 (1943).

[2] J. H. OORT u. M. SCHMIDT: Bull. Astronom. Inst. Netherl. **11**, Nr. 419 (1951). — V. VANYSEK: Contrib. Astronom. Inst. Univ. Brno **1**, Nr. 9 (1952). — A. H. DELSEMME u. P. SWINGS: Ann. d'Astrophys. **15**, 1 (1952).

IV. Zur Deutung der Kometenformen.

11. Die mechanische Theorie von BESSEL und BREDICHIN. Die am Kern des Kometen Halley 1835 beobachteten Ausströmungserscheinungen von Materie und die damit verbundene Schweifentwicklung veranlaßten BESSEL[1] zum Entwurf der Elemente einer Theorie der Kometenformen. Nach den Vorstellungen BESSELs verlassen die Materieteilchen der Kometenatmosphären den Kometenkern mit einer gewissen Geschwindigkeit g (bevorzugt in der Richtung zur Sonne) und stehen von da ab nur noch unter dem Einfluß der Sonne, die außer mit der Schwereattraktion noch mit einer Repulsionskraft auf die Teilchen wirkt. BESSEL nimmt der Einfachheit halber und mangels irgendeiner Kenntnis der Natur der Repulsionskraft an, daß diese ebenso wie die Schwere proportional r^{-2} mit der Distanz r von der Sonne variiert. An die Stelle der Sonnenmasse (gleich 1 gesetzt) tritt die hypothetische Masse $\mu = 1 - R$ mit positivem R. Jedes Teilchen beschreibt also eine selbständige Bahn um die Sonne, bestimmt durch $1 - \mu$, die Anfangslage und Anfangsgeschwindigkeit. Wir können folgende drei Möglichkeiten unterscheiden:

$0 < \mu < 1$. Das Teilchen bewegt sich unter verminderter Attraktion. Die Bahn ist ein Hyperbelast, der konkav zur Sonne ist.

$\mu = 0$. Gravitation und Repulsionskraft heben sich gegenseitig auf. Es resultiert eine geradlinige Bewegung.

$\mu < 0$. Die Repulsionskraft übersteigt die Sonnenattraktion. Es ergibt sich als Bahnkurve eine zur Sonne konvexe Hyperbel.

Bei der Beobachtung eines Kometen wird die Gesamtheit aller leuchtenden Teilchen wahrgenommen, die innerhalb eines längeren Zeitintervalls in den verschiedensten Richtungen vom Kern aus ausgeströmt sind. Beschränkt man sich auf die nähere Umgebung des Kometenkopfes und den Anfang des Schweifes, so ist dieses Zeitintervall von der Größenordnung von Stunden. Beim gesamten Schweif wächst dasselbe auf Tage an. BESSEL behandelt die Schweife als flache Gebilde, die in der Ebene der Kometenbahn liegen. Wie weit diese Annahme berechtigt ist, läßt sich bis heute noch nicht sicher entscheiden. Es ist allerdings vielfach beobachtet worden, daß bei Annäherung der Erde an die Ebene einer Kometenbahn die Breite des Schweifes sich verringert.

Das wesentlichste Element der Besselschen Theorie ist die analytische Darstellung der rechtwinkligen kometozentrischen Koordinaten ξ, η für ein beliebiges Teilchen zu irgendeinem Zeitpunkt t, das zum Zeitpunkt $t - \tau$ den Kometenkern mit der Geschwindigkeit g und unter dem Richtungswinkel G gegen die Richtung zur Sonne verlassen hat. ξ und η liegen in der Kometenbahnebene, die positive ξ-Achse fällt mit dem *verlängerten* Radiusvektor zusammen, die positive η-Achse zeigt *entgegengesetzt* zur Bewegungsrichtung des Kernes. Bezeichnet man mit x, y die rechtwinkligen Koordinaten eines heliozentrischen Koordinatensystems mit der positiven x-Achse zum Perihel gerichtet und mit der positiven y-Achse in Richtung der Bahnbewegung im Perihel, so bestehen zwischen den kometozentrischen Koordinaten ξ, η und den heliozentrischen Koordinaten x', y' der Teilchen die Beziehungen:

$$\left.\begin{aligned} r\xi &= x x' + y y' - r^2, \\ r\eta &= y x' - x y'. \end{aligned}\right\} \tag{11.1}$$

x und y bezeichnen die heliozentrischen Koordinaten des Kometenkernes.

[1] F.W. BESSEL: Astronom. Nachr. **13**, 185 (1836). BESSEL zeichnet in seiner Mitteilung ähnliche Bilder des Kometen Halley 1835 auf, wie sie in der Fig. 5 gegeben sind.

Die heliozentrischen Koordinaten x', y' des Teilchens werden bei BESSEL nach dem Zeitintervall τ entwickelt:

$$\left.\begin{aligned} x' &= x'_0 + \frac{dx'_0}{dt}\tau + \frac{d^2x'_0}{dt^2}\frac{\tau^2}{2} + \frac{d^3x'_0}{dt^3}\frac{\tau^3}{6}\cdots, \\ y' &= y'_0 + \frac{dy'_0}{dt}\tau + \frac{d^2y'_0}{dt^2}\frac{\tau^2}{2} + \frac{d^3y'_0}{dt^3}\frac{\tau^3}{6}\cdots. \end{aligned}\right\} \tag{11.2}$$

x'_o, y'_o bezeichnen die Positionen des Teilchens zur Zeit $t-\tau$. Die Differentialquotienten $d^n x_0/dt^n$, $d^n y_0/dt^n$ ergeben sich aus den Bewegungsgleichungen für Teilchen und Kern. Als Zeiteinheit wird der reziproke Wert der Gaußschen Konstanten $1/k = 58{,}132$ Tage benützt, als Längeneinheit die astronomische Einheit $1{,}494 \times 10^{13}$ cm. Die Durchführung der Entwicklung führt unter der Voraussetzung, daß zur Zeit $t-\tau$ die Position des Teilchens mit der Position des Kernes übereinstimmt und bei Abbruch der Entwicklung mit τ^3 zu folgenden Ausdrücken:

$$\left.\begin{aligned} \xi &= -g\cos G\cdot\tau + \left[\frac{1-\mu}{r^2} - 2g\sin G\frac{\sqrt{p}}{r^2}\right]\frac{\tau^2}{2} + \\ &\quad + \left[\frac{1-\mu}{r^3}\frac{4e\sin v}{\sqrt{p}} - g\cos G\left(\frac{2\mu}{r^3} - \frac{3p}{r^4}\right) - g\sin G\cdot\frac{6e\sin v}{r^3}\right]\frac{\tau^3}{6}, \\ \eta &= +g\sin G\cdot\tau - g\cos G\left(\frac{2\sqrt{p}}{r^2}\right)\frac{\tau^2}{2} + \\ &\quad + \left[\frac{1-\mu}{r^4}\cdot 2\sqrt{p} - g\sin G\left(\frac{\mu}{r^3} + \frac{3p}{r^4}\right) - g\cos G\frac{6e\sin v}{r^3}\right]\frac{\tau^3}{6}. \end{aligned}\right\} \tag{11.3}$$

e, p und v in (11.3) bedeuten die Exzentrität, den Parameter und die wahre Anomalie der Kometenbahn zum Zeitpunkt t.

Die Gln. (11.3) erlauben die Lage eines Schweifteilchens relativ zum Kern zu bestimmen, wenn die Anfangsbedingungen und die Repulsivkraft bekannt sind. Gestatten die Beobachtungen die Ermittlung der Bewegung der Schweifteilchen, so führen die Gleichungen zur Bestimmung der Repulsivkraft.

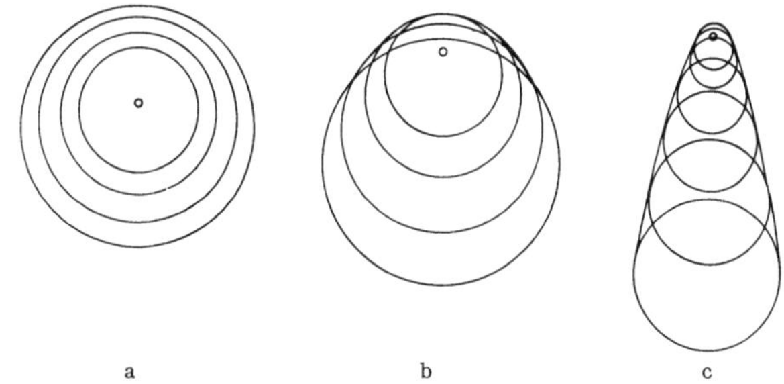

Fig. 20a—c. Ausbreitung der Gase vom Kometenkern bei symmetrischer Ausströmung. Die Reihenfolge a, b, c entspricht bei gleicher Ausströmungsgeschwindigkeit v zunehmender Repulsion $1-\mu$.

BREDICHIN hat später die Methode BESSELs weiter ausgebaut und auch strengere Formeln der Bewegung der Schweifteilchen aufgestellt, auf deren Wiedergabe wir jedoch hier verzichten. Die Schwierigkeiten der Theorie liegen weniger in der mathematischen Formulierung als in der richtigen Deutung der Beobachtungen.

12. Form des Kometenkopfes. Bei einer Beschränkung auf den Kometenkopf bleibt τ eine sehr kleine Größe, so daß wir in (11.3) die Glieder mit τ^3 und ebenso die mit g multiplizierten Glieder in τ^2 verwerfen können. Wir haben damit

$$\left.\begin{aligned} \xi &= -g\cos G\cdot\tau + \frac{1-\mu}{r^2}\frac{\tau^2}{2}, \\ \eta &= +g\sin G\cdot\tau. \end{aligned}\right\} \tag{12.1}$$

Ist g für alle Richtungen gleich groß, so erhält man durch Elimination von G aus (12.1) diejenige Kurve, auf der sich alle jene Teilchen befinden, die in demselben Augenblick $(t-\tau)$ vom Kern ausgeströmt sind. Die Gleichung dieser Kurve ergibt sich so zu

$$\left(\xi - \frac{1-\mu}{r^2}\,\frac{\tau^2}{2}\right)^2 + \eta^2 = g^2\,\tau^2. \tag{12.2}$$

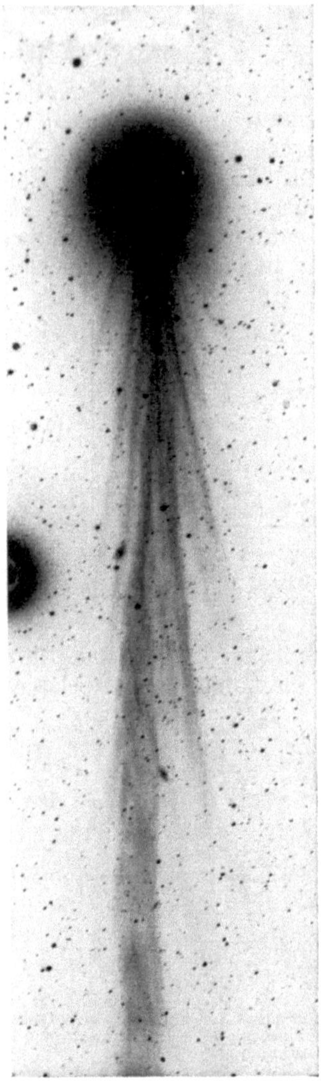

Fig. 21. Komet Whipple-Fedtke (1942g). Der rundliche Kometenkopf entsteht im wesentlichen durch die Emission in der CN-Bande λ3883. (P. AHNERT, Sonneberg.)

Das ist die Gleichung eines Kreises mit dem zeitlich variablen Radius $g \cdot \tau$, dessen Mittelpunkt sich im Laufe der Zeit um $\frac{1-\mu}{r^2}\,\tau^2$ auf dem verlängerten Radiusvektor verschiebt (s. Fig. 20a bis c). Die äußere Begrenzungskurve der Gesamtheit dieser Kreise hat die Gleichung

$$\eta^2 = \frac{2g^2 \cdot r^2}{1-\mu}\,\xi + \frac{g^4\,r^4}{(1-\mu)^2}\,. \tag{12.3}$$

Dies ist eine Parabel mit dem Kern als Fokus.

Die Materie in den vom Kern sich ablösenden Kreisringen wird mit wachsenden Radien mehr und mehr verdünnt. An den Rändern des Kopfes (Umhüllenden) tritt wieder eine Verdichtung infolge der Überlagerung der Ringe ein.

Die Entfernung des Scheitels der umhüllenden Parabel vom Kern ist

$$\varepsilon = \frac{g^2\,\pi^2}{2(1-\mu)}\,, \tag{12.4}$$

welcher Ausdruck eine einfache Relation zwischen der Ausströmungsgeschwindigkeit, der Repulsionskraft und dem Radiusvektor ergibt.

Bei den vorwiegend gasförmigen Kometen werden die rundlichen Kometenköpfe (Fig. 21) bis zu den Rändern vorwiegend von den leuchtenden CN Molekülen gebildet. Die monochromatischen Bilder des CN ähneln der Fig. 20a. Zu einer merklichen Verlängerung dieser Bilder in Richtung des Schweifes kommt es nicht, da die CN Moleküle vorzeitig durch Dissoziation oder Ionisation vernichtet werden. Genauere Werte von g und $1-\mu$ konnten bisher noch nicht ermittelt werden. Erstere scheinen in der Höhe von 1 km/sec zu liegen, die Repulsivkraft $1-\mu$ ist wahrscheinlich von der Ordnung 1.

13. Heller Schweifrand. Es ist fast die Regel, daß bei den stärker gekrümmten, diffusen Schweifen der im Sinne der Bahnbewegung vordere Schweifrand von erhöhter Helligkeit ist (s. Fig. 22 und 23). BESSEL gab dafür im Anschluß an die Formeln (11.3) folgende Erklärung.

Betrachten wir zunächst die Bewegung der Materie, die den Kometenkern genau in Richtung zur Sonne verläßt. Der Ausströmungswinkel G ist in diesem

Fig. 22. Komet Donati, 5. Oktober 1858. (Zeichnung von H. A. PICKERING nach visuellem Anblick.)

Fig. 23. Großer September-Komet von 1882. (Aufnahme D. GILL, Cape of Good Hope, Oktober 1882.)

Falle gleich Null. Wir haben somit für die Koordinaten ξ und η:

$$\left.\begin{aligned} \xi &= -g\cdot\tau + \left[\frac{1-\mu}{r^2}\right]\frac{\tau^2}{2}\cdots, \\ \eta &= -g\frac{2\sqrt{p}}{r^2}\frac{\tau^2}{2} + \left[\frac{1-\mu}{r^4}2\sqrt{p} - g\frac{6e\sin v}{r^3}\right]\frac{\tau^3}{6}. \end{aligned}\right\} \tag{13.1}$$

Für nicht zu kleine Ausströmungsgeschwindigkeiten g überwiegt in dem Ausdruck für η zunächst immer das erste negative Glied. Alle genau in Richtung zur Sonne ausgeströmten Teilchen gehen somit dem Kometenkern in der Bewegung voran und zwar so lange, bis das zweite Glied in τ^3 das erste zu überwiegen beginnt. Dies wird im allgemeinen bei einem um so größeren τ eintreten je größer die Geschwindigkeit g ist. Bei größeren g erfolgt eine Bewegung zu negativem η auch noch für solche Teilchen, deren Ausströmungsrichtungen einen gewissen positiven Winkel G mit der Richtung zur Sonne bilden. Wenn also die Materie in gleicher Quantität zu beiden Seiten des Radiusvektors den Kern verläßt, so wird sie sich zum größeren Teil auf die vordere Seite des Kometen ergießen; der entsprechende Rand des Schweifes wird aufgehellt erscheinen.

Damit die in der Bewegung nachfolgende Seite des Schweifes eine Aufhellung erfährt, muß die Achse des Ausströmungskegels zurück geneigt, G also stärker positiv sein. Auch solche Fälle sind gelegentlich, wenn auch seltener beobachtet worden (Komet[1] 1860 III, Juni 25 und 26).

14. Syndynamen. Eliminiert man aus (11.3) die Zeit τ, so erhält man die Gleichung der Kurve für die Gesamtheit der Teilchen, welche in dem Zeitintervall $(t-\tau)$ bis t unter demselben Winkel G und mit derselben Geschwindigkeit g dem Kern entwichen sind und unter derselben Repulsionskraft $1-\mu$ standen. Um diese Elimination durchzuführen ist es nach BESSEL zweckmäßig, die folgende Substitution einzuführen:

$$\tau = \tau' - \frac{2}{3}\frac{e\sin v}{r\sqrt{p}}\tau'^2. \tag{14.1}$$

Verzichtet man auf die Einbeziehung von sehr kleinen Werten der Koordinate ξ, so gelangt man durch die Elimination von τ' zu dem Ausdruck:

$$\eta = g\sin G\left[\frac{r\sqrt{2\xi}}{\sqrt{1-\mu}} - \frac{4er\sin v}{3(1-\mu)\sqrt{p}}\xi\right] + \frac{2\sqrt{2p}}{3r\sqrt{1-\mu}}\xi^{\frac{3}{2}}. \tag{14.2}$$

Die durch (14.2) dargestellten Kurven werden nach BREDICHIN als *Syndynamen* bezeichnet.

Wie man leicht erkennt, bilden in dem Falle, daß g für alle Richtungen gleich und konstant ist, die beiden Schweifränder Syndynamen, nämlich für die Ausströmungsrichtungen $G = -90°$ (vorderer Rand) und $G = +90°$ (nachfolgender Rand).

Bei Kenntnis einer Syndyname gestattet deren Gl. (14.2) die Ermittlung der Repulsivkraft $1-\mu$ und des Produktes $g\sin G$. Zur Bestimmung von $1-\mu$ und $g\sin G$ wird im allgemeinen folgendermaßen vorgegangen. Dividiert man die Koordinate η nach (14.2) durch die zugehörige Koordinate ξ, so erhält man die trigonometrische Tangente des Winkels φ, welche die Verbindungslinie vom Kern zu dem Punkte ξ, η mit dem verlängerten Radiusvektor bildet:

$$\tan\varphi = g\sin G\left[\frac{r\sqrt{2}}{\sqrt{1-\mu}\sqrt{\xi}} - \frac{4er\sin v}{3(1-\mu)\sqrt{p}}\right] + \frac{2\sqrt{2p}}{3r\sqrt{1-\mu}}\sqrt{\xi}. \tag{14.3}$$

[1] Siehe R. JÄGERMANN: Kometenformen, loc. cit., S. 181.

Um tan φ zu errechnen hat man natürlich die an der Sphäre beobachteten Schweifpunkte auf die Ebene der Kometenbahn zu projizieren. Die Bestimmung von tan φ für zwei ausreichend getrennt liegende Punkte ξ, η ergibt $1-\mu$ und $g \sin G$.

Eine der ersten sorgfältigen Anwendungen von (14.3) wurde von C. F. PAPE[1] auf den vollentwickelten Hauptschweif des Kometen Donati 1858 gemacht und zwar für fünf Nächte zwischen September 28 und Oktober 8 unter Ausmessung der beiden Schweifränder. Die Kurven der Schweifränder ergaben übereinstimmend für die fünf Nächte und die beiden Schweifränder

$$\mu = 0{,}6 \quad \text{und} \quad g \sin G = \mp\, 0{,}14 .$$

Ist g unabhängig von G, wie der Autor annimmt, so erhalten wir $g = 0{,}14$, was 4,2 km/sec entspricht. Die Repulsionskraft $R = 1 - \mu$ beträgt also das 0,4fache der Massenanziehung der Sonne. Es braucht wohl kaum erwähnt zu werden, daß Beobachtung und Analyse nicht von einer solchen Genauigkeit waren, daß die Ergebnisse mehr als die Größenordnungen von μ und g garantieren.

Tabelle 10.

	I. Typus	II. Typus	III. Typus
$1-\mu$	18	2,2 −0,5	0,3 −0,1
g	0,34−0,1	0,07−0,03	0,02−0,01

Auf der von BESSEL vorgezeichneten Linie hat in der zweiten Hälfte des vorigen Jahrhunderts BREDICHIN nicht weniger als 50 verschiedene Kometen mit dem Ziel einer Bestimmung der Repulsivkraft und der Ausströmungsgeschwindigkeiten bearbeitet[2], wobei die Besselsche Methode, wie schon bemerkt wurde, gewisse Erweiterungen gefunden hat, die allerdings im Prinzip nichts Neues bieten. Der Vieldeutigkeit, die sich fast immer bei der Darstellung einer Syndyname in bezug auf das Wertsystem (μ, G) ergibt, wird soweit wie möglich durch ein Studium der Ausströmungsvorgänge im Kometenkopf unter Benutzung der Formeln der Ziff. 12 zu begegnen versucht.

Abgesehen von hin und wieder auftretenden anomalen Schweifformen kam BREDICHIN zu dem allgemeinen Resultat, daß bei den Kometen drei verschiedene, immer wiederkehrende Haupttypen existieren, die sich sowohl im Betrage der Repulsivkräfte $1-\mu$ wie auch der Ausströmungsgeschwindigkeit g größenordnungsmäßig unterscheiden (s. Tabelle 10). Der Typus I wird repräsentiert durch langgestreckte, sehr schwach gekrümmte Schweife, die sich eng an den verlängerten Radiusvektor anschließen. Typus II ist mäßig und Typus III sehr stark gekrümmt. In vielen Fällen treten nach BREDICHIN zwei oder sämtliche drei Typen gleichzeitig auf.

Bei den Schweifen vom Typus I läßt sich die Repulsivkraft besser auf Grund der Verfolgung der Bewegung isolierter Schweifwolken (Fig. 24) ableiten als aus der Schweifform. Seit der Einführung der Photographie hat man reichlich Gelegenheit gehabt, dieses zuverlässigere Verfahren anzuwenden. Es hat sich gezeigt, daß die Repulsivkräfte in den Schweifen vom Typus I weit höher sind als es von BREDICHIN nach seinen Ergebnissen angenommen wurde (s. Ziff. 16).

Die stark gekrümmten Staubschweife zeigen gelegentlich eine Aufgliederung in nahezu parallel verlaufende, gerade Streifen (s. Fig. 31). Dieses Phänomen bedarf ebenso wie die Strahlenstruktur der Gasschweife noch der Aufklärung.

[1] C. F. PAPE: Astronom. Nachr. **49**, 309 (1859). Als Material dienten nach freiem Auge in Sternkarten eingetragene Zeichnungen.

[2] Eine Zusammenfassung dieser Arbeiten mit Literaturangaben findet man bei R. JÄGERMANN: Kometenformen loc. cit.

15. Anomale Schweife. Der erste helle Komet des Jahres 1957 Arend-Roland zeigte bei seinem Wiedererscheinen nach der Perihelpassage ein überraschendes Phänomen: Neben dem von der Sonne abgewandten Hauptschweif bildete sich am 21.—22. April ein *Gegenschweif* aus, der in Richtung zur Sonne zeigte (Fig. 25 und 26). Anfänglich, in der Form ähnlich einem schwächeren, schiefen Spiegelbild

Fig. 24. Abgerissener Schweif beim Kometen Morehouse 1908. (Aufnahme MAX WOLF, Heidelberg.)

des Hauptschweifes, bildete er mit diesem (im Uhrzeigersinn vom Hauptschweif zum Gegenschweif gerechnet) einem Winkel von etwa 165°. In den nächsten drei Tagen wuchs dieser Winkel auf 180° an, wobei der Gegenschweif gleichzeitig an Schärfe und Länge zunahm. In der Nacht vom 25. auf den 26. April erschien er wie ein scharfer Lichtstrahl von 15° Länge, der direkt auf die Position der Sonne gerichtet war. Bei weiterem Anwachsen des Drehwinkels nahm die Schärfe des anomalen Schweifes wieder rasch ab, bis er Ende April bei einem Winkel von 190° kaum noch wahrzunehmen war. Der Komet Arend-Roland zeigte im Spektrum ein intensives Kontinuum, dem im Kopfgebiet die bekannten Emissionen von C_2, CN und C_3 überlagert waren (s. Fig. 27).

Fig. 25.

Fig. 26.

Fig. 25. Komet Arend-Roland. (Aufnahme R. FOLGELQUIST, Uppsala, Schweden.) 22. bis 23. August 1957, $21^{h}30^{m}$ U.T. [Entnommen aus Sky and Telescope **16**, Nr. 9 (1957).]

Fig. 26. Komet Arend-Roland (Aufnahme E. MENDOZA und M. KREBS, McDonald Obs. Texas.) 25.—26. August, 1957, $3^{h}5^{m}$ U.T. [Entnommen Sky and Telescope **16**, Nr. 9 (1957).]

Die Erklärung des beobachteten Gegenschweif-Phänomens erfolgte unmittelbar und unabhängig voneinander durch mehrere Autoren[1], wobei dann auch gleichzeitig darauf aufmerksam gemacht wurde, daß die Astronomen des vorigen Jahrhunderts mit einer solchen Erscheinung und deren Deutung schon wohlvertraut waren.

Es handelt sich bei diesem Gegenschweif, wie wir noch weiter erläutern werden, um feste Teilchen, die vom Kometenkern ausgeschleudert worden sind und sich vorwiegend flach über die Kometenbahnebene um den Kometen ausgebreitet haben. Um den 25. April, der Zeit der Sichtbarkeit des Gegenschweifes, stand die Erde nahe der Bahnebene des Kometen, d.h. die flache, weitverbreitete Teilchenwolke wurde zu dieser Zeit mehr oder weniger von der Kante gesehen. Die höchste Schärfe und größte Länge des Gegenschweifes fällt exakt mit dem Durchgang der Erde durch die Ebene der Kometenbahn zusammen.

OLBERS[2] erwähnt bereits folgendes in einer kurzen Abhandlung aus dem Jahre 1830:

1. Geschwänzte Kometen zeigen zuweilen außer ihrem gewöhnlichen einen schwächeren, gegen die Sonne gerichteten Schweif.

2. Diese wunderbare Erscheinung dauert nur einige Tage.

3. Bisher hat man sie nur nach dem Perihel und ziemlich lange nach dem Perihel gesehen.

Wie R. JÄGERMANN[3] bemerkt, wurde mit dem Kometen 1862 III, dem Erzeuger des periodischen Meteorstromes den Perseiden, die letzte Feststellung von OLBERS ungültig. Dieser Komet zeigte einen anomalen Schweif vor dem Perihel. Insgesamt werden von JÄGERMANN im Jahre 1903 nicht weniger als zehn bekannte Objekte mit anomalen Schweifen aufgeführt. Es wird dort nicht ausdrücklich erwähnt, daß zu den Zeiten der Sichtbarkeit der anomalen Schweife die Erde nahe den Kometenbahnebenen stand. Wie sich leicht nachprüfen läßt, trifft dies jedoch zumindest für sechs der zehn aufgezählten Fälle zu, und zwar sind dies gerade die eindrucksvollsten.

Nach BREDICHIN entstehen die anomalen Schweife durch vom Kern abgestoßene, feste Partikeln, für die infolge ihrer Größe die Repulsionskraft $1-\mu$ gleich Null oder doch verschwindend klein ist. Sie unterliegen in ihrer Bewegung also nur der gewöhnlichen Sonnenattraktion. Der Autor hat die Bewegungsverhältnisse für solche Teilchen, die in irgendeiner beliebigen Richtung und mit beliebigen Geschwindigkeiten vom Kern sich trennen, sowohl mit den genäherten Besselschen Formeln (11.3) wie auch mit den strengen Formeln einer Zentralbewegung untersucht. Wir beschränken uns hier auf einige qualitative Erläuterungen an Hand der noch weiter vereinfachten Besselschen Formeln. Setzt man in (11.3) die Größe $\mu=1$ und vernachlässigt die Glieder mit τ^n größer als τ^2, so ergibt sich

$$\left.\begin{aligned} \xi &= -g\cos G\cdot\tau - \frac{g\sin G\, 2\sqrt{p}}{r^2}\,\frac{\tau^2}{2}, \\ \eta &= +g\sin G\cdot\tau - \frac{g\cos G\, 2\sqrt{p}}{r^2}\,\frac{\tau^2}{2}. \end{aligned}\right\} \tag{15.1}$$

Die vorstehende Koordinatendarstellung gilt natürlich nur für kleine ξ und η. Für Teilchen, die sich direkt zur Sonne bewegen ($G=0$) geht (15.1) über in:

$$\xi = -g\cdot\tau; \qquad \eta = -g\,\frac{2\sqrt{p}}{r^2}\,\frac{\tau^2}{2} \tag{15.2}$$

[1] GIOVANNI u. ANGELO BERNASCONI, dem Verfasser zugesandtes, ausführliches Manuskript. FRED L. WHIPPLE: Sky a. Telescope **16**, Nr. 9, 426 (1957). — G. LARSSON-LEANDRE: Observatory **77**, Nr. 899, 128 (1957). — E.B. ARMSTRONG: Observatory **77**, Nr. 899, 135 (1957).

[2] Astronom. Nachr. **1830**, Nr. 192.

[3] R. JÄGERMANN: Kometenformen, S. 19.

und für Teilchen mit einer Anfangsgeschwindigkeit in der genau entgegengesetzten Richtung ($G = 180°$):

$$\xi = + g \cdot \tau; \qquad \eta = + g \frac{2\sqrt{p}}{r^2} \frac{\tau^2}{2}. \tag{15.3}$$

Im ersten Falle bleiben also die Koordinaten ξ, η ausschließlich negativ; das entstehende Schweifbüschel liegt innerhalb der Kurve der Kometenbahn und geht dem Kometenkern in der Bewegung voraus. Die Schweifkrümmung ist *konvex* zur Sonne. Der zweite Fall liefert ausschließlich positive ξ und η, das Schweifbüschel liegt hinter dem Radiusvektor, seine Krümmung ist *konkav* zur Sonne.

Der Gegenschweif des Kometen Arend-Roland muß vom zweiten Typus gewesen sein, er erstreckte sich noch im Quadranten positiver ξ und positiver η. Er folgte dem Kern und dem Hauptschweif in seiner Bewegung nach, und diese relative Position blieb auch nach dem Durchgang der Erde durch die Bahnebene erhalten. Man beachte, daß der Komet zur Zeit der Existenz des Gegenschweifes zwischen Sonne und Erde stand; wie schon bemerkt wurde.

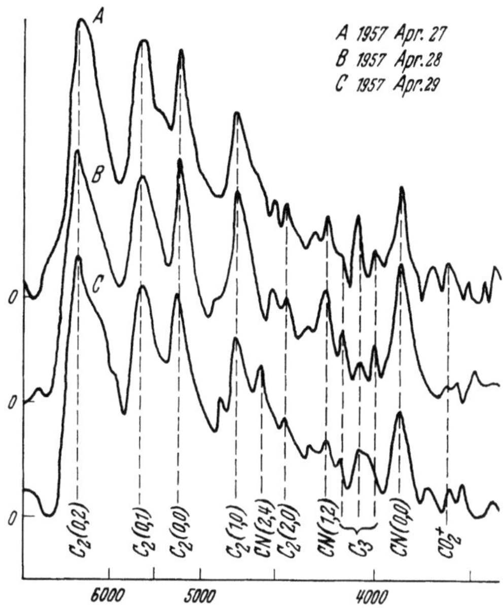

Fig. 27. Registrierkurven des Spektrums des Kometen Arend-Roland. (J. Bouška, Prag.)

Beispiele für den Kometen vorangehende anomale Schweife liefern die Kometen 1844 III und 1862 III. Bei solchen muß beim Durchgang der Erde durch die Kometenbahn-Ebene die Drehung um den Kopf entgegengesetzt dem Uhrzeigersinn erfolgen.

Genauere Angaben über die Ausbreitung der Materie der anomalen Schweife über die Bahnebene sind sehr schwierig zu gewinnen, da man dieselben stets nur unter starker perspektivischer Verkürzung beobachten kann.

16. Die Natur der Repulsivkräfte. Die Erklärung der Existenz der stark gekrümmten Schweife (Typus II und III nach Bredichin) bietet keine prinzipiellen Schwierigkeiten. Die Repulsivkräfte sind in diesem Falle von der Ordnung 1 und kleiner und lassen sich als *Lichtdruckkräfte* auf kleine Staubteilchen erklären.

Dagegen ist es nicht möglich, die beobachteten Beschleunigungen in den CO^+-Schweifen (Typus I) durch den solaren Lichtdruck zu deuten. In Tabelle 11 haben wir die bei einer Reihe von Kometen aus der Bewegung von Schweifwolken ermittelten $1-\mu$ aufgeführt[1]. Man sieht einmal, daß sie das mehr als Hundertfache der Sonnenattraktion erreichen können und weiterhin, daß sie beträchtlichen Schwankungen unterliegen.

[1] Gelegentlich sind auch von einigen Autoren Werte von $1-\mu$ in der Höhe von einigen Tausend publiziert worden. Ob diesen sehr hohen Werten eine reelle Bedeutung zukommt, wäre noch zu prüfen.

Tabelle 11.

Komet	$1-\mu$	Autor	Literatur
1892 I	39,3	JÄGERMANN	Astronom. Nachr. **171**, 1 (1906)
	35,1	JÄGERMANN	Astronom. Nachr. **171**, 1 (1906)
	71,0	KOPFF	Publ. Astrophys. Inst. Königstuhl-Heidelberg, Bd. III, Nr. 2. 1907
1903 IV	89,1	JÄGERMANN	Astronom. Nachr. **166**, 279 (1904)
	90,8	KOPFF	Astronom. Nachr. **176**, 149 (1907)
1908 III	62,0	A. ORLOFF	Publ. Sternwarte Jurjew **21**, 1 (1910)
	72,0	S.V. ORLOW	Nachr. russ. astronom. Ges. **16**, 60 (1910)
	162.0	S.V. ORLOW	Nachr. russ. astronom. Ges. **16**, 60 (1910)
	105,3	PROKOWSKI	Astronom. Nachr. **184**, 3 (1910)
	151,0	BOBROVNIKOFF	Lick Bull. **13**, 161 (1928)
	88,0	BOBROVNIKOFF	Lick Bull. **13**, 161 (1928)
1910 X	70,0	S.V. ORLOW	Publ. Sternwarte Jurjew **21**, 43 (1910)
	90,0	GONDOLATSCH	
1942	bis	AHNERT u.	Z. Astrophys. **22**, 286 (1943)
	1940	HOFFMEISTER	Z. Astrophys. **23**, 1 (1943)

Das Verhältnis μ der Lichtdruckbeschleunigung eines absorbierenden Moleküls zur Gravitation errechnet sich aus[1]

$$\mu = \frac{Z/c}{m\,G} \tag{16.1}$$

mit

$$Z = \frac{\pi e^2 f}{m_e} \varrho'_\nu . \tag{16.2}$$

Z ist die pro Sekunde vom Molekül gestreute Energie [s. Gl. (7.1)], c die Lichtgeschwindigkeit, G die Gravitationsbeschleunigung der Sonne und m die Masse des Moleküls. Bei Einsetzung numerischer Werte in (16.1) erhält man für das CO^+-Molekül

$$\mu = 46{,}9 f . \tag{16.3}$$

Der f-Wert der Resonanzbanden des CO^+ ist nicht genauer bekannt, doch kann es keinem Zweifel unterliegen, daß ein Wert $f = 10^{-1}$ eine obere Grenze für den möglichen Wert darstellt. Der Lichtdruck auf die Schweifionen ist also viel zu gering um deren hohe Beschleunigungen zu erklären.

L. BIERMANN[2] hat die Idee geäußert und quantitativ erläutert, daß die Beschleunigung der Schweifionen durch solare Korpuskelströme zustande kommt. Die Existenz von solchen Strömen ist lange bekannt, insbesondere auf Grund des Entstehens von erdmagnetischen Störungen und der damit häufig verbundenen Nordlichter. Schwieriger ist es, über die Ausbreitung und die Dichten in diesen Strömen ein klares Bild zu gewinnen. Bei den von BIERMANN postulierten Ionendichten der Korpuskularstrahlung ($n_i = 10^3$ bis $10^4\,\mathrm{cm}^{-3}$) ergibt sich ziemlich überzeugend, daß allein infolge der „Reibungskräfte" zwischen den solaren und den kometarischen Ionen die beobachteten hohen Werte von μ eine Erklärung finden können. Im Gegensatz zu BIERMANN ist H. ALFVÉN[3] der Meinung, daß die postulierten hohen Ionendichten äußerst unwahrscheinlich sind. Der letzte

[1] W. BAADE u. W. PAULI: Naturwiss. **15**, 49 (1927). Die Baade-Paulische Formel enthält ein Versehen, vgl. dazu K. WURM, Mitt. Hamburger Sternwarte **8**, Nr. 51 (1943).
[2] L. BIERMANN: Z. Astrophys. **29**, 274 (1951). — Symposium Liège 1952, S. 255.
[3] H. ALFVÉN: Tellus **9**, 92 (1957).

Autor schlägt einen neuen Mechanismus vor. Dieser stützt sich zwar auch auf solare Korpuskelströme als Ursache der Beschleunigung, jedoch abweichend von der Biermannschen Theorie allein auf die mit den Strömen mitwandernden magnetischen Felder, d.h. auf einen magneto-hydrodynamischen Mechanismus. Die von ALFVÉN vorausgesetzten Korpuskeldichten von nur $n_i = 10^{-1}$ bis 10^{-2} cm³ sind zweifellos leichter akzeptabel. Wie vom Autor vorher an anderer Stelle ausgeführt worden ist, führen solare Ströme dieser Dichte magnetische Felder der Ordnung 10^{-5} Gauß mit sich.

Da beide Theorien gegenwärtig zur Diskussion stehen, so sei anschließend noch ein wenig mehr dazu gesagt.

Bei der Biermannschen Theorie handelt es sich wesentlich um eine Impulsübertragung zwischen den solaren Elektronen und den Schweifionen. Wegen der höheren Beweglichkeit der Elektronen stoßen diese wesentlich häufiger auf die CO^+-Ionen als die solaren Ionen. Letztere sind zur Hauptsache Protonen. Die Energieverluste der Elektronen werden durch die Zusammenstöße mit den Protonen wieder ausgeglichen. Die ungeordnete mittlere Geschwindigkeit im solaren Korpuskelstrom soll einer Temperatur von 10000° entsprechen. Die Dichte der Schweifionen ist größenordnungsmäßig wesentlich kleiner als die Dichte der solaren Ionen. Die gerichtete Geschwindigkeit der letzteren ist von der Ordnung 10^8.

Der Autor geht aus von den Gleichungen für den Impulsaustausch in einem Plasma, wie sie von A. SCHLÜTER[1] aufgestellt wurden, und die für 3 Partikelsorten wie folgt lauten:

$$\left.\begin{aligned}
m_m \frac{d\mathfrak{V}_m}{dt} &+ \frac{m_m m_p}{m_m + m_p} \gamma_{mp} (\mathfrak{V}_m - \mathfrak{V}_p) + \frac{m_m m_e}{m_m + m_e} \gamma_{me} (\mathfrak{V}_m - \mathfrak{V}_e) \\
&= e\,\mathfrak{E} + \frac{e}{c} [\mathfrak{V}_m \mathfrak{H}], \\
m_p \frac{d\mathfrak{V}_p}{dt} &+ \frac{m_p m_m}{m_m + m_p} \gamma_{pm} (\mathfrak{V}_p - \mathfrak{V}_m) + \frac{m_p m_e}{m_p + m_e} \gamma_{pe} (\mathfrak{V}_p - \mathfrak{V}_e) \\
&= e\,\mathfrak{E} + \frac{e}{c} [\mathfrak{V}_p \mathfrak{H}], \\
m_e \frac{d\mathfrak{V}_e}{dt} &+ \frac{m_m m_e}{m_m + m_e} \gamma_{em} (\mathfrak{V}_e - \mathfrak{V}_m) + \frac{m_p m_e}{m_p + m_e} \gamma_{ep} (\mathfrak{V}_e - \mathfrak{V}_p) \\
&= -e\,\mathfrak{E} - \frac{e}{c} [\mathfrak{V}_e \mathfrak{H}].
\end{aligned}\right\} \tag{16.4}$$

Im vorliegenden Falle bezieht sich der Index m auf die kometarischen Ionen (CO^+), der Index p auf die solaren Ionen (wesentlich Protonen) und der Index e auf die freien Elektronen. m_m, m_p, m_e sind die Massen der Teilchen, $\mathfrak{V}_m$, $\mathfrak{V}_p$, $\mathfrak{V}_e$ deren makroskopische Geschwindigkeiten, $\mathfrak{E}$ die elektrische und $\mathfrak{H}$ die magnetische Feldstärke. Die γ definieren die Stoßwahrscheinlichkeiten pro Sekunde und pro Volumeneinheit, beispielsweise γ_{em} die Wahrscheinlichkeit eines Stoßes von einem Elektron auf ein kometarisches Ion.

Da die ungeordneten Geschwindigkeiten der Ionen m und p sehr klein sind gegenüber der ungeordneten Geschwindigkeit der freien Elektronen, und $m_e \ll m_m$ sowie $n_m \ll n_e$, so gilt $\gamma_{mp} \ll \gamma_{me}$ und $\gamma_{pm} \ll \gamma_{pe}$. Die Glieder mit γ_{mp} und γ_{pm} in (16.4) können deshalb gestrichen werden. Die Tatsache, daß die ungeordnete Geschwindigkeit der Elektronen ($T_e \approx 10^4$ Grad) von derselben Ordnung ist wie die makroskopische Geschwindigkeit der Elektronen und Protonen, erlaubt es

[1] A. SCHLÜTER: Z. Naturforsch. 5a, 72 (1950).

nun, γ_{em} und γ_{ep} mit der normalen Leitfähigkeit σ eines Plasmas in Beziehung zu setzen. Es gilt genähert

$$\gamma_{em} + \gamma_{ep} = \frac{e^2 n_e}{\sigma m_e} = \frac{10^{8.4}}{\sigma} n_e \tag{16.5}$$

und, da

$$\gamma_{em} : \gamma_{ep} = n_m : n_p, \tag{16.6}$$

weiterhin

$$\left.\begin{aligned} \gamma_{em} &\approx \frac{e^2}{\sigma m_e} n_m, & \gamma_{me} &\approx \frac{e^2}{\sigma m_e} n_e, \\ \gamma_{ep} &\approx \frac{e^2}{\sigma m_e} n_p, & \gamma_{pe} &\approx \frac{e^2}{\sigma m_i} n_e. \end{aligned}\right\} \tag{16.7}$$

Da es sich darum handelt, zunächst einmal einen ungefähren Anhaltspunkt betreffs der Größe der möglichen Beschleunigungen der CO^+-Ionen zu erhalten, werden noch weiterhin folgende Vereinfachungen eingeführt. Sämtliche Geschwindigkeiten und Dichten sollen nur von einer Koordinate X abhängig sein, Magnetfelder im solaren Korpuskelstrom werden ausgeschlossen, und weiterhin soll die Dichte der kometarischen Ionen sehr klein sein gegenüber der Dichte der solaren Ionen, $m_m \ll m_p$. Letzteres wird den wahren Verhältnissen in den Schweifen entsprechen, falls n_p von der Ordnung $10^3\,\mathrm{cm}^{-3}$ oder größer ist. Für $t = 0$ wird weiterhin $v_m = 0$ vorausgesetzt, und zusätzlich gelten die Dichte n_p und alle Geschwindigkeiten als ortsunabhängig.

Die Gln. (16.4) gehen damit über in

$$\left.\begin{aligned} m_m \frac{d v_m}{d t} - \frac{e^2}{\sigma} n_e (v_e - v_m) &= e\,\mathfrak{E}, \\ m_p \frac{d v_p}{d t} + \frac{e^2}{\sigma} n_e (v_p - v_e) &= e\,\mathfrak{E}, \\ m_e \frac{d v_e}{d t} - \frac{e^2}{\sigma} (n_p v_p + n_m v_m - n_e v_e) &= -e\,\mathfrak{E}. \end{aligned}\right\} \tag{16.8}$$

Beim Übergang von (16.4) auf (16.8) ist noch weiter angenommen, daß die durch Ladungstrennungen verursachten elektrischen Ströme durch das damit entstehende Feld $\mathfrak{E}$ so klein gehalten werden, daß kein Magnetfeld entsteht, was nachträglich verifiziert werden kann.

Die weitere Diskussion der Gln. (16.8), die hier nicht wiedergegeben werden soll, wird mit festen, numerischen Werten für $n_p (= 10^4\,\mathrm{cm}^{-3})$, $v_p (= 10^8\,\mathrm{cm\,sec}^{-1})$ und $n_m (= 10^1\,\mathrm{cm}^3)$ durchgeführt. Für die in erster Linie interessierende Beschleunigung der Ionen m ergibt sich schließlich in ausreichender Näherung

$$\left.\begin{aligned} &\frac{d v_m}{d t} = \frac{e^2}{\sigma m_m} n_e v_e \approx 10^3\,\mathrm{cm\,sec}^{-2} \\ &(\sigma = 10^{12,5},\; e^2/\sigma = 10^{-31,3},\; m_m = 10^{-22,3},\; n_e = n_p). \end{aligned}\right\} \tag{16.9}$$

Eine Beschleunigung von $10^3\,\mathrm{cm\,sec}^{-2}$ ist gleich dem 2000fachen der Sonnenatraktion für $r = 1$, letztere hat den Wert $0{,}56\,\mathrm{cm\,sec}^{-2}$. Wie schon bemerkt, sind solche Werte von $1-\mu$ gelegentlich von Autoren genannt worden[1]. Falls sie in dieser Höhe überhaupt wirklich auftreten sollten, so scheinen sie jedoch Ausnahmefälle darzustellen. Die in Tabelle 11 aufgeführten Fälle, die uns zuverlässiger erscheinen, gehen nicht über $1-\mu = 200$ hinaus. Sie wären nach der Biermannschen Theorie mit Korpuskeldichten der Ordnung 10^3 zu verstehen.

[1] Vgl. N. T. BOBROVNIKOFF, in Astrophysics (Ed. J. A. HYNEK), S. 326f. New York: McGraw-Hill 1951.

Die neue Theorie von ALFVÉN hat soweit noch keine mathematische Ausarbeitung erfahren und beschränkt sich bisher darauf, auf einige mögliche Phänomene hinzuweisen. Die Fig. 28a—d, die von ALFVÉN gezeichnet sind, sollen veranschaulichen, was sich möglicherweise ereignen kann, wenn ein ausgedehnter Korpuskelstrom, der ein Magnetfeld H mit sich führt, auf einen Kometen auftrifft. Der Einfachheit halber ist das Feld H als homogen angenommen. Beim Auftreffen des Korpuskelstromes auf den Kometenkopf tritt nach dem Autor eine Stoßwelle auf, die eine intensive Erhitzung der Kometenkopf-Atmosphäre und damit eine Steigerung der Ionisation der Gase erzeugt. In dem ionisierten Gas des Kopfes werden die magnetischen Kraftlinien festgehalten, und da der breite Strom sich weiterbewegt, erfahren die magnetischen Kraftlinien hinter dem Kometenkopf eine Deformation, wie sie in den Fig. 28 dargestellt ist. Das am Kometenkopf aufgehängte, divergierende und im weiterlaufenden Strom endende Magnetfeld zwingt die ionisierten Schweifteilchen zu einer Bewegung entlang den magnetischen Kraftlinien. Nach ALFVÉN reichen die in der Schweifrichtung auftretenden elektromagnetischen Kräfte aus, um die höchsten beobachteten Beschleunigungen der Schweifionen zu erklären. Die postulierten Ionendichten im solaren Strom sind nur von der Ordnung $n = 0{,}1\,\mathrm{cm}^{-3}$, deren Geschwindigkeiten von der Ordnung 10^8 cm sec^{-1}.

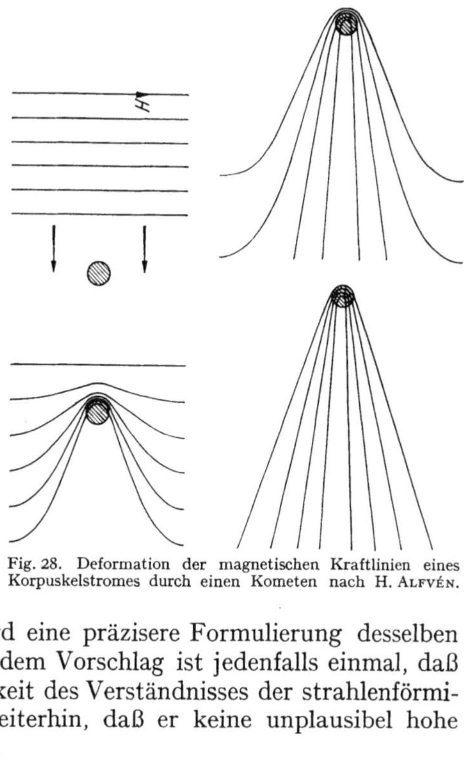

Fig. 28. Deformation der magnetischen Kraftlinien eines Korpuskelstromes durch einen Kometen nach H. ALFVÉN.

Über den Wert des Alfvénschen Gedankens läßt sich gegenwärtig kaum ein sicheres Urteil fällen, und man wird eine präzisere Formulierung desselben abzuwarten haben. Beachtenswert an dem Vorschlag ist jedenfalls einmal, daß derselbe zum ersten Male eine Möglichkeit des Verständnisses der strahlenförmigen Schweifstrukturen aufzeigt und weiterhin, daß er keine unplausibel hohe Korpuskeldichten zu verlangen scheint.

Nach den Vorstellungen des Autors wird nicht nur die Struktur der Schweife, sondern auch die Schweifaktivität (Bildung der Ionen) auf die solaren Korpuskelströme zurückgeführt. Verfasser scheint es sehr zweifelhaft, ob der letzte Gedanke sich akzeptieren lassen wird. Wenn die Ionisierung beim Auftreffen der solaren Korpuskeln den ganzen Kometenkopf ergreift, so müßte mit einer Vermehrung der ionisierten Teilchen gleichzeitig ein Verschwinden oder eine Verminderung der neutralen eintreten, was sicherlich nicht der Wirklichkeit entspricht. Es ist außerdem eine lange bekannte Tatsache, daß die CO^+-Schweifionen durchaus nicht aus dem ganzen Kometenkopf abströmen, sondern nur aus einer relativ eng begrenzten Zone um den Kometenkern.

V. Kosmogonie der Kometen.

17. Bemerkungen zur Herkunft der Kometen. Bei einer Diskussion der Herkunft der Kometen ist es nach OORT bei dem gegenwärtigen Stand des Problems zweckmäßig, folgende zwei Fragen auseinander zu halten: 1. Wo haben die Kometenkerne ihren Ursprung? 2. Aus welchen Räumen kommen sie heute, soweit sie neu auftauchen, zu uns?

In bezug auf die erste Frage ist man bis heute noch immer nicht über Vermutungen hinaus. LAPLACE[1] war der erste, der an einen *interstellaren Ursprung* dachte, an einen Einfang interstellarer Materiebrocken durch das Gravitationsfeld der Sonne. Unter der Annahme einer bestimmten Verteilungsfunktion $y(v)$ der Geschwindigkeiten v der Kernkörper versuchte er die Frage zu beantworten, welche Verteilung der Hauptachsen für die uns sichtbar werdenden Kometenbahnen zu erwarten ist. LAPLACE fand für die von ihm angenommene (aber ganz unwahrscheinliche) Verteilungsfunktion $y(v) = \text{const}$ für alle v, daß die Anzahl hyperbolischer Bahnen äußerst gering sein müsse gegenüber den elliptischen und parabolischen (Verhältnis etwa 1:6000). SCHIAPARELLI[2] zeigte später, daß das Resultat sich ändert, wenn man die Raumbewegung der Sonne gegen den Kometenkern-Schwarm berücksichtigt. Wie ohne weiteres verständlich, muß die relative Zahl der hyperbolischen Bahnen zunehmen, und dies um so mehr, je höher die Raumgeschwindigkeit der Sonne angenommen wird. FABRY[3] hat später eine andere Verteilungsfunktion behandelt, die den wahren Verhältnissen schon etwas näher kommen könnte als die von LAPLACE [$y(v) = 0$ für v größer als eine obere Grenze], obwohl auch diese sich nicht begründen läßt. FABRY findet, daß unter Zugrundelegung seiner Verteilungsfunktion (oder ähnlicher) die Abwesenheit stark hyperbolischer Bahnen sich nicht in Einklang mit dem interstellaren Ursprung bringen läßt.

VAN WOERKOM[4] untersuchte die Frage unter der Annahme einer Maxwell-Verteilung der Geschwindigkeiten mit einer Streuung von $\pm 9{,}4$ km/sec und einer Asymmetrie der Sonnenbewegung von 20 km/sec. Er kommt damit zu der Forderung von 1% hyperbolischer Bahnen, und zwar mit zum Teil stark hyperbolischer Bahnen. Daß solche nie beobachtet worden sind, wäre nur verständlich, wenn die Sonne keine Bewegung gegen die Kernwolke hätte, was ganz unwahrscheinlich ist. VAN WOERKOMs Resultat spricht mit größter Wahrscheinlichkeit gegen einen interstellaren Ursprung.

Von R.A. LYTTLETON[5] ist neuerdings eine Theorie vorgetragen worden, nach welcher die Sonne beim Durchqueren interstellarer Materiewolken Verdichtungen von Staub erzeugt und diese gleichzeitig in Kometenbahnen zwingt. Solange der Ursprung der Kometenkerne noch ganz ungeklärt bleibt, wäre es unangebracht, diese neue Theorie vollständig zu übersehen, in ihrem jetzigen Zustand scheint sie uns jedoch wenig anziehend, weshalb wir es bei diesem Hinweis bewenden lassen.

Nach einer zweiten Hypothese, die zuerst von LAGRANGE[6] geäußert wurde, sind die Kometenkerne *Auswurfprodukte eines Planeten* oder Splitter eines ehemals existierenden, zerborstenen Planeten. Die Bruchstücke eines zerfallenen Planeten sollten Bahnen verfolgen, die alle einen gemeinsamen Kreuzungspunkt besitzen, nämlich den Ort der Katastrophe. Da dies bei den Kometen nicht beobachtet wird, so hat man diesen zweiten Gedanken bald wieder fallen gelassen. Allerdings muß gesagt werden, daß das Fehlen eines gemeinsamen Bahnpunktes

[1] P. S. DE LAPLACE: Oeuvres Complètes **13**, 88 (1904).

[2] G.V. SCHIAPARELLI: Entwurf einer astronomischen Theorie der Sternschnuppen und Meteore. Stettin 1871.

[3] L. FABRY: Etudes sur la probabilité des comètes hyperboliques et l'origine des comètes. Thèse, Marseille 1893.

[4] A. J. J. VAN WOERKOM: Bull. Astronom. Inst. Netherl. **10**, Nr. 399 (1948).

[5] R.A. LYTTLETON: Monthly Notices Roy. Astronom. Soc. London **108**, 465 (1948); **111**, 268 (1951). — Symposium, Liège 1953.

[6] J.L. LAGRANGE: Add. Com. d. Temps 1814.

nicht unbedingt gegen die Explosionshypothese spricht. In einem Zeitabschnitt, vergleichbar mit der Existenzzeit des Sonnensystems, könnte durch planetarische Störungen der Bahnen eine solche Bahnverwandtschaft vollkommen beseitigt worden sein.

Die neueren Befürworter der Auswurfhypothese stützen sich in erster Linie auf die Jupiterfamilie der Kometen[1]. Trotzdem diese Hypothese die auffälligsten Eigentümlichkeiten der Jupitergruppe ziemlich zwanglos zu erklären vermag, so erweist sich jedoch, wie VAN WOERKOM[2] gezeigt hat, die Einfanghypothese der Auswurfhypothese überlegen (vgl. weiter unten). Derselbe Autor kommt in seinen Untersuchungen auch zu dem Resultat, daß die langperiodischen Kometen keinesfalls durch Eruptionen der Planeten entstanden sein können.

Weiterhin ist dann kürzlich in bezug auf den Entstehungsprozeß der Kometenkerne der Gedanke geäußert worden, daß derselbe mit der *Entstehung der Planeten* zusammenfällt[3]. Die Bildung des Sonnensystems ist bis heute noch keineswegs klar aufgedeckt, vieles spricht jedoch dafür, daß sich die Planeten gleichzeitig mit der Sonne aus einer ausgebreiteten, flachen, rotierenden Gasmasse gebildet haben. Im Zentrum formte sich die Sonne, und um diese herum kondensierten sich größere und kleinere Massen. Es wäre denkbar, daß im Verlaufe dieses Prozesses sich an den Rändern des Nebels zahllose kleinere Körper als selbständige „Kometenkerne" bildeten.

Zusammenfassend muß leider gesagt werden, daß wir über die Entstehung der Kometenkerne noch sehr im Ungewissen bleiben. In einer kurzen Monographie scheint es kaum gerechtfertigt, den einzelnen, sich entgegenstehenden Hypothesen noch mehr Raum zu widmen.

In bezug auf die oben aufgeworfene zweite Frage befinden wir uns in einer etwas günstigeren Situation. Hier liefert das bisher gesammelte Material an berechneten Kometenbahnen bestimmte Hinweise. Das Studium der Bahnverhältnisse, insbesondere der Häufigkeitsverteilung der halben großen Bahnachsen a führt, wie OORT[4] gezeigt hat, zu dem Schluß, daß die uns sichtbar gewordenen und noch neu sichtbar werdenden Kometen aus einem großen Reservoir von Kometenkernen stammen, welches die Sonne in der Distanz von 50000 bis 150000 A.E. umgibt. Die Bahnen der Kerne dieses Reservoirs bleiben in allen Punkten weit außerhalb des Systems der Planeten. Störungen der Bahnen der Kerne dieser Kernwolke können nur durch den Vorbeigang benachbarter Sterne hervorgerufen werden. Infolge dieser Sternstörungen entstehen zu einem kleinen Teil auch stark exzentrische Bahnen, die im Perihel in die Nähe der Sonne bzw. Erde gelangen. Wir kommen in Ziff. 18 hierauf zurück.

18. Die langperiodischen Kometen. Es ist seit längerem bekannt, daß unter den langperiodischen Kometen die sehr langen Perioden sehr stark überwiegen. Ordnet man die rund 220 Objekte aus dieser Gruppe, für die genauere Bahnelemente vorliegen, nach steigenden Werten der reziproken Halbachsen $1/a$ mit einer Intervallbreite von 0,002 A.E.$^{-1}$, so erhält man das Bild der Tabelle 12. Über 80% der Gesamtheit fällt in das erste Intervall von 0,00 bis 0,002.

J. OORT hat gezeigt, daß unter den 220 in der Tabelle 12 benützten Kometenbahnen 41 vorhanden sind, bei denen die Genauigkeit, mit der das Bahnelement a bekannt ist, eine feinere Intervallbildung gestattet. Diese 41 Kometenbahnen

[1] S. VSESSVIATSKY: Astronom. Nachr. **240**, 273 (1930); **243**, 281 (1931). — Russ. Astronom. J. **10**, 18 (1934); **14**, 63 (1951).
[2] A. J. J. VAN WOERKOM: loc. cit.
[3] Vgl. dazu G. P. KUIPER: Symposium, Liège 1952.
[4] J. OORT: Bull. Astronom. Inst. Netherl. **11**, Nr. 408 (1950). Vgl. auch Symposium, Liège 1953.

sind bei der Aufstellung der nächsten Tabelle 13 mit den Intervallbreiten von $5 \cdot 10^{-5}$ A.E.$^{-1}$ benutzt worden[1]. Es handelt sich in allen Fällen um die reziproken Halbachsen von ursprünglichen Bahnen, wie sie also vor dem Eintritt der Kometenkerne in das System der Planeten vorlagen. Die n der zweiten Spalte bezeichnen die Anzahlen zu den links davon aufgeführten Intervallen. Die $\bar{n}$ der Spalte 3 entsprechen den Anzahlen der Kometen pro einheitlichem Intervall $5 \cdot 10^{-5}$ A.E.$^{-1}$. Die vier am Anfang stehenden Objekte mit negativen $1/a$ beruhen aller Wahrscheinlichkeit nach auf Beobachtungsfehlern, und es ist anzunehmen, daß sie in Wahrheit in das dritte Intervall gehören.

Tabelle 12.

$1/a$	Anzahl n
0,000—0,002	177
0,002—0,004	10
0,004—0,006	8
0,006—0,008	7
0,008—0,010	2,5
0,010—0,012	6,5
0,012—0,014	1
0,014—0,016	1,5
0,016—0,018	1,5
0,018—0,020	4
0,020—0,022	2
0,022—0,024	0

Nach VAN WOERKOM (Fußnote 4 auf S. 510).

Die Tabelle 13 zeigt auf jeden Fall, daß eine ausgesprochene Häufung der Kometen für Werte $1/a$ kleiner als $5 \cdot 10^{-5}$ existiert. Ein beträchtlicher Bruchteil aller Kometen mit längerer Periode hat also die Aphele bei heliozentrischen Distanzen oberhalb von 40000 A.E. Der Komet Delavon 1914 V, der zu den Objekten in der dritten Gruppe der Tabelle 13 gehört, hat nach VAN BIESBROECK eine Apheldistanz von 170000 A.E. und eine Umlaufperiode von 24 Millionen Jahren.

Tabelle 13.

$1/a \cdot 10^5$ A.E.$^{-1}$	Anzahl n	Anzahl pro Interval $5 \cdot 10^{-5}$ $\bar{n}$	Mittlere Periode Jahre
-8	1	1	
-3 bis 0	3	3	
0 bis $+5$	9	9	$\approx 10^7$
$+5$ bis 10	4	4	$1{,}5 \cdot 10^6$
10 bis 15	2	2	$7{,}5 \cdot 10^5$
15 bis 20	2	2	$4{,}5 \cdot 10^5$
20 bis 25	1	1	$3{,}0 \cdot 10^5$
25 bis 50	3	0,6	$1{,}7 \cdot 10^5$
50 bis 75	2	0,4	$7{,}0 \cdot 10^4$
75 bis 100	1	0,2	$4{,}0 \cdot 10^4$
100 bis 200	3	0,15	$2{,}0 \cdot 10^4$
200 bis 400	4	0,10	$2{,}0 \cdot 10^3$
400 bis 600	1	0,02	$3{,}0 \cdot 10^3$
600 bis 800	4	0,10	$1{,}7 \cdot 10^3$
800 bis 1000	0	0,00	$1{,}1 \cdot 10^3$
1000 bis 2000	1	0,005	$6{,}5 \cdot 10^2$
2000 bis 3000	0	0,000	$2{,}5 \cdot 10^2$
3000 bis 4000	0	0,000	$1{,}5 \cdot 10^2$
Summe	41		

Nach J. OORT: Symposium, Liège 1952.

OORT macht nun darauf aufmerksam, daß die Kometen im oberen Teil der Tabelle 13 mit Werten von $1/a$ kleiner als etwa $10 \cdot 10^{-5}$ bei ihrer Beobachtung zum ersten Male in das System der Planeten eingedrungen sein müssen. Dieser Schluß stützt sich auf folgenden Umstand. Der Durchgang eines Kometen durch das System der Planeten ist stets mit Bahnstörungen verbunden. FAYET[2] und VAN WOERKOM[3] haben errechnet, daß unter Berücksichtigung aller möglichen Bahnformen bei einer einzelnen Passage allein durch Jupiter[4] im Durchschnitt eine Änderung des reziproken Wertes der Bahnachse um den Betrag $\Delta(1/a) = 0{,}0005$ A.E.$^{-1}$ erfolgt. Die Verteilung der $\Delta(1/a)$ ist symmetrisch zu $\Delta(1/a) = 0$, gleiche positive und negative Werte sind gleich wahrscheinlich. Die Größe dieser Jupiterstörungen besagt, daß die im Maximum der n (oder $\bar{n}$) liegenden Kometen größtenteils „neue“ Kometen sein müssen.

[1] Man beachte, daß die Größen der Intervalle der ersten Spalte wechseln und mit wachsendem $1/a$ größer werden.

[2] G. FAYET: Publ. Obs. Paris, Mém. **26**, A (1910).

[3] A. J. J. VAN WOERKOM: loc. cit.

[4] Die Jupiterstörungen sind im allgemeinen ausschlaggebend.

Es liegt nahe, anzunehmen, daß alle Kometen aus diesen großen Distanzen gekommen sind und die Einwirkung der Planeten eine stetige Diffusion zu größeren $1/a$ bewerkstelligt hat und noch weiter bewerkstelligt[1].

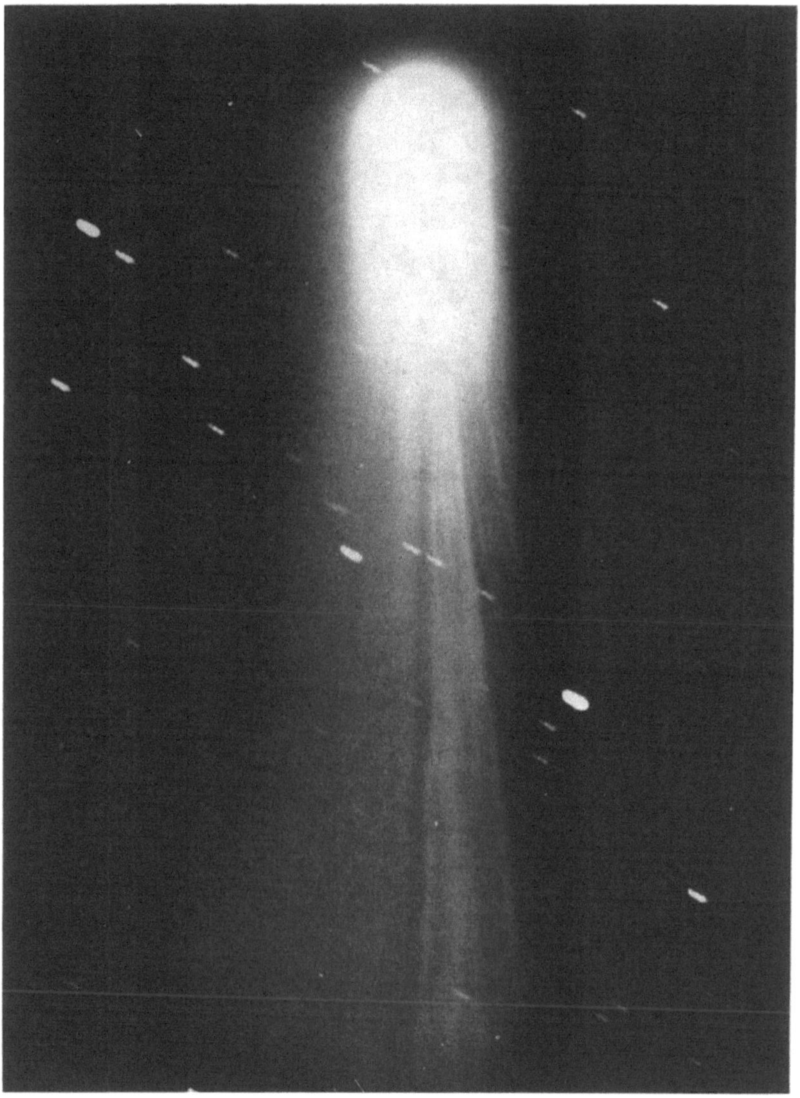

Fig. 29. Komet Mrkos (1957d), 28. August 1957. (Aufnahme G. MANNINO, Asiago.)

Das Auftauchen neuer Kometen aus Distanzen von 40000 bis 150000 A.E. erklärt sich — wiederum nach OORT — in folgender Weise. Es kann keinesfalls angenommen werden, daß sich die Kometenkerne in diesen Entfernungen, wo

[1] Die Berechnungen von FAYET und VAN WOERKOM sind statistischer Art. Der Mittelwert der Störung in $1/a$ kann ebenfalls aus den Störungen der 21 individuellen Bahnen der Tabelle 2 (Ziff. 2) berechnet werden. Nach den Ergebnissen dieser Tabelle erhält man $\Delta(1/a) = 0{,}00048$ A.E.$^{-1}$. Daß $\Delta(1/a)$ bei den Objekten der Tabelle 2 fast ausschließlich positiv ausfällt, ist natürlich darin begründet, daß die Ausgangsbahnen vorwiegend oskulierende hyperbolische Bahnen sind.

keine anderen Körper vorliegen, ständig neu bilden. Sie müssen bereits als fertige Kerne in diesen entfernten Regionen vorhanden sein und in Bahnen laufen, die nicht in das innere des Sonnensystems eindringen. Diese Kometenkern-Wolke bleibt vollständig unbeeinflußt durch die Planeten. Da jedoch ihre Distanz

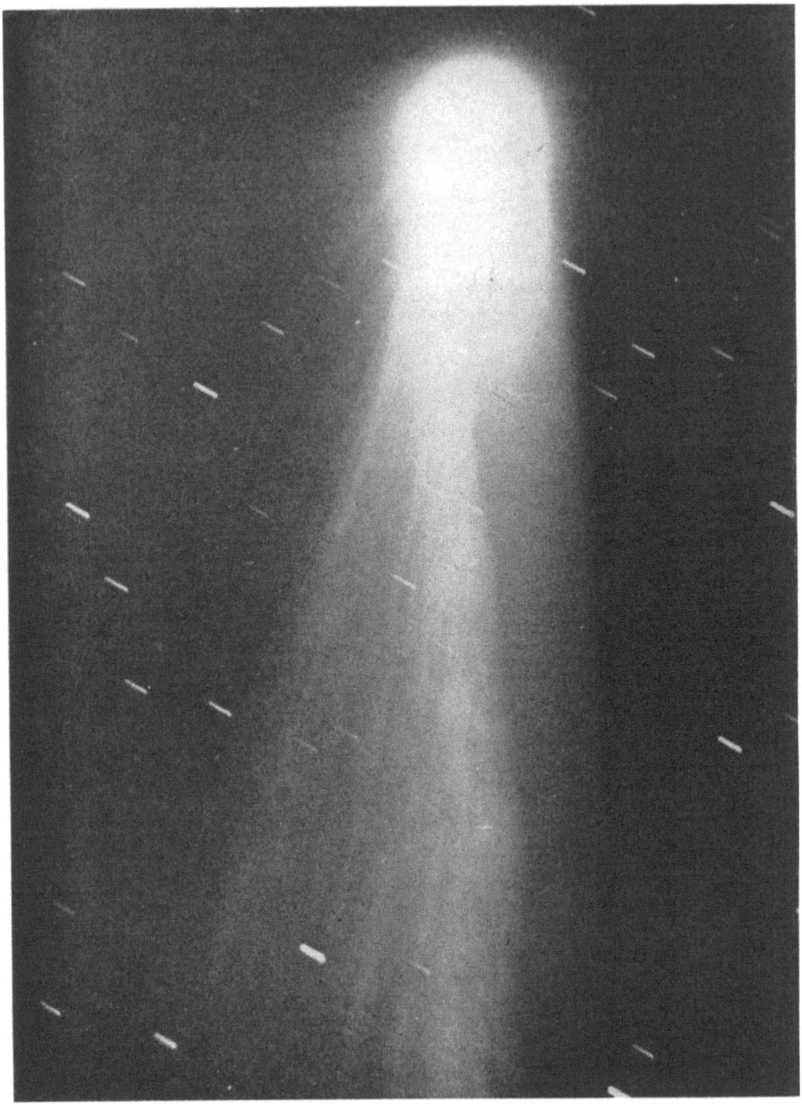

Fig. 30. Komet Mrkos (1957d), 30. August 1957. (Aufnahme G. MANNINO, Asiago.)

von der Sonne von derselben Ordnung ist wie die Distanzen zwischen benachbarten Sternen, so unterliegt sie hin und wieder dem störenden Einfluß vorbeiziehender Sterne. Unter den zahlreichen Bahnumwandlungen, die ein solcher Vorbeigang eines Sternes hervorruft, entstehen auch einige Bahnen positiver Exzentrizitäten, deren Perihele nahe der Erdbahn liegen, und deren Kerne also nach entsprechenden Zeiten für uns sichtbar werden[1].

[1] Die Gesamtheit der Kerne in der Kernwolke wird auf 10^{10} bis 10^{11} Einzelkerne geschätzt.

Nach VAN WOERKOM[1], der den Einfluß der Jupiterstörungen auf den nach innen gehenden Strom der Kometenkerne untersucht hat, sollte unter quasistationären Bedingungen eine solche Verteilung der Hauptachsen sich ergeben,

Fig. 31. Komet Mrkos, 13. August 1957. (Aufnahme A. McCLURE, Los Angeles.)

daß gleiche Intervalle $1/a$ eine gleiche und konstante Anzahl n von Kometenkernen enthalten[2]. Ein stationärer Zustand dieser Art ist offensichtlich nach den Tabellen 12 und 13 nicht vorhanden, obwohl doch die Zeit der Existenz

[1] A. J. J. WOERKOM: loc. cit.

[2] Man beachte, daß auch ein nach außen gerichteter Strom hyperbolischer Bahnen existiert, dessen Mitglieder dem System verloren gehen.

des Sonnensystems ($3 \cdot 10^9$ Jahre) sehr groß ist im Vergleich zu den längsten Perioden. Dieser Umstand wird ohne weiteres verständlich. Die sich der Sonne nähernden Kerne unterliegen einem fortschreitenden Zerfall, der im Durchschnitt um so schneller erfolgt, je kürzer die Periode ist.

19. Die kurzperiodischen Kometen. Der 1770 von MESSIER entdeckte Komet Lexell lief zur Zeit seiner Entdeckung in einer Bahn mit einer Umlaufsperiode von 5,6 Jahren und einem Perihelabstand von 0,67 A.E. Während der nächsten Perihelannäherung waren die Beobachtungsbedingungen zu ungünstig, um ihn auffinden zu können. Bei der darauffolgenden Perihelpassage blieb er wiederum unauffindbar, trotz sehr günstiger Stellungen von Erde und Komet. LAPLACE[1] zeigte, daß der Komet im Jahre 1779 sich Jupiter auf 0,01 A.E. genähert hatte und in eine Bahn mit einer Periode von 16,2 Jahren und einer Periheldistanz von 3,33 A.E. übergegangen war. Es ergab sich außerdem, daß die Bahn, in welcher er 1770 entdeckt wurde, ebenfalls das Resultat einer kurz vorangegangenen Jupiterannäherung war. Vor 1767 besaß der Komet eine Periode von 11,4 Jahren und eine Periheldistanz von 2,96 A.E. Aus dieser Erfahrung schloß LAPLACE auf eine Entstehung der kurzperiodischen Kometen infolge starker Jupiterstörungen aus langperiodischen Bahnen.

Das Problem der Störungen der Kometenbahnen durch nahe Vorbeigänge an den Planeten ist später am ausführlichsten und erfolgreichsten durch den amerikanischen Astronomen H. A. NEWTON[2] behandelt worden, neuere Beiträge von Wichtigkeit zu diesem Thema stammen von H. N. RUSSELL[3] und A. J. J. WOERKOM[4]. NEWTONs Publikation liefert die Grundlagen einer statistischen Theorie starker planetarischer Störungen. Die Voraussetzungen der Theorie sind die folgenden. Es existiert im Abstand von Sonne gleich dem Bahnradius r des betrachteten Planeten ein homogenes Feld parabolischer Kometen. Die Bahn des Planeten wird als kreisförmig angenommen. Die Geschwindigkeiten der Kometenkerne sind also in der Entfernung r überall dieselben und gleich der parabolischen Geschwindigkeit, die Richtungen der Bewegung der Kerne sind nach dem Zufall verteilt. Um den durch das Feld streichenden Planeten gibt es im Sinne von LAPLACE eine Aktionszone, innerhalb welcher die Schwereattraktion des Planeten die der Sonne so stark überwiegt, daß letztere vernachlässigt werden kann. Außerhalb der Aktionszone wird die Attraktion des Planeten vernachlässigt. Die Bahnkurve innerhalb der Aktionszone ist eine Hyperbel mit dem Mittelpunkt des Planeten als Fokus; sie ist bestimmt durch die Geschwindigkeit, deren Richtung und den Ort beim Eintritt in die Zone.

Wir müssen davon absehen, die nicht wenig verwickelte Theorie hier darzulegen und beschränken uns auf die Erwähnung einiger der wesentlichsten Resultate[5].

Der Autor leitet unter anderem eine Formel ab, welche die Wahrscheinlichkeit zu berechnen gestattet, daß eine einzelne Passage eines Kometenkerns durch den Raum innerhalb der Bahn des Planeten mit dem Radius r zur Entstehung einer elliptischen Bahn mit einer Halbachse kleiner als ein bestimmter Wert a führt. Bezeichnen wir diese Wahrscheinlichkeit mit F, so lautet der entsprechende

[1] P. S. DE LAPLACE: Méc. Cél. IV., Livre 9, Chap. 3. 1805.
[2] H. A. NEWTON: Mem. Nat. Acad. Sci. **6**, 7 (1893).
[3] H. N. RUSSELL: Astronom. J. **23**, 49 (1920).
[4] A. J. J. WOERKOM: loc. cit.
[5] Die Lektüre der Newtonschen Publikation ist ziemlich schwierig. Man findet die wesentlichsten Elemente der Theorie noch einmal und zwar leichter verständlich dargestellt bei M. J. BOSLER, J. Observateurs **14**, 117 (1931).

Ausdruck:

$$F = \frac{1}{4}\frac{m^2}{r^2}\int\left\{4a^2 - \left(\frac{a - r - a s^2}{s}\right)^2\right\} ds. \tag{19.1}$$

In dieser Formel bezeichnet m die Masse des Planeten in der Einheit der Sonnenmasse, s ist das Verhältnis der planetozentrischen Geschwindigkeit des Kometen zur heliozentrischen des Planeten. Die Integration ist über alle $s > \sqrt{2} - 1$ zu erstrecken, die dem Integranden einen positiven Wert lassen. Ausgehend von einer Gesamtzahl von 10^9 parabolischen Kometen, welche die Distanz der Jupiterbahn nach innen zur Sonne hin überschreiten, findet NEWTON, daß diese nach (19.1) 126 Kometen mit einer Periode kleiner als die halbe Jupiterperiode P_J ergeben, 839 kleiner als P_J, 1701 kleiner 1,5 P_J und 2670 kleiner als 2 P_J.

Nach NEWTON werden die eingefangenen Kometen vorwiegend in direkte Bahnen übergehen, also denselben Umlaufssinn haben wie der Planet. Bezeichnen

Tabelle 14.

ψ (in °)	n	ψ (in °)	n	ψ (in °)	n
0— 15	6	60— 75	124	120—135	37
15— 30	91	75— 90	101	135—150	23
30— 45	170	90—105	67	150—165	11
45— 60	152	105—120	53	165—180	4

wir mit ψ den Winkel, welchen die heliozentrische Geschwindigkeit des Planeten (Jupiter) mit der heliozentrischen Geschwindigkeit des Kometen in der umgewandelten Bahn bildet, so zeigen beispielsweise die umgewandelten Bahnen von 839 Kometen mit einer Periode kleiner als zwölf Jahren eine Verteilung über zwölf gleiche Intervalle von 15°, wie sie in der Tabelle 14 aufgezeigt ist.

$\psi < 90°$ bedeutet direkte, $\psi > 90°$ retrograde Bahnen[1]. Aus der Verteilung über ψ kann die Verteilung über die Neigungswinkel i gegen die Ekliptik nicht ohne weiteres entnommen werden; es ergibt sich jedoch, wie man sich leicht klar macht bei der Verteilung über i eine stärkere Zusammendrängung zu kleineren Winkeln hin. Der Autor hat zum Vergleich die Besetzungszahlen i für die Winkel von 0 bis 30° und von 150 bis 180° errechnet und findet 257 im ersten Intervall gegen nur 51 im Endintervall.

Die Kometen der Jupitergruppe mit Perioden unter 12 Jahren zeigen ausschließlich Bahnneigungen $i < 45°$; die mittlere Neigung beträgt überhaupt nur 14°. Diese extreme Bevorzugung der kleinen Neigungen zeigt, daß die Mitglieder dieser Gruppe höchstens zu einem kleinen Teil das Produkt nur *einer* Jupiterbegegnung sein können, sie sind vielmehr das Ergebnis einer weiteren Aussonderung nach der ersten Begegnung. Nach der Newtonschen Theorie ergibt sich[2], daß für Bahnen, deren ψ nach dem Einfang 45° übersteigt, bei nachfolgenden Begegnungen die Chance für eine Vergrößerung der Periode größer ist als für eine weitere Verkürzung.

Das entscheidende Kriterium für einen „kürzlich" vor sich gegangenen Einfang eines Kometen durch einen Planeten muß darin gesehen werden, daß die Kometenbahn in einem Stück sehr dicht an die Bahnkurve des Planeten heranführt, der für den Einfang verantwortlich gemacht wird. Eine Anwendung dieses Kriteriums auf die typischen Mitglieder der Jupitergruppe durch RUSSELL[3]

[1] Wir sehen hier von der kleinen Bahnneigung des Planeten ab.
[2] Vgl. dazu A. J. J. VAN WOERKOM: loc. cit., S. 470f.
[3] H. N. RUSSELL: loc. cit., S. 56.

ergab, daß es voll erfüllt wird. Die fraglichen Minimaldistanzen d liegen unter 0,2 A.E. mit einer Häufung in dem Intervall von 0,00 bis 0,05 [1].

VAN WOERKOM hat noch auf eine weitere statistische Relation bei den Jupiterkometen aufmerksam gemacht, die in der Tabelle 15 aufgezeigt wird.. Die bekannten Bahnelemente der Kometen gestatten unter Benutzung des Jacobischen Integrals der Dreikörperbewegung (Sonne, Jupiter, Komet) die Beantwortung der Frage, mit welchen Geschwindigkeiten s die verschiedenen Kerne nach der Jupiterbegegnung das Schwerefeld des Planeten (Aktionsradius) verlassen haben. In Tabelle 15 ist s in der Einheit der Jupiter-Bahngeschwindigkeit ausgedrückt. Wir bemerken, daß eine Häufung der Objekte bei der halben Jupitergeschwindigkeit existiert, was als ein sehr bemerkenswertes Charakteristikum der Bahnen angesehen werden muß. VAN WOERKOM konnte zeigen, daß die Relation der Tabelle 15 automatisch aus der Newtonschen Theorie folgt. Für $i \approx 0{,}5$ besteht

Tabelle 15.

s	Anzahl	s	Anzahl	s	Anzahl
0,00—0,12	0	0,36—0,48	11	0,73—0,80	3
0,12—0,24	3	0,48—0,61	13	0,80—0,97	3
0,24—0,36	9	0,61—0,73	9		

für einen Kometen die größte Chance, innerhalb der Jupitergruppe zu bleiben die Wahrscheinlichkeit dafür fällt dann aber sowohl zu kleineren wie zu größeren s allmählich ab.

[1] Der Komet Encke macht hier eine Ausnahme und nimmt eine Sonderstellung ein, was sehr wahrscheinlich auf die schon in Ziff. 3 erwähnten Bewegungsanomalien zurückzuführen ist.

Meteors.

By

Fred L. Whipple and Gerald S. Hawkins.

With 17 Figures.

1. Introduction. In recent years the subject of meteors and meteoritics has emerged from its isolation as a purely academic subject. We now know that meteors affect conditions in the lower ionospheric regions by producing columns of ionization which have recently found a practical use in the propagation of radio waves beyond the optical horizon. Meteoritic dust has even been ascribed the property of engendering heavy rainfall on a world-wide scale, while the diffusion of the dust before reaching the Earth is largely responsible for the zodiacal light or scattered sunlight near the plane of the Earth's orbit. Furthermore, the microscopic dust particles are a source of considerable concern to the engineers who are constructing artificial satellites or man-made vehicles which are invading the silent realms of space. The phenomena of the meteoric processes represent an ultimate in ultra-sonic projectiles, and meteors provide critical data in this area of ballistics. The extreme of terminal ballistic processes is represented by the large meteoritic craters that pit the surface of the Earth. As our understanding of the Earth's crust increases, we can recognize more and more of these imprints made by great meteoritic bodies in the past.

As a consequence of the increasing demand for basic knowledge of meteoritic phenomena, as it relates both to pure science and to technology, research covering the entire field, from meteoritic dust to lunar crater formation, has received an impetus that few would have predicted a few decades ago, and has greatly increased. The study of meteors has passed from the hands of the amateurs into those of the professional astronomers, the ballisticians, the electronics experts and other groups of scientists and technicians who are interested in various phases of the problem. Many of the basic academic questions have now been answered, but the new fields of interest that have been opened are presenting, in turn, questions of a far-reaching nature and thus continually extend and enlarge our original basic concepts.

A vital question, "What percentages of meteors are of cometary origin, of asteroidal origin, and of interstellar origin?" has now been answered. No real evidence exists for hyperbolic meteors with origin outside the gravitational limits of the solar system. Ordinary visual meteors and the fainter radio meteors appear to be almost entirely of cometary origin, while asteroidal debris must certainly account for most of the brighter fireballs and certainly for all of the sizeable meteorites recovered on the surface of the Earth. The evidence for these conclusions will be presented later in this paper.

Today we face important problems concerning the detailed physical processes of ablation, or loss of matter, as the atmosphere interacts with and destroys the surface of a meteoroid (meteoritic body) at velocities from 11 to 72 km/sec.

What fraction of this available energy goes into the production of electrons and observable photons? What are the masses and densities of meteoroids? What physical processes affect meteoritic material from the time that it leaves a cometary nucleus until it finally plunges into the Earth's atmosphere?

To answer these and other questions, we must deal with meteors of all sizes, with dimensions ranging from less than one micron to many meters. The methods of study chosen and the basic nature of the phenomenon depend markedly upon the size of the bodies, as indicated in Table 1. Here the dimensions given are to be considered as only indicative, and not as exact limits. Furthermore, the methods of detection listed for each class of meteoroid are limited to those that appear to be most useful in each size range. For example, all meteoroids in space must contribute to the zodiacal light, but it has become evident that only the smaller ones contribute enough scattering to make their study profitable by this technique.

Table 1. *Meteoritic phenomena.*

Class of phenomena or bodies	Rough dimensions	Possible methods of detection
Micrometeorites	$< 1\ \mu$	Artificial satellites and rockets
Micrometeorites	1 to 100 μ	Light scattering, in space (zodiacal light) Light scattering in atmosphere Meteoritic dust collection Deep sea sediments Artificial satellites Rockets
Faint meteors	100 μ to 0.1 cm	Radio Zodiacal light Artificial satellites
Meteors	0.1 to 10 cm	Photographic Radio Visual
Fireballs and bolides	10 to 300 cm	Visual Meteorites Radio
Great fireballs and bolides	> 300 cm	Meteorites Meteoritic craters and geological structures Visual, radio

In the subsequent sections of this article, we shall deal first with a basic theory of the meteoritic processes in the atmosphere and describe the celestial mechanics which enables us to compare the orbital characteristics of meteors, comets and asteroids. We shall then deal with the various meteoric phenomena in the order of the dimensions of the bodies, stressing in each size range the methods of study that have proved most fruitful, and discussing the results of such studies as are now available. Although this method of presentation involves some overlapping of size range and methods, it seems unavoidable and would also occur in any other method of presentation.

2. Basic theory of the meteoric processes. The basic theory presented here for meteoric phenomena was first used practically in the problem by one of the authors[1]. It is a modification of a theory presented by HOPPE[2], and is based also on fundamental thinking concerning the problem by a number of authors

[1] F.L. WHIPPLE: Proc. Amer. Philos. Soc. **79**, 499 (1938). — Rev. Mod. Phys. **15**, 246 (1943).

[2] J. HOPPE: Astronom. Nachr. **262**, 169 (1937).

including MARIS[1], SPARROW[2], and particularly ÖPIK[3]. A general summary of this work has been presented by LEVIN[4]. The equations describe the interaction of the meteoroid with the atmosphere and relate the quantities observable by visual and photographic techniques,—velocity, deceleration and light emission. A modification of these equations for their application in the production of free electrons, which produce radio scattering and reflection, is simple and direct without change in principle, and was first presented by HERLOFSON[5] (see A.C.B. LOVELL, Vol. XLVIII of this Encyclopedia).

Let us suppose that a meteoroid of mass m, irregular dimensions, and internal density ϱ_m, enters the upper atmosphere with velocity v and its direction of motion inclines at an angle z to the direction of the zenith. In the practical problem the velocity v is high enough so that the kinetic energy of the incoming body greatly exceeds the internal energies necessary to pulverize, melt, or vaporize the material.

Air particles are trapped momentarily on or near the surface of the meteoroid, imparting to it some of the energy of collision and decreasing its forward momentum. The air particles are given a forward velocity, while the meteoroid suffers a deceleration dv/dt, so that for the momentum exchange we may write

$$m\frac{dv}{dt} = -\Gamma S \varrho v^2, \tag{2.1}$$

where ϱ is the density of the atmosphere, Γ is the drag coefficient and S is the effective cross-sectional area of the meteoroid. It is convenient to eliminate one variable from this equation by expressing the cross section as a function of the mass m. If, for example, the meteoroid were a sphere of density ϱ_m, the cross-sectional area S could be written

$$S = \left(\frac{9\pi}{16}\right)^{\frac{1}{3}} \varrho_m^{-\frac{2}{3}} m^{\frac{2}{3}}. \tag{2.2}$$

In general, the meteoroid will be an irregularly shaped object and we must replace the $(9\pi/16)^{\frac{1}{3}}$ by a general dimensionless shape factor, A, which for bodies of ordinary shape, such as cones, ellipsoids and short cylinders, is of the order of unity. With the transformation we can establish the first fundamental equation in the meteor theory, the *drag equation:*

$$\frac{dv}{dt} = -\Gamma A \varrho_m^{-\frac{2}{3}} m^{-\frac{1}{3}} \varrho v^2. \tag{2.3}$$

From the principle of the conservation of energy, we may equate the energy imparted by the colliding particles to the energy required for ablation (vaporization, melting or fragmentation) of the meteoroid, since the kinetic energy lost by deceleration of the meteoroid is negligible. Thus

$$\zeta\frac{dm}{dt} = -\frac{1}{2}\Lambda A \varrho_m^{-\frac{2}{3}} m^{\frac{2}{3}} \varrho v^3, \tag{2.4}$$

where ζ is the energy required to ablate one gram of the meteoroid and Λ is the efficiency of the energy exchange. The second fundamental equation, the *mass*

[1] H.B. MARIS: J. Terr. Magn. **34**, 309 (1929).
[2] C.M. SPARROW: Astrophys. Journ. **63**, 90 (1926).
[3] E. ÖPIK: Veröff. Dorpat **25**, 1, 32 (1922).
[4] B.I. LEVIN: The physical theory of meteors and meteoric matter in the solar system. This appeared in Dokl. Akad. Nauk SSSR. **1956**, 293 pp., and is available in English translation, ASTIA Document No. AD 110091, published by the American Meteorological Society.
[5] N. HERLOFSON: Phys. Soc. Rep. Progr. Phys. **11**, 444 (1948).

equation, is obtained by dividing (2.4) by (2.3), which gives

$$\frac{1}{m}\frac{dm}{dt} = \frac{\Lambda}{2\Gamma\zeta} v \frac{dv}{dt}. \tag{2.5}$$

The material that is ablated from the surface of the meteoroid collides with the particles in the atmosphere with a forward velocity v, and hence each meteor atom has a kinetic energy of from 100 to 1000 electron volts. This energy is sufficient to ionize and excite the air particles with consequent emission of light. If the fraction of the *kinetic energy converted into light* is τ, then

$$I = -\frac{1}{2}\left(\frac{dm}{dt}\right) v^2 \tau. \tag{2.6}$$

Similarly, if β is the probability that an evaporated meteor atom will *produce an ion pair*, then the number of ion pairs, q, per cm of track is given by the expression:

$$q = -\frac{dm}{dv}\cdot\frac{1}{v}\cdot\frac{\beta}{\mu}, \tag{2.7}$$

where μ is the mean mass of an evaporated atom.

Thus we have four basic equations in the meteor theory, Eqs. (2.3), (2.5) to (2.7), with m, v, and I or q as variables, and six unknown quantities, Γ, Λ, ϱ_m, ζ, τ and β, which we hope are approximately constant. Photographically we can measure v, height, dv/dt and I as functions of t, and by radio techniques we can measure v, height, and q. From four basic equations we can evaluate only three quantities; this fact represents the crucial problem facing meteor theory at present. We choose to evaluate the following groups of constants: $\tau/\varrho_m^2(\Gamma A)^3$, $\Lambda/2\Gamma\zeta$, τ/β. To evaluate $\tau/\varrho_m^2(\Gamma A)^3$, we make use of the fact that the mass at time t is given by integrating Eq. (2.6) from t to t_e (the end of the meteor) and, on substituting this value of m in the drag equation (2.3), we find

$$\frac{\tau}{\varrho_m^2}(\Gamma A)^3 = 2\left(\frac{1}{\varrho v^2}\frac{dv}{dt}\right)^3 \int_t^{t_e} \frac{I}{v^2}\,dt. \tag{2.8}$$

In practice, v is found to be approximately constant and may be brought outside the integral sign. The intensity I is found from the estimated visual magnitude M_v of the meteor from the expression of ÖPIK:

$$M_v = 24.6 - 2.5\log_{10} I. \tag{2.9}$$

The quantity $\sigma = \Lambda/2\Gamma\zeta$ is found in a similar way by inserting m and dm/dt in the mass equation (2.5):

$$\frac{1}{\sigma} = \frac{2\Gamma\zeta}{\Lambda} = -\frac{v^3}{I}\frac{dv}{dt}\int_t^{t_e}\frac{I}{v^2}\,dt. \tag{2.10}$$

To express τ/β in terms of measured quantities, we eliminate the mass between Eqs. (2.6) and (2.7) and substitute for I in Eq. (2.9). This yields the radar magnitude relationship[1,2],

$$M_v = 80.35 - 2.5\left(3\log_{10} v + \log_{10}\frac{\tau}{\beta} + \log_{10} q\right) \tag{2.11}$$

(in c.g.s. units), from which τ/β can be found.

[1] J. S. GREENHOW and G. S. HAWKINS: Nature, Lond. **170**, 355 (1952).
[2] G. S. HAWKINS: Astrophys. Journ. **124**, 311 (1956).

For computational work and for theoretical reasons it is useful to develop subsidiary equations from the basic theory. For example, we may eliminate dm/dt between Eqs. (2.6) and (2.4), thus expressing the intensity of light as a function of air density and meteor velocity, as follows:

$$I = \frac{\Lambda A m_m^{\frac{2}{3}}}{4\zeta \varrho_m^{\frac{2}{3}}} \tau \varrho v^5 . \tag{2.12}$$

By logarithmic differentiation we may then determine the air density $\varrho_{\max}$, where the meteoroid attains its maximum luminosity:

$$\varrho_{\max} = \frac{\cos Z_R \cdot \varrho_m^{\frac{2}{3}} m_{\max}^{\frac{1}{3}}}{5 H \Gamma A \left\{ \frac{\Lambda v^2}{15 \Gamma \zeta} + 1 \right\}} . \tag{2.13}$$

It will be noted that the differential $\frac{1}{\varrho} \frac{d\varrho}{dt}$ has been evaluated by assuming an isothermal atmosphere with a scale height H, and that the meteor enters the atmosphere at an angle Z_R to the vertical with negligible deceleration. Since $\Lambda v^2/15\Gamma\zeta \gg 1$, Eq. (2.13) simplifies to the form

$$\varrho_{\max} = \frac{3 \varrho_m^{\frac{2}{3}} m_{\max}^{\frac{1}{3}} \zeta \cos Z_R}{\Lambda H A v^2} . \tag{2.14}$$

We may express the mass of the meteor in terms of its height h in the atmosphere and its mass m_∞ outside the atmosphere by integrating Eq. (2.4);

$$m^{\frac{1}{3}} = m_\infty^{\frac{1}{3}} + \frac{\Lambda A v^2}{6 \zeta \varrho_m^{-\frac{2}{3}} \cos Z_R} \int_{+\infty}^{h} \varrho \, dh . \tag{2.15}$$

Evaluating the integral by assuming an isothermal atmosphere and substituting for $\varrho_{\max}$ from Eq. (2.14), we may deduce the simple relationship:

$$m_{\max} = \tfrac{8}{27} m_\infty . \tag{2.16}$$

The intensity of a meteor is simply formulated if an isothermal atmosphere is again assumed and $\varrho_{\max}$ is substituted in the intensity equation. It can be shown that

$$\frac{I}{I_{\max}} \frac{q}{q_{\max}} = \frac{9}{4} \frac{\varrho}{\varrho_{\max}} \left\{ 1 - \left[\frac{\varrho}{3 \varrho_{\max}} \right]^2 \right\} . \tag{2.17}$$

Thus the normalized light curves of all the meteors are identical when plotted as a function of height, and are independent of velocity, angle of approach, and any physical characteristics of the meteoroid. The integrations involved from Eqs. (2.12) to (2.17), of course, are not valid if the meteoroid explodes or produces flares of light along its path, which are frequent occurrences in the brighter photographic meteors, or where the atmospheric scale height varies with height.

The light curve of bright meteors is dependent on velocity; this fact was used to determine the dependence of atmospheric density on height[1], a procedure that provided an excellent check on the basic principles of the theory.

The brighter photographic meteors have light curves that agree remarkably well with the simplified theory. Fig. 1, from the work of JACCHIA[2], shows typical examples of bright meteors for which the measured light curve is compared with

[1] F. L. WHIPPLE: Pop. Astron. **47**, 419 (1939).
[2] L. G. JACCHIA: Harvard Coll. Obs. Techn. Rep. 3, 1949.

that computed from Eq. (2.17). The phenomenon of flares along the path of the meteor is illustrated in the same figure, and it can be seen that the meteor obeys the simple theory until the flare occurs. After the flare the meteoroid disintegrates rapidly.

The Super-Schmidt cameras photograph meteors that are some three or four magnitudes fainter than the meteors photographed in the earlier programs. Some 300 meteors have been reduced in the Short Trail Reduction Program[1,2], so that Eq. (2.17) can be tested statistically. If we substitute for the isothermal atmosphere and for the relation between intensity and magnitude from Eq. (2.9), we find that

$$\left.\begin{aligned} &\frac{2.303}{2.5}(M_{\mathrm{max}}-M) \\ &= \log_e \frac{9}{4} + \frac{h_{\mathrm{max}}-h}{H} + \\ &+ 2\log_e\left[1-\frac{1}{3}\exp\left\{\frac{h_{\mathrm{max}}-h}{H}\right\}\right]. \end{aligned}\right\} \quad (2$$

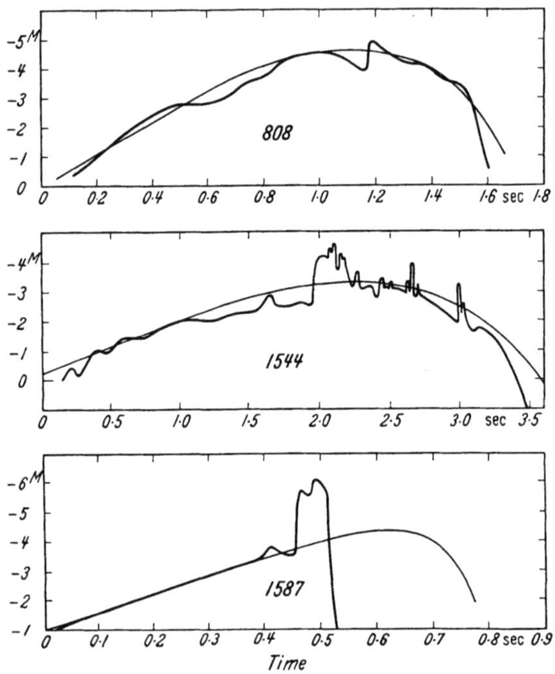

Fig. 1. Light curves of typical bright meteors, observed and theoretical.

Continuous light curves have not been measured for these meteors, but the magnitude has been measured at three points: at the position of maximum light, at the beginning, and at the end of the trail, when the magnitude of the meteor equalled the limiting magnitude of the plate, M_{lim}. Thus,

at the beginning of the trail

$$h = h_b, \qquad M_{\mathrm{lim}} - M_{\mathrm{max}} = \Delta M;$$

at the end of the trail

$$h = h_e, \qquad M_{\mathrm{lim}} - M_{\mathrm{max}} = \Delta M.$$

We can therefore test the theory by plotting ΔM against $h_{\mathrm{max}} - h_b$, and ΔM against $h_{\mathrm{max}} - h_e$, as shown in Fig. 2 where the theoretical curve as given by Eq. (2.18) has been drawn for comparison. Almost all of the meteor measurements lie within the theoretical light curve, indicating that the observed length of meteor trails is considerably shorter than that predicted, and departs from the theory in a random fashion. It has been suggested by JACCHIA[3] that this discrepancy can be accounted for by the phenomenon of fragmentation of the meteoroid. It is hoped that the theory may be extended to cover the case of fragmentation when this process is more fully understood. It is probably correct to state that the above theory applies to each individual fragment and can be extended to cover the case of a fragmenting meteor if account can be taken of all the individual particles.

3. Meteor spectra. A spectral analysis of the light emitted by a meteor gives an essential clue to the physical processes that occur in the coma and trail of a

[1] G. S. HAWKINS: Smithson. Contrib. Astrophys. **1**, 207 (1957).
[2] G. S. HAWKINS and R. B. SOUTHWORTH: Smithson. Contrib. Astrophys. **2**, 349 (1958).
[3] L. G. JACCHIA: Astrophys. Journ. **121**, 521 (1955).

meteor. Unfortunately it has proved to be extremely difficult to obtain adequate numbers of meteor spectra because the sensitivity of a camera is necessarily reduced when an objective prism or diffraction grating is placed in front of the lens. Early visual work indicated that meteors produced a faint continuum of radiation on which was superimposed a line spectrum. Two lines identified with reasonable certainty were the *D*-line of neutral sodium and a line from neutral magnesium. Since 1900 the collection of meteor spectra obtained photographically has been increasing, and the World List of all known photographed spectra contained 148 events at the beginning of 1955[1].

The emission lines of the spectra are usually produced by neutral atoms which presumably have been excited by inelastic collision between evaporating meteor

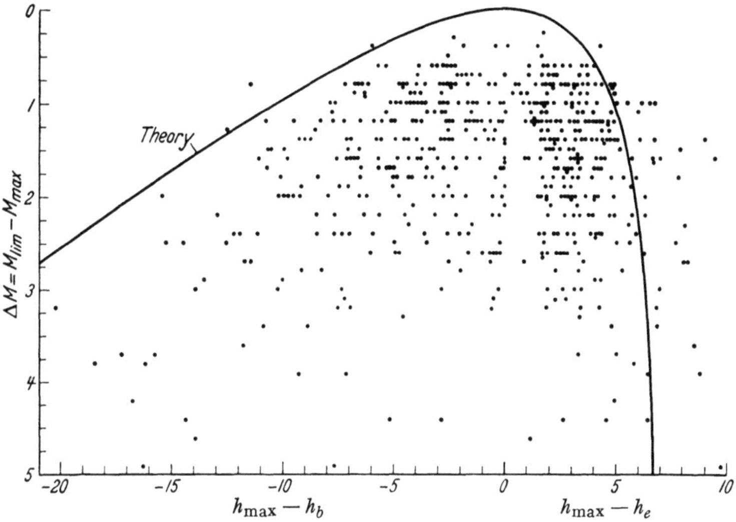

Fig. 2. Points from the light curves of faint meteors, compared with nonfragmentation theory.

atoms and atmospheric particles. In some cases, emission from singly ionized atoms supplies a great deal of the radiated energy. The neutral atoms which have so far been detected are H, N, O, Na, Mg, Al, Si, Ca, Cr, Mn, Fe, and Ni, while the ionized atoms are Mg, Si, Ca, Fe. The fact that all of the identified lines are produced by radiation from atoms is a confirmation of the hypothesis that the meteor is vaporized into single atoms, because the elements listed above, except nitrogen and perhaps a small part of the oxygen, are supplied by the meteoroid and not by the atmosphere. There is one observation of radiation from a molecule[2], N_2, but in this instance the nitrogen molecule was probably supplied by the Earth's atmosphere. It is interesting to note that although the energy of the evaporated meteor atom is of the order of 500 ev, the energy levels most frequently involved are centered at 4.5 ev with extreme values ranging from 1.9 to 13.9 ev. Thus the most probable transitions are from upper levels, with energies that are about 1% of the kinetic energy of the evaporated meteor atoms.

The level of excitation and ionization depends on the velocity of the meteor. For example, no emission lines from ionized atoms appear in slow meteors with velocities below about 30 km/sec. As the velocity increases, Ca II is the first ionized atom to be detected and occurs in the spectra of Geminid meteors, with

[1] P. M. Millman: J. Roy. Astronom. Soc. Canada **49**, 169 (1955).
[2] A. F. Cook and P. M. Millman: Astrophys. Journ. **121**, 250 (1955).

a velocity of 35 km/sec. In high-velocity meteors such as the Perseids (velocity 60 km/sec), the ionized lines become the strongest feature[1]. This is consistent with the dependence of ionization on velocity which has been found by radio methods, and is described in a later section.

For bright meteors a progressive change in the character of the spectrum has been noted along the path. The neutral atoms and ions are excited to higher energy levels as the meteor penetrates the atmosphere and the ratio of positive ions to neutral atoms increases. Thus the spectrum of the beginning portion is predominantly emission from neutral atoms, while the end portion is from excited ions. It has been suggested by COOK[2] that excitation and ionization increase with the ratio of meteoroid size to mean free path of evaporated atoms. When the meteoroid is relatively small, the atoms are colliding with air particles, but as it penetrates a region in which its diameter is large compared with the mean free path, collisions between the meteor atoms in the coma become more frequent and the transfer of energy more efficient. The light curve of individual spectral lines departs considerably from the form of Eq. (2.17), yet the integrated light intensity follows the simple theory remarkably well.

It must be observed that the interpretation of the spectrum given in the preceding paragraph applies to the coma near the head of the meteoroid, and does not necessarily apply to the tail extending well behind. For example, the height of maximum ionization determined by radio methods does not occur below the position of maximum light (see Sect. 7) and there is no evidence for a progressive increase in ionizing efficiency along the meteor path.

The continuous spectrum noted by early visual observers has not been studied adequately by photographic techniques, partly because of the difficulty of calibrating the spectral response of the photographic plate. COOK and MILLMAN[3], however, have analyzed the continuous spectrum for a single Perseid meteor and have explained the continuum in terms of the first and second positive groups of the neutral nitrogen molecule, combined with black body radiation of characteristic temperature 7000° K. The latter temperature was adopted to account in a formal way for the continuum, but must not be regarded as the actual temperature of the meteoroid surface which is probably at about 3000° K[4].

The elements observed in meteor spectra are those expected on the basis of general cosmic abundances. Unfortunately, abundance determinations from meteor spectra have not yet been attempted because of uncertainties in the theory of the state of excitation and the physical circumstances of the meteor phenomena.

A bright meteor usually leaves a faint column of light, or train, which may persist for many seconds. MILLMAN[5] has obtained the only photograph of a train spectrum (between rotating shutter breaks) at present available, although a considerable number of visual observations exist[6]. The conspicuous characteristic in both the visual and photographic spectra is the complete absence of lines from ionized atoms. Photographically the elements Fe, Mg, Ca and Na are identified, all in the neutral state. This is interpreted by MILLMAN as an indication that the light from the train is produced by a recombination process, but this explanation has been shown to be inadequate by COOK and HAWKINS[7]. There seems to be some mechanism that stores energy for considerable periods of time

[1] P.M. MILLMAN: Astronom. J. **54**, 177 (1949).
[2] A.F. COOK: J. Atmos. and Terrest. Phys., Spec. Suppl. **2**, 8 (1955).
[3] See footnote 2, p. 525.
[4] A.F. COOK, H. EYRING and R.N. THOMAS: Astrophys. Journ. **113**, 475 (1951).
[5] P.M. MILLMAN: J. Roy. Astronom. Soc. Canada **44**, 209 (1950).
[6] C.C. TROWBRIDGE: Astrophys. Journ. **26**, 95 (1907).
[7] A.F. COOK and G.S. HAWKINS: Astrophys. Journ. **124**, 605 (1956).

in the train; and this energy, on its release, excites the neutral atoms. It should be pointed out that the Millman train spectrum may arise largely from light emitted only small fraction of a second (<0.01 sec) after the meteor. Thus this spectrum may differ systematically from train spectra observed visually a half second or more after the meteor.

4. Meteor dynamics and orbits of comets, asteroids, and meteors. The velocity of escape from the Earth is 11.3 km/sec, the lowest velocity with which a meteoric body from space can strike the upper atmosphere. The Earth's orbital motion about the Sun is approximately 30 km/sec in a nearly circular orbit, while the velocity of escape from the solar system at the distance of the Earth corresponds to 42 km/sec. Hence the maximum velocity with which a meteoric body can strike the Earth occurs when a meteoroid at its perihelion strikes the Earth head-on. Under these conditions, the velocity components total about 72 km/sec.

The vectorial difference v_G between the Earth's velocity and the velocity of the body is increased to v_∞, the geocentric velocity outside the Earth's atmosphere, because of the attraction of the Earth. These velocities are related as follows:

$$v_\infty^2 = v_G^2 + \frac{2 G M_e}{R}, \tag{4.1}$$

where G is the constant of gravity, R the radius of the Earth, and M_e the mass of the Earth. More precise equations involving the observed shape and distribution of mass in the Earth are not required in the usual meteor studies. The Earth's attraction adds materially to the atmospheric velocity of slow incoming meteors, but contributes little to faster ones.

Atmospheric resistance, of course, reduces v as the meteoroid is ablated. Only among the brightest and the slowest photographed meteors has this reduction in velocity ever attained 50% of the initial velocity, before ablation had destroyed the meteor. Extremely bright fireballs and detonating bolides undoubtedly undergo reductions of velocity to the terminal velocity of free-fall at moderate altitudes, a few kilometers, in the atmosphere. These terminal velocities may be of the order of 1 km/sec, while for micrometeorites the velocities are reduced to negligible values at altitudes of the order of 100 km above the Earth's surface.

In determining the nature of interplanetary material, we are limited in our sampling procedures by the fact that meteoric bodies must cross the Earth's orbit rather precisely, if we are to observe them. Among visual, photographic, and radio meteors, the possible sample is further restricted by the efficiency of light production, of electron production, and of atmospheric diffusion in case of radio meteors, as well as by the general sensitivity of our observing equipment. The sample collected by each observing technique is thus biased in favor of those meteors appearing within a particular range of height and velocity. Hence our measurements of meteoritic bodies, to be representative of the interplanetary content, must be corrected not only for the relative sensitivity of the detection equipment, but also for losses of material in the atmosphere, velocity effects, atmospheric effects, and the probability that the bodies will physically encounter the Earth.

Öpik[1] has formulated the probability for collision with the Earth by a particle in an orbit of known a, e and i, as given in the following equation:

$$\frac{1}{P} = \frac{\pi v_G \sin i}{R^2 v_\infty^2}\left(2 - \frac{1}{a} - p\right)^{\frac{1}{2}}, \tag{4.2}$$

[1] E. Öpik: Proc. Roy. Irish Acad., Sect. A **54**, No. 12 (1951).

where $p = a(1-e^2)$, the parameter of the orbit, and R is the radius of the Earth plus relevant atmosphere. In this equation the unit of length is the astronomical unit, and velocities are expressed in terms of the Earth's orbital velocity. The equation includes the effects of perturbative motions in the node and the argument of perihelion, as long as the inclination of the orbit remains essentially unchanged. The equation fails for very small i or v_G. A perusal of Eq. (4.2) shows that our observable sample of interplanetary matter favors the selection of particles moving with inclinations close to the plane of the Earth's orbit, particularly those having retrograde motions and orbits with perihelion or aphelion nearly equal to one astronomical unit.

One of the major factors that influence the observation of the interplanetary particles is their velocity in the Earth's atmosphere. If the velocity distribution of meteors observed photographically is compared with that observed by radio methods, the distribution of radio meteors is found to contain a preponderance of high-velocity meteors, with speeds up to about 50 km/sec [1]. For meteors detected at a constant line density of electrons the limiting visual magnitude is related to the velocity of the meteor in the manner shown in Table 2 [2]. The actual

Table 2. *Limiting magnitude as a function of velocity.*

Limiting visual magnitude of	Velocity (km/sec)				
	20	30	40	50	60
Radio equipment . .	0.0	0.7	1.2	(1.6)	(1.9)
Camera	0.0	−0.3	−0.8	−1.0	−1.2

limiting magnitudes depend, of course, on the sensitivity of the equipment, and Table 2 quotes a relative scale on the assumption that mag 0.0 is detected by radar at a velocity of 20 km/sec. On the other hand, the photographic intensity of the meteor image decreases as the trailing velocity across the plate increases. The dependence of limiting visual magnitude on velocity is given in Table 2 on the assumption that there is no reciprocity failure in the photographic plate. Since meteors become more numerous at the fainter magnitudes this difference in sensitivity adequately accounts for the different velocity distributions found by radio and photographic methods. At $v \gtrsim 50$ km/sec the number of detected radio meteors decreases, probably because diffusion of the ion column reduces the echo amplitude [3]. The limiting magnitudes of meteors affected by diffusion are indicated by parentheses in Table 2.

A further complication arises if we consider the limiting mass that can be detected. It has been shown that the visual brightness is proportional to $m v^3$. Let us define the sensitivity of the various pieces of equipment to be proportional to $1/m_{\text{lim}}$, where m_{lim} is the mass of a meteor at the limit of detection. The sensitivity of a visual observer is then proportional to v^3. Table 2 shows that the corresponding sensitivities of a camera and of radio equipment vary as v^2 and $v^{5.5}$, if we neglect the effect of diffusion for the radio meteors of high velocities.

Visual and photographic methods are particularly favorable for the detection of interstellar meteors whose velocities exceed the elliptical limits of solar-system objects. Although the higher velocities augment the opportunity for visual, photographic or radio observation of meteors, at the same time they reduce the

[1] F. L. WHIPPLE: Astrophys. Journ. **121**, 241 (1955).
[2] G. S. HAWKINS: Astrophys. Journ. **124**, 311 (1956).
[3] G. S. HAWKINS: Proc. Inst. Rad. Engrs. **44**, 1192 (1956).

probability of recovery on the ground because ablation increases with velocity in the atmosphere.

An asteroidal source for meteors is suggested by the nature of meteorites (see Sect. 13) but this theory is not supported by the known orbits of the asteroids, almost all of which have perihelia placed well beyond the Earth's orbit. Less than 0.5%, 7 in all, have been observed to pass within the Earth's orbit. These are listed in Table 3. All show a second peculiarity, that all of the aphelia are beyond Mars' orbit. ÖPIK[1] has shown that the sweeping effect of the Earth on asteroidal material crossing the Earth's orbit would have been essentially complete over a period of a billion years, a period shorter than the time elapsed since the origin of the solar system; thus we should expect to find no remnants of such initial material. The sweeping effect of Mars, however, would by no

Table 3. *Orbits of asteroids crossing the Earth's orbit.*

Object	Discoverer	Orbital elements[2]							
		ω	Ω	i	e	a (a.u.)	q (a.u.)	q' (a.u.)	P (yr.)
Icarus	BAADE	31	88	23	0.83	1.08	0.19	1.97	1.1
Adonis	DELPORTE	39	353	2	0.78	1.97	0.44	3.50	2.5
Apollo	REINMUTH	285	36	6	0.56	1.48	0.65	2.30	1.8
Hermes	REINMUTH	91	35	5	0.48	1.29	0.68	1.90	2.1
1948 OA	WIRTANEN	126	274	10	0.44	1.38	0.77	1.98	1.61
1951 RA	WILSON and MINKOWSKI	276	337	13	0.34	1.24	0.83	1.66	1.4
1950 DA	WIRTANEN	223	356	12	0.51	1.69	0.84	2.55	2.1
1948 EA	WIRTANEN	265	349	18	0.69	2.26	0.89	3.63	3.41

means be complete, and consequently an appreciable fraction of asteroidal material that initially crossed the orbit of Mars might still remain. Those larger pieces of asteroidal debris, observable in space, must have been perturbed by Mars into orbits that crossed the Earth's orbit during more recent astronomical times. The smaller fragments, on the other hand, may have been broken from the asteroids by collisions, some of which may conceivably have caused orbital changes of a kind to make the material reach the Earth's surface, as meteorites. PIOTROWSKY[3] has investigated the production of asteroidal debris by continual collisions and suggests the possibility that such material might be adequate for the production of meteors.

The elongated and highly diversified character of comet orbits, and the known association between meteoric showers and comets, constitute a strong basis for the theory that cometary debris may contribute extensively to meteors and micrometeorites. Long-period comets with perihelia less than 1 a.u. show a nearly random distribution of orbital inclinations. Comets in short-period orbits, whose existence is attributable to the perturbative effects of Jupiter, move mostly in orbits inclined less than 30° to the plane of the ecliptic with aphelion distances not greatly beyond Jupiter's orbit. Here we have an abundant and continuing source of meteoritic material moving in orbits appropriate for encountering the Earth.

The vicissitudes undergone by cometary particles, after their ejection by escaping vapor pressure from the nuclei of comets, have been discussed in some

[1] E. ÖPIK: Proc. Roy. Irish Acad., Sect. A **54**, No. 12 (1951).

[2] ω, argument perihelion; Ω, longitude, ascending node; i, inclination; e, eccentricity; a, semi-major axis; q, q', perihelion, aphelion distance; P, period.

[3] S. L. PIOTROWSKY: Acta Astronomica, Ser. A **5**, 115 (1953).

detail by WHIPPLE[1]. In comets of extremely long period, some particles will be lost immediately by their initial velocity of ejection even though it is very small. Others will be blown away from solar system attraction by the interstellar wind, produced by interstellar gas at distances of thousands of astronomical units from the Sun. The perturbations by Jupiter will produce a "random walk" in the changes of extremely elongated orbits in the manner described by J. OORT[2]. Some particles will be lost to the solar system by near encounters with the planets (particularly Jupiter) which result in atmospheric capture, ejection from the system, or destruction by the Sun. All particles remaining in orbits will have their aphelion distances reduced by the Poynting-Robertson effect of momentum transfer from sunlight, by a similar but probably more important effect arising from corpuscular radiation from the Sun and by the resistance given by the solar corona[3]. The corpuscular radiation and the corona, furthermore, will slowly etch away the surfaces of the meteoroids while their orbits are being reduced. As the aphelion distances of larger particles are contracted to the neighborhood of Jupiter's orbit, the relatively close encounters with Jupiter will produce the "Jupiter Barrier", a reflecting gravitational barrier that greatly retards further reduction of orbital size until it has been penetrated.

When we allow for the various losses and destructive forces that affect the meteoritic material, we see that only about 1% of the debris from comets may become effective as particles in the zodiacal cloud. An estimated contribution of 100 tons or more per second from comets appears to be adequate to account for the intensity of the scattered sunlight in the zodiacal light and Fraunhofer corona. Furthermore, some 80% of the injection orbits are probably of low inclination, similar to the orbits of the zodiacal particles; hence it is natural to visualize a continuous process that dissipates comets and replenishes the zodiacal cloud. However, a precise description of the processes and more detailed calculations and estimates of masses will depend upon a better knowledge of the amount and nature of solar corpuscular radiation and the extended solar corona, which appear to be dominating factors in reducing the orbits of meteoritic material. Collisions among the meteoric particles themselves will further reduce the percentage of particles greater than about 0.1 cm in radius.

A similar phenomenon of dissipation might be expected to occur for asteroidal dust produced by collisions among the asteroids. To determine the relative contributions made by cometary and asteroidal sources to the meteoric population, we must extrapolate from the precise observations which were obtained from somewhat larger particles, or we must make direct observations on asteroidal dust.

5. Meteoritic dust. The term micrometeorites is a convenient designation for meteorites that are originally so small that they are not markedly altered when they strike the Earth's atmosphere. It is difficult to set exact size limits, but micrometeorites are too small to be detected by radio or optical methods. They are to be distinguished from other objects included by the broader term, meteoritic dust, which also includes fragmentation products from larger meteoroids. The presence of meteoritic dust in interplanetary space is indicated by the scattered sunlight of the zodiacal light and in the Fraunhofer corona, the part of the coronal radiation which consists of scattered sunlight without a great change in the energy distribution and without appreciable line broadening. Rockets and

[1] F.L. WHIPPLE: Astrophys. Journ. **121**, 750 (1955).
[2] J. OORT: Bull. Astronom. Inst. Netherl. **11**, 91 (1950).
[3] E. ÖPIK: Irish Astronom. J. **4**, 84 (1956).

satellites make possible more direct observations: (a) the pitting of polished surfaces; (b) acoustic effects on the surface of a rocket nose-cone, which may be detected at frequencies of the order of 60000 c/sec[1]; (c) faint light flashes on a surface exposed in a rocket above the atmosphere[2]; (d) possible puncturing of the surface of satellites and rockets exposed a sufficient time above the Earth's atmosphere, and detectable by leaks in a pressurized volume; (e) the etching away of thin conducting or radioactive surfaces on satellite skins, which possibly could be confused with etching caused by solar corpuscular radiation and the solar corona; (f) crater formation on thin, opaque surfaces exposed to space and detectable by transmission of light through the resultant holes or by reaction on a photographic emulsion; (g) ionization or light produced in rocket-borne gas chambers that are sealed with exceedingly thin covers.

The classic, independent investigations of the zodiacal light in terms of meteoritic material in space, by VAN DE HULST[3] and ALLEN[4], demonstrate beyond reasonable doubt that diffraction of sunlight by micrometeorites in space between us and the Sun, contributes markedly to the Fraunhofer solar corona, while at greater angular distances from the Sun ordinary scattering occurs to produce the zodiacal light. Relationship with the Gegenschein has not yet been established although it has been suspected by numerous investigators. Very probably the Gegenschein arises from high back scatter from rough meteoritic surfaces as in the Heiligenschein[5]. The diffraction phenomenon at small angles from the Sun proves conclusively that most of the material is rather small, with dimensions of the order of 10^{-3} cm according to ALLEN; it is distributed over a range in particle radius s following an $s^{-2.7}$ law, according to VAN DE HULST, with a possible cut-off at a dimension of the order of 0.03 cm.

On the assumption that ordinary meteoritic densities are of the order of 5 gm per cm^{-3}, mean space densities of meteoritic material at the Earth's distance from the Sun are found to be of the order of 10^{-21} and 10^{-23} gm/cm^3, by VAN DE HULST and ALLEN, respectively. The unknown density of the particles themselves adds to the uncertainty. The total mass of such material encountering the Earth's atmosphere in a unit of time depends not only upon the quantities of the micrometeorites in space but also on their orbital characteristics, which are not known at present. VAN DE HULST assumes that the orbital distribution is comparable to that of ordinary meteors, and so estimates a total influx of some 10000 tons per day on the entire Earth. We shall see that this volume is large when compared with earlier estimates obtained by extrapolating visual observations to smaller meteoritic dimensions. It is important to note that a fair approximation to the orbital distribution of these small particles in space might be obtained from a careful study of the distribution of light in the zodiacal cloud. A calculation of this distribution from a theory of the orbital changes and destruction of small particles by corpuscular radiation, the solar corona and the Poynting-Robertson effect, as mentioned in the previous section, will some day constitute an important check on this basic theory.

The few observations made from rockets are not, so far, conclusive with regard to the actual quantity of micrometeorites in space. Satellite observations, however, should provide basic and perhaps definite information concerning the distribution, both in space and in size, of these particles.

[1] J.L. BOHN and F.H. NADIG: Res. Inst. of Temple Univ., Report No. 18, 1950.
[2] O.E. BERG and L.H. MEREDITH: J. Geophys. Res. **61**, 751 (1956).
[3] H.C. VAN DE HULST: Astrophys. Journ. **105**, 471 (1947).
[4] C.W. ALLEN: Monthly Notices Roy. Astronom. Soc. London **106**, 137 (1947).
[5] F.L. WHIPPLE: Trans. Internat. Astronom. Union **9**, 321 (1955).

The meteoritic process involved when small particles pass through the atmosphere differs from that for larger bodies; with sufficiently small particles, the heat of interaction with the atmosphere can be radiated away rapidly enough to prevent melting or ablation. The upper limit of the radius, s, for a sphere of velocity v_∞ is given by the following formula[1]:

$$s = \frac{18\,\mathrm{e}\,\beta\,\sigma\,\Gamma H(T_m^4 - T_0^4)}{\alpha\,\varrho_m \cos Z\, v_\infty^3},$$

where e is the base of natural logarithm, β is an emissivity factor, α is the accommodation coefficient ($\beta/\alpha \sim 1$), σ is the Stefan-Boltzmann constant, Γ is the drag coefficient [Eq. (2.1)], T_m is the maximum permissible temperature of the meteoroid with an initial temperature T_0, H is the scale height in an isothermal atmosphere, ϱ_m is the density of the meteoroid, and Z is the zenith distance of the radiant.

Table 4. *Numerical values for micrometeorites.*

v_∞ (km/sec)	Air density at $T_m = 1600°$ K	Radius in microns	
		$\varrho_m = 3$	$\varrho_m = 0.3$
11.3	5.6×10^{-9}	41.4	414
15.0	2.4×10^{-9}	19.2	192
20.0	1.0×10^{-9}	8.8	88
23.1	6.5×10^{-10}	6.0	60
25.0	5.2×10^{-10}	4.8	48
30.0	3.0×10^{-10}	2.94	29.4
40.0	1.3×10^{-10}	1.36	13.6
50.0	6.5×10^{-11}	0.74	7.4
60.0	3.7×10^{-11}	0.46	4.6
70.0	2.4×10^{-11}	0.30	3.0

Table 4 gives such dimensions for various velocities and assumed densities of the micrometeorites (Γ is taken as unity). It is improbable, according to the work of Wood[2], that the sputtering effect of atmospheric molecules, atoms and ions on the surface of the micrometeorites causes appreciable disintegration while they are being stopped in the atmosphere. The velocities are reduced to the extremely slow terminal velocities of free fall at altitudes in the neighbor hood of 100 km above the Earth's surface. Unless the micrometeorites are extremely fragile, or some basic error exists in the theories presented above, then particles within the dimensional ranges indicated in Table 4 should enter the atmosphere without being destroyed and probably without much change in shape, except possibly a minor melting down of the sharp corners and projections. Whether any glassy film should be formed on the surface is difficult to guess; it seems unlikely for particles well below the limiting dimensions indicated.

The most comprehensive discussion of the collection and analysis of meteoritic dust is that by Buddhue[3]. A bibliography of the subject to 1952 was compiled by Hoffleit[4]. The unavoidable contamination of meteoric dust by terrestrial materials presents obvious problems. Since most meteorites contain an appreciable percentage of iron, which is accompanied by an unusually high percentage of nickel as compared to that in normal terrestrial materials, and since meteor spectra show both nickel and iron, investigators of meteoric dust usually solve the problem by studying only magnetic material that contains appreciable nickel. Thus the value of collections made on sticky surfaces, in basins of liquid, on roofs and tree crotches in semiarid regions, and in high-flying aircraft, is limited by the inherent difficulty of proving that the small particles not easily identified as of terrestrial origin are indeed of interplanetary origin. On sticky surfaces the customary procedure is to use a microscope to identify as possible meteoric

[1] F.L. Whipple: Proc. Nat. Acad. Sci. U.S.A. **36**, 687 (1950); **37**, 19 (1951).
[2] J.A. Wood jr.: Astronom. J. **62**, 40 (1957).
[3] J.D. Buddhue: Meteoritic Dust. Albuquerque: Univ. of New Mexico Press 1950.
[4] D. Hoffleit: Harvard Obs. Reprint Ser. II, No. 43, 1952.

dust the small, shiny spherules that are opaque to visual light. Thus LANDSBERG[1], for example, found fairly convincing evidence for the fall of an unusually large number of these small spherules, with diameters ranging from a few microns to a hundred microns, a few days to a month after the large meteor stream of the Giacobini-Zinner comet on October 9, 1946. In more recent studies of sticky surfaces exposed at the Table Mountain Observatory of the Smithsonian Institution in California, at the University of Alaska at Fairbanks, Alaska, and at one or two other sites, HODGE[2] has found that, on the average, an opaque, shiny spherule averaging perhaps 4 microns in diameter, is deposited per day on a square centimeter surface exposed horizontally. If such particles have the density of terrestrial stones, of the order of 3.5 gm/cm^3, the total fall on the Earth of material from this small size range alone would amount to around 150 tons per day[3].

Until we have better methods for determining the chemical composition of particles with dimensions of only a few microns, and can learn more of their physical or crystalline structure, it will be difficult to place the study of meteoritic dust on a sound observational basis. All possible methods for studying such particles should of course be investigated.

The magnetic spherules detected in deep sea oozes by PETTERSSON and ROTSCHI[4] are subject to the same difficulty of interpretation as are the meteoritic dust collections. The quantity of such particles is appreciable, in terms of the van de Hulst estimate of 10000 tons per day falling on the entire Earth. The problem is not simplified by the possible effects of disintegration suffered by these particles during the long, extremely slow fall through ocean waters to a final, sedimentary growth on the ocean bottoms.

BOWEN[5] has introduced a new aspect to the subject, in suggesting that micrometeorites may serve as freezing nuclei in clouds, and thus cause a super-cooled cloud to precipitate extremely heavy rainfall at various points on the Earth's surface. BOWEN has presented evidence that such deluges occur about 30 days after recognized meteoritic bombardments. Among calendar dates of exceptionally heavy rainfall all over the world, the period of January 11 to 13 is particularly conspicuous, and occurs approximately 30 days after the great Geminid meteor shower. The meteorological statistics and physics have been a matter of considerable discussion among the experts in these fields; for example, MARTYN[6]. From the viewpoint of meteoric astronomy, WHIPPLE and HAWKINS[7] have shown that BOWEN's hypothesis requires that meteor streams be accompanied by micrometeoritic activity at a level that is not detectable by radio, photographic or visual observations; that in some (or possibly all) cases such activity must arise from streams that are not recognized in terms of larger particles by these methods. Although the Bowen hypothesis cannot be disproved by the arguments and data presently available, most meteoriticists are reserving their opinions; it is extremely important that extended observations of the various relevant kinds be made as soon as possible, to clarify the situation.

---

[1] H.E. LANDSBERG: Popular Astr. **55**, 322 (1947).

[2] P. HODGE: Nature, Lond. **178**, 1251 (1956).

[3] See also I. HASEGAWA: Collecting Meteoritic Dust in Japan. Oriental Astronomical Assn. Memoir No. 378, 1956.

[4] H. PETTERSSON and H. ROTSCHI: Nature, Lond. **166**, 308 (1950). — Geochim. et Cosmochim. Acta **2**, 81 (1952).

[5] E.G. BOWEN: Austral. J. Phys. **6**, 490 (1953). — J. Meteorology **13**, 142 (1956).

[6] D.F. MARTYN: Austral. J. Phys. **7**, 358 (1954).

[7] F.L. WHIPPLE and G.S. HAWKINS: J. Meteorology **13**, 236 (1956).

At present, actual numbers, orbits, physical structure, chemical structure, and current rates of micrometeorites must be considered as at best uncertain if, indeed, direct evidence concerning them in the lower atmosphere and on the surface of the Earth is available at all.

6. Visual meteor studies. Until the advent of World War II the most sensitive device for detecting meteors was the human eye. Visual observers have been invaluable in establishing the science of Meteor Astronomy and have made discoveries that would have been beyond the reach of other methods of observation; in many instances, however, the visual technique has been pushed beyond its limits of accuracy and has produced some misleading or erroneous results.

The visual work prior to 1900 established the fact that the light from a meteor is caused by the collision of an extraterrestrial particle with the upper atmosphere; and the radiant point was correctly interpreted as the perspective vanishing point of parallel paths. Furthermore, most of the associations between the orbits of comets and those of meteor streams were established through visual observations; the complete list is presented in Table 7, Sect. 10. An account of this work has recently been given by PORTER[1] and LOVELL[2]. The discovery of the radiant point in the spectacular Leonid meteor shower of 1833 gave an impetus to routine visual observing and a search was made for radiant points throughout the year by many noted observers, the culmination of this work being the catalogue of 3035 radiants published by DENNING[3]. However, a radiant point determined from the intersection of the tracks made by a group of possibly unrelated meteors has been criticized by MALTZEV[4], and the work of DENNING and other observers is, in this respect, almost worthless. These catalogues are interesting, however, because they do indicate in a statistical sense the general distribution of sporadic radiants[5]. A concentration of radiant points found towards the plane of the ecliptic confirms the results obtained by modern techniques (see Sects. 10 and 11). In addition there is a concentration at the declination corresponding to the zenith of the observer, which is a spurious effect arising from the fact that most meteors have a downward component in their angular motion across the sky. This secondary peak may be used as an index of reliability for a catalogue of radiants, and it is interesting to note that it is entirely absent in the work of MCINTOSH[6] and occurs only to a small extent in the catalogue of ÖPIK[7].

ÖPIK'S work in the Arizona meteor expedition conducted by HARVARD in 1932 was a very thorough and systematic effort to utilize visual observations in the determination of radiants and orbits for sporadic meteors. ÖPIK combined the parallactic observations made from two stations simultaneously with a method for determining angular velocity by means of a rocking mirror. The trail appeared as a cycloid and the shape and separation of the individual cycloids enabled the velocity of the meteor to be determined. This measurement of angular velocity coupled with the parallactic determination of heights and radiants should lead theoretically to a determination of the velocity vectors of the meteorite motion, which, when coupled with the Earth's motion in space, should lead to a complete heliocentric coordinate and velocity vector for the body and, finally, to its heliocentric orbit. Unfortunately, photographic and radio reflection techniques have not verified ÖPIK'S conclusion that a large percentage of the fainter visual meteors

[1] J. G. PORTER: Comets and Meteor Streams. London: Chapman & Hall 1952.
[2] A. C. B. LOVELL: Meteor Astronomy. Oxford 1954.
[3] W. F. DENNING: Monthly Notices Roy. Astronom. Soc. London **47**, 38 (1886).
[4] V. A. MALTZEV: Astronom. Nachr. **234**, 243 (1928).
[5] G. S. HAWKINS: Smithson. Contrib. Astrophys. **3**, 7 (1958).
[6] R. A. MCINTOSH: Monthly Notices Roy. Astronom. Soc. London **95**, 8 (1935).
[7] E. ÖPIK: Harvard Coll. Obs. Circ. No. 388, 1934.

move in hyperbolic orbits about the Sun and hence originate in interstellar space The cause of the discrepancy is still not understood but appears to be due to a subjective error in the visual observer. On the other hand, ÖPIK's careful statistical evaluation of a criterion for determining a group radiant, and the analysis of the observations in these terms have been highly valuable in furthering our knowledge of shower meteors.

A considerable controversy has arisen concerning the method of reducing heights by statistical means, utilized by ÖPIK, which involves the averaging of inverse heights instead of linear heights in order to reduce the statistical error introduced by gross measuring errors in direction. This type of error, regardless of the method of reduction, has seriously weakened the value of the statistical information available from visual observations from two or more stations with regard to meteor heights, although the general conclusions agree qualitatively with those made by the more precise photographic and radio techniques.

Table 5. *Beginning and end heights.*

v km/sec	H_1 km	H_2 km
10.2	85.7	76.4
14.5	92.1	78.7
17.5	98.5	80.4
20.5	95.7	83.2
24.0	100.6	84.8
29.7	106.9	89.4
36.1	107.1	89.7
45.0	112.2	94.7
71.9	119.7	96.2

When the visual observing technique is pushed to its limit it is possible to estimate[1] the direction of flight of a meteor with a twisting error of 1°, and the position of the beginning and end points can be found to within ±2°. Two observers separated by about 50 km can then determine the beginning and end heights with an error of approximately ±4 km. PORTER[2] has analyzed some 770 meteors observed in Britain, as shown in Table 5. The mean values of beginning and end heights, H_1, H_2, are given as a function of observed velocity v. It must be pointed out that the observed velocities are subject to considerable error, and there is clear evidence that the velocity is systematically underestimated by a factor of from 10 to 20%. The mean absolute visual magnitude of these meteors was between 0 and +0.5.

A great deal of visual effort has been expended to determine the diurnal variation of the occurrence rate of meteors; the most comprehensive of such studies is that made by HOFFMEISTER[3]. The basic object of these studies was to determine the percentage of interstellar meteors that were falling on the Earth, for by assuming that the distribution of directions of meteors in space was random, their mean velocity could be obtained as a function of the number of meteors that were approaching from the apex, and the number approaching from the antapex. A preliminary determination of the radiant distribution of sporadic meteors by HAWKINS[4], has shown that the distribution of directions is not in fact random and that there is a preponderance of directly moving meteors. This vitiates the determination of meteor velocity from the diurnal-rate curves but, nevertheless, the original observations remain a substantial contribution to the subject.

Early estimates of the total number of meteors falling on the Earth per day were obtained by visual telescopic observations[5]; these established the well-known law that between magnitudes +5 and +9, the number of meteors between

[1] J. P. M. PRENTICE: Rep. Progr. Phys. **11**, 389 (1948).

[2] J. G. PORTER: Monthly Notices Roy. Astronom. Soc. London **104**, 257 (1944).

[3] See A. C. B. LOVELL: Meteor Astronomy, Oxford 1954 and C. HOFFMEISTER: Ergebn. exakt. Naturw. **24**, 1 (1951).

[4] G. S. HAWKINS: Monthly Notices Roy. Astronom. Soc. London **116**, 92 (1956). — Astronom. J. **61**, 386 (1956).

[5] F. G. WATSON: Proc. Amer. Phil. Soc. **81**, 493 (1939). See also Sect. 11.

magnitudes m and $m+1$ is greater than the number between m and $m-1$ by the factor 2.5. This work has been extended by Millman[1] who finds that for absolute magnitudes between 0 and -6, the factor is 3.7. It is therefore clear that the number of meteors incident on the Earth cannot be expressed as a simple function of magnitude, as was originally supposed. Systematic visual observations are now being carried out by amateurs in England, Canada, Czechoslovakia, and the United States; in the latter, the work is ably directed by C. P. Olivier. Visual observations are also used in conjunction with photographic and radio studies, and, at present, there is no technique that can compete with visual observations when information is required on the structure of a meteor train[2,3], or on the rare occassions when meteorites rend the sky and strike the Earth.

7. Radio meteor techniques. Our knowledge of meteors has been greatly extended during the past few years through the techniques of radar, where pulses of radio energy are received after being reflected by a target. A line of q electrons per meter and length L will act as a scattering target of cross section $L^2 q^2 \sigma_e$ if the electrons scatter in phase. The quantity σ_e is the back-scattering cross section for a single electron and has a value 9.8×10^{-29} square meters. The power p returned to the transmitter at range R is given by the radar equation

$$p = \frac{P G^2 \lambda^2}{64\pi^3 R^4} \cdot L^2 q^2 \sigma_e \tag{7.1}$$

where the antenna gain is G and the transmitter operates at wavelength λ with peak power P. Lovell and Clegg[4] have shown that a meteor trail of infinite length is reduced to an effective scattering length L of $\sqrt{\lambda R/2}$ by Fresnel diffraction. The power reflected from a meteor of line density q is given by the Lovell-Clegg formula;

$$p = 2.54\times 10^{-32} \frac{P G^2 \lambda^3 q^2}{R^3} \text{ watts/meter}^2. \tag{7.2}$$

It will be recognized that $\sqrt{\lambda R/2}$ is equal to half the length of the first Fresnel zone. The beamwidth of the scattered radiation is $\sim\lambda/L$ which, for a meteor at a range of 200 km observed with a wavelength of 5 m, is equal to $0\overset{\circ}{.}4$. Thus a meteor produces an echo when its trail is perpendicular to the line of sight; this property of specular reflection has enabled the direction of meteors to be studied and their radiant points deduced.

The Lovell-Clegg formula holds, provided that the length of the meteor is somewhat greater than that of the Fresnel zone, and that the electrons are contained within a cylinder of radius $\lambda/2\pi$. It is further assumed that the electron density within the cylinder is not greater than the critical electron density corresponding to the wavelength of the radar. In practice, the cylinder of ionization diffuses radially and the echo amplitude decays exponentially with time. For short wavelength radars the effects of diffusion become serious, and the effective scattering length is reduced to $v\lambda^2/16\pi^2 D$. Under these conditions the echo power is given by the formula of Hawkins[5,6];

$$p = 2.54\times 10^{-32} \frac{P G^2 \lambda^3 q^2}{R^3} \cdot \frac{2v\,\lambda^{\frac{3}{2}}}{(16\pi^2 D R^{\frac{1}{2}})^2}\,. \tag{7.3}$$

[1] P. M. Millman: Sky and Telescope **16**, 222 (1957).
[2] C. Hoffmeister: Astronom. Nachr. **197**, 98 (1914).
[3] G. S. Hawkins: Sky and Telescope **16**, 168 (1957).
[4] A. C. B. Lovell and J. A. Clegg: Proc. Phys. Soc. Lond. **60**, 491 (1948).
[5] G. S. Hawkins: Proc. Inst. Rad. Engrs. **44**, 1192 (1956).
[6] Recent work by G. S. Hawkins and D. F. Winter [Proc. Inst. Radio Engrs. **45**, 1290 (1957)] has shown that Eq. (7.3) also applies when the electron density is greater than critical.

It can be seen that there is an attenuation term introduced into Eq. (7.2) which is a function of the ambipolar diffusion coefficient of the atmosphere D and the velocity v of the meteor. When diffusion is severe, $\lambda \sim 1$ m, the effective length of the meteor may be as small as 20 m. Consequently, the beamwidth of the scattered radiation is of the order of 5° and echoes are possible for a considerable distance on either side of the minimum range position.

Early methods of radiant determination are now of historic interest only. Originally, HEY and STEWART[1] used three radar stations at three locations in south-eastern England and deduced the direction of a meteor stream from the variation in echo rate at each station. McKINLEY and MILLMAN[2] have described a statistical method of determining the radiant point of an isolated meteor stream, but this method has not been used extensively as it depends on an estimation of the mean height of the meteor trails. They[3] have also described a method which is useful for determining the radiant point of an individual meteor. A bright meteor does not follow Eq. (7.2) but appears to produce a large coma of ionization in the vicinity of the meteoroid, so that it is possible to obtain radar reflections when the meteoroid is at great distances from the minimum range point. By measuring the range from the three stations the trajectory of the meteor can be found and the change of range with time enables the velocity to be determined. This method is, however, severely limited as it is applicable only to bright meteors that are moving nearly parallel to the Earth's surface.

Later CLEGG[4] used a single antenna that could be pointed at different azimuths, and deduced the position of a radiant point from the variation of echo range with time on two different azimuths. For survey work in the northern hemisphere a system using two fixed antennas was developed by ASPINALL, CLEGG, and HAWKINS[5], and similar equipment has been built in the southern hemisphere[6]. With the twin beam antennas, echoes are recorded throughout the 24-hour period by photographing an intensity-modulated cathode ray tube with a continuously moving film. General surveys of the radiant distributions of stream and sporadic meteors were begun in 1949. Recently GILL and DAVIES[7] described a three-station radar system which measured the radiant point of an individual meteor. They measured the meteor velocity by the Fresnel technique and also the time delay between echoes at three stations. Thus the direction cosines of the meteor trajectory could be found from the known lengths of the baselines and their projections on the meteor track.

The range-time method described above was the first radio method used in determining meteor velocities, but it has been superseded by amplitude-time methods. DAVIES and ELLYETT[8] have shown that the power p received from a meteor takes the form

$$\left.\begin{aligned} p &= p_\infty \left[\int_{-\infty}^{x_1} \cos\frac{\pi x^2}{2}\,dx\right]^2 + p_\infty \left[\int_{-\infty}^{x_1} \sin\frac{\pi x^2}{2}\,dx\right]^2, \\ x^2 &= \frac{4 v^2 t^2}{R\lambda} \end{aligned}\right\} \quad (7.4)$$

---

[1] J. S. HEY and G. S. STEWART: Proc. Phys. Soc. Lond. **59**, 858 (1947).
[2] D.W.R. McKINLEY and P.M. MILLMAN: Proc. Inst. Radio Engrs. **37**, 364 (1949).
[3] D.W.R. McKINLEY and P.M. MILLMAN: Canad. J. Res. **27**, 53 (1949).
[4] J.A. CLEGG: Phil. Mag. **39**, 577 (1948).
[5] A. ASPINALL, J.A. CLEGG and G.S. HAWKINS: Phil. Mag. **42**, 504 (1951).
[6] C.D. ELLYETT and C.S.L. KEAY: Austral. J. Phys. **9**, 471 (1956).
[7] J.C. GILL and J.G. DAVIES: Monthly Notices Roy. Astronom. Soc. London **116**, 105 (1956).
[8] J.G. DAVIES and C.D. ELLYETT: Phil. Mag. **40**, 614 (1949).

when expressed as a function of time t. The asymptotic value p_∞ is the value given by the Lovell-Clegg formula. The integrals are the standard "Fresnel integrals" encountered in optics, and the radar scattering is analogous to the problem of diffraction at a straight edge. As the meteoroid passes the minimum range point, zones are added to the electron column which are successively in phase and out of phase with the main signal so that the echo amplitude varies with time as shown in Fig. 3. The time interval between maxima and minima in the Fresnel pattern gives a measure of the velocity of the meteor. It is possible to observe the pattern as the envelope of a pulse train or as the audio component of a continuous wave signal which has been reflected from the meteor (see MANNING, VILLARD and PETERSON[1] and McKINLEY[2]).

An accuracy approaching 1% can be attained in measuring meteor velocities with the Fresnel technique, the fundamental limit to the accuracy being set by atmospheric turbulence which distorts the ion column during the formation of the Fresnel zones. When the velocity measures are used to compute the orbit of the meteor an additional uncertainty is introduced through the deceleration of the meteor in the atmosphere. When we use velocity measures to compute the orbit of a meteor we do not know the exact correction for deceleration in the atmosphere. We measure the meteor's velocity during its passage through the Fresnel zones. This represents only a small segment of the trail and so deceleration is difficult to measure. Nevertheless, the Fresnel technique has been invaluable in the search for interstellar meteors, and in determining approximate orbital elements, as in the case of the daytime meteor streams.

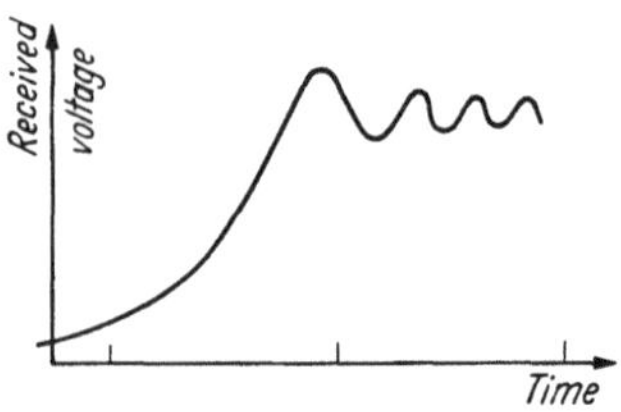

Fig. 3. Radio echo amplitudes from a meteor trail.

Radio observations are, of course, independent of adverse weather conditions and do not appear to be affected by solar radiation during the daytime. The figures obtained on the number of meteors striking the Earth are therefore extremely accurate, and are affected only by changes in the sensitivity of radar equipment which can be calibrated to almost any required degree of accuracy. Some difficulty remains, however, in interpreting the statistics and in relating the radio results to the photographic and visual measures, because of the selection effects described in Sect. 4.

8. Photographic meteor techniques. The first systematic attempt to photograph meteors simultaneously from two stations, with a rotating shutter to break the photographic trail into measurable segments, was made by ELKIN, at Yale University, over a period of 16 years from 1893 to 1909 (see summary by OLIVIER[3]). This program failed to yield results of great significance because of technological difficulties; for example, the slowness of the older photographic cameras, and particularly the shortness of the baseline used.

At the two Massachussetts stations of the Harvard College Observatory, double-station meteor photography developed further in the period 1936 to 1942 under a program directed by WHIPPLE[4]. Using Ross Express $F/4$ lenses of wide fields, the cameras were separated by 37 km and equipped with a rotating shutter powered by a small synchronous motor. The program yielded one doubly-photographed

[1] L.A. MANNING, O.G. VILLARD and A.M. PETERSON: J. Appl. Phys. **20**, 475 (1949).
[2] D.W.R. McKINLEY: Astrophys. Journ. **113**, 225 (1951).
[3] C.P. OLIVIER: Astronom. J. **46**, 41 (1937).
[4] F.L. WHIPPLE: Proc. Amer. Phil. Soc. **79**, 499 (1938); **83**, 711 (1940).

meteor for approximately 100 hours of overall exposure time. The program netted four or five photographed meteors per year, and, by the early 1940's, the accumulation of data had become significant, if not extensive.

By 1948 the program had gained substantial support from the U.S. Naval Bureau of Ordnance and later from the U.S. Air Force, Cambridge Research Center. The older meteor cameras, supplemented by newer aerial mapping cameras, were moved to two isolated stations near Las Cruces in New Mexico that provided an optical base-line of 27 km. Here the rate of meteor photography was increased markedly by the utilization of newer photographic emulsions with increased sensitivity.

With such equipment, by 1954, the number of doubly photographed meteors had been increased by some fifty-two by the Harvard program in Massachusetts, ninety-two in New Mexico, and many more by the Russian, Czechoslovakian, and Japanese astronomers.

In 1949 JAMES G. BAKER designed a new type of Schmidt optics admirably suited for the direct photography of meteors, the Super-Schmidt Meteor Camera (Figs. 4 and 5). The chief innovation consists of two spherical shells to correct the major portion of the spherical aberration inherent in the Schmidt system. A two-element correcting plate aids in the achromatization. The system operates at $f/0.65$ with an aperture of 31 cm, a focal length of 20.3 cm, and a field of 54°. Since the field is curved with a radius of curvature equal to the focal length, the molding and processing of fast photographic films presented a new problem, which was solved at Harvard[1] with the aid of advice from the Eastman Kodak Company.

Six Super-Schmidt meteor cameras were produced by the Perkin-Elmer Corporation and two were put into operation in New Mexico in late 1951, two by P.M. MILLMAN in stations north of Edmonton, Canada, and two more, for the photography of persistent meteor trains, in New Mexico in 1953. Fig. 5 shows the general external character of this revolutionary type of camera. Because of the small amount of space between the spherical focal surface and the inner surface of the middle correcting hemisphere, the rotating shutter is driven by a motor located at the base of the spherical mirror. The drive shaft passes through the mirror and is supported by a second bearing located in the glass of the correcting plate. Of necessity, this shaft, which drives the shutter at 1800 rpm, blocks the light from the center of the film and hence produces an unexposed circular area as appears in Fig. 6.

Because of the enormous speed of the optical system and the time involved in loading spherical film, 18 cm in chordal diameter, the two-sectored shutter is made to occult the light three quarters of the time, to permit the use of exposure times in the range from eight to twelve minutes. Otherwise the exposure time of the full system with fast photographic emulsions would be limited by night-sky brightness to two to three minutes.

In operation, both cameras for the direct photography of meteors are directed at an identical volume of space centered at a height of approximately 90 km; the cameras are exposed simultaneously on a prearranged time schedule. The shutters on the two cameras operate independently but with synchronous motors controlled in rate to better than one part in a thousand. Observers at both stations watch the region of the sky at which their camera is directed, and record by an automatic printing clock the time that each meteor crosses the field, to

[1] P. CARROLL, R.E. MCCROSKY, R.C. WELLS and F.L. WHIPPLE: Harvard Repr. Ser. II, No. 39, 1951.

0.01 min. They record also the general path of the meteor's motion, its apparent brightness, and color.

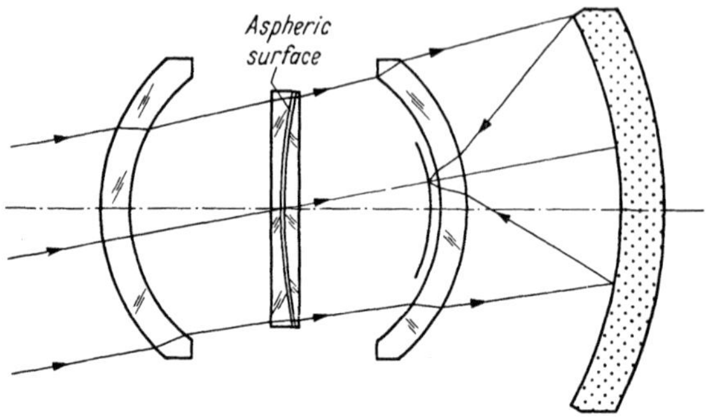

Fig. 4. The Super-Schmidt Meteor Camera of J. G. Baker, general design.

Fig. 5. The Super-Schmidt Meteor Camera of J. G. Baker, external view.

The Super-Schmidt cameras are most sensitive with fast blue-sensitive emulsions. Over three years they averaged 3.4 doubly photographed meteors per hour of exposure time, or one meteor every eighteen minutes. Thus through 1957, more than 6000 meteors have been doubly photographed from the New Mexico pair of stations alone.

Precision reduction of the longer meteor trails yields velocities with an accuracy of the order of 0.3%, significantly accurate decelerations at one or more points

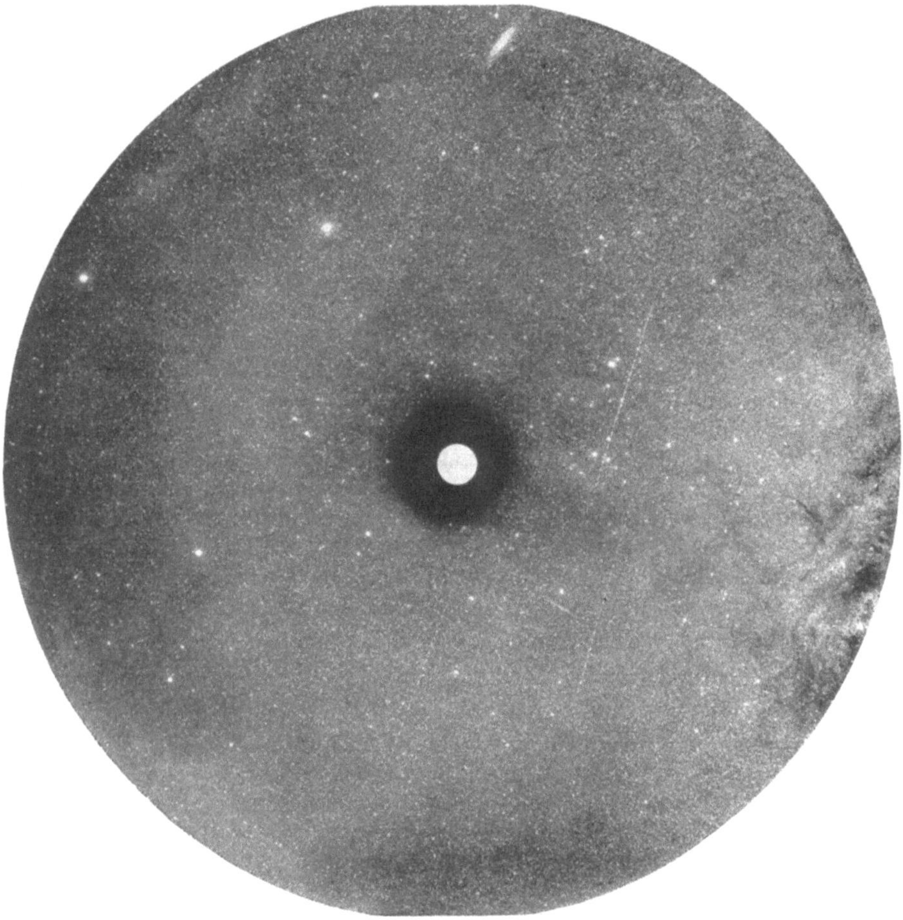

Fig. 6. Typical meteor photograph by the Super-Schmidt Meteor Camera.

along the trail, heights accurate to a few meters and apparent radiants accurate to about one minute of arc. The curved film is photographed onto a flat glass plate by a precise one-to-one copying camera designed by BAKER. Measurements are made with a precision measuring engine. The final projection of the sky is almost perfectly gnomonic[1].

The precision reduction technique described by WHIPPLE and JACCHIA[2] has been carried out for some 500 meteors of the Harvard Program. Since the method

[1] Techniques for reduction were developed by one of the authors and improved by JACCHIA, who has directed the reduction program for some ten years, first at the Massachusetts Institute of Technology, later at Harvard College Observatory and more recently at the Smithsonian Astrophysical Observatory. The work has been supported by the U.S. Office of Naval Research, the U.S. Army Office of Ordnance Research and the U.S. Air Force.

[2] L. G. JACCHIA and F. L. WHIPPLE: Smithson. Contrib. Astrophys. 1, 183 (1957).

is time-consuming, an abridged version has been developed by HAWKINS[1] in which deceleration is not measured. This method is being used on a comparable number of trails selected at random. McCROSKY[2] has developed a rapid graphical method for the approximate reduction of meteor trails and has measured some 2000 velocities and orbits for statistical studies. A graphical method, utilizing the sky charts of the Bonn-Durchmusterung and Córdoba catalogues, has been devised for the reduction of singly or doubly photographed meteors for the determination of group radiants[3].

The rough reduction techniques (which require a total time of less than one hour per meteor pair) developed by McCROSKY have an accuracy about an order of magnitude less than that of the most precise techniques; for meteor velocity, the error is roughly 3%, and for height, of the order of 1 km. Radiants are determined to an accuracy of the order of 1° when the geometrical circumstances are favorable. Perihelion distances and orbital inclinations are quite accurately determined by precise reduction methods and by the intermediate techniques. In the later methods, of course, a precise correction for atmospheric resistance cannot be applied, but it is possible to apply a general correction determined statistically from the more accurate data. Aphelion distances by the precise techniques are probably accurate to the order of half an astronomical unit at Jupiter's orbit, and relatively less precise for longer orbits. The McCrosky method yields a fair result on aphelion distances within the orbit of Mars. Beyond Mars, the data are useful only in statistical analysis. The graphical method of WHIPPLE and WRIGHT[3] for studying meteor radiants allows one to draw the great circle of the meteor trail on star charts so that it passes the true radiant with an accuracy of the order of six minutes of arc, which is roughly the observed scatter in the most closely packed meteor streams of shortest duration.

In addition to orbital results, the doubly photographed meteors provide physical data concerning the upper atmosphere and the meteoric process. Results from the photographic meteor studies are presented in various appropriate sections of this article.

9. The masses and densities of meteoroids. The most direct method of determining the masses of meteoroids stems from measures of atmospheric resistance to their motion. In the double-station photography of meteors, we observe meteoric deceleration directly. A rather satisfactory knowledge of the atmospheric density in this region of the atmosphere is now available from high-altitude sounding measures such as those by the Rocket Panel[4]. The drag Eq. (2.3) couples the observed deceleration, dv/dt, with the atmosphere density, ϱ, as follows:

$$\frac{1}{m^{\frac{1}{3}}\,\varrho_m^{\frac{2}{3}}} = -\frac{dv/dt}{\Gamma A\,\varrho\, v^2} \tag{9.1}$$

where Γ and A are the dimensionless *drag coefficient* and the shape factor, respectively.

At a point in the trajectory where dv/dt is measured, the value of $m^{\frac{1}{3}}\varrho_m^{\frac{2}{3}}$ for the meteoroid can be determined observationally, if a value of ΓA is assumed. Mass and density cannot be separated numerically, however. A more general expression relating ϱ_m and luminous efficiency, τ, can be derived. For bright

[1] G. S. HAWKINS: Smithson. Contrib. Astrophys. **1**, 207 (1957).
[2] R. E. McCROSKY: Smithson. Contrib. Astrophys. **1**, 215 (1957).
[3] F. L. WHIPPLE and F. W. WRIGHT: Smithson. Contrib. Astrophys. **1**, 239 (1957).
[4] Rocket Panel: Phys. Rev. **88**, 1027 (1952).

photographic meteors JACCHIA[1] has shown that τ varies linearly with velocity, or that

$$\tau = \tau_0 v, \tag{9.2}$$

where τ_0 is a constant for $v > 15$ km/sec.

Thus Eq. (2.8) becomes

$$\frac{\tau_0}{\varrho_m^2}(\Gamma A)^3 = 2\left(\frac{1}{\varrho v^2}\frac{dv}{dt}\right)^3 \int_t^{t_e} \frac{I}{v^3}\,dt, \tag{9.3}$$

where all of the quantities on the right-hand side of the equation are directly observable.

Thus in the regime of very bright meteors with $v > 15$ km/sec, the ratio τ_0/ϱ_m^2 is a determinable constant if ΓA is known.

For steel pellets moving at velocities up to 2.5 km/sec and aluminum pellets up to 5 km/sec, RINEHART, ALLEN and WHITE[2] at the U.S. Naval Test Station at Inyokern have measured both Γ and A. In the case of the steel pellets they find that $\Gamma = 0.418$ and $\Gamma A = 0.17/\varrho_m^{\frac{2}{3}}$. Similar results are found by HANSCHE and RINEHART[3] with steel spheres at velocities up to Mach number 3. For a cube the average value of $A\,\varrho_m^{\frac{2}{3}}$ is 1.50 and for a brick $(2\times 3\times 6)$, 1.66. At high altitudes where the effect of the air cap is reduced we expect Γ to increase from the sea-level value of RINEHART *et al.*[2]. Hence we adopt as a representative value, $\Gamma A = 1$.

For many bright $(M_v < 0)$ photographic meteors JACCHIA[4] finds that the expression

$$\frac{(\Gamma A)^3\,\tau_0}{\varrho_m^2} = 2.79\times 10^{-19}\ \text{(c.g.s.)} \tag{9.4}$$

is consistent with the known density of the upper atmosphere. Since ΓA is close to unity, the ratio τ_0/ϱ_m^2 equals 2.79×10^{-19} (c.g.s.).

ÖPIK[5] has estimated theoretically that

$$``\tau_0" = 6.452\times 10^{-19}\ \text{c.g.s. units} \tag{9.5}$$

and this value was assumed by JACCHIA in his derivation of mass from the integrated form of (2.6):

$$m_\infty = \frac{2}{``\tau_0"} \int_t^{t_e} \frac{I}{v^3}\,dt. \tag{9.6}$$

We use quotation marks for the symbol "τ_0" to indicate that it is a tentative value. Eq. (9.6) yields a mass of 0.15 gms for a meteor of visual magnitude 0 and velocity 30 km/sec.

Fresh, independent data are required to determine unambiguously the mass m_∞, density ϱ_m, and luminous efficiency. It is possible to determine the mass of a meteoroid by studying the motion of the glowing column or train. A double-station photograph has been obtained of a train that peristed for several seconds and exhibited a downward velocity component, i.e., the train was "coasting" behind the meteoroid. This coasting velocity gives a measure of the momentum

[1] L. G. JACCHIA: Harvard Reprint II, 31, 1949.
[2] J. S. RINEHART, W. A. ALLEN and W. C. WHITE: Techn. Mem. Naval Ordnance Test Station, Inyokern, No. 338, 1951.
[3] G. E. HANSCHE and J. S. RINEHART: J. Aero. Sci. **19**, 83 (1952).
[4] L. G. JACCHIA: Harvard Coll. Obs. Reprint II, No. 26, 1948.
[5] E. ÖPIK: Harvard Coll. Obs. Reprint I, No. 100, 1933.

transferred to the glowing train and hence enables the original mass to be computed since the velocity of the meteoroid is known. A preliminary analysis by COOK and WHIPPLE[1] shows that the mass scale is a factor of about 200 too low. This implies that the value deduced by ÖPIK for "τ_0" must be decreased by 200 to a value of about 3×10^{-21}. From Eq. (9.4) we note that τ_0/ϱ_m^2 is an observed mean constant. Thus the value of ϱ_m^2 must be reduced by 200. If $\Gamma A = 1.0$, Eq. (9.4) yields a density for the meteoroid of 0.12 gm/cm³. Eq. (9.1) leads to a similar result ($\varrho_m \sim 0.1$). This low value is not surprising in view of the fragile nature of meteoroids as exhibited by the phenomenon of fragmentation. Furthermore, evaporation of ices from the "cosmic mix" that might be expected in comets would leave a solid structure of approximately the above density; the maximum value is $\varrho_m \sim 0.3$ gm/cm³.

10. Shower meteors. Meteor showers and the falling of stones from the sky are recorded in the ancient HEBREW, CHINESE, GREEK, ROMAN and other literatures. Some of this fascinating history can be found in the books by OLIVIER[2] and WATSON[3], while BIOT'S *Catalogue Générale des Étoiles Filantes* has remained, since 1841, a standard reference to the ancient records[4]. The periodic nature of meteor streams was first noticed by BOGUSLAWSKI, who attributed the great Leonid meteor storms of 1799 and 1833 to the collision of the earth with a cloud of debris moving in an elliptical orbit. In some meteor streams the debris is dispersed uniformly around the orbit and a meteor shower is observed on the same date each year. Extensive and painstaking visual observations, as described in Sect. 6, have established the basic facts about the annually recurring streams with radiants located in the night-time sky, but the visual measurements have been gradually superseded by the results obtained from the photographic and radio techniques described in Sects. 7 and 8. The best results obtained from all three techniques have been combined in Tables 6 and 7.

Table 6 gives the basic data relating to the streams. The visual rate refers to the number of meteors visible to one observer; the photographic rate, to the number of trails recorded by a Baker Super-Schmidt camera; and the radio rate, the number of echoes detected by a radar with characteristics as described by ASPINALL, CLEGG and HAWKINS[5].

Table 7 gives the orbital elements of the streams and indicates the parent comet. All major streams are given and minor streams have been included when the orbit has been determined by precise photographic or radio techniques. The number of minor streams could be increased almost indefinitely if wider selection tolerances were allowed in defining a meteor stream. The more precise photographic information indicates that among the fainter meteors a surprisingly high percentage, of the order of 30 to 50%, can still be identified with more and more diffuse streams of great duration. The radiants of these streams tend to be concentrated towards the ecliptic[6] but the overlap is so great that even the addition of another measurable criterion, velocity, does not serve to distinguish membership of an individual meteor in very diffuse streams. At this level of uncertainty it is convenient to describe the meteors as "sporadic" (see Sect. 11) and deal with the orbital characteristics on a statistical basis. There can be little doubt that we observe in this phenomenon the dynamic effects of age on the streams

---

[1] A. F. COOK and F. L. WHIPPLE: Unpublished.
[2] C. P. OLIVIER: Meteors. Baltimore: Williams & Wilkins Co. 1925.
[3] F. G. WATSON: Between the Planets. Cambridge: Harvard University Press 1956.
[4] See also IMATO and HASEGAWA: Smithson. Contrib. Astrophys. **2**, No. 6 (1958).
[5] A. ASPINALL, J. A. CLEGG and G. S. HAWKINS: Phil. Mag. **42**, 504 (1951).
[6] G. S. HAWKINS: Monthly Notices Roy. Astronom. Soc. London **116**, 92 (1956).

Table 6. *Meteor streams: observational quantities.*

Stream	U.T. Date at maximum	Extreme limits	Radiant 1950 R.A. Dec.	V_∞ km/sec	Hourly rate at maximum			Radiant transit (local time) midnight = 00h
					visual	photo	radio	
Quadrantids	Jan. 3	Jan. 1 4	230 + 48	42.7	30	1.9	95	08h 28m
Virginids	Mar. 13	Mar. 5 21	183 + 4	30.8	1	—	< 5	00 49
II (WHIPPLE[1])		Mar. 13 Apr. 21	157 + 56	15.2	—	—	—	20 49
Lyrids	Apr. 21	Apr. 20 23	270 + 33	48.4	5	—	11	03 59
η Aquarids	May 4	May 2 6	336 + 0	64	5	—	15	07 36
Daytime Arietids	June 8	May 29 June 18	44 + 23	39	—	—	66	09 51
Daytime ζ Perseids	June 9	June 1 16	62 + 23	29	—	—	42	10 59
Daytime β Taurids	June 30	June 24 July 6	86 + 19	32	—	—	27	11 12
Southern δ Aquarids	July 30	July 21 Aug. 15	339 − 17	43.0	—	—	—	02 14
					10		34	
Northern δ Aquarids		July 14 Aug. 19	339 − 5	42.3		—		02 08
Southern ι Aquarids		July 16 Aug. 25	338 − 14	35.8	—	—	—	02 04
Northern ι Aquarids		July 16 Aug. 25	331 − 5	31.2	—	—	—	01 36
α Capricornids	Aug. 1	July 17 Aug. 21	309 − 10	25.5	—	—	10	00 00
Perseids	Aug. 12	July 29 Aug. 17	46 + 58	60.4	37	2.5	49	05 43
$\varkappa$ Cygnids		Aug. 19 22	289 + 56	26.6	—	—	< 5	21 25
Draconids	Oct. 10	Oct. 10	264 + 54	23.1	Periodic, next return 1959			16 13
Orionids	Oct. 22	Oct. 18 26	94 + 16	66.5	13	2.9	18	04 12
Southern Taurids	Nov. 1	Sept. 15 Dec. 15	51 + 14	30.2	5	—	—	00 42
							< 15	
Northern Taurids	Nov. 1	Oct. 17 Dec. 2	52 + 21	31.3	5	—		00 46
Andromedids	Nov. 7	Nov. 7	22 + 27	21.3	—	—	< 5	22 23
Leonids	Nov. 17	Nov. 14 20	152 + 22	72.0	6	—	< 10	06 22
Geminids	Dec. 14	Dec. 7 15	113 + 32	36.5	55	5.6	80	02 01
χ Orionids		Dec. 9 14	87 + 21	30.6	—	—	—	00 25
Monocerotids		Dec. 13 15	103 + 8	44.0	—	—	—	01 21
Ursids	Dec. 22	Dec. 17 24	206 + 80	35.2	15	—	13	08 24

[1] F. L. WHIPPLE: Astronom. J. **59**, 201 (1954).

Table 7. *Meteor streams: orbital elements.*

Stream	ω	Ω	ι	π	a (a.u.)	e	q (a.u.)	Parent Comet
Quadrantids	167°.9	282°.6	73°.8	90°.1	3.42	0.715	0.974	—
Virginids	285.8	353.7	5.2	279.5	2.82	0.857	0.403	—
II	187.1	27.3	11.0	214.4	2.67	0.626	0.999	—
Lyrids	213.9	31.8	79.9	245.6	29.6	0.969	0.918	1861 I
η Aquarids	83.0	43.1	160.0	126.1	5.0	0.91	0.47	Halley?
Daytime Arietids	29	76.8	21	106	1.6	0.94	0.09	(Second crossing of δ Aquarids)
Daytime ζ Perseids	59	77.8	0.4	137	1.6	0.79	0.34	—
Daytime β Taurids	246	276.4	6	162	2.2	0.85	0.34	Encke
Southern δ Aquarids	154	302	29.3	97.6	2.60	0.976	0.062	—
Northern δ Aquarids	332.6	138.9	20.4	111.4	2.62	0.973	0.070	—
Southern ι Aquarids	127.5	311.0	6.0	78.5	2.88	0.920	0.230	—
Northern ι Aquarids	308.0	150.9	4.7	98.9	1.75	0.842	0.265	—
α Capricornids	270.5	132.8	4.0	43.3	2.57	0.779	0.568	1948 n
Perseids	151.2	138.1	113.7	289.3	20.8	0.955	0.936	1862 III
$\varkappa$ Cygnids	204.2	144.3	37.0	348.4	4.09	0.762	0.973	—
Draconids	171.8	196.3	30.7	8.1	3.51	0.717	0.996	Giacobini-Zinner
Orionids	86.8	29.8	163.2	116.5	7.70	0.930	0.539	Halley?
Southern Taurids	111.9	45.1	5.4	156.9	2.30	0.835	0.380	Encke
Northern Taurids	298.4	221.8	3.2	160.2	2.14	0.849	0.323	Encke
Andromedids	242.4	224.4	6.0	107.8	3.34	0.776	0.748	Biela
Leonids	173.7	235.0	162.5	48.7	12.76	0.924	0.970	Temple
Geminids	324.3	261.2	24.0	225.6	1.39	0.899	0.140	—
χ Orionids	105.4	79.8	0.8	185.2	2.92	0.859	0.412	—
Monocerotids	128.2	81.6	35.2	209.9		1.002	0.186	—
Ursids	212.2	264.6	52.5	116.8	5.91	0.845	0.916	Tuttle

produced largely by the perturbations of the major planets, which act differently upon particles having slightly different initial orbits.

The finest example of a highly perturbed stream that is still clearly distinguishable is the Taurid stream complex, whose identity with the comet of shortest period, Comet Encke, was proved by one of the authors[1]. The nodes of some of these orbits have moved westwardly, under the perturbing effect of Jupiter, for some 1700 revolutions or about 5000 years. The inclinations vary over the enormous range from 3 to 17 degrees in half this period. A large toroidal volume of space has been populated by these meteoroids from Comet Encke; individual members are observed to strike the earth both at the ascending and descending nodes in October and November. As a consequence, the stream radiants are distributed along two great circles nearly symmetrically placed about the ecliptic, and persist for some two months. The simple perturbational theory for this

[1] F. L. WHIPPLE: Proc. Amer. Phil. Soc. **83**, 711 (1940).

motion has been greatly developed and extended by BROUWER[1] while calculations by WHIPPLE and HAMID[2] show details of the motions and history. Debris that is being scattered by Comet Encke at the present time cannot strike the earth until far in the future; thus we cannot measure the current loss of meteoroids by the comet. A striking confirmation of the great volume occupied by the Taurid meteorites was afforded through the radio discovery by CLEGG, HUGHES and LOVELL[3] of the same stream at a second crossing in July. The existence of summer Taurids, in the form of fireballs from the general direction of the Sun, had earlier been predicted on the basis of the dynamical structure of the stream. A second stream of this type is quite probably represented by the night-time δ-Aquarid meteors[4] and the daytime Arietids[5].

HAMID[6] has studied the perturbational motions of the night-time Perseid stream in August, and found that the observed scatter in radiants and other elements of these meteors is consistent with the original projection of these particles from the Comet 1862-III some 250 revolutions ago when its orbit crossed that of Jupiter. PLAVEC[7] finds that the perturbations of the Geminid stream are so rapid that the observable shower should have a relatively short life of perhaps the order of only a hundred years. WRIGHT, JACCHIA and WHIPPLE[8] have studied the radiant motions of a number of major streams from singly and doubly photographed Super-Schmidt meteor trails, and a similar study has been made from radio echoes by HAWKINS, BULLOUGH and DAVIDSON[9]. Both groups of investigators find that the daily motion of the radiant is always appreciable and generally amounts to the order of 1 degree per day eastwardly with a direction not too far from parallelism to the plane of the ecliptic. A notable exception is the radiant of the daytime Arietids which moves at an angle of about 30° to the ecliptic.

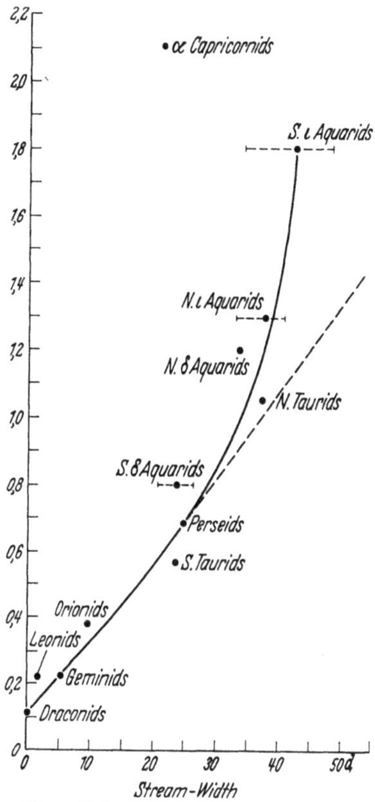

Fig. 7. Relation between duration of meteor streams and radiant diameter.

A statistical correlation is found between the duration of the stream and the diffuseness of the radiant, as shown in Fig. 7. No theoretical explanation for

[1] D. BROUWER: Astronom. J. **52**, 190 (1947).
[2] F. L. WHIPPLE and S. E. HAMID: Helwan Obs. Bull. No. 41, 1951.
[3] J. A. CLEGG, V. A. HUGHES and A. C. B. LOVELL: Monthly Notices Roy. Astronom. Soc. London **107**, 369 (1947).
[4] F. W. WRIGHT, F. L. WHIPPLE and L. G. JACCHIA: Astronom. J. **59**, 400 (1954).
[5] A. ASPINALL, G. S. HAWKINS et al.: Monthly Notices Roy. Astronom. Soc. London **111**, 18 (1951).
[6] S. E. HAMID: Formation and Evolution of the Perseid Meteor Stream. Doctoral Thesis, Harvard University 1950.
[7] M. PLAVEC: Nature, Lond. **165**, 362 (1950).
[8] F. W. WRIGHT, L. G. JACCHIA and F. L. WHIPPLE: Harvard Coll. Obs. Reprints II Nos. 16, 35, 38, 47, 67, 80, 108 (1947—1957).
[9] T. W. DAVIDSON: Jodrell Bank Ann. **1**, 116 (1956).

the relationship has been derived. Nevertheless, it is apparent that as a stream ages, it spreads out over a greater volume in space and the individual meteors crossing the earth's orbit at any point move in orbits that diverge increasingly in time. Such a phenomenon should of course be expected, because of the effects of dynamical aging, although a formulation of the complex perturbational relationships may present considerable theoretical difficulties. Dispersive effects due to physical causes are indicated by the radio results. HAWKINS and ALMOND[1] have shown that the percentage of bright meteors increases at the maximum of the Perseids and Geminids and that these streams contain a "core" of large, undisturbed particles. LINDBLAD[2] has presented evidence indicating that faint radio meteors of the δ-Aquarid stream are observed at an earlier date than the bright photographic meteors; he suggests that the separation may be due to the Poynting-Robertson effect.

On the basis of his icy model for comets, WHIPPLE[3] has suggested that several causes act to change the orbits of meteoroids in space and also to produce their physical deterioration and destruction. The most prominent cause of deterioration is probably mutual collisions, although atomic sputtering by corpuscular radiation from the Sun may play an important role. Other causes of deterioration include the photoelectric effect, encounters with ions or electrons in the interplanetary medium, and heating by solar radiation which may drive off residual gases as well as the primitive ices. ÖPIK[4] has investigated the sweeping effects of the planets in removing solid debris from the solar system; he finds that whereas insufficient time has elapsed for Mars to eliminate entirely material crossing its orbit, the earth would reduce this material by a factor of 1/e in some 10^8 years. Jupiter, of course, is much more effective in removing meteoric debris from the solar system. The gravitational perturbations of this planet are also severe, and Jupiter is largely responsible for the spreading of orbits in meteor streams.

It is possible that the general complex of sporadic orbits may represent the final stages of the dissolution of meteor streams. On the other hand, we must not neglect the destructive effects described above. For example, the correlation of meteor showers with heavy rainfall, as found by BOWEN, implies that small meteors are annihilated during the dynamic lifetime of the stream[5]. To understand in detail the relation between stream and sporadic meteors will be a difficult, and yet perhaps our most interesting, problem for future research.

11. Sporadic meteors and the total influx. Visual observations by ÖPIK, HOFFMEISTER, WATSON and others (see MILLMAN and BURLAND[6]) have established that the number of meteors, dN, appearing in the earth's atmosphere with visual magnitudes between M and dM is given by the incremental law

$$dN = a\,r^M\,dM \tag{11.1}$$

where a is a constant and r is the ratio of increase in the number of meteors between magnitudes M and $M+1$. In this paper dN will be used to define an influx rate of meteors passing through an area of 1 km² parallel to the earth's surface during a period of 1 hour, and M is the magnitude at maximum light.

[1] G. S. HAWKINS and M. ALMOND: Monthly Notices Roy. Astronom. Soc. London **112**, 219 (1952).
[2] B. LINDBLAD: Trans. Chalmers Univ. of Technology, Gothenburg, Sweden, No. 129, 3, 1952.
[3] F. L. WHIPPLE: Astrophys. Journ. **121**, 750 (1955).
[4] E. ÖPIK: Proc. Roy. Irish Acad. **54**, 165 (1951).
[5] F. L. WHIPPLE and G. S. HAWKINS: Amer. J. Meteorol. **13**, 236 (1956).
[6] P. M. MILLMAN and S. BURLAND: Sky and Telescope **16**, 222 (1957).

The value of r appears to be constant over a wide range of magnitudes. A similar law has been shown to hold for the photographic magnitude scale. The cumulative law, giving the total number of meteors brighter than magnitude M which pass through 1 km² per hour, is found by integrating Eq. (11.1) between the limits of $-\infty$ and M:

$$N = a\, r^M / \log_e r \quad (\text{km}^{-2}\ \text{hour}^{-1}). \tag{11.2}$$

Thus

$$\left.\begin{aligned} \log_{10} N &= \log_{10} a - \log_{10}(\log_e r) + M \log_{10} r, \\ \log_{10} N &= N_0 + M \log_{10} r. \end{aligned}\right\} \tag{11.3}$$

N_0 is the logarithm of the influx rate for meteors with $M \leqq 0$ and r is the ratio of increase given by the incremental law. Eq. (11.3) can be written with the variable M replaced by photographic magnitude M_{pg}, the meteoroid mass m, or the electron line density in the meteor trail q. In observational work there are two problems; to determine the ratio of increase r, and the absolute rate N_0.

There is considerable uncertainty in the ratio r, determined for sporadic meteors; published values vary between 2.0 and 4.1 (MILLMAN and BURLAND)[1]. Part of this variation may be attributed to an actual decrease in the ratio for faint meteors, and indeed r must finally tend to unity since the number of meteors in the solar system is finite. Part of the variation, on the other hand, is due to uncertainties in the observations and in the correction factors involved. The ratio, r, has been determined from radar observations by MCKINLEY and by BROWNE, BULLOUGH, EVANS and KAISER[2]. Both visual and radar work provide evidence that r has different values for sporadic meteors and for meteors belonging to well-defined streams. The absolute influx rate, N_0, has been determined visually by WATSON[3] who found that 75×10^6 meteors with $M \leqq +5$ were incident on the earth during a period of 24 hours. MILLMAN and BURLAND[1] revised the number to 200×10^6 but HAWKINS[4] by radar methods finds a value in good agreement with that of WATSON.

The photographic method is not subject to the uncertainties introduced by the variation of sensitivity across the observing field, and HAWKINS and UPTON[5] have deduced the following influx rates from the Super-Schmidt data of the Harvard Meteor Project:

$$\log_{10} N = 0.538\, M_{pg} - 4.34, \tag{11.4}$$

$$\log_{10} N \approx 0.538\, M_v - 5.17, \tag{11.5}$$

$$\log_{10} N = 15.95 - 1.34 \log_{10} q, \tag{11.6}$$

with N in km⁻² hour⁻¹, where M_{pg} is the photographic magnitude, M_v is the visual magnitude and q is the electron line density (cm⁻¹) in the trail. Eqs. (11.4) and (11.5) indicate an increase in numbers of 3.44 per magnitude, and the uncertainty in this value is ± 0.2. The observations show that 90 million meteors brighter than 5th magnitude visual hit the atmosphere of the earth in 24 hours.

The meteor streams described in Sect. 10 represent only a fraction of the meteoric debris falling on the Earth. For example, the percentage of meteors

---

[1] See footnote 6, p. 548.

[2] I. E. BROWNE, K. BULLOUGH, S. EVANS and T. R. KAISER: Proc. Phys. Soc. Lond. B **69**, 83 (1956).

[3] F. G. WATSON: Between the Planets. Cambridge: Harvard University Press 1956.

[4] G. S. HAWKINS: Astronom. J. **61**, 386 (1956).

[5] G. S. HAWKINS and E. K. L. UPTON: Astrophys. Journ. **128**, 727 (1958).

that do not belong to the major streams as determined by various techniques is shown in Table 8. Even when the streams are well defined and the meteors

Table 8. *Percentage of meteors not belonging to major streams.*

Technique	Limiting visual magnitude	Percentage	Authority
Visual	+5.5	63%	Denning[1]
Photographic . . .	+0	58%	Whipple[2]
Photographic . . .	+4.5	83%	Hawkins and Southworth[3]
Radio	+4.2	95%	Hawkins[4]

relatively easy to identify this fraction varies considerably. A possible explanation of the variation is that the distribution of masses is different within the streams and that each technique acts selectively on a particular range of masses. Historically, the policy has been to classify any meteor that could not be identified with a stream as "sporadic", and Table 8 gives the percentage of sporadic meteors on this basis. However, the more accurate photographic technique shows that a considerable fraction of sporadic meteors may be associated in small groups which may be conveniently termed minor streams. At the time of writing, the unravelling of the minor streams has only just begun, and it is therefore expedient to classify all meteors outside the major streams as sporadic and adopt a statistical approach with the reservation that our studies must inevitably include many minor streams.

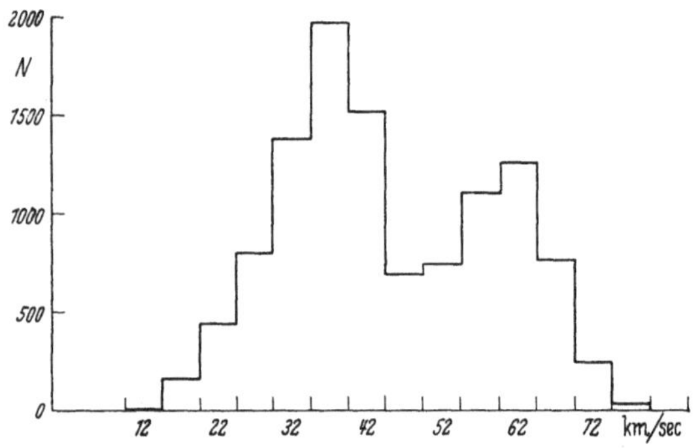

Fig. 8. Velocity distribution of 11073 meteors.

For many years it was considered that sporadic meteors were interstellar in origin and travelled through the solar system in hyperbolic orbits. A simple test of this thesis is to measure the velocity of sporadic meteors, since the velocity of escape from the Sun at a distance of 1 a.u. is known (42.2 km/sec). Observations by Whipple[2] of several bright sporadic meteors at the commencement of the photographic program showed that these meteors, at least, were moving in short-period elliptical orbits. Of some 500 well determined photographic meteor orbits none have yet been found to be certainly hyperbolic. A detailed study of the velocities of sporadic meteors was made by Almond, Davies and Lovell[5] using the radio technique in a series of six experiments in which the sensitivity of the equipment was successively increased from a limiting magnitude of +6 to +9. Over 1000 velocities were measured and it was shown that the number of hyperbolic meteors must be considerably less than 1% of the total. In Fig. 8 the velocity distribution for 11073 meteors is given, as determined by McKinley[6]

[1] W. F. Denning: Obs. Handbook, Brit. Ast. Ass. 1922.
[2] F. L. Whipple: Astronom. J. **59**, 201 (1954).
[3] G. S. Hawkins and R. B. Southworth: Smithson. Contrib. Astrophys. **2**, 349 (1958).
[4] G. S. Hawkins: Monthly Notices Roy. Astronom. Soc. London **116**, 96 (1956).
[5] M. Almond, J. G. Davies and A. C. B. Lovell: Monthly Notices Roy. Astronom. Soc. London **113**, 411 (1953).
[6] D. W. R. McKinley: Astrophys. Journ. **113**, 255 (1951).

who used a cw transmitter in contrast to the pulsed radar of ALMOND *et al.*[1]. It can be seen that very few meteors were observed with velocities greater than 72 km/sec, which is the geocentric velocity of a parabolic meteor with radiant at the apex. On this evidence we may conclude that interstellar meteors are

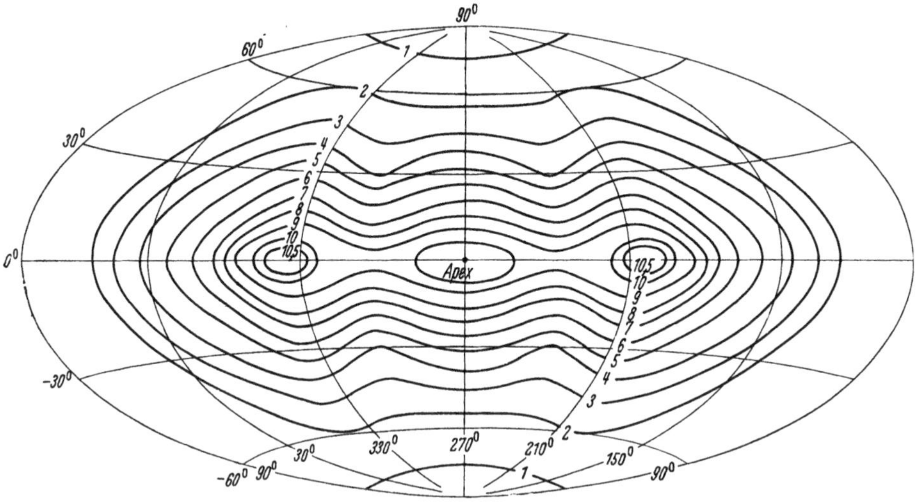

Fig. 9. Contours of equal radiant density for sporadic meteors. Radar observations.

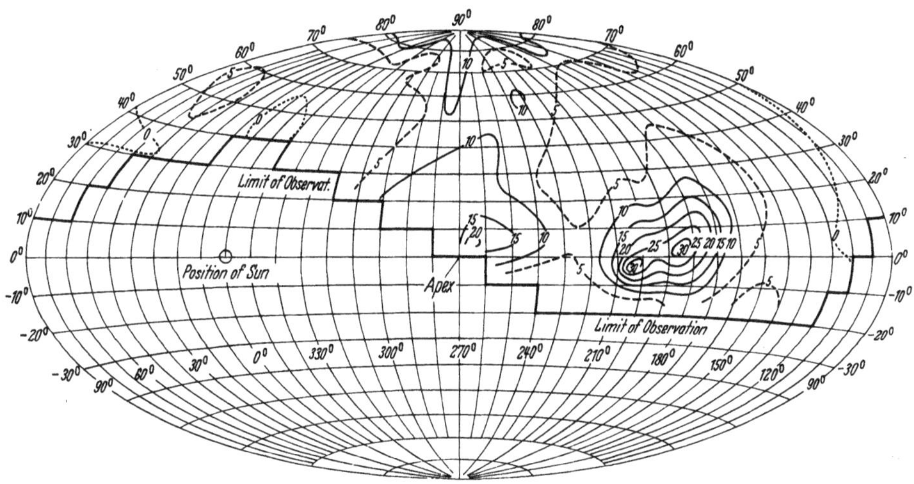

Fig. 10. Same as Fig. 9. Visual observations.

extremely rare within the present range of detection. There is still the possibility of detecting hyperbolic meteors below a magnitude of $+9$ where the dimensions of the meteor are of the same order as the grains of interstellar dust, and, of course, we may expect to find a few larger meteors which have been thrown into hyperbolic orbits by planetary perturbations.

A statistical study of the radiant distribution of sporadic meteors has been made by one of the authors[2] using a radar technique. The number of sporadic

[1] See footnote 5, p. 550.

[2] G. S. HAWKINS: Monthly Notices Roy. Astronom. Soc. London **116**, 92 (1956). — Astronom. J. **61**, 386 (1956).

radiants was determined in a strip lying along the prime meridian of the observing station and the distribution of radiants was sampled once during a period of 24 hours as the earth rotated. An analysis of 240000 echoes showed a concentration of radiant points towards the plane of the ecliptic, and a marked asymmetry in the distribution of meteor orbits. The mean radiant distribution determined between the period October 1949 to September 1951 is shown in Fig. 9, where contours of equal radiant density are plotted on an equal-area projection of the sky centered at the apex of the Earth's way. The horizontal coordinates are circles at ecliptic latitudes 0°, ±30° and ±60°. The vertical circles are drawn at 60° intervals in ecliptic longitude with the position of the Sun taken as 0°. Thus the radiant pattern is fixed with respect to the Sun and there are three concentrations of radiant points, — at the apex, near the Sun and near the

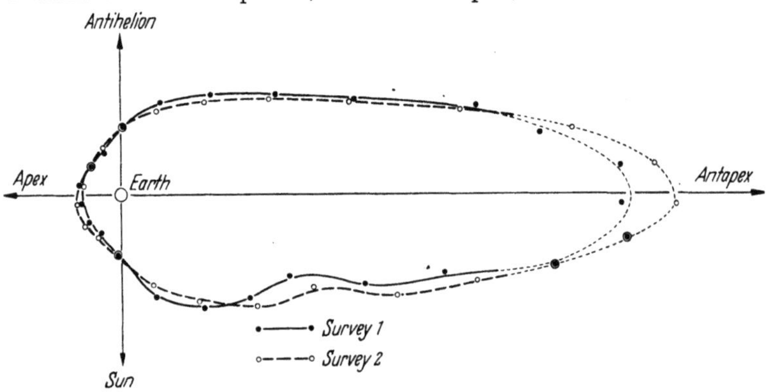

Fig. 11. Distribution of directions of meteors in space, calculated from the results of two radio echo surveys.

antihelion point. The concentration at the apex corresponds to the one predicted by classical theory and is produced by the sweeping action of the Earth as it moves round its orbit. The other concentrations correspond to meteors moving in short-period direct orbits which are aligned closely to the plane of the ecliptic.

A similar measurement of the distribution of sporadic radiants has been made from visual observations by HAWKINS and PRENTICE[1], who determined the radiant point of a meteor by triangulation with an accuracy of approximately ±1°. The results of this analysis are given in Fig. 10 where contours of equal radiant density have again been plotted. It can be seen that the visual results confirm the grouping of radiants at the antihelion point and also show a more diffuse concentration of radiants at the apex. The observations are incomplete, of course, because of the limitations of daylight, the group of radiants near the Sun being unobservable. It must be emphasized that both the radio and visual radiant distributions represent an integration taken over a period of one year, and may show marked variations from night to night. Such irregularities are undoubtedly present and their study will undoubtedly lead into the realm of minor streams.

It is interesting to deduce the distribution of directions of meteors in space from the observed distribution of radiant points. When corrections are applied to allow for the velocity of the Earth relative to the Sun, the radio echo results produce the distribution shown in Fig. 11. In this figure the Earth is shown at the center of a polar diagram and the length of the radius vector is proportional to the number of meteors crossing the Earth's orbit at a given polar angle. Clearly there is a preponderance of meteors overtaking the Earth, the ratio being close

[1] G. S. HAWKINS and J. P. M. PRENTICE: Astronom. J. 62, 234 (1957).

to 30:1. A similar correction may be applied to the visual data. With these data each radiant point can be treated separately and it is permissible to assume that the heliocentric velocity of the meteor was 42.2 km/sec. We can then construct a direction diagram for meteors with orbits at various inclinations i to the plane of the ecliptic as shown in Fig. 12. The visual results confirm the preponderance of directly moving meteors for orbital inclinations less than 50°, the ratio of direct to retrograde orbits being 50:1. Orbits of high inclination, on the other hand, are uniformly distributed in direction, the cardioid pattern in the polar diagram being due to the variation in the area of the sampling element.

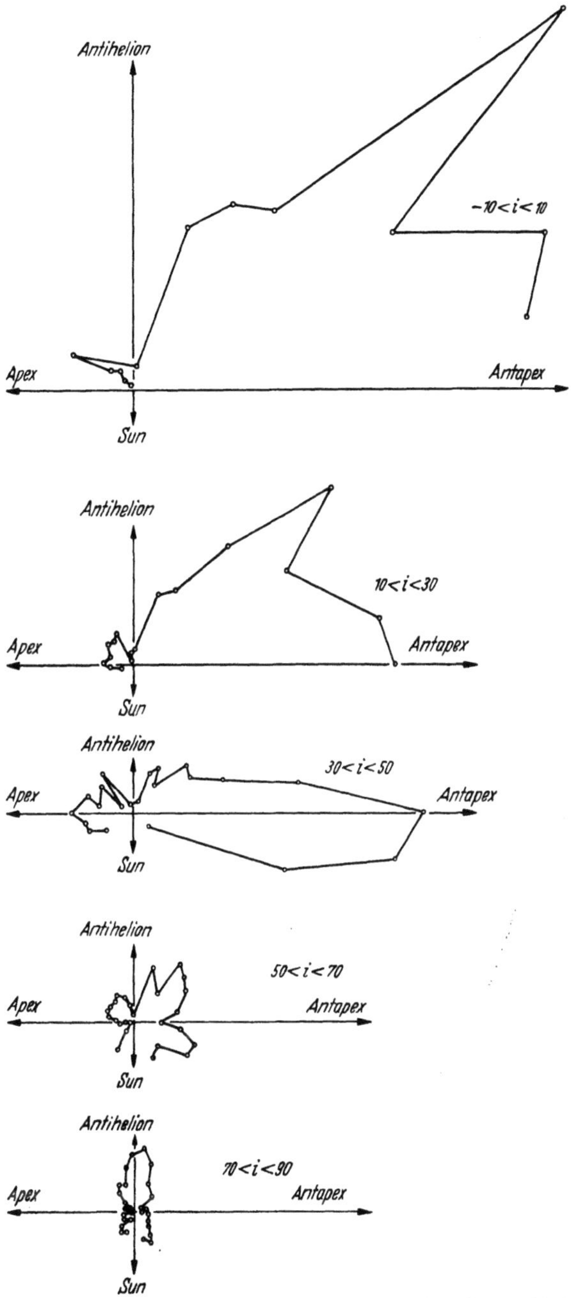

Fig. 12. Distribution of directions of meteors in space for different orbital inclinations.

The photographic technique provides us with accurate data that may enable us to discover the life history of many individual sporadic meteors. WHIPPLE[1] has published detailed orbits for 144 meteors, more than half of which fall into the category of sporadic or minor stream meteors, and WHIPPLE and JACCHIA have obtained further data on 308 meteors which has not yet been published. They find that a distinction can be drawn between directly moving and retrograde orbits. The direct orbits tend to have short periods, high eccentricities and low inclinations, resembling very closely the orbits of short period comets. On the other hand, retrograde orbits have a longer period and a more uniform distribution of inclinations, resembling the distribution of long period comets. Almost every

[1] F. L. WHIPPLE: Astronom. J. **59**, 201 (1954).

meteor orbit has a positive K criterion and we may infer from this evidence that the bulk of sporadic meteors have a cometary origin. This will be discussed at greater length in Sect. 12.

12. Origin of meteors, fireballs and meteorites. In Sect. 10 and 11 we have seen that hyperbolic orbits, if present at all, occur at rates below the 1% level among photographic meteors and perhaps below the 0.5% level among radio meteors, the latter extending down to magnitudes of the order of +8. Thus, from magnitudes −3 to +8 the existence of *any* interstellar meteors requires additional proof. In this range our problem is to distinguish, if possible, between contributions by cometary debris and those by asteroidal chips. In the latter part of this section the general problem will be considered for brighter fireballs.

Among the shower meteors some twelve streams are definitely associated with comets so that their origin is clearly delineated (see Table 7, Sect. 10). A question arises as to whether any of the showers, particularly those with asteroidal-like orbits, can be of asteroidal origin. This suggestion has been implied by Hoffmeister[1], on the basis of his visual estimates of velocities and orbits for a number of such showers. Unfortunately, Hoffmeister's extremely ingenious method of determining velocities by the visual nature of the trails tends to yield velocities that are too small, and hence only 3 streams remain within this category on the basis of the precise photographic studies: the Geminids, $q'=2.6$, the Virginids, $q'=5.2$, and the Taurids, $q'=4.0$. It seems safe to assume that stream orbits with aphelion in the neighborhood of Jupiter's orbit or beyond it are of cometary origin. Hoffmeister's suggestion might be somewhat more tenable were the Taurid meteors not clearly associated with Comet Encke. Meteoroid density and physical characteristics in the atmosphere appear to provide the best criteria for settling this question.

If, indeed, meteoroids of cometary origin have densities as low as the results of Sect. 9 indicate ($\sim$0.1 gm/cm^{-3}), then the densities for asteroidal particles (3.5 or 8.0 gm/cm^3) should produce conspicuous effects in measurements of drag and calculated atmospheric densities among the doubly-photographed meteors. Only for the Geminids do we have sufficient data to approach this problem with confidence. Here we do find that the Geminid meteors deviate significantly from the average. Jacchia[2] has shown that their specific densities are roughly 2.4 ϱ_m where ϱ_m is the density of average meteors; or alternately, that their luminous efficiencies are about an order of magnitude greater than those of average meteors. We should expect that the density ratio should be greater than 2.4 were the Geminids asteroidal, but we cannot be certain about luminous efficiency.

A density of 2.4 ϱ_m and the peculiarly small perihelion distance, 0.14 a.u., of the Geminid orbit are not incompatible with the general concept of a cometary origin. If the Geminids originated from a comet with an aphelion point which has slowly contracted from Jupiter's orbit, the comet must originally have been of relatively great dimensions to have survived. Geminid meteors must then represent material from the central nucleus of an unusually large comet, quite possibly greater than that of the original Encke comet. We should not be surprised, then, that the fragments should have a density somewhat greater than is typical for cometary meteors. In addition, excessive heating by the Sun may have compressed the original dust balls, although this effect is not evident for the δ Aquarids, $q=0.06$ a.u.

[1] C. Hoffmeister: Meteorströme. Weimar and Leipzig 1948.
[2] L. G. Jacchia: Unpublished.

It is interesting to note that JACCHIA[1] finds the Taurid meteoroids to be of unusually tough, non-fragmentable material although their densities are typical of average meteors. Data for the Virginid meteors are not adequate to establish the density relationship, although the few available cases show no marked abnormalities. We may conclude, therefore, that the evidence strongly indicates, although it does not quite prove, the hypothesis that all shower meteors are indeed of cometary origin.

Among the orbits of sporadic photographic meteors[2] some 20% show the characteristic of long-period comet orbits, that is, great aphelion distance and a random distribution of orbital inclinations. A cometary origin for these orbits seems obvious.

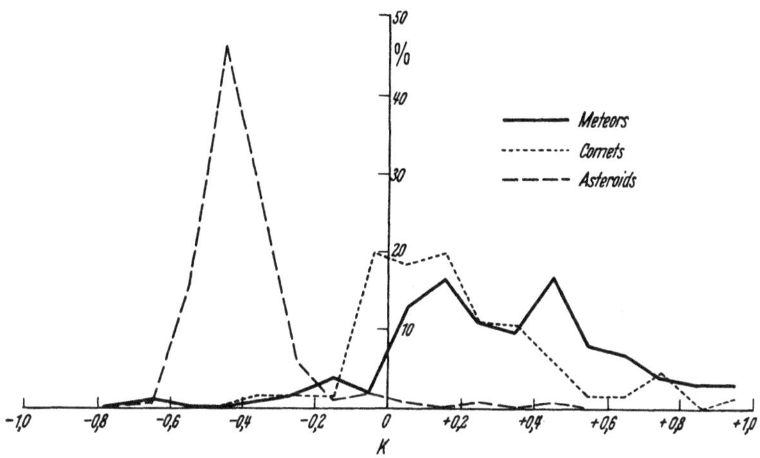

Fig. 13. Frequency of the K criterion for meteors, comets and asteroids.

Among the remaining sporadic meteors, the orbits are only moderately inclined to the plane of the ecliptic, much like those of short-period comets ($i < 30°$ for 80%). Some 65%, however, have aphelion distance well within Jupiter's orbit. This makes the hypothesis of cometary origin less certain because only some 5% of short-period comets have such small aphelion distances. It is probably significant, however, that only three of the first 308 Super-Schmidt orbits studied by WHIPPLE and JACCHIA[3] have aphelia within the orbit of Mars and two of these practically at the Earth's orbit. All seven asteroids with perihelion within the Earth's orbit move beyond the orbit of Mars (see Table 3 on p. 529).

To help solve this problem, WHIPPLE[2] set up an arbitrary K criterion as a single quantity to distinguish between asteroidal and cometary orbits. The criterion is defined as follows:

$$K = \log_{10}\left[\frac{a(1+e)}{1-e}\right] - 1,$$

where the bracketed quantity is proportional to the inverse square of aphelion velocity. Fig. 13 presents the distribution function of K in the range of interest for asteroids, photographic meteors, and the short-period comets with $q < 1$ a.u.

It is immediately evident from Fig. 13 that the photographic meteors show a distribution of the K criterion that is more characteristic of the periodic comets with perihelion distance of $q < 1$ a.u. than of the asteroids. There is, however,

[1] L. G. JACCHIA: Astrophys. Journ. **121**, 521 (1955).
[2] F. L. WHIPPLE: Astronom. J. **59**, 201 (1954).
[3] F. L. WHIPPLE and L. G. JACCHIA: Astronom. J. **62**, 37 (1957).

a region of overlap, and one must admit that the K criterion has no basic physical significance. On the other hand, it is a quantitative statement of the general orbital characteristics that are known to distinguish asteroids from comets.

It may be argued that among the photographic meteors are a great many of asteroidal origin whose orbits are so like those of the short-period comets that there is no method of distinguishing them. JACCHIA[1] has segregated from the Super-Schmidt photographic meteors those whose orbits were of particularly asteroidal character. For this class of orbit, the atmospheric densities have been calculated from the observed decelerations and light curves and compared with the values obtained from the remaining meteors. There is no significant difference in these values, whereas a difference in density of even 20 or 30%, or a similar difference in the luminous efficiency, would have shown up strongly as measured by the interaction of the bodies with the upper atmosphere. A thorough search of differences between the physical characteristics of meteors in this "asteroidal" group as compared to the remaining meteors of more clearly cometary origin indicates no difference at the measurable level. If asteroidal meteors are in fact present among the sample of photographed meteors, they number at most a very few per cent, probably fewer than 3%.

It has been shown by JACCHIA[2] that significantly different characteristics are indicated for the major meteor showers as compared to the average, each shower showing a tendency towards a peculiar density (or luminous efficiency) or fragmentation index. We must conclude, therefore, that the photographic meteors, when separated according to orbital characteristics, form homogeneous groups in physical character and, since most of the streams have been identified with comets, the remaining streams probably have a similar origin.

Very bright fireballs and stone-dropping fireballs, on the other hand, show no tendency to move with radiants of known showers nor do their occurrence frequencies throughout the year tend to show peaks coincident with the known showers. Hence, there is no *a priori* reason to associate extremely bright fireballs and meteorites with cometary streams.

Visual observations of velocities, radiants and orbits, however, have tended to indicate that the brighter fireballs and meteorites have heliocentric velocities very frequently of hyperbolic character[3]. VON NIESSL and HOFFMEISTER in their famous catalog of fireballs, find 79% with hyperbolic heliocentric velocities. Other investigators such as ÖPIK[4] and ASTAPOVICH[5] have confirmed this conclusion. Of contrary mind has been WYLIE[6], who made an extraordinary effort to remove the systematic errors from visual observations of fireball velocities. By joining with the observers in reenacting the circumstances of the phenomenon and making on-the-spot measurements of angles, trajectories and time intervals, he discovered the rather surprising result that angular measurements are grossly overestimated while time measurements are fairly reliable when made by unprepared observers. As a consequence, he concludes that the orbits of meteorites and bright fireballs are essentially asteroidal in character, the atmospheric velocities lying well below the hyperbolic limit.

---

[1] L. G. JACCHIA: Unpublished.

[2] L. G. JACCHIA: Astronom. J. **61**, 6 (1956).

[3] VON NIESSL and C. HOFFMEISTER: Denkschr. Akad. Wiss. Wien, math.-naturwiss. Kl. 1925, 100.

[4] E. ÖPIK: Harvard Coll. Obs. Circ. No. 389 and 391, 1934.

[5] I. S. ASTAPOVICH: Russ. Astronom. J. **16**, 41 (1939).

[6] C. C. WYLIE: Univ. Iowa Obs. Contr. No. 8, 1939. — Science, Lancaster, Pa. **90**, 264 (1939).

Long ago, NEWTON[1] studied the radiants and their distribution with respect to the apex of the Earth's motion; he concluded for 116 stone falls that the orbits were of solar-system origin, with inclinations less than 90°.

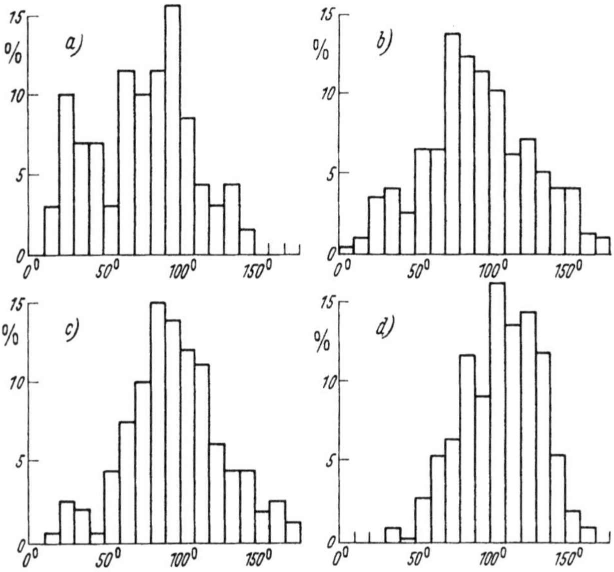

Fig. 14a—d. Distribution of elongations of geocentric radiant from the Earth's apex. (a) 70 photographic meteors (b) 325 fainter fireballs, (c) 160 great fireballs, (d) 89 detonating bolides.

WHIPPLE and HUGHES[2] reinvestigated the problem with more recent radiant data for meteorite falls and found the distribution of elongations from the apex as seen in Figs. 14a—d and Fig. 15a and b for the sequence: photographic

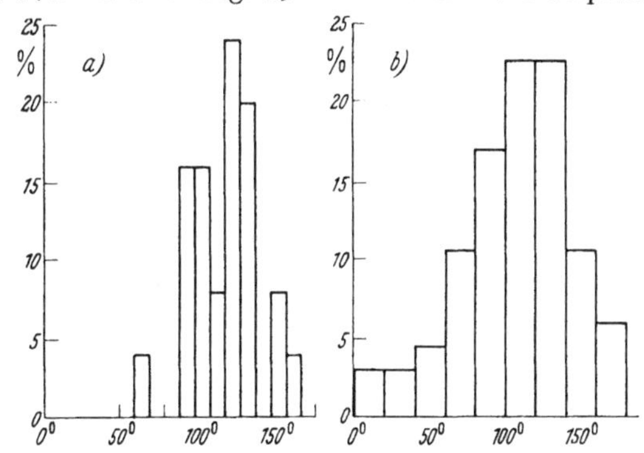

Fig. 15a and b. Same as Fig. 14. (a) 25 meteorites, (after WHIPPLE and HUGHES) (b) 66 meteorites (after ASTAPOVICH).

meteors, fainter fireballs, great fireballs, detonating bolides, and meteorite falls. The elongations given are from the apex of the Earth's motion to the apparent radiant, without correction for the zenith attraction. By comparison with the velocity-elongation diagram (Fig. 16) for photographic meteors, they deduce the mode, mean, and root-mean-square velocities for bolides and fireballs as given in Table 9. They adopt a mean meteorite velocity of 16.5 km/sec and 17 km/sec

[1] H. A. NEWTON: Amer. J. Sci. **36**, 1 (1888).

[2] F. L. WHIPPLE and R. F. HUGHES: J. Atmos. Terr. Phys., Spec. Suppl. **2** (1955).

for the root-mean-square velocity in the Earth's upper atmosphere. Wylie[1] from his corrected observations for meteorites derived a mean velocity of 16.0 km/sec, in good agreement.

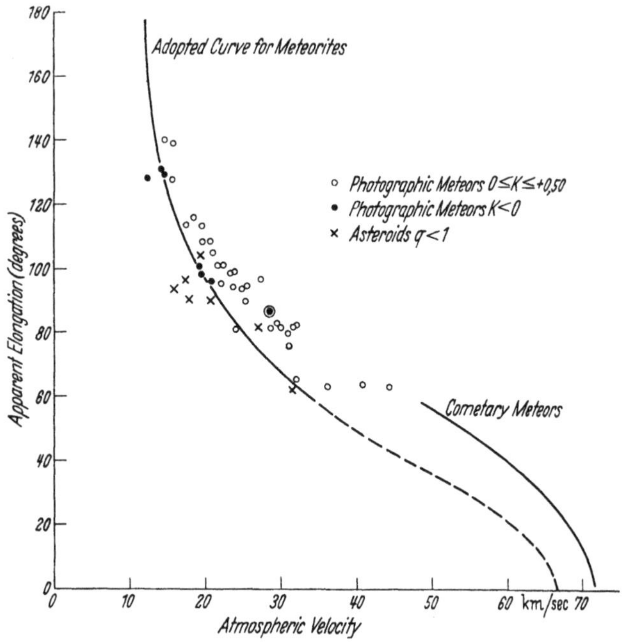

Fig. 16. Velocity—elongation diagram for photographic meteors.

Since the radiants of bolides and meteorites are highly concentrated to the plane of the ecliptic as shown by Watson[2], it is quite impossible to disagree with Newton's early conclusion that meteorites are of solar-system origin. Their

Table 9. *Atmospheric velocities (km/sec).*

Phenomenon	Mode	Mean	Root-mean-square
89 Bolides	18.0	19.3	20.3
25 Meteorite falls	14.8	16.5	16.9
66 Meteorite falls	15.4	21.3	24.3

orbits are distinctly asteroidal in character and hence practically no doubt remains that meteorites are samples of asteroidal material; whether they are typical or not requires more consideration.

In arranging the sequence from meteorites to photographic meteors in terms of brightness, one asks at what brightness-level occurs the transition from cometary meteors to asteroidal fireballs. Probably the transition occurs within a few magnitudes of -8 although the data are inadequate to separate the two groups well in magnitude. The ratio, r, has been defined (Sect. 11) as the increase in the number of meteors from an increase of 1 magnitude. Thus if the proportion of cometary meteors decreases as we proceed to the region of fireballs and meteo-

[1] C. C. Wylie: Unpublished. Private communication.
[2] F. G. Watson: Proc. Amer. Phil. Soc. **81**, 473 (1939).

rites we would expect the value of r to decrease. MILLMAN and BURLAND[1] have presented evidence to show that r is 3.7 for meteors with visual magnitude M_v given by the expression $0 < M_v < -6$ and for fainter meteors, $M_v < 5$, $r = 2.5$. This evidence is not supported by an analysis of photographic images[2] which yields $r = 3.4$ over the range $-0.5 < M_v < 4$. Nor is it consistent with the results of an analysis of 100000 meteors of the American Meteor Society[3] which gives $r = 2.5$ for $0 < M_v < -6$. No data are available to estimate this ratio for the bright fireballs and meteorite falls because magnitudes are difficult to determine. If the visual[3] and photographic[2] work is correct the ratio r changes from 3.4 to 2.5 between $M_v = 2$ and $M_v = -3$. This would indicate that somewhere in the neighborhood of $M_v = -3$ we are observing a falloff in the percentage of cometary meteors in the influx and that relatively few contribute to extremely bright fireballs and none to meteorite falls.

We may conclude, therefore, that radio and visual meteors are almost entirely of cometary origin and that detonating bolides and meteorites are perhaps entirely of asteroidal origin. We note that the frequency distribution in terms of meteoroid radius follows approximately an inverse fourth power for radio, photographic and visual meteors (to 3rd magnitude) while PIOTROWSKY[4] calculates an inverse third power law for asteroidal particle production by grinding among the asteroids. This law holds for the very brightest asteroids but cannot be confirmed observationally over a large range in dimension. If we assume that these two relationships are indeed representative over large ranges, we see then that the asteroidal source compared to the cometary source should fall off with decreasing radius, in keeping with our conclusion above regarding the origin. If we extrapolate to the range of micrometeorites or meteoritic dust in interplanetary space, we can conclude that for all practical purposes this fine material is entirely of cometary origin, in keeping with WHIPPLE's[5] hypothesis concerning the origin of the zodiacal cloud materials (Sect. 4). This argument for the micrometeorites, however, must be taken as only qualitative in character until we have more evidence to demonstrate the physical and orbital characteristics of the micrometeorites.

13. Meteorites. Since the year 1803, when the French Academy was forced to acknowledge the cosmical origin of meteorites by BIOT's study of a fall of stones at L'Aigle, France, the subject has been an accredited scientific endeavor. The literature is enormous, particularly with regard to detailed descriptions, and includes many studies of some 1600 meteorite falls and finds. Of these about half were seen to "fall", and the other half have been "finds" of meteorites whose fall was not observed. The meteorites of a fall or find, usually single but frequently multiple, are identified by the name of the geographic locality in which they were discovered. A symbolic system of identification has been developed by LEONARD[6,7]. The present account of meteorites will be quite abbreviated, but the reader is referred to a few of the major accounts and more recent summaries. Meteorites are readily classified into three major groups: *irons* or *siderites*; *stony irons* or *siderolites*; and *stones* or *aerolites*. A fourth group, *tektites*, glassy in structure and frequently button or tearshaped, are found irregularly scattered

[1] P. M. MILLMAN and S. BURLAND: Sky and Telescope **16**, 222 (1957).
[2] G. S. HAWKINS and E. K. L. UPTON: Astrophys. Journ. **128**, 727 (1958).
[3] C. OLIVIER and G. S. HAWKINS: Unpublished.
[4] S. L. PIOTROWSKY: Acta Astronomica, Ser. A **5**, 115 (1953).
[5] F. L. WHIPPLE: Astrophys. Journ. **121**, 750 (1955).
[6] F. C. LEONARD and R. D. VIOLINI: Univ. Calif. Publ. Astr. **2**, No. 1, 1 (1956).
[7] F. C. LEONARD and G. L. ROWLAND: Meteoritics **1**, 440 (1956).

over the surface of the Earth, and are not certainly of cosmic origin. SUESS[1] has recently shown from the argon content that tektites generally have been solid for less than 10^7 years. Since this significant observation still does not clarify the problem of cosmic or terrestrial origin for tektites, we shall not discuss them further in this article.

Meteoritic stones weather rapidly and are often indistinguishable superficially from terrestrial rocks. Hence among the unobserved meteoritic finds, stones are relatively rare but predominate among the observed falls. They are generally estimated to constitute some 90% of the meteoritic mass striking the Earth. The stony irons are relatively rare, constituting some 2.5% of the total finds and falls, and often give the appearance in cross-section of raisin pudding, the raisins being stony inclusions in an iron-nickel matrix. The irons are composed almost completely of iron-nickel alloy with occasional inclusions of stony material. The three major types of meteorites will be discussed briefly from a chemical and metallographic point of view in terms of general composition and structure in the sequence, irons, stones and stony irons.

Besides iron and nickel, the average iron meteorite contains about 0.6% cobalt, 0.2% phosphorus, a few hundredths of a percent of sulphur, carbon, copper and chromium with only traces of the other elements which altogether add to less than half a percent. When an iron meteorite is sliced, polished and etched, it usually shows a fairly coarse and often times artistic surface structure known as the Widmanstätten figures. The nature of these figures is highly dependent upon the nickel content as noted by PFANN[2] and elaborated by PERRY[3]. The different crystalline structures within the pattern differ markedly in composition (see Fig. 17). The component of low nickel content, approximately 6%, is commonly known as *kamacite*, a body-centered cubic crystalline structure in the alpha phase of the iron-nickel alloy. The high nickel component, *taenite*, is of variable content but averages about 30% nickel and is a face-centered cubic structure in the gamma phase of the alloy. A mixture of finely divided kamacite and taenite, called *plessite*, constitutes a third crystalline structure in iron meteorites. The irons of low nickel content, about 5.5%, include as a major group the relatively uncommon hexahedrites manifesting a hexagonal crystalline structure. The most common variety of iron is the octahedrite ranging from 6 to 13% nickel.

Two other structures frequently observable in the polished and etched surfaces of meteorites are the stony inclusions and the Neumann bands or lamellae. Frequent inclusions are iron carbide, *cohenite*, iron sulfide, *troilite* and iron-nickel phosphide, *schreibersite*. Occasional graphite nodules and rare microscopic diamonds are found. The Neumann bands are of an entirely different character; they give the appearance of straight lines or parallel scratches over large areas of the meteoritic sample. Microscopic examination shows the bands to be twin thin lines, which can be formed in body-centered cubic iron crystals as a consequence of severe shock or rapid external stress. The mode of formation by mechanical stress is quite temperature sensitive, ranging from relatively slow deformation at liquid air temperatures, to projectile-type shocks at room temperatures, to increasingly high shocks at higher temperatures, until finally at sufficiently high temperatures the phenomena cannot be produced. In a general account of

---

[1] H. E. SUESS: Geochim. et Cosmochim. Acta **2**, 76 (1951).
[2] E. PFANN: Int. Z. Metallogr. **9**, No. 2, 65 (1917).
[3] S. H. PERRY: The metallography of meteoric iron. Bull. U.S. Nat. Museum, No. 184, 1, 1944.

this phenomenon, UHLIG[1] concludes that the Neumann bands "were formed previous to the time the meteorite entered the Earth's atmosphere".

Since the crystalline structure of the Widmanstätten figures cannot be duplicated by artificial means, or found naturally in the Earth except on a microscopic scale, it is generally concluded that the complex structures found in the

Fig. 17. Typical surface patterns on meteorites.

iron-nickel meteorites must have been induced by slow cooling, possibly accompanied by extremely great pressure. With prolonged heating at temperatures of the order of 900 to 1000° C, the structures disappear in meteoritic irons and cannot be restored in the laboratory. The difficulties in the laboratory of reproducing the time scales and pressures likely to have existed when the meteorites were formed, and the difficulties of the theory have made reconstruction of the circumstances of pressure and cooling extremely uncertain. In the summary of the subject, UHLIG[1] concludes that the Widmanstätten pattern was perhaps

[1] H. H. UHLIG: Geochim. et Cosmochim. Acta **6**, 282 (1954); **7**, 34 (1955).

achieved at temperatures as low as 300 to 400° C and that both temperature rises and rapid changes of pressure may also play a role in the more detailed structure formation.

The stony meteorites present such a complicated internal structure, both chemically and physically, that no detailed description can be given here. A minority of about 10%, called *achondrites*, are silicates not too dissimilar from terrestrial basic rocks. Although the chemical composition is complex and varies markedly from region to region within an achondrite and from one to another, the material is 90 or more percent silicate and more than one half silica (SiO_2).

The *chondritic* stones are the most prevalent and are characterized by the presence of chondrules, inclusions which are somewhat similar in chemical composition to the achondritic stones. The chondrites, however, contain varying degrees of iron-nickel in the metallic phase up to as much as 35%, and the background structure of broken irregular crystals is so chaotic and complex that detailed mineralogical analysis becomes practically impossible. With regard to the systematic composition of stones, PRIOR[1] found that the percentage of iron in the silicates varied in an inverse fashion with the percentage of iron in the metallic phase. The metal phase in the stones shows much the same composition and range as in the irons. UREY and CRAIG[2] and WAHL[3] find that although the internal structure and composition of chondritic meteorites is extremely heterogeneous, the average composition is surprisingly uniform. Table 10 shows the atomic abundances of the elements in chondrites as compiled by UREY and CRAIG for two groups, high and low in iron content and the average of all. In view of the amazing irregularity in the chondritic structure, the fact that the main compositions of extreme groups are so nearly uniform represents a mixing phenomenon of extraordinary power.

Table 10. *Atomic abundances of the elements in chondrites (silicon = 10000).*

	Low iron	High iron	Δ %	Total average
Oxygen	34760	33882	− 2.6	34386
Sodium	509	490	− 3.8	500
Magnesium	9261	9444	+ 2.0	9335
Aluminum	778	769	− 1.2	774
Silicon	10000	10000	—	10000
Phosphorus	47	42	− 12.6	45
Sulphur	1000	1100	+ 10.0	1041
Potassium	58	59	+ 1.1	58
Calcium	532	595	+ 12.0	558
Titanium	21	23	+ 8.8	22
Chromium	87	63	− 37.9	76
Manganese	57	54	− 6.0	56
Iron	6084	8498	+ 39.7	7091
Cobalt	17	29	+ 68.6	23
Nickel	274	476	+ 73.3	360
Fe/Ni (weight %)	22.07	17.01	− 23.9	18.71

The stony meteorites are a sample of the solar system which may be more representative of the original composition than other available sources; variations, however, must have been introduced by chemical and physical circumstances. A discussion of such matters is beyond the scope of this article. Those interested

[1] G. T. PRIOR: Miner. Mag. **18**, 26 (1916); **19**, 51 (1920).
[2] H. C. UREY and H. CRAIG: Geochim. et Cosmochim. Acta **4**, 36 (1953).
[3] W. WAHL: Geochim. et Cosmochim. Acta **2**, 1 (1951).

in the abundances of the elements may refer to the discussions, e.g., by SUESS and UREY[1], BROWN and PATTERSON[2], or GERLING[3]. Similarities between the abundances in meteorites and in other cosmic sources such as the Sun are striking in many cases.

The analysis of noble gas and radioactive isotopes in meteorites is becoming increasingly important. The absence of krypton and xenon supports the assumption that no primordial noble gas occurred in meteorites. The isotope He^4 is found in meteorites and can result from the decay of uranium and thorium; A^{40} can result from potassium decay. In stony meteorites uranium, thorium, and potassium are measurable; in iron meteorites they are almost entirely absent.

BAUER[4] and HUNTLEY[5] suggested that the helium measured in iron meteorites by PANETH[6] was produced by cosmic ray spallation interactions and not by radioactive decay. The existence of spallation isotopes in meteorites is now established by their high He^3 contents[7–11], their high He^3 to He^4 ratios[7–10], H^3 in recent meteorite falls[10–12], the proportionality of A^{36}, A^{38}, Ne^{21}, and Ne^{22} to He^3 [9,10,13] and the lack of proportionality of Li^6 to He^3 [10,12]. The study of spallation isotopes in meteorites combined with the companion study of the isotopes produced in iron targets by high energy protons[14,15] has led to the following tentative conclusions: The intensity of the cosmic rays striking some meteorites has averaged three times that of the rays striking the Earth[10], and the exposure time of meteorites to cosmic rays varies from 0.3×10^9 years to 1.5×10^9 years[10,12]. The results also show that the slow neutron flux in space is less than 1 neutron/cm^2 sec [10] and the average energy of the bombarding cosmic ray for the meteorites is many billion volts. Furthermore the shape of meteorites in space may now be approximated; for example, the Carbo meteorite was an egg-shaped object in space[16].

The number of different spallation isotopes measured is rapidly increasing and each isotope adds a new chapter to the history of meteorites and of cosmic rays. The isotope A^{39} which has a half-life of 265 years is presently being measured. A^{39} can be detected with a much greater sensitivity than H^3 because there is much less argon than hydrogen in meteorites and argon is a better counter fill than hydrogen. For dated falls the A^{39} combined with the A^{38} gives a more accurate radiation age than the H^3—He^3 age, because the relative production rates of A^{39} and A^{38} are nearly equal and independent of energy while the H^3 and He^3 production rates are somewhat different and energy dependent. At low bombarding energies more He^3 than H^3 is produced. The rate of diffusion loss of A^{39} and A^{38} from materials is also low compared to that of He^3 or H^3. The relative production rate of H^3 to A^{39} is six to one. The most recent iron fall is

[1] H. E. SUESS and H. C. UREY: Rev. Mod. Phys. **28**, 53 (1956).
[2] H. BROWN and C. PATTERSON: J. Geology **55**, 405, 508 (1947).
[3] E. K. GERLING: Dokl. Akad. Nauk. USSR. **77**, No. 1, 85 (1951).
[4] C. A. BAUER: Phys. Rev. **72**, 354 (1947).
[5] H. E. HUNTLEY: Nature, Lond. **161**, 356 (1948).
[6] W. J. ARROL, R. B. JACOBI and F. A. PANETH: Nature, Lond. **149**, 235 (1942).
[7] F. A. PANETH, P. REASBECK and K. I. MAYNE: Geochim. et Cosmochim. Acta **2**, 300 (1951).
[8] J. C. DALTON, F. A. PANETH, P. REASBECK, S. J. THOMSON and K. I. MAYNE: Nature, Lond. **172**, 1168 (1953).
[9] W. GENTNER and J. ZÄHRINGER: Z. Naturforsch. **10**a, 498 (1955).
[10] E. L. FIREMAN and D. SCHWARZER: Geochim. et Cosmochim. Acta **11**, 252 (1957).
[11] E. L. FIREMAN: Bull. Amer. Phys. Soc., Ser. II **7**, 343 (1956).
[12] E. L. FIREMAN: Nuclear Processes in Geology. Nuclear Science Series. 1958 (in press).
[13] P. REASBECK and K. I. MAYNE: Nature, Lond. **176**, 733 (1955).
[14] E. L. FIREMAN: Phys. Rev. **97**, 1303 (1955).
[15] E. L. FIREMAN and J. ZÄHRINGER: Phys. Rev. **107**, 1695 (1957).
[16] Private communication from E. L. FIREMAN.

that of Sikhote-Alin, in 1947. Half of the tritium in this meteorite has decayed but the A^{39} content remains relatively unchanged. The degree of the cosmic ray bombardment of the Sikhote-Alin meteorite, as indicated by its He^3 content of 3×10^{-7} cm³/g, is not very great. This content indicates an age of about 6×10^8 years. The A^{39} is not limited to recently fallen meteorites since practically all known falls have occurred within the past 265 years. The A^{39} is not affected by terrestrial water which can spoil the H^3 measurements. A further advantage in the determination of age by the argon method is that the A^{39} can be measured in both stone and iron meteorites. Thus by this method an "argon" age can be obtained for practically every known fall. Where the date of fall is not known, the A^{39} content still provides interesting information. For example, the lack of detectable A^{39} in Carbo indicates that it must have fallen more than 2000 years ago[1].

In the early determination of ages from the decay products of natural uranium, thorium, and potassium we meet difficulties. For iron meteorites the helium ages based on uranium and thorium decay[1] are of dubious value because of the effect of cosmic ray spallation and because the exact content of uranium and thorium is difficult to measure. The diffusion loss of helium from stones is serious. A spectacular advance in the technique of measuring potassium and argon by neutron activation has just opened the door for the measurement of radiogenic ages of iron meteorites[2]. With regard to the older measurements of the lead ages of stony meteorites, a serious discrepancy exists between the amount of lead in the stone and the amount of radiogenic lead expected from its uranium and thorium content. There is only one stone meteorite, Nuevo Laredo, where the discrepancy is small enough to permit a reliable lead age. The uranium and lead age of this meteorite is 4.6×10^9 years[3]. Nuevo Laredo has the largest uranium content of any known meteorite. The argon-potassium ages of six stone meteorites have also been determined. These ages are between 1.0 and 4.2×10^9 years, with the chondrites showing a tendency to be older than the achondrites. It should be noted that the argon-potassium age refers to the period of time that the meteorite has been solid, i.e., it gives the date of solidification. The measurements of the argon-potassium age of iron meteorites[2] indicate very old ages, all more than 6×10^9 years. It therefore appears that the achondrites, chondrites, and irons solidified as groups at different times. The analysis of isotopes in meteorites will certainly progress greatly during the next decade.

Acknowledgements.

This account of the astronomy and physics of meteors is based on the researches of many authors who in most instances are directly referred to in the text. In particular we are indebted to Dr. L. G. JACCHIA for major contributions to the concept of fragmentation and to Dr. E. L. FIREMAN for the account of isotope studies in meteorites.

[1] See footnote 16, p. 563.
[2] R. W. STOENNER and J. ZÄHRINGER: Geochim. et Cosmochim. Acta (in press).
[3] C. C. PATTERSON: Geochim. et Cosmochim. Acta 7, 151 (1955).

Sachverzeichnis.

(Deutsch-Englisch.)

Bei gleicher Schreibweise in beiden Sprachen sind die Stichwörter nur einmal aufgeführt.

Subject Index.

(English-German.)

Where English and German spelling of a word is identical the German version is omitted.

FSC
www.fsc.org
MIX
Papier aus verantwortungsvollen Quellen
Paper from responsible sources
FSC® C105338

If you have any concerns about our products, you can contact us on
ProductSafety@springernature.com

In case Publisher is established outside the EU, the EU authorized representative is:
Springer Nature Customer Service Center GmbH
Europaplatz 3, 69115 Heidelberg, Germany

Printed by Libri Plureos GmbH
in Hamburg, Germany